PREFIXES

Prefix	Abbreviation	Multiplier
Atto	a	10^{-18}
Femto	f	10^{-15}
Pico	p	10^{-12}
Nano	n	10^{-9}
Micro	μ	10^{-6}
Milli	m	10^{-3}
Centi	c	10^{-2}
Deci	d	10^{-1}
Deka	da	10^{1}
Hecto	h	10^{2}
Kilo	k	10^{3}
Mega	M	10^{6}
Giga	G	10^{9}
Tera	T	10^{12}

BASIC CIRCUIT ANALYSIS

BASIC CIRCUIT ANALYSIS

DAVID R. CUNNINGHAM
University of Missouri — Rolla

JOHN A. STULLER
University of Missouri — Rolla

HOUGHTON MIFFLIN COMPANY   BOSTON
Dallas   Geneva, Illinois   Palo Alto   Princeton, New Jersey

BRIEF CONTENTS

CONTENTS

PREFACE

Basic Circuit Analysis is written for the two-semester electric circuits course typically offered as the first course in the standard electrical engineering curriculum. Knowledge of first-year calculus is assumed. A strong high school physics course or a college physics course is desirable, but not absolutely necessary. This book does not assume a familiarity with complex arithmetic, matrices, or differential equations.

An electric circuits text is often a student's first engineering book. The book should convey a sense of the thoroughness, innovation, and skill that characterize good engineering. Our reason for writing this book is to convey this sense to our readers and point to the direction they should take to excel in engineering.

Pedagogy: Functional Aids to Enhance Student Learning

In a first circuits book it is essential to present each topic in a pedagogically sound way. Each section of this book has been analyzed for the appropriate level, motivation, supporting examples, and exercises. In addition, the material was class-tested over a six-year period at the University of Missouri-Rolla.

We emphasize problem solving based on fundamental concepts that underlie circuit and system theory. Each of the 227 worked-out examples emphasizes the importance of setting up a problem in a clear and organized way before solving it. To be successful engineers, students must be able to do much more than just plug numbers into equations. They must know the fundamentals and how to apply them. Wherever possible, the practical importance and applicability of a given topic is discussed. We frequently refer to practical systems and devices to help students relate theory to real problems. This emphasis and the section on the history of electrical science and engineering give students a perspective missing from many texts.

Learning comes through practice and reinforcement. To aid in this, we have developed an extensive program of pedagogical features for this text.

Perspective

♦   **Chapter introductions** highlight the content and basic concepts presented.

♦   **New material** is presented in the context of preceding work.

- **Frequent reference to practical systems and devices** motivates students and helps them relate theory to real problems.

- **A brief history of the development of electrical science and engineering** conveys appreciation for the achievements of many people over hundreds of years.

Reinforcement

- **Key definitions** are concisely stated and highlighted for easy reference.

- **Remember statements** that identify principal or noteworthy ideas conclude individual sections.

- **More than 200 carefully chosen examples** are included. The solutions to these examples are organized to serve as models for solving exercises and problems in the book.

- **Summary and Key Facts** sections conclude each chapter.

Problems and Exercises

- **More than 650 exercises** appear within chapters, designed for immediate reinforcement of the material.

- **More than 1100 end-of-chapter problems keyed to individual sections** are included.

- **Over 250 unkeyed comprehensive problems** also appear at the ends of chapters, calling on the student to combine skills learned in various chapters.

- **Many shorter, less time-consuming problems** are included to encourage students.

- **Answers** to exercises and selected problems are provided at the back of the book.

Symbols

- **Optional sections, marked with an asterisk (*),** can be omitted without loss of continuity.

- **A firecracker ( )** precedes sections and problems that will challenge students. These sections and problems are optional.

- **A marginal floppy disk ( ) and reference** refers students to particular sections and problems in the PSpice® manual, which is described below.

- **A floppy disk and problem number (5.3)** following a text problem refers students to problems in the PSpice manual.

Figures and Photographs

- **More than 1500 figures** illustrate important concepts.

- **Photographs** help students relate theoretical concepts to concrete systems or devices.

PSpice

Computer-aided circuit analysis is widely used in industry, and its introduction into the undergraduate engineering curriculum has been steadily growing. SPICE appears to be the most popular circuit analysis program in both industry and schools. We have selected PSpice, the personal computer version of SPICE, because it is freely available and is a de facto standard at universities across the United States.

Many schools prefer to postpone the use of computer-aided circuit analysis until the electronics courses. We have incorporated PSpice in a way that encourages its use without requiring it. We have also incorporated the PSpice material in a manner that allows the instructor to make minor changes to adapt the material to other versions of SPICE.

For those who wish to provide a more extensive coverage of PSpice, a complete manual entitled *Introduction to PSpice® with Student Exercise Disk* accompanies this text. The manual contains a **student exercise disk** and gives students direct, hands-on instruction with PSpice. The manual teaches students how to use PSpice as a tool to learn more about circuits and to see circuit design principles in action.

The manual begins with installation of PSpice. Students then work through an introductory hands-on **tutorial** and eight worked-out **examples** of PSpice programs. A separate chapter containing dozens of **problems**, graduated in difficulty from simple to complex, directly extend the presentation of circuit design principles provided in the text. **Cross references** keyed with a floppy disk symbol appear in the text to refer the student to particular sections and problems in the PSpice manual.

Ancillaries

♦ **A solutions manual**, which includes transparency masters for more than 100 text figures, provides complete solutions.

♦ **A student problem supplement** provides hundreds of extra problems. Each problem is followed by its solution.

♦ **The solutions manual and problem supplement can be obtained** by contacting Houghton Mifflin Company.

Coverage and Organization

To help students recognize the interrelation of major ideas, we divided the topics into five major parts: Part One, Fundamentals; Part Two, Time-Domain Circuit Analysis; Part Three, Frequency-Domain Circuit Analysis; Part Four, Series and Transform Methods; and Part Five, Selected Topics. Some chapters can be covered in an alternative sequence. The flowchart in Fig. P.1 shows the prerequisites for each chapter.

The pace of the technical development is slow initially and increases gradually to accommodate students with differing backgrounds. Part One, Fundamentals, begins with an introduction to electrical current, voltage, and power. It then presents Kirchhoff's current law (KCL) and Kirchhoff's voltage law (KVL), giving examples of each law before introducing the network components. We then introduce sources, resistance R, capacitance C, and inductance L.

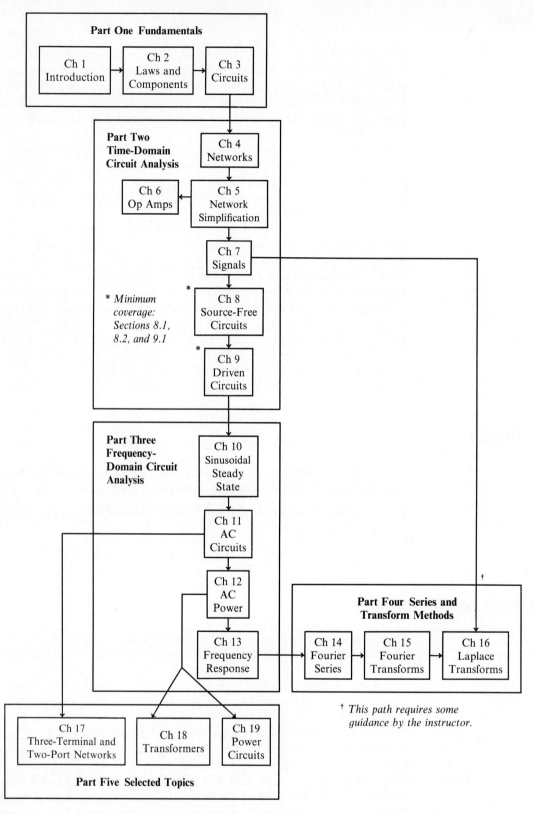

FIGURE P.1
Prerequisites flowchart

Circuits composed of only sources and *R*s are emphasized at first. However, the early development includes very simple examples of circuits also containing *L*s and *C*s. As the development progresses, the number of examples of circuits containing all components increases gradually.

We also introduce matrix notation early. The goal is simply to familiarize students with matrix notation. The inverse of a 2×2 matrix is derived and used in Part Five.

Part Two, Time-Domain Circuit Analysis, contains all the fundamental techniques of linear circuit analysis, except graph theory, beginning with node-voltage analysis. Here, and throughout the book, we present new material in the context of preceding work. For example, in Chapter 4 we present circuit analysis by node voltages as a generalization of the analysis of a parallel circuit. Similarly, we present mesh-current analysis as a generalization of the analysis of a series circuit.

We conclude Part Two with the classical method for solving the differential equations arising in linear circuit theory. We divide transient circuit analysis between source-free circuits and driven-circuits, rather than between first-order and second-order circuits. This division promotes an appreciation of the way in which a circuit's complete response depends on the interaction of the input and the circuit's own dynamics.

Part Three, Frequency-Domain Circuit Analysis, develops phasor circuit analysis beginning with basic definitions and concluding with poles, zeros, and Bode plots. We describe the mathematical transformations from time domain to frequency domain in greater detail and depth than many texts do. This thorough treatment is designed to provide students with a particularly strong foundation for future courses in linear systems.

In Part Four, Series and Transform Methods, we follow the traditional development of the Fourier series to the Fourier transform and the Fourier transform to the Laplace transform. Circuit analysis by Fourier series is presented as a straightforward extension of the methods of Part Three.

The material in Part Five, Selected Topics, can be used to tailor the book to different curriculum requirements or to feed into subsequent courses. These topics can be covered in any of the sequences shown in Fig. P.1.

Concluding appendixes provide clear, concise introductions to elementary matrix manipulations and the algebra of complex numbers.

Terminology

Circuit theory abstracts three elementary electrical properties, *resistance, inductance,* and *capacitance* from distributed electromagnetic phenomena and models these characteristics individually as lumped *elements.* Many books, including this one, use the terms *resistance, inductance,* and *capacitance* to denote the elements of circuit theory. Some of this book's reviewers preferred the terms *resistor, inductor,* and *capacitor* for the theoretical elements. These terms are also used in many texts and are, of course, universally accepted names of physical devices. It is widely acknowledged that these physical devices should never be confused with the elements of the theory. We have retained the original terminology because it clearly distinguishes between the elements of the theory and the physical devices.

Acknowledgments

We thank the University of Missouri-Rolla for providing an ideal teaching environment in which we could develop this book. This text benefited immeasurably from the feedback we received from instructors and students here. We thank many of our colleagues for their suggestions. Suggestions from John Oliver, Jim Paunicka, Steve Saliga, Kumar Shiv, Professor Gordon Carlson, Professor Frank Kern, Professor Charles McDowell, Professor Tom VanDoren, Professor Mehmet Saridereli, Professor George McPherson, and Professor Earl Richards led to improvements to the original manuscript. A student, Terry Trippe, provided an especially insightful suggestion.

We also wish to acknowledge our own circuit theory teachers. An outstanding circuit theory course taught at the Massachusetts Institute of Technology by Amar G. Bose (lecturer) and Harry L. VanTrees (recitation instructor) was a turning point in the education of one of the authors.

We thank the reviewers for their constructive criticisms.

Jonny Andersen, University of Washington
Douglas B. Brumm, Michigan Technological University
Arthur R. Butz, Northwestern University
Roger C. Conant, University of Illinois at Chicago
James F. Delansky, The Pennsylvania State University
Chris DeMarco, University of Wisconsin-Madison
Lawrence Eisenberg, University of Pennsylvania
John A. Fleming, Texas A & M University
Gary E. Ford, University of California, Davis
Eddie R. Fowler, Kansas State University
Douglas Frey, Lehigh University
Richard R. Gallagher, Kansas State University
Glen C. Gerhard, University of New Hampshire
James G. Gottling, The Ohio State University
Walter L. Green, University of Tennessee, Knoxville
Natale J. Ianno, University of Nebraska-Lincoln
Kenneth A. James, California State University, Long Beach
Yogendra P. Kakad, University of North Carolina at Charlotte
Robert H. Miller, Virginia Polytechnic Institute and State University
Robert W. Newcomb, University of Maryland, College Park
William D. O'Brien, Jr., University of Illinois at Urbana-Champaign
John R. O'Malley, University of Florida, Gainesville
Arthur L. Pike, Tufts University
Arthur D. van Rheenen, University of Minnesota, Twin Cities
Susan A. Riedel, Marquette University
Ronald J. Roedel, Arizona State University
Lee Rosenthal, Fairleigh Dickinson University
F. William Stephenson, Virginia Polytechnic Institute and State University
Don L. Steuhm, North Dakota State University
James Svoboda, Clarkson University
M. D. Wvong, The University of British Columbia

This book would not have been possible without its typists, who include Linda Boswell Bramel, Krista Fester, and Lorrie Evans Hamilton. Special thanks are due to Kathy Collins and Janice Spurgeon for their patience and typing skill in preparing the final manuscript.

Finally, we convey our warmest thanks to our wives, Mary Cunningham and Sandra Lanning Stuller, and our children, Linda and Mark Cunningham, and David, Michael, Peter, and John K. Stuller, for their patience and love throughout the lengthy development of this book.

D. R. C.
J. A. S.

THE BEGINNINGS

The development of modern electrical engineering has been intimately intertwined with the scientific discovery and understanding of electric and magnetic phenomena. The history of this development is a fascinating story. In some instances, the application of observed phenomena encouraged a quest for understanding; in other instances, the search for understanding pointed the way to new phenomena and applications.

Surely the first practical, although uncontrolled, use of electricity occurred when a fire started by lightning was used for heating or cooking. The first reported application of magnetism was by Emperor Huan-ti of China, who used a magnetic compass as early as 2637 B.C. Static electricity, developed by friction on amber (the Greek name for amber is *electron*), is said to have been first observed over twenty centuries later by Thales of Miletus, the Greek philosopher.

Dr. William Gilbert, court physician to Queen Elizabeth of England, published six books in 1600 containing the results of seventeen years of intense research on electricity and magnetism. His second book is considered the first published work on electricity, and for this work he is called the founding father of modern electricity.

In 1660 Otto von Guericke, a German physicist, constructed the first machine for generating static electricity. A sulfur sphere revolved around its axis while a cloth was held against the sphere.

The Leyden jar, a form of capacitor that could store electrical energy, was developed in 1745. Although several people claimed the development, Heinrich Bernt Wilhelm von Kleist published his discovery first. As early as 1767, Joseph Bozolus proposed using Leyden jars and long wires to transmit coded messages by creating sparks.

The quantitative study of electricity began with Charles A. Coulomb, who was the founder of electrostatics and the school of experimental physics in France. In 1785, with the aid of a small torsional balance electroscope, he established Coulomb's law by showing that the force of attraction or repulsion between two charged spheres varied inversely with the square of the distance between their centers.

Alessandro Volta

Alessandro Volta developed the voltaic pile, an early form of the electric battery in 1796, and the publication of his work in 1800 provided the greatest boost for electrical experimentation since the Leyden jar. The battery provided the first continuous supply of electrical energy.

André Marie Ampère

Michael Faraday

Georg Simon Ohm

Jean Baptiste Joseph Fourier

The next truly great discovery occurred in 1820 when the Danish physicist and experimenter Hans Christian Oersted found that a wire connecting two battery terminals would influence the needle of a compass placed under it. This demonstrated that an electric current creates a magnetic field.

Scientists were extremely active over the next few years. Perhaps the most significant work was performed by André Marie Ampère, who discovered that a force was exerted between wires carrying current. This helped establish the elementary laws of electrodynamics.

In England in 1821, Michael Faraday showed that electricity can produce continuous mechanical motion. His greatest discovery occurred in 1831 when he verified his prediction that magnetism could induce electric current. This opened the door for practical applications of electricity, and electrical engineering was born.

The German mathematician and physicist Georg Simon Ohm published *The Galvanic Theory of Electricity* in Berlin in 1827 and established Ohm's law: The current through a conductor is proportional to the voltage across it. Not to be overlooked are the contributions of two famous experimentalists and mathematicians of the late eighteenth and early nineteenth centuries: Jean Baptiste Joseph Fourier and Marquis Pierre Simon de Laplace, who developed mathematical techniques now used in circuit analysis.

Joseph Henry, a professor in Albany, New York, conducted many experiments on electromagnetism and constructed an electric motor in 1829. While others, such as the German mathematician Karl Friedrich Gauss, contributed to electromagnetic theory, James Clerk Maxwell unified this theory with the concept of displacement current. This completed what are now known as Maxwell's equations and contributed to the 1864 prediction of electromagnetic wave propagation.

By the end of 1831 Hypolite Pixii of Paris demonstrated a rotating electromagnetic generator that could supply a constant voltage and current. In 1839 Professor Jacobi of Russia demonstrated a battery-powered boat, and in 1840 Robert Davidson, a Scot, equipped a car to run on rails using a solenoid-type motor.

Gustav Robert Kirchhoff, a Prussian working in Germany, published some of his work between 1845 and 1882. This work gave us Kirchhoff's voltage law and Kirchhoff's current law, which form the basis of electric circuit analysis.

Following the development of the Morse code by Samuel F. B. Morse and Alfred Vail, a public telegraph was opened in the United States in 1844 and in England in 1845; the modern era of communications began. In 1861 the New York-to-San Francisco telegraph line was completed and established high-speed transcontinental communications. A major breakthrough occurred in 1875 when Alexander Graham Bell developed the telephone.

Comparable advances were being made in electric power. By 1851 a few electromechanical generators (dynamos) were in use for commercial electroplating, and other new applications of electricity evolved rapidly. The South Foreland lighthouse in England first shined an arc lamp to sea in 1858 and established electric lighting as practical. In 1879 Thomas A. Edison invented the incandescent light, which opened the way to electric lights for the home. The world was introduced to electrically powered transportation by an electric train with a 3-horsepower motor, which was constructed by the Siemens and Halske Company in 1879 for a trade fair in Berlin.

Marquis Pierre Simon
de Laplace

Joseph Henry

Gustav Robert Kirchhoff

William Siemens

In the post–Civil War period, static electricity was taught as science in the physics department, while galvanic electricity* was taught in the chemistry department. The telegraph as well as the dynamo and the incandescent lamp created a greater demand for trained electrical workers than these educational programs could provide. This demand was recognized prior to 1880, but the first formal electrical engineering programs were introduced at the Massachusetts Institute of Technology and Cornell in 1882. By the end of the decade most universities had introduced a formal program in electrical science or engineering, and in 1893 Ohio State University graduated Bertha Lamme, the first woman electrical engineer. The programs differed, but all were built on a strong foundation of mathematics, physics, and mechanical engineering.

With many people trained in electrical science and engineering, progress was rapid. Nikola Tesla produced a two-phase alternator in 1887 and soon developed the first practical ac motor. This established the feasibility of ac power, which is the basis of our power systems.

In 1894 Marchese Guglielmo Marconi successfully transmitted feeble electrical signals through space. This introduced wireless communication, which was well established by the beginning of World War One.

The products developed since 1918 that rely on electrical phenomena are too numerous to mention. Some developments could best be described as revolutionary, whereas others were more evolutionary. Developments of both types were required for practical products.

More recent developments include television broadcasting (1936), radar (World War Two), and the electronic digital computer (1945). In 1948 John Bardeen, Walter Brattain, and William Shockley invented the transistor, which made the modern era of communications possible. The first communications satellite was orbited in 1970.

The pocket calculator and desktop computer are perhaps the most conspicuous products of the modern era, but many less visible advances have been made. Navigation satellites guide shipping, solar cells supply power to remote weather stations, and the newer home appliances are more efficient.

This brief history is obviously incomplete, but we hope that it will stimulate you to read more about this fascinating subject. As for future developments, well, these are partly up to you.

---

* Electricity supplied at constant voltage was called galvanic electricity, because it was first generated by galvanic action in a battery.

BASIC CIRCUIT
ANALYSIS

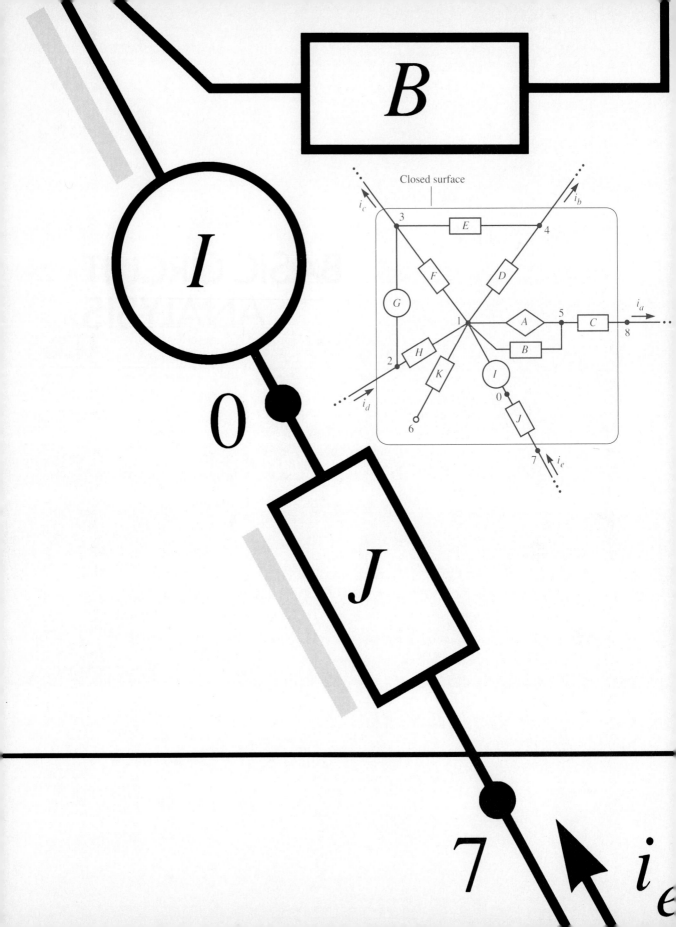

FUNDAMENTALS

We begin with the physical units and foundations of circuit analysis. Circuit analysis is presented from an axiomatic point of view by the introduction of Kirchhoff's current law, Kirchhoff's voltage law, and the component equations. We then apply these axioms and definitions to the analysis of simple circuits.

CHAPTER
1

Introduction

Electricity and magnetism are very broad terms encompassing many interesting and useful phenomena. As a consequence, electrical engineering has many facets. This textbook considers circuit theory and methods of circuit analysis, which are the first topics usually learned in electrical engineering. Circuit theory is concerned with the analysis of currents and voltages in electrical systems such as stereo amplifiers, computers, and power systems.

Although we discuss the physical foundations of circuit analysis, circuit theory begins with definitions and axioms—not with physics. We shall form mathematical models of physical electrical systems and analyze these models.

When we move on from the study of this textbook into the practice of electrical engineering (or into an electric circuits lab), we must remember that like other theories, the one presented here has limitations. For example, circuit theory tells us nothing of radio-wave propagation. For that we need Maxwell's equations of electromagnetic field theory.

1.1 Units

In keeping with international practice, we will use the International System of Units, abbreviated SI and popularly called the metric system.* The six basic units of the SI system are shown in Table 1.1.

TABLE 1.1
The International
System of Units

QUANTITY	UNIT	SYMBOL
Length	Meter	m
Mass	Kilogram	kg
Time	Second	s
Electric current	Ampere	A
Temperature	Degree kelvin	K
Luminous intensity	Candela	cd

All other units are derived from these. For example, work (energy) is force times distance and has the derived dimension of newton-meters or joules (J). The SI system uses the prefixes listed in Table 1.2 to specify powers of 10.

TABLE 1.2
Prefixes

PREFIX	ABBREVIATION	MULTIPLIER
Atto	a	10^{-18}
Femto	f	10^{-15}
Pico	p	10^{-12}
Nano	n	10^{-9}
Micro	μ	10^{-6}
Milli	m	10^{-3}
Centi	c	10^{-2}
Deci	d	10^{-1}
Deka	da	10^{1}
Hecto	h	10^{2}
Kilo	k	10^{3}
Mega	M	10^{6}
Giga	G	10^{9}
Tera	T	10^{12}

For example, the following are equivalent values of power in watts (W):

$$1000 \text{ nW} \qquad 1 \text{ } \mu\text{W} \qquad 0.001 \text{ mW} \qquad 0.000001 \text{ W}$$

We will use an uppercase letter to imply that a quantity cannot be a function of time. For example, an average value is constant, and

$$P = 2 \text{ W} \tag{1.1}$$

* The abbreviation SI comes from the French name for the system, Système International des Unités.

implies that the average power P has a constant value of 2 W. A lowercase letter implies that the quantity can be, but is not necessarily, a function of time. For example, instantaneous power p can be a functon of time, such as

$$p = p(t)$$
$$= 12 \cos 50t \text{ W} \tag{1.2}$$

We will often omit the explicit dependence on time unless it is needed for emphasis or clarity.

EXERCISES

Write the following quantities in terms of prefixed values that contain no more than three digits. (For example, 360,000 m is equivalent to 360 km.)

1. 0.002 m
2. 15,000 m
3. 0.05 s
4. 0.0004 ms
5. 27,000 kW
6. 0.007 s

With voltages given in volts (V) and currents in amperes (A), which of the following are consistent with the notation defined in Eqs. (1.1) and (1.2)?

7. $v = 120 \sqrt{2} \cos 2\pi 60t$
8. $I = 5t^2$
9. $v = 12$
10. $V = 12 \sin 2t$
11. $I = 6$

1.2 Charge

All electrical phenomena are manifestations of *electric charge*. Anyone who has removed a wool sweater and had it stick to his or her shirt has experienced an effect of electric charge.

The atomic particles that carry electric charge are protons and electrons. They have charges of equal but opposite sign. The charge on a proton is 1.602×10^{-19} coulomb (C), where the coulomb is the derived unit of charge in the SI system. The charge on an electron is -1.602×10^{-19} C.

The quantity of charge involved in simple electrical processes can be very large. For example, it requires 3,036 C of charge to electroplate 1 gram (g) of copper.

Charges of opposite sign attract each other, and charges of like sign repel each other. Two 1-C charges separated by 1 m would exert a force on each other of

$$F = 8.99 \times 10^9 \text{ newtons (N)}$$

which is approximately 2×10^9 pounds or one million tons. Fortunately an object normally has approximately equal quantities of positive and negative charge, whose effects cancel each other with respect to forces exerted on other objects.

1.3 Current

Electric circuit theory deals with the effects of charge. Charge flow is called *electric current* (refer to Fig. 1.1).

FIGURE 1.1
Electric current due to charge
flow in a fluorescent lamp

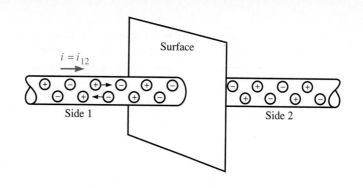

DEFINITION

Electric **current** $i = i_{12}$ that passes through a surface from side 1 to side 2* is the net positive charge per unit of time that passes through the surface in this direction.

The SI unit for charge is the *ampere* (A). An ampere is equivalent to one coulomb per second (C/s).

As indicated by the two small arrows on two charges in Fig. 1.1, a positive value of current i_{12} implies that positive $(+)$ charge passes through the surface from side 1 to side 2, or negative $(-)$ charge passes through the surface in the opposite direction, or both.

An equivalent statement of the definition of current is given by the following equation:[†]

Current

$$i = \frac{dq}{dt} \qquad (1.3)$$

where i is current in amperes
        q is charge in coulombs
        t is time in seconds

We can determine the charge that passes through the surface in the time interval $-\infty$ to t if we integrate Eq. (1.3) with respect to time. With the assumption that $q(-\infty)$ is zero, this gives[‡]

Charge

$$q(t) = \int_{-\infty}^{t} i(\tau)\, d\tau \qquad (1.4)$$

* Current i_{12} is read i one-two, not i twelve.

[†] Lowercase q and i are used to imply that these quantities can be functions of time. The time dependence will be shown explicitly when necessary to avoid ambiguity.

[‡] The variable of integration in Eq. (1.4) is changed to τ because t appears in the upper limit of the integral.

In practice, current is measured with an instrument called an ammeter, as shown in Fig. 1.2a. The meter will indicate the current through the meter, and device A, in the direction from 1 to 2. For convenience the circuit drawing is simplified as shown in Fig. 1.2b.

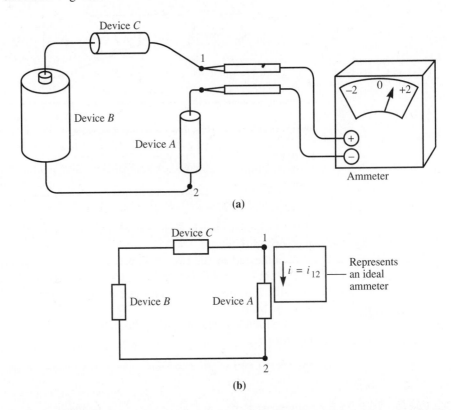

The symbol i represents the ammeter reading, and the arrow represents the location and relative connection of the meter. By convention the arrow points from the point where the $(+)$ terminal of the meter is connected toward the point where the $(-)$ terminal of the meter is connected, as shown in Fig. 1.2. The value of i may be positive or negative.

Subscripts can be used to indicate the current reference direction. The equivalent arrow points from the location of the first subscript toward the location of the second subscript. For current i as shown in Fig. 1.2b,

$$i = i_{12} \tag{1.5}$$

The representation of Fig. 1.2b implies an ideal current measurement that does not affect any part of the circuit.

REMEMBER

A current symbol i represents the current measured by an ideal ammeter. The reference arrow indicates the relative meter connections. The arrow points from the point where the positive $(+)$ terminal of the meter is connected toward the point where the negative $(-)$ terminal of the meter is connected. With the double-subscript notation for current (i_{12}), the reference arrow points from the point labeled with the first subscript (1) toward the point labeled with the second subscript (2). We must always include *both a current symbol and a reference arrow (or double subscripts)* on a circuit diagram to define a current.

EXERCISES

12. A current of 1 μA flows through a device from terminal 1 to terminal 2. Assume that all current is caused by electron flow. How many electrons enter device terminal 1 in a time interval of 1 ns? This is a relatively small current and a short time interval. Do you see why we can usually ignore the discrete nature of charge in circuit analysis?

Charge q passes through a surface from side 1 to side 2 along a wire. Sketch $q(t)$ and $i_{12}(t)$ for $0 \leq t \leq 20$ s, if $q(t)$, in coulombs, is defined by the functions in Exercises 13 and 14. *Hint*: Use Eq. (1.3).

13. $q = 10e^{-2t}$ 14. $q = 10 \cos 2t$

A current of i_{12} amperes passes through a surface from 1 to 2 along a wire. Sketch current $i_{12}(t)$ and the charge $q(t)$ that passes through the surface between time 0 and t, for the functions in Exercises 15 and 16. *Hint*: Use Eq. (1.4).

15. $i_{12} = 10e^{-2t}$ 16. $i_{12} = 10 \cos 2t$

1.4   Voltage

The electrical *energy* per unit charge required to move charge from one point to another is called voltage.

Multimeter (a combination ammeter, voltmeter, and ohmmeter). Courtesy of John Fluke Mfg. Co., Inc.

DEFINITION

The **voltage** $v = v_{12}$ of point 1 with respect to point 2 is the electrical energy expended per unit of positive charge as it moves from point 1 to point 2.

We often call the voltage v_{12} the *voltage drop* going from point 1 to point 2. The SI unit for voltage is the *volt* (V). A volt is equivalent to one joule per coulomb (J/C).

An equivalent definition of voltage is given by the following equation:

Voltage

$$v = \frac{dw}{dq} \tag{1.6}$$

where v is voltage in volts
        w is energy in joules
        q is charge in coulombs

In practice, the voltage of one point with respect to another is usually measured with an instrument called a voltmeter, as shown in Fig. 1.3a. The voltmeter indicates the voltage of point 1 with respect to point 2. For convenience the drawing is simplified as shown in Fig. 1.3b. The symbol v represents the voltmeter reading. The $(+)$ and $(-)$ signs correspond to the points where the similarly marked meter terminals are connected. Subscripts can be used to indicate the meter connection. The first subscript indicates where the $(+)$ terminal is connected, and the second indicates where the $(-)$ terminal is connected. For voltage v as shown in Fig. 1.3b,

$$v = v_{12} \tag{1.7}$$

The quantity v may be positive or negative. The representation of Fig. 1.3b implies an ideal voltage measurement that does not affect any part of the circuit.

FIGURE 1.3
Voltage measurement. (*a*) Measuring voltage v_{12} with a voltmeter; (*b*) symbolic representation of voltage measurement (box shown to emphasize that the letter v and the polarity marks correspond to the meter reading and terminal connections of the voltmeter)

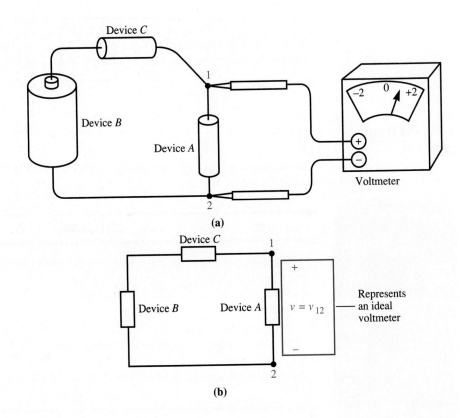

REMEMBER

A voltage symbol v represents the voltage measured by an ideal voltmeter. The location of the $(+)$ and $(-)$ signs corresponds to the points where the like-marked terminals of the meter are connected. With the double-subscript notation for voltage (v_{12}), the first subscript indicates the point in the circuit where the $(+)$ terminal of the voltmeter is connected, and the second subscript indicates the point where the $(-)$ terminal is connected. We must always include *both a symbol and polarity marks (or double subscripts) on a circuit diagram to define a voltage.*

17. A device converts all electrical energy delivered to it directly into heat. Temperature measurements indicate that the energy input was 50 J when a charge of 4 C passed through the device. If the voltage across the device was constant, what was its value?

18. The voltage across a device v_{12} is a constant 12 V. Temperature measurements indicate that 48 J of energy was absorbed by the device in a time of 2 s. What charge passed through the device in this time? If the current i_{12} was constant, what was its value?

1.5 Work and Power

Electricity has many uses, but its principal function is to perform a task that requires energy, whether it is to heat a house, run a refrigerator, or deliver a message by telephone that might take hours to deliver by hand.

DEFINITION

Power p absorbed by a device is the energy per unit of time delivered to the device.

The SI unit for power is the *watt* (W). A watt is equivalent to one joule per second (J/s).

An equivalent definition of power is given by the following equation:

Power

$$p = \frac{dw}{dt} \tag{1.8}$$

where p is power in watts
w is energy (work) in joules
t is time in seconds

Voltage v, as shown in Fig. 1.4, is the energy in joules per coulomb absorbed from positive charge moving through device X, from the $(+)$-marked end to the $(-)$-marked end. Current i is the charge in coulombs per second of positive charge that moves in this direction. Therefore the product of voltage v and current i is the instantaneous electric power absorbed by (delivered to) device X in joules per second, or watts:

$$p = \frac{dw}{dt} = \left(\frac{dw}{dq}\right)\left(\frac{dq}{dt}\right) = vi \tag{1.9}$$

FIGURE 1.4
Relative reference directions
for the passive sign
convention

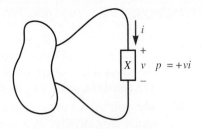

DEFINITION

Passive Sign Convention

The relative reference directions for v and i shown in Fig. 1.4—the reference arrow for i entering the end of the device with the $(+)$ reference mark for v and leaving at the end with the $(-)$ mark—are called the **passive sign convention**.

This gives the following:

Power Absorbed—Passive Sign Convention

$$p = +vi \qquad\qquad (1.10)$$

The instantaneous power absorbed, p, may be either a positive or a negative quantity. If the power absorbed is negative, the device supplies power to the rest of the circuit.

FIGURE 1.5
Relative reference directions
for the active sign convention

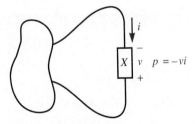

If the reference direction (physically, the meter connection) for either v or i, but not both, is reversed as shown in Fig. 1.5, the electric power absorbed by element X is given by the following equation:

Power Absorbed—Active Sign Convention

$$p = -vi \qquad\qquad (1.11)$$

The passive sign convention is normally used and is implied unless otherwise stated.

The electrical energy absorbed by a device in the interval t_1 to t_2 is

$$W = \int_{t_1}^{t_2} p \, dt \qquad (1.12)$$

With the references assigned as in Fig. 1.4 (passive sign convention), the electrical energy absorbed by device X in the interval t_1 to t_2 is given by the following equation:

Energy Absorbed

$$W = \int_{t_1}^{t_2} vi \, dt \qquad (1.13)$$

In general, power p is a function of time and is referred to as the *instantaneous power*. The *average power* absorbed by a device in the interval t_1 to t_2 is

Average Power

$$P = \frac{1}{t_2 - t_1} \int_{t_1}^{t_2} p \, dt \qquad (1.14)$$

Conservation of energy requires that the power absorbed by an element be supplied from some other source: perhaps chemical energy from a battery or mechanical energy driving a generator. For this reason, the algebraic sum of the instantaneous electric power absorbed by all components in a network must be zero (power supplied is treated as negative power absorbed). The following equation is equivalent to this statement:

Power Balance Equation

The algebraic sum of the instantaneous electric power absorbed by all components in a network is zero:

$$\sum p(t)_{\text{absorbed}} = 0 \qquad (1.15)$$

where the sum is over all components in the network.

Integration of Eq. (1.15) gives the result that the algebraic sum of the *average* electric power absorbed by all components in a network also must be zero.

EXAMPLE 1.1

The voltage and current for the device shown in Fig. 1.4 are $v = 12$ V and $i = 0.5t$ A, for $0 \leq t \leq 10$ s.

(a) Find the instantaneous power absorbed by the device X.

(b) Determine the average power absorbed in the interval $0 \leq t \leq 10$ s.

The voltage and current references in Fig. 1.4 are assigned according to the passive sign convention. Therefore:

(a) $p = vi = (12)(0.5t) = 6t$ W

(b) $P = \dfrac{1}{t_2 - t_1} \displaystyle\int_{t_1}^{t_2} p\,dt = \dfrac{1}{10} \int_0^{10} 6t\,dt = 30$ W. ■

EXAMPLE 1.2

Given the system shown in Fig. 1.6, determine the power p_B that must be absorbed by component B at some instant in time, if the power absorbed by the other elements is $p_1 = -1$ W, $p_2 = 2$ W, $p_3 = 3$ W, and $p_4 = 4$ W at the same instant in time.

FIGURE 1.6
Circuit for Example 1.2

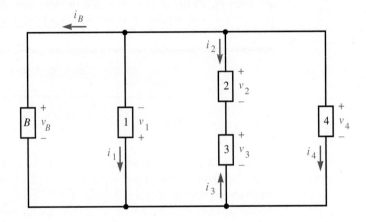

■ **SOLUTION**

From Eq. (1.15),

$$\sum p_{\text{absorbed}} = 0$$

$$p_B + p_1 + p_2 + p_3 + p_4 = 0$$

$$p_B + (-1) + (2) + (3) + (4) = 0$$

$$p_B = -8 \text{ W}$$

Component B must absorb -8 W or supply $+8$ W. ■

EXERCISES

Determine which of the devices shown in Fig. 1.6 have the voltage and current references selected according to the passive sign convention.

19. Device 1
20. Device 2
21. Device 3
22. Device 4
23. Device B

The power absorbed by each device in the circuit of Fig. 1.6 is given in Example 1.2. Determine the specified current if the device voltage is as stated.

24. $v_1 = -12$ V; determine i_1.

25. $v_2 = \frac{24}{5}$ V; determine i_2.

26. $i_3 = -\frac{5}{12}$ A; determine v_3.

27. $v_4 = 12$ V; determine i_4.

28. $v_B = 12$ V; determine i_B.

Voltage and current measurements are made according to the passive sign convention for a device X.

29. The voltage is $v = 50$ V, and the current is $i = 4$ A.
    (a) Determine the instantaneous power p.
    (b) Calculate the average power P over the time interval $0 \le t \le T$.

30. The voltage is $v = 50 \cos 2\pi 60t$ V, and the current is $i = 4 \cos (2\pi 60t - \frac{1}{3}\pi)$ A.
    (a) Determine the instantaneous power p. Sketch p for $0 \le t \le 1/60$ s.
    (b) Calculate the average power P over the time interval $0 \le t \le 1/60$ s. (The argument of the cosine is in radians.)

An electrical system with five devices is shown in the following diagram. Determine the power absorbed by the specified device.

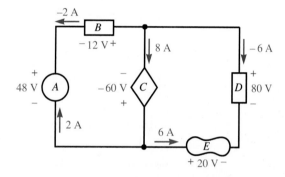

31. Device A

32. Device B

33. Device C

34. Device D

35. Device E

36. Sum the powers absorbed by the devices. Is the power balance equation satisfied?

1.6 Summary

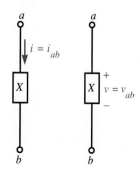

In this chapter we introduced the fundamental units of the SI system and two derived electrical units, charge and voltage. Current i, in amperes, through a circuit component from end a to end b ($i = i_{ab}$) is the net rate of positive charge, in coulombs per second, passing through the device from end a to end b. We can interpret the numerical value of i_{ab} as the current measured by an ideal ammeter when the current i_{ab} enters the (+)-marked terminal of the ammeter and exits at the (−)-marked terminal. Voltage v, in volts, of point a with respect to point b ($v = v_{ab}$) is defined as the energy in joules per coulomb lost by a positive charge as it moves from point a to point b in a circuit. The numerical value of v_{ab} is the voltage indicated by an ideal voltmeter with the (+) terminal connected to the point indicated

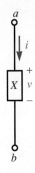

by the first subscript (a) and the ($-$) terminal connected to the point indicated by the second subscript (b).

From the definitions of voltage and current, we showed that the power p absorbed by (delivered to) a device is given by the product of the voltage v across the device and the current i through the device ($p = vi$).

KEY FACTS

♦ The SI system of units is used.

♦ The unit of current is the ampere (coulomb per second):

$$i = \frac{dq}{dt}$$

♦ The unit of charge is the coulomb (ampere · second):

$$q = \int_{-\infty}^{t} i(\tau)\,d\tau$$

♦ The unit of voltage is the volt (joule per coulomb):

$$v = \frac{dw}{dq}$$

♦ The unit of energy is the joule (newton · meter):

$$w = \text{force} \cdot \text{distance}$$

♦ The unit of power is the watt (joule per second):

$$p = \frac{dw}{dt}$$

♦ The passive sign convention implies that the current reference arrow enters the end of the component with the ($+$) voltage reference. The power absorbed by device X is

$$p = vi$$
$$= v_{ab}i_{ab}$$

♦ Conservation of energy gives the power balance equation:

$$\sum p(t)_{\text{absorbed}} = 0$$

and for average power,

$$\sum P_{\text{absorbed}} = 0$$

where the sums are over all components in the circuit.

PROBLEMS

Section 1.3

Charge q passes through a surface from point 1 to point 2 along a wire. Sketch $q(t)$ and $i_{12}(t)$ for $-20 \le t \le 20$ s if $q(t)$, in coulombs, is defined by the following functions:

1. $q = 4 + 2t$
2. $q = |t|$
3. $q = e^{-|t|}$
4.
$$q = \begin{cases} t, & -20 \le t < 5 \text{ s} \\ 5, & 5 \le t < 10 \text{ s} \\ -2(t - 12.5), & 10 \le t < 15 \text{ s} \\ -5, & 15 \le t < 20 \text{ s} \end{cases}$$

A current i_{12} passes through a surface along a wire. Sketch $i_{12}(t)$ and $q(t)$, the charge that passes through the surface between -2 and t, for $-2 \le t \le 2$ s, if $i_{12}(t)$, in amperes, is defined by the following functions:

5. $i = \dfrac{1}{(1 + t^2)}$
6. $i = e^{2t}$
7. $i = e^{-|t|}$
8.
$$i = \begin{cases} t, & -20 \le t < 5 \text{ s} \\ 5, & 5 \le t < 10 \text{ s} \\ -2(t - 12.5), & 10 \le t < 15 \text{ s} \\ -5, & 15 \le t < 20 \text{ s} \end{cases}$$

9. A photodetector releases an average of one electron for every ten photons that strike its surface. If 5×10^{12} photons per second strike the surface, how many coulombs of charge are released in one second? What is the average current through the photodetector?

10. A copper bar with a square cross section of 1 cm × 1 cm is conducting 400 A. The current is entirely due to the motion of free electrons. In copper the free electron density is approximately 10^{29} electrons/m^3.
    (a) What is the average velocity (in the opposite direction to the current) of the electrons due to this current?
    (b) How long would it take an electron to travel 1 km at this velocity?
    (c) Does this contradict the common statement, "Electricity travels at the speed of light"?

11. Silver is a univalent element, so one electron is required to electroplate each atom of silver. There are 6.023×10^{23} atoms in a gram atomic weight (Avogadro's number). The atomic weight of silver is 107.88, and the charge on an electron is -1.602×10^{-19} C.
    (a) How many coulombs are required to electroplate a gram atomic weight of silver? (This is the quantity of electric charge often called a faraday.)
    (b) How many coulombs are required to electroplate 1 g of silver?

12. A power supply must be designed to electroplate 20 g of copper when the current is constant during a time period of 30 min. What current must the power supply provide? (Copper is a bivalent element and requires two electrons per atom of plated material.)

Section 1.4

13. Temperature measurements indicate that a heater has provided 1200 J of energy to a container of water in a time interval of 1 min. The charge that has passed through the heater in this time interval is 100 C. What is the constant voltage across the heater?

An electrical system consists of three parts, A, B, and C, as indicated in the following illustration. It is known that 2 J is absorbed by device A for each coulomb of positive charge that moves through part A from terminal a to terminal b. At the same time, if 1 C of positive charge moves through part B from terminal b to terminal c, 4 J is absorbed by element B.

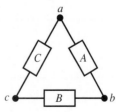

14. What is v_{ab}?
15. What is v_{bc}?
16. Based on the fact that charge will not accumulate inside a device and on conservation of energy, what would be the energy absorbed by

device C for each coulomb of charge that passes through device C from c to a?

17. What is v_{ca}?

18. Find the sum of v_{ab}, v_{bc}, and v_{ca}.

Section 1.5

Given meter connections as shown in Fig. 1.4 (passive sign convention), find and plot the instantaneous power absorbed by device X as a function of time and the total energy absorbed by X from time 0 to t, and calculate the average power absorbed by X over the interval $0 < t < \infty$. The voltage and current are in volts and amperes, respectively.

19. $v = 10$ and $i = 5$, $t > 0$
20. $v = 10$ and $i = 5e^{-4t}$, $t > 0$
21. $v = 10e^{-4t}$ and $i = 5e^{-4t}$, $t > 0$
22. $v = 10$ and $i = 20 \cos 2\pi t$, $t > 0$
23. $v = 10 \cos 2\pi t$ and $i = 20 \cos 2\pi t$, $t > 0$
24. $v = 10 \cos 2\pi t$ and $i = 20 \sin 2\pi t$, $t > 0$
25. $v = 10 \cos 2\pi t$ and $i = 20 \cos (2\pi t + \pi/4)$, $t > 0$
26. $v = 10 \cos 2\pi t$ and $i = 20 \cos 4\pi t$, $t > 0$
27. Given meter connections as shown in Fig. 1.4 (passive sign convention) and voltage (top) and current (bottom) as shown in the following illustration, find and plot the instantaneous power absorbed by device X as a function of time for $-1 < t < 4$. Find the average power absorbed over this time interval.

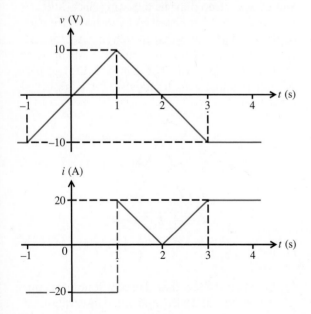

Determine the power absorbed by each device in the specified circuit below. Check your result with the power balance equation.

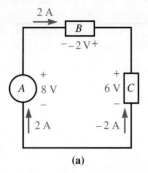

(a)

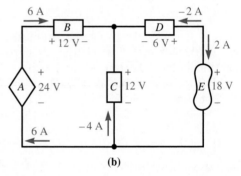

(b)

28. Circuit a 29. Circuit b

Calculate the power absorbed by the specified device in the circuit below if the following voltages and currents are known: $v_{12} = -5$ V, $i_{12} = -6$ A, $v_{23} = -6$ V, $i_{32} = 6$ A, $v_{30} = 21$ V, $i_{30} = -4$ A, $v_{10} = 10$ V, $i_{01} = -6$ A, $v_{43} = 7$ V, $i_{43} = 2$ A, $v_{45} = -9$ V, $i_{54} = 2$ A, $v_{50} = 37$ V, and $i_{05} = 2$A.

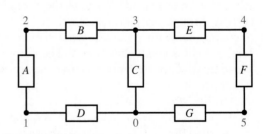

30. Device A 31. Device B
32. Device C 33. Device D
34. Device E 35. Device F
36. Device G

Comprehensive Problems

37. The instantaneous power p absorbed by a two-terminal component is $60e^{-4t}$ watts for $t \geq 0$. If $v_{ab}(t) = 3e^{-2t}$ volts, $t > 0$, calculate the total charge input to terminal a over the interval $0 \leq t < \infty$.

Under starting conditions, the terminal voltage of an automobile starter is 10 V, and the current through the starter is 200 A. Assume that the engine starts after 5 s.

38. What is the power absorbed by the starter?
39. What is the energy absorbed by the starter?
40. What is the charge that has passed through the starter?

The famous bicycle racer Eddy Merckz once held the 1-h closed-course distance record of just over 30 mi. It has been estimated that during this hour his power output was approximately seven-tenths of one horsepower. (One horsepower is approximately 745.7 W.)

41. What was his power output in watts?
42. What was his energy output in joules?
43. If his pedals had been connected to an electrical generating system with an overall efficiency of 80 percent and a terminal voltage of 12 V, what would be the electric power output of the system?

44. What would be the output current of the generator in Problem 43?
45. A family of energy wasters lives in a house with six rooms and two baths. Each room and bath has 150 W of lighting, which is left on an average of 6 h per day. If electricity costs 10 cents per kWh, how much a week does it cost for the energy used by the lights?
46. You wish to design an electric teapot for a motor home. It must heat 1 liter (L) of water, and raise the temperature from 20°C to 100°C in 10 min. If heat losses are neglected, what must be the power rating of the heater? The motor home has a 12-V electrical system. What current is required? (It requires approximately 4.2×10^3 J to raise the temperature of 1 L of water one Celsius degree.)

An automobile battery has a terminal voltage of approximately 12 V when the engine is not running and the starter motor is not engaged. A car is parked at a picnic with the stereo playing, which requires 60 W, and some lights on, which require 120 W. With a load of this type, the battery will supply approximately 1.2 MJ of energy before it will no longer start the car.

47. What power must the battery supply?
48. What current must the battery supply?
49. What is the approximate amount of time that the car can remain parked with the lights on and the stereo playing and still start?

Network Laws and Components

As engineers we need to predict the performance of *electrical
systems* composed of *physical devices*. We will analyze a system
with the aid of a model called a *network* diagram or *circuit*
diagram. A circuit diagram is an interconnection of idealized
devices called network *components*. The model is based on the
assumption that all electrical energy is confined to the network
components. This lets us characterize the interaction of the
network components by two axioms: *Kirchhoff's current law*
(KCL), which implies conservation of charge, and *Kirchhoff's
voltage law* (KVL), which implies conservation of energy. These
laws are the only axioms of network theory and do not depend
on the type of network components making up the circuit.

The network components model the idealized electrical
properties of physical devices in terms of a *component equation*.
In this chapter, we will introduce models for sources of electrical
energy, the conversion of electrical energy into heat, and the
storage of energy in electric and magnetic fields. We must
remember that we may require more than one network
component to model a simple physical device.

2.1 Kirchhoff's Current Law

In Chapter 1, we commented that the net charge of an object is ordinarily zero, and experiments show that electric charge is neither created nor destroyed in a network. Because of this, if we were to visualize a closed surface, such as an imaginary soap bubble, enclosing any portion of a circuit, we would expect that any charge entering the closed surface would be matched by a like quantity of charge leaving along some other path. This statement of the conservation of charge is formalized in Kirchhoff's current law (KCL), which is one of the two axioms of circuit theory.

AXIOM

Kirchhoff's Current Law

At any instant in time, the algebraic sum of all currents leaving any closed surface is zero.

An equivalent statement of KCL is given by the following equation:

KCL Equation

$$\sum_{k=1}^{N} i_k = 0 \tag{2.1}$$

where i_k is the kth current of the N currents leaving the closed surface.

When applying KCL, we must treat a current leaving the closed surface along a path as the negative of a current entering along that path. We can then rephrase KCL to state that *the sum of the currents entering a closed surface must be equal to the sum of the currents leaving the closed surface*. We will now apply KCL to an example circuit.

EXAMPLE 2.1

Apply KCL to the closed surface indicated in Fig. 2.1. Use this result to determine current i_a, if $i_b = 2$ A, $i_c = -3$ A, $i_d = 5$ A, and $i_e = -12$ A.

■ **SOLUTION**

The sum of all currents leaving the indicated closed surface is

$$i_a + i_b + i_c - i_d - i_e = 0$$

from KCL. A minus sign appears on i_d and i_e in the above equation, because the associated reference arrows *enter* the closed surface. Substitution of the given current values into this KCL equation gives

$$i_a + (2) + (-3) - (5) - (-12) = 0$$

We can easily solve this to obtain

$$i_a = -6 \text{ A}$$

This implies that an ammeter connected as indicated by the reference arrow for i_a would indicate -6 A. ■

FIGURE 2.1
Application of KCL to a
closed surface

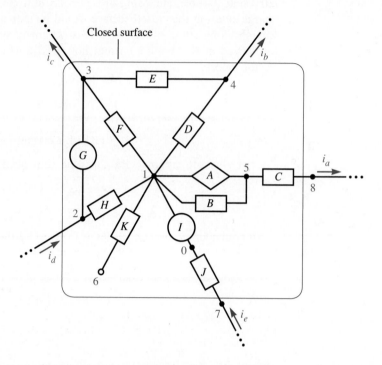

Observe that KCL holds for any closed surface and is *not* restricted to one enclosing a single point or component. By choosing the closed surface wisely, we can often save considerable work when solving a problem.

The symbol for a network component and the associated line segments form a *branch*. For example, in Fig. 2.1 the path from point 1 to point 2 consists of a single branch (branch H), as does the path from point 1 to point 6 (branch K). The path from point 1 to point 7 contains two series branches (branches I and J).* Two parallel branches (branches A and B) connect node 1 to node 5.

No current can flow directly from node 6 to node 0, because no branch connects these two points. A path through which no current can flow, regardless of the voltage across it, is an *open circuit*.

The termination of one or more branches is a *node*. Thus points 0 through 8 of Fig. 2.1 form nodes.† A *junction* is a node where three or more branches terminate. Therefore nodes 1 through 5 of Fig. 2.1 are also junctions. The straight line segment between points x and y of Fig. 2.2 represents an ideal wire connecting

---

* Some books would consider that the combination of branches I and J also formed a branch.

† Some texts define a node as a point where two or more branches terminate. With this definition, point 6 of Fig. 2.1 would be a *terminal* (a point where an electrical connection can be made), but not a node.

these two points. This implies that v_{xy} is zero regardless of the value of current i_{xy}. We say that there is a *short circuit* between points x and y. We consider these points as a single node (which we have labeled as one node 0). In general, all points joined to each other by short circuits form a single node.

EXAMPLE 2.2

For the circuit shown in Fig. 2.2, determine current i_x.

FIGURE 2.2
Network for Example 2.2

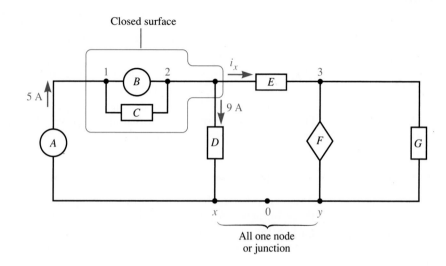

■ SOLUTION

If we select the closed surface as shown, the only unknown in the corresponding KCL equation is current i_x. The sum of the currents leaving the closed surface is

$$-5 + 9 + i_x = 0$$

which gives

$$i_x = -4 \text{ A}$$

We can see that there was no need to know the current through component B or C. These currents lie totally within the selected closed surface. In a similar manner we do not need to know the currents through branches F and G, because these currents lie totally outside the closed surface. ■

We say that a network is *connected* if there is at least one path of branches joining any two specified nodes. The network of Fig. 2.2 is an example of a connected network. If a network contains two or more nodes that are not connected by a path of branches, the network has two or more *separate parts*.

The closed surface shown in Fig. 2.2 cuts the branches containing components A, D, and E. Branches A, D, and E form a *cutset*.* The removal of all branches in a cutset will cut the original *connected network* into two *separate parts* as shown

---

* The concept of a cutset is used in Chapter 4 when we discuss network analysis with the use of the computer program PSpice®.

FIGURE 2.3
The network of Fig. 2.2 after
removal of the branches in
a cutset

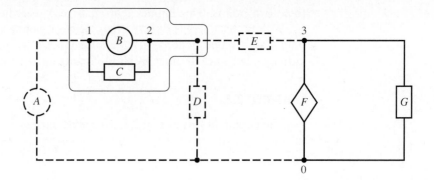

in Fig. 2.3. One part consists of the connected set of nodes and branches left inside the closed surface after removal of the branches in the cutset. The other part consists of the connected set of nodes and branches remaining outside the closed surface after removal of the branches in the cutset. The replacement of any branch in a cutset always reconnects the parts.

REMEMBER

The sum of all currents leaving any closed surface is zero. The currents through those branches that lie totally inside or totally outside a closed surface play no part in the writing of the KCL equation for that surface. We should select our closed surfaces wisely to make the problem as simple as possible.

EXERCISES

Refer to the network shown below.

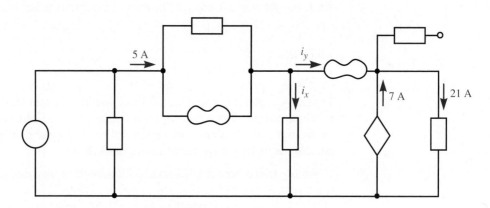

1. How many nodes are in this network?
2. How many junctions are in this network?
3. Determine current i_x.
4. Determine current i_y.

For the network shown on page 23, $i_1 = 3$ A, $i_2 = -4$ A, and $i_3 = 5$ A.

5. How many nodes are in the network?

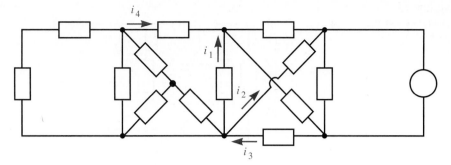

6. How many junctions are in the network?

7. Determine i_4 from a single KCL equation.

2.2 Kirchhoff's Voltage Law

Voltage v_{12} is the work in joules per coulomb *performed* by charge moving along any path from point 1 to point 2. An energy source must supply the same energy per coulomb to move charge from point 2 to point 1 along any path. This statement of the conservation of energy is formalized in Kirchhoff's voltage law (KVL), which is the second and last axiom of circuit theory.

AXIOM

Kirchhoff's Voltage Law

At any instant in time, the algebraic sum of all the voltage drops taken around any closed path is zero.

An equivalent statement of KVL is given by the following equation:

KVL Equation

$$\sum_{k=1}^{N} v_k = 0 \tag{2.2}$$

where v_k is the voltage drop, taken in the direction of the path, along the kth segment of the N segments in the closed path.

When writing a KVL equation, we will use $+v_k$ if the path takes us from the $(+)$ to the $(-)$ reference mark for v_k. If the path is from the $(-)$ to the $(+)$ reference mark, we will use the negative of the value of v_k. A restatement of KVL is *the voltage difference of one point with respect to another is independent of the path.* We will now apply KVL to an example circuit.

EXAMPLE 2.3

Apply KVL to the closed path A indicated in Fig. 2.4. Use this result to determine v_a, if $v_b = 2$ V, $v_c = -3$ V, $v_d = 5$ V, and $v_e = -12$ V.

FIGURE 2.4
Application of KVL to closed
paths

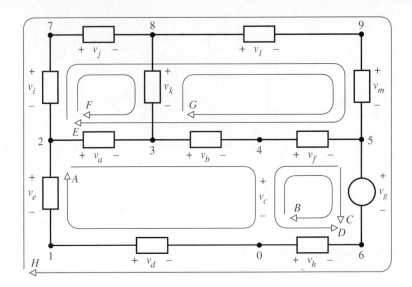

■ SOLUTION

In the circuit of Fig. 2.4 we choose to apply KCL to paths in the clockwise direction. The sum of the voltage drops around closed path A is

$$v_a + v_b + v_c - v_d - v_e = 0$$

from KVL. A minus sign appears on v_d and v_e, because for the chosen direction, the path goes from the $(-)$ to the $(+)$ reference for these two voltages. Substitution of the given voltage values into this KVL equation gives

$$v_a + (2) + (-3) - (5) - (-12) = 0$$

which can be solved to give

$$v_a = -6 \text{ V}$$

This implies that an ideal voltmeter connected as shown by the reference marks for v_a would indicate -6 V. ■

Observe that the *closed path* we chose in the previous example *did not* follow along a single component as we went from node 4 to node 0. The voltage v_c is simply voltage v_{40}.

EXAMPLE 2.4

The voltage $v_c = -3$ V, $v_f = 12$ V, and $v_h = 30$ V are known for the network shown in Fig. 2.4. Determine v_g.

■ SOLUTION

Application of KVL to closed path B gives

$$-v_c + v_f + v_g - v_h = 0$$

Substitution of the numerical values gives

$$-(-3) + (12) + v_g - (30) = 0$$

which yields
$$v_g = 15 \text{ V}$$

We could skip a bit of algebra by using the alternative statement of KVL: The voltage difference between two points is independent of the path. The voltage of node 5 with respect to node 6 along path C must equal the sum of the voltage drops from node 5 to node 6 along path D:

$$v_g = -(12) + (-3) + (30)$$
$$= 15 \text{ V} \blacksquare$$

EXAMPLE 2.5

Apply KVL to closed path E indicated in Fig. 2.4.

■ SOLUTION

The sum of the voltage drops around closed path E is

$$-v_i + v_j + v_l + v_m - v_f - v_b - v_a = 0 \quad \blacksquare$$

Observe that closed path E does not jump across part of the circuit as path A did when it went from node 4 to node 0. A closed path of circuit components that does not pass through the same node twice, such as path E, is called a *loop*. We call loop H, which encloses the entire circuit, the *outer loop*.

EXAMPLE 2.6

Apply KCL to loop F and then around loop G for the circuit of Fig. 2.4. Then add the two equations.

■ SOLUTION

The sum of the voltage drops around loop F is

$$-v_i + v_j + v_k - v_a = 0$$

The sum of the voltage drops around loop G is

$$-v_k + v_l + v_m - v_f - v_b = 0$$

Addition of these two equations gives

$$-v_i + v_j + v_l + v_m - v_f - v_b - v_a = 0$$

which is the same KVL equation we obtained for loop E in Example 2.5. ■

Loop E encloses two other loops, loops F and G. Neither loop F nor loop G encloses another loop. A loop, such as loop F or loop G, that does not enclose another loop, is called a *mesh*. We define meshes only for planar networks. A *planar network* is one that can be drawn so that no lines cross, as in Fig. 2.4.* We observed that addition of the KVL equations for loops F and G yielded the KVL equation for loop E. In general, the KVL equation around a loop will be linearly dependent on the KVL equations for the meshes contained within the loop. (Addition of the KVL equations for the meshes contained in a loop will yield the KVL equation for the loop, if all KVL equations are written clockwise or counterclockwise.)

We applied KVL to closed paths that were not necessarily loops or meshes. We can define a closed path in any way that is convenient. By choosing the path wisely, we can often save considerable effort when analyzing a circuit.

* See Problems 8 and 9 at the end of the chapter for an example of a nonplanar network.

EXAMPLE 2.7

For the circuit shown in Fig. 2.5, determine voltage v_x.

FIGURE 2.5
Circuit for Example 2.7

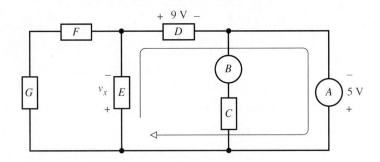

■ SOLUTION

Select a loop as indicated in Fig. 2.5. The sum of the voltages is

$$v_x + 9 - 5 = 0$$

which yields

$$v_x = -4 \text{ V}$$

We can see that there was no need to know the voltage across component B or C. Our selected path did not follow these branches. For the same reason, the voltages across components F and G did not enter into our KVL equation. ■

REMEMBER

The sum of the voltage drops around any closed path is zero. Voltages that are not on a closed path play no part in writing a KVL equation for that path. We should select our closed path wisely to make the problem as simple as possible.

EXERCISES

The network shown below has voltage v_{10} equal to 5 V.

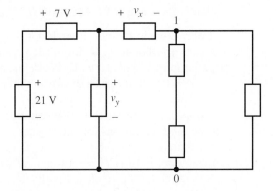

8. How many nodes are in the network?
9. How many junctions are in the network?

10. How many meshes are in the network?

11. How many loops can be found in the network?

12. Determine voltage v_x.

13. Determine voltage v_y.

For the circuit shown below, $v_{10} = 5$ V, $v_{20} = 17$ V, and $v_{30} = 21$ V.

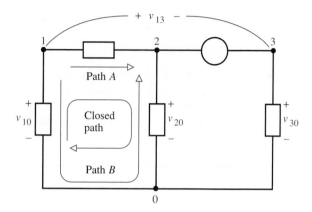

14. Write a KVL equation around the indicated closed path and solve for v_{12}.

15. Use the fact that KVL says that the voltage difference is independent of the path, and equate the voltage difference between node 1 and node 2 along path A to the sum of the differences along path B. This gives v_{12} directly.

16. Use KVL to find v_{13}.

17. For the circuit shown below, $v_{10} = 2$ V, $v_{20} = 4$ V, $v_{30} = 7$ V, $v_{40} = 11$ V, $v_{50} = 16$ V, $v_{60} = 22$ V, and $v_{70} = 29$ V. Find v_x, v_{21}, v_{12}, v_{53}, v_{16}, v_{67}, v_y, and v_{14}.

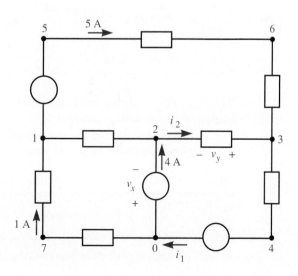

2.3 Independent Sources

An electrical system can perform no useful function without a source of electrical energy. Sources of electrical energy include generators or alternators that convert mechanical energy into electrical energy, but other devices are also used. For example, a flashlight relies on a battery to convert chemical energy into electrical energy, and a solar cell converts light (electromagnetic energy) directly into a conveniently used form of electrical energy. In each case we can model the availability of electrical energy with network components called *sources*.

Voltage Source

Under normal operating conditions a practical energy source, such as the battery in an automobile, can maintain a voltage across its terminals that is relatively independent of the current required by the accessories. We idealize such a device with a network component called an *independent voltage source*, or more simply a *voltage source*.

DEFINITION

An **independent voltage source** is a two-terminal network component with terminal voltage v_{ab} specified by a time function $v(t)$ that is independent of the terminal current i_{ab}.

The network symbol we use to represent an independent voltage source is shown in Fig. 2.6.

COMPONENT SYMBOL

Independent Voltage Source

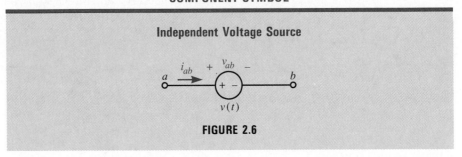

FIGURE 2.6

The $(+)$ and $(-)$ marks inside the circle of the voltage source symbol identify the component as a voltage source.

An equivalent definition of an independent voltage source is given by the following *component equation* or *source equation*:

SOURCE EQUATION

Independent Voltage Source

$$v_{ab} = v(t) \tag{2.3}$$

FIGURE 2.7
Terminal characteristics of
an independent voltage
source

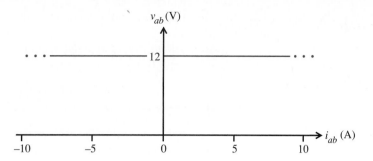

Observe that current i_{ab} has no effect on voltage v_{ab} as given by Eq. (2.3). In Fig. 2.7 we graphically depict the component equation for an independent voltage source with voltage $v_{ab} = 12$ V to emphasize that the terminal voltage is independent of the terminal current. The definition of a voltage source implies that a *short circuit* can be considered to be a voltage source of value zero. Because an independent voltage source represents an idealization of a practical device, such as a battery, many engineers refer to an independent voltage source as an *ideal voltage source*.

Current Source

A few practical devices generate a current that is relatively independent of the terminal voltage over the normal range of operating voltages. A constant-current transformer, once widely used to supply power to incandescent street lights, is a good example. Another example is an automobile alternator, which provides a relatively constant current under certain operating conditions. We idealize such a device with a network component called an *independent current source*, or more simply a *current source*.

DEFINITION

An **independent current source** is a two-terminal network component with terminal current i_{ab} specified by a time function $i(t)$ that is independent of the terminal voltage v_{ab}.

The network symbol we use to represent an independent current source is shown in Fig. 2.8.

COMPONENT SYMBOL

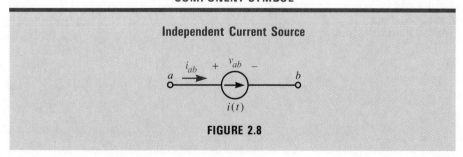

Independent Current Source

FIGURE 2.8

The arrow inside the circle of the current source symbol identifies the component as a current source.

An equivalent definition of an independent current source is given by the following component equation or source equation:

SOURCE EQUATION

Independent Current Source

$$i_{ab} = i(t) \tag{2.4}$$

Observe that voltage v_{ab} has no effect on current i_{ab} as given by Eq. (2.4).

FIGURE 2.9
Terminal characteristics of an independent current source

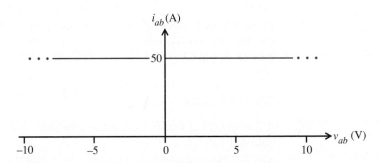

In Fig. 2.9 we graphically depict the component equation of an independent current source with current $i_{ab} = 50$ A to emphasize that the terminal current is independent of the terminal voltage. Engineers frequently refer to an independent current source as an *ideal current source*, because it is an idealization of a practical device. The definition of a current source implies that an *open circuit* can be considered to be a current source of value zero.

We can now begin to form a model for a practical electrical system.

EXAMPLE 2.8

A relatively crude model of an automobile electrical system is shown in Fig. 2.10. The voltage source represents the lead-acid storage battery, and the current source represents the alternator. We have lumped the automobile accessories together in one box, called the load, which requires a current of $i_L = 30$ A. Determine the load voltage v_L, the battery current i_B, and the power absorbed by the battery.

FIGURE 2.10
A crude model of an automobile electrical system

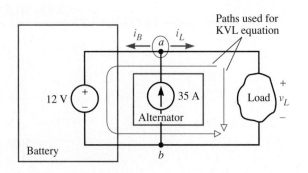

Application of KVL to the paths indicated by arrows gives the load voltage:

$$v_L = 12 \text{ V}$$

We can obtain the battery current by summing the currents leaving the indicated closed surface:

$$i_B - 35 + 30 = 0$$

which gives

$$i_B = 5 \text{ A}$$

The power absorbed by the battery is

$$p = 12i_B = 12(5) = 60 \text{ W}$$

Therefore we are storing chemical energy in the battery that will be returned in the form of electrical energy at a later time. For this reason we say that the battery is being "charged." ■

We will require more types of network components to refine this model.

REMEMBER

The voltage across a voltage source is independent of the current through it, and the current through a current source is independent of the voltage across it.

EXERCISES

Do the connections of components shown in Exercises 18 through 21 satisfy KVL? If so, use KVL to determine v_{ab} in the network.

18.

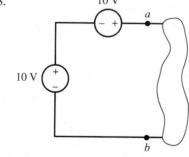

19.

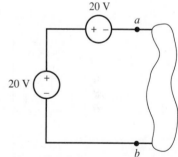

20.

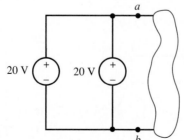

21.

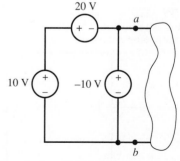

Do the connections of components shown in Exercises 22 through 25 satisfy KCL? If so, use KCL to determine i_{ab} in the network.

22.

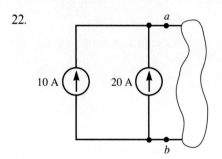

23.

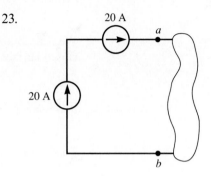

24.

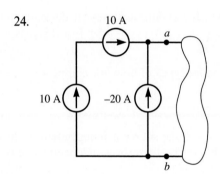

25.

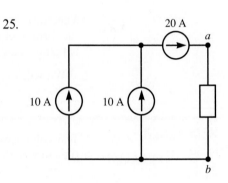

For the networks shown in Exercises 26 through 29, find i_x, v_{ab}, and the power absorbed by the right-hand source.

26.

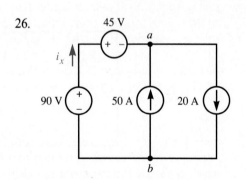

27.

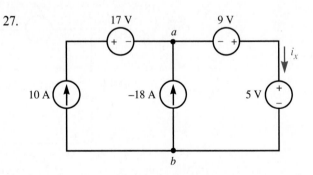

28.

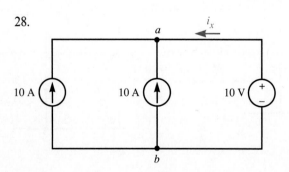

29.

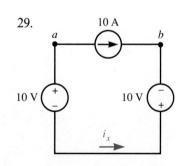

Each interconnection of sources shown in Exercises 30 through 33 violates KCL or KVL and cannot form a network. Determine the contradiction in each case.

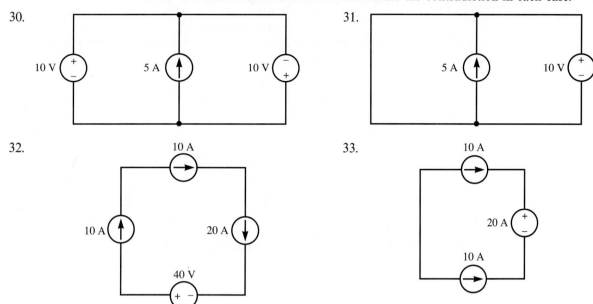

30.

31.

32.

33.

2.4 Dependent Sources

Many electrical systems of interest take a small amount of electric power from one source and use this to control the delivery of a large amount of electric power from another source. A common example is an amplifier used in a tape player. The magnetized tape moving past the tape head, which consists of magnetic material and a small coil of wire, generates a very small voltage signal that delivers a small fraction of a watt to the amplifier. This low-power signal ultimately controls the delivery of tens of watts to the speaker system. The power to the speaker is obtained from the amplifier power source: a car battery or an electrical outlet. We model this ability to control the delivery of power by the introduction of four-terminal network components called *dependent sources* or *controlled sources*.

Voltage-Controlled Voltage Source

Some practical voltage amplifiers generate a large voltage across two terminals that is intended to be proportional to a smaller voltage established across two other terminals. A tape player amplifier is one such device. We idealize a practical voltage amplifier with a four-terminal network component called a *voltage-controlled voltage source* (VCVS).

DEFINITION

A **voltage-controlled voltage source** is a four-terminal network component that establishes a voltage v_{cd} between two points c and d in the circuit that is proportional to a voltage v_{ab} between two points a and b.

We use the network symbol shown in Fig. 2.11 to represent a VCVS.

COMPONENT SYMBOL

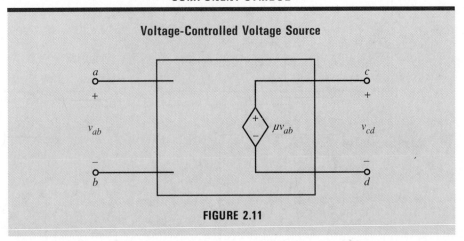

Voltage-Controlled Voltage Source

FIGURE 2.11

The $(+)$ and $(-)$ marks inside the diamond of the component symbol identify the component as a voltage source.

An equivalent definition is given by the following component equation or *control equation*:

CONTROL EQUATION

Voltage-Controlled Voltage Source

$$v_{cd} = \mu v_{ab} \qquad (2.5)$$

Observe that voltage v_{cd} depends only on the constant μ, a dimensionless constant called the *voltage gain*, and the *control voltage* v_{ab}. Current i_{cd} can affect voltage v_{cd} only if it affects voltage v_{ab}. Although we show the VCVS symbol with four terminals, *the control voltage v_{ab} is typically the voltage across some other network component*, as shown in Fig. 2.12.

FIGURE 2.12
Example with a VCVS

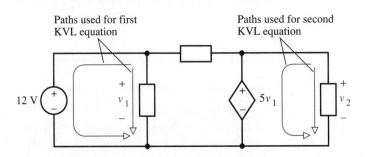

EXAMPLE 2.9

Determine voltage v_2 in the circuit of Fig. 2.12.

■ **SOLUTION**

Kirchhoff's voltage law easily establishes that

$$v_1 = 12 \text{ V}$$

and

$$v_2 = 5v_1 = 5(12) = 60 \text{ V} \ \blacksquare$$

Voltage-Controlled Current Source

Some devices permit the control of a current directly with a voltage. One such electronic device is a field-effect transistor. We idealize such a device with a four-terminal network component called a *voltage-controlled current source* (VCCS).

DEFINITION

A **voltage-controlled current source** is a four-terminal network component that establishes a current i_{cd} in a branch of the circuit that is proportional to the voltage v_{ab} between two points a and b.

We use the network symbol shown in Fig. 2.13 to represent a VCCS.

COMPONENT SYMBOL

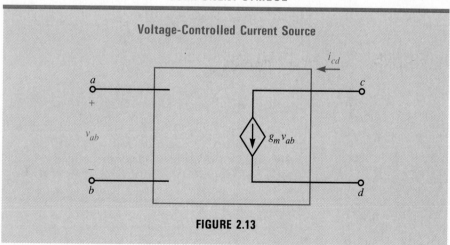

Voltage-Controlled Current Source

FIGURE 2.13

The arrow inside the diamond of the component symbol identifies the component as a current source.

An equivalent definition is given by the following component equation or control equation:

$$i_{cd} = g_m v_{ab} \qquad\qquad (2.6)$$

Observe that current i_{cd} depends only on the control voltage v_{ab} and the constant g_m, called the *transconductance* or mutual conductance. Constant g_m has dimensions of ampere per volt, or siemens (S).

FIGURE 2.14
Example with a VCCS

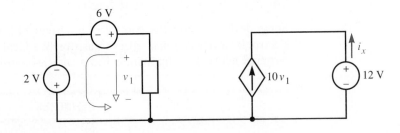

EXAMPLE 2.10

Determine current i_x in the circuit of Fig. 2.14.

■ **SOLUTION**

We can use KVL to show that

$$v_1 = 6 - 2 = 4 \text{ V}$$

and KCL to determine that

$$i_x = -10v_1 = -10(4) = -40 \text{ A} \quad \blacksquare$$

Current-Controlled Voltage Source

Electronic devices that generate a voltage that is almost proportional to a current can be constructed. We idealize such a device with a four-terminal network component called a *current-controlled voltage source* (CCVS).

DEFINITION

A **current-controlled voltage source** is a four-terminal network component that establishes a voltage v_{cd} between two points c and d in the circuit that is proportional to current i_{ab} in some branch of the circuit.

We use the network symbol shown in Fig. 2.15 to represent a CCVS.

COMPONENT SYMBOL

Current-Controlled Voltage Source

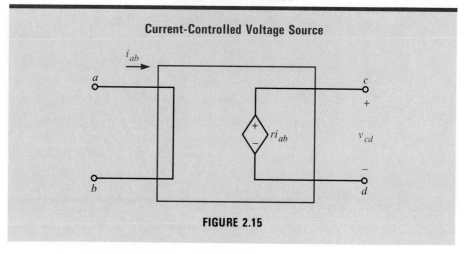

FIGURE 2.15

The $(+)$ and $(-)$ marks inside the diamond of the component symbol identify the component as a voltage source.

An equivalent definition is given by the following component equation or control equation:

CONTROL EQUATION

Current-Controlled Voltage Source

$$v_{cd} = ri_{ab} \tag{2.7}$$

Observe that voltage v_{cd} depends only on the *control current* i_{ab} and the constant r, called the *transresistance* or mutual resistance. Constant r has the dimension of volt per ampere, or ohm (Ω). Although we show the CCVS symbol with four terminals, terminals a and b and the lines joining them are not usually shown explicitly. *The control current i_{ab} is typically the current through some other network component,* as shown in Fig. 2.16.

FIGURE 2.16
Example with a CCVS

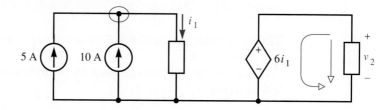

EXAMPLE 2.11

Determine voltage v_2 for the network shown in Fig. 2.16.

■ SOLUTION

We can use KCL to establish that

$$i_1 = 15 \text{ A}$$

and KVL gives us $\qquad v_2 = 6i_1 = 6(15) = 90 \text{ V} \quad ■$

Current-Controlled Current Source

Some devices permit the control of one current with another. An example is the electronic device called a bipolar junction transistor. We idealize a practical current amplifier with a four-terminal network component called a *current-controlled current source* (CCCS).

DEFINITION

A **current-controlled current source** is a four-terminal network component that establishes a current i_{cd} in one branch of a circuit that is proportional to current i_{ab} in some branch of the network.

We use the network symbol shown in Fig. 2.17 to represent a CCCS.

COMPONENT SYMBOL

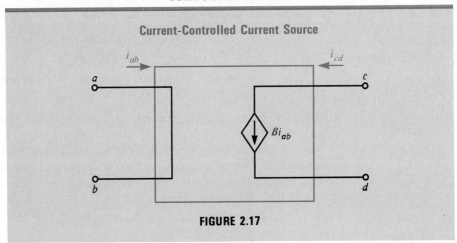

Current-Controlled Current Source

FIGURE 2.17

The arrow inside the diamond of the component equation identifies the component as a current source.

An equivalent definition is given by the following component equation or control equation:

CONTROL EQUATION

Current-Controlled Current Source

$$i_{cd} = \beta i_{ab} \qquad (2.8)$$

Observe that current i_{cd} depends only on the control current i_{ab} and the dimension-less constant β, called the *current gain*.

EXAMPLE 2.12

Determine current i_2 for the circuit shown in Fig. 2.18.

FIGURE 2.18
Example with a CCCS

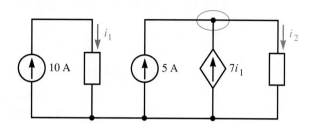

■ **SOLUTION**

We can obtain the solution from two KCL equations,

$$i_1 = 10 \text{ A}$$

and
$$i_2 = 5 + 7i_1 = 5 + 7(10) = 75 \text{ A} \quad ■$$

We conclude with a more complicated example.

We say that an open circuit exists between the two terminals a and b in Fig. 2.19, because the current i_{ab} must be zero, regardless of the value of v_{ab}. As we noted earlier, an open circuit can be thought of as a branch with a current source of value zero.

FIGURE 2.19
Dependent source example

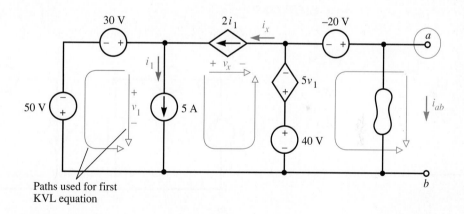

Paths used for first
KVL equation

EXAMPLE 2.13

For the network of Fig. 2.19, determine i_1, v_1, i_x, v_x, and v_{ab}, if no other compo-nents are connected to terminals a and b.

■ **SOLUTION**

The independent current source gives

$$i_1 = 5 \text{ A}$$

(This can be considered to be a very simple KCL equation.)

Application of KVL to the paths indicated on the left-hand mesh yields

$$v_1 = 30 - 50 = -20 \text{ V}$$

The control equation gives

$$i_x = 2i_1 = 2(5) = 10 \text{ A}$$

A KVL equation for the paths shown on the center mesh gives

$$v_x = v_1 - 40 + 5v_1$$
$$= -20 - 40 - 100 = -160 \text{ V}$$

Application of KVL to the paths indicated on the right gives

$$v_{ab} = +(-20) - 5v_1 + 40$$
$$= -20 + 100 + 40 = 120 \text{ V} \blacksquare$$

A network component that can supply positive average power (absorb negative energy) is an *active component*. Independent and dependent sources are active components. In the following sections we will introduce four components that cannot supply positive average power (the absorbed energy is never negative). These are *passive components*. A network that contains one or more active components is an *active network*. A *passive network* contains only passive components.

In addition to providing sources of electrical energy, physical devices, unavoidably or by design, convert electrical energy into heat or store energy in electric or magnetic fields. We model these properties of devices with network components called *elements*. In the next section, we will consider the conversion of electrical energy into heat.

EXERCISES

For the networks in Exercises 34 through 37, determine v_{ab} and i_y.

34.

35.

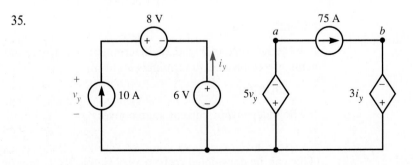

36.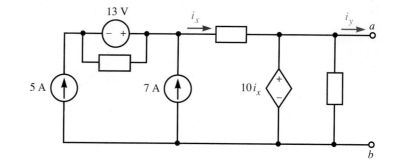

37.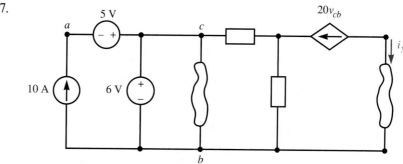

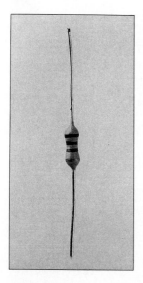

Resistor.
Photograph by
James Scherer

2.5   Resistance

We are all familiar with devices, such as an electric furnace, that are intentionally designed to convert electrical energy into heat. Other devices, such as a television receiver, unavoidably convert some fraction of the electrical energy they absorb into heat. We call a device that converts electrical energy into heat a *dissipative device*, because it dissipates electrical energy. Even a simple piece of wire exhibits this characteristic. Perhaps you have noticed that the cord to an electric toaster was warm when you unplugged it. Electric charge moving through a material, such as metal, causes the molecules of the material to vibrate and thus converts electrical energy into heat.

We call a physical device designed to convert electrical energy into heat a *resistor*. If we were to experimentally measure the relationship between the voltage across and the current through a physical resistor, we might obtain a characteristic such as that given by curve *a* or curve *b* in Fig. 2.20. In general the voltage is a function of current:

$$v_{ab} = f(i_{ab}) \qquad (2.9)$$

A resistor constructed of a semiconducting material, such as silicon carbide, could have a characteristic such as curve *b* of Fig. 2.20. We say that the device exhibits *nonlinear resistance* and refer to the device as a *nonlinear resistor*.

Experiments show that for a metallic conductor at constant temperature, the voltage across the conductor is directly proportional to the current through it. This proportional relationship is referred to as *Ohm's law*. A resistor constructed of metal would obey Ohm's law and would have a characteristic that follows a straight line, as given by curve *a* of Fig. 2.20. We say that the device exhibits *linear resistance*

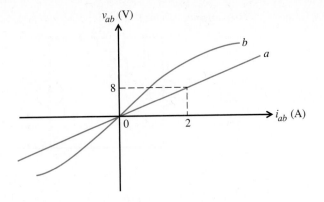

and is a *linear resistor*. A linear resistor is the basis for the network component that we call a *resistance* or *ideal resistor*.*

DEFINITION

A **resistance** is a two-terminal network component with terminal voltage v_{ab} directly proportional to current i_{ab}. The constant of proportionality is also called resistance.

The network symbol we use to represent a resistance is shown in Fig. 2.21.

COMPONENT SYMBOL

Resistance

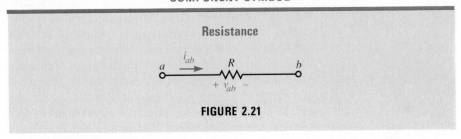

FIGURE 2.21

An equivalent definition of resistance is given by the following component equation or *terminal equation*:

TERMINAL EQUATION

Resistance

$$v_{ab} = Ri_{ab} \qquad (2.10)$$

Resistance R has the dimension of volt per ampere, or *ohm* (Ω). Observe that Eq. (2.10) *uses the passive sign convention*. That is, the current arrow for $i = i_{ab}$ enters the end of the resistance with the ($+$) reference mark for $v = v_{ab}$. With the

---

* Some texts refer to both the physical device and the network component as a resistor and use the word resistance to denote only the constant of proportionality R.

active sign convention, the current arrow for $i = i_{ba}$ enters the end of the resistance with the $(-)$ reference mark for $v = v_{ab}$ and

$$v_{ab} = -Ri_{ba} \qquad (2.11)$$

We often find it convenient to work with the reciprocal of resistance, which is called *conductance*.

DEFINITION

Conductance

$$i_{ab} = Gv_{ab} \qquad (2.12)$$

where

$$G = \frac{1}{R}$$

Conductance G has the dimension of ampere per volt or *siemens* (S). [An archaic term for siemens is the *mho* (℧), which is ohm spelled backward.] We refer to Eqs. (2.10) and (2.12) as Ohm's law equations.

EXAMPLE 2.14

Determine the resistance of the resistor with terminal characteristics given by curve *a* of Fig. 2.20.

■ **SOLUTION**

The slope of line *a* in Fig. 2.20 is

$$R = \frac{8}{2} = 4\ \Omega$$

Within the measured current range, we can describe the device by the equation

$$v_{ab} = 4i_{ab}$$

and we would model the device by a 4-Ω resistance. ■

We can easily calculate the power absorbed by a resistance if the component voltage and current are known.

$$p = v_{ab}i_{ab} \qquad (2.13)$$

where the voltage and current references satisfy the passive sign convention. Substitution from Ohm's law gives

$$p = (Ri_{ab})i_{ab} = Ri_{ab}^2 \qquad (2.14)$$

or

$$p = (v_{ab})\frac{1}{R}v_{ab} = \frac{1}{R}v_{ab}^2 \qquad (2.15)$$

We can see from either of these two equations that the power absorbed by a resistance is never negative:

$$p \geq 0 \qquad (2.16)$$

Therefore resistance is a passive component.

We show an example of the relation between the voltage across, the current through, and the power and energy absorbed by a resistance in Fig. 2.22. Note that the voltage and current waveforms have exactly the same shape, and that the power and energy absorbed are never negative.

FIGURE 2.22
An example of the relation between current, voltage, power, and energy absorbed by a resistance.
(a) Resistance terminal equation; (b) current; (c) voltage; (d) power; (e) energy

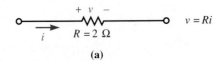

(a)

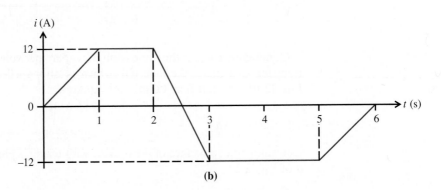

(b)

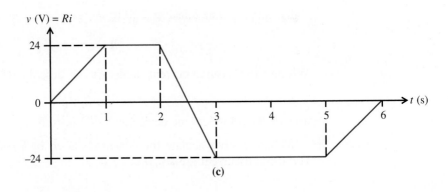

(c)

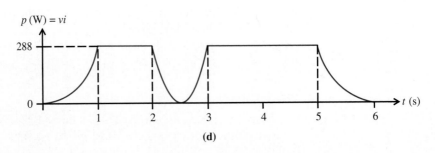

(d)

FIGURE 2.22
Continued

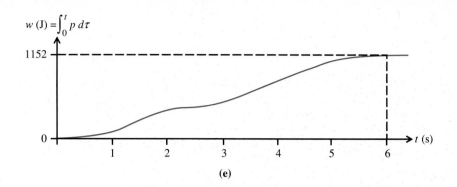

$$w\ (\text{J}) = \int_0^t p\, d\tau$$

(e)

EXAMPLE 2.15

We can improve the model we used for a storage battery in Example 2.8 by including a resistance in series with the voltage source, as shown in Fig. 2.23. This resistance is a model for the *internal resistance* of the battery. Assume that the load current remains unchanged. Determine the load voltage v_{ab}.

FIGURE 2.23
An improved model of an automobile electrical system

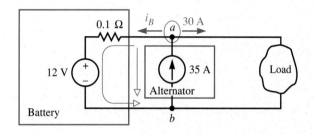

■ SOLUTION

We obtain current i_B by the use of KCL:

$$i_B - 35 + 30 = 0$$

which gives

$$i_B = 5\ \text{A}$$

We can now obtain voltage v_{ab} by the use of KVL and Ohm's law:

$$v_{ab} = 0.1i_B + 12 = 0.1(5) + 12 = 12.5\ \text{V}$$

The alternator is supplying more current than is required by the load, so the battery terminal voltage is greater than 12 V. ■

REMEMBER

Resistance is the network model that accounts for the conversion of electrical energy into heat and for voltage that is proportional to current.

EXERCISES

38. The volt-ampere measurements of a device are shown in the following figure. What value of resistance would have this terminal equation?

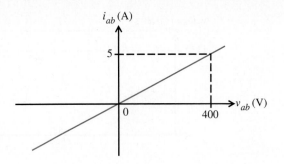

39. The terminal characteristics of a *practical voltage source* were measured and found to be as shown below. Determine a series connection of a resistance and an independent voltage source that would have the same terminal characteristics.

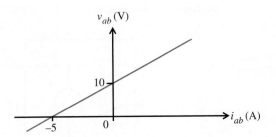

For each of the circuits shown in Exercises 40 through 48, find v_{ab}, i_{ab}, and the power absorbed by the resistance.

40.

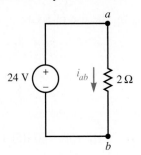

41.

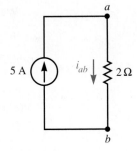

42.

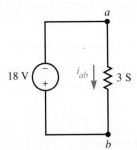

43.

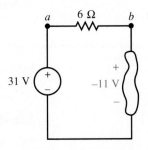

44.

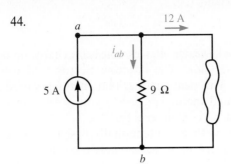

45.

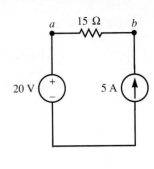

46.

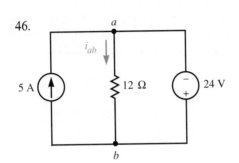

47.

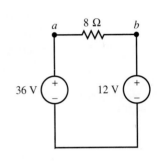

48.

49. Determine current i_2 in the following circuit.

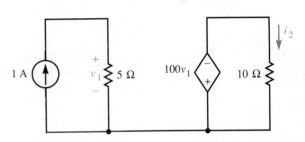

50. Determine voltage v_2 in the following circuit.

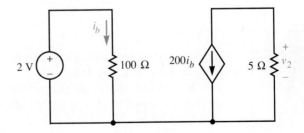

51. The alternator modeled as the current source in Example 2.15 is no longer charging, so the current source becomes zero (an open circuit). What is the battery terminal voltage v_{ab}?

2.6 Capacitance

Removal of a wool sweater often establishes a charge on the sweater and an opposite charge on our shirt. The resulting electric field will pull the sweater back toward the shirt. This demonstrates that energy is stored in the electric field. The electric field around the power line to your school also stores electrical energy. Electric fields always contain energy.

A physical device that is intentionally designed to store energy in an electric field is called a *capacitor*, or occasionally a condenser. We make extensive use of these devices in electronic and power systems. For instance, a radio tuner uses a capacitor of adjustable value. A capacitor is physically made from two conducting plates (often aluminum foil) separated by a dielectric (nonconducting or insulating) medium, such as plastic, waxed paper, or even air. The physical device is depicted as connected to an electrical system in Fig. 2.24.

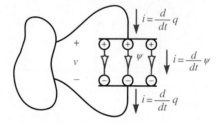

Charge of equal magnitude but opposite sign accumulates on the opposing capacitor plates. When the voltage v is not zero, the resulting electric field causes an *electric flux* ψ to exist in the region between the two plates. The total electric flux ψ will equal the excess charge q on the $(+)$ plate. As long as ψ is constant, i will be zero, but as v changes, ψ will change and i will be nonzero.

Conduction current

$$i = \frac{d}{dt} q \qquad (2.17)$$

in the wire is the result of moving charge. Although no charge passes through the insulating region between the two plates, the changing electric flux causes a *displacement current*

$$i = \frac{d}{dt} \psi \qquad (2.18)$$

in the dielectric region. This displacement current precisely equals the conduction current, so KCL is not violated.

The ratio of flux to voltage is approximately constant for most dielectrics. Because flux ψ is equal to charge q, the charge-to-voltage ratio is also a constant C:

$$C = \frac{q}{v} \qquad (2.19)$$

This lets us write current i in terms of voltage v:

$$i = \frac{d}{dt} q = \frac{d}{dt}\left[\left(\frac{q}{v}\right)v\right]$$

Ceramic capacitor.
Photograph by
James Scherer

$$= \frac{d}{dt}[Cv]$$

$$= C\frac{d}{dt}v \qquad (2.20)$$

The constant of proportionality C is the *capacitance* of the plates.

We model energy storage in an electric field with a network component called a *capacitance* or *ideal capacitor*.* All currents induced by changing voltage are assumed to flow through such components.

DEFINITION

A **capacitance** is a two-terminal network component with current i_{ab} directly proportional to the time derivative of the voltage v_{ab}. The constant of proportionality C is also called the capacitance.

The network symbol we use to represent capacitance is shown in Fig. 2.25.

COMPONENT SYMBOL

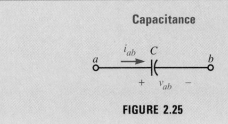

Capacitance

FIGURE 2.25

An equivalent definition of capacitance is given by the following component equation or terminal equation:

TERMINAL EQUATION

Capacitance

$$i_{ab} = C\frac{d}{dt}v_{ab} \qquad (2.21)$$

Capacitance C has the dimension of coulomb per volt, or farad (F). Observe that Eq. (2.21) implies the passive sign convention. With the *active sign convention*,

$$i_{ab} = -C\frac{d}{dt}v_{ba} \qquad (2.22)$$

Electrolytic capacitor.
Photograph by
James Scherer

Variable capacitor.
Photograph by
James Scherer

* Some texts refer to both the practical device and the network component as a capacitor and use the word capacitance to denote only the constant of proportionality C.

EXAMPLE 2.16

Voltages and currents are assigned according to the passive sign convention. The voltage across a 3-μF capacitance is

$$v(t) = 24e^{40t} \text{ V}$$

Calculate the capacitance current i.

■ **SOLUTION**

$$i = C\frac{d}{dt}v = 3 \times 10^{-6}\frac{d}{dt}(24e^{40t})$$

$$= 0.00288e^{40t} \text{ A} \quad ■$$

Integration of Eq. (2.21) with respect to t gives*

$$\int_{v_{ab}(-\infty)}^{v_{ab}(t)} dv_{ab} = \frac{1}{C}\int_{-\infty}^{t} i_{ab}\,d\tau \tag{2.23}$$

Completion of the integration on the left-hand side gives us

$$v_{ab}(t) - v_{ab}(-\infty) = \frac{1}{C}\int_{-\infty}^{t} i_{ab}\,d\tau \tag{2.24}$$

With the assumption that the capacitance voltage is zero at time minus infinity,

$$v_{ab}(t) = \frac{1}{C}\int_{-\infty}^{t} i_{ab}\,d\tau \tag{2.25}$$

Equation (2.25) can be written as

$$v_{ab}(t) = \frac{1}{C}\int_{-\infty}^{t_0} i_{ab}\,d\tau + \frac{1}{C}\int_{t_0}^{t} i_{ab}\,d\tau \tag{2.26}$$

This equation can be written in the form

$$v_{ab}(t) = v_{ab}(t_0) + \frac{1}{C}\int_{t_0}^{t} i_{ab}\,d\tau \tag{2.27}$$

We often call the voltage

$$v_{ab}(t_0) = \frac{1}{C}\int_{-\infty}^{t_0} i_{ab}\,d\tau \tag{2.28}$$

the *initial condition* at time t_0.

EXAMPLE 2.17

Voltages and currents are assigned according to the passive sign convention. The voltage across a 3-μF capacitance at $t = 0$ is $v(0) = 4$ V. The current is

$$i(t) = 24e^{-40t} \text{ mA}$$

for $t \geq 0$. Determine $v(t)$ for $t \geq 0$.

---

* The variable of integration is changed from t to τ because t appears in the upper limit.

$$v(t) = v(0) + \frac{1}{C} \int_0^t i \, d\tau$$

$$= 4 + \frac{1}{3 \times 10^{-6}} \int_0^t 24 \times 10^{-3} \, e^{-40\tau} \, d\tau$$

$$= 204 - 200e^{-40t} \text{ V } \blacksquare$$

A capacitance does not dissipate electrical energy in the form of heat as does a resistance. It *stores energy* in an electric field as the magnitude of the voltage increases. It returns this energy to the rest of the circuit as the magnitude of the voltage decreases. The stored energy in joules (J) is

$$w(t) = \int_{-\infty}^t p(\tau) \, d\tau = \int_{-\infty}^t v(\tau) i(\tau) \, d\tau$$

$$= \int_{-\infty}^t v(\tau) C \left[\frac{d}{d\tau} v(\tau) \right] d\tau$$

$$= \int_{v(-\infty)}^{v(t)} Cv(\tau) \, dv(\tau) = \frac{1}{2} \, Cv^2(\tau) \Big|_{v(-\infty)=0}^{v(t)} \qquad (2.29)$$

which gives

$$w(t) = \frac{1}{2} \, Cv^2(t) \qquad (2.30)$$

Because C is greater than zero, it follows that

$$w \geq 0 \qquad (2.31)$$

Therefore capacitance is a passive component.

An example of the relation between the current through, the voltage across, the power supplied to, and the energy stored in a capacitance is shown in Fig. 2.26. (This figure assumes that voltage v equals zero when t equals zero.) Note that the voltage and current waveforms for the capacitance do not have the same shape. The instantaneous power supplied to the capacitance can be either positive or negative. Power supplied is negative when the capacitance returns stored energy to the rest of the circuit. The energy stored can never be negative.

FIGURE 2.26
An example of the relation between current, voltage, power supplied to, and energy stored by capacitance. (*a*) Capacitance terminal equation; (*b*) current; (*c*) voltage; (*d*) power; (*e*) energy

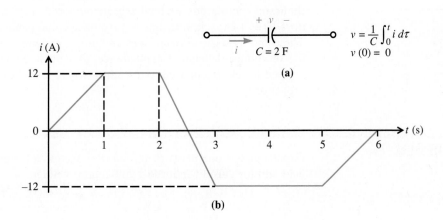

FIGURE 2.26
Continued

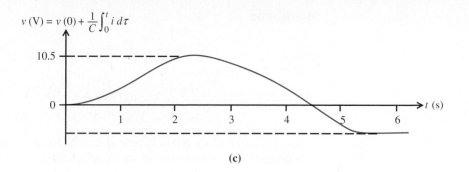

$$v\,(\text{V}) = v\,(0) + \frac{1}{C}\int_0^t i\,d\tau$$

(c)

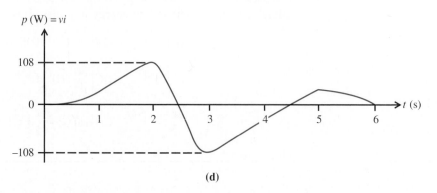

$$p\,(\text{W}) = vi$$

(d)

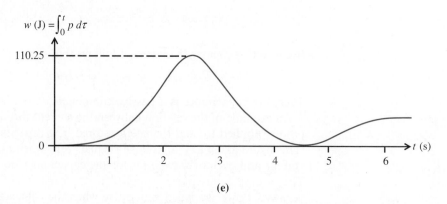

$$w\,(\text{J}) = \int_0^t p\,d\tau$$

(e)

Capacitors with values of capacitance between a few picofarads and a few thousand microfarads are regularly encountered in electronic systems. Most capacitors have a capacitance that is very nearly independent of voltage and current. These find wide use in electronics and power systems. Some practical devices make use of voltage-dependent capacitance. One such device, the varactor diode, can be used for electronic tuning in radios, where it replaces a mechanically adjusted capacitor.

REMEMBER

Capacitance is our network model that accounts for energy stored in an electric field and for current induced by changing voltage.

Assume the passive sign convention for the relation between v and i. If $C = 0.01$ F and $v(t) = 100 \cos 2\pi 60t$ V, determine and sketch the following quantities. (The quantity $2\pi 60t$ is in radians.)

52. $i(t)$ 53. $p(t)$

54. $w(t)$

For the network shown below, $v_a = 24e^{-3t}$ V and $v_b = 2e^{-3t}$ V
Determine the following quantities for $t > 0$.

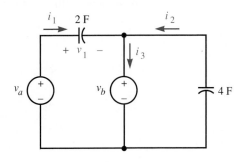

55. v_1 56. i_1

57. i_2 58. i_3

59. Integrate both sides of Eq. (2.22) to show that for the active sign convention,

$$v_{ab}(t) = v_{ab}(t_0) - \frac{1}{C} \int_{t_0}^{t} i_{ba} \, d\tau$$

2.7   Inductance

When we pick up a paper clip with a permanent magnet, we use the energy stored in the magnetic field to lift the paper clip. Current flowing through a wire also creates a magnetic field, and therefore magnetic flux, ϕ, around the wire. Magnetic fields always contain energy.

    A physical device designed to store energy in a magnetic field is called an *inductor*, or occasionally a coil. We make extensive use of these devices in such diverse equipment as transformers, radios, and radar. An automobile ignition coil, for example, uses an inductor. We can physically make an inductor by winding a coil of

FIGURE 2.27
An *N*-turn inductor or coil

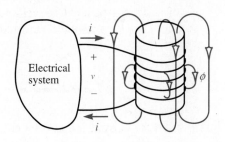

Inductor.
Photograph by James Scherer

wire around some supporting structure or form, such as plastic or iron. The physical device is depicted as connected to an electrical system in Fig. 2.27. Current i causes magnetic flux to link (pass through) each turn of the coil. As indicated in the figure, not all flux will link every turn. With resistance neglected, Faraday's law gives

$$v = N \frac{d}{dt} \phi$$

where ϕ is the average magnetic flux linking a turn and N is the number of turns. For coil forms constructed of linear magnetic materials, such as plastic, the ratio of current i to magnetic flux ϕ is constant. We define the constant L:

$$L = \frac{N\phi}{i} \tag{2.32}$$

This permits us to write voltage v in terms of current i:

$$v = N \frac{d}{dt} \phi = \frac{d}{dt} \left[\frac{N\phi}{i} i \right]$$

$$= \frac{d}{dt} (Li)$$

$$= L \frac{d}{dt} i \tag{2.33}$$

The constant of proportionality L is the *inductance* of the coil.

We model energy storage in a magnetic field with a network component called an *inductance* or *ideal inductor*.* All voltages created by a changing current are assumed to appear across such components.

DEFINITION

An **inductance** is a two-terminal network component with voltage v_{ab} proportional to the time derivative of the current i_{ab}. The constant of proportionality L is also called the inductance.

The network symbol we use to represent inductance is shown in Fig. 2.28.

COMPONENT SYMBOL

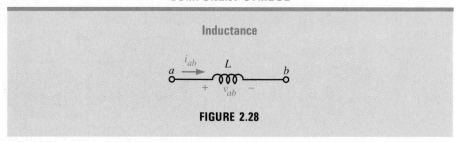

FIGURE 2.28

---

* Some texts refer to both the practical device and the network component as an inductor and use the word inductance to denote only the constant of proportionality L.

An equivalent definition of inductance is given by the following equation:

TERMINAL EQUATION

Inductance

$$v_{ab} = L \frac{d}{dt} i_{ab} \qquad (2.34)$$

Inductance L has the dimension of volt-second per ampere, or henry (H). Observe that Eq. (2.34) implies the passive sign convention. With the active sign convention,

$$v_{ab} = -L \frac{d}{dt} i_{ba} \qquad (2.35)$$

EXAMPLE 2.18

Voltages and currents are assigned according to the passive sign convention. The current through a 3-mH inductance is

$$i(t) = 24 \cos 40t \text{ A}$$

Calculate the inductance voltage v.

■ **SOLUTION**

$$v = L \frac{d}{dt} i = 3 \times 10^{-3} \frac{d}{dt} 24 \cos 40t$$

$$= -2.88 \sin 40t \text{ V} \quad ■$$

Integration of Eq. (2.34) with respect to t gives

$$\int_{i_{ab}(-\infty)}^{i_{ab}(t)} di_{ab} = \frac{1}{L} \int_{-\infty}^{t} v_{ab} \, d\tau \qquad (2.36)$$

Completion of the integration on the left-hand side gives us

$$i_{ab}(t) - i_{ab}(-\infty) = \frac{1}{L} \int_{-\infty}^{t} v_{ab} \, d\tau \qquad (2.37)$$

With the assumption that inductance current is zero at time minus infinity,

$$i_{ab}(t) = \frac{1}{L} \int_{-\infty}^{t} v_{ab} \, d\tau \qquad (2.38)$$

Equation (2.38) can be written as

$$i_{ab}(t) = \frac{1}{L} \int_{-\infty}^{t_0} v_{ab} \, d\tau + \frac{1}{L} \int_{t_0}^{t} v_{ab} \, d\tau \qquad (2.39)$$

This equation can be written in the form

$$i_{ab}(t) = i_{ab}(t_0) + \frac{1}{L} \int_{t_0}^{t} v_{ab} \, d\tau \qquad (2.40)$$

We often call the current

$$i_{ab}(t_0) = \frac{1}{L} \int_{-\infty}^{t_0} v_{ab} \, d\tau \qquad (2.41)$$

the *initial condition* at time t_0.

EXAMPLE 2.19

Voltages and currents are assigned according to the passive sign convention. The current through a 3-mH inductance at $t = 0$ is $i(0) = 4$ A. The inductance voltage is

$$v(t) = 24 \cos 4000t \text{ V} \qquad \text{for } t \geq 0$$

Determine $i(t)$ for $t \geq 0$.

■ **SOLUTION**

$$i(t) = i(0) + \frac{1}{L} \int_{0}^{t} v \, d\tau$$

$$= 4 + \frac{1}{3 \times 10^{-3}} \int_{0}^{t} 24 \cos 4000\tau \, d\tau$$

$$= 4 + 2 \sin 4000t \text{ A} \qquad \text{for } t \geq 0 \quad ■$$

As contrasted with capacitance, which accounts for energy stored in an electric field, inductance is our network model that accounts for energy stored in a magnetic field. An inductance does not dissipate electrical energy in the form of heat as does a resistance. It only stores energy in a magnetic field as the magnitude of the current increases. It returns this energy to the rest of the circuit as the magnitude of the current decreases. The energy stored is

$$w(t) = \int_{-\infty}^{t} p(\tau) \, d\tau = \int_{-\infty}^{t} i(\tau)v(\tau) \, d\tau$$

$$= \int_{-\infty}^{t} i(\tau)L\left[\frac{d}{d\tau} i(\tau)\right] d\tau$$

$$= \int_{i(-\infty)}^{i(t)} Li(\tau) \, di(\tau) = \frac{1}{2} Li^2(\tau)\Big|_{i(-\infty)=0}^{i(t)} \qquad (2.42)$$

which gives

$$w(t) = \frac{1}{2} Li^2(t) \qquad (2.43)$$

Because inductance L is greater than zero, it follows that

$$w \geq 0 \qquad (2.44)$$

Therefore inductance is a passive component.

FIGURE 2.29
An example of the relation
between current, voltage,
power supplied to, and
energy stored by inductance.
(*a*) Inductance terminal
equation; (*b*) current;
(*c*) voltage; (*d*) power;
(*e*) energy

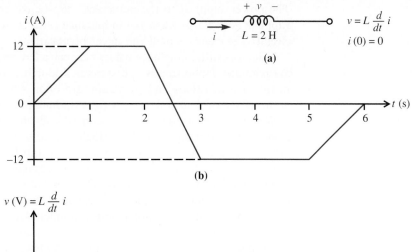

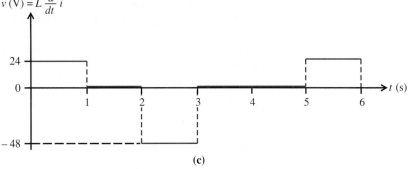

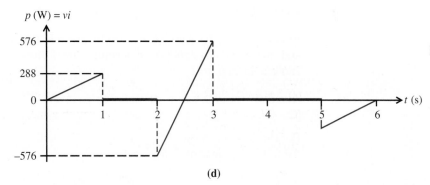

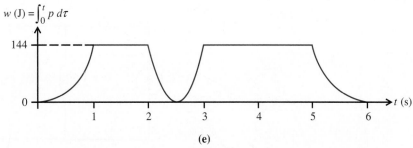

An example of the relation between the voltage across, the current through, the power supplied to, and the energy stored in an inductance is shown in Fig. 2.29. (This figure assumes that current *i* equals zero for *t* equal to zero.) Note that the voltage and current waveforms for an inductance do not have the same shape.

The power supplied to the inductance can be either positive or negative. Power supplied is negative when the inductance returns stored energy to the rest of the circuit. The energy stored can never be negative.

Inductors with inductance values between a few microhenries and several henries are regularly found in physical systems. Inductors used in communication equipment, such as radios and television, are often wound on plastic forms. We can accurately model such devices with an inductance that is independent of current.

We cannot accurately model a coil wound on ferromagnetic material, such as iron, by a fixed inductance. Nevertheless, a constant inductance model often provides a useful approximation for iron-cored transformers and some other devices.

Examples of devices that use nonlinear magnetic material are the iron-core transformer and the electric motor. Devices that intentionally exploit the nonlinear magnetization of ferromagnetic material are the magnetic core memory of a computer and a device called a magnetic amplifier. Although they were once common, magnetic core memories and magnetic amplifiers are seldom encountered in modern equipment, having been replaced by less expensive semiconductor devices such as transistors and thyristors.

REMEMBER

Inductance is our network model that accounts for energy stored in a magnetic field and for voltage induced by a changing current.

EXERCISES

Assume the passive sign convention for the relation between v and i. If $L = 10$ mH and $i(t) = 100 \cos 2\pi60t$ A, determine the following quantities. (The quantity $2\pi60t$ is in radians.)

60. $v(t)$ 61. $p(t)$

62. $w(t)$

For the network shown below, $i_a = 24e^{-3t}$ A and $i_b = 2e^{-3t}$ A for $t \geq 0$. Determine the following quantities for $t > 0$.

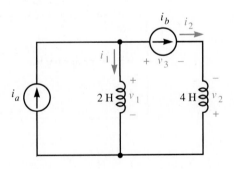

63. i_1 64. i_2

65. v_1 66. v_2

67. Integrate both sides of Eq. (2.35) to show that for the active sign convention,

$$i_{ab}(t) = i_{ab}(t_0) - \frac{1}{L} \int_{t_0}^{t} v_{ba}\, d\tau$$

2.8 Mutual Inductance*

Faraday established experimentally that the changing magnetic flux due to a changing current in one conductor induces a voltage in another conductor linked by the flux. Such conductors are said to be *magnetically coupled*. This is an undesirable effect when power lines induce a voltage in adjacent telephone lines, but magnetic coupling can also be beneficial. Magnetically coupled inductors (coils) are widely used in electronics and power systems. They are the basis for *transformers*. We use transformers to convert from high voltage and low current to low voltage and high current, or back.

Coils of wire placed in close proximity form coupled inductors. We usually place one coil inside the other, so that a large fraction of the magnetic flux that links (is enclosed by) one coil will also link the other.

In Fig. 2.30 we have drawn the coils side by side to permit a clearer picture. A careful look at this figure is important. Observe that we have assigned a dot mark (•) to the end of the coils at terminals a and c. This is because current into

FIGURE 2.30
A simplified model for coupled coils. The dot marks (•) by terminals a and c of coils 1 and 2 indicate that currents into terminals a and c cause magnetic flux to pass through the two coils in the same direction

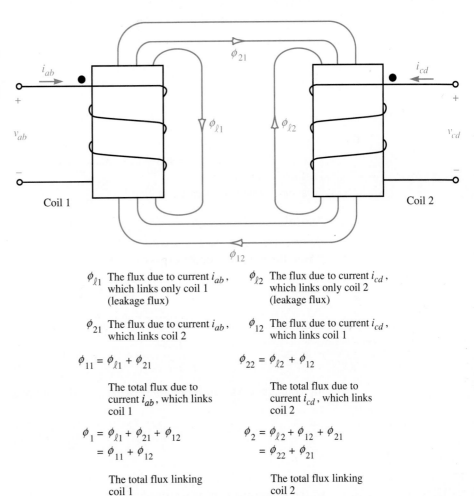

$\phi_{\ell 1}$ The flux due to current i_{ab}, which links only coil 1 (leakage flux)

$\phi_{\ell 2}$ The flux due to current i_{cd}, which links only coil 2 (leakage flux)

ϕ_{21} The flux due to current i_{ab}, which links coil 2

ϕ_{12} The flux due to current i_{cd}, which links coil 1

$\phi_{11} = \phi_{\ell 1} + \phi_{21}$

   The total flux due to current i_{ab}, which links coil 1

$\phi_{22} = \phi_{\ell 2} + \phi_{12}$

   The total flux due to current i_{cd}, which links coil 2

$\phi_1 = \phi_{\ell 1} + \phi_{21} + \phi_{12}$
$\quad = \phi_{11} + \phi_{12}$

   The total flux linking coil 1

$\phi_2 = \phi_{\ell 2} + \phi_{12} + \phi_{21}$
$\quad = \phi_{22} + \phi_{21}$

   The total flux linking coil 2

* This section can be postponed until needed in Chapter 4. If the mutual inductance topics in Chapter 4 are omitted, this material may be postponed until Chapter 18.

the dot-marked end of each coil causes magnetic flux to pass through the coils in the same direction. Not all flux linking (passing through) one turn of a coil will link the others, so the fluxes shown are the average flux per turn. This simplified model yields the same result as one that considers the individual turns of the coils. With the wire resistance neglected, Faraday's law gives the voltage v_{ab} across the coil with N_1 turns:

$$v_{ab} = N_1 \frac{d}{dt}\phi_1 = N_1 \frac{d}{dt}\phi_{11} + N_1 \frac{d}{dt}\phi_{12} \tag{2.45}$$

For a linear magnetic material, the ratio of flux to current is constant. This implies that the *self-inductance* of coil 1,

$$L_1 = N_1 \frac{\phi_{11}}{i_{ab}} \tag{2.46}$$

and the *mutual inductance* between coil 1 and coil 2,

$$M_{12} = N_1 \frac{\phi_{12}}{i_{cd}} \tag{2.47}$$

are constants. These definitions let us write voltage v_{ab} in terms of current:

$$
\begin{aligned}
v_{ab} &= \frac{d}{dt}\left(\frac{N_1\phi_{11}}{i_{ab}}i_{ab}\right) + \frac{d}{dt}\left(\frac{N_1\phi_{12}}{i_{cd}}i_{cd}\right) \\
&= \frac{d}{dt}(L_1 i_{ab}) + \frac{d}{dt}(M_{12}i_{cd}) \\
&= L_1\frac{d}{dt}i_{ab} + M_{12}\frac{d}{dt}i_{cd} \tag{2.48}
\end{aligned}
$$

In a similar way we can show that

$$v_{cd} = L_2\frac{d}{dt}i_{cd} + M_{21}\frac{d}{dt}i_{ab} \tag{2.49}$$

where the self-inductance of coil 2,

$$L_2 = N_2 \frac{\phi_{22}}{i_{cd}} \tag{2.50}$$

and the mutual inductance between coils 2 and 1,

$$M_{21} = N_2 \frac{\phi_{21}}{i_{ab}} \tag{2.51}$$

are constants. As will be established later in this chapter,

$$M = M_{12} = M_{21} \tag{2.52}$$

so the subscripts can be dropped.

We modeled inductors with network components called inductances. We now extend this model to coupled coils by defining the network component mutual inductance.

We represent mutual inductance and coupled inductances with the symbol shown in Fig. 2.31.

COMPONENT SYMBOL

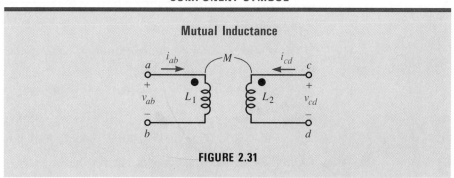

FIGURE 2.31

The component terminal equation also defines mutual inductance:

TERMINAL EQUATION

Mutual Inductance

$$v_{ab} = L_1 \frac{d}{dt} i_{ab} + M \frac{d}{dt} i_{cd} \qquad (2.53)$$

$$v_{cd} = L_2 \frac{d}{dt} i_{cd} + M \frac{d}{dt} i_{ab} \qquad (2.54)$$

The unit of both self-inductance and mutual inductance is the henry (H). Equations (2.53) and (2.54) imply that currents i_{ab} and i_{cd} cause flux through the coils that is in the same direction (additive), as shown in Fig. 2.30. The relative directions in which the coils are wound are implied by the dot marks shown in Fig. 2.30 and 2.31. Current into the dot-marked end of each coil causes flux to pass through the coils in the same direction.

The beginner often experiences difficulty in assigning the proper sign to the terms in the equations. Just remember that the sign on the self-induced voltage term is positive for the passive sign convention. The sign on the mutually induced voltage term is the same as the sign on the self-induced term, if the current reference arrows are oriented the same way relative to the corresponding dot marks.

This is true because currents that are oriented the same way relative to the corresponding dot marks cause flux through the coils in the same direction.

EXAMPLE 2.20

For the circuit shown in Fig. 2.32, find v_1 and v_2 for $t > 0$.

FIGURE 2.32
First coupled inductance
example

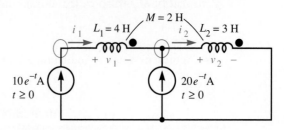

■ **SOLUTION**

The source equations and application of KCL to the indicated closed surfaces give

$$i_1 = 10e^{-t} \text{ A} \qquad \text{for } t \geq 0$$

and

$$i_2 = 10e^{-t} + 20e^{-t} = 30e^{-t} \text{ A} \qquad \text{for } t \geq 0$$

From the terminal equations for coupled inductances,

$$v_1 = L_1 \frac{d}{dt} i_1 + M \frac{d}{dt} i_2$$

$$= 4 \frac{d}{dt} (10e^{-t}) + 2 \frac{d}{dt} (30e^{-t})$$

$$= -100e^{-t} \text{ V} \qquad \text{for } t > 0$$

and

$$v_2 = L_2 \frac{d}{dt} i_2 + M \frac{d}{dt} i_1$$

$$= 3 \frac{d}{dt} (30e^{-t}) + 2 \frac{d}{dt} (10e^{-t})$$

$$= -110e^{-t} \text{ V} \qquad \text{for } t > 0 \quad ■$$

We will next work through an example in which one current enters the end of one inductance with a dot mark, and the other current enters the end of the second inductance without a dot mark.

EXAMPLE 2.21

Assume that the coils were wound so that the dot mark is moved from the right end to the left end of inductance L_1 of Fig. 2.32. Solve for voltages v_1 and v_2.

■ **SOLUTION**

Currents i_1 and i_2 are as found in Example 2.20. Voltages v_1 and v_2 are

$$v_1 = L_1 \frac{d}{dt} i_1 + M \frac{d}{dt} (-i_2)$$

$$= 4 \frac{d}{dt} (10e^{-t}) + 2 \frac{d}{dt} (-30e^{-t})$$

$$= 20e^{-t} \text{ V} \qquad \text{for } t > 0$$

and

$$v_2 = 3 \frac{d}{dt} i_2 + 2 \frac{d}{dt} (-i_1)$$

$$= 3 \frac{d}{dt} (30e^{-t}) + 2 \frac{d}{dt} (-10e^{-t})$$

$$= -70e^{-t} \text{ V} \qquad \text{for } t > 0 \quad \blacksquare$$

We can obtain an integral form of the terminal equations by solving Eq. (2.54) for

$$\frac{d}{dt} i_{cd} = \frac{1}{L_2} v_{cd} - \frac{M}{L_2} \frac{d}{dt} i_{ab} \qquad (2.55)$$

and substituting this into Eq. (2.53), to obtain

$$v_{ab} = L_1 \frac{d}{dt} i_{ab} + M \left[\frac{1}{L_2} v_{cd} - \frac{M}{L_2} \frac{d}{dt} i_{ab} \right] \qquad (2.56)$$

A little more algebra and integration with respect to time yields current i_{ab} in terms of the integral of v_{ab} and v_{cd}. We can similarly solve for current i_{cd} in terms of the integral of v_{cd} and v_{ab}. The results are given below in Eqs. (2.57) and (2.58).

$$i_{ab} = \frac{L_2}{L_1 L_2 - M^2} \int_{-\infty}^{t} v_{ab} \, d\tau - \frac{M}{L_1 L_2 - M^2} \int_{-\infty}^{t} v_{cd} \, d\tau \qquad (2.57)$$

$$i_{cd} = \frac{L_1}{L_1 L_2 - M^2} \int_{-\infty}^{t} v_{cd} \, d\tau - \frac{M}{L_1 L_2 - M^2} \int_{-\infty}^{t} v_{ab} \, d\tau \qquad (2.58)$$

With the integral form of the equations, a rule for the sign is that the sign on the self-induced current term is positive for the passive sign convention. The sign on the mutually induced term is the negative of the sign on the self-induced term, if the (+) marks are oriented the same way relative to the corresponding dot marks for both inductances.

EXAMPLE 2.22

For the circuit shown in Fig. 2.33, find i_1 and i_2 for $t \geq 0$ when the currents at time zero are $i_1(0) = 1$ A and $i_2(0) = 2$ A.

FIGURE 2.33
Third coupled inductance example

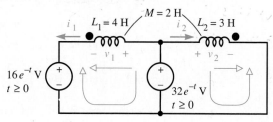

■ SOLUTION

The source equations and application of KVL to the indicated paths give

$$v_1 = 32^{-t} - 16e^{-t} = 16e^{-t} \text{ V} \qquad \text{for } t \geq 0$$

and

$$v_2 = 32e^{-t} \text{ V} \qquad \text{for } t \geq 0$$

Then

$$i_1 = \frac{L_2}{L_1 L_2 - M^2} \int_{-\infty}^{t} v_1 \, d\tau - \frac{M}{L_1 L_2 - M^2} \int_{-\infty}^{t} v_2 \, d\tau$$

$$= i_1(0) + \frac{L_2}{L_1 L_2 - M^2} \int_{0}^{t} v_1 \, d\tau - \frac{M}{L_1 L_2 - M^2} \int_{0}^{t} v_2 \, d\tau$$

$$= 1 + \frac{3}{8} \int_{0}^{t} 16e^{-\tau} \, d\tau - \frac{2}{8} \int_{0}^{t} 32e^{-\tau} \, d\tau$$

$$= 1 + \frac{3}{8}(16 - 16e^{-t}) - \frac{2}{8}(32 - 32e^{-t})$$

$$= 2e^{-t} - 1 \text{ A} \qquad \text{for } t \geq 0$$

and

$$i_2 = i_2(0) + \frac{L_1}{L_1 L_2 - M^2} \int_{0}^{t} v_2 \, d\tau - \frac{M}{L_1 L_2 - M^2} \int_{0}^{t} v_1 \, d\tau$$

$$= 2 + \frac{4}{8} \int_{0}^{t} 32e^{-\tau} \, d\tau - \frac{2}{8} \int_{0}^{t} 16e^{-\tau} \, d\tau$$

$$= 2 + \frac{4}{8}(32 - 32e^{-t}) - \frac{2}{8}(16 - 16e^{-t})$$

$$= 14 - 12e^{-t} \text{ A} \qquad \text{for } t \geq 0 \quad ■$$

We will next work through an example in which one current enters the end of the inductance with a dot mark and the other current enters the end of the inductance without a dot mark.

EXAMPLE 2.23

On Fig. 2.33 the coil has been wound so that the dot mark is moved from the left end to the right end of inductance L_1. Repeat Example 2.22.

■ SOLUTION

Voltages v_1 and v_2 are as found in Example 2.22. Currents i_1 and i_2 are

$$i_1 = i_1(0) + \frac{L_2}{L_1 L_2 - M^2} \int_{0}^{t} v_1 \, d\tau - \frac{M}{L_1 L_2 - M^2} \int_{0}^{t} -v_2 \, d\tau$$

$$= 1 + \frac{3}{8} \int_{0}^{t} 16e^{-\tau} \, d\tau - \frac{2}{8} \int_{0}^{t} -32e^{-\tau} \, d\tau$$

$$= 1 + \frac{3}{8}(16 - 16e^{-t}) - \frac{2}{8}(32e^{-t} - 32)$$

$$= 15 - 14e^{-t} \text{ A} \qquad \text{for } t \geq 0$$

and

$$i_2 = i_2(0) + \frac{L_1}{L_1 L_2 - M^2} \int_0^t v_2 \, d\tau - \frac{M}{L_1 L_2 - M^2} \int_0^t -v_1 \, d\tau$$

$$= 2 + \frac{4}{8} \int_0^t 32 e^{-\tau} \, d\tau - \frac{2}{8} \int_0^t -16 e^{-\tau} \, d\tau$$

$$= 2 + \frac{4}{8} (32 - 32 e^{-t}) - \frac{2}{8} (16 e^{-t} - 16)$$

$$= 22 - 20 e^{-t} \, \text{A} \qquad \text{for } t \geq 0 \quad \blacksquare$$

Mutually coupled coils with three or more windings are often used in transformers to supply more than one output voltage. The assignment of more than one mark may be required as shown in Fig. 2.34. Current entering like-marked ends for any two coils will cause flux through these two coils in the same direction.

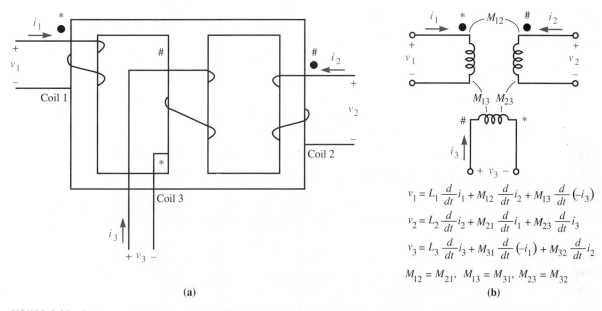

(a)   **(b)**

FIGURE 2.34 Three coupled inductors and circuit model using coupled inductances. (a) Three coupled inductors; (b) circuit model

Coupled inductors have many uses. The ignition coil for an automobile is one example of coupled inductors. Electrical transformers used in power systems and in electronics are another example. Many transformers used in electronics are air-cored and linear. Although power transformers have nonlinear magnetic cores, linear approximations are often sufficient.

We did not introduce a model for mutual capacitance. Mutual capacitance is not required in circuit analysis to describe the properties of any physical device.

REMEMBER

Mutual inductance is our network model that accounts for the transfer of electrical energy from one inductor to another by magnetic coupling. Mutual inductance accounts for a voltage induced across one coil as a result of a change in current through another coil. Current into like-marked ends of coils (dotted or undotted) will cause flux through the coils in the same direction.

EXERCISES

For the circuit shown below, $i_s = 100 \cos 3t$ A.

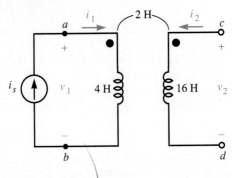

68. Calculate v_1 and v_2.
69. Short-circuit terminals c and d. (A short circuit is represented by a solid line and makes $v_{cd} = 0$.) Determine v_1, v_2, and i_2.
70. Replace the current source in the circuit with a voltage source, having the (+) reference mark at the top, of value $v_s = 0$ V, $t < 0$, and $v_s = 10 \cos 3t$ V, $t \geq 0$. Find i_1 and v_2 for $t > 0$.
71. Repeat Exercise 70 if terminals c and d are short-circuited.

A network with two coupled inductances is shown below.

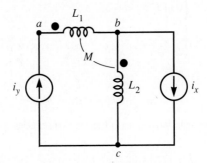

72. Use derivatives to express v_{ab} and v_{cb} as functions of i_x and i_y.
73. In the preceding circuit, replace the current sources i_x and i_y by corresponding voltage sources v_x and v_y, both with (+) reference marks at the top. Express i_{ba} and i_{bc} as functions of v_x and v_y. Use definite integrals with the limits of integration from $-\infty$ to t.

Magnetic Coefficient of Coupling

The *magnetic coefficient of coupling* k is defined by the equation

$$k = \frac{M}{\sqrt{L_1 L_2}} \qquad (2.59)$$

With the magnetic flux variables defined as in Fig. 2.30, this can be written as

$$k = \frac{\sqrt{M_{21}}\sqrt{M_{12}}}{\sqrt{L_1 L_2}} = \frac{\sqrt{N_2 \dfrac{\phi_{21}}{i_{ab}}}\sqrt{N_1 \dfrac{\phi_{12}}{i_{cd}}}}{\sqrt{N_1 \dfrac{\phi_{11}}{i_{ab}}}\sqrt{N_2 \dfrac{\phi_{22}}{i_{cd}}}} = \sqrt{k_1 k_2} \qquad (2.60)$$

which is the geometric mean of the *coupling coefficients* k_1 and k_2. The coupling coefficient for coil 1, defined by

$$k_1 = \frac{\phi_{21}}{\phi_{11}} \qquad (2.61)$$

is the fraction of the average flux linking the turns of coil 1, due to current i_{ab}, which also links coil 2. The coupling coefficient for coil 2 is similarly defined:

$$k_2 = \frac{\phi_{12}}{\phi_{22}} \qquad (2.62)$$

For these definitions it follows that k_1, k_2, and k are always greater than or equal to 0 and less than or equal to 1:

$$0 \le k \le 1 \qquad (2.63)$$

This result can also be justified from energy considerations (see Problem 70). If both coils have a similar geometry, then they will have nearly the same coupling coefficient.

EXERCISES

74. Determine the coupling coefficient for two inductances if $L_1 = 16$ H, $L_2 = 9$ H, and $M = 5$ H.

75. Two coupled inductances have self-inductances of $L_1 = 25$ H and $L_2 = 9$ H. If the coupling coefficient is 0.4, determine the value of the mutual inductance M.

Energy Stored in Coupled Inductances

The total power p absorbed by the coupled inductances of Fig. 2.31 is the sum of p_1, the power input to coil 1, and p_2, the power input to coil 2. Thus

$$
\begin{aligned}
p &= p_1 + p_2 \\
&= i_{ab} v_{ab} + i_{cd} v_{cd} \\
&= i_{ab}\left(L_1 \frac{d}{dt} i_{ab} + M_{12} \frac{d}{dt} i_{cd}\right) + i_{cd}\left(L_2 \frac{d}{dt} i_{cd} + M_{21} \frac{d}{dt} i_{ab}\right) \qquad (2.64)
\end{aligned}
$$

Because $M = M_{12} = M_{21}$, Eq. (2.64) can be written as

$$p = L_1 i_{ab} \frac{d}{dt} i_{ab} + M \frac{d}{dt}(i_{ab} i_{cd}) + L_2 i_{cd} \frac{d}{dt} i_{cd} \qquad (2.65)$$

The total energy stored in the coupled inductances at time t is

$$w = \int_{-\infty}^{t} p \, d\tau$$

$$= L_1 \int_{\tau=-\infty}^{\tau=t} i_{ab}(\tau) \, di_{ab}(\tau) + M \int_{\tau=-\infty}^{\tau=t} d[i_{ab}(\tau)i_{cd}(\tau)] + L_2 \int_{\tau=-\infty}^{\tau=t} i_{cd} di_{cd}(\tau) \quad (2.66)$$

With $i_1(-\infty) = i_2(-\infty) = 0$, assumed, this becomes

$$w = \frac{1}{2} L_1 i_{ab}^2 + M i_{ab} i_{cd} + \frac{1}{2} L_2 i_{cd}^2 \qquad (2.67)$$

for the current reference arrows oriented the same way with respect to the corresponding dot mark.

If the current reference arrows were oriented oppositely with respect to the dot marks, so that the currents would cause flux through the coils that opposed each other, the stored energy would be given by

$$w = \frac{1}{2} L_1 i_{ab}^2 - M i_{ab} i_{dc} + \frac{1}{2} L_2 i_{dc}^2 \qquad (2.68)$$

Coupled coils are passive because energy is stored in a magnetic field as the magnetic flux increases and returned as the flux decreases.*

EXERCISES

76. Refer to the circuit used in Exercise 72. Determine the energy stored in the coupled inductances if $L_1 = 16$ H, $L_2 = 9$ H, and $M = 5$ H, and if $i_x = 4 \cos 2t + 10 \sin 2t$ A and $l_y = 10 \sin 2t$ A.

77. The location of the dot mark is reversed on inductance L_2 of the circuit used in Exercise 76. Determine the energy stored in the coupled inductances.

Symmetry of Mutual Inductance†

The energy stored by coupled inductances as given by integration of Eq. (2.64) is

$$w(t) = \int_{-\infty}^{t} p \, d\tau$$

$$= \frac{1}{2} L_1 i_{ab}^2 + M_{12} \int_{\tau=-\infty}^{\tau=t} i_{ab}(\tau) \, di_{cd}(\tau) + M_{21} \int_{\tau=-\infty}^{\tau=t} i_{cd}(\tau) \, di_{ab}(\tau) + \frac{1}{2} L_2 i_{cd}^2 \quad (2.69)$$

This equation can be used, in conjunction with the fact that the energy stored in the coupled inductances must be zero when i_1 and i_2 are zero, to prove that $M = M_{12} = M_{21}$.

If M_{12} can be shown to be equal to M_{21} for some currents i_{ab} and i_{cd}, then

---

* Symmetry of mutual inductance ($M_{12} = M_{21} = M$) implies that coupled coils are passive. See Problem 72.

† This section can be omitted without loss of continuity.

M_{12} and M_{21} must be equal for all currents i_{ab} and i_{cd}, because M_{12} and M_{21} are constant. Assume that

$$i_{ab}(t) = t^3(t-1) \qquad \text{for } t \geq 0 \qquad (2.70)$$

$$i_{cd}(t) = t^2(t-1) \qquad \text{for } t \geq 0 \qquad (2.71)$$

and i_{ab} and i_{cd} are zero for all time less than zero. This gives

$$i_{ab}(0) = i_{cd}(0) = i_{ab}(1) = i_{cd}(1) = 0 \qquad (2.72)$$

and

$$w(1) = 0 \qquad (2.73)$$

The integrals

$$\int_{\tau=-\infty}^{\tau=1} i_{cd}(\tau)\, di_{ab}(\tau) = \int_{\tau=0}^{\tau=1} \tau^2(\tau-1)\, d[\tau^3(\tau-1)] = \frac{1}{210} \qquad (2.74)$$

and

$$\int_{\tau=-\infty}^{\tau=1} i_{ab}(\tau)\, di_{cd}(\tau)\, d\tau = \int_{\tau=0}^{\tau=1} \tau^3(\tau-1)\, d[\tau^2(\tau-1)] = -\frac{1}{210} \qquad (2.75)$$

can be substituted into Eq. (2.69) to give

$$w(1) = \frac{1}{210} M_{12} - \frac{1}{210} M_{21} = \frac{1}{210}(M_{12} - M_{21}) = 0 \qquad (2.76)$$

or

$$M_{12} = M_{21} = M \qquad (2.77)$$

2.9   Summary

We introduced Kirchhoff's current law and Kirchhoff's voltage law as the two axioms of network theory:

KCL: The sum of the currents leaving any closed surface is zero.

KVL: The sum of the voltage drops around any closed path is zero.

These axioms are based on the assumption that the network does not radiate electromagnetic energy.

We presented definitions relating to network theory and introduced the network components as ideal models for physical properties of practical devices. Table 2.1 contains physical properties modeled by network components. The equations

TABLE 2.1
Physical properties modeled by network components

NETWORK COMPONENT	PHYSICAL PROPERTY
Independent source	Source of electrical energy
Dependent source	Control of electrical energy
Resistance	Conversion of electrical energy to heat
Capacitance	Storage of energy in an electric field
Inductance	Storage of energy in a magnetic field
Mutual inductance	Transfer of energy stored in a magnetic field from one coil to another

describing these components, in conjunction with KVL and KCL, are the basis of network analysis (the concept of "mutual capacitance" is not required in circuit analysis to describe any physical property).

KEY FACTS

◆ Kirchhoff's current law:

$$\sum_{k=1}^{N} i_k = 0$$

◆ Kirchhoff's voltage law:

$$\sum_{k=1}^{N} v_k = 0$$

◆ Voltage across an independent voltage source is not a function of the current through the source:

$$v_{ab} = v(t)$$

◆ Current through an independent current source is not a function of the voltage across the source:

$$i_{ab} = i(t)$$

◆ Voltage across a dependent voltage source is determined by the control voltage or current:

$$v_{cd} = \mu v_{ab} \qquad \text{or} \qquad v_{cd} = r i_{ab}$$

◆ Current through a dependent current source is determined by the control voltage or current:

$$i_{cd} = g_m v_{ab} \qquad \text{or} \qquad i_{cd} = \beta i_{ab}$$

◆ Voltage across a resistance is proportional to the current through it:

$$v_{ab} = R i_{ab} \qquad \text{and} \qquad i_{ab} = G v_{ab}$$

◆ Current through a capacitance is proportional to the time derivative of the voltage across it:

$$i_{ab} = C \frac{d}{dt} v_{ab}$$

◆ Voltage across an inductance is proportional to the time derivative of the current through it:

$$v_{ab} = L \frac{d}{dt} i_{ab}$$

◆ Current into like-marked ends (both dotted or both undotted) of coupled inductances causes magnetic flux that passes through the coils in the same direction:

$$v_{ab} = L_1 \frac{d}{dt} i_{ab} + M \frac{d}{dt} i_{cd}$$

$$v_{cd} = L_2 \frac{d}{dt} i_{cd} + M \frac{d}{dt} i_{ab}$$

Section 2.1

For the network shown below, $i_1 = 1$ A, $i_2 = 8$ A, and $i_3 = 5$ A. Determine the specified currents.

Determine the requested currents in the network shown below.

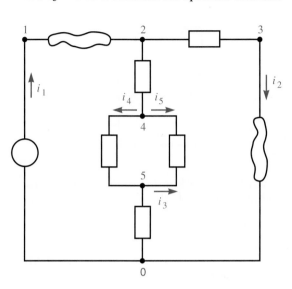

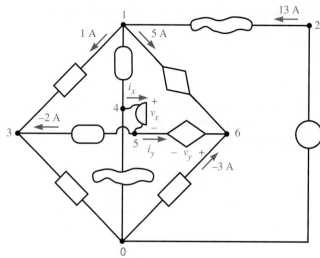

1. i_{12}
2. i_{21}
3. i_{32}
4. i_{24}
5. i_{05}
6. i_4
7. i_5

8. Currents i_x and i_y
9. Currents i_{31}, i_{14}, i_{41}, and i_{40}
10. Is the circuit shown above planar?

11. Determine i_x, i_y, and i_z for the circuit shown below.

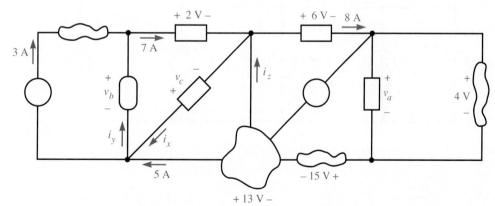

Section 2.2

For the network shown at the left, $v_{10} = 1$ V, $v_{20} = 2$ V, $v_{30} = 3$ V, $v_{40} = 4$ V, $v_{50} = 5$ V, and $v_{60} = 6$ V. Determine the specified voltage.

12. v_{12}
13. v_{21}
14. v_{34}
15. v_{43}
16. v_{41}
17. v_{15}
18. v_{45}
19. v_{16}

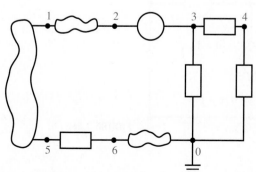

20. v_{62}

21. v_{53}

22. v_{64}

23. v_{03}

24. For the network used in Problem 11, find v_a, v_b, and v_c.

Determine the requested voltages for the network used in Problem 8, if $v_{10} = 6$ V, $v_{20} = 8$ V, $v_{30} = 11$ V, $v_{40} = 15$ V, $v_{50} = 20$ V, and $v_{60} = 26$ V.

25. Voltages v_{13}, v_{15}, v_{52}, and v_{24}

26. Voltages v_x and v_y

27. The following three KVL equations describe an electrical network. Can the equations be solved for voltages v_1, v_2, and v_3?

$$v_1 + v_2 + v_3 = 6$$

$$2v_1 + 0v_2 + 4v_3 = 16$$

$$3v_1 + v_2 + 5v_3 = 22$$

Section 2.4

28. For the network in Fig. 2.35, find i_x, v_{ab}, i_{ab}, and the power absorbed by the 15-V voltage source.

Section 2.5

29. For the network below, find the total energy absorbed by the load when $i = 0$ for $t < 0$, $i = 5e^{-2t}$ A for $t \geq 0$, and $v_s = 10$ V.

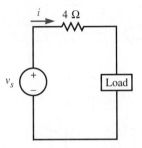

30. Given the network in Fig. 2.36, find i_x, v_1, i_{ba}, and the power absorbed by the -30-V source.

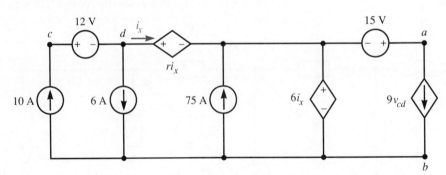

FIGURE 2.35

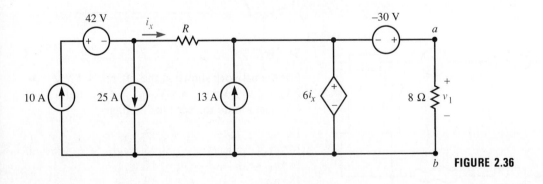

FIGURE 2.36

31. The voltage $v_x = 9$ V and the current $i_y = -1$ A have been measured for the network below. Find $v_{10}, v_{20}, v_{30}, v_{40}, i_1, i_2, i_3,$ and i_4.

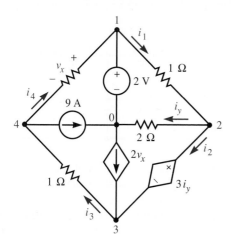

32. For the network above, currents $i_1 = -2$ A, $i_2 = 0$ A, $i_3 = -1$ A, and $i_4 = 7$ A have been measured. Find $v_{10}, v_{20}, v_{30},$ and v_{40}.

33. Part of a network is shown below. In the most convenient order for you, determine (a) v_x; (b) v_z; (c) i_y; (d) i_a; and (e) R.

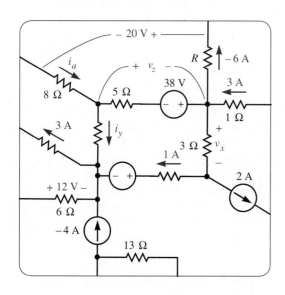

34. Part of a network follows. In the most convenient order for you, determine (a) v_z; (b) v_x; (c) i_y; (d) i_a; and (e) R.

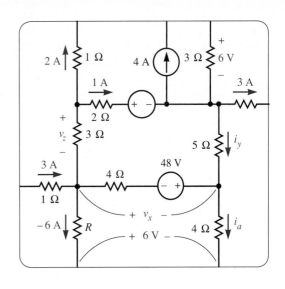

Section 2.6

Determine the capacitance current i_{ab} for $t > 0$ if the capacitance is 2 μF and the voltage v_{ab} in volts is as given below for $t \geq 0$.

35. $12e^{-4000t}$ 36. $50 \cos 300t$

Determine the capacitance voltage v_{ab} for $t \geq 0$ if the capacitance is 12 μF, $v_{ab}(0)$ is 6 V, and current i_{ab} in amperes is as given below for $t \geq 0$.

37. $0.36e^{-8000t}$ 38. $0.48 \cos 2000t$

39. Capacitance current i_{ab} for a $\frac{1}{2}$-F capacitance is as shown below. If $v(0) = -4$ V, plot v_{ab} for $0 \leq t \leq 4$ s.

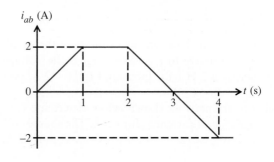

40. If the ordinate in the above graph is changed to v_{ab} in volts, where v_{ab} is the voltage across a $\frac{1}{2}$-F capacitance, plot the current i_{ab} through the capacitance for $0 \leq t \leq 4$ s.

41. Given the network below, determine current i_{ab} and voltage v_{cd} for $t > 0$. The source voltage is $v_s = 12 \cos 200t$ V for $t \geq 0$, and $v_{cd}(0)$ is 3 V.

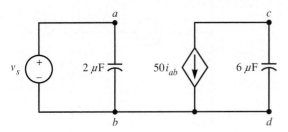

Section 2.7

Determine the inductance voltage v_{ab} for $t > 0$ if the inductance is 2 mH and the current i_{ab} in amperes is as given below for $t \geq 0$.

42. $13e^{-200t}$

43. $20 \cos (2\pi 60t)$

Determine the inductance current i_{ab} for $t \geq 0$ if the inductance is 36 mH, $i_{ab}(0)$ is 2 A, and voltage v_{ab} in volts is as given below for $t \geq 0$.

44. $18e^{-400t}$

45. $40 \cos 90t$

46. The current i_{ab} through a 2-H inductance is given below. Plot the voltage v_{ab} across the inductance for $0 \leq t \leq 8$ s.

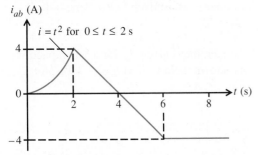

47. Change the label on the ordinate of the graph shown above to v_{ab} volts. If v_{ab} is the voltage across a 2-H inductance and $i_{ab}(0) = 2$ A, plot i_{ab} for $0 \leq t \leq 8$ s.

48. For the network shown below, determine voltage v_{ab} and current i_{cd} for $t > 0$. The source current is $i_s = 12e^{-2t}$ A for $t \geq 0$, and $i_{cd}(0)$ is 2 A.

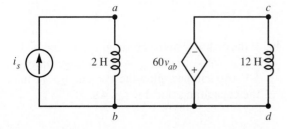

Section 2.8

49. Determine voltages v_1, v_2, and v_3 for $t > 0$ in the network shown below. The source currents are $i_a = 3e^{-2t}$ A and $i_b = 7e^{-2t}$ A.

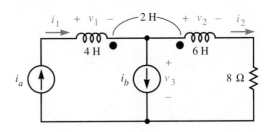

50. Repeat Problem 49 if the dot mark on the 6-H inductance is relocated to the right-hand side of the inductance.

51. Develop the integral form of the terminal equations for mutual inductance as given in Eqs. (2.57) and (2.58) from the differential form given in Eqs. (2.53) and (2.54).

52. Determine voltages v_1 and v_2 and currents i_1 and i_2 for $t \geq 0$ in the following circuit if $v_a = 100e^{-2t}$ V, $t \geq 0$; $i_b = 4e^{-2t}$ A, $t \geq 0$; and $i_1(0) = 5$ A.

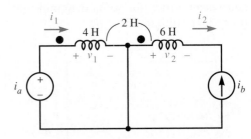

53. In the circuit above, relocate the dot mark on the 4-H inductance to the right-hand end of the inductance. Then repeat Problem 52.

54. For the network shown below, find v_1, v_2, v_3, v_a, and v_b, if $L_1 = 2$ H; $L_2 = 4$ H; $L_3 = 8$ H; $M_{12} = M_{21} = 1$ H; $M_{13} = M_{31} = 2$ H; $M_{23} = M_{32} = 3$ H; $i_a = 0$, $t < 0$; $i_a = 2[1 - e^{-t}]$ A, $t \geq 0$; $i_b = 0$, $t < 0$; and $i_b = 3[1 - e^{-t}]$ A, $t \geq 0$.

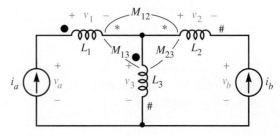

55. Start with the equation for three coupled inductances as given in Fig. 2.34, and develop equations that give the inductance currents as integrals of inductance voltages.

56. What is the coefficient of coupling for the coupled inductances of the circuit used in Problem 52?

57. Two coupled inductances have terminals labeled as shown below. Current $i_{cd} = 0$ and current i_{ab} is measured to be $2 \cos 100t$ A. If $v_{ab} = -800 \sin 100t$ V and $v_{cd} = 400 \sin 100t$ V, assign a suitable set of dot marks to the inductances and find L_1 and M. The manufacturer states that the coefficient of coupling is $\frac{1}{18}$. What is L_2?

58. Two coupled inductances have terminals labeled as shown above. $L_1 = 3$ H, $L_2 = 12$ H, and $M = 4$ H. The dots are at terminals a and d. Terminals b and c are connected, and $i_{ad} = 2 \cos 10t$ A. Find v_{ab}, v_{cd}, the instantaneous power input to the series combination, and the instantaneous energy that is stored in the series combination.

59. Repeat Problem 58 if the dots are at terminals a and c.

60. Two coupled inductances are as shown in the circuit of Problem 57. The dot mark on L_1 is on terminal a, but the dot mark on L_2 has been lost. A voltmeter is connected from terminal c to d with the $(+)$ reference mark on c. The $(-)$ terminal of a D cell (flashlight battery) is connected to terminal b, and the $(+)$ terminal is momentarily connected to terminal a. When the connection is made, the voltmeter momentarily reads positive. When the connection is broken, the meter momentarily reads negative. Where should the dot mark be placed on L_2?

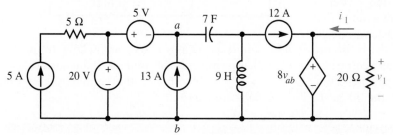

Comprehensive Problems

61. For the network in Fig. 2.37, determine v_{ab}, v_1, i_1, and the power absorbed by the dependent voltage source.

62. For the network below, determine i_{45}, i_{30}, i_{01}, i_{12}, i_{23}, v_{10}, v_{12}, v_{20}, v_{43}, and v_{53} for $t > 0$. For $t \geq 0$ the source values in amperes are $i_a = 10t^2$, $i_b = 5t^2$, and $i_c = 2t^2$. The initial capacitance voltage is $v_{12} = 5$ V.

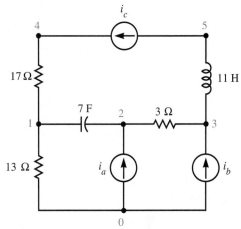

63. For the network below, determine v_{10}, v_{20}, v_{30}, v_{12}, v_{23}, v_{34}, i_x, i_y, i_z, and i_w for $t > 0$. The initial condition is $i_z(0) = 5$ A. The source voltages for $t \geq 0$ are $v_a = 2e^{-2t}$ V, $v_b = 7e^{-2t}$ V, and $v_c = 3e^{-2t}$ V.

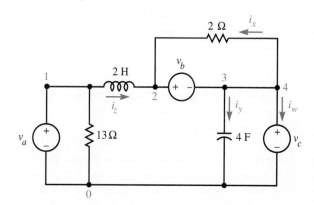

FIGURE 2.37

64. Determine i_{10}, i_{20}, i_{23}, v_{20}, v_{12}, v_{30}, and v_{34} for $t > 0$ for the network shown below. The sources in amperes for $t \geq 0$ are $i_a = 6e^{-2t}$, $i_b = 16e^{-2t}$, and $i_c = 7e^{-2t}$. The initial capacitance voltage is $v_c(0) = 10$ V.

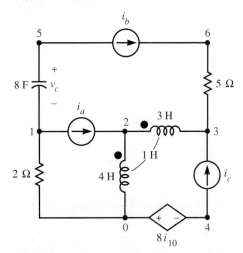

65. Measurements on a two-terminal network device indicate that

$$v_{ab} = 10 + 5i_{ab}$$

Construct a network model for the device. Use two network components in series.

66. Construct a network model with two network components in parallel that has the terminal equation given in Problem 65.

67. A two-terminal device is found to have the volt-ampere characteristics

$$v_{ab} = 5i_{ab} + 10\frac{d}{dt}i_{ab}$$

Construct a network model for the device. Use two network components in series.

68. A four-terminal device has the characteristics $i_{ab} = 0$, $v_{bd} = 0$, and $v_{cd} = -10v_{ab} + 2i_{cd}$. Use network components to construct a model.

69. An electric water heater must raise the temperature of 30 gal of water from 10°C to 70°C in 1 h. The supply voltage to the heater is 240 V. Neglect all heat losses and assume that the heater is a resistance. Determine the power rating of the heater and the resistance of the heating element. What is the electric current through the heater? At 10 cents per kWh, how much does it cost to heat the 30 gal of

water? (Approximately 15,800 J is required to raise the temperature of one gallon of water one degree Celsius.)

70. The upper bound of $k = 1$ for the coefficient of coupling can be established if you calculate the energy stored in coupled inductances when one inductance is short-circuited. Refer to Fig. 2.31. With inductance 2 short-circuited, show that

$$v_{ab} = \frac{L_1 L_2 - M^2}{L_2}\frac{d}{dt}i_{ab}$$

Use this to calculate

$$w(t_0) = \int_{-\infty}^{t_0} i_{ab}v_{ab}\, dt$$

with $i_1(-\infty) = 0$ assumed. Use the fact that

$$L_1 \geq 0 \qquad L_2 \geq 0 \qquad \text{and} \qquad w(t_0) \geq 0$$

to show that

$$L_1 L_2 = M^2 \geq 0$$

Write this inequality in terms of k by the use of

$$k = \frac{M}{\sqrt{L_1 L_2}}$$

to show that $k^2 \leq 1$

Because mutual inductance is nonnegative, this implies

$$0 \leq k \leq 1$$

71. The unknown devices shown in the figure below absorb the indicated powers. Determine the unknown current i_x.

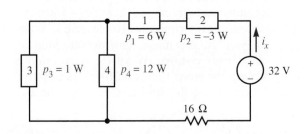

72. Use Eqs. (2.59), (2.63), and (2.67) to prove that symmetry of mutual inductance ($M_{12} = M_{21} = M$) implies that mutually coupled coils are passive.

Introduction to Circuit Analysis

In this chapter we use KCL, KVL, and the component terminal equations to determine the voltages and currents in some simple, but highly important, circuits. We first apply KCL to the analysis of the general *two-node* or *parallel circuit*. This analysis provides us with our first two-terminal *equivalent circuits*. The analysis also yields the *current divider* equation, which shows how the sum of currents through parallel conductances divides proportionally to the individual conductances. We next use KVL to analyze the general *single-loop* or *series circuit*. Analysis of the series circuit yields additional two-terminal equivalent circuits. This analysis also shows that the voltage across series resistances divides proportionally to the individual resistances. This is the *voltage divider* relation. We use the equivalent circuits we have developed to simplify and analyze circuits that initially appear complicated. The use of equivalent circuits is extended to three-terminal equivalent circuits in a final section that considers three resistances connected in delta or wye.

3.1 Two-Node or Parallel Circuits

Most equipment for industry, appliances for the house, and automotive accessories are designed to run on a specified voltage. For instance, automotive accessories are designed so that approximately 12 V should appear across each unit. Ideally they should be connected in *parallel*. An example of parallel-connected components is shown in Fig. 3.1.

FIGURE 3.1
A parallel circuit

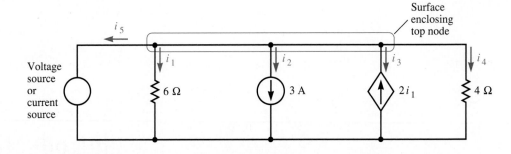

DEFINITION

Parallel Components

Two or more components are in parallel if their terminals are connected to two common (shared) nodes.

As an alternative definition, *A set of components is in parallel if every combination of two components in the set forms a loop.* Application of KVL to a loop of any two of the components in parallel gives the following result:

At any specified time the same voltage v must appear across each component in parallel.

The *general procedure* for solving a two-node or parallel circuit problem is to *first find the voltage v between the nodes.* If this is specified by a voltage source, the problem is simple. If the voltage is not known, apply KCL to a surface enclosing one node, write all unknown currents in terms of the voltage v with the use of terminal or control equations, and solve for v. Once v is known, all currents are easily calculated. We will now demonstrate the procedure with three simple examples.

EXAMPLE 3.1

For the network of Fig. 3.1, the left-hand source is a 12-V voltage source with the (+) reference mark at the top of the figure. Determine the indicated currents.

■ SOLUTION

The component equations are

$$i_1 = \frac{1}{6}v = \frac{1}{6}(12) = 2A \quad \text{(terminal equation)} \tag{3.1}$$

$$i_2 = 3 \text{ A} \quad \text{(source equation)} \tag{3.2}$$

$$i_3 = -2i_1 = -2(2) = -4 \text{ A} \quad \text{(control equation)} \tag{3.3}$$

$$i_4 = \frac{1}{4}v = \frac{1}{4}(12) = 3 \text{ A} \quad \text{(terminal equation)} \tag{3.4}$$

We can now find the current through the voltage source from a KCL equation for a surface enclosing the top node (or the bottom node).

$$i_5 + i_1 + i_2 + i_3 + i_4 = 0 \tag{3.5}$$

This KCL equation can be written directly in terms of v with the use of the component equations.

$$i_5 + \frac{1}{6}v + 3 - 2\left(\frac{1}{6}v\right) + \frac{1}{4}v = 0 \tag{3.6}$$

With $v = 12$ V,

$$i_5 = -4 \text{ A} \quad ■ \tag{3.7}$$

If the voltage between the nodes is not specified by a voltage source, we can readily find the voltage by the use of KCL and terminal, source, and control equations.

EXAMPLE 3.2

For the circuit of Fig. 3.1, the left-hand source is a 12-A current source (reference arrow pointing up). Determine the voltage v.

■ SOLUTION

Use the component equations to write a KCL equation in terms of the voltage v for a surface enclosing the top node. This is the same as Eq. (3.6), except that current i_5 is now known and voltage v is unknown:

$$-12 + \frac{1}{6}v + 3 - 2\left(\frac{1}{6}v\right) + \frac{1}{4}v = 0 \tag{3.8}$$

This gives

$$v = 108 \text{ V} \tag{3.9}$$

Once voltage v is known, the currents not already specified by the independent current sources are easily found. ■

It is no more difficult to write the necessary equations for a two-node circuit with resistance, capacitance, and inductance than it is for the resistive case.

EXAMPLE 3.3

Find the indicated currents for the network shown in Fig. 3.2 if (a) the left-hand source is a voltage source; (b) the left-hand source is a current source.

FIGURE 3.2
A parallel *RLC* circuit

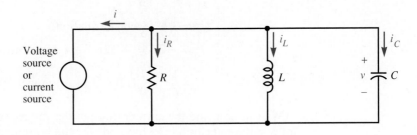

■ SOLUTION

(a) The element currents are given by the terminal equations (the dependence of v and i on t is implicit):

$$i_R = \frac{1}{R} v \tag{3.10}$$

$$i_L = \frac{1}{L} \int_{-\infty}^{t} v \, d\tau \tag{3.11}$$

$$i_C = C \frac{d}{dt} v \tag{3.12}$$

where voltage v is specified by the voltage source. Current i through the source can now be found from a KCL equation for a surface enclosing the top node:

$$i + i_R + i_L + i_C = 0 \tag{3.13}$$

Substitution from the component equations gives us

$$i + \frac{1}{R} v + \frac{1}{L} \int_{-\infty}^{t} v \, d\tau + C \frac{d}{dt} v = 0 \tag{3.14}$$

where voltage v is known (we can skip the first step when we write the KCL equation and write the last equation directly).

(b) If the voltage source in the network of Fig. 3.2 is replaced by a current source, we cannot find the element currents immediately. The KCL equation for a surface enclosing the top node, as given in Eq. (3.14), is an *integro-differential equation* (includes both integration and differentiation) and must first be solved for v in terms of i. The solution of equations of this type is described in Chapters 8 and 9. ■

REMEMBER

The voltage is the same across each component in parallel. For a parallel circuit driven by a voltage source, this voltage is known. The current through the resistance, capacitance, or inductance is easily found with the use of this voltage and the terminal equation. The current through the voltage source must be found from a KCL equation. If a parallel circuit is not driven by a voltage source, the voltage across the components is found from a KCL equation.

EXERCISES

1. For the circuit shown below, find currents i_1, i_2, i_3, i_4, and i_5, and the powers p_1, p_2, p_3, p_4, and p_5 absorbed by each component.

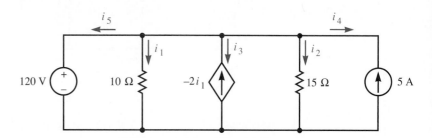

2. Replace the voltage source in the network shown above with a current source, with the reference arrow pointing toward the bottom of the page, of value 49 A. Find voltage v and currents i_1, i_2, i_3, i_4, and i_5.

3. For the circuit shown below, find currents i_1, i_2, i_3, i_4, i_5, and i_6 for $t > 0$ if $i_3(0) = -5$ A.

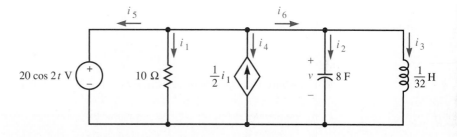

4. Replace the voltage source in the circuit shown above with a current source (reference arrow pointing up) of value $i_s = 2 \cos 2t$ A for $t \geq 0$. Write a KCL equation that gives an integro-differential equation that can be solved for voltage v. The initial conditions are $v(0) = 20$ V and $i_3(0) = 10$ A. Do not solve for voltage v.

3.2 Parallel Elements

The analysis of a network containing parallel elements can often be simplified if we replace the parallel elements with an *equivalent element*. Consider the N parallel

FIGURE 3.3
The equivalent resistance of N resistances in parallel: (a) parallel resistances; (b) the equivalent resistance

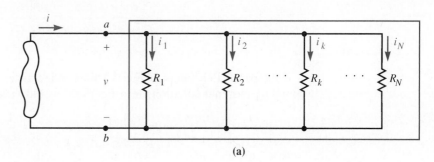

(a)

FIGURE 3.3
Continued

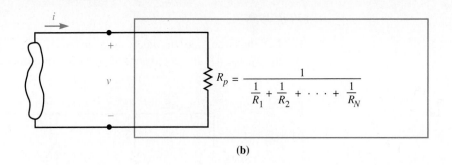

(b)

resistances of Fig. 3.3a. We can relate the current i to the individual element currents by a KCL equation:

$$i = i_1 + i_2 + \cdots + i_N \tag{3.15}$$

Substitution of the terminal equations for resistance into Eq. (3.15) yields

$$\begin{aligned} i &= \frac{1}{R_1} v + \frac{1}{R_2} v + \cdots + \frac{1}{R_N} v \\ &= \left(\frac{1}{R_1} + \frac{1}{R_2} + \cdots + \frac{1}{R_N} \right) v \\ &= \frac{1}{R_p} v \end{aligned} \tag{3.16}$$

where the *equivalent resistance* R_p *for N resistances in parallel* is

---

Resistances in Parallel

$$R_p = \frac{1}{\dfrac{1}{R_1} + \dfrac{1}{R_2} + \cdots + \dfrac{1}{R_N}} \tag{3.17}$$

Thus the *reciprocals of resistances in parallel are added* to give the *reciprocal* of the equivalent resistance. By defining the conductances $G_k = 1/R_k$, we can see that *conductances in parallel are added.*

Conductances in Parallel

$$G_p = G_1 + G_2 + \cdots + G_N \tag{3.18}$$

If the parallel elements of Fig. 3.3 were inductances, substitution of the corresponding terminal equations into Eq. (3.15) would yield

$$i = \frac{1}{L_1} \int_{-\infty}^{t} v \, d\tau + \frac{1}{L_2} \int_{-\infty}^{t} v \, d\tau + \cdots + \frac{1}{L_N} \int_{-\infty}^{t} v \, d\tau = 0$$

$$= \left(\frac{1}{L_1} + \frac{1}{L_2} + \cdots + \frac{1}{L_N} \right) \int_{-\infty}^{t} v \, d\tau$$

$$= \frac{1}{L_p} \int_{-\infty}^{t} v \, d\tau \tag{3.19}$$

where the *equivalent inductance L_p of N inductances in parallel* is

Inductances in Parallel

$$L_p = \frac{1}{\dfrac{1}{L_1} + \dfrac{1}{L_2} + \cdots + \dfrac{1}{L_N}} \tag{3.20}$$

Thus the *reciprocals of inductances in parallel are added* to give the *reciprocal* of the equivalent inductance.

Similarly, if the parallel elements of Fig. 3.3 were capacitances, substitution of the corresponding terminal equations into Eq. (3.15) would yield

$$i = C_1 \frac{d}{dt} v + C_2 \frac{d}{dt} v + \cdots + C_N \frac{d}{dt} v$$

$$= (C_1 + C_2 + \cdots + C_N) \frac{d}{dt} v$$

$$= C_p \frac{d}{dt} v \tag{3.21}$$

where the *equivalent capacitance for N capacitances in parallel* is

Capacitances in Parallel

$$C_p = C_1 + C_2 + \cdots + C_N \tag{3.22}$$

Thus *capacitances in parallel are added* to give the equivalent capacitance.

Equations (3.17), (3.20), and (3.22) give us our first two-terminal *equivalent circuits*. The circuits are equivalent in the sense that *the relationship between the voltage v of terminal a with respect to terminal b and the current i into terminal a is the same for both the original parallel elements and the equivalent parallel element*. Equivalent networks play an important role in the design and analysis of all types of electric circuits. We will now use an equivalent parallel resistance to analyze a simple circuit.

EXAMPLE 3.4

We model three lights on a boat by three parallel resistances, as shown in Fig. 3.4a. Find the equivalent resistance and the current i for the parallel combination of resistances. The voltage $v = 12$ V is given.

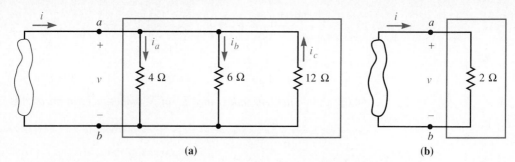

FIGURE 3.4 Equivalent parallel resistance example: (*a*) parallel resistances; (*b*) equivalent parallel resistance

■ SOLUTION

$$R_p = \frac{1}{(1/4) + (1/6) + (1/12)} = \frac{12}{3 + 2 + 1} = 2\,\Omega$$

$$i = \frac{1}{R_p}v = \frac{12}{2} = 6\text{ A} \quad ■$$

Note that it really saves us no time to combine resistances to find R_p if the desired result is i rather than R_p. It is just as easy to write KCL for the top node to obtain i directly:

$$i = \frac{12}{4} + \frac{12}{6} + \frac{12}{12} = 6\text{ A} \tag{3.23}$$

Further note that the identities of all resistances and individual resistance currents, i_a, i_b, and i_c, in the original circuit are lost in the equivalent circuit of Fig. 3.4b. Nevertheless the use of equivalent parallel resistances is often a valuable aid in the simplification of a complex network.

REMEMBER

We add reciprocals of resistances in parallel to obtain the reciprocal of the equivalent resistance (conductances in parallel are added). We add reciprocals of inductances in parallel to obtain the reciprocal of the equivalent inductance. We add capacitances in parallel to obtain the equivalent capacitance.

EXERCISES

Replace the network to the right of terminals *a* and *b* in the circuits of Exercises 5 through 8 by a single equivalent element. Use this equivalent value to calculate the voltage v_{ab} if the current *i* into terminal *a* is e^{2t} A.

5.

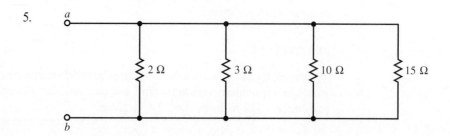

6.

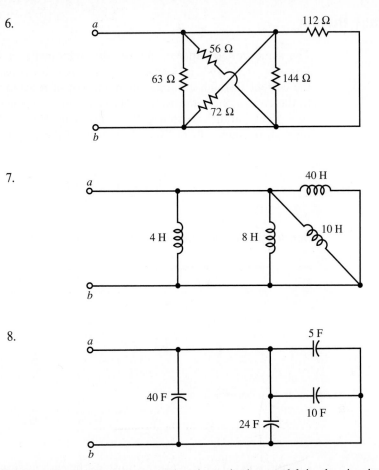

7.

8.

Replace the network to the right of terminals a and b in the circuits of Exercises 9 and 10 by an equivalent network consisting of only two elements. Use the equivalent circuit to determine the current into terminal a, if v_{ab} is $360e^{2t}$ V.

9.

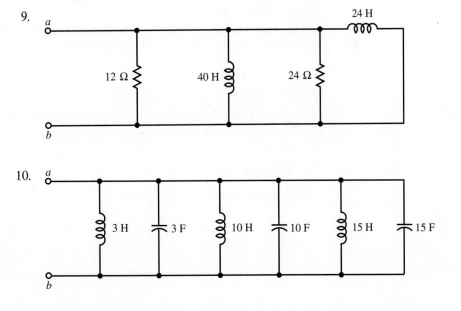

10.

3.3   Current Divider

The same voltage appears across all components in parallel, and the current through a resistance is inversely proportional to the resistance. This implies that the current through a parallel combination of resistances divides proportionally to the inverse of each resistance. We can express this mathematically by using the equivalent parallel resistance R_p as given by Eq. (3.17) to write the current i_k through resistance R_k of Fig. 3.3 in terms of the current i entering node a.

$$i_k = \frac{1}{R_k} v = \frac{1}{R_k}(R_p i) = \frac{\dfrac{1}{R_k}}{\dfrac{1}{R_p}} i \tag{3.24}$$

Substitution for R_p from Eq. (3.17) gives us the following:

Current Divider
$$i_k = \frac{\dfrac{1}{R_k}}{\dfrac{1}{R_1} + \dfrac{1}{R_2} + \cdots + \dfrac{1}{R_N}} i \tag{3.25}$$
or $\qquad$ $$i_k = \frac{G_k}{G_1 + G_2 + \cdots + G_N} i \tag{3.26}$$

Equations (3.25) and (3.26) are known as the *current divider relation*. This says that *the current through a parallel combination of conductances divides proportionally to the conductance.* Although this relation is not of the same fundamental importance as KVL and KCL, it is frequently a time-saver and provides a thought pattern that is very useful in electronic circuit design. We must use care in applying the current divider relation because the resistances must be in parallel, and if the current i enters the node, current i_k must leave the node or the sign on the equation will change. This is illustrated by the following example.

EXAMPLE 3.5

For the example of parallel resistances shown in Fig. 3.4, suppose that voltage v is unknown, but current i has been measured and is 6 A. Determine the values of i_b and i_c.

■ **SOLUTION**

The current divider relation gives

$$i_b = \frac{1/6}{(1/4) + (1/6) + (1/12)} 6 = 2 \text{ A}$$

and $\qquad$ $$i_c = -\frac{1/12}{(1/4) + (1/6) + (1/12)} 6 = -1 \text{ A} \qquad ■$$

The current divider relation should become a part of an engineer's analysis tools, since it is a valuable aid in analysis of more complex networks.

REMEMBER

The current through a parallel combination of conductances divides proportionally to the individual conductances.

EXERCISES

For the networks in Exercises 11 and 12, determine currents i_x and i_y with the use of the current divider relation. (Be careful with the current reference directions.)

11.

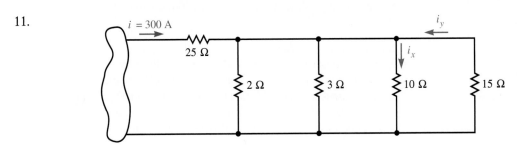

12.

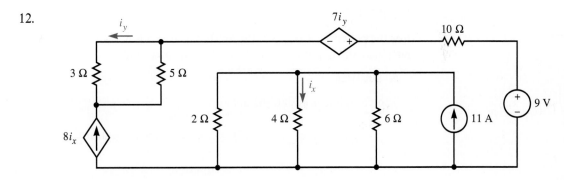

3.4 Single-Loop or Series Circuits

Although most electrical equipment is connected to the power source in parallel, internally there are often many devices connected in *series*, or devices modeled by components in series. An example of series-connected components is shown in Fig. 3.5.

FIGURE 3.5
A series circuit

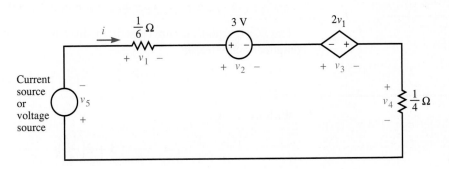

Series Components

Two or more components are in **series** if they are connected end to end with only two components connected at each of the intervening nodes.

Application of KCL to a closed surface that cuts any two branches of a series circuit gives the following result:

At any specified time the same current flows through each component in series.

The *general procedure* to solve a single-loop circuit problem is to first find the current i through the elements. If this is specified by a current source, current i is known. If the current is not known, apply KVL around the loop. Write all unknown voltages in terms of the current i with the use of terminal or control equations and solve for i. Once i is found, the individual component voltages are easily calculated. The following three examples demonstrate the procedure.

EXAMPLE 3.6

For the network of Fig. 3.5, the left-hand source is a 12-A current source with the reference arrow pointing up. Find the indicated voltages.

■ **SOLUTION**

The component equations are

$$v_1 = \frac{1}{6}i = \frac{1}{6}(12) = 2 \text{ V} \quad \text{(terminal equation)} \tag{3.27}$$

$$v_2 = 3 \text{ V} \quad \text{(source equation)} \tag{3.28}$$

$$v_3 = -2v_1 = -2(2) = -4 \text{ V} \quad \text{(control equation)} \tag{3.29}$$

$$v_4 = \frac{1}{4}i = \frac{1}{4}(12) = 3 \text{ V} \quad \text{(terminal equation)} \tag{3.30}$$

We can now find the voltage across the current source from a KVL equation written around the loop:

$$v_5 + v_1 + v_2 + v_3 + v_4 = 0 \tag{3.31}$$

This KVL equation can be written directly in terms of i with the use of the component equations.

$$v_5 + \frac{1}{6}i + 3 - 2\left(\frac{1}{6}i\right) + \frac{1}{4}i = 0 \tag{3.32}$$

With $i = 12$ A,

$$v_5 = -\frac{12}{6} - 3 + 2\left(\frac{12}{6}\right) - \frac{12}{4} = -4 \text{ V} \quad ■ \tag{3.33}$$

If the current through the loop is not specified by a current source, we can easily find the current by the use of KVL and terminal, source, and control equations.

EXAMPLE 3.7

For the circuit of Fig. 3.5, the left-hand source is a 12-V source with the (+) reference mark at the top of the page, so $v_s = -12$ V. Determine current i.

■ **SOLUTION**

Use the component equations to write KVL around the loop in terms of current i. This is the same as Eq. (3.32), except that voltage v_s is now known and current i is unknown:

$$-12 + \frac{1}{6}i + 3 - 2\left(\frac{1}{6}i\right) + \frac{1}{4}i = 0 \tag{3.34}$$

This gives
$$i = 108 \text{ A} \tag{3.35}$$

Now that we know current i, the voltages not already specified by voltage sources are easily found. ■

It is no more difficult to write the necessary equations for a single-loop circuit with resistance, inductance, and capacitance than it is for the resistive case.

EXAMPLE 3.8

Find the indicated voltages for the circuit shown in Fig. 3.6 if (a) the left-hand source is a current source, and (b) the left-hand source is a voltage source. (This would be a model for an unloaded power transmission line. R and L are the resistance and inductance of the wire, and C is the capacitance between wires.)

FIGURE 3.6
A series *RLC* circuit

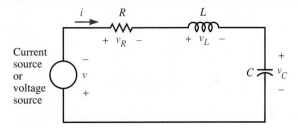

■ **SOLUTION**

(a) The terminal equations give the element voltages:

$$v_R = Ri \tag{3.36}$$

$$v_L = L\frac{d}{dt}i \tag{3.37}$$

$$v_C = \frac{1}{C}\int_{-\infty}^{t} i\, d\tau \tag{3.38}$$

where current i is specified by the current source. We can now find the voltage v across the current source from a KVL equation around the loop:

$$v + v_R + v_L + v_C = 0 \tag{3.39}$$

Substitution from the component equations gives us

$$v + Ri + L\frac{d}{dt}i + \frac{1}{C}\int_{-\infty}^{t} i\,d\tau = 0 \qquad (3.40)$$

(We can skip the first step when we write the KVL equations and write the KVL equation directly in terms of i.) With current i known, this KVL equation gives voltage v.

(b) If the current source in the network is replaced by a voltage source, the element voltages cannot be found immediately. The KVL equation given in Eq. (3.40) is an integro-differential equation and must be solved for current i in terms of voltage v, which is specified by the voltage source. ∎

REMEMBER

The current through each component in series is the same. This current is known if the series circuit is driven by a current source. The voltage across the resistance, inductance, or capacitance is found with the use of this current and the terminal equation. The voltage across the current source is found from a KVL equation. If a series circuit is not driven by a current source, the current through the components is found from a KVL equation.

EXERCISES

13. For the network shown below, determine voltages v_1, v_2, v_3, v_4, and v_5, and the powers p_1, p_2, p_3, p_4, and p_5 absorbed by each component.

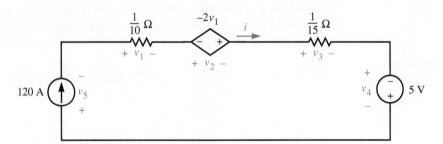

14. Replace the current source in the circuit above with a voltage source, $(+)$ reference mark at the bottom, of value 49 V. Find voltages v_1, v_2, v_3, v_4, and v_5.

15. For the circuit shown below, find voltages v_1, v_2, v_3, v_4, v_5, and v_6 for $t > 0$, if $v_3(0) = -5$ V.

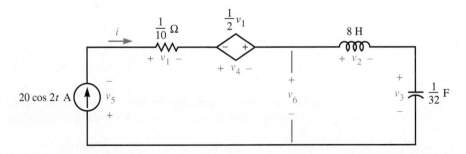

16. Replace the current source in the network above with a voltage source, $(+)$ reference mark at the top, of value $v_s = 2 \cos 2t$ V for $t \geq 0$. Write a KVL equation that gives a differential equation that can be solved for i. The initial conditions are $i(0) = 20$ A and $v_3(0) = 10$ V. Do not solve for current i.

3.5 Series Elements

The analysis of a network containing series elements can often be simplified if we replace the series elements with an equivalent element. Consider the N series resistances of Fig. 3.7a.

FIGURE 3.7
The equivalent resistance of N resistances in series:
(a) series resistances;
(b) the equivalent resistance

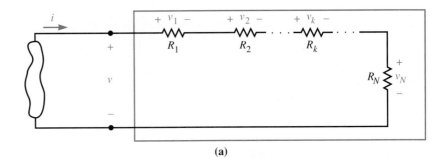

(a)

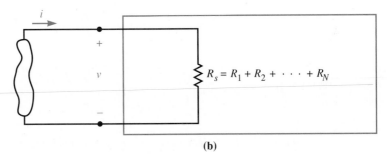

(b)

We can relate the voltage v to the individual element voltages by a KVL equation:

$$v = v_1 + v_2 + \cdots + v_N \tag{3.41}$$

Substitution of the terminal equations for resistance into Eq. (3.41) yields

$$\begin{aligned} v &= R_1 i + R_2 i + \cdots + R_N i \\ &= (R_1 + R_2 + \cdots + R_N)i \\ &= R_s i \end{aligned} \tag{3.42}$$

where the *equivalent resistance* R_s *for N resistances in series* is

---

Resistances in Series

$$R_s = R_1 + R_2 + \cdots + R_N \tag{3.43}$$

We see that *resistances in series are added* to give the equivalent resistance.

If the series elements of Fig. 3.7 were inductances, substitution of the corresponding terminal equations into Eq. (3.41) would yield

$$v = L_1 \frac{d}{dt} i + L_2 \frac{d}{dt} i + \cdots + L_N \frac{d}{dt} i$$

$$= (L_1 + L_2 + \cdots + L_N) \frac{d}{dt} i$$

$$= L_s \frac{d}{dt} i \qquad (3.44)$$

where the *equivalent inductance of N inductances in series* is

Inductances in Series

$$L_s = L_1 + L_2 + \cdots + L_N \qquad (3.45)$$

Thus *inductances in series are added* to give the equivalent inductance.

Similarly, if the series elements of Fig. 3.7 were capacitances, substitution of the corresponding terminal equations into Eq. (3.41) would yield

$$v = \frac{1}{C_1} \int_{-\infty}^{t} i \, d\tau + \frac{1}{C_2} \int_{-\infty}^{t} i \, d\tau + \cdots + \frac{1}{C_N} \int_{-\infty}^{t} i \, d\tau$$

$$= \left(\frac{1}{C_1} + \frac{1}{C_2} + \cdots + \frac{1}{C_N} \right) \int_{-\infty}^{t} i \, d\tau$$

$$= \frac{1}{C_s} \int_{-\infty}^{t} i \, d\tau \qquad (3.46)$$

where the *equivalent capacitance of N capacitances in series* is

Capacitances in Series

$$C_s = \frac{1}{\dfrac{1}{C_1} + \dfrac{1}{C_2} + \cdots + \dfrac{1}{C_N}} \qquad (3.47)$$

Thus the *reciprocals of capacitances in series are added* to give the *reciprocal* of the equivalent capacitance.

As with parallel circuits, the equivalence is with respect to the terminal voltage and current, v and i only. The individual element voltages do not appear in the equivalent circuit.

EXAMPLE 3.9

Find the equivalent resistance for the series combination of resistances shown in Fig. 3.8a. Then determine voltage v. Current $i = 2$ A is known.

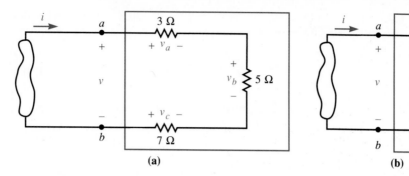

(a)   **(b)**

FIGURE 3.8 Equivalent series resistance example: (*a*) series resistances; (*b*) equivalent resistance

■ **SOLUTION**

$$R_s = 3 + 5 + 7 = 15\,\Omega \tag{3.48}$$

$$v = R_s i = 15(2) = 30 \text{ V} \quad ■ \tag{3.49}$$

We should note that it really saved no time to find the equivalent series resistance since the desired result was to find v rather than R_s. The voltage v is just as easily found from KVL as follows:

$$v = 3(2) + 5(2) + 7(2) = 30 \text{ V} \tag{3.50}$$

Further note that the identities of all the original resistances and the original resistance voltages, v_a, v_b, and v_c, are lost in the equivalent circuit of Fig. 3.8b. Nevertheless the use of an equivalent series resistance is often a valuable aid in the simplification of more complex networks.

REMEMBER

We add resistances in series to obtain the equivalent resistance. We add inductances in series to obtain the equivalent inductance. We add reciprocals of capacitances in series to obtain the reciprocal of the equivalent capacitance.

EXERCISES

What is the equivalent element looking into terminals a and b of the networks in Exercises 17 through 22? Use this equivalent value to calculate the current i if v_{ab} is $148e^{3t}$ V.

17.

18.

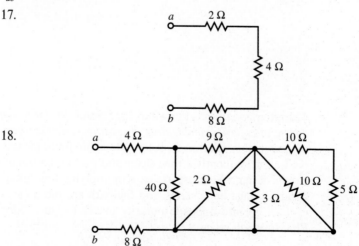

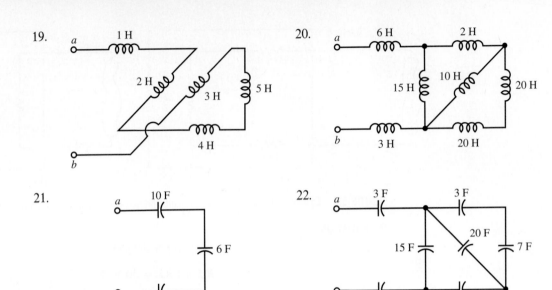

3.6 Voltage Divider

The current is the same through all components in series, and the voltage across a resistance is directly proportional to the resistance. This implies that the voltage across a series connection of resistances divides proportionally to each resistance. The equivalent series resistance R_s given by Eq. (3.43) can be used to give the voltage v_k across resistance R_k of Fig. 3.7 in terms of the voltage v across the series combination:

$$v_k = R_k i = R_k \left(\frac{1}{R_s} v \right) \tag{3.51}$$

Substitution for R_s from Eq. (3.43) gives us

Voltage Divider

$$v_k = \frac{R_k}{R_1 + R_2 + \cdots + R_N} v \tag{3.52}$$

The latter equation is known as the *voltage divider relation*. This says that *the voltage across a series combination of resistances divides proportionally to the resistance*. Although this relation is not of the same fundamental importance as KVL and KCL, it is frequently a time saver and provides a useful way for us to think about circuit design. We must use care in applying the voltage divider relation. The resistances must be in series, and the path around a loop containing voltages v and v_k must go into opposite polarity marks ($+$ or $-$) for v and v_k, or the sign on Eq. (3.52) will change.

EXAMPLE 3.10

For the previous example of series resistances shown in Fig. 3.8, suppose that current i is unknown, but voltage v is measured to be 30 V. Determine the value of v_b and v_c.

■ **SOLUTION**

From the voltage divider relation,

$$v_b = \frac{5}{3 + 5 + 7} \, 30 = 10 \text{ V} \tag{3.53}$$

and

$$v_c = -\frac{7}{3 + 5 + 7} \, 30 = -14 \text{ V} \quad ■ \tag{3.54}$$

The voltage divider relation should become a part of an engineer's analysis tools, since it is a valuable aid in analysis of more complex networks.

REMEMBER

The voltage across a series combination of resistances divides proportionally to the resistance.

EXERCISES

23. The storage battery for a boat is modeled as a 12-V voltage source in series with a 0.1-Ω resistance. We will use a resistance of 0.2 Ω as a crude model for the boat's starter motor. Use the voltage divider relation to determine the voltage across the starter motor when it is starting the engine. (The two resistances and the source are in series.)

For the networks shown in Exercises 24 and 25, calculate voltage v_x and v_y.

24.

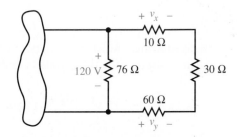

25.

3.7 Network Analysis by Simplification

We have seen how a group of parallel or series elements of the same type can be replaced by a single equivalent element without affecting the remainder of the circuit. The equivalent element *simplifies* the circuit. Repeated replacement of series or parallel elements by an equivalent element may permit us to greatly simplify a complicated circuit and solve for some voltage or current value we need. The following example demonstrates network analysis by simplification.

EXAMPLE 3.11

The circuit shown in Fig. 3.9 is a model for the lighting system on a small boat. Determine the terminal voltage v_b of the battery and the voltage v_L for the 10-Ω lamp.

FIGURE 3.9
Lighting system for a small boat

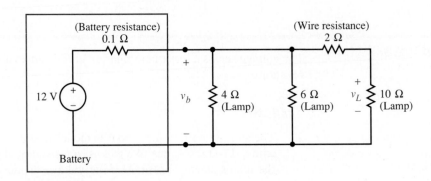

■ SOLUTION

The equivalent resistance for the series combination of the 2-Ω and 10-Ω resistances is

$$R_s = 2 + 10 = 12 \ \Omega \tag{3.55}$$

This equivalent resistance is in parallel with the 4-Ω and 6-Ω resistances. The equivalent resistance for this parallel combination is

$$R_p = \frac{1}{(1/4) + (1/6) + (1/12)} = 2 \ \Omega \tag{3.56}$$

This gives us the simple equivalent circuit (Fig. 3.10).

FIGURE 3.10
An equivalent lighting circuit

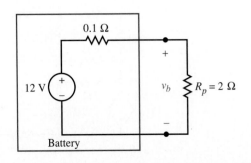

We can obtain voltage v_b from the voltage divider relation:

$$v_b = \frac{2}{2 + 0.1} \, 12 = 11.43 \text{ V}$$

Because the battery resistance is much lower than the load resistance, which consists of the lights, most of the 12-V internal voltage of the battery appears across the battery terminals. Use of the voltage divider relation with the original circuit gives

$$v_L = \frac{10}{10 + 2} \, v_b = \frac{10}{12} (11.43) = 9.52 \text{ V} \tag{3.57}$$

We can see that the 2-Ω resistance of the wire further reduces the voltage for the 10-Ω lamp. ■

A physical system is only approximated by a network model. For example, the circuit of Fig. 3.9 neglected the resistance of the wire between the 4-Ω and 6-Ω lamps. We will now analyze a more complicated circuit.

EXAMPLE 3.12

Use series and parallel resistance combinations to find an equivalent resistance for the portion of the network to the right of terminals a and b in Fig. 3.11a. Solve for voltage v_{ab} with one KCL equation.

FIGURE 3.11
Network simplification used to determine the voltage v_{ab}: (a) network with series and parallel elements; (b) network after making one equivalent series and one equivalent parallel resistance; (c) after a second series and parallel equivalent; (d) a parallel equivalent used; (e) a series equivalent used

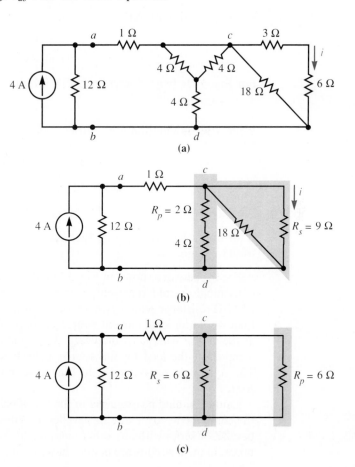

FIGURE 3.11
Continued

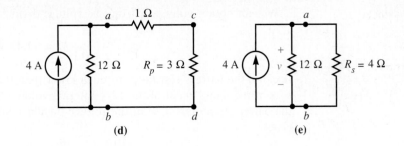

(d)   **(e)**

■ SOLUTION

For Fig. 3.11b (with reference to Fig. 3.11a),

$$R_s = 3 + 6 = 9\ \Omega \tag{3.58}$$

$$R_p = \frac{1}{(1/4) + (1/4)} = 2\ \Omega \tag{3.59}$$

For Fig. 3.11c (with reference to Fig. 3.11b),

$$R_s = 2 + 4 = 6\ \Omega \tag{3.60}$$

$$R_p = \frac{1}{(1/9) + (1/18)} = 6\ \Omega \tag{3.61}$$

For Fig. 3.11d (with reference to Fig. 3.11c),

$$R_p = \frac{1}{(1/6) + (1/6)} = 3\ \Omega \tag{3.62}$$

For Fig. 3.11e (with reference to Fig. 3.11d),

$$R_s = 3 + 1 = 4\ \Omega \tag{3.63}$$

Write a KCL equation for a surface enclosing the top node.

$$-4 + \frac{1}{12}\ v_{ab} + \frac{1}{4}\ v_{ab} = 0 \tag{3.64}$$

This gives

$$v_{ab} = 12\ \text{V} \tag{3.65}$$

If the object had been to find the current i through the 6-Ω resistance, some work would remain. ■

For the network shown in Fig. 3.11a, the current source and resistance to the left of terminals a and b represent a *practical current source*. The inclusion of the parallel 12-Ω resistance would give a more accurate model for a physical current source than would an ideal current source in that the output current i_{ab} would decrease as the output voltage v_{ab} increased. The resistances to the right of terminals a and b represent the *load* on the source. (The load absorbs power from the source.) The equivalent resistance R_s in Fig. 3.11e is the *equivalent load* on the practical source.

Caution should be exercised in the use of parallel and series equivalents. Be sure the components are really in parallel or series. Some components may not be in parallel or series with any other component. For example, see Fig. 3.12. The resistances in these circuits are neither in series nor in parallel. In Chapter 2, we defined

FIGURE 3.12
Networks with no series
or parallel components:
(a) planar network;
(b) nonplanar network

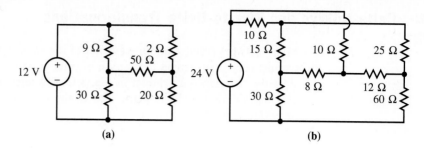

(a)   (b)

a network such as that in Fig. 3.12a, which can be drawn so that no lines cross, to be a *planar network*. The network of Fig. 3.12b cannot be drawn so that no lines cross and is a *nonplanar network*.

REMEMBER

Check to be sure that components are in parallel before calculation of a parallel equivalent or use of the current divider equation. Be sure that components are in series before calculation of a series equivalent or use of the voltage divider equation.

EXERCISES

Calculate v_x and v_y for the networks shown in Exercises 26 through 28.

26.

27.

28.

3.8 Delta-to-Wye and Wye-to-Delta Transformations

Many networks that contain elements of the same type that are not in series or parallel can be simplified by the use of three-terminal equivalent networks. Consider the two networks shown in Fig. 3.13. We refer to the network of Fig. 3.13a as a wye (Y) or tee (T) network, and to the network of Fig. 3.13b as a delta (Δ) or pi (Π) network.

FIGURE 3.13
Equivalent Δ and Y networks: (*a*) Y-connected resistances (T-connected resistances); (*b*) Δ-connected resistances (Π-connected resistances)

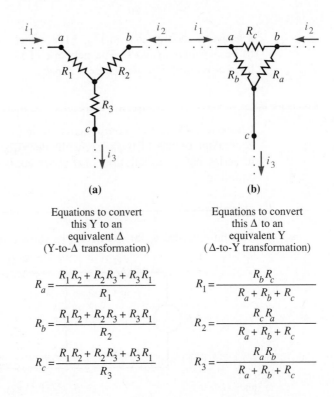

(a)

(b)

Equations to convert
this Y to an
equivalent Δ
(Y-to-Δ transformation)

$$R_a = \frac{R_1 R_2 + R_2 R_3 + R_3 R_1}{R_1}$$

$$R_b = \frac{R_1 R_2 + R_2 R_3 + R_3 R_1}{R_2}$$

$$R_c = \frac{R_1 R_2 + R_2 R_3 + R_3 R_1}{R_3}$$

Equations to convert
this Δ to an
equivalent Y
(Δ-to-Y transformation)

$$R_1 = \frac{R_b R_c}{R_a + R_b + R_c}$$

$$R_2 = \frac{R_c R_a}{R_a + R_b + R_c}$$

$$R_3 = \frac{R_a R_b}{R_a + R_b + R_c}$$

The Δ and Y networks will be equivalent with respect to the effect on any other network if the numerical value of the coefficients in the equations that relate v_{ac}, v_{bc}, i_1, and i_2 are the same for the Δ and Y networks. (From KVL and KCL, this also includes v_{ab} and i_3.) Equivalence holds if the resistances are related in the manner given by the equations of Fig. 3.13.

The equations for conversion between Δ- and Y-connected inductances are identical in form to those for resistances given in the equations of Fig. 3.13. Equations for conversion between Δ- and Y-connected capacitances can be formed by replacement of the resistances in the equations of Fig. 3.13 by the reciprocals of the corresponding capacitances. Remember that, as with all equivalent circuits, the equivalence is with respect to the external voltages and currents.

A derivation of the resistive Δ-to-Y and Y-to-Δ transformations follows Example 3.14.

EXAMPLE 3.13

Determine a Y that is equivalent to the Δ shown in Fig. 3.14 (the Y will have the same terminal characteristics as the Δ).

FIGURE 3.14
A Δ-Y transformation: (a)
The original Δ or Π; (b)
the equivalent Y or T

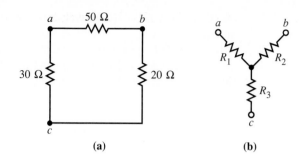

(a)   **(b)**

■ SOLUTION

We can obtain the resistance values from the equations given in Fig. 3.13:

$$R_1 = \frac{(30)(50)}{20 + 30 + 50} = \frac{1500}{100} = 15 \ \Omega$$

$$R_2 = \frac{(50)(20)}{100} = 10 \ \Omega$$

$$R_3 = \frac{(20)(30)}{100} = 6 \ \Omega \ \blacksquare$$

EXAMPLE 3.14

Determine the equivalent resistance for the network to the right of terminals a and b in Fig. 3.15. Use this simplified circuit to calculate current i.

FIGURE 3.15
Network with resistances in
Δ

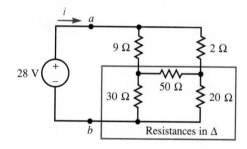

■ SOLUTION

The lower three resistances in Fig. 3.15 are connected in a Δ configuration and can be replaced with an equivalent Y. This will not alter the current through any

FIGURE 3.16
The first step in simplifying
the network of Fig. 3.15

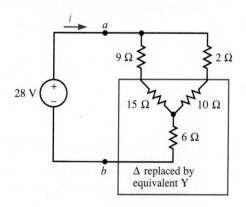

other component. (Alternatively, the upper three resistances can be replaced by an equivalent Y.)

In the preceding example we calculated the resistance values for a Y that is equivalent to the Δ formed by the lower three resistances. Replacement of the bottom three resistances with this equivalent Δ gives the equivalent circuit shown in Fig. 3.16.

The 9-Ω and 15-Ω resistances are in series, for an equivalent resistance of 24 Ω, and the 10-Ω and 2-Ω resistances are in series, for an equivalent resistance of 12 Ω. The equivalent resistance of these 24-Ω and 12-Ω resistances in parallel is

$$R_p = \frac{1}{(1/24) + (1/12)} = 8 \ \Omega$$

Thus the resistive portion of the network is equivalent to

$$R = 8 + 6 = 14 \ \Omega$$

and

$$i = \frac{28}{R} = \frac{28}{14} = 2 \text{ A} \quad \blacksquare$$

Equivalent Δ-Y circuits can sometimes be used to reduce a nonplanar network to a planar network (see Problem 115).

Derivation of the Δ-to-Y and Y-to-Δ Transformations*

Ohm's law and KVL applied to the network of Fig. 3.13a give

$$v_{ac} = R_1 i_1 + R_3 i_3 = R_1 i_1 + R_3(i_1 + i_2)$$
$$= (R_1 + R_3)i_1 + R_3 i_2 \tag{3.66}$$

where the second step required the use of KCL. In a similar manner,

$$v_{bc} = R_3 i_1 + (R_2 + R_3)i_2 \tag{3.67}$$

We can solve Eqs. (3.66) and (3.67) for currents i_1 and i_2 by the use of Cramer's rule.[†] This gives

$$i_1 = \frac{R_2 + R_3}{R_1 R_2 + R_2 R_3 + R_3 R_1} v_{ac} - \frac{R_3}{R_1 R_2 + R_2 R_3 + R_3 R_1} v_{bc} \tag{3.68}$$

$$i_2 = -\frac{R_3}{R_1 R_2 + R_2 R_3 + R_3 R_1} v_{ac} + \frac{R_1 + R_3}{R_1 R_2 + R_2 R_3 + R_3 R_1} v_{bc} \tag{3.69}$$

We can easily analyze the network of Fig. 3.13b by the use of KCL and Ohm's law to give

$$i_1 = \frac{1}{R_b} v_{ac} + \frac{1}{R_c} v_{ab} = \frac{1}{R_b} v_{ac} + \frac{1}{R_c} (v_{ac} - v_{bc})$$

$$= \left(\frac{1}{R_b} + \frac{1}{R_c} \right) v_{ac} - \frac{1}{R_c} v_{bc} \tag{3.70}$$

where the second step required the use of KVL. In a similar manner,

$$i_2 = -\frac{1}{R_c} v_{ac} + \left(\frac{1}{R_a} + \frac{1}{R_c} \right) v_{bc} \tag{3.71}$$

---

* This derivation may be omitted without loss of continuity.

† Cramer's rule is presented in Appendix A.

We can solve Eqs. (3.70) and (3.71) for voltages v_{ac} and v_{bc} by the use of Cramer's rule to give

$$v_{ac} = \frac{R_aR_b + R_bR_c}{R_a + R_b + R_c}i_1 + \frac{R_aR_b}{R_a + R_b + R_c}i_2 \qquad (3.72)$$

and

$$v_{bc} = \frac{R_aR_b}{R_a + R_b + R_c}i_1 + \frac{R_aR_b + R_cR_a}{R_a + R_b + R_c}i_2 \qquad (3.73)$$

The Δ-to-Y transformation of Fig. 3.13 can be established if we equate the co-efficients for i_1 and i_2 in Eqs. (3.66) and (3.67) to the corresponding coefficients in Eqs. (3.72) and (3.73). If we equate the coefficients on v_{ac} and v_{bc} in Eqs. (3.70) and (3.71) to the corresponding coefficients in Eqs. (3.68) and (3.69), we establish the Y-to-Δ transformation of Fig. 3.13.

EXERCISES

Transform each of the following Δ circuits to a Y circuit. Be sure to label the terminals for the equivalent circuit.

29.

30.

Transform each of the following Y circuits to a Δ circuit. Label the terminals for the equivalent circuit.

31.

32.

33. In the circuit below, use a Y-Δ transformation to calculate currents i_1 and i_2.

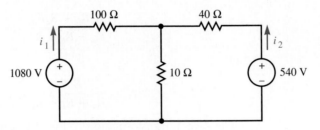

34. Find the equivalent resistance for the resistive network connected between terminals *a* and *b* in the circuit below. Use this to calculate current *i*.

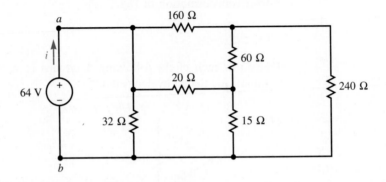

3.9 Summary

We began this chapter by analyzing circuits composed of components that were all in parallel. This analysis led to the concept of an equivalent component or circuit, and to the current divider relation. A similar analysis of series circuits yielded additional equivalent components and the voltage divider relation. We used these equivalent components to reduce more complicated networks to equivalent series or parallel circuits, which are easily analyzed. Finally we extended our use of equivalent circuits by the introduction of Δ-Y transformations.

KEY FACTS

◆ The voltage is the same across components in parallel.

◆ Equivalent values for elements in parallel:

$$R_p = \frac{1}{\dfrac{1}{R_1} + \dfrac{1}{R_2} + \cdots + \dfrac{1}{R_N}}$$

$$L_p = \frac{1}{\dfrac{1}{L_1} + \dfrac{1}{L_2} + \cdots + \dfrac{1}{L_N}}$$

$$C_p = C_1 + C_2 + \cdots + C_N$$

- Current divider relation for parallel resistances:

$$i_k = \frac{\dfrac{1}{R_k}}{\dfrac{1}{R_1} + \dfrac{1}{R_2} + \cdots + \dfrac{1}{R_N}} \, i$$

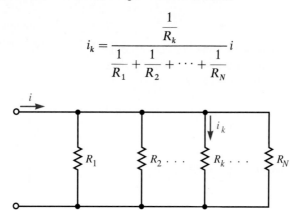

- The current is the same through components in series.

- Equivalent values for elements in series:

$$R_s = R_1 + R_2 + \cdots + R_N$$

$$L_s = L_1 + L_2 + \cdots + L_N$$

$$C_s = \frac{1}{\dfrac{1}{C_1} + \dfrac{1}{C_2} + \cdots + \dfrac{1}{C_N}}$$

- Voltage divider relation for series resistances:

$$v_k = \frac{R_k}{R_1 + R_2 + \cdots + R_N} \, v$$

- Δ and Y equivalents:

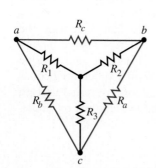

Y-to-Δ TRANSFORMATION	**Δ-to-Y TRANSFORMATION**
$R_a = \dfrac{R_1 R_2 + R_2 R_3 + R_3 R_1}{R_1}$	$R_1 = \dfrac{R_b R_c}{R_a + R_b + R_c}$
$R_b = \dfrac{R_1 R_2 + R_2 R_3 + R_3 R_1}{R_2}$	$R_2 = \dfrac{R_c R_a}{R_a + R_b + R_c}$
$R_c = \dfrac{R_1 R_2 + R_2 R_3 + R_3 R_1}{R_3}$	$R_3 = \dfrac{R_a R_b}{R_a + R_b + R_c}$

Section 3.1

For each circuit below, determine voltage v_{ab}; currents i_a, i_b, and i_c; and the instantaneous power absorbed by each component. The sources are $v_s = 60 \cos 2t$ V and $i_s = 50 \cos 2t$ A. Would the ratio v_{ab}/i_a represent an equivalent resistance connected to the left-hand source? Explain.

1.

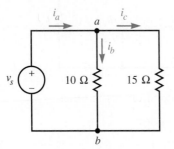

2.

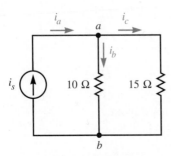

3.

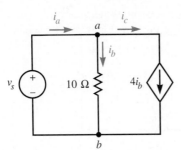

4.

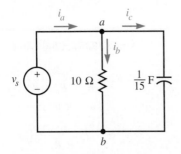

5. Calculate voltages v_{ab}, v_{cb}, and v_{ac} and currents i_1, i_2, and i_3 for the network shown below.

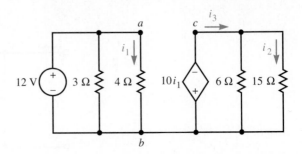

6. Calculate voltages v_{ab} and v_{cd} and currents i_1 and i_2 for the network shown below.

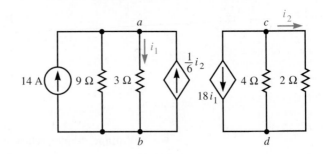

7. Solve for current i in the network shown below if $i_1(0) = i_2(0) = 0$.

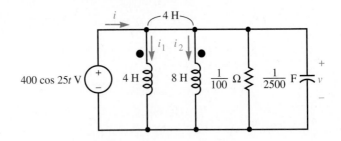

8. Repeat Problem 7 if the voltage source value is changed to $400 \cos 50t$ V.
9. For the network of Problem 7, replace the voltage source with a current source having the reference arrow pointing up and a value of $2 \cos 25t$ A, $t \geq 0$. Write a differential equation that can be solved for v. Assume that no energy is stored in the inductances or capacitance at $t = 0$.

10. Repeat Problem 7 if the dot mark on the 8-H inductance is located on the opposite end.

Section 3.2

Determine a single equivalent component with respect to terminals a and b for each network shown below.

11.

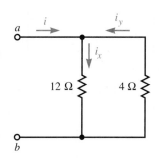

12.

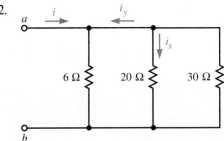

13.

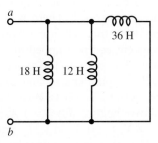

14.

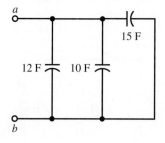

15. Replace the network to the right of terminals a and b in the circuit below with a single resistance, inductance, and capacitance.

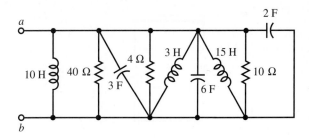

16. Use Eq. (3.17) and algebra to show that the equivalent resistance of two resistances R_1 and R_2 in parallel is given by

$$R_p = \frac{R_1 R_2}{R_1 + R_2}$$

From Eq. (3.17), develop a relation similar to the above equation for three resistances in parallel, then for four resistances in parallel.

17. A circuit consists of N resistances $R_1, R_2, \ldots, R_N$ in parallel. What is the equivalent resistance if $R_n = n^2 R$, $n = 1, 2, \ldots, N$? What is the equivalent resistance as N becomes infinite?

18. For the network below, $i_1(0) = 8$ A and $i_2(0) = 2$ A. Calculate $i(0)$ and the energy stored in L_1 and L_2 at $t = 0$.

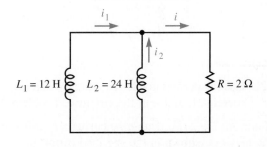

Replace the parallel inductances in the circuit above by an equivalent inductance as shown below. What is $i(0)$ for this network? Calculate the energy stored in L_p at $t = 0$. Is this equal to the sum of the energies stored in L_1 and

L_2 at $t = 0$? If not, explain the difference. What is $i(\infty)$? Use this and the initial value of stored energy to find $i_1(\infty)$ and $i_2(\infty)$.

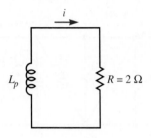

Calculate the value of a single inductance that is equivalent to the coupled inductances connected between terminals a and b.

19.

20.

Section 3.3

21. Use the current divider equation to determine currents i_b and i_c in the circuit of Problem 2.

If current $i = 24$ A, determine currents i_x and i_y in the network, shown in the specified problem.

22. Problem 11   23. Problem 12

24. You have an ammeter that reaches full scale when the current through it is 1 mA. The meter resistance is known to be 50 Ω. What resistance must you place in parallel with this meter so that you can use the meter with this *shunt resistance* in an application that requires a meter that measures 1 A at full scale? Refer to the illustration below.

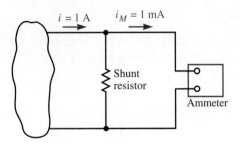

25. Determine voltage v for the network shown below.

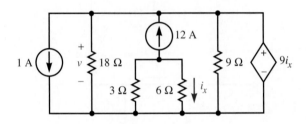

26. For the dependent voltage source shown above, substitute a dependent current source with the reference arrow pointing up and a control equation giving this current a value of $2i_x$. Determine v.

27. Develop a current divider relation for N inductances in parallel. There is a theoretical problem with this relation because a circulating current can be trapped in a loop of inductances. The nonzero resistance of practical coils will dissipate energy and cause the circulating current to go to zero. For the network shown below, $i_1(0) = -1$ A, $i_2(0) = 1$ A, and $v = 8$ V, $t \geq 0$. Find i_1, i_2, and i for $t > 0$ and show that these particular initial conditions void the current divider relation for these ideal coils.

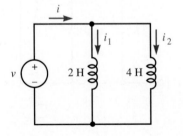

28. Develop a current divider relation for N capacitances in parallel.

Section 3.4

For each circuit shown below, determine current i; voltages v_{ac}, v_{ab}, and v_{bc}; and the instantaneous power absorbed by each component. The sources are $i_s = 60 \cos 2t$ A and $v_s = 50 \cos 2t$ V. Would the ratio v_{ac}/i represent an equivalent resistance connected to the left-hand source? Explain.

29.

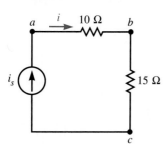

30.

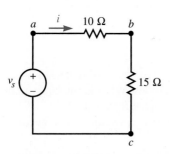

31.

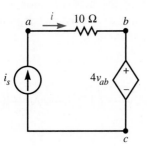

32.

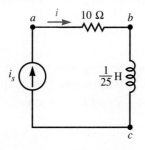

33. Find the voltage v, the current i, and the power absorbed by each component of the network given below.

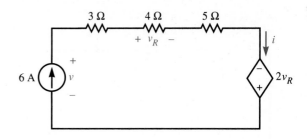

34. Replace the 6-A source in the circuit above with a 24-V voltage source having the (+) reference mark toward the top of the page. Determine voltage v and current i.

35. Calculate voltages v_1, v_2, and v_3 in the network shown below.

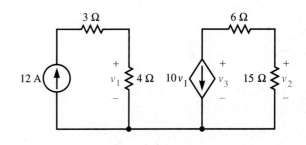

For the networks shown below, find v_o or i_o.

36.

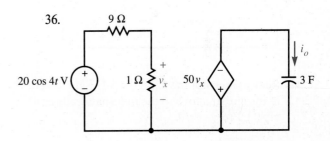

37.

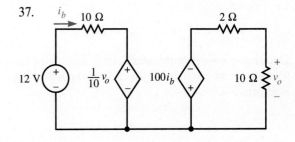

38. Determine v_1, v_2, v_3, v_4, and v_5 for the circuit shown below, if $v_3(0) = 0$ and (a) $i_s = 20 \cos 2t$ A; (b) $i_s = 20 \cos 4t$ A.

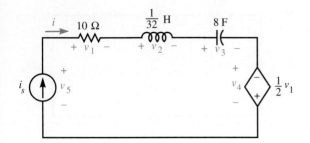

39. Replace the current source in the network above with a voltage source, (+) reference mark at the top, of value $v_s = 100 \cos 2t$ V for $t \geq 0$. Write a KVL equation in terms of current i for the circuit. The initial conditions are $i(0) = 2.0$ A and $v_3(0) = 10$ V. Do not solve the equation for i.

40. A series circuit contains coupled inductances as shown below. Find voltage v for $t > 0$ if $v_o(0) = 0$ V and $i = 4 \cos 2t$ A, $t \geq 0$.

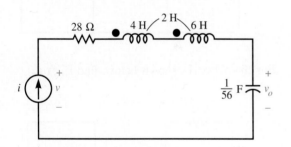

41. Relocate the dot mark on the 6-H inductance in the circuit above to the right-hand side and repeat Problem 40.

Section 3.5

Determine a single equivalent component with respect to terminals a and b for each of the following networks.

42.

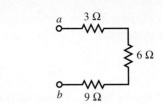

43.

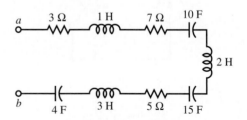

44.

45. A circuit consists of N resistances R_1, R_2, ..., R_N in series. What is the equivalent resistance if $R_n = (1/n)R$, $n = 1, 2, \ldots, N$? What is the equivalent resistance as N becomes infinite?

46. Replace the network to the right of terminals a and b in the network shown below by an equivalent series resistance R_s, inductance L_s, and capacitance C_s.

47. For the following network, $v_1(0) = 8$ V and $v_2(0) = 2$ V. Calculate $v(0)$ and the energy stored in C_1 and C_2 at $t = 0$.

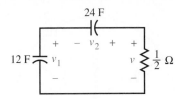

Replace the series capacitances by an equivalent capacitance as shown below. What is $v(0)$ for this network? Calculate the energy stored in C_s at $t = 0$. Is this equal to the sum of the energies stored in C_1 and C_2 at $t = 0$? If not, explain the difference. What is $v(\infty)$? Use this and the results from above to find $v_1(\infty)$ and $v_2(\infty)$.

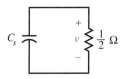

Calculate the value of a single inductance that is equivalent to the coupled inductances connected between terminals a and b.

48.

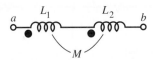

49.

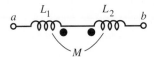

Section 3.6

50. Use the voltage divider equation to calculate voltages v_{ab} and v_{cb} in the network of Problem 30.

51. A practical voltmeter can often be modeled as a resistance, and the meter reading, although calibrated in volts, is proportional to the current the voltage causes to flow through this resistance. A 10-V meter requires 1 mA of current for a full-scale reading. What resistance must be placed in series with the voltmeter so that 100 V is required for a full-scale reading?

52. Determine current i for the network shown below.

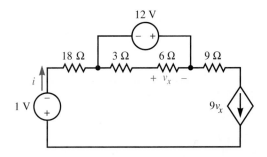

53. An unknown resistance R_x can be measured in terms of known resistances R_a, R_b, and R_c with an instrument called a Wheatstone bridge. Refer to the illustration below. We say that the bridge is balanced when $v_{12} = 0$. For the balanced condition, find R_x as a function of R_a, R_b, and R_c.

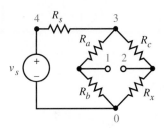

54. Develop a voltage divider relation for N inductances in series.

55. Develop a voltage divider relation for N capacitances in series. There are theoretical and practical problems with this relation because a voltage can be trapped on the capacitors as a result of capacitor leakage (modeled by a resistance in parallel with the capacitance) or as a result of previously connected components. These voltages can be trapped for minutes or even days on practical capacitors. For the network shown below, $v_1(0) = -1$ V,

$v_2(0) = 1$ V, and $i = 1$ A, $t \geq 0$. Find v_1, v_2, and v for $t \geq 0$ and show that these particular initial conditions void the voltage divider relation.

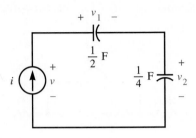

Section 3.7

Find the equivalent resistance when connections to the network shown below are made to the terminals specified in Problems 56 through 62.

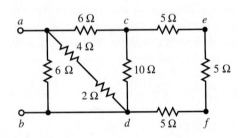

56. a and b
57. c and d
58. e and f
59. e and d
60. b and d
61. c and e
62. a and c

For the networks of Problems 63 through 65, determine the equivalent resistance connected between terminals a and b.

63.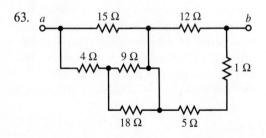

64.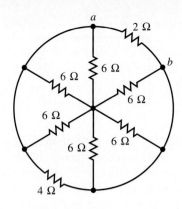

65.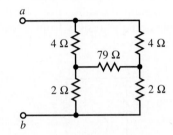

66. For the network shown below, find the resistance measured between terminals a and b as the number N of "L" sections goes to infinity.

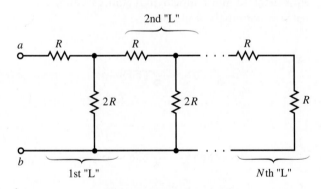

Twelve resistances, each with the same value R, are connected to form the edges of a cube as shown in the following illustration. What resistance would be measured between the indicated terminals?

67. a and b
68. b and c
69. a and d

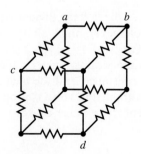

70. Determine the simplest *RLC* network possible that will exhibit the same overall relationship between v and i as in Fig. 3.17.

71. Find the voltage v across the current source for the following network.

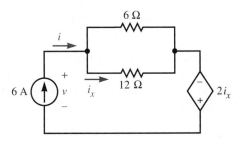

72. If the 6-A current source shown above is replaced by a 100-V voltage source, find the power absorbed (a) by the independent voltage source, and (b) by the dependent voltage source.

Determine the voltage v and the power absorbed by each component in the networks of Problems 73 and 74.

73.

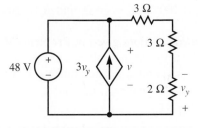

74.

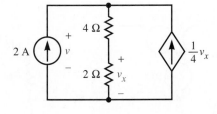

75. For the network that follows, determine voltage v, current i, and the power absorbed by the independent source.

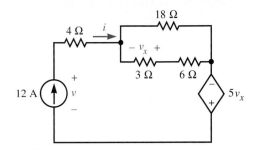

76. Find the voltage v and the current i in the network given below.

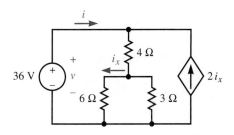

77. For the network shown below, find current i.

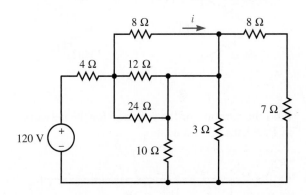

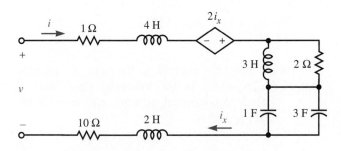

FIGURE 3.17

Section 3.8

Determine an equivalent Δ for each Y circuit and an equivalent Y for each Δ circuit. (Be sure to label the terminals on the equivalent circuit.)

78.

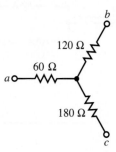

79.

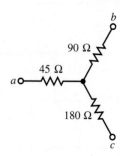

80.   81.

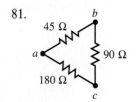

Find the current i or the voltage v for the networks in Problems 82 and 83.

82.

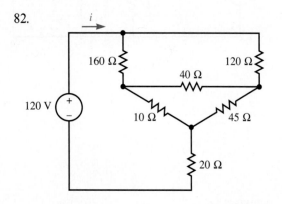

83.

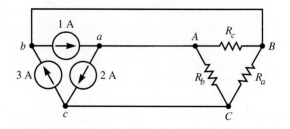

For Problems 84 and 85, use the values given for the network shown below and KCL to determine i_{aA}, i_{bB}, and i_{cC}. Then make a Δ-Y transformation and calculate v_{AB}, v_{BC}, and v_{CA}.

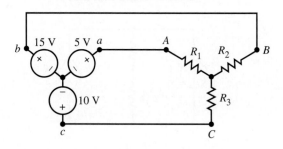

84. $R_a = R_b = R_c = 9\ \Omega$

85. $R_a = 100\ \Omega$, $R_b = 80\ \Omega$, and $R_c = 20\ \Omega$

The resistance values given in Problems 86 and 87 are for the network that follows. Use KVL to determine v_{AB}, v_{BC}, and v_{CA}. Then make a Y-Δ transformation and calculate i_{aA}, i_{bB}, and i_{cC}.

86. $R_1 = R_2 = R_3 = 15\ \Omega$

87. $R_1 = 15\ \Omega$, $R_2 = 20\ \Omega$, and $R_3 = 60\ \Omega$

88. Replace the network to the right of terminals a, b, and c in the following figure with a single Δ-connected network and then calculate i_1 and i_2.

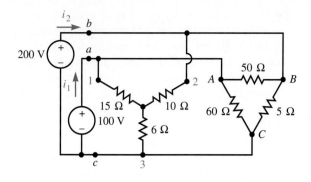

(a) Determine v_{GS}, i, i_1, i_2, and v_o.

(b) You wish to deliver the same current to both R_1 and R_2. What resistance would you place in series with R_2?

(c) You wish to deliver the same power to both R_1 and R_2. What resistance would you place in series with R_2?

92. Semiconductor devices, such as a field-effect transistor, operate properly only within a certain range of voltages and must be *biased* to operate within that range. The equivalent circuit shown below represents the bias problem for a field-effect transistor (FET). The large-signal transconductance is $g = 0.001$ S when the drain current i_D is in the range $0.8 < i_D < 1.2$ mA.

(a) Use the value of $R_2 = 100$ kΩ, and select R_1 so that the current i_D is 1 mA.

(b) Select R_1 and R_2 so that the series combination is 100 kΩ and v_{SO} is 1 V.

Comprehensive Problems

89. Only the relation between v and i is important in the following network. Use series and parallel combinations to reduce the network to as few components as possible.

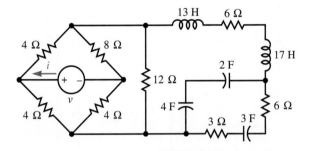

90. If the 8-Ω resistance in the network of Problem 89 is replaced by a 4-Ω resistance, symmetry can be used to easily find the relation between v and i. What is the relationship?

91. A small-signal equivalent circuit for a field-effect transistor (FET) amplifier is shown in Fig. 3.18. Terminals G, S, and D represent the gate, source, and drain, respectively.

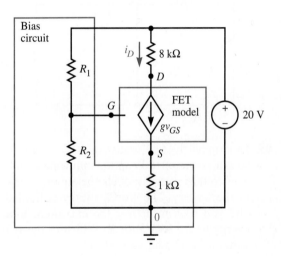

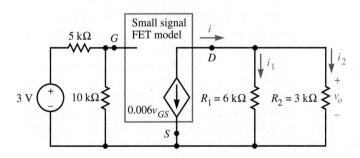

FIGURE 3.18

93. A practical ammeter can often be modeled as a resistance. The meter reading is proportional to the current through the resistance. A meter has a full-scale reading of 10 A when the voltage across the meter is 50 mV. What resistance must be placed in parallel with the meter as a *shunt* so that the combination will yield a full-scale reading of 100 A?

94. Two devices, one modeled as a resistance of 4 Ω and the other modeled as a resistance of 12 Ω, are to be operated from a 12-V source as shown in the following illustration. The devices are designed to operate from a 6-V source.
    (a) What value of R should you use to supply 6 V to the two devices as shown?
    (b) Determine the power absorbed by each component.

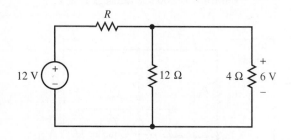

95. Determine the two values of R that will absorb a power of 16 W in the following circuit. The voltage source models the stored energy in the battery. The stored energy is limited, and you wish to deliver the maximum total energy to the resistance R. Which of the two values of R should you select?

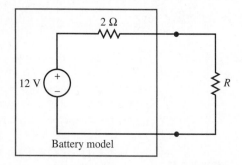

96. Semiconductor devices operate properly only within a certain range of voltages and currents, and must be *biased* to operate within that range. The following figure represents the bias problem for a bipolar transistor, where terminals B, C, and E are the base, collector, and emitter connections, respectively. You must bias for a collector current i_C of 1 mA. Determine (a) the currents i_B and i_E; (b) the value of v_{CO}, v_{EO}, and v_{BO}; (c) the value of i_1 and i_2 if $R_1 = 10$ kΩ; and (d) the value of R_2.

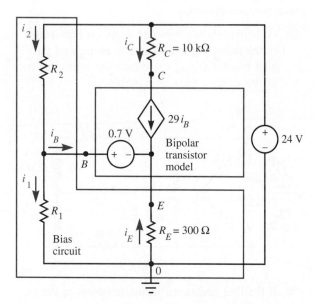

97. A battery must supply 100 W of power to a small heater. The battery is modeled as a 100-V source in series with a 10-Ω resistance. You must design the heater, which can be modeled as a resistance. What two values of resistance can the heater have? Which value should you use? Why?

A low-current ammeter is used to measure resistance in an instrument called an ohmmeter. One of the possible configurations is as shown on page 117. We can simplify this circuit model by use of the equivalent series resistance $R_2 = R_z + R_M$. The scale of the ohmmeter is determined by R_1, and R_z is adjusted so that when terminals a and b are short-circuited, the meter reads a full-scale value. The ammeter reads a full-scale value of 50 μA when the voltage across the meter is 1 V.

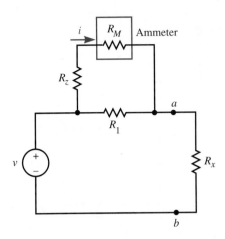

98. What is the resistance R_M of the ammeter?

99. Resistance R_2 is the series combination of R_M and R_z. Find an equation that gives i as a function of R_1, R_2, R_x, and v.

100. Find an equation that gives R_x as a function of R_1, R_2, v, and i.

101. With voltage $v = 3$ V, the ammeter reading must be 25 μA (one-half scale) when $R_x = 30\ \Omega$ and 50 μA when $R_x = 0$. What must be the value of R_1, R_2, and R_z?

102. If v changes and the meter is set to read full scale for $R_x = 0$ by adjusting R_z, will the half-scale reading of 25μA still correspond to $R_x = 30\ \Omega$?

103. A resistance R_s is included in series with the voltage source v to yield a more accurate model. With R_1 as previously calculated, will adjustment of R_z for a full-scale reading when $R_x = 0$ yield a proper reading when $R_x = 30\ \Omega$?

Find the equivalent resistance looking into terminals a and b of the following networks if $v_{ab} = 200$ V. Is the answer unique?

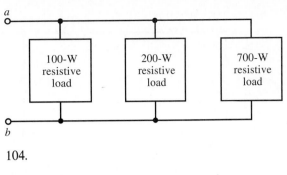

104.

105.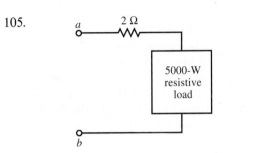

106.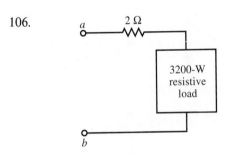

107.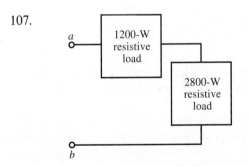

108. The following is a direct consequence of Ohm's law (current is proportional to voltage). If a voltage source of value v_s (or a current source of value i_s), which drives a network of resistances, is replaced by one of value av_s (or ai_s), all voltages and currents caused by the source are also multiplied by a. That is,

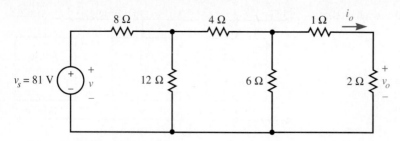

FIGURE 3.19

the voltages and currents are proportional to the source value. (This idea is extended in Chapter 6 by the formal introduction of linearity and the superposition theorem.) This proportionality provides a useful method to solve for the output voltage v_o of a *ladder network* as shown in Fig. 3.19. Assign a convenient value v_o' to v_o or i_o' to i_o, then determine the other from Ohm's law. Use the value of i_o' to find the voltage across the 3-Ω equivalent resistance. Then find the current through the 6-Ω resistance by the use of Ohm's law. Ohm's law and KCL can then be used to find the voltage across the 12-Ω resistance. The procedure can be repeated until the value v' for v that would correspond to v_o' is obtained. The output voltage will then be

$$v_o = \frac{v}{v'} v_o'$$

where v is the actual value of v, $v = v_s = 81$ V. Let $i_o' = 1$ A, and use this procedure to determine the actual value of v_o in the circuit of Fig. 3.19.

109. Use the method presented in Problem 108 to determine current i in the network of Fig. 3.11.

For each of the networks in Problems 110 and 111, replace the sources with a single equivalent source that will have the same effect on the load that is connected to terminals a and b.

110.

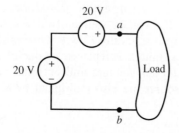

111.

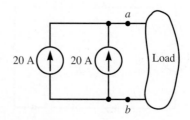

112. This problem refers to the Wheatstone bridge, described in Problem 53. With $R_s = 0$ and balance detected by measuring v_{12}, what relation between R_a, R_b, R_c, and R_x results in maximum sensitivity? That is, what relation maximizes

$$\frac{\partial}{\partial R_x} v_{12} \bigg|_{v_{12}=0}$$

113. A series circuit consists of a voltage source, resistance, and nonlinear device, as shown below. If the characteristics of the nonlinear device are as given, what are the values of v and i? (A graphical solution is necessary. Both KCL and KVL must be satisfied.)

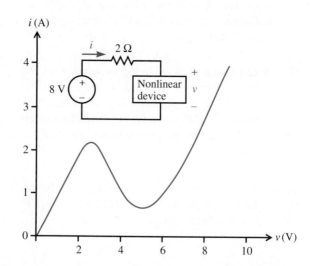

114. A permanent-magnet dc motor is operated from a 24-V source. The motor can be modeled as a resistance of 2 Ω in series with a dependent voltage source with value dependent on motor speed n, as shown in the following figure. The power absorbed by this dependent source is equal to the mechanical power developed by the motor. Calculate this power and the motor current i as a function of n, where n is in revolutions per minute.

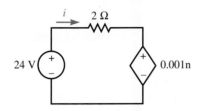

115. Determine the current through the voltage source of Fig. 3.12b.

116. Current i for the left-hand circuit shown below can be determined from the equivalent circuit shown at the right. Determine the value of voltage gain A for the equivalent circuit, and use this to calculate current i.

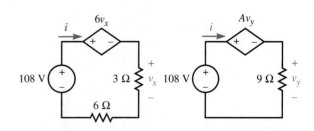

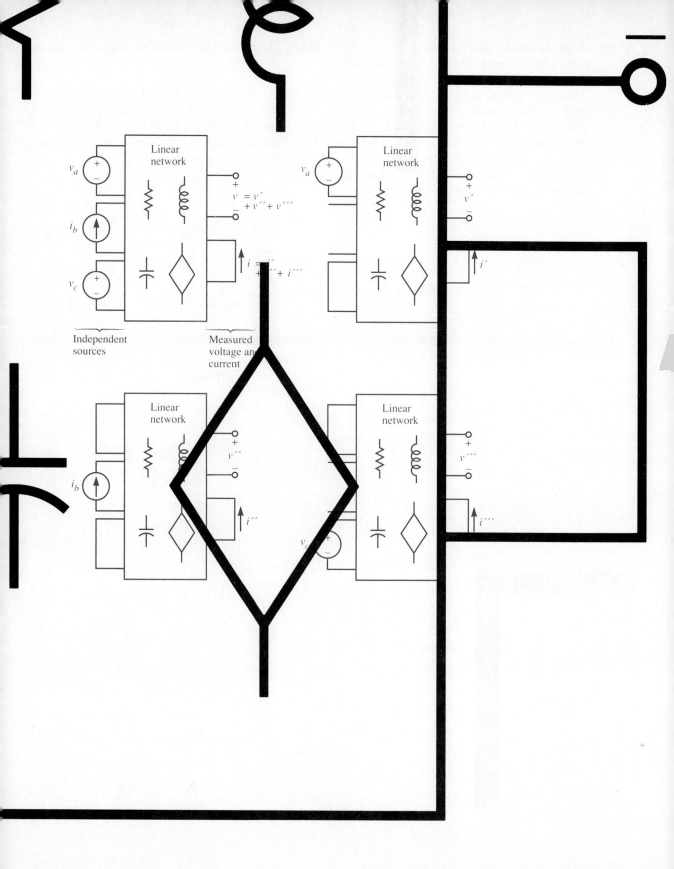

Linear
network

$v = v'$
$+ v'' + v'''$

$i = i''$
$+ i''' + i''''$

v_a

i_b

v_c

Independent
sources

Measured
voltage and
current

Linear
network

v_a

$+$
v'
$-$

i'

Linear
network

$+$
v''
$-$

i'''

i_b

Linear
network

$+$
v'''
$-$

i''''

v_c

TIME-DOMAIN
CIRCUIT ANALYSIS

In this part of the text we generalize the analysis of simple parallel and series circuits to obtain node-voltage equations and mesh-current equations. A number of network simplification techniques, including Thévenin's theorem, are also introduced. We describe the mathematical functions used to model voltages and currents that typically occur in electric circuits. We conclude Part Two with a technique to solve the differential equations that arise in circuit analysis.

CHAPTER
4

Network Analysis

In this chapter we develop two systematic approaches to network analysis. The first method, called node-voltage analysis, is applicable to *any* electrical network. Node-voltage analysis is a generalization of the method we used in Chapter 3 to analyze simple parallel circuits. The second method, called mesh-current analysis, is applicable to any *planar* network. Mesh-current analysis is a generalization of the method we used in Chapter 3 to analyze simple series circuits. Both methods yield a set of linearly independent equations that must be solved to obtain the *node voltages* or *mesh currents.*

These systematic approaches used to combine KCL, KVL, and the component equations provide two efficient ways to arrive at a complete specification of all component voltages and currents in the network. In addition, the methods described here are also useful to establish network theorems presented in later chapters.

We conclude the chapter with an introduction to PSpice® for computer-aided network analysis.

4.1   Node Voltages

A standard method to measure device voltages is to connect one terminal of a voltmeter or oscilloscope to a node of an electrical system and measure the voltages of all other nodes with respect to this *reference node*. This method is illustrated in Fig. 4.1.

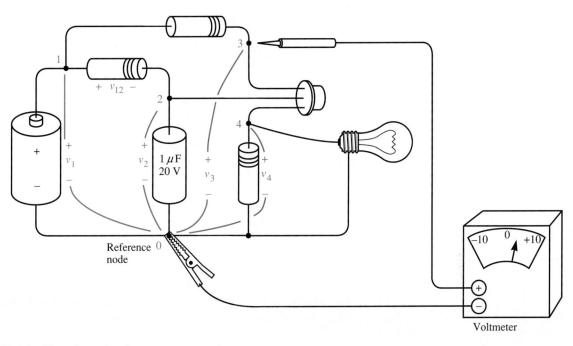

FIGURE 4.1   Measuring node voltages v_1, v_2, v_3, and v_4

Five nodes are accessible in the system illustrated. Therefore, four node voltages can be measured with respect to the reference node. We define node voltage v_k as the voltage measured with the ($+$) terminal of the voltmeter connected to node k and the ($-$) terminal of the voltmeter connected to the reference node, node 0. That is, for an $N + 1$-node circuit (see Fig. 4.1 for $N + 1 = 5$),

---

DEFINITION

Node Voltages

$$v_k = v_{k0}, \qquad k = 0, 1, 2, \ldots, N \tag{4.1}$$

Once we have measured all node voltages, we know the voltage across any device that is connected directly to the reference node. Moreover, we can calculate the voltage across any device that is not directly connected to the reference node as the difference of two node voltages. For example, in Fig. 4.1, KVL gives us

$$v_{12} = v_1 - v_2 \tag{4.2}$$

In general, for a connected circuit with $N + 1$ nodes, the voltage across any branch is

---

Branch Voltages

$$v_{kj} = v_k - v_j, \qquad k, j = 0, 1, 2, \ldots, N \tag{4.3}$$

We can readily adapt this experimental procedure to the analysis of a network model. The first step is to draw the network diagram and select node 0, the *reference node*. The choice of a reference node is completely arbitrary, but the node with the largest number of components or voltage sources connected to it is usually most convenient.

After we choose a reference node, the next step is to number the remaining nodes $1, 2, \ldots, N$. The choice of a numbering scheme is arbitrary.

The third step is to write the *node-voltage equations*. We must write one node-voltage equation for each node other than the reference node. For most circuits the number of node-voltage equations that are KVL equations is equal to the number of voltage sources,* and the remaining node-voltage equations are KCL equations.

The next step is to solve the equations for the node voltages. After solving the equations for the node voltages, we can write any component voltage (branch voltage) either as a node voltage or as the difference between two node voltages. Terminal equations then give us the element currents. The source or control equation determines the current through each current source. Finally, we determine the current through each voltage source from a KCL equation.

The procedure for writing node-voltage equations is *easily* and best learned by studying examples. We will begin by looking at a circuit in which the node-voltage equations are KVL equations.

Node-Voltage Equations that Are KVL Equations

We must determine only one node voltage for a parallel or two-node circuit. If the circuit is driven by a voltage source, a KVL equation gives the node voltage immediately. In a more complicated circuit, a KVL equation[†] also gives us the node voltage for any node connected to the reference node by a path of voltage sources. We will demonstrate this with the following example.

EXAMPLE 4.1

Write a set of node-voltage equations for the network shown in Fig. 4.2.

■ **SOLUTION**

Since there are four nodes and three voltage sources in the network of Fig. 4.2, we must write three KVL equations to obtain the node voltages. The node-voltage equation for node 1 is found from KVL by equating voltages along the paths

---

* This assumes that no loops of voltage sources are present. Loops of voltage sources, as in Problem 91, are an unusual case.

[†] These KVL equations are often called *constraint equations*, because voltages between nodes are constrained by the voltage sources.

FIGURE 4.2
Network with node-voltage
equations that are KVL
equations

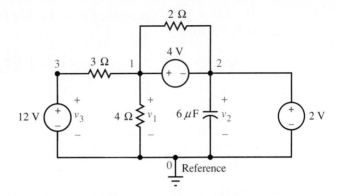

indicated in the illustrations. (Only the components that influence the equations are shown explicitly; the other components are represented by dotted lines in the following illustrations.)

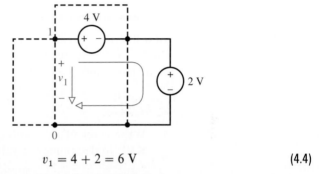

$$v_1 = 4 + 2 = 6 \text{ V} \qquad (4.4)$$

We quickly obtain the node-voltage equations for nodes 2 and 3 from KVL:

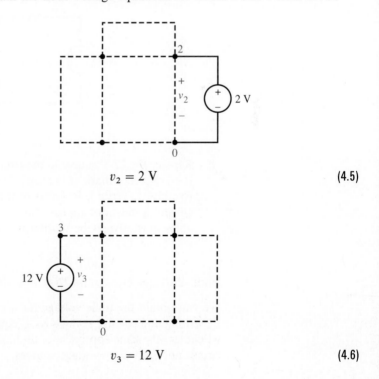

$$v_2 = 2 \text{ V} \qquad (4.5)$$

$$v_3 = 12 \text{ V} \qquad (4.6)$$

The node-voltage equations, Eqs. (4.4), (4.5), and (4.6), can be collected in matrix form to yield

$$\begin{bmatrix} v_1 \\ v_2 \\ v_3 \end{bmatrix} = \begin{bmatrix} 6 \\ 2 \\ 12 \end{bmatrix} V$$

The element currents are obtained directly from the terminal equations, but we must use KCL equations to find the currents through the voltage sources. Notice that the values of resistance and capacitance have no effect on the node voltages in this circuit. ∎

In this section we examined a circuit for which all node-voltage equations were KVL equations. In the next section we consider a circuit for which all node-voltage equations are KCL equations.

We will write a KVL equation for each node connected to the reference node by a path of voltage sources. The KVL equation is a node-voltage equation and must be written in terms of known source values and node voltages.

EXERCISES

1. Write a set of node-voltage equations for the network shown below. Use KVL to determine the value of voltage v_{12}, and calculate current i_b.

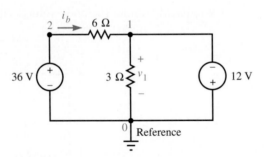

2. Replace the 12-V source in the circuit above with a dependent voltage source [(+) reference mark at the top]. The control equation is $v = 6i_b$. Then repeat Exercise 1. Write i_b in terms of the node voltages v_1 and v_2. The unknown current i_b must not appear in the final node-voltage equations. (Control variables must always be written in terms of node voltages or known voltages and currents.)

Node-Voltage Equations that Are KCL Equations

We can obtain the node voltage for a parallel circuit driven by a current source by writing and solving a single KCL equation. The following example shows how we can use the same approach for finding the node voltages for a more complicated circuit not excited by voltage sources.

EXAMPLE 4.2

(a) Write a set of node-voltage equations for the network for Fig. 4.3 and solve for all node voltages.

(b) Replace the 6-Ω resistance with a 6-F capacitance and the 10-Ω resistance with a 10-H inductance. Write the node-voltage equations.

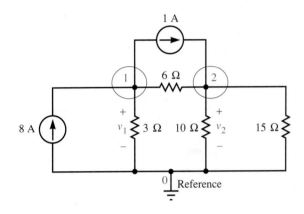

■ SOLUTION

(a) The node numbers for the network of Fig. 4.3 were arbitrarily assigned. We obtain the node-voltage equations by the use of KCL. The KCL equation for a surface enclosing node 1 is

$$-8 + 1 + i_{10} + i_{12} = 0 \tag{4.7}$$

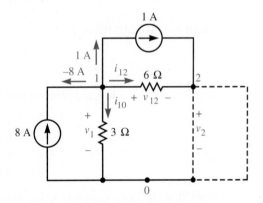

The terminal equation for the 3-Ω resistance gives

$$i_{10} = \frac{1}{3} v_1$$

and (with the use of KVL) the terminal equation for the 6-Ω resistance gives

$$i_{12} = \frac{1}{6} v_{12} = \frac{1}{6} (v_1 - v_2)$$

Substitution from these terminal equations permits Eq. (4.7) to be written as

$$-8 + 1 + \frac{1}{3} v_1 + \frac{1}{6} (v_1 - v_2) = 0 \tag{4.7a}$$

The previous steps could have been bypassed and Eq. (4.7a) written directly. Just remember that the current i_{kj} through an element is obtained by substitution of $v_k - v_j$ for v_{kj} in the terminal equation.

We can rearrange Eq. (4.7a) to give

$$\frac{1}{2} v_1 - \frac{1}{6} v_2 = 7 \qquad\qquad (4.7b)$$

This is the first node-voltage equation. We obtain the second node-voltage equation from a KCL equation for a surface enclosing node 2:

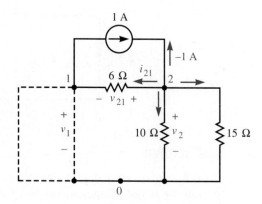

$$-1 + \frac{1}{6}(v_2 - v_1) + \frac{1}{10} v_2 + \frac{1}{15} v_2 = 0 \qquad\qquad (4.8)$$

which can be rearranged to give

$$-\frac{1}{6} v_1 + \frac{1}{3} v_2 = 1 \qquad\qquad (4.8a)$$

These two node-voltage equations written as a single matrix equation are

$$\begin{bmatrix} \dfrac{1}{2} & -\dfrac{1}{6} \\[2mm] -\dfrac{1}{6} & \dfrac{1}{3} \end{bmatrix} \cdot \begin{bmatrix} v_1 \\ v_2 \end{bmatrix} = \begin{bmatrix} 7 \\ 1 \end{bmatrix}$$

Note that the only unknowns that appear in this equation are node voltages.

We can solve this matrix equation by Cramer's rule to give the node voltages:

$$\Delta = \begin{vmatrix} \dfrac{1}{2} & -\dfrac{1}{6} \\[2mm] -\dfrac{1}{6} & \dfrac{1}{3} \end{vmatrix} = \left(\frac{1}{2}\right)\left(\frac{1}{3}\right) - \left(-\frac{1}{6}\right)\left(-\frac{1}{6}\right) = \frac{5}{36}$$

$$v_1 = \frac{\begin{vmatrix} 7 & -\dfrac{1}{6} \\[2mm] 1 & \dfrac{1}{3} \end{vmatrix}}{\Delta} = \frac{(7)(1/3) - (1)(-1/6)}{5/36} = 18 \text{ V}$$

$$v_2 = \frac{\begin{vmatrix} \frac{1}{2} & 7 \\ -\frac{1}{6} & 1 \end{vmatrix}}{\Delta} = \frac{(1/2)(1) - (7)(-1/6)}{5/36} = 12 \text{ V}$$

The voltage across any component is given by either a node voltage or the difference in node voltages. For example,

$$v_{12} = v_1 - v_2 = 18 - 12 = 6 \text{ V}$$

Thus
$$i_{12} = \frac{1}{6}v_{12} = \frac{1}{6}(18 - 12) = 1 \text{ A}$$

(b) With the two resistances replaced by a capacitance and inductance as requested, the first KCL equation is

$$-8 + 1 + \frac{1}{3}v_1 + 6\frac{d}{dt}(v_1 - v_2) = 0$$

and the second KCL equation is

$$-1 + 6\frac{d}{dt}(v_2 - v_1) + \frac{1}{10}\int_{-\infty}^{t} v_2 \, d\tau + \frac{1}{15}v_2 = 0 \quad \blacksquare$$

The inclusion of inductance and capacitance does not significantly complicate writing the two node-voltage equations, but the resulting integro-differential equations are more difficult to solve than algebraic equations. We will postpone the solution of such equations until Chapters 8 and 9.

This section has dealt with a circuit with node-voltage equations that are all KCL equations. We will introduce the concept of supernodes in the next section. These require both KCL and KVL equations.

REMEMBER

We will write a KCL equation for each node (except the reference node) that has no voltage sources connected to it. The KCL equation is a node-voltage equation and must be written in terms of node voltages and known source values.

EXERCISES

3. Write a set of node-voltage equations for the network shown below, and solve for the node voltages. Determine the value of v_{12}.

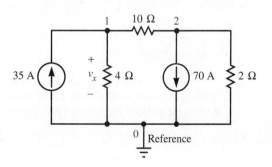

4. Write a KCL equation for a surface enclosing the reference node. Observe that this is the negative of the sum of the two KCL equations written in Exercise 3, and is thus linearly dependent on them. This is why we will not use a KCL equation for the reference node as a node-voltage equation.

5. Replace the 70-A source in the above network with a dependent current source (reference arrow pointing up). The control equation is $i = \frac{1}{4} v_x$. Repeat Exercise 3. Write v_x in terms of the node voltage. The unknown voltage v_x must not appear in the final equations.

6. Replace the 10-Ω resistance in the above figure with a 10-H inductance and the 2-Ω resistance with a 2-F capacitance. Write the two node-voltage equations, but do not solve them.

Supernodes Require KCL and KVL Equations

Nodes that are connected to each other by voltage sources, but not to the reference node by a path of voltage sources, form a *supernode*. We include in the supernode all components connected in parallel with these sources. A supernode requires one equation for each node contained within the supernode. We must write a KCL equation for a surface enclosing the supernode. The remaining equations are KVL equations.

We also demonstrate a shortcut. We select one node within the supernode as a *principal node*, and use KVL to write the other node voltages in the supernode as a sum of this principal node voltage and voltage source values. We then write a KCL equation for a surface enclosing the supernode, just as we would for a node of a parallel circuit. The following example demonstrates the procedure.

EXAMPLE 4.3

Write a set of node-voltage equations for the network shown in Fig. 4.4.

FIGURE 4.4
Network with one supernode that requires both a KCL and a KVL equation

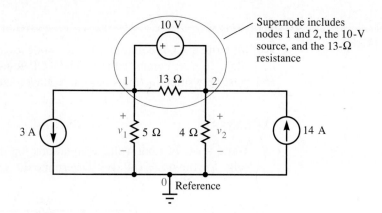

■ SOLUTION

With the nodes numbered as in Fig. 4.4, nodes 1 and 2, together with the connecting voltage source and 13-Ω resistance, form one supernode. A KCL equation for a surface enclosing the supernode is

$$3 + \frac{1}{5} v_1 + \frac{1}{4} v_2 - 14 = 0 \qquad (4.9)$$

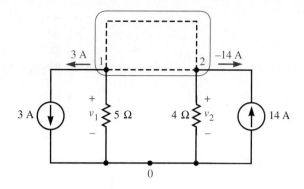

which is the first node-voltage equation. Use of KVL to equate the voltages along the paths indicated in the illustration gives

$$v_1 - v_2 = 10 \qquad (4.10)$$

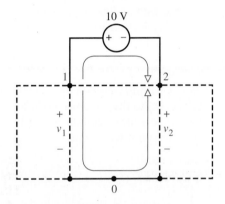

which is the second node-voltage equation, and relates the voltage source within the supernode to the node voltages. Equations (4.9) and (4.10) can be written as

$$\begin{array}{l}\text{From KCL} \left\{ \\ \text{From KVL} \left\{ \right. \end{array} \begin{bmatrix} \dfrac{1}{5} & \dfrac{1}{4} \\ 1 & -1 \end{bmatrix} \begin{bmatrix} v_1 \\ v_2 \end{bmatrix} = \begin{bmatrix} 11 \\ 10 \end{bmatrix} \right\} \begin{array}{l} \text{Current due to} \\ \text{current source} \\ \text{Voltage due to} \\ \text{voltage source} \end{array}$$

We can solve this matrix node-voltage equation for the node voltages.

The labor necessary to write and solve the node-voltage equations can be reduced if we use KVL to write

$$v_2 = v_1 - 10 \qquad (4.10a)$$

directly on the network diagram before we write the KCL equation. Because we have written v_2 in terms of v_1, we call node 1 a *principal node*. We can now write the KCL equation for a surface enclosing the supernode as

$$3 + \frac{1}{5}v_1 + \frac{1}{4}(v_1 - 10) - 14 = 0 \qquad (4.9a)$$

which immediately yields

$$v_1 = 30 \text{ V}$$

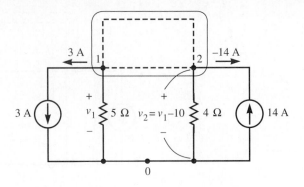

Observe how this shortcut reduced the analysis procedure to one similar to that for a simple parallel circuit. The second node voltage is given by the KVL equation as

$$v_2 = 30 - 10 = 20 \text{ V } \blacksquare \qquad (4.10b)$$

This section has investigated a group of nodes we call a supernode. In the following section we will examine the general circuit.

REMEMBER

Nodes within a supernode are connected by a path of voltage sources, but are not connected to the reference node by a path of voltage sources. A supernode requires one node-voltage equation that is a KCL equation. The remaining node-voltage equations are KVL equations.

EXERCISES

7. Write a set of node-voltage equations for the network shown in the circuit below (v_x must be written in terms of the node voltages). First, write four node-voltage equations (two KVL and two KCL equations). Next, use the KVL equation to write v_3 and v_4 in terms of v_1 and v_2 directly on the network diagram. Then write a KCL equation for each supernode. The resulting two node-voltage equations should contain v_1 and v_2 as the only unknowns. Solve for all node voltages.

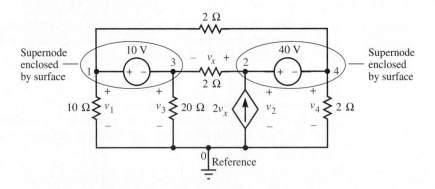

8. Use the node voltages found in Exercise 7 to solve for i_{10} and i_{14}. Now write a KCL equation for a surface enclosing node 1 and solve for i_{13}. Observe

that we could not write i_{13} in terms of v_1 and v_3 alone, but that $i_{13} = i_{30} + i_{32}$ can be written in terms of v_2 and v_3. This is why we wrote the node-voltage equations for surfaces enclosing the supernodes.

4.2 Network Analysis by Node Voltages

In many networks, some nodes are connected to the reference node by a path of voltage sources, some nodes are not connected to voltage sources, and some nodes form supernodes. We shall see that the methods we developed in Section 4.1 still apply.

EXAMPLE 4.4

Use the method of node voltages to analyze the circuit shown in Fig. 4.5.

FIGURE 4.5
Network with all three types of nodes

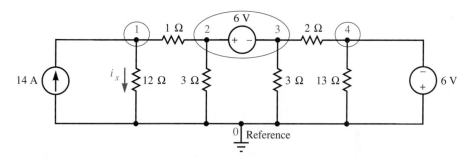

■ SOLUTION

Select the reference node as shown. Node 1 requires a KCL equation. Nodes 2 and 3 form a supernode and require both KCL and KVL equations. Node 4 requires only a KVL equation. Write a KCL equation for a surface that encloses node 1:

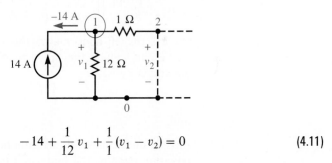

$$-14 + \frac{1}{12} v_1 + \frac{1}{1} (v_1 - v_2) = 0 \qquad (4.11)$$

Write a KCL equation for a surface enclosing the supernode formed by nodes 2 and 3:

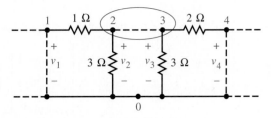

$$\frac{1}{1}(v_2 - v_1) + \frac{1}{3}v_2 + \frac{1}{3}v_3 + \frac{1}{2}(v_3 - v_4) = 0 \qquad (4.12)$$

The third and fourth node-voltage equations are obtained from KVL:

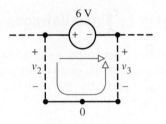

$$v_2 - v_3 = 6 \qquad (4.13)$$

and

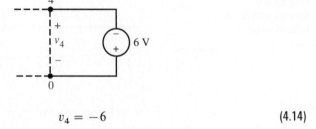

$$v_4 = -6 \qquad (4.14)$$

These four node-voltage equations form a single matrix equation:

$$\begin{matrix} \text{From} \\ \text{KCL} \end{matrix} \left\{ \begin{bmatrix} \dfrac{13}{12} & -1 & 0 & 0 \\ -1 & \dfrac{4}{3} & \dfrac{5}{6} & -\dfrac{1}{2} \\ 0 & 1 & -1 & 0 \\ 0 & 0 & 0 & 1 \end{bmatrix} \cdot \begin{bmatrix} v_1 \\ v_2 \\ v_3 \\ v_4 \end{bmatrix} = \begin{bmatrix} 14 \\ 0 \\ 6 \\ -6 \end{bmatrix} \right. \begin{matrix} \text{Currents due to} \\ \text{current sources} \\ \\ \text{Voltages due to} \\ \text{voltage sources} \end{matrix}$$

$$\begin{matrix} \text{From} \\ \text{KVL} \end{matrix} \left\{ \right.$$

We call the left-hand matrix the *node transformation matrix.*

We can reduce the labor involved in writing and solving the node-voltage equations by writing

$$v_3 = v_2 - 6 \qquad (4.13a)$$

and
$$v_4 = -6 \qquad (4.14)$$

directly on the network diagram before writing the KCL equations. The first KCL equation remains the same, and the second KCL equation becomes

$$\frac{1}{1}(v_2 - v_1) + \frac{1}{3}v_2 + \frac{1}{3}(v_2 - 6) + \frac{1}{2}[(v_2 - 6) + 6] = 0 \qquad (4.12a)$$

in which case the node-voltage equations reduce to

$$\begin{bmatrix} \dfrac{13}{12} & -1 \\ -1 & \dfrac{13}{6} \end{bmatrix} \cdot \begin{bmatrix} v_1 \\ v_2 \end{bmatrix} = \begin{bmatrix} 14 \\ 2 \end{bmatrix}$$

The left-hand matrix is the *reduced node transformation matrix*. We can solve this matrix equation for the *principal-node* voltages, v_1 and v_2:

$$\begin{bmatrix} v_1 \\ v_2 \end{bmatrix} = \begin{bmatrix} 24 \\ 12 \end{bmatrix} V$$

We obtain the two remaining node voltages v_3 and v_4 from Eqs. (4.13a) and (4.14). ∎

A summary of the shortcut procedure we have developed follows. We present the list for easy reference, not for memorization.

Shortcut Method ▶

Analysis of Node Voltages

1. Draw the network diagram and select the reference node. Usually the node with the largest number of voltage sources or components connected to it is the most convenient choice for the reference node.

2. Number the remaining nodes in sequence.

3. Write the node-voltage equations.
   (a) Identify each node that is connected to the reference node by a path of voltage sources. Use KVL to write each of these node voltages as a sum of the source voltages. Write these relationships on the network diagram.
   (b) Identify each group of nodes and components that form a supernode. Select one node within each group as a principal node. Use KVL to write the voltage of the other nodes within this group as the sum of the principal-node voltage and the source voltages. Write these relationships on the network diagram.
   (c) The remaining nodes, other than the reference node, are also principal nodes.
   (d) Now write a KCL equation for surfaces that enclose each node identified in (c) and each supernode.
   Remember that all unknowns must be written in terms of principal-node voltages.

4. Solve the KCL equations for the principal-node voltages. Each remaining node voltage is then obtained from a KVL equation. ◀

We will now apply the shortcut procedure to a five-node network.

EXAMPLE 4.5

Use the shortcut procedure to write a set of node-voltage equations for the network shown in Fig. 4.6.

■ SOLUTION

1. Select the reference node as shown.
2. Number the remaining nodes 1, 2, 3, and 4 as indicated.

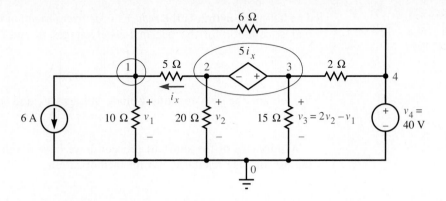

FIGURE 4.6
Example circuit for the shortcut procedure for node-voltage analysis

3. Write the node-voltage equations.
   (a) Node 4 is connected to the reference node by a voltage source. Application of KVL gives us

$$v_4 = 40 \text{ V} \qquad (4.15)$$

   Write $v_4 = 40$ V on the network diagram as shown.
   (b) Nodes 2 and 3 form a supernode. Select node 2 as the principal node. We again use KVL and Ohm's law to obtain

$$v_3 = 5i_x + v_2 = 5\left[\frac{1}{5}(v_2 - v_1)\right] + v_2 = 2v_2 - v_1 \qquad (4.16)$$

   Write $v_3 = 2v_2 - v_1$ on the network diagram as shown.
   (c) Node 1 is also a principal node, because no voltage sources are connected to it.
   (d) Write a KCL equation for a surface enclosing node 1:

$$6 + \frac{1}{10}v_1 + \frac{1}{5}(v_1 - v_2) + \frac{1}{6}(v_1 - 40) = 0 \qquad (4.17)$$

   Next write a KCL equation for a surface enclosing the supernode:

$$\frac{1}{5}(v_2 - v_1) + \frac{1}{20}v_2 + \frac{1}{15}(2v_2 - v_1) + \frac{1}{2}[(2v_2 - v_1) - 40] = 0 \quad (4.18)$$

4. These two KCL equations, written as a single matrix equation, give us

$$\begin{bmatrix} \dfrac{7}{15} & -\dfrac{1}{5} \\ -\dfrac{23}{30} & \dfrac{83}{60} \end{bmatrix} \cdot \begin{bmatrix} v_1 \\ v_2 \end{bmatrix} = \begin{bmatrix} \dfrac{2}{3} \\ 20 \end{bmatrix}$$

We can easily solve this matrix equation for $v_1 = 10$ V and $v_2 = 20$ V. From Eq. (4.16), $v_3 = 30$ V. Equation (4.15) gives v_4. ∎

The following example illustrates how easy it is to apply the shortcut procedure to even a complicated-looking network.

EXAMPLE 4.6

Use the shortcut procedure to write a set of node-voltage equations for the network shown in Fig. 4.7.

FIGURE 4.7
Node-voltage example with
inductance and capacitance

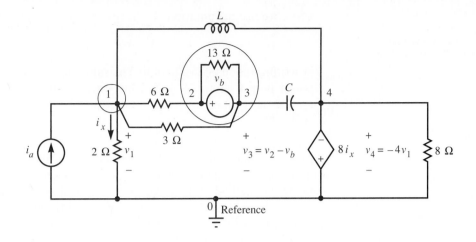

■ **SOLUTION**

1. Select the reference node as shown.
2. Number the remaining nodes 1, 2, 3, and 4.
3. Write the node-voltage equations.
   (a) Node 4 is connected to the reference node by the dependent voltage source. From KVL,

   $$v_4 = -8i_x = -8\left(\frac{1}{2}v_1\right) = -4v_1 \qquad (4.19)$$

   Write this on the network diagram.
   (b) Nodes 2 and 3 form a supernode. Select node 2 as the principal node. From KVL,

   $$v_3 = v_2 - v_b \qquad (4.20)$$

   Write this on the network diagram.
   (c) Node 1 will also be a principal node, because no voltage sources are connected to it.
   (d) Write a KCL equation for a surface enclosing node 1:

   $$-i_a + \frac{1}{2}v_1 + \frac{1}{3}\left[v_1 - (v_2 - v_b)\right] + \frac{1}{6}(v_1 - v_2) + \frac{1}{L}\int_{-\infty}^{t}\left[v_1 - (-4v_1)\right]d\tau = 0$$
   $$(4.21)$$

   Next apply KCL to a surface enclosing the supernode:

   $$\frac{1}{6}(v_2 - v_1) + \frac{1}{3}\left[(v_2 - v_b) - v_1\right] + C\frac{d}{dt}\left[(v_2 - v_b) - (-4v_1)\right] = 0 \quad (4.22)$$

4. We can write the two KCL equations, Eqs. (4.21) and (4.22), as a single matrix differential equation:

$$
\begin{bmatrix}
1 + \dfrac{5}{L}\displaystyle\int_{-\infty}^{t} d\tau & -\dfrac{1}{2} \\[3mm]
4C\dfrac{d}{dt} - \dfrac{1}{2} & C\dfrac{d}{dt} + \dfrac{1}{2}
\end{bmatrix}
\cdot
\begin{bmatrix} v_1 \\ v_2 \end{bmatrix}
=
\begin{bmatrix}
i_a - \dfrac{1}{3}v_b \\[3mm]
C\dfrac{d}{dt}v_b + \dfrac{1}{3}v_b
\end{bmatrix}
$$

where we use the notation

$$\int_{-\infty}^{t} v \, d\tau = \int_{-\infty}^{t} d\tau \cdot v$$

in writing the matrix equation. The solution of differential equations is postponed until Chapters 8 and 9, but once the principal-node voltages, v_1 and v_2, are determined, v_3 and v_4 can easily be obtained from Eqs. (4.19) and (4.20). ∎

The networks of Figs. 4.2 through 4.7 are all one-part (connected) networks. Occasionally the need arises to analyze a network like that shown in Fig. 4.8. This is an unconnected network consisting of *two separate parts*. To analyze this network by the method of node voltages, one must select a reference node for each part, as shown. Notice that $v_{00'}$, and thus v_{12}, is indeterminant, because no line connects the two parts of this circuit. More generally, a network with S separate parts will require S reference nodes, one for each part.

FIGURE 4.8
Two-part network with two reference nodes

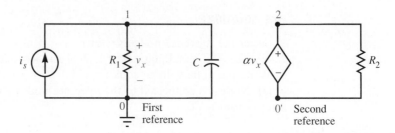

REMEMBER

The number of node-voltage equations required for a network is equal to the number of nodes minus the number of separate parts. If there are no loops of voltage sources, the number of node-voltage equations that are KVL equations is equal to the number of voltage sources. The remainder of the node-voltage equations are KCL equations.

EXERCISES

9. Use the method of analysis by node voltages to calculate the node voltages of the network shown in the following circuit. Then calculate the power absorbed by the dependent source.

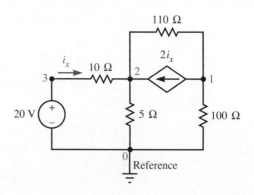

10. Replace the 20-V source in the above circuit with a short circuit and the 5-Ω resistance with the 20-V source. Keep the $(+)$ reference mark at the top. Then repeat Exercise 9.

11. Replace the dependent current source of the circuit used in Exercise 9 with a dependent voltage source $[(+)$ reference to the right] having a control equation of $v = 110i_x$. Then repeat Exercise 9.

12. Replace the right-hand 2-Ω resistance of the circuit used in Exercise 7 with a voltage source of value 100 V [the $(+)$ reference mark at the top]. Write the four node-voltage equations.

13. Use the shortcut procedure and repeat Exercise 12. This should give one KCL equation that contains v_1 as the only unknown. Solve for all node voltages.

14. Write the node-voltage equations for the network of Fig. 4.8.

Replace the upper voltage source of Fig. 4.5 with a dependent voltage source $[(+)$ reference mark on the left] having a control equation of $v = 3i_x$.

15. Write the four node-voltage equations and arrange in matrix form. You should have two KCL equations and two KVL equations.

16. Use the two KVL equations written in Exercise 15 to write v_3 and v_4 in terms of the specified voltage and current sources and node voltages v_1 and v_2. Write these on the network diagram. Directly from the network diagram, write two KCL equations that contain v_1 and v_2 as the only unknowns.

17. Solve for all node voltages and the power absorbed by the dependent voltage source.

4.3 Node-Voltage Equations by Inspection*

After elimination of the KVL equations by substitution, the reduced node transformation matrix in Example 4.4 was symmetric about the diagonal. The reduced node transformation matrix of a network containing no dependent sources can always be arranged to be symmetric. If all sources are independent current sources and all elements are resistances, the equations can be written by inspection. The resulting form for a circuit with N node voltages is

$$\begin{bmatrix} G_{11} & G_{12} & \cdots & G_{1N} \\ G_{21} & G_{22} & \cdots & G_{2N} \\ \vdots & \vdots & & \vdots \\ G_{N1} & G_{N2} & \cdots & G_{NN} \end{bmatrix} \cdot \begin{bmatrix} v_1 \\ v_2 \\ \vdots \\ v_N \end{bmatrix} = \begin{bmatrix} i_1 \\ i_2 \\ \vdots \\ i_N \end{bmatrix} \qquad (4.23)$$

where G_{kk} = sum of all conductances connected to node k
$G_{jk} = G_{kj}$ = the negative of the sum of all conductances connecting node k to node j, $k \neq j$
v_k = node voltage
i_k = sum of all currents entering node k from independent current sources

Remember that this is valid only for a network consisting of linear resistances and independent current sources. A generalization to include all linear elements and dependent current sources is easily developed.

* This section can be omitted without loss of continuity.

EXAMPLE 4.7

Write a set of node-voltage equations for the circuit shown in Fig. 4.9: (a) first by writing the KCL equations, (b) then by inspection.

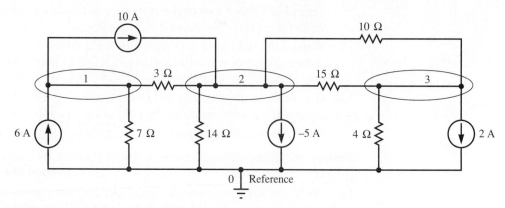

FIGURE 4.9 A node-voltage example with only current sources

■ SOLUTION

(a) Select as the reference node the node with the most components connected to it, and number the nodes as shown. Write KCL equations for surfaces enclosing each node except the reference. This yields

$$-6 + 10 + \frac{1}{7}v_1 + \frac{1}{3}(v_1 - v_2) = 0 \tag{4.24}$$

$$-10 + (-5) + \frac{1}{3}(v_2 - v_1) + \frac{1}{14}v_2 + \frac{1}{15}(v_2 - v_3) + \frac{1}{10}(v_2 - v_3) = 0 \tag{4.25}$$

$$2 + \frac{1}{4}v_3 + \frac{1}{15}(v_3 - v_2) + \frac{1}{10}(v_3 - v_2) = 0 \tag{4.26}$$

Collection of the equations without addition of the conductances yields

$$\begin{bmatrix} \frac{1}{7} + \frac{1}{3} & -\frac{1}{3} & 0 \\ -\frac{1}{3} & \frac{1}{3} + \frac{1}{14} + \frac{1}{10} + \frac{1}{15} & -\frac{1}{10} - \frac{1}{15} \\ 0 & -\frac{1}{10} - \frac{1}{15} & \frac{1}{10} + \frac{1}{15} + \frac{1}{4} \end{bmatrix} \cdot \begin{bmatrix} v_1 \\ v_2 \\ v_3 \end{bmatrix} = \begin{bmatrix} 6 - 10 \\ 10 - (-5) \\ -2 \end{bmatrix}$$

The solution for the node voltages is

$$\mathbf{v} = \begin{bmatrix} v_1 \\ v_2 \\ v_3 \end{bmatrix} = \begin{bmatrix} 21 \\ 42 \\ 12 \end{bmatrix} \text{V}$$

(b) You should verify that the procedure outlined by Eq. (4.23) yields the same result. ■

Use the method presented in this section to write a set of node-voltage equations for the network indicated.

18. The network of Fig. 4.3

19. The network used in Exercise 3

4.4 Mutual Inductance and Node-Voltage Equations

Node-voltage equations are as easy to write for networks containing coupled inductances as they are for networks containing uncoupled inductances. The most straightforward approach is to use the integral form of the terminal equations presented in Chapter 2 and repeated below. Coupled inductances are shown in Fig. 4.10.

FIGURE 4.10
Coupled inductances

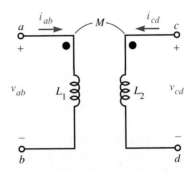

$$i_{ab} = \frac{L_2}{L_1 L_2 - M^2} \int_{-\infty}^{t} v_{ab} \, d\tau - \frac{M}{L_1 L_2 - M^2} \int_{-\infty}^{t} v_{cd} \, d\tau$$

$$i_{cd} = \frac{L_1}{L_1 L_2 - M^2} \int_{-\infty}^{t} v_{cd} \, d\tau - \frac{M}{L_1 L_2 - M^2} \int_{-\infty}^{t} v_{ab} \, d\tau$$

EXAMPLE 4.8

Write a set of node-voltage equations for the network of Fig. 4.11.

FIGURE 4.11
Node-voltage example with mutual inductance

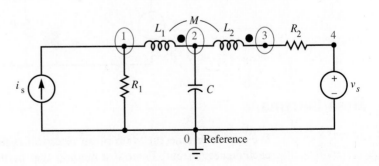

■ **SOLUTION**

Use KCL and sum the currents leaving a surface enclosing node 1:

$$-i_s + \frac{1}{R_1} v_1 + \frac{L_2}{L_1 L_2 - M^2} \int_{-\infty}^{t} (v_1 - v_2)\, d\tau - \frac{M}{L_1 L_2 - M^2} \int_{-\infty}^{t} (v_2 - v_3)\, d\tau = 0$$

(4.27)

Write a KCL equation for a surface enclosing node 2:

$$C \frac{d}{dt} v_2 + \frac{L_2}{L_1 L_2 - M^2} \int_{-\infty}^{t} (v_2 - v_1)\, d\tau - \frac{M}{L_1 L_2 - M^2} \int_{-\infty}^{t} (v_3 - v_2)\, d\tau$$

$$+ \frac{L_1}{L_1 L_2 - M^2} \int_{-\infty}^{t} (v_2 - v_3)\, d\tau - \frac{M}{L_1 L_2 - M^2} \int_{-\infty}^{t} (v_1 - v_2)\, d\tau = 0 \quad (4.28)$$

Sum the currents leaving a surface enclosing node 3. Make use of the known node voltage, $v_4 = v_s$.

$$\frac{L_1}{L_1 L_2 - M^2} \int_{-\infty}^{t} (v_3 - v_2)\, d\tau - \frac{M}{L_1 L_2 - M^2} \int_{-\infty}^{t} (v_2 - v_1)\, d\tau + \frac{1}{R_2} (v_3 - v_s) = 0$$

(4.29)

The fourth node voltage is specified by a voltage source:

$$v_4 = v_s \quad ■$$

(4.30)

Although the magnetic coefficient of coupling [see Eq. (2.59)] is always less than 1 in a physical device it may exceed 0.99. In such a case the coefficient of coupling is often approximated by unity. With this approximation, $L_1 L_2 - M^2 = 0$, and the terms in the equations of Fig. 4.10 are undefined. A unity coupling coefficient introduces no special problem if the method of mesh-current analysis, which is presented in Section 4.8, is used. The problem can also be solved by the introduction of ideal transformers, as discussed in Chapter 18.

EXERCISE

20. Write a set of node-voltage equations for the network in the circuit shown below.

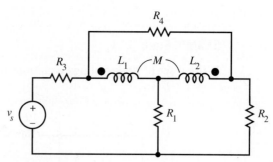

4.5 Mesh Currents

We can determine the state of an electrical system by calculating or measuring each device current. There is a method that permits a solution for all device currents in terms of a smaller number of currents, called *mesh currents*. The method of analysis by mesh currents applies only to planar networks, that is, networks that can be drawn so that no lines cross.* Circuit analysis by mesh currents is very popular among students and some electrical engineers, even though it has no experimental equivalent.

We calculate the mesh currents by writing and solving a set of KVL and KCL equations, called *mesh-current equations*.

We can understand the definition of mesh currents by considering the system of electrical devices shown in Fig. 4.12. We assign currents i_1, i_2, and i_3 clockwise around each mesh as if it were an isolated single loop or series circuit. These are the mesh currents. A mesh current can be measured by a single ammeter reading only if that mesh current passes through a branch in the outer loop, as do mesh currents i_1 and i_2. We can see that mesh current i_3 cannot be measured directly by a single ammeter reading, because i_3 does not constitute the total current through any single device.

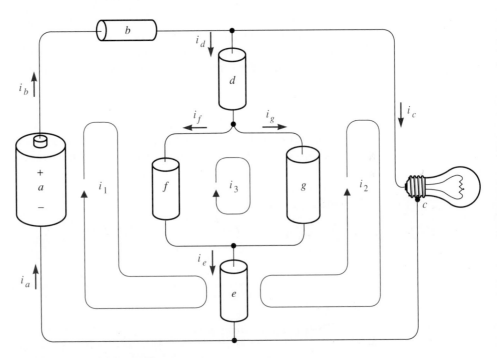

FIGURE 4.12   Circulating currents i_1, i_2, and i_3 are mesh currents

There are seven device currents to find. KCL directly implies $i_a = i_b$, and a KCL equation for a surface enclosing the top node yields $i_d = i_b - i_c = i_a - i_c$. It is not too hard to see that the device currents can be represented by a smaller set of currents with the aid of KCL. A clever way to do this is to write all currents in terms of mesh currents. Clearly

$$i_a = i_1 \tag{4.31}$$

$$i_b = i_1 \tag{4.32}$$

$$i_c = i_2 \tag{4.33}$$

* A generalization, called *link-current analysis* or *loop analysis*, applies to any network. Link-current analysis is not discussed in this chapter.

and in general

Mesh Currents

The current i_x through any device x in the outer loop is

$$i_x = i_j \tag{4.34}$$

where current i_j is the mesh current through the device x in the same direction as current i_x.

Application of KCL to a surface enclosing the top node yields

$$i_d = i_1 - i_2 \tag{4.35}$$

Another way to look at this latter equation is that device current i_d is composed of two components: i_1 in the direction of i_d, and i_2 in the opposite direction. (This is analogous to representing the voltage across a device connecting two nodes as the difference of two node voltages.) From this line of reasoning, device current i_e is

$$i_e = i_1 - i_2 \tag{4.36}$$

We can easily verify this by applying KCL to a surface enclosing the bottom node of the network. Repeated application of KCL will verify that the remaining device currents can be represented in terms of the mesh currents as

$$i_f = i_1 - i_3 \tag{4.37}$$

and

$$i_g = i_3 - i_2 \tag{4.38}$$

In general,

Branch Currents

The current i_x through any device x not in the outer loop is

$$i_x = i_j - i_k \tag{4.39}$$

where current i_j is the mesh current through device x in the direction of i_x, and current i_k is the mesh current through device x in the opposite direction. (This assumes that all mesh currents are assigned in a clockwise or all in a counter-clockwise direction.)

The first step in the analysis by mesh currents is to draw the network diagram so that no unconnected lines cross. (If this is not possible, the circuit is nonplanar, so we cannot use this method.)

The second step is to assign a mesh current and number to each mesh. (Mesh currents are usually assigned clockwise, but this is not a necessity.) Any numbering scheme will work.

The third step is to write the *mesh-current equations*. There is one mesh-current equation for each mesh. For most circuits the number of mesh-current equations that are KCL equations is equal to the number of current sources,* and the remaining mesh-current equations are KVL equations.

The next step is to solve the equations for the mesh currents. After these equations are solved for the mesh currents, we can write any component current as either a mesh current or the difference of two mesh currents. Element voltages are then found by the use of the terminal equations, the voltage across each voltage source is determined by the source or control equation, and current source voltages are determined from KVL equations.

The procedure to write mesh-current equations is *easily* and best learned by studying examples. We begin with a network for which all mesh-current equations are KCL equations.

Mesh-Current Equations that Are KCL Equations

A series circuit has only one mesh and one mesh current. Kirchhoff's current law gives this current if the circuit is driven by a current source. In some more complicated networks, mesh currents can also be determined from KCL equations.[†] We use KCL if we can go from the interior of a mesh to the exterior of the circuit along a path that cuts across only branches with current sources (see the dashed paths *A*, *B*, and *C* shown on Fig. 4.13). We will demonstrate the procedure by example.

FIGURE 4.13
Network with mesh-current equations that are KCL equations

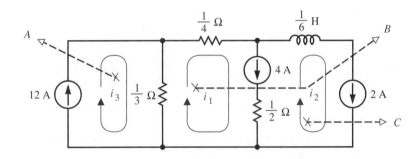

EXAMPLE 4.9

Write a set of mesh-current equations for the network shown in Fig. 4.13, and solve for all mesh currents.

■ SOLUTION

Since there are three meshes in the network of Fig. 4.13, three mesh currents must be found. We can always write the current through any current source in terms of at most two mesh currents. Thus from KCL (only the components that influence the equation are shown explicitly in the following illustrations),

---

* A description of the rare exceptions depends on the concept of a *cutset*, which will not be considered at this time, but is examined in Problem 93.

† These KCL equations are often called *constraint equations*, because a mesh current or the difference between two mesh currents is constrained by a current source.

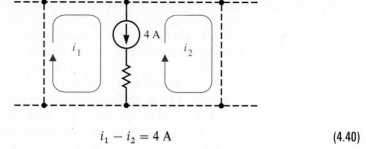

$$i_1 - i_2 = 4 \text{ A} \tag{4.40}$$

and KCL also gives

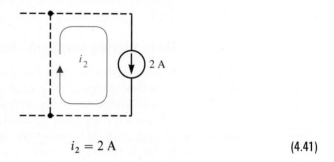

$$i_2 = 2 \text{ A} \tag{4.41}$$

and

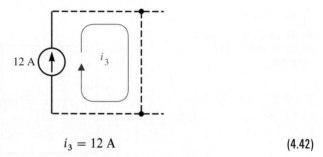

$$i_3 = 12 \text{ A} \tag{4.42}$$

The mesh-current equations, Eqs. (4.40), (4.41), and (4.42), yield

$$\begin{bmatrix} i_1 \\ i_2 \\ i_3 \end{bmatrix} = \begin{bmatrix} 6 \\ 2 \\ 12 \end{bmatrix} \text{A}$$

The element voltages are easily found by use of the terminal equations, but KVL equations are required to find the voltage across current sources. ∎

In this section we have investigated a circuit for which all mesh-current equation are KCL equations. In the next section we examine a circuit for which all mesh-current equations are KVL equations.

REMEMBER

We will write a KCL equation for a mesh if we can go from the interior of the mesh to the exterior of the circuit along a path that cuts across only branches with current sources. This KCL equation is a mesh-current equation and must be written in terms of known sources and mesh currents.

21. Write a set of mesh-current equations for the network shown below. Use KCL to determine the value of i_{12} and v_b.

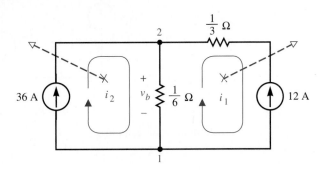

22. Replace the 12-A source in the circuit above with a dependent current source having the reference arrow pointing down. The control equation is $i = 6v_b$. Repeat Exercise 21. Write v_b in terms of mesh currents i_1 and i_2. The unknown current i_b must not appear in the mesh-current equations.

Mesh-Current Equations that Are KVL Equations

A KVL equation is used to determine the mesh current for a series circuit without a current source. We shall show that this same technique gives us the mesh currents for more complicated networks not excited by current sources.

EXAMPLE 4.10

(a) Write a set of mesh-current equations for the network shown in Fig. 4.14, and solve for all mesh currents.

(b) Replace the $\frac{1}{6}$-Ω resistance with a $\frac{1}{6}$-H inductance and the $\frac{1}{10}$-Ω resistance with a $\frac{1}{10}$-F capacitance. Write the mesh-current equations.

FIGURE 4.14
Network with mesh-current equations that are KVL equations

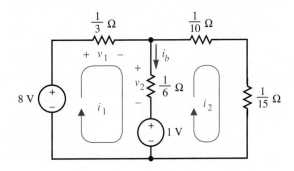

■ **SOLUTION**

(a) Select the mesh currents as shown in the network of Fig. 4.14. The mesh-current equations are obtained from KVL. Application of KVL to the closed path around mesh 1 gives

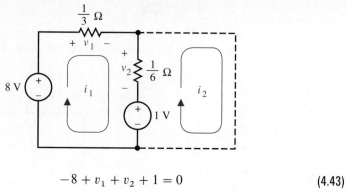

$$-8 + v_1 + v_2 + 1 = 0 \tag{4.43}$$

The element equations are

$$v_1 = \frac{1}{3} i_1$$

and

$$v_2 = \frac{1}{6} (i_1 - i_2)$$

Substitution from these terminal equations into Eq. (4.43) gives the first mesh-current equation:

$$-8 + \frac{1}{3} i_1 + \frac{1}{6} (i_1 - i_2) + 1 = 0 \tag{4.43a}$$

The previous steps can be bypassed and Eq. (4.43a) written directly. Just remember that the total current through a branch, which is in common with mesh k and mesh j, in the direction of mesh current i_k is $i = i_k - i_j$.

The second mesh-current equation is obtained from KVL applied around mesh 2.

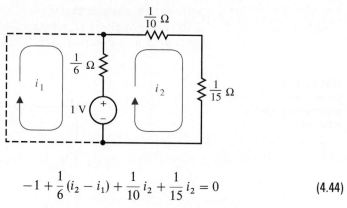

$$-1 + \frac{1}{6} (i_2 - i_1) + \frac{1}{10} i_2 + \frac{1}{15} i_2 = 0 \tag{4.44}$$

We can write these two mesh-current equations as a single matrix equation:

$$\begin{bmatrix} \dfrac{1}{2} & -\dfrac{1}{6} \\ -\dfrac{1}{6} & \dfrac{1}{3} \end{bmatrix} \cdot \begin{bmatrix} i_1 \\ i_2 \end{bmatrix} = \begin{bmatrix} 7 \\ 1 \end{bmatrix}$$

Note that the only unknowns that appear in this equation are mesh currents.

We can easily solve these equations by Cramer's rule or substitution to give

$$\begin{bmatrix} i_1 \\ i_2 \end{bmatrix} = \begin{bmatrix} 18 \\ 12 \end{bmatrix} \text{A}$$

The current through any component is given by either a mesh current or the difference between two mesh currents. For example,

$$i_b = i_1 - i_2$$
$$= 18 - 12 = 6 \text{ A}$$

Thus

$$v_2 = \frac{1}{6} i_b = \frac{1}{6}(6) = 1 \text{ V}$$

(b) With the two resistances replaced by an inductance and a capacitance, the first KVL equation becomes

$$-8 + \frac{1}{3} i_1 + \frac{1}{6} \frac{d}{dt}(i_1 - i_2) + 1 = 0$$

and the second KVL equation is

$$-1 + \frac{1}{6} \frac{d}{dt}(i_2 - i_1) + 10 \int_{-\infty}^{t} i_2 \, d\tau + \frac{1}{15} i_2 = 0$$

The solution of integro-differential equations will be covered in Chapters 8 and 9. ∎

We have considered a circuit for which all mesh-current equations are KVL equations. We introduce the concept of a supermesh in the next section. This requires both KVL and KCL equations.

REMEMBER

We write a KVL equation for each mesh that has no current source in any branch. This KVL equation is a mesh-current equation and must be written in terms of mesh currents and known source values.

EXERCISES

23. Write a set of mesh-current equations for the network shown below and solve for the mesh currents. Determine the value of i_{12}.

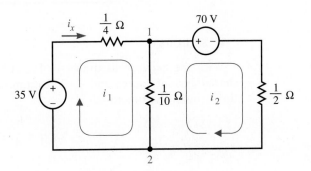

24. Write a KVL equation around the outer loop for the above circuit. Observe that this is the sum of the two KVL equations written in Exercise 23. This is why we will not use a KVL equation around loops that are not meshes when we write mesh-current equations.

25. Replace the 70-V source in the network shown above with a dependent voltage source having the (+) reference mark at the right-hand side. The control equation is $v = \frac{1}{4}i_x$. Repeat Exercise 23. Write i_x in terms of the mesh currents. The unknown current i_x must not appear in the mesh-current equations.

26. For the network used in Exercise 23, replace the $\frac{1}{10}$-Ω resistance with a $\frac{1}{10}$-F capacitance and the $\frac{1}{2}$-Ω resistance with a $\frac{1}{2}$-H inductance. Write the two mesh-current equations, but do not solve them.

Supermeshes Require KVL and KCL Equations

Meshes that share a current source with other meshes, none of which contains a current source in the outer loop, form a *supermesh*. A path around a supermesh does not pass through a current source. A path around each mesh contained within a supermesh passes through a current source. The total number of equations required for a supermesh is equal to the number of meshes contained in the supermesh. We must write a KVL equation around the supermesh. The remaining equations are KCL equations.

We also demonstrate a shortcut. We designate one mesh within a supermesh as a *principal mesh*, and use KCL to write the other mesh currents in the supermesh as a sum of this principal-mesh current and current source values. We then write a KVL equation for the loop around the supermesh. This last step is similar to the analysis of a series circuit that does not have a current source.

EXAMPLE 4.11

Write a set of mesh-current equations for the network shown in Fig. 4.15.

FIGURE 4.15
Network with one supermesh that requires both a KVL and a KCL equation

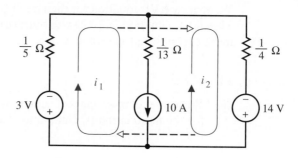

■ SOLUTION

With the network as drawn in Fig. 4.15, meshes 1 and 2, together with the shared current source and $\frac{1}{13}$-Ω resistance, form one supermesh. A KVL equation around the supermesh is

$$3 + \frac{1}{5}i_1 + \frac{1}{4}i_2 - 14 = 0 \qquad (4.45)$$

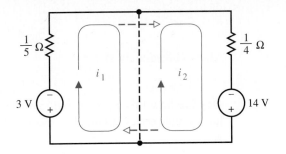

which is the first mesh-current equation. The use of KCL permits us to write the 10-A current source current in terms of mesh-current components:

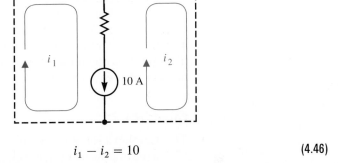

$$i_1 - i_2 = 10 \qquad (4.46)$$

This is the second mesh-current equation. Equations (4.45) and (4.46) can be written as

$$\begin{matrix} \text{From KVL} \\ \\ \text{From KCL} \end{matrix} \left\{ \begin{bmatrix} \dfrac{1}{5} & \dfrac{1}{4} \\ 1 & -1 \end{bmatrix} \cdot \begin{bmatrix} i_1 \\ i_2 \end{bmatrix} = \begin{bmatrix} 11 \\ 10 \end{bmatrix} \right\} \begin{matrix} \text{Voltage due to} \\ \text{voltage sources} \\ \text{Current due to} \\ \text{current source} \end{matrix}$$

This matrix mesh-current equation can be solved for the mesh currents.

The labor needed to write and solve the mesh-current equations can be reduced if we use KCL to write

$$i_2 = i_1 - 10 \qquad (4.46a)$$

directly on the network diagram before writing the KVL equation. Because we have written i_2 in terms of i_1, we call mesh 1 a *principal mesh*.

The KVL equation around the supermesh can be written as

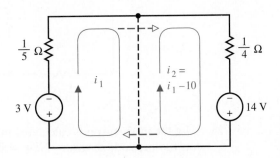

$$3 + \frac{1}{5}i_1 + \frac{1}{4}(i_1 - 10) - 14 = 0 \qquad \text{(4.45a)}$$

which immediately yields

$$i_1 = 30 \text{ A}$$

Observe how this shortcut reduced the analysis procedure to one similar to that for a simple series circuit. The second mesh current is given by the KCL equation as

$$i_2 = 30 - 10 = 20 \text{ A} \quad \blacksquare \qquad \text{(4.46b)}$$

REMEMBER

We can go from the interior of any mesh within a supermesh to the interior of any other mesh within the same supermesh by cutting across branches that contain current sources. You cannot reach the exterior of the network by cutting across a branch containing a current source (see the dashed paths in Fig. 4.15). A supermesh requires one mesh-current equation that is a KVL equation. The remaining mesh-current equations are KCL equations.

EXERCISES

27. Develop a set of mesh-current equations for the network shown below. First, write four mesh-current equations (two KVL and two KCL equations). Next, express i_3 and i_4 in terms of i_1 and i_2, and write them directly on the network diagram. Then write a KVL equation for each supermesh. The resulting two mesh-current equations should contain i_1 and i_2 as the only unknowns. Solve for all four mesh currents.

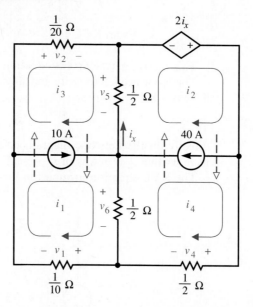

28. Use the mesh currents calculated in Exercise 27 to solve for v_1 and v_6 shown on the network diagram. Now write a KVL equation around mesh 1 and solve for the voltage of the left end of the 10-A source with respect to the

right end. Observe that we could not write this voltage in terms of i_1 and i_3, but that this voltage is $v_2 + v_5$, which can be written in terms of i_2 and i_3. This is why we wrote the mesh-current equations around the supermeshes.

4.6   Network Analysis by Mesh Currents

A more complex network may have current sources in the outer loop, meshes without current sources, and meshes that form a supermesh. The following example demonstrate that the analysis methods we developed still apply.

EXAMPLE 4.12

Analyze the circuit of Fig. 4.16 by the method of mesh currents.

FIGURE 4.16
A more general mesh-current example

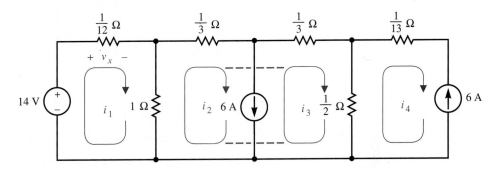

■ **SOLUTION**

Assign the mesh currents as shown in Fig. 4.16. Mesh 1 requires a KVL equation. Meshes 2 and 3 form a supermesh that requires one KVL equation and one KCL equation. Mesh 4 requires only a KCL equation. Write a KVL equation around mesh 1:

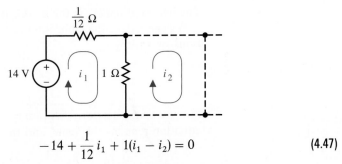

$$-14 + \frac{1}{12}i_1 + 1(i_1 - i_2) = 0 \qquad (4.47)$$

Write a KVL equation around meshes 2 and 3 (a supermesh):

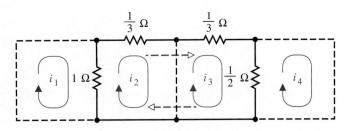

$$1(i_2 - i_1) + \frac{1}{3}i_2 + \frac{1}{3}i_3 + \frac{1}{2}(i_3 - i_4) = 0 \qquad \text{(4.48)}$$

The third and fourth mesh equations are KCL equations:

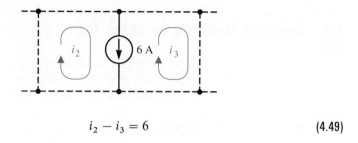

$$i_2 - i_3 = 6 \qquad \text{(4.49)}$$

and

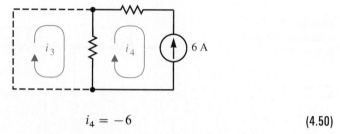

$$i_4 = -6 \qquad \text{(4.50)}$$

We can write these four mesh-current equations as a single matrix equation:

$$
\left\{
\begin{matrix}
\text{From} \\
\text{KVL} \\
\\
\text{From} \\
\text{KCL}
\end{matrix}
\right.
\begin{bmatrix}
\dfrac{13}{12} & -1 & 0 & 0 \\
-1 & \dfrac{4}{3} & \dfrac{5}{6} & -\dfrac{1}{2} \\
0 & 1 & -1 & 0 \\
0 & 0 & 0 & 1
\end{bmatrix}
\cdot
\begin{bmatrix}
i_1 \\
i_2 \\
i_3 \\
i_4
\end{bmatrix}
=
\begin{bmatrix}
14 \\
0 \\
6 \\
-6
\end{bmatrix}
\left.
\begin{matrix}
\\
\\
\\
\\
\end{matrix}
\right\}
\begin{matrix}
\text{Voltages due to} \\
\text{voltage sources} \\
\\
\text{Currents due to} \\
\text{current sources}
\end{matrix}
$$

The left-hand matrix is the *mesh transformation matrix*.

We can reduce the labor involved in writing and solving the mesh-current equations by writing

$$i_3 = i_2 - 6$$

and

$$i_4 = -6$$

directly on the network diagram before writing the KVL equations. The first KVL equation remains the same, and the second KVL equation is

$$1(i_2 - i_1) + \frac{1}{3}i_2 + \frac{1}{3}(i_2 - 6) + \frac{1}{2}\big[(i_2 - 6) + 6\big] = 0 \qquad \text{(4.48a)}$$

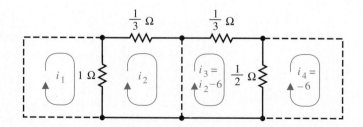

in which case the mesh-current equations reduce to

$$\begin{bmatrix} \dfrac{13}{12} & -1 \\ -1 & \dfrac{13}{6} \end{bmatrix} \begin{bmatrix} i_1 \\ i_2 \end{bmatrix} = \begin{bmatrix} 14 \\ 2 \end{bmatrix}$$

We refer to the left-hand matrix as the *reduced mesh transformation matrix*. When we solve for the *principal-mesh* currents, i_1 and i_2, we get

$$\begin{bmatrix} i_1 \\ i_2 \end{bmatrix} = \begin{bmatrix} 24 \\ 12 \end{bmatrix} \text{ A}$$

We obtain currents i_3 and i_4 from Eqs. (4.49) and (4.50). ∎

A summary of the shortcut procedure we have developed follows. The list is presented for easy reference, not for memorization.

Shortcut Method ▶

Analysis of Mesh Currents

1. Given a network diagram, draw the network so that no lines cross without being connected. If this cannot be done, the network is not planar and the method of mesh currents is not applicable.

2. Number each mesh beginning with 1, and assign a mesh current clockwise to each mesh.

3. Write the mesh-current equations.
   (a) Identify each mesh that has its mesh current defined by current sources. Use KCL to write each of these mesh currents as the difference of source currents. Write these relationships on the network diagram.
   (b) Identify each group of meshes that form a supermesh. Select one mesh within each supermesh as a principal mesh. Use KCL to write each other mesh current within the group in terms of the principal-mesh current and source currents. Write these relationships on the network diagram.
   (c) The remaining meshes are also principal meshes.
   (d) Now write a KVL equation around each mesh identified in (c) and around each supermesh. Remember that all unknowns must be written in terms of principal-mesh currents.

4. Solve the KVL equations for the principal-mesh currents. Each remaining mesh current is then obtained from a KCL equation.

We will now apply the shortcut procedure to a simple network.

EXAMPLE 4.13

Use the shortcut procedure to write a set of mesh-current equations for the network shown in Fig. 4.17.

FIGURE 4.17
Example circuit for the
shortcut procedure for
mesh-current analysis

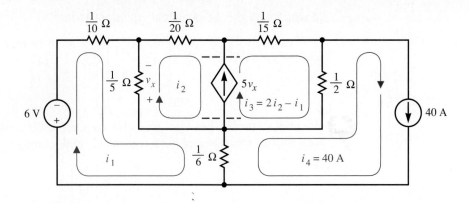

■ SOLUTION

1. The network is drawn so that no lines cross without being connected.
2. Assign the mesh currents as shown.
3. Write a set of mesh-current equations.
   (a) Mesh current i_4 is determined by the current source. From KCL,

$$i_4 = 40 \text{ A} \tag{4.51}$$

   (b) Meshes 2 and 3 form a supermesh. Select mesh 2 as the principal mesh. From KCL,

$$i_3 - i_2 = 5v_x = 5\left[\frac{1}{5}(i_2 - i_1)\right]$$

   This gives

$$i_3 = 2i_2 - i_1 \tag{4.52}$$

   Write this on the network diagram.
   (c) Mesh 1 is also a principal mesh, because it contains no current sources.
   (d) Write a KVL equation around mesh 1:

$$6 + \frac{1}{10}i_1 + \frac{1}{5}(i_1 - i_2) + \frac{1}{6}(i_1 - 40) = 0 \tag{4.53}$$

   Next apply KVL around the supermesh:

$$\frac{1}{5}(i_2 - i_1) + \frac{1}{20}i_2 + \frac{1}{15}(2i_2 - i_1) + \frac{1}{2}[(2i_2 - i_1) - 40] = 0 \tag{4.54}$$

4. These two KVL equations can be written as a single matrix equation:

$$\begin{bmatrix} \dfrac{7}{15} & -\dfrac{1}{5} \\ -\dfrac{23}{30} & \dfrac{83}{60} \end{bmatrix} \begin{bmatrix} i_1 \\ i_2 \end{bmatrix} = \begin{bmatrix} \dfrac{2}{3} \\ 20 \end{bmatrix}$$

This is easily solved to give $i_1 = 10$ A and $i_2 = 20$ A. From Eq. (4.52), $i_3 = 30$ A. Equation (4.51) gives i_4. ■

The following example illustrates how easy it is to apply the shortcut procedure to even a complicated-looking network.

EXAMPLE 4.14

Use the shortcut procedure to write a set of mesh-current equations for the network shown in Fig. 4.18.

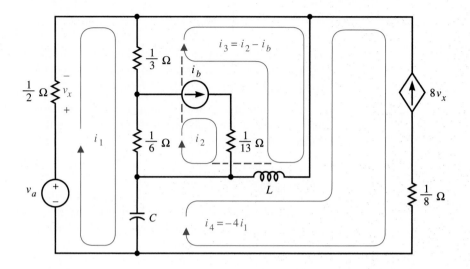

■ SOLUTION

1. The network is drawn so that no lines cross without being connected.
2. Assign the mesh currents as shown.
3. Write the mesh-current equations.
   (a) Mesh current i_4 is determined by the dependent current source. From KCL,

$$i_4 = -8v_x = -8\left(\frac{1}{2}i_1\right) = -4i_1 \qquad (4.55)$$

   Write this on the network diagram.
   (b) Meshes 2 and 3 form a supermesh. Select mesh 2 as the principal mesh. From KCL,

$$i_b = i_2 - i_3$$

   which gives

$$i_3 = i_2 - i_b \qquad (4.56)$$

   Write this on the network diagram.
   (c) Mesh 1 is also a principal mesh, because it contains no current sources.
   (d) Write a KVL equation around mesh 1:

$$-v_a + \frac{1}{2}i_1 + \frac{1}{3}[i_1 - (i_2 - i_b)] + \frac{1}{6}(i_1 - i_2) + \frac{1}{C}\int_{-\infty}^{t}[i_1 - (-4i_1)]\,d\tau = 0$$

$$(4.57)$$

   Next apply KVL around the supermesh:

$$\frac{1}{6}(i_2 - i_1) + \frac{1}{3}[(i_2 - i_b) - i_1] + L\frac{d}{dt}[(i_2 - i_b) - (-4i_1)] = 0 \quad (4.58)$$

4. The two KVL equations, Eqs. (4.57) and (4.58), can be written as a single matrix equation:

$$
\begin{bmatrix}
1 + \dfrac{5}{C}\displaystyle\int_{-\infty}^{t} d\tau & -\dfrac{1}{2} \\[2ex]
4L\dfrac{d}{dt} - \dfrac{1}{2} & L\dfrac{d}{dt} + \dfrac{1}{2}
\end{bmatrix}
\cdot
\begin{bmatrix} i_1 \\[1ex] i_2 \end{bmatrix}
=
\begin{bmatrix}
v_a - \dfrac{1}{3} i_b \\[2ex]
L\dfrac{d}{dt} i_b + \dfrac{1}{3} i_b
\end{bmatrix}
$$

The solution of a matrix differential equation of this type will be postponed until Chapters 8 and 9, but once the principal-mesh currents, i_1 and i_2, are determined, i_3 and i_4 can easily be obtained from Eqs. (4.55) and (4.56). ■

Mesh equations are also suitable for planar networks with two or more separate parts, as in Fig. 4.8. Each separate part has an outer loop, but this introduces no added complication.

EXERCISES

29. Use the method of analysis by mesh currents to calculate the mesh currents for the network in the following figure. Then calculate the power absorbed by the dependent source.

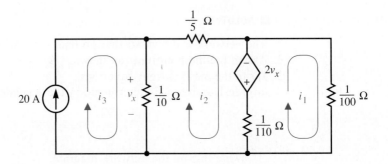

30. Replace the 20-A source in the network shown above with an open circuit and the $\frac{1}{5}$-Ω resistance with a 20-A source (reference arrow pointing to the right). Then repeat Exercise 29.

31. Replace the dependent voltage source of the preceding network with a dependent current source (current reference arrow pointing down) with a control equation of $i = 110v_x$. Then repeat Exercise 29.

32. Replace the lower $\frac{1}{2}$-Ω resistance of the circuit used in Exercise 27 with a current source of value 100 A (reference arrow pointing toward the left). Use the shortcut procedure to write the mesh-current equations. This should leave one KVL equation that contains i_1. Solve for all mesh currents.

For Exercises 33 through 35, replace the center current source of Fig. 4.16 with a dependent current source, reference arrow pointing down, with a control equation $i = 3v_x$.

33. Write the four mesh-current equations, and arrange in matrix form. You should have two KVL equations and two KCL equations.

34. Use the two KCL equations to write i_3 and i_4 in terms of i_1 and i_2. Write these on the network diagram. Directly from the network diagram, write two KVL equations that contain i_1 and i_2 as the only unknowns.

35. Solve for all mesh currents and the power absorbed by the dependent source.

4.7 Mesh-Current Equations by Inspection*

After elimination of the KCL equations by substitution in Example 4.12, the reduced mesh transformation matrix was symmetric about the diagonal. A network with no dependent sources is described by a symmetric reduced mesh transformation matrix. If all sources are independent voltage sources and all elements resistances, the equations for a circuit with N meshes can be written by inspection.

$$\begin{bmatrix} R_{11} & R_{12} & \cdots & R_{1N} \\ R_{21} & R_{22} & \cdots & R_{2N} \\ \vdots & \vdots & & \vdots \\ R_{N1} & R_{N2} & \cdots & R_{NN} \end{bmatrix} \cdot \begin{bmatrix} i_1 \\ i_2 \\ \vdots \\ i_N \end{bmatrix} = \begin{bmatrix} v_1 \\ v_2 \\ \vdots \\ v_N \end{bmatrix} \qquad (4.59)$$

where R_{kk} = sum of all resistances in mesh k

$R_{jk} = R_{kj}$ = the negative of the sum of all resistances in common with mesh k and mesh j, $k \neq j$

i_k = mesh current chosen clockwise

v_k = sum taken counterclockwise of the voltages across all voltage sources in mesh k

Remember that this is valid only for a planar network consisting of linear resistances and independent voltage sources when all mesh currents are assigned a clockwise direction. A generalization to include all linear elements and dependent voltage sources is easily developed.

EXAMPLE 4.15

Write a set of mesh-current equations for the circuit for Fig. 4.19, first by writing KVL equations and then by inspection.

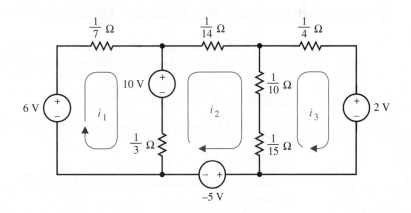

* This section can be omitted without loss of continuity.

■ SOLUTION

Assign mesh currents clockwise as shown. Write KVL equations around each mesh:

$$-6 + \frac{1}{7}i_1 + 10 + \frac{1}{3}(i_1 - i_2) = 0 \tag{4.60}$$

$$\frac{1}{3}(i_2 - i_1) - 10 + \frac{1}{14}i_2 + \frac{1}{10}(i_2 - i_3) + \frac{1}{15}(i_2 - i_3) + (-5) = 0 \tag{4.61}$$

$$\frac{1}{15}(i_3 - i_2) + \frac{1}{10}(i_3 - i_2) + \frac{1}{4}i_3 + 2 = 0 \tag{4.62}$$

Collection of the equations without addition of the resistances yields

$$\begin{bmatrix} \frac{1}{7} + \frac{1}{3} & -\frac{1}{3} & 0 \\ -\frac{1}{3} & \frac{1}{3} + \frac{1}{14} + \frac{1}{10} + \frac{1}{15} & -\frac{1}{10} - \frac{1}{15} \\ 0 & -\frac{1}{10} - \frac{1}{15} & \frac{1}{10} + \frac{1}{15} + \frac{1}{4} \end{bmatrix} \cdot \begin{bmatrix} i_1 \\ i_2 \\ i_3 \end{bmatrix} = \begin{bmatrix} 6 - 10 \\ 10 - (-5) \\ -2 \end{bmatrix}$$

The solution for the mesh currents is

$$\mathbf{i} = \begin{bmatrix} i_1 \\ i_2 \\ i_3 \end{bmatrix} = \begin{bmatrix} 21 \\ 42 \\ 12 \end{bmatrix} \text{A}$$

You should verify that the procedure outlined by Eq. (4.59) yields the same result. ■

EXERCISE

36. Use the method presented in this section to write a set of mesh-current equations for the network used in Exercise 23.

4.8 Mutual Inductance and Mesh-Current Equations*

Mutual inductance does not significantly complicate the writing of mesh equations. An example that includes mutual inductance follows.

EXAMPLE 4.16

Write a set of mesh-current equations for the network of Fig. 4.20.

■ SOLUTION

Assign the mesh currents as shown. Write a KVL equation around mesh 1 and around mesh 2:

___

* This section can be postponed without loss of continuity.

FIGURE 4.20
Mesh-current example with
mutual inductance

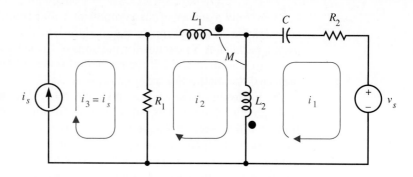

$$L_2 \frac{d}{dt}(i_1 - i_2) + M \frac{d}{dt}(-i_2) + \frac{1}{C} \int_{-\infty}^{t} i_1 \, d\tau + R_2 i_1 + v_s = 0 \qquad (4.63)$$

$$R_1(i_2 - i_s) + L_1 \frac{d}{dt} i_2 + M \frac{d}{dt}(i_2 - i_1) + L_2 \frac{d}{dt}(i_2 - i_1) + M \frac{d}{dt} i_2 = 0 \quad (4.64)$$

where the third mesh equation,

$$i_3 = i_s \qquad (4.65)$$

was used to eliminate i_3 from the second mesh equation when it was written. ∎

EXERCISE

37. Write a matrix mesh-current equation for the network shown in the circuit
    below.

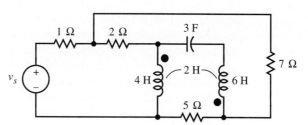

4.9   The Best Method

The best network analysis method to use depends not only on the network to be
analyzed but also on the information required. If we require the voltage across
and the current through each component, direct application of the method of node
voltages or mesh currents may be the best method. This is especially true when
a computer program is available to solve the resulting equations. However, even
when a computer is available, it is wise to pick the method that results in the
smallest set of equations. The set of node-voltage equations can easily be reduced
to the number of nodes minus the number of reference nodes, minus the number
of voltage sources.* The set of mesh-current equations can easily be reduced to
the number of meshes minus the number of current sources.

* See Problems 91 through 94 for the unusual cases of circuits that contain loops of voltage sources
and cutsets of current sources.

Often, only a subset of the component voltages and currents is required. Here considerable simplification of the analysis can result from the use of equivalent (series, parallel, Δ-Y) circuits for portions of the network, as described in Chapter 3. Try to eliminate nodes if the method of node voltages is used. Try to eliminate meshes if the method of mesh currents is used.

The inclusion of inductance and capacitance does not increase the difficulty of writing network equations, but it may significantly increase the difficulty of solution, as will be seen in Chapters 8 and 9. However, in certain very important circumstances, the increased complexity is only that of going from the algebra of real numbers to that of complex numbers, as will be shown in Chapter 10. In the next section we will introduce a computer program, based on node-voltage analysis, that performs circuit analysis for us.

4.10 Network Analysis with PSpice®*

Digital computers are routinely used to solve systems of simultaneous algebraic equations, and you may have solved node-voltage and mesh-current equations in this manner. There are also programs to solve systems of simultaneous differential equations. Computer-aided analysis takes this a step further. You simply input the component or device specifications and their node connections. The program then generates and solves the equations for specified voltages and currents. The most popular programs are the various versions of SPICE, which stands for Simulation Program with Integrated Circuit Emphasis. SPICE, which was developed at the University of California, Berkeley, is based on node-voltage analysis. All node voltages and the currents through the independent voltage sources are calculated, as well as the total power supplied by these sources. With the use of the proper solution control statement, any voltage and current can be included in the output. Outputs are available in either tabular or graphical form. SPICE is an incredibly versatile circuit analysis tool that can analyze networks of linear or nonlinear devices. In this introductory treatment, we will limit our discussion to networks containing only independent sources and the standard linear elements and dependent sources. You can refer to the PSpice supplement that accompanies this textbook or other textbooks for a more advanced treatment.[†]

SPICE can be used in several modes, including dc (sources constant for all time), transient (sources applied at time $t = 0$), and ac (sources that are a sine or cosine function for all time) analysis. We shall consider dc analysis here, transient analysis in Chapters 8 and 9, and ac analysis in Chapters 11 and 13.

DC Analysis

In the dc analysis mode, all sources are assumed constant since the beginning of time, so that the time derivatives of all voltages and currents are zero. This results in zero capacitance current, zero inductance voltage, and zero voltage due to mutual inductance:

---

* PSpice® is a registered trademark of MicroSim Corporation. With the exception of the .OPT statement and the .PROBE command, the PSpice material presented in this textbook is compatible with most versions of SPICE.

† Paolo Antognetti and Giuseppe Massobrio, *Semiconductor Device Modeling with SPICE*. New York: McGraw-Hill, 1988; Paul W. Tuinenga, *SPICE: A Guide to Circuit Simulation and Analysis Using PSpice*. Englewood Cliffs, N.J.: Prentice-Hall, 1988.

$$i_C = C \frac{d}{dt} v_C = 0 \tag{4.66}$$

$$v_L = L \frac{d}{dt} i_L = 0 \tag{4.67}$$

$$v_M = M \frac{d}{dt} i_M = 0 \tag{4.68}$$

Thus dc analysis proceeds with all capacitances replaced by open circuits, all inductances replaced by short circuits, and all mutual inductances set equal to zero.

■ *Restrictions; see Sec. 3.2.1 of the PSpice manual.*

The following restrictions are always placed on the circuit to be analyzed:

1. The reference node must be numbered 0 (zero).
2. All other nodes must be assigned a unique nonnegative integer identification number.
3. The circuit cannot contain a loop of voltage sources and/or inductances.
4. The circuit cannot contain a cutset* of current sources and/or capacitances.
5. Each node in the circuit must have a dc path to the reference node (a path through sources, resistances, and/or inductances).
6. Every node must have at least two components connected to it (no "dangling nodes" allowed).

PSpice

■ *How to load PSpice; see Sec. 1.1 of the PSpice manual.*

We will assume that the personal computer version of SPICE, known as PSpice, which is available from MicroSim Corporation, has been loaded on our computer. This program analyzes a circuit described by an input file or circuit file that contains the following:

1. Title and comment statements
2. Circuit description (data statements)
3. Solution control statements
4. Output control statements
5. End statement

We will demonstrate the dc analysis of a circuit with the following example.

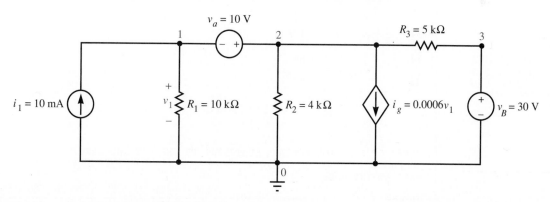

FIGURE 4.21 Circuit for the first dc PSpice example

* A cutset is defined on page 21.

EXAMPLE 4.17

Use PSpice to calculate the node voltages of the circuit shown in Fig. 4.21.

■ SOLUTION

PSpice text editor; see Sec. 2.1.3 of the PSpice manual.

We must first log on to our computer. The procedure depends on the particular system and often consists of simply turning on the power. We next insert a formatted floppy disk in drive A, call our particular ASCII text editor[†] (EDLIN was used for the file presented below). We can then type in the circuit file (input file).

We will name our file EX4-17.CIR and store it on the floppy disk in drive A by typing

    EDLIN A: EX4-17.CIR

This file, which follows, contains the five types of statements in the list preceding this example. The comment statements, which begin with an asterisk (*), are for explanation and we need not include them. (For a more detailed explanation of the statements, refer to Section 4.11.)

```
INPUT FILE FOR EXAMPLE 4.17

*DESCRIBE THE CIRCUIT. THE FIRST FIELD IDENTIFIES THE

*COMPONENT. THE NEXT TWO IDENTIFY THE NODE CONNECTIONS

*(THE VCVS REQUIRES TWO MORE NODE NUMBERS TO SPECIFY

*THE CONTROL VOLTAGE). THE FINAL FIELD(S) SPECIFY THE

*COMPONENT PARAMETERS. THE PASSIVE SIGN CONVENTION IS USED

*FOR ALL VOLTAGES AND CURRENTS.

*

I1      0       1       DC      10M

VA      2       1       DC      10

VB      3       0       DC      30

R1      1       0       10K

R2      2       0       4K

R3      2       3       5K

G1      2       0       1       0       0.0006

*

*THE .OP SOLUTION CONTROL STATEMENT IS DEFAULT AND NEED

*NOT BE INCLUDED.
```

[†] The most recent versions of PSpice include a simple text editor that can be used as an alternative to your text editor.

```
*

.OP

*

*THE .OP STATEMENT CAUSES ALL NODE VOLTAGES, AND THE

*VOLTAGE SOURCE CURRENTS TO BE CALCULATED AND ENTERED IN

*THE OUTPUT FILE. NO OUTPUT CONTROL STATEMENT IS REQUIRED

*FOR THE .OP CONTROL STATEMENT.

.END
```

Running PSpice in the batch mode; see Sec. 1.3.2 of the PSpice manual.

We use the EDLIN command e to terminate the editing. This will save our file on the floppy disk in drive A (your system may require a slightly different command) and exit the editor. Then type

    PSPICE

and press the return key. The program will request an input (circuit) file. The default ending of the input file is .CIR, so we only need to type

    A:EX4-17

and press the return key. The program next requests a name for the output file. We will use the default, so press the return key. This initiates the program and puts the output in a file named EX4-17.OUT on the floppy disk in drive A.

Our output file now contains our circuit description, the calculated node voltages, and the current through the voltage sources (passive sign convention). The output file can be viewed on the screen. Type

    TYPE A:EX4-17.OUT

and press the return. If a copy is required, type

    PRINT A:EX4-17.OUT

and press the return. At this point, some systems require the specification of an output device, or simply a second return. In any case, the output will contain the original circuit description and the following information:

```
NODE    VOLTAGE     NODE    VOLTAGE     NODE    VOLTAGE

( 1)    10.0000     ( 2)    20.0000     ( 3)    30.0000


VOLTAGE SOURCE CURRENTS

NAME         CURRENT

VA          -9.000E-03

VB          -2.000E-03

TOTAL POWER DISSIPATION    1.50E-01   WATTS
```

The total power dissipation listed is the power supplied by the independent voltage sources. ∎

We now have all node voltages and the current through each voltage source. At this point we can easily calculate any additional voltages or currents by hand, but manual calculations are not necessary, as the following example demonstrates.

EXAMPLE 4.18

Use PSpice to analyze the small-signal equivalent circuit of the transistor amplifier shown in Fig. 4.22a. Print out the node voltages, v_{12}, v_{43}, the current through R_C, the current through R_1, and the current through R_L.

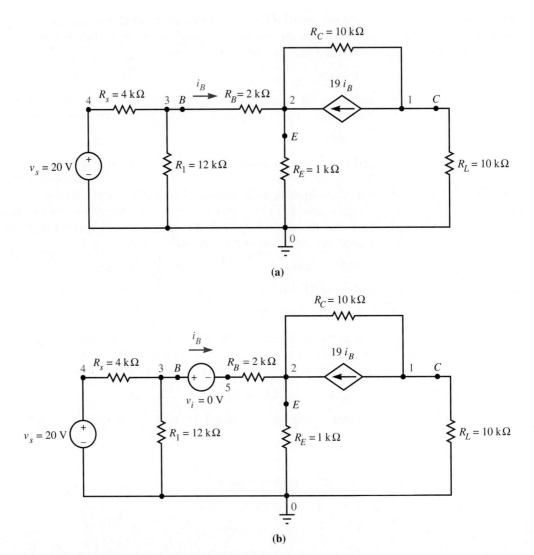

(a)

(b)

FIGURE 4.22 Amplifier for PSpice analysis, Example 4.18: (*a*) amplifier equivalent circuit; (*b*) dummy voltage source inserted to force calculation of control current i_B

■ SOLUTION

First build an input file. Only currents through voltage sources are initially calculated, so a dummy voltage source must be introduced to force the program to calculate the control current. (Refer to Fig. 4.22b and observe that this introduces another node, so that one node connection for R_B is changed.) The polarity of the source is determined by the passive sign convention. Outputs other than node voltages are required, so the control statement .DC must be used instead of .OP. Although not initially calculated, currents other than voltage source currents will be calculated for printout if requested. The beginning and end voltages for v_s will be set at 20 V, and the step size at 1. This will force a calculation for $v_s = 20$ V only. A print statement is required with .DC. The input file is as follows:

```
INPUT FILE FOR EXAMPLE 4.18

*COMPONENT DATA STATEMENTS

VS      4      0      DC      20

VI      3      5      DC      0

RS      4      3      4K

R1      3      0      12K

RB      5      2      2K

RC      2      1      10K

RE      2      0      1K

F1      1      2      VI      19

RL      1      0      10K

*SOLUTION CONTROL STATEMENT .DC IS USED.

*THIS MAKES VS EQUAL TO 20 VOLTS.

.DC    VS    20    20    1

*USE .OPT TO REDUCE PAGE COUNT.

.OPT NOPAGE

*.DC REQUIRES A PRINT STATEMENT.

.PRINT DC  V(1)  V(2)  V(3)

.PRINT DC  V(4)  V(1,2)  V(4,3)

.PRINT DC  I(RC)  I(R1)  I(RL)

*END STATEMENT IS REQUIRED.

.END
```

The information in the output file includes (along with some unneeded output):

VS	V(1)	V(2)	V(3)
2.000E+01	-9.000E+01	1.000E+01	1.200E+01

VS	V(4)	V(1,2)	V(4,3)
2.000E+01	2.000E+01	-1.000E+02	8.000E+00

VS	I(RC)	I(R1)	I(RL)
2.000E+01	1.000E-02	1.000E-03	-9.000E-03

Observe that the calculated currents through resistances R_C, R_1, and R_L are determined from the passive sign convention and the specified polarities of the resistance connections. ∎

For our final example, we will consider the dc analysis of a circuit containing resistance, inductance, and capacitance. PSpice assumes that the sources have been constant for a very long time and replaces all inductances by short circuits and all capacitances by open circuits. (The transient case for a suddenly applied voltage is considered in Chapter 9.)

EXAMPLE 4.19

Use PSpice to calculate the node voltages for the circuit shown in Fig. 4.23. Assume that the voltage source has been constant for a very long time.

FIGURE 4.23
PSpice example with inductance, capacitance, resistance, and a constant source

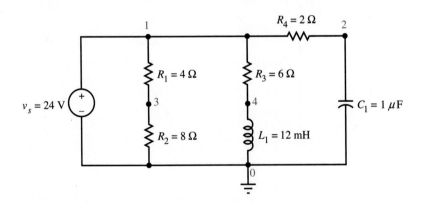

∎ SOLUTION

Use the default .OP for solution control. This will replace all inductances by a short circuit and all capacitances by an open circuit before the node voltages are calculated. The input file is

```
INPUT FILE FOR EXAMPLE 4.19

*COMPONENT DATA STATEMENTS

VS      1    0    24

R1      1    3    4

R2      3    0    8

R3      1    4    6

R4      1    2    2

L1      4    0    12M

C1      2    0    1U

*USE DEFAULT .OP FOR SOLUTION CONTROL.

*USE .OPT TO REDUCE PAGE COUNT.

.OPT NOPAGE

*END STATEMENT IS REQUIRED.

.END
```

The desired node voltages are found in the output file (this file was edited to reduce
the printed width).

```
NODE    VOLTAGE    NODE    VOLTAGE    NODE    VOLTAGE

( 1)    24.0000    ( 2)    24.0000    ( 3)    16.0000


NODE    VOLTAGE

( 4)    0.0000



VOLTAGE SOURCE CURRENTS

NAME         CURRENT

VS           -6.000E+00


TOTAL POWER DISSIPATION   1.44E+02  WATTS
```

You should replace the inductance by a short circuit and the capacitance by
an open circuit and use hand calculations to check the results of the previous
example. ■

4.11   PSpice Statements for DC Analysis

We have used five types of PSpice statements in our circuit files:

1. Title and comment statements
2. Circuit description (data statements)
3. Solution control statements
4. Output control statements
5. End statement

We will now describe these statements in more detail.

Title and Comment Statements

Expanded coverage of PSpice statements; see Sec. A.1 of the PSpice manual.

The first statement of the circuit file is a *title statement*. This is a *single line* that identifies the analysis problem—for example,

    DC ANALYSIS OF CIRCUIT ONE

Any number of comment lines can be inserted in the program that follows, but each comment line must begin with an asterisk (*).

Circuit Description

PSpice component descriptions; see Sections A.1.1 and A.1.3 of the PSpice manual.

We describe the circuit by a data statement for each component. Each statement is composed of three fields:

$$\underbrace{\texttt{XNAME}}_{\substack{\text{Component}\\\text{name}}} \qquad \underbrace{\texttt{N+ \ N-} \ \cdots}_{\substack{\text{Node}\\\text{connections}}} \qquad \underbrace{\texttt{VALUE} \ \cdots}_{\substack{\text{Parameter}\\\text{values}}}$$

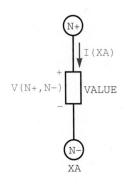

1. The *component name* contains from one to eight alphanumeric characters. (PSpice permits names longer than eight characters.) The first letter of the name determines the component type; for example, a resistance name must begin with R.
2. The *node connections* for the component are specified by the identification numbers N+ for the (+) reference node and N− for the (−) reference node. *The passive sign convention is used*, so the current I(XA) through component XA would be from the first node listed, N+, to the second node listed, N−, as shown in the illustration.
3. The *parameter values* can be listed in integer, floating-point, or exponential format:

Format	Examples
Integer	47, −13
Floating-point	3.14, −0.17
Exponential	4.7E3, 3.3E-6

PSpice circuit element values; See Sec. A.1.2 of the PSpice manual.

As an additional convenience we can also use scale factors with these numbers.

$$F = 10^{-15} \qquad U = 10^{-6} \qquad MEG = 10^{6}$$
$$P = 10^{-12} \qquad M = 10^{-3} \qquad G = 10^{9}$$
$$N = 10^{-9} \qquad K = 10^{3} \qquad T = 10^{12}$$

For example, the following are equivalent statements:

$$3.3E\text{-}12 \qquad 3.3E\text{-}9M \qquad 3.3E\text{-}6U \qquad 3.3P$$

Some components have one or more parameters; some of these are optional and will be shown enclosed in brackets [ ], but the brackets are not entered in the control statement. Some optional parameters and nonlinear components are not considered in the elementary treatment of this text.

We now present a detailed description and example of the data statement for each component type.

Resistance; see Sec. A.2.1 of the PSpice manual.

DATA STATEMENT FOR RESISTANCE   We completely specify a resistance element with its name, node connections, and value:

```
RNAME     N+     N-     VALUE
```

RNAME is a unique resistance name beginning with R, N+ and N− represent the node connections (passive sign convention), and VALUE is the resistance in ohms.

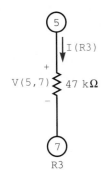

■ RESISTANCE EXAMPLE

A resistance $R3$ that is connected from node 5 [(+) reference mark] to node 7 [(−) reference mark] and has a resistance of 47 kΩ is specified by

```
R3     5     7     47K
```

Current I(R3) is current i_{57}. ■

Capacitance; see Sec. A.2.2 of the PSpice manual.

DATA STATEMENT FOR CAPACITANCE   In addition to the name, node connections, and value, we can also specify an initial condition (voltage) for a capacitance:

```
CNAME     N+     N-     VALUE     [IC = VOLTS]
```

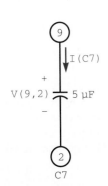

CNAME is a unique capacitance name beginning with C, N+ and N− are the node connections (passive sign convention), and VALUE is the capacitance in farads. IC = VOLTS is the value of capacitance voltage V(N+, N−) measured in volts, at time zero. This entire field is omitted for dc analysis (its omission would give a default value of zero for transient analysis).

■ CAPACITANCE EXAMPLE

A capacitance C7 is connected between node 9[(+) reference mark] and node 2[(−) reference mark] and has a value of 5 μF. The voltage $v_{92}(0)$ is 0. This is specified by

```
C7     9     2     5U
```
■

Inductance; see Sec. A.2.3 of the PSpice manual.

DATA STATEMENT FOR INDUCTANCE   Like capacitance, inductance is specified by a name, node connections, value, and an initial condition (current):

```
LNAME     N+     N-     VALUE     [IC = AMPS]
```

LNAME is a unique inductance name beginning with L, N+ and N− are the node connections (passive sign convention), and VALUE is the inductance in henries. IC = AMPS is the value of inductance current I(LNAME) in amperes at time zero. This entire field is omitted for dc analysis (its omission would give a default value of zero for transient analysis).

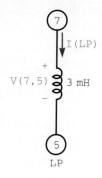

■ INDUCTANCE EXAMPLE

An inductance LP is connected from node 7 [(+) reference mark] to node 5 [(−) reference mark] and has a value of 3 mH. The current $i_{75}(0)$ is 0. This is specified by

```
LP      7      5      3M ■
```

DATA STATEMENT FOR MUTUAL INDUCTANCE
We must specify each coupled coil with its own data statement and then specify the coupling coefficient with this data statement:

```
KNAME      LNAMEA      LNAMEB      VALUE
```

KNAME is a unique coupling coefficient name beginning with K. LNAMEA is the name of one of the coupled inductances. LNAMEA must be defined by its own data statement. The N+ node of LNAMEA must correspond to the end with the dot mark. LNAMEB is the name of the second coupled inductance. LNAMEB must be defined by its own data statement. The N+ node of LNAMEB must correspond to the end with the dot mark. VALUE is the numerical value of the coupling coefficient and lies between 0 and 1.

⬛ *Mutual inductance; see Sec. A.2.4 of the PSpice manual.*

■ MUTUAL INDUCTANCE EXAMPLE

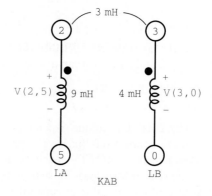

A mutual inductance named LA is connected from node 2 to node 5 and has a value of 9 mH. A second mutual inductance named LB is connected from node 3 to node 0 and has a value of 4 mH. The inductances are coupled, and the dot marks are on nodes 2 and 3. The mutual inductance is 3 mH (this gives KAB = 0.5). This situation is specified by the three statements

```
LA      2      5      9M

LB      3      0      4M

KAB     LA     LB     0.5 ■
```

⬛ *Independent sources; see Sec. A.3 of the PSpice manual.*

DATA STATEMENT FOR INDEPENDENT SOURCES
Constant sources are specified by a name, node connections, type, and value:

```
SNAME      N+      N-      [TYPE]      VALUE
```

SNAME is a unique source name. The first letter must be V for an independent voltage source and I for an independent current source. N+ and N− identify

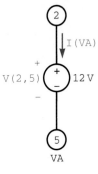

VA

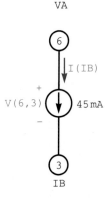

IB

the node connections (passive sign convention). TYPE is DC for dc analysis and AC for ac analysis (omission of this field gives DC as a default). Other types are considered in Chapter 9. VALUE is the dc value in volts or amperes for DC (for AC, both the magnitude and phase, in degrees, must be specified).

■ VOLTAGE SOURCE EXAMPLE

A voltage source named VA is connected with the (+) reference mark at node 2 and the (−) reference mark at node 5, and has a value of 12 V. This is specified by the statement

VA   2   5   12 ■

■ CURRENT SOURCE EXAMPLE

A current source named IB is connected from node 6 to node 3 with the reference arrow pointing toward node 3. The source has a value of 45 mA. This is specified by the statement

IB   6   3   45M ■

DATA STATEMENTS FOR VOLTAGE-CONTROLLED DEPENDENT SOURCES   We completely specify a voltage-controlled source with a name, node connections, control voltage, and control parameter:

DNAME   N+   N−   NC+   NC−   VALUE

DNAME is a unique source name beginning with E (voltage-controlled voltage source) or G (voltage-controlled current source). N+ and N− identify the node connections (passive sign convention) for the dependent source. NC+ and NC− identify the node connections (passive sign convention) for the controlling voltage. VALUE is the numerical value of the control parameter (dimensionless or in siemens), which can be positive or negative.

⬚ *VCVS; see Sec. A.4.1 of the PSpice manual.*

■ VOLTAGE-CONTROLLED VOLTAGE SOURCE (VCVS) EXAMPLE

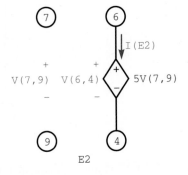

E2

A VCVS named E2 is connected from node 6 [(+) reference] to node 4 [(−) reference]. The control voltage is v_{79}, and the voltage gain is 5. This component is described by the statement

E2   6   4   7   9   5 ■

⏷ VCCS; see Sec. A.4.3
of the PSpice manual.

■ **VOLTAGE-CONTROLLED CURRENT SOURCE (VCCS) EXAMPLE**

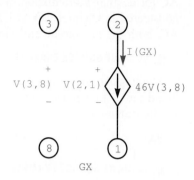

A VCCS named GX is connected from node 2 to node 1 with the reference arrow pointing toward node 1. The control voltage is v_{38}, and the transconductance is 46 S. The component specification statement is

```
GX    2    1    3    8    46 ■
```

DATA STATEMENT FOR CURRENT-CONTROLLED DEPENDENT SOURCES We specify a current-controlled source with a name, terminal connections, control current, and control parameter:

```
DNAME    N+    N-    VX    VALUE
```

DNAME is a unique source name beginning with H (current-controlled voltage source) or F (current-controlled current source). N+ and N− identify the node connections (passive sign convention) for the dependent source. VX is the name of a voltage source through which the controlling current passes (passive sign convention). It may be necessary to introduce a dummy voltage source of 0 V into the circuit to obtain this current. VALUE is the numerical value of the control parameter (dimensionless or in ohms), which can be positive or negative.

⏷ CCVS; see Sec. A.4.4
of the PSpice manual.

■ **CURRENT-CONTROLLED VOLTAGE SOURCE (CCVS) EXAMPLE**

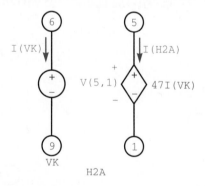

A CCVS named H2A is connected from node 5 [(+) reference] to node 1 [(−) reference]. The control current is the current (passive sign convention) I(VK) through a voltage source named VK connected between node 6 [(+) reference] and node 9 [(−) reference]. The mutual resistance is 47 Ω. The statement that describes this component is

```
H2A    5    1    VK    47 ■
```

CCCS; see Sec. A.4.2 of the PSpice manual.

■ CURRENT-CONTROLLED CURRENT SOURCE (CCCS) EXAMPLE

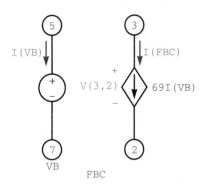

A CCCS named FBC is connected from node 3 to node 2, reference arrow pointing toward node 2. The control current is the current (passive sign convention) I(VB) through a voltage source named VB connected between node 5 [(+) reference] and node 7 [(−) reference]. The current gain is 69. The statement that describes the component is

    FBC   3   2   VB   69 ■

Solution Control Statements

There are several types of solution control statements, but in this chapter we will only consider two statements that apply to dc circuit analysis. We will introduce other control statements in later chapters.

Operating point analysis; see Sec. B.2.6 of the PSpice manual.

OPERATING (BIAS) POINT DC ANALYSIS   The .OP solution control statement is specifically designed to calculate the bias point in electronic circuits, but this statement also permits us to perform dc analysis:

    .OP

This simple statement specifies calculation of all node voltages and the current through all independent voltage sources. (These values will automatically be printed out, so no .PRINT statement will be required.) .OP is the *default* if no control statement is used.

DC analysis; see Sec. B.2.2 of the PSpice manual.

DC ANALYSIS   If we wish to calculate voltages that are not node voltages or currents through components other than voltage sources, we can use the .DC control statement. For example, this permits us to print out the voltage V(1, 2), which is v_{12}, or the current I(R), which is the current through resistance R (passive sign convention).

    .DC   SVAR   SSTART   SSTOP   SSTEP

SVAR is the name of an independent source that is to be stepped over a range of values, and the circuit analyzed at each step. SSTART is the beginning value of the source. SSTOP is the final value of the source. SSTEP is the incremental increase in source value with each step. SSTART can be greater or less than SSTOP, but SSTEP must always be positive. With the DC option a print statement is required, but any voltage or current can be printed in the output.

■ DC SOURCE EXAMPLE

A current source I3 (which is specified by a data statement) has a value of 20 A. We need to calculate voltages that are not node voltages or currents other than those through voltage sources. This is accomplished with the solution control statement

```
.DC    I3    20    20    1
```

The difference between the start value of current (20 A) and the stop value of current (20 A) is 0. This is less than the step size of 1 A, so only the value of 20 A is used for current source I3. ■

Output Control Statements

We must supply an output control statement unless the .OP solution control statement is used. The three output control statements are .PRINT, .OPT, and .PLOT.

The .PRINT statement; see Sec. B.2.9 of the PSpice manual.

THE .PRINT STATEMENT   When used with the .DC control statement, the .PRINT statement causes the program to calculate and print any requested voltage or current.

```
.PRINT    TYPE    OUTVARS
```

TYPE must correspond to the solution type used: DC, AC, or TRANS. OUTVARS is a list of the desired output variables. There is no limit to the number of output variables. The output format is controlled with the .WIDTH statement, or alternatively with the .OPT statement.

THE .WIDTH STATEMENT   The .WIDTH statement permits us to control the width of our printer or printer plot output:

```
.WIDTH    OUT = VALUE
```

VALUE is the number of columns to be printed or plotted. For PSpice, one .WIDTH statement is required for each .PRINT or .PLOT statement if other than default is used. The default is 80 if the .WIDTH statement is omitted. The only acceptable values are 80 and 132.

The .OPT statement; see Sec. B.2.7 of the PSpice manual.

THE .OPT STATEMENT   PSpice has a number of options that are controlled by the .OPT statement:

```
.OPT    PARAMETER
```

At this point we will use only NOPAGE, which will suppress page ejection and save paper.

The .PLOT statement; see Sec. B.2.8 of the PSpice manual.

THE .PLOT STATEMENT   We can obtain a printer plot of our output if we use the .DC control statement to step over a range of voltages. In later chapters we will use this statement to plot voltages and currents as a function of time.

```
.PLOT    TYPE    OUTVARS    [LOWLIM, UPLIM]
```

TYPE must correspond to the solution type used: DC, AC, or TRANS. OUTVARS is a list of the desired output variables. The x axis range is fixed by the variables plotted. [LOWLIM, UPLIM] specifies the lower and upper limits, respectively, for the y axis and will be used to plot all variables listed on the left of this limit. If the limits are omitted, the program calculates the limits. Additional variables and limits may follow. A maximum of eight variables can be plotted with a single .PLOT statement.

The End Statement

Every program must include an .END statement:

```
. END
```

This marks the end of the program.

REMEMBER

🔲 *PSpice circuit restrictions; see Sec. 3.2.1 of the PSpice manual.*

The following restrictions are placed on all circuits to be analyzed with PSpice.

1. The network cannot contain a loop of voltage sources and/or inductances.
2. The network cannot contain a cutset of current sources and/or capacitances.
3. Each node must have a dc path to the reference node.
4. No dangling nodes are allowed.
5. A control current must be the current through an independent voltage source (this may require the introduction of a dummy voltage source of value zero volts).
6. The input file contains five types of statements:
   (a) Title and comment statements
   (b) Circuit description (data statements)
   (c) Solution control statements
   (d) Output control statements
   (e) End statement

The following conventions are used in the input file.

1. The reference node is numbered zero.
2. All other node numbers must be unique positive numbers.
3. The passive sign convention is used.

EXERCISES

38. Use PSpice to calculate the node voltages in the network for Fig. 4.5. Use the .OP control statement.

39. Use the .DC control statement with PSpice to calculate current i_{12} in the network of Fig. 4.5.

40. Calculate the node voltages in the network of Fig. 4.6 with the use of PSpice.

41. Calculate the node voltages and current i_{12} in the network of Fig. 4.7, if $i_a = 10$ A and $v_b = 20$ V for all time. Use PSpice with the .DC control statement.

4.12   Summary

In this chapter we developed two systematic methods of circuit analysis. The first method, node-voltage analysis, was obtained through generalization of the methods of Chapter 3 to analyze a parallel circuit. First we analyzed networks that contained only one of three classes of nodes. We then saw that these techniques were applicable to networks containing all three classes of nodes. The method is applicable to any electrical network. We next generalized the method used to analyze a series circuit to obtain the second systematic method, called mesh-current analysis. This method is applicable only to planar networks. We followed with a brief discussion of which analysis technique is best for a given circuit. We concluded the chapter with an introduction to computer-aided analysis of circuits with the use of PSpice.

KEY FACTS

◆ Node-voltage analysis can be used for any network.

◆ One node-voltage equation is required for each node except the reference node.

◆ Voltage sources introduce KVL equations for node-voltage analysis. We can easily eliminate these by substitution to simplify the analysis. (A shortcut method to eliminate these KVL equations is summarized on page 135.)

◆ When we use network simplification with node-voltage analysis, we should try to reduce the number of nodes.

◆ Mesh-current analysis can be used only for planar networks.

◆ One mesh-current equation is required for each mesh.

◆ Current sources introduce KCL equations for mesh-current analysis. We can easily eliminate these KCL equations by substitution and simplify the analysis. (A shortcut method to eliminate these KCL equations is summarized on page 155.)

◆ We can use network simplification with mesh-current analysis. We should eliminate meshes.

◆ When numerical values are given for all network components, we can solve for any voltage or current with the computer program PSpice.

■ P R O B L E M S ■

Section 4.1

In Problems 1 and 2, write a set of node-voltage equations for the network in the circuit shown. Solve for all component currents for $t > 0$. ◧ 4.4

1.

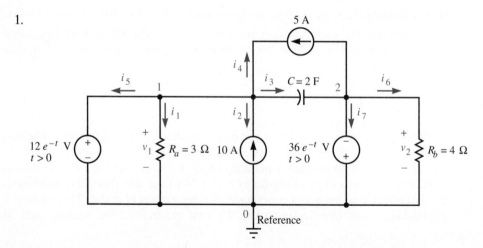

2.

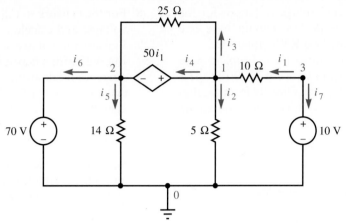

Use the method of node voltages to solve for voltage v_x or current i_x in Problems 3 and 4. ▣ 4.1, 4.2

3.

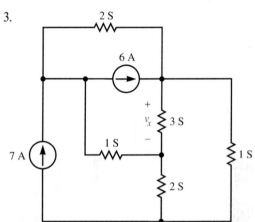

4.

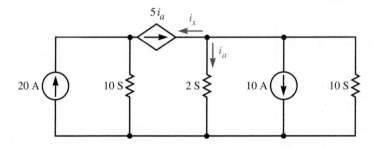

5. Write a set of node-voltage equations for the network shown below. ▣ 4.3

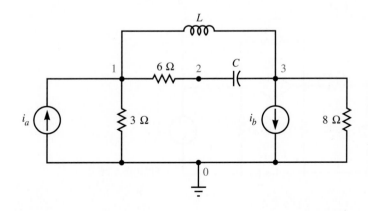

6. Write a set of node-voltage equations for the network in Fig. 4.24. You should have one KCL equation and two KVL equations.
7. Use the KVL equations from Problem 6 to solve for v_2 and v_3 in terms of v_1. Write these on the network diagram. Directly from the diagram, write a single KCL equation containing v_1 as the only unknown.

8. For the network in Fig. 4.25, solve for the node voltages and calculate current i_x. ☒ 4.5
9. Replace the voltage source in the network in Fig. 4.25 with a dependent voltage source [(+) reference mark at the right] that has a control equation of $v = 1.5v_1$. Solve for the node voltages. ☒ 4.6

Section 4.2

10. Write a set of node-voltage equations in matrix form for the circuit in Fig. 4.26. Reduce this set of equations to two KCL equations by substitution, and solve for the node voltages with the use of determinants.
11. Use the shortcut procedure to solve for the node voltages in the circuit in Fig. 4.26.

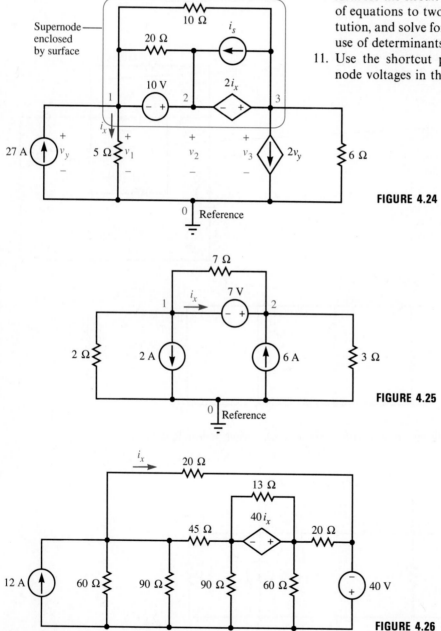

FIGURE 4.24

FIGURE 4.25

FIGURE 4.26

12. Analyze the network in Fig. 4.27 by the method of node voltages. Solve for current i_{23}. ▣ 4.7

13. (a) Write a set of node-voltage equations for the network in Fig. 4.28.
    (b) Solve for the node voltages.
    (c) Find i_{12} and i_{43}.

Use the reference node and node numbers given in Problems 14 and 15 and write a matrix node-voltage equation that includes both the KCL and KVL equations. Next, use the shortcut procedure to write a matrix node-voltage equation that does not contain any KVL equations.

14. See Fig. 4.29.   15. See Fig. 4.30.

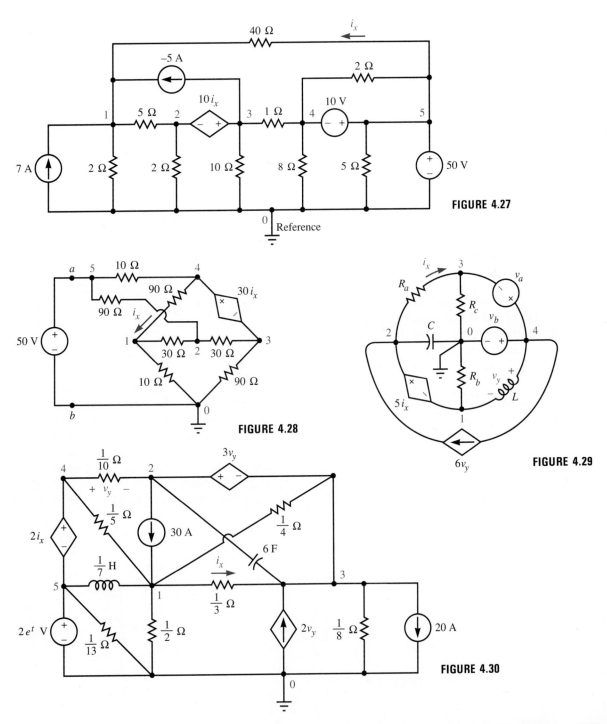

FIGURE 4.27

FIGURE 4.28

FIGURE 4.29

FIGURE 4.30

16. Use the shortcut procedure to solve for the node voltages in the following circuit. Find v_e/v_s and v_c/v_s. These are *voltage gains*. Find the *current gains* i_e/i_b and i_c/i_b. Also find the *power gains* $v_e i_e/v_s i_b$ and $-v_c i_c/v_s i_b$. Find the ratio v_s/i_b, which is the *input resistance* seen by the independent voltage source, because $i_b = 0$ when $v_s = 0$. This is a simplified small-signal equivalent circuit for a one-transistor amplifier.

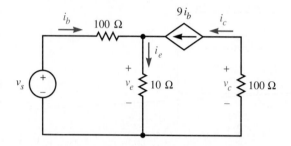

17. The network shown below represents a simplified model of the small-signal equivalent circuit of a field-effect transistor amplifier in the grounded-source configuration.
    (a) Use the method of analysis by node voltages to calculate v_o and the voltage gain v_o/v_i of the amplifier.
    (b) Connect a 100-Ω resistance between terminals G and D and repeat (a). Also calculate i_i. The ratio v_i/i_i is the input resistance of the amplifier.

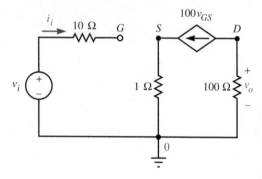

18. Replace the voltage source in the circuit used in Problem 17 with a short circuit, and replace the 1-Ω resistance with this voltage source, leaving the $(+)$ reference mark at the top. The input current i_i will now be the current leaving the $(+)$ reference mark of this source. This would represent an amplifier using a field-effect transistor in the grounded-gate configuration. Repeat Problem 17 for this configuration.

19. Transistors are often used in the Darlington connection (compound amplifier). A simplified small-signal equivalent circuit for this connection is shown below. Analyze the circuit by the method of node voltages and solve for the gains v_e/v_b, v_c/v_b, i_e/i_b, and i_c/i_b. Calculate the ratio v_b/i_b, which is the input resistance seen by the voltage source because $i_b = 0$ when $v_b = 0$.

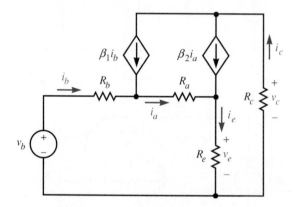

20. A simplified small-signal model for a cascode-connected two-transistor amplifier is shown in Fig. 4.31. Use the method of node voltages to find the voltage gain, v_L/v_b. Calculate the ratio v_b/i_b. This is the input resistance seen by the voltage source, because $i_b = 0$ when $v_b = 0$.

21. Repeat Problem 20 if the short circuit between terminals a and 0 is replaced by a resistance of value R_e.

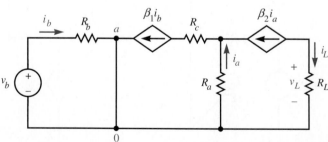

FIGURE 4.31

22. Use some of the analysis tools introduced in Chapter 3 in conjunction with the method of node voltages to find i_x in the network in Fig. 4.32. ▣ 4.8

Section 4.3

23. Use the method described in Section 4.3 to write the node-voltage equations for the network of Problem 3 by inspection.
24. Refer to the network used in Problem 10. Replace the dependent voltage source with a 20-A current source (reference arrow pointing to the right). Replace the 40-V source with a 40-A source (reference arrow pointing down). Now write a set of node-voltage equations by inspection.

Section 4.4

Use the shortcut procedure to write a set of node-voltage equations for the networks shown in Problems 25 and 26.

25. See Fig. 4.33.

26. See Fig. 4.34.

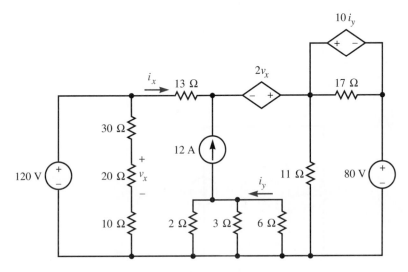

FIGURE 4.32

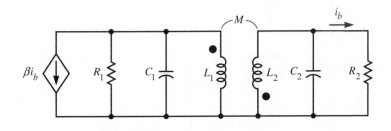

FIGURE 4.33

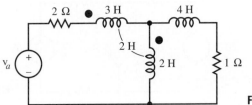

FIGURE 4.34

Write a set of mesh-current equations for the networks shown in Problems 27 and 28. Solve for all component voltages.

27. See Fig. 4.35.   28. See Fig. 4.36.

Use the method of mesh currents to solve for i_x in each network shown in Problems 29 and 30.

29. See Fig. 4.37.   30. See Fig. 4.38.

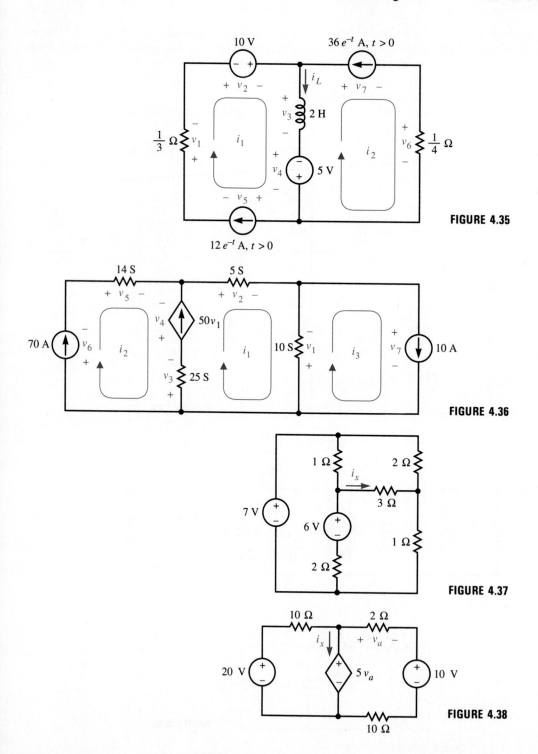

FIGURE 4.35

FIGURE 4.36

FIGURE 4.37

FIGURE 4.38

31. Write a set of mesh-current equations for the network shown in Fig. 4.39.
32. Write a set of mesh-current equations for the network in Fig. 4.40. You should have one KVL equation and two KCL equations.
33. Use the KCL equations from Problem 32 to solve for i_2 and i_3 terms of i_1. Write these on the network diagram. Directly from the diagram, write a single KVL equation containing i_1 as the only unknown.
34. Solve for the mesh currents and then calculate voltage v_x for the following network.

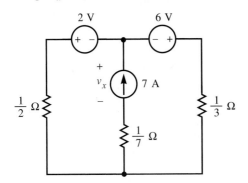

Section 4.6

35. Write a set of mesh-current equations in matrix form for the following circuit. Reduce this set of equations to two KVL equations by substitution, and solve for the mesh currents with the use of determinants.

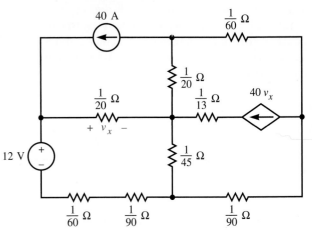

36. Use the shortcut procedure to solve for mesh currents in the preceding circuit.

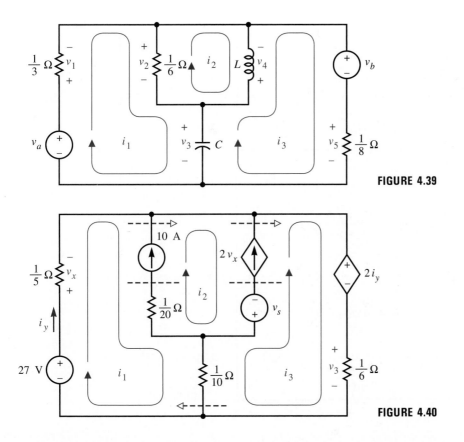

FIGURE 4.39

FIGURE 4.40

37. Analyze the network in Fig. 4.41 by the method of mesh currents. Solve for voltage v_{23}.

38. (a) Write a set of mesh-current equations in matrix form for the network in Fig. 4.42.
   (b) Solve for the mesh currents.
   (c) Find i_y and v_{ab}.

In the networks of Problems 39 and 40, use the indicated mesh currents and write a matrix mesh-current equation that includes both KVL and KCL equations. Then use the shortcut procedure to write a matrix mesh-current equation that does not contain any KCL equations.

39. See Fig. 4.43.   40. See Fig. 4.44.

41. Solve for the mesh currents in the circuit in Fig. 4.45. Find i_y/i_x and i_y/i_s. These are current gains.

42. Use mesh currents to analyze the amplifier circuit given in Problem 16, and calculate the gains and input resistance indicated in Problem 16.

43. Use some of the analysis tools introduced in Chapter 3 in conjunction with the method of mesh currents to find v_x indicated in the network in Fig. 4.46..

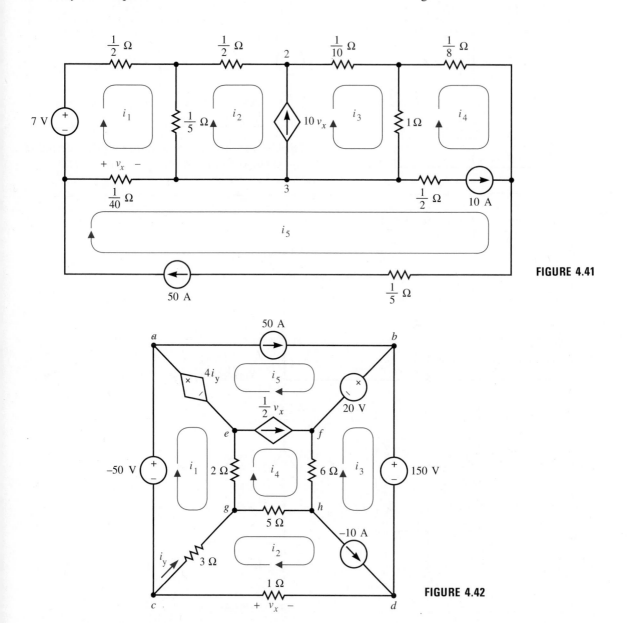

FIGURE 4.41

FIGURE 4.42

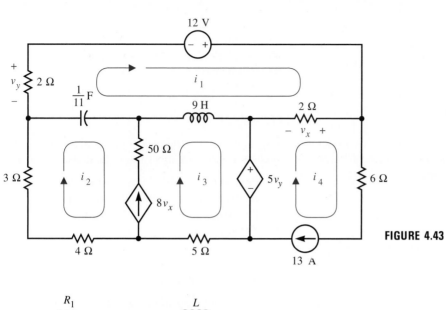

FIGURE 4.43

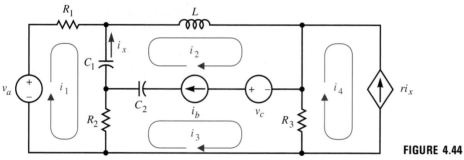

FIGURE 4.44

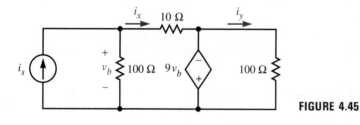

FIGURE 4.45

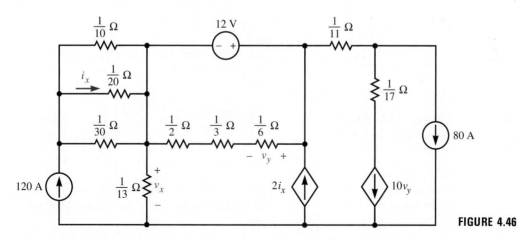

FIGURE 4.46

Section 4.7

44. Use the procedure of Section 4.7 to write a set of mesh-current equations for the network of Problem 29 by inspection.
45. Replace the dependent current source of the network used in Problem 35 with a 20-V voltage source [(+) reference at the left]. Replace the 40-A source with a 40-V source [(+) reference at the left]. Now write a set of mesh-current equations by inspection.

Section 4.8

Write a set of mesh-current equations in matrix form for the networks in Problems 46 and 47.

46.

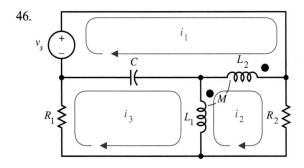

47.

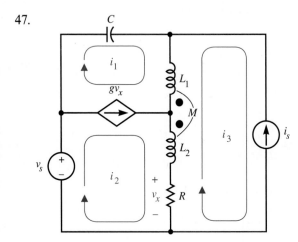

Section 4.9

Indicate which method of analysis, node voltages or mesh currents, would be most appropriate for the networks in the indicated problems. Why?

48. Problem 1
49. Problem 2
50. Problem 3
51. Problem 4
52. Problem 5
53. Problem 6
54. Problem 12
55. Problem 13

56. Problem 27
57. Problem 28
58. Problem 29
59. Problem 30
60. Problem 31
61. Problem 32
62. Problem 37
63. Problem 39
64. Problem 40
65. Problem 47

Section 4.10

Use a computer-aided circuit analysis program, such as PSpice, to solve the indicated problems.

66. Problem 8
67. Problem 9
68. Problem 10
69. Problem 12
70. Problem 13
71. Problem 4
72. Problem 16, with $v_s = 2$ V.

Comprehensive Problems

Draw a circuit diagram for a network with the constraints stated in Problems 73 through 78.

73. Seven nodes and the following set of mesh-current equations (x_1, x_2, and x_3 are mesh currents).

$$\begin{bmatrix} 6 & -2 & -3 \\ -2 & 11 & -4 \\ -3 & -4 & 12 \end{bmatrix} \cdot \begin{bmatrix} x_1 \\ x_2 \\ x_3 \end{bmatrix} = \begin{bmatrix} 10 \\ -20 \\ 30 \end{bmatrix}$$

74. Six meshes and the preceding set of node-voltage equations (x_1, x_2, and x_3 are node voltages).
75. Three nodes and the following set of mesh-current equations (x_1 and x_2 are mesh currents).

$$\begin{bmatrix} 100 & 10 \\ 30 & -1 \end{bmatrix} \cdot \begin{bmatrix} x_1 \\ x_2 \end{bmatrix} = \begin{bmatrix} 50 \\ 0 \end{bmatrix}$$

76. Two meshes and the preceding set of node-voltage equations (x_1 and x_2 are node voltages).
77. Five nodes and the following set of mesh-current equations (x_1, x_2, and x_3 are mesh currents).

$$\begin{bmatrix} 8 + 4\dfrac{d}{dt} & -3 & -4\dfrac{d}{dt} \\ 0 & -1 & 1 \\ 0 & 0 & 1 \end{bmatrix} \cdot \begin{bmatrix} x_1 \\ x_2 \\ x_3 \end{bmatrix} = \begin{bmatrix} 20 \\ 14 \\ 24 \end{bmatrix}$$

78. Four meshes and the preceding set of node-voltage equations (x_1, x_2, and x_3 are node voltages).

79. Simplify the network shown in Fig. 4.47 and make use of any specified voltages so that voltage v can be found from a single node-voltage (KCL) equation.
80. Simplify the network in Fig. 4.47 and make use of any specified currents so that voltage v can be found with a single mesh-current (KVL) equation.

Given the network in Fig. 4.48, find v_2 and i_1 by the method specified in Problems 81 through 83.

81. The use of node-voltage equations.
82. The use of mesh-current equations.
83. The use of node-voltage equations for the left-hand part of the circuit and a mesh-current equation for the right-hand part of the circuit.

84. Use the method of your choice to calculate current i_x in the circuit in Fig. 4.49.
85. Use the most convenient method to determine voltage v_x in the network in Fig. 4.50.

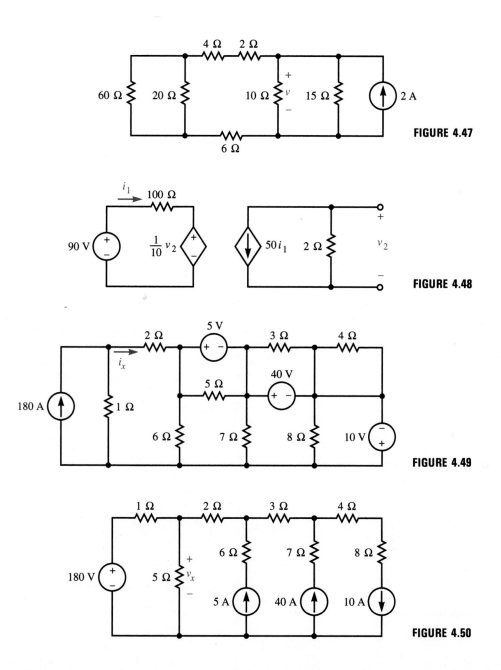

FIGURE 4.47

FIGURE 4.48

FIGURE 4.49

FIGURE 4.50

86. Determine i_o for the network shown below.

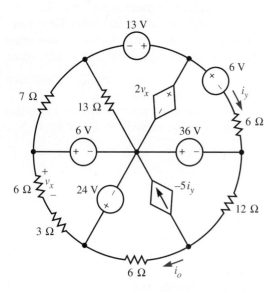

87. Calculate v_o for the network shown below.

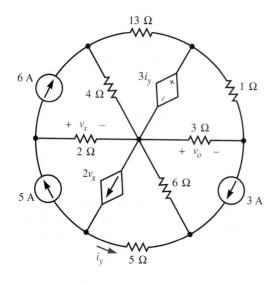

Analyze the contrived networks shown in Problems 88 through 90.

88.

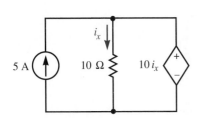

89.

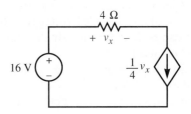

90.

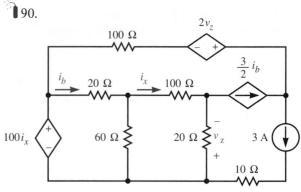

91. The network that follows contains a loop of voltage sources. Note that the KVL equations

$$v_1 - v_2 = 10$$
$$v_2 - v_3 = 5$$
$$v_3 - v_1 = -15$$

are linearly dependent. More voltage sources are present than are required to define the node voltages. The result is that although the node voltages are uniquely determined, the current through the voltage sources is not. Replace the 15-V source by an open circuit, and solve for the node voltages in this network. Next, determine all component currents. Verify that these satisfy KVL and KCL for the original circuit.

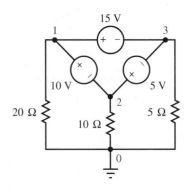

92. The network that follows contains a cutset of current sources. Select the center node as the reference, and verify that the four KCL equations written for the remaining nodes are linearly dependent. Next, replace the 1-A current source by a short circuit. Write three node-voltage equations and solve for the node voltages and component currents. Verify that these satisfy KVL and KCL for the original network. The node voltages were not uniquely defined in the original network.

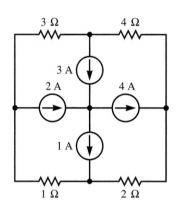

93. The network that follows contains a cutset of current sources. Note that the equations

$$i_1 = 2$$

$$i_1 - i_2 = 5$$

$$i_2 = -3$$

are linearly dependent. More current sources are present than are needed to define the mesh currents. The result is that although the mesh currents are uniquely determined, the voltage across the current sources is not. Replace the 5-A current source by a short circuit, and solve

for the mesh currents in this network. Next, determine all component voltages. Verify that these satisfy KCL and KVL for the original circuit.

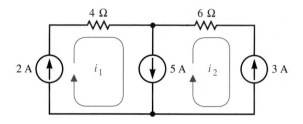

94. The network shown below contains a loop of voltage sources. Verify that the three mesh-current equations are linearly dependent. Next, replace the 15-V source by an open circuit. Write two mesh-current equations, and solve for the mesh currents and component voltages. Verify that these satisfy KCL and KVL for the original network. The mesh currents were not uniquely defined in the original network.

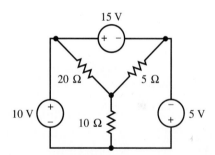

95. Try to solve for the node voltages of the contrived networks of Problems 88 through 90 by the use of a computer-aided circuit analysis program such as PSpice, and see what happens.

CHAPTER
5

Network Properties and Simplification

Up to this point we have emphasized networks in which resistance, capacitance, inductance, and the control parameters for dependent sources are constants, independent of voltage, current, and time. In this chapter we will show that the input-output equations for these networks have important general properties, known as linearity and time invariance. The linearity property implies that if the input is multiplied by a constant, then the output is multiplied by the same constant. Linearity further implies that the response to two or more simultaneous inputs is simply the sum of the responses to each input taken separately. Time invariance implies that a delay in the input just delays the response by the same amount of time.

This chapter will introduce the concepts of linearity and time invariance and some additional network properties. We will use these concepts to simplify network analysis. These techniques will be used throughout the remainder of the book.

5.1 Linearity and Time Invariance

Almost all networks that we will study in this book have constant values for resistance, inductance, and capacitance. In this section we will see that these networks are linear and time-invariant. To understand linearity and time invariance, we will need to introduce the idea of a transformation.

The terminal equation for a resistance,

$$v(t) = Ri(t) \tag{5.1}$$

relates the input i to the output v. We say that voltage v is a function of current i:

$$v(t) = f[i(t)] \tag{5.2}$$

An alternative way to say this is that the output v is related to the input i by a *transformation T*:

$$v(t) = T[i(t)] \tag{5.3}$$

For example, the transformation that describes a constant resistance is multiplication of the input i by the resistance value R to obtain the output v.

If we plot voltage v versus current i for a constant resistance, we obtain a *straight line through the origin*. As a result, if we multiply the input current by some constant a, the output voltage is multiplied by the same constant:

$$v(t) = T[ai(t)] = Rai(t) = aRi(t) = aT[i(t)] \tag{5.4}$$

We say that such a transformation is *homogeneous*.

The *linear relationship* between voltage and current for a constant resistance also says that the output voltage due to two input currents is simply the sum of the outputs due to each current applied separately:

$$\begin{aligned} v(t) &= T[i_1(t) + i_2(t)] = R[i_1(t) + i_2(t)] \\ &= Ri_1(t) + Ri_2(t) = T[i_1(t)] + T[i_2(t)] \end{aligned} \tag{5.5}$$

We say that *superposition* holds for the transformation.

The output current for a capacitance is also related to the input voltage by a transformation (the terminal equation):

$$i = T[v(t)] = C\frac{d}{dt}v(t) \tag{5.6}$$

We can easily see that this transformation is homogeneous (the derivative of a constant times any function is the constant times the derivative of the function). Superposition also holds (the derivative of the sum of two functions is the sum of the derivatives of the two functions). We will now formalize the idea of a linear relationship.

If a transformation is homogeneous and superposition holds, we say that the transformation is *linear*. From this we see that a constant resistance and a constant capacitance are represented by linear transformations. We now give the formal definition of a linear transformation.

Linear Transformation

The transformation T is **linear** if and only if for arbitrary inputs $x_1(t)$ and $x_2(t)$ with corresponding outputs

$$y_1(t) = T[x_1(t)] \qquad (5.7)$$

and

$$y_2(t) = T[x_2(t)] \qquad (5.8)$$

It follows that

$$\begin{aligned} y(t) &= T[a_1 x_1(t) + a_2 x_2(t)] \\ &= a_1 T[x_1(t)] + a_2 T[x_2(t)] \\ &= a_1 y_1(t) + a_2 y_2(t) \end{aligned} \qquad (5.9)$$

where a_1 and a_2 are arbitrary constants.

An example of the response of a circuit with input current i and output voltage v that is related by a linear transformation is shown in Fig. 5.1.

FIGURE 5.1
The input current and output voltage of a parallel resistance and capacitance are related by a linear transformation: (a) input current $i = i_1$, and response $v = v_1$; (b) input current $i = ai_1$, and response $v = av_1$

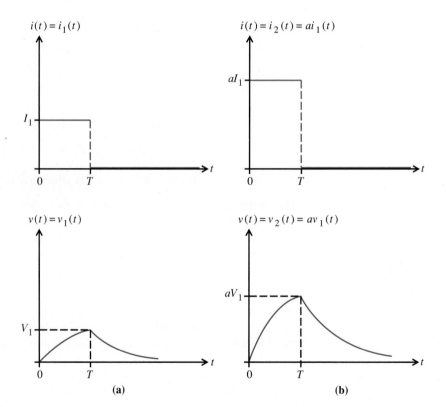

We use the definition of a linear transformation to define a linear component.

Linear Component

We say a circuit component is **linear** if and only if the component input (voltage or current) and response (current or voltage) are related by a linear transformation.

We have assumed that resistance, inductance, capacitance, and the control parameters for dependent sources do not change with time. This implies that a delay in the input simply delays the output by the same amount. For example, for a resistance,

$$v_1(t) = T[i_1(t)] = Ri_1(t) \tag{5.10}$$

Delaying the input current by time t_0 gives

$$\begin{aligned} v_2(t) = T[i_1(t - t_0)] &= Ri_1(t - t_0) \\ &= v_1(t - t_0) \end{aligned} \tag{5.11}$$

Therefore the output is also delayed by time t_0, as expected. The transformation is *time-invariant*. The formal definition of a time-invariant transformation is

DEFINITION

Time Invariance

The transformation T is **time-invariant** if and only if for every input $x(t)$ with response

$$y(t) = T[x(t)] \tag{5.12}$$

it follows that

$$y(t - t_0) = T[x(t - t_0)] \tag{5.13}$$

for every t_0, where $-\infty < t_0 < \infty$.

We now define a time-invariant component.

DEFINITION

Time-Invariant Component

A network component is **time-invariant** if and only if the input and output are related by a time-invariant transformation.

We will make use of the implications of time invariance beginning with Chapter 8.

We will see that our standard network components, with the exception of independent sources, are both linear and time-invariant, and are called *linear time-invariant (LTI) components*.

We define a linear time-invariant circuit in the following way:

DEFINITION

Linear Time-Invariant Circuit

A circuit composed of only linear time-invariant components is a **linear time-invariant (LTI) circuit**. A circuit that contains one or more nonlinear components is a nonlinear circuit, and a circuit that contains one or more time-varying components is a time-varying circuit.

Figure 5.2 gives an example of the response of a circuit with input and output related by an LTI transformation. Although this book considers, with few exceptions, the response of only LTI networks driven by sources, this LTI restriction *is not* required by KVL and KCL.

We showed that a constant resistance was linear and time-invariant as the definitions were presented. We will now examine capacitance.

FIGURE 5.2
The input current and output voltage of a parallel resistance and capacitance are related by a linear time-invariant transformation: (a) input current $i = i_1(t)$, and response $v = v_1(t)$; (b) time-shifted input current $i = i_1(t - t_0)$ causes the same shift in the output voltage, $v = v_1(t - t_0)$

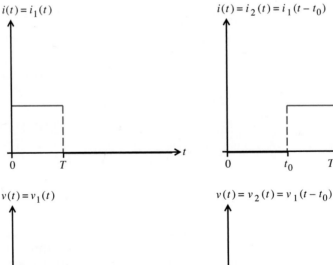

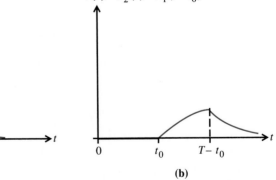

EXAMPLE 5.1

Is a constant capacitance linear and time-invariant? Assume an input voltage.

■ SOLUTION

Assume that the reference marks are as given by the passive sign convention. For input voltages $v_1(t)$ and $v_2(t)$, the responses are

$$i_1(t) = C \frac{d}{dt} v_1$$

and

$$i_2(t) = C \frac{d}{dt} v_2$$

If some multiple of both voltages is applied simultaneously,

$$i = C \frac{d}{dt} (a_1 v_1 + a_2 v_2) = a_1 C \frac{d}{dt} v_1 + a_2 C \frac{d}{dt} v_2$$

$$= a_1 i_1 + a_2 i_2$$

Thus a capacitance is linear.

With an input $v_1(t)$,

$$i_1(t) = C \frac{d}{dt} v_1(t)$$

If the input voltage is delayed by t_0 seconds,

$$v_2(t) = v_1(t - t_0)$$

Then

$$i_2(t) = C \frac{d}{dt} [v_1(t - t_0)] = i_1(t - t_0)$$

Therefore a capacitance is time-invariant. ■

EXAMPLE 5.2

Is a capacitance linear? Assume an input current.

■ SOLUTION

Assume the passive sign convention. For input currents $i_1(t)$ and $i_2(t)$, the responses are

$$v_1(t) = \frac{1}{C} \int_{-\infty}^{t} i_1(\tau) \, d\tau$$

and

$$v_2(t) = \frac{1}{C} \int_{-\infty}^{t} i_2(\tau) \, d\tau$$

If some multiple of both currents is applied simultaneously,

$$v(t) = \frac{1}{C} \int_{-\infty}^{t} \left[a_1 i_1(\tau) + a_2 i_2(\tau) \right] d\tau$$

$$= a_1 \frac{1}{C} \int_{-\infty}^{t} i_1(\tau) d\tau + a_2 \frac{1}{C} \int_{-\infty}^{t} i_2(\tau) d\tau$$

$$= a_1 v_1(t) + a_2 v_2(t)$$

Therefore a capacitance is linear regardless of whether the input is voltage and the response current or the input is current and the response voltage.

Be careful when you use the integral form of the terminal equation with initial values of voltage:

$$v(t) = \frac{1}{C} \int_{-\infty}^{t} i(\tau) d\tau$$

$$= \frac{1}{C} \int_{-\infty}^{t_0} i(\tau) d\tau + \frac{1}{C} \int_{t_0}^{t} i(\tau) d\tau$$

which we can write as

$$v(t) = v(t_0) + \frac{1}{C} \int_{t_0}^{t} i(\tau) d\tau$$

Voltage v and current i, as given by the last equation, are not related by a linear transformation if $v(t_0)$ is specified other than by the integral of i. This does not imply that capacitance is nonlinear, but that we have chosen to represent it in a manner that introduces a nonlinear transformation. ∎

From the previous two examples, it follows that inductance is also an LTI element. The same caution applies to the integral form of the terminal equation for an inductance.

REMEMBER

Resistances, inductances, and capacitances of constant value are LTI components. Dependent sources with a constant parameter are LTI components (see Exercise 1). Independent sources of nonzero value are not linear (see Exercises 4 and 5). Superposition holds for linear transformations and therefore for linear components.

EXERCISES

1. Prove that a dependent source with a control equation of the form $y = kx$ is linear and time-invariant.

2. Is a component whose terminal equation is $v = 5i^2$ linear? Apply the definition of a linear transformation to verify your answer.

3. Use the integral form of the capacitance terminal equation, as in Example 5.2, and show that a capacitance is time-invariant.

4. Show that an independent voltage source of nonzero value is a nonlinear component (use current as the input and voltage as the response).

5. Show that an independent current source of nonzero value is a nonlinear component (use voltage as the input and current as the response).

5.2 Linearity and Superposition

In the previous section we saw that the voltage across and the current through each component (except independent sources) of a linear network is related by a *linear transformation*. A linear transformation implies that *superposition* applies. Superposition also holds for networks of linear components. That is, the voltage and current in any part of a circuit are the sums of the voltages and currents due to each independent source acting alone.

THEOREM 1

Superposition Theorem

The voltage and current response of a linear network to a number of independent sources is the *sum* of the responses obtained by applying each independent source once with other independent sources set equal to zero.

See Problems 5.1 and 5.2 in the PSpice manual.

We will find it useful to observe that a 0-V voltage source is equivalent to a short circuit, and a 0-A current source is equivalent to an open circuit. To apply superposition, *dependent sources are left intact and are not set equal to zero.* Any analysis technique can be used to calculate each response.

A pictorial interpretation of the superposition theorem for a network with three independent sources is shown in Fig. 5.3. The values of a voltage v and a current i are to be calculated for the linear network depicted by Fig. 5.3a. The voltage and current are calculated by use of the superposition theorem as follows:

1. Set all independent sources except the voltage source of value v_a equal to zero. This yields the network depicted in Fig. 5.3b. The responses v' and i' due to v_a acting alone are now calculated by any convenient method.
2. Set all independent sources except the current source of value i_b equal to zero. This yields the network depicted in Fig. 5.3c. The responses v'' and i'' due to i_b acting alone are now calculated by any convenient method.

FIGURE 5.3
Pictorial interpretation of the superposition theorem for a network with three independent sources.
(*a*) Original network with three independent sources;
(*b*) network with all independent sources except v_a set equal to zero;
(*c*) network with all independent sources except i_b set equal to zero;
(*d*) network with all independent sources except v_c set equal to zero

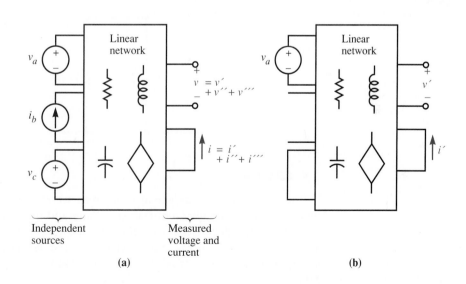

FIGURE 5.3
Continued

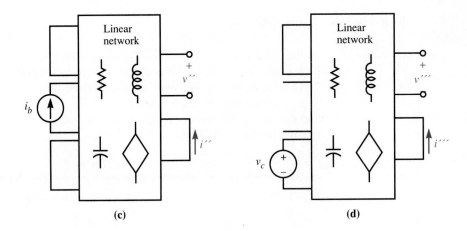

(c)   (d)

3. Set all independent sources except the voltage source of value v_c equal to zero. This yields the network depicted in Fig. 5.3d. The responses v''' and i''' due to v_c acting alone are now calculated by any convenient method.

4. The responses v and i due to all sources acting simultaneously in the original circuit of Fig. 5.3a are found as the sum of the responses due to each source acting alone:

$$v = v' + v'' + v''' \tag{5.14}$$

and

$$i = i' + i'' + i''' \tag{5.15}$$

The following two examples illustrate the use of the superposition theorem for a network containing three independent sources as in Fig. 5.3a.

EXAMPLE 5.3

Use superposition to determine voltage v_x in the network of Fig. 5.4.

FIGURE 5.4
Original network with three independent sources

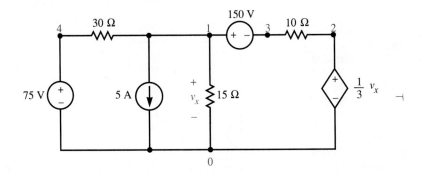

■ SOLUTION

Since we have three independent sources, we should use them one at a time, find the corresponding component of v_x, and add the three components.

First set all independent sources except the 75-V voltage source equal to zero. The result is shown in Fig. 5.5. Observe that the dependent source of value $(\frac{1}{3})v_x$ is *not* set equal to zero. (The reason for this will be clear after you work Example 5.4.)

FIGURE 5.5
All independent sources except the 75-V voltage source set equal to zero

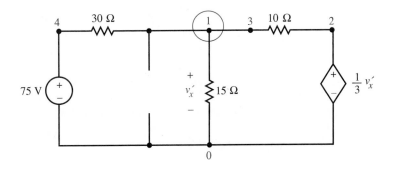

The KCL equation for the closed surface indicated in Fig. 5.5 is

$$\frac{1}{30}(v'_x - 75) + \frac{1}{15}v'_x + \frac{1}{10}\left(v'_x - \frac{1}{3}v'_x\right) = 0$$

which gives the component of v_x that corresponds to the 75-V source as

$$v'_x = 15 \text{ V}$$

Next, set all independent sources except the 5-A current source equal to zero. The result is shown in Fig. 5.6. The KCL equation for the indicated closed surface is

$$\frac{1}{30}v''_x + 5 + \frac{1}{15}v''_x + \frac{1}{10}\left(v''_x - \frac{1}{3}v''_x\right) = 0$$

FIGURE 5.6
All independent sources except the 5-A current source set equal to zero

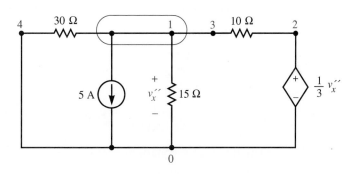

which gives the component of v_x that corresponds to the 5-A source:

$$v''_x = -30 \text{ V}$$

Now retain the last independent source, as shown in Fig. 5.7.

FIGURE 5.7
All independent sources except the 150-V voltage source set equal to zero

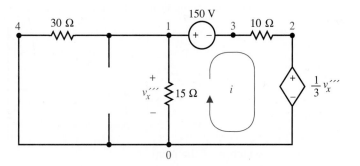

The equivalent resistance for the 30-Ω and 15-Ω parallel resistances is

$$R_p = \frac{1}{(1/30) + (1/15)} = 10\ \Omega$$

A KVL equation around the indicated loop can then be written as

$$10i + 150 + 10i + \frac{1}{3}(-10i) = 0$$

which gives

$$i = -9\ \text{A}$$

From Ohm's law, the component of v_x that corresponds to the 150-V source is

$$v_x''' = -10(-9) = 90\ \text{V}$$

Addition of the three components of v_x as required by the superposition theorem gives us

$$v_x = v_x' + v_x'' + v_x''' = 15 - 30 + 90 = 75\ \text{V} \quad \blacksquare$$

We will now solve for all node voltages in the network of Fig. 5.4 in a manner that will demonstrate how a general proof of the superposition theorem can be obtained.

EXAMPLE 5.4

Use superposition to determine the node voltages for the network of Fig. 5.4.

■ **SOLUTION**

Write the node-voltage equations for the network of Fig. 5.4 without the use of any shortcuts. A KCL equation for the supernode composed of nodes 1 and 3 is

$$\frac{1}{30}(v_1 - v_4) + 5 + \frac{1}{15}v_1 + \frac{1}{10}(v_3 - v_2) = 0 \tag{5.16}$$

The three KVL equations are

$$v_2 = \frac{1}{3}v_x = \frac{1}{3}v_1 \tag{5.17}$$

$$v_1 - v_3 = 150 \tag{5.18}$$

$$v_4 = 75 \tag{5.19}$$

These equations written in matrix form are

$$\begin{bmatrix} \dfrac{1}{10} & -\dfrac{1}{10} & \dfrac{1}{10} & -\dfrac{1}{30} \\ -\dfrac{1}{3} & 1 & 0 & 0 \\ 1 & 0 & -1 & 0 \\ 0 & 0 & 0 & 1 \end{bmatrix} \cdot \begin{bmatrix} v_1 \\ v_2 \\ v_3 \\ v_4 \end{bmatrix} = \begin{bmatrix} -5 \\ 0 \\ 150 \\ 75 \end{bmatrix} \tag{5.20}$$

which is of the form

$$\mathbf{T} \cdot \mathbf{v} = \mathbf{s} \tag{5.21}$$

Observe that the effect of the dependent source appears in the node transformation matrix $\mathbf{T}$ (row 2, column 1), and the independent sources appear in the source column vector $\mathbf{s}$. If, in turn, all independent sources except one were set equal to zero, we would obtain three matrix equations,

$$\mathbf{T} \cdot \mathbf{v}' = \mathbf{s}' \tag{5.22}$$

$$\mathbf{T} \cdot \mathbf{v}'' = \mathbf{s}'' \tag{5.23}$$

$$\mathbf{T} \cdot \mathbf{v}''' = \mathbf{s}''' \tag{5.24}$$

in which the node transformation matrix $\mathbf{T}$ is unchanged,

$$\mathbf{s}' = \begin{bmatrix} 0 \\ 0 \\ 0 \\ 75 \end{bmatrix} \tag{5.25}$$

$$\mathbf{s}'' = \begin{bmatrix} -5 \\ 0 \\ 0 \\ 0 \end{bmatrix} \tag{5.26}$$

$$\mathbf{s}''' = \begin{bmatrix} 0 \\ 0 \\ 150 \\ 0 \end{bmatrix} \tag{5.27}$$

and $\mathbf{v}'$, $\mathbf{v}''$, and $\mathbf{v}'''$ are the components of the node-voltage column vector $\mathbf{v}$ that are due to the 75-V, 5-A, and 150-V sources, respectively. You should write node-voltage equations for the networks shown in Figs. 5.5, 5.6, and 5.7. (When a voltage source is set equal to zero, a node is eliminated, but treat the nodes as if they remained distinct.) These will be as shown in Eqs. (5.22), (5.23), and (5.24).

Addition of these three equations gives

$$\mathbf{T} \cdot \mathbf{v}' + \mathbf{T} \cdot \mathbf{v}'' + \mathbf{T} \cdot \mathbf{v}''' = \mathbf{s}' + \mathbf{s}'' + \mathbf{s}''' \tag{5.28}$$

Because matrix multiplication is distributive, this can be written as

$$\mathbf{T} \cdot [\mathbf{v}' + \mathbf{v}'' + \mathbf{v}'''] = \mathbf{s}' + \mathbf{s}'' + \mathbf{s}''' \tag{5.29}$$

which is of the form

$$\mathbf{T} \cdot \mathbf{v} = \mathbf{s} \tag{5.30}$$

where

$$\mathbf{s} = \mathbf{s}' + \mathbf{s}'' + \mathbf{s}''' \tag{5.31}$$

and

$$\mathbf{v} = \mathbf{v}' + \mathbf{v}'' + \mathbf{v}''' \tag{5.32}$$

are the solution to Eq. (5.20). Observe that it was necessary that the $\mathbf{T}$ matrix be the same in all cases, and, therefore, the dependent source could not be set to zero. This procedure can easily be generalized to prove the superposition theorem. ∎

As we will see, in many parts of this text, the concept of superposition provides an exceedingly useful way of thinking about network analysis and design.

REMEMBER

Superposition applies to linear circuits. Each independent source must be used once while the unused independent sources are set equal to zero. A voltage source set equal to zero is equivalent to a short circuit. A current source set equal to zero is equivalent to an open circuit. In all cases, dependent sources are left in the circuit and not changed.

EXERCISES

Use the principle of superposition to find voltage v_x in the networks shown in Exercises 6 and 7.

6.

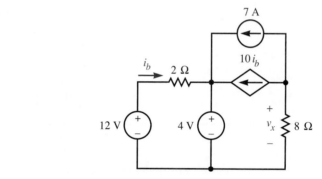

7.

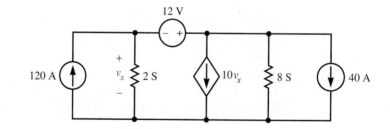

5.3 Equivalent Sources

In this section we introduce a new technique for network simplification. Many *practical sources* of electrical energy, such as a storage battery, are adequately modeled for some applications by an ideal voltage source and series resistance,* as shown in Fig. 5.8a. A KVL equation quickly provides the terminal equation

---

* An ideal voltage source with zero series resistance can supply infinite current and thus infinite power. This is physically impossible.

FIGURE 5.8
Practical sources and their
terminal characteristics.
(*a*) Practical voltage source;
(*b*) practical current source

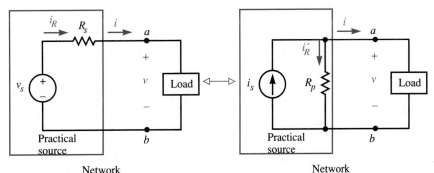

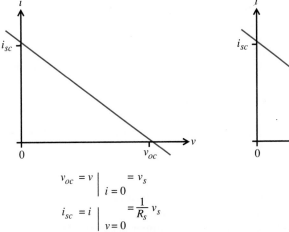

$$v_{oc} = v \Big|_{i=0} = v_s$$

$$i_{sc} = i \Big|_{v=0} = \frac{1}{R_s} v_s$$

Volt-ampere characteristic

(a)

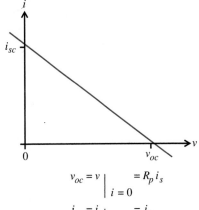

$$v_{oc} = v \Big|_{i=0} = R_p i_s$$

$$i_{sc} = i \Big|_{v=0} = i_s$$

Volt-ampere characteristic

(b)

given in Fig. 5.8a. This is the equation of a straight line that gives the volt-ampere characteristic shown. The intercepts with the voltage and current axes are the open-circuit voltage v_{oc} and the short-circuit current i_{sc}.

A *practical* current source is modeled by an ideal current source and parallel resistance,* as shown in Fig. 5.8b. A KCL equation provides the terminal equation shown in the figure. Comparison reveals that the two terminal equations in Fig. 5.8 are identical if

$$R_p = R_s \tag{5.33}$$

and R_s, v_s, and i_s are related by Ohm's law, that is,

$$v_s = R_s i_s = R_p i_s \tag{5.34}$$

The conclusion is

---

* An ideal current source without a parallel resistance can supply current to an infinite resistance and thus supply infinite power. This is physically impossible.

Source Transformations

See Problem 5.4 in the PSpice manual.

A voltage source of value v_s with series resistance of value R_s is equivalent, with respect to the terminal equation relating v to i, to a current source of value

$$i_s = \frac{1}{R_s} v_s \qquad (5.35)$$

and parallel resistance of value $R_p = R_s$.

A current source of value i_s with parallel resistance of value R_p is equivalent, with respect to the terminal equation relating v to i, to a voltage source of value

$$v_s = R_p i_s \qquad (5.36)$$

and series resistance $R_s = R_p$.

The *equivalence is only with respect to the terminal characteristics* that relate v to i. Typically the currents through R_s and R_p are different. (We can easily verify this by calculating these resistance currents when the terminals a and b and a' and b' are open-circuited.)

A voltage source with zero series resistance has a terminal voltage that is independent of the output current and therefore cannot be replaced with an equivalent current source and a parallel resistance. A current source with zero parallel conductance (infinite parallel resistance) has an output current that is independent of the terminal voltage and therefore cannot be replaced with an equivalent voltage source and series resistance.

The following gives a numerical example of source transformations.

EXAMPLE 5.5

(a) Develop an equivalent source, with respect to terminals a and b, for the network shown in Fig. 5.9a.

(b) Perform a source transformation on the network of Fig. 5.9b.

FIGURE 5.9
Equivalent source example

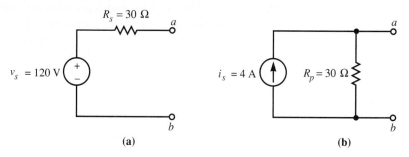

(a)   (b)

■ **SOLUTION**

(a) The resistance in the equivalent circuit must be the same as in the original circuit:

$$R_p = R_s = 30 \ \Omega$$

The current source required for equivalence can be obtained from Ohm's law:

$$i_s = \frac{1}{R_s} v_s = \frac{1}{30}(120) = 4 \text{ A}$$

The equivalent source is as shown in Fig. 5.9b.

(b) The resistance of the equivalent source must be the same as in the original circuit:

$$R_s = R_p = 30 \ \Omega$$

The voltage source required for equivalence is given by Ohm's law:

$$v_s = R_p i_s = 30(4) = 120 \text{ V}$$

The equivalent source is as shown in Fig. 5.9a. ∎

⊡ *See Problem 5.4 in the PSpice manual.*

Source transformations are useful to simplify a circuit for analysis. Conversion from a practical voltage source to a practical current source eliminates a node. Conversion from a practical current source to a practical voltage source eliminates a mesh. A source transformation may also permit further simplification by series and parallel equivalents. This is illustrated in the following example.

EXAMPLE 5.6

Find voltage v shown in Fig. 5.10a with the aid of repeated source transformations.

FIGURE 5.10
An example in which source transformations are convenient. (*a*) The original circuit; (*b*) after the first source transformation; (*c*) after a resistance combination and a second source transformation; (*d*) after a resistance combination and a third source transformation; (*e*) equivalent current source; (*f*) equivalent voltage source

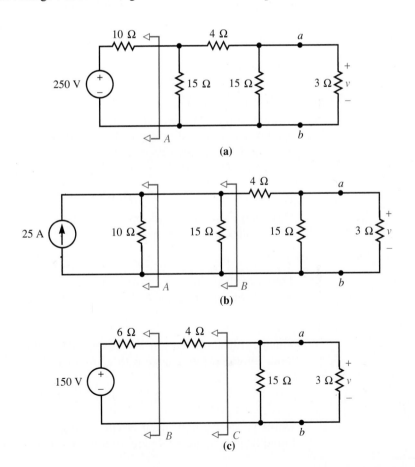

FIGURE 5.10
Continued

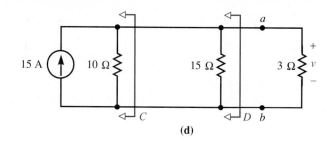

(d)

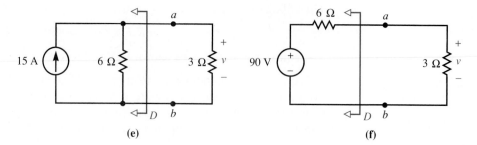

(e)   **(f)**

■ **SOLUTION**

Begin at the left of the circuit in Fig. 5.10a, and make repeated source transformations until only a two-node circuit remains.

Transformation of the 250-V voltage source and 10-Ω series resistance gives a current source

$$i_s = \frac{250}{10} = 25 \text{ A}$$

with a 10-Ω parallel resistance, as shown in Fig. 5.10b. (Remember that the current through this 10-Ω resistance is not the same as that through the original 10-Ω resistance.) Now find an equivalent resistance for the 10-Ω resistance in parallel with the 15-Ω resistance:

$$R_p = \frac{1}{(1/10) + (1/15)} = \frac{150}{25} = 6 \text{ Ω}$$

Next, make a source transformation on the 25-A current source with the equivalent 6-Ω parallel resistance:

$$v_s = 25 \times 6 = 150 \text{ V}$$

The equivalent circuit is shown in Fig. 5.10c. Find an equivalent resistance for the series combination of the 6-Ω and 4-Ω resistances:

$$R_s = 6 + 4 = 10 \text{ Ω}$$

Finally, transform the 150-V voltage source and the 10-Ω equivalent series resistance into a current source with a 10-Ω parallel resistance:

$$i_s = \frac{150}{10} = 15 \text{ A}$$

which yields the equivalent circuit of Fig. 5.10d. At this point, v can easily be found from a single KCL equation written for a surface enclosing the top node:

$$-15 + \frac{1}{10}v + \frac{1}{15}v + \frac{1}{3}v = 0$$

or
$$v = 30 \text{ V}$$

To emphasize the concept of the 3-Ω load resistance being connected to an equivalent current source with a parallel internal resistance, we can combine the 10-Ω and 15-Ω resistances as shown in Fig. 5.10e. An additional source transformation yields the network of Fig. 5.10f. ∎

Source transformations are often used when node-voltage equations are to be written for networks containing voltage sources with series resistances. Each voltage source with a series resistance is replaced with an equivalent current source and parallel resistance. This reduces the number of nodes and thus the number of equations that must be solved (Exercise 10). Again, remember that the current through the parallel resistance will not be the same as that through the original series resistance. Several other source equivalences are shown in Fig. 5.11. The equivalence between the circuit pairs is with respect to v_{ab} and i_{ab} for any connected load.

FIGURE 5.11
Sources with equivalent terminal characteristics.
(a) Series voltage sources;
(b) parallel voltage sources;
(c) parallel current sources;
(d) series current sources;
(e) voltage source and parallel resistance;
(f) current source and series resistance

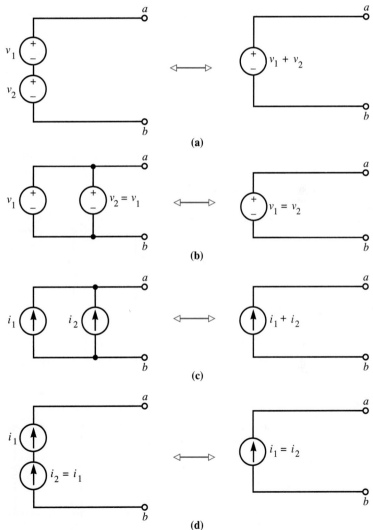

FIGURE 5.11
Continued

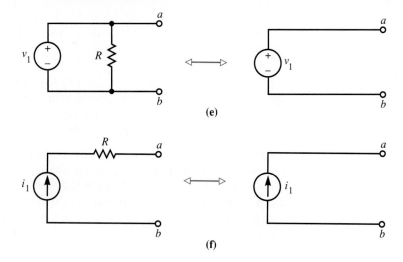

(e)

(f)

REMEMBER

Conversion from a practical voltage source to a practical current source eliminates a node. Conversion from a practical current source to a practical voltage source eliminates a mesh. Either conversion typically alters the current through the source resistance. The equivalence is only with respect to the terminal characteristics.

EXERCISES

8. Use repeated source transformations, and combine parallel and series resistances to find voltage v in the network below by writing a single KCL equation.

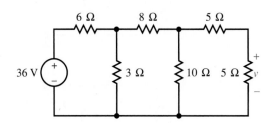

9. Use source transformations to determine currents i_x and i_y for the network that follows.

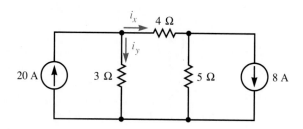

10. Use source transformations to eliminate two nodes in the following network, and write node-voltage equations for the remaining two nodes. This is a useful technique.

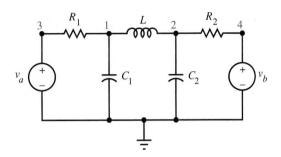

Use source transformations to find two-component networks that are equivalent with respect to the relation between v_{ab} and i_{ab} in the portion of the network shown to the left of terminals a and b in Exercises 11 and 12.

11.

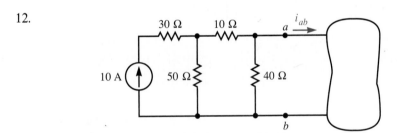

12.

5.4 Thévenin's and Norton's Theorems

Frequently, one wants to connect two terminals of one network, such as an audio amplifier, to two terminals of another, such as a speaker. Here the pertinent electrical quantities are the speaker voltage and current, not the various voltages and currents in the internal components of the amplifier. The concept of equivalent sources, introduced in the previous section, can be substantially generalized to apply to this type of problem.

The problem is depicted in Fig. 5.12, and the generalization is stated in the following two theorems.

FIGURE 5.12
Replacing a network of
resistances and independent
and dependent sources by
an equivalent network: (*a*)
networks coupled at only
two terminals; (*b*) network
A replaced by a Thévenin
equivalent; (*c*) network *A*
replaced by a Norton
equivalent

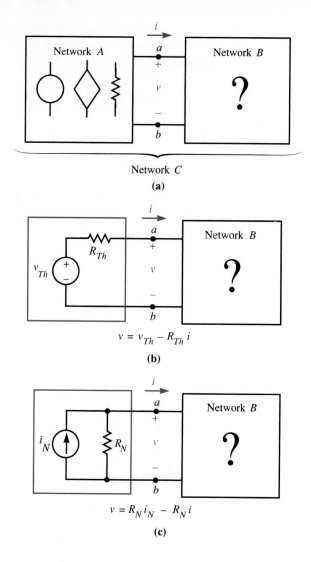

THEOREM 3

Thévenin's Theorem

Given two networks *A* and *B* that satisfy the following conditions:

1. Networks *A* and *B* are connected at only two terminals *a* and *b*, as shown in Fig. 5.12a. No control voltage or current for a dependent source in one network is found in the other. No other restrictions, such as linearity, are required for network *B*.
2. Network *A* contains only linear resistances and independent and dependent sources.
3. The magnitude of the open-circuit voltage of network *A* is

$$|v_{oc}| = |v|\big|_{i=0} < \infty \qquad (5.37)$$

Network A can be replaced, with respect to its effect on network B, by an equivalent circuit consisting of a voltage source and a series resistance. The voltage source has a value v_{Th} equal to the open-circuit voltage of network A. The series resistance R_{Th} is the resistance of network A measured between terminals a and b when network B is not connected and all independent sources in network A are set equal to zero. (A voltage source set equal to zero is a short circuit; a current source set equal to zero is an open circuit.) Dependent sources are unchanged.

A source transformation yields the following theorem.

THEOREM 4

Norton's Theorem

Given two networks A and B that satisfy conditions 1 and 2 given in Theorem 3 and where condition 3 is replaced by

3(a) The magnitude of the short-circuit current of network A is

$$|i_{sc}| = |i|\big|_{v=0} < \infty \tag{5.38}$$

Network A can be replaced, with respect to its effect on network B, by an equivalent circuit consisting of a current source and a parallel resistance. The current source has a value i_N equal to the short-circuit of network A. The parallel resistance R_N is the same as R_{Th} of Theorem 3.

If

$$|v_{oc}|, |i_{sc}| < \infty$$

either Thévenin's or Norton's equivalent network can be used. Application of KVL and Ohm's law to the network of Fig. 5.12b gives

$$R_{Th} = \frac{v_{oc}}{i_{sc}} \tag{5.39}$$

which provides an alternative method to calculate or measure the Thévenin or Norton equivalent resistance.

Obviously any two of the three parameters v_{oc}, i_{sc}, and R_{Th} can be measured and the third calculated from Ohm's law. Use the method that requires the least work.

See Problems 5.5 and 5.6 in the PSpice manual.

We use Thévenin's theorem if we wish to minimize the number of meshes and Norton's theorem if we wish to minimize the number of nodes.

The procedure will now be clarified by several examples.

EXAMPLE 5.7

⬛ *See Problem 5.5 in the PSpice manual.*

Find a Thévenin equivalent network for the network to the left of terminals *a* and *b* in Fig. 5.13.

FIGURE 5.13
Network for Thévenin equivalent, Example 5.7

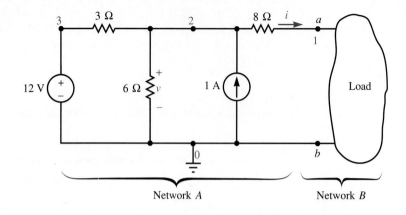

■ SOLUTION

First find the open-circuit voltage. Apply KCL to a surface enclosing the nodes indicated in Fig. 5.14:

FIGURE 5.14
Diagram for calculating v_{oc} for the network of Fig. 5.13

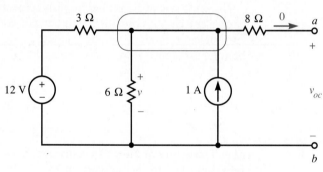

$$\frac{1}{3}(v - 12) + \frac{1}{6}v - 1 + 0 = 0$$

which gives

$$v = 10 \text{ V}$$

A simple KVL equation gives

$$v_{Th} = v_{oc} = 8(0) + v = 10 \text{ V}$$

Next set all independent sources to zero, as shown in Fig. 5.15.

FIGURE 5.15
Diagram for calculating the Thévenin resistance with respect to terminals *a* and *b* of the network of Fig. 5.13

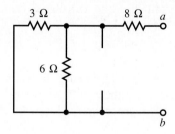

The 8-Ω resistance is in series with the parallel combination of the 3-Ω and 6-Ω resistances:

$$R_{Th} = 8 + \frac{1}{(1/3) + (1/6)} = 10 \, \Omega$$

This gives the Thévenin equivalent shown in Fig. 5.16. ■

FIGURE 5.16
Network *A* of Fig. 5.13 replaced by its Thévenin equivalent

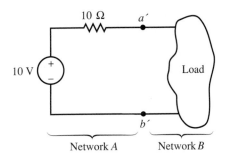

Network *A*   Network *B*

We can apply a source transformation to the network of Fig. 5.16 to obtain the Norton equivalent circuit for the network to the left of terminals *a* and *b* in Fig. 5.13, but for practice we will apply Norton's theorem directly to the circuit of Fig. 5.13.

EXAMPLE 5.8

Construct a Norton equivalent for the network to the left of terminals *a* and *b* in Fig. 5.13.

FIGURE 5.17
Diagram for calculating the short-circuit current of the network of Fig. 5.13

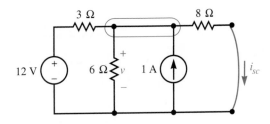

■ **SOLUTION**

Determine the short-circuit current as indicated in Fig. 5.17. Application of KCL to the indicated closed surface yields

$$\frac{1}{3}(v - 12) + \frac{1}{6}v - 1 + \frac{1}{8}v = 0$$

which gives

$$v = 8 \text{ V}$$

From KVL,

$$-v + 8i_{sc} = 0$$

which gives

$$i_{sc} = 1 \text{ A}$$

The Norton resistance is found as in Example 5.7:

$$R_N = R_{Th} = 10 \ \Omega$$

The resulting Norton equivalent is shown in Fig. 5.18. This Norton equivalent could have been obtained directly from a source transformation on the Thévenin equivalent found in Example 5.7. ■

FIGURE 5.18
Network *A* of Fig. 5.13 replaced by its Norton equivalent

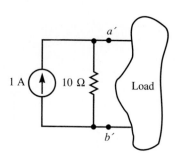

Transistor.
Photograph by
James Scherer

We will now examine the input characteristics of a simple amplifier.

EXAMPLE 5.9

FIGURE 5.19
A signal source and amplifier supplying power to a load

Figure 5.19 represents a signal source (for example, a microphone) driving an emitter-follower amplifier, which in turn drives the input of a power amplifier (the load), modeled as a 10-Ω resistance. The amplifier model is somewhat simplified, and the numerical values are chosen to simplify the arithmetic. Determine the Thévenin equivalent to the right of terminals *a* and *b* with the 10-Ω load connected. (This Thévenin resistance, R_{ab}, is called the *input resistance* of the amplifier.)

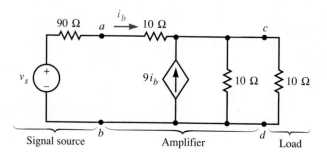

🖳 *See Problem 5.6 of the PSpice manual.*

■ **SOLUTION**

Disconnect the signal source from terminals *a* and *b*. Node-voltage or mesh-current analysis will verify that both the open-circuit voltage and the short-circuit current for the amplifier input terminals are zero:

$$v_{oc} = v_{ab}\Big|_{i_{ab}=0} = 0$$

and

$$i_{sc} = -i_b\Big|_{v_{ab}=0} = 0$$

This is always true for a resistive network that does not contain any independent sources. The Thévenin resistance can be found if we connect a source of power

FIGURE 5.20
Measuring the input
resistance of the amplifier
of Fig. 5.19

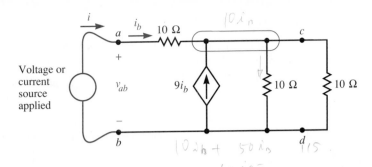

to terminals a and b, as shown in Fig. 5.20. Although specification of the type of source (voltage or current) is not required, it is often easiest to specify it as a 1-V or 1-A source. For this example, let the source remain unspecified. Apply KCL to the indicated node.

$$\frac{1}{10}(v_{cd} - v_{ab}) - 9\left[\frac{1}{10}(v_{ab} - v_{cd})\right] + \frac{1}{10}v_{cd} + \frac{1}{10}v_{cd} = 0$$

This gives

$$v_{cd} = \frac{5}{6}v_{ab}$$

and the input current is

$$i = \frac{1}{10}(v_{ab} - v_{cd}) = \frac{1}{10}\left(v_{ab} - \frac{5}{6}v_{ab}\right) = \frac{1}{60}v_{ab}$$

Ohm's law yields the input resistance

$$R_{ab} = \frac{v_{ab}}{i} = \frac{v_{ab}}{v_{ab}/60} = 60 \ \Omega$$

The signal source is connected to an equivalent load of 60 Ω, as shown in Fig. 5.21. ∎

FIGURE 5.21
Equivalent load on the
signal source obtained by
Thévenin's theorem

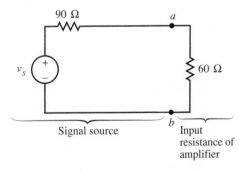

We frequently call the Thévenin resistance measured between two terminals (terminals a and b in the preceding example) the *driving-point resistance* looking into terminals a and b, because this is the resistance a source must drive when connected between the two terminals.

We will now examine the output characteristics of the amplifier of Fig. 5.19.

EXAMPLE 5.10

Determine the Thévenin equivalent circuit to the left of terminals c and d in Fig. 5.19 with the source connected. (The Thévenin resistance, R_{cd}, is called the *output resistance* of the amplifier with the source connected.)

FIGURE 5.22
Signal source and amplifier with terminals c and d open-circuited

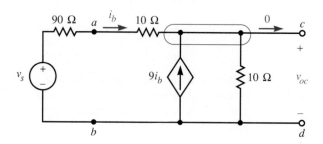

■ SOLUTION

Open circuit terminals c and d as shown in Fig. 5.22. KCL applied to the indicated closed surface yields

$$\frac{1}{100}(v_{oc} - v_s) - 9\left[\frac{1}{100}(v_s - v_{oc})\right] + \frac{1}{10}v_{oc} = 0$$

which gives

$$v_{oc} = \frac{1}{2}v_s$$

which is the Thévenin equivalent voltage looking into terminals c and d.

FIGURE 5.23
Source voltage v_s set equal to zero so the output resistance of the amplifier of Fig. 5.19 can be measured

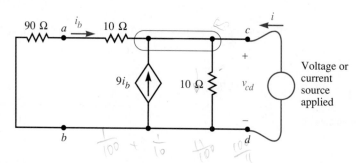

We can find the *output resistance*, or the Thévenin resistance looking into terminals c and d, by setting all *independent sources* to zero and applying a source to terminals c and d, as shown in Fig. 5.23. KCL applied to the indicated node yields

$$\frac{1}{100}v_{cd} - 9\left[\frac{1}{100}(-v_{cd})\right] + \frac{1}{10}v_{cd} - i = 0$$

or

$$i = \frac{1}{5}v_{cd}$$

From Ohm's law,

$$R_{cd} = \frac{v_{cd}}{v_{cd}/5} = 5\ \Omega$$

The load is connected to or "sees" an equivalent voltage source and series resistance as shown in Fig. 5.24. The equivalent circuits shown in Figs. 5.21 and 5.24 easily yield the current required from the source and the current delivered to the load. ■

FIGURE 5.24
Equivalent source seen by the load

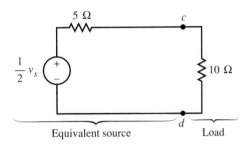

Equivalent source   Load

REMEMBER

Thévenin's and Norton's theorems apply to circuits with independent sources, linear resistances, and linear dependent sources. (This restriction does not apply to the load.)

The Thévenin voltage is the open-circuit voltage. The Norton current is the short-circuit current. If the network is without an independent source, the open-circuit voltage and short-circuit current are zero.

The ratio of open-circuit voltage to short-circuit current is the Thévenin resistance (also the Norton resistance).

The Thévenin resistance can also be measured by setting all independent sources to zero (any dependent sources are left as they are) and applying a voltage or current to the two terminals considered. The ratio of the terminal voltage to the input current is the Thévenin resistance.

EXERCISES

⬇ *PSpice .TF command; see Sec. B.2.12 of the PSpice manual.*

Determine the Thévenin and Norton equivalent networks with respect to terminals a and b for the networks shown in Exercises 13 through 15.

13.

14.

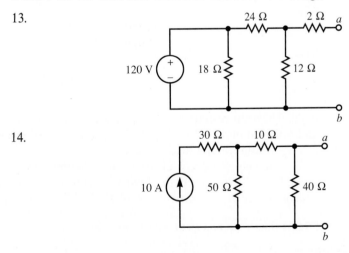

15.

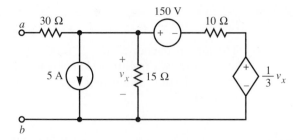

5.5 Derivation of Thévenin's Theorem*

Thévenin's theorem can be established by the use of node-voltage equations and Cramer's rule. Refer to Fig. 5.12a. Select node b as the reference node and node a as node 1. The node-voltage equations can be written in the form

$$
\begin{matrix}
\text{KCL} \\
\text{equations}
\end{matrix}
\left\{
\begin{matrix}
\text{KVL} \\
\text{equations}
\end{matrix}
\right.
\begin{bmatrix}
T_{11} & T_{12} & \cdots & T_{1N} \\
\vdots & \vdots & & \vdots \\
T_{M1} & T_{M2} & & T_{MN} \\
\vdots & \vdots & & \vdots \\
T_{N1} & T_{N2} & \cdots & T_{NN}
\end{bmatrix}
\begin{bmatrix}
v_1 \\
\vdots \\
v_M \\
\vdots \\
v_N
\end{bmatrix}
=
\begin{bmatrix}
i_1 - i \\
\vdots \\
v_{sM} \\
\vdots \\
v_{sN}
\end{bmatrix}
\begin{matrix}
\text{Load current} \\
\text{Due to} \\
\text{current sources} \\
\\
\text{Due to} \\
\text{voltage sources}
\end{matrix}
\tag{5.40}
$$

Observe that the node transformation matrix **T** is for network A alone. The contribution of network B to the equations appears as the load current i.

Voltage v_{ab} can be found with the use of Cramer's rule:

$$
v_{ab} = v_1
$$

$$
= \underbrace{\frac{\Delta_{11}}{\Delta} i_1 + \cdots + \frac{\Delta_{N1}}{\Delta} v_{sN}}_{\begin{matrix}\text{Open-circuit} \\ \text{voltage}\end{matrix}} - \underbrace{\frac{\Delta_{11}}{\Delta}}_{\begin{matrix}\text{Equivalent} \\ \text{resistance}\end{matrix}} \overset{\text{Load current}}{i}
\tag{5.41}
$$

where Δ is the determinant of the node transformation matrix and Δ_{ij} is the cofactor for row i and column j (refer to Apppendix A). Equation (5.41) is of the form

$$
v_{ab} = v_{oc} - R_{ab}i
\tag{5.42}
$$

where

$$
R_{ab} = \frac{\Delta_{11}}{\Delta}
\tag{5.43}
$$

is the Thévenin resistance or *driving-point* resistance with respect to terminals a and b. Equation (5.42) is the terminal equation for an ideal voltage source with a series resistance and load current i.

To complete the proof of Thévenin's theorem, we must consider two special cases. First, if there is a path of voltage sources from node 1 (node a) to the reference node (node b in network A of Fig. 5.12), the node-voltage equation corre-

---

* This derivation can be omitted without loss of continuity.

sponding to node 1 would be a KVL equation. Voltage v_{ab} would be independent of the load current, and for this case, the Thévenin equivalent circuit is an ideal voltage source with zero series resistance. Second, if the only component (in network A of Fig. 5.12) that is connected to node a is a current source, the determinant Δ is zero. This gives an infinite open-circuit voltage, and Thévenin's theorem does not apply. The equivalent circuit is an ideal current source with infinite parallel resistance. This completes the proof of Thévenin's theorem.

EXAMPLE 5.11

Find the Thévenin equivalent circuit for the network to the left of terminals a and b in Fig. 5.13. Use node-voltage equations as in the proof of Thévenin's theorem.

■ **SOLUTION**

The network is described by the node-voltage equations:

$$
\begin{bmatrix}
\dfrac{1}{8} & -\dfrac{1}{8} & 0 \\[2mm]
-\dfrac{1}{8} & \dfrac{5}{8} & -\dfrac{1}{3} \\[2mm]
0 & 0 & 1
\end{bmatrix}
\cdot
\begin{bmatrix} v_1 \\ v_2 \\ v_3 \end{bmatrix}
=
\begin{bmatrix} -i \\ 1 \\ 12 \end{bmatrix}
$$

The determinant of the node transformation matrix is

$$
\Delta =
\begin{vmatrix}
\dfrac{1}{8} & -\dfrac{1}{8} & 0 \\[2mm]
-\dfrac{1}{8} & \dfrac{5}{8} & -\dfrac{1}{3} \\[2mm]
0 & 0 & 1
\end{vmatrix}
= \dfrac{1}{16}
$$

The cofactors are

$$
\Delta_{11} =
\begin{vmatrix}
\dfrac{5}{8} & -\dfrac{1}{3} \\[2mm]
0 & 1
\end{vmatrix}
= \dfrac{5}{8}
\qquad
\Delta_{21} = (-1)
\begin{vmatrix}
-\dfrac{1}{8} & 0 \\[2mm]
0 & 1
\end{vmatrix}
= \dfrac{1}{8}
\qquad
\Delta_{31} =
\begin{vmatrix}
-\dfrac{1}{8} & 0 \\[2mm]
\dfrac{5}{8} & -\dfrac{1}{3}
\end{vmatrix}
= \dfrac{1}{24}
$$

Cramer's rule gives

$$
v_1 = \dfrac{\Delta_{11}}{\Delta}(-i) + \dfrac{\Delta_{21}}{\Delta}(1) + \dfrac{\Delta_{31}}{\Delta}(12)
$$

Comparing this equation to Eq. (5.42), we have

$$
v_{ab} = v_1 = -R_{Th}i + v_{oc}
$$

with

$$
R_{ab} = R_{Th} = \dfrac{\Delta_{11}}{\Delta} = 10\ \Omega
$$

and

$$
v_{oc} = \dfrac{\Delta_{21}}{\Delta}(1) + \dfrac{\Delta_{31}}{\Delta}(12) = 10\ \text{V}
$$

This last equation clearly shows the contribution of each independent source to the open circuit voltage (superposition). ∎

EXERCISES

Use node-voltage equations, as in Example 5.11, to determine the Thévenin equivalent circuit with respect to terminals a and b of the network in the circuit of the indicated exercise.

16. Exercise 13
17. Exercise 14
18. Exercise 15

5.6 Maximum Power Transfer

A principal function of many electrical systems is to deliver power to a load. Given a specified open-circuit voltage, how can the power absorbed by the load be maximized? There are two possible cases—either the source resistance or the load resistance can be selected to maximize the power delivered to the load.

FIGURE 5.25
Load power when the source resistance R_s is adjustable. (*a*) Circuit with adjustable source resistance; (*b*) load power

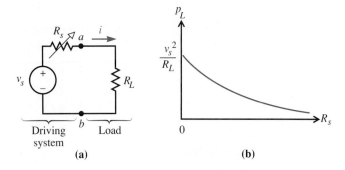

(a)   (b)

Assume that the load resistance R_L and open-circuit voltage v_s are specified, and the output resistance R_s of the driving system can be adjusted. Refer to Fig. 5.25a. The current is

$$i = \frac{v_s}{R_s + R_L} \tag{5.44}$$

and the power absorbed by the load is

$$p_L = R_L i^2 = \frac{R_L}{(R_s + R_L)^2} v_s^2 \tag{5.45}$$

Load power p_L as a function of source resistances is sketched in Fig. 5.25b, and is obviously maximized when

$$R_s = 0 \tag{5.46}$$

The second, more interesting, case occurs when the open-circuit voltage v_s and output resistance R_s of the driving network are specified, and the designer is free to choose the load resistance R_L, as indicated in Fig. 5.26a. As before, the power absorbed by the load is given by Eq. (5.45).

FIGURE 5.26
Load power when the load
resistance is adjustable.
(*a*) Circuit with adjustable
load; (*b*) load power

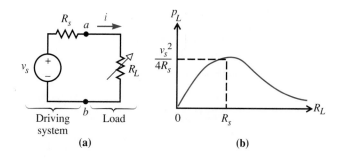

A maximum for p_L obviously occurs interior to the interval $0 \le R_L \le \infty$ and can be found if we set the derivative of p_L with respect to R_L equal to zero. Differentiation of Eq. (5.45) gives

$$\frac{dp_L}{dR_L} = \frac{(R_s + R_L)^2 - 2R_L(R_s + R_L)}{(R_s + R_L)^4} v_s^2 = 0 \qquad (5.47)$$

This equation is satisfied by

$$R_L = R_s \qquad (5.48)$$

See Problems 5.7 and 5.8 in the PSpice manual.

The result can be stated in the following theorem.

THEOREM 5

Maximum Power Transfer Theorem

When the source resistance is fixed and the load resistance can be selected, maximum power is absorbed by the load when the source and load resistance are equal or *matched*.

Substitution of $R_L = R_s$ into Eq. (5.45) gives the load power maximized with respect to R_L for R_s and v_s fixed:

$$\max_{R_L} \{p_L\} = \frac{v_s^2}{4R_s} \qquad (5.49)$$

An electric utility would never match the source and load resistances for its system because the energy lost in the Thévenin resistance of the system would equal the energy delivered to the load. The utility would design the system to minimize the cost of supplying the load power. Resistance (or more generally impedance, as we will see in later chapters) matching is common in electronics, particularly when short-duration pulses are transmitted over cables. The reason for this is explained in texts on electromagnetic fields.

EXAMPLE 5.12

Refer to the network of Fig. 5.13.

(a) Determine the value of load resistance connected between terminals *a* and *b* that will absorb the maximum power from network *A*.

(b) What power is absorbed by the load resistance found in (a)?

(a) The equivalent resistance seen by the load is the Thévenin resistance with respect to terminals a and b of network A. The maximum power is absorbed by the load when

$$R_L = R_{Th} = 10 \, \Omega$$

(b) The load current will be

$$i = \frac{V_{Th}}{R_{Th} + R_L} = \frac{10}{10 + 10} = \frac{1}{2} \, A$$

with a resulting load power of

$$P_L = Ri^2 = 10\left(\frac{1}{2}\right)^2 = 2.5 \, W \quad ■$$

REMEMBER

If you are free to select only the source resistance, make it as small as possible to maximize the power delivered to the load. If you are free to select only the load resistance, make it equal to the source resistance to maximize the power delivered to the load. This latter case causes the power absorbed by the Thévenin resistance of the source to be equal to the power delivered to the load.

EXERCISES

19. For the network shown below, select R_1 so that the maximum power is absorbed by R_2. Find P_1 and P_2.

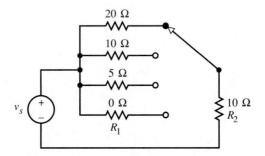

20. For the network shown above, select R_1 so that the maximum power is absorbed by R_1. Find P_1 and P_2.

21. Select R_L so that the power absorbed by R_L is a maximum, and find P_L for the following circuit. (Use Thévenin's theorem.)

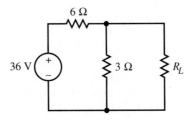

22. Find a load resistance R_L that, when connected to terminals a and b of the network described in Exercise 15, will absorb the maximum power.

5.7 Source Mobility*

As mentioned earlier, a voltage source with *zero* series resistance cannot be replaced by a current source and a parallel resistance. Nevertheless, a preliminary step will often permit source transformations to be made. To see why, consider the network fragment shown in Fig. 5.27a.

FIGURE 5.27
An example of how a voltage source is "pushed through" a node to form an equivalent circuit. (*a*) Original circuit; (*b*) circuit after voltage source has been pushed through node *a*

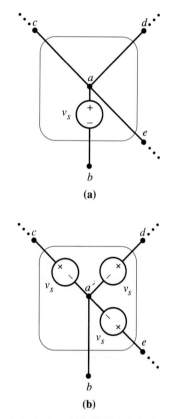

A network with terminal characteristics equivalent to those of Fig. 5.27a can be formed if we replace the voltage source by three voltage sources, one in series with each of the other branches connected to node a, as is shown in Fig. 5.27b. We often say that the voltage source was "pushed through" node a. KVL can easily be used to verify that we have selected the source polarities so that the voltages v_{cb}, v_{db}, and v_{eb} are preserved. The branch currents are also preserved, but the identity of node a has been changed.

Thus the network fragment in Fig. 5.27b is equivalent with respect to terminals b, c, d, and e to the network fragment of Fig. 5.27a. If any of the branches now contains a voltage source and a series resistance, they can be replaced by an equivalent current source and parallel resistance. This technique is occasionally useful to simplify a network.

* This section can be omitted with no loss of continuity.

FIGURE 5.28
An example of how a
current source is "pulled
around" a loop to form
an equivalent circuit. (a)
Original circuit; (b) circuit
after the current source has
been pulled around the
loop

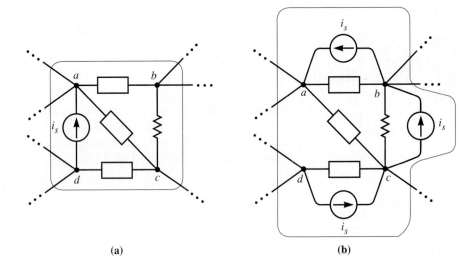

(a)   (b)

A manipulation that is similar to that just discussed for voltage sources can be performed on a current source. Consider the network fragment of Fig. 5.28a.

A current source in a loop can be replaced by a current source in parallel with each of the other branches in the loop, as is shown in Fig. 5.28b. The current source has been "pulled around" the loop. KCL can easily be used to verify that we have selected the source reference directions so that the sum of the currents leaving nodes a, b, c, and d is preserved. Thus the network fragment shown in Fig. 5.28b is equivalent to the network fragment of Fig. 5.28a. The equivalence is with respect to the voltages across and the currents through all components, except the current sources, shown. Any parallel combination of a current source and resistance can now be replaced by an equivalent voltage source and series resistance. This is also a useful technique in network simplification.

EXAMPLE 5.13

Use network reduction techniques, including source mobility, to solve for node voltages in the network of Fig. 5.29a.

■ SOLUTION

The network reduction is illustrated in Fig. 5.29b–e. Observe that the identities of only the reference node and nodes 1 and 5 have survived.

Ohm's law and KVL for the network of Fig. 5.29e quickly give

$$i = \frac{54 - 20}{17} = 2 \text{ A}$$

and

$$v_1 = -(2)(9) = -18 \text{ V}$$

From KVL in the same network,

$$v_5 = -54 + 6(2) = -42 \text{ V}$$

Application of KVL to the original network gives

$$v_4 = 30 + v_5 = -12 \text{ V}$$

FIGURE 5.29
Node voltage v_1 found by the use of source mobility. (a) Original network; (b) current source pulled around a loop, and voltage source pushed through node 4; (c) source transformations; (d) series resistances and voltage sources combined and source transformations; (e) parallel resistance combinations and source transformations

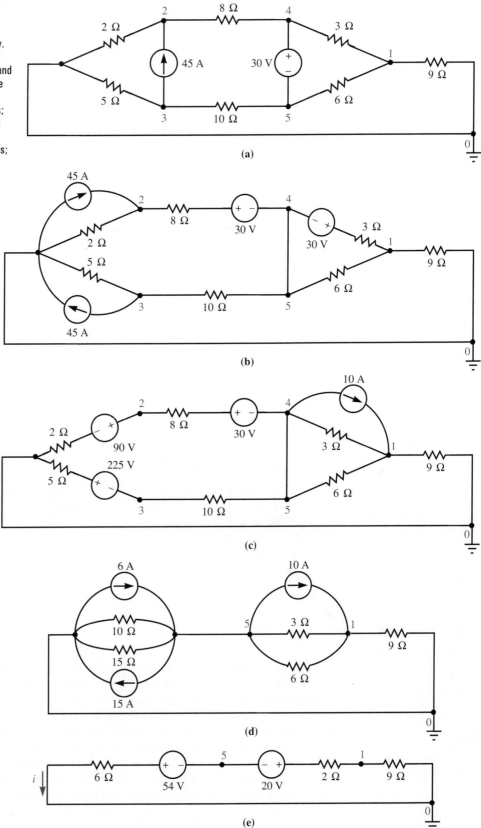

(a)

(b)

(c)

(d)

(e)

The KCL equation for a surface enclosing node 2 is

$$\frac{1}{2} v_2 + \frac{1}{8}(v_2 + 12) - 45 = 0$$

which gives

$$v_2 = 69.6 \text{ V}$$

and a KCL equation for a surface enclosing node 3 gives

$$\frac{1}{5} v_3 + \frac{1}{10}(v_3 + 42) + 45 = 0$$

with the result

$$v_3 = -164 \text{ V} \blacksquare$$

REMEMBER

When you push a voltage through a node, be sure the polarities of your sources preserve the relationship between the voltages of the other nodes. When you pull a current source around a loop, be sure the reference directions for the sources preserve the sum of the currents leaving the nodes.

EXERCISES

Use source mobility to calculate voltage v_{ab} in Exercises 23 and 24.

23.

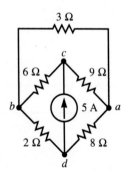

24.

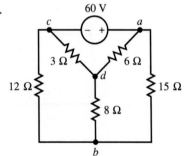

5.8 Duality

Perhaps you noticed that the node-voltage equations for many examples in Sections 4.1 and 4.2 were similar in form to the mesh-current equations for examples in Sections 4.5 and 4.6, but with current and voltage interchanged. For two circuits, if the node-voltage equations for each circuit can be found when we interchange voltage and current in the mesh-current equations for the other, we say the circuits are *duals*. The examples in Chapter 4 were intentionally selected to illustrate the concept of *duality*. Duality is a consequence of the similarity in component equations used in KCL and KVL equations. This is illustrated in Fig. 5.30. Some dual relationships for networks are given in Table 5.1. These are mutual

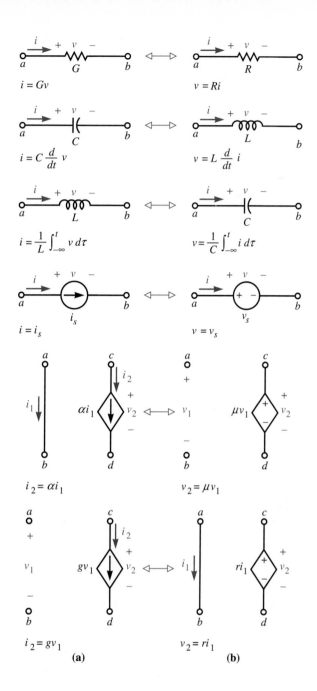

FIGURE 5.30
The dual nature of network components: (*a*) used in KCL equations; (*b*) used in KVL equations

$$i = Gv \qquad\qquad v = Ri$$

$$i = C\frac{d}{dt}v \qquad\qquad v = L\frac{d}{dt}i$$

$$i = \frac{1}{L}\int_{-\infty}^{t} v\,d\tau \qquad\qquad v = \frac{1}{C}\int_{-\infty}^{t} i\,d\tau$$

$$i = i_s \qquad\qquad v = v_s$$

$$i_2 = \alpha i_1 \qquad\qquad v_2 = \mu v_1$$

$$i_2 = g v_1 \qquad\qquad v_2 = r i_1$$

(a)   **(b)**

relationships, and the columns in which the terms of a dual pair appear can be interchanged.

The principal application of duality is that once you have the solution to a problem, you also know the solution to the dual problem. This will be enlightening when we examine parallel and series *RLC* circuits in Chapter 8. Duality is also useful in standardized network design. If you have a table of networks with the desired properties and the network is driven by a voltage source, the dual of that network has the dual properties when driven by a current source.

TABLE 5.1
Dual pairs

Conductance	Resistance
Capacitance	Inductance
Current source	Voltage source
Current-controlled current source	Voltage-controlled voltage source
Voltage-controlled current source	Current-controlled voltage source
Voltage	Current
Branch voltage	Branch current
Node	Mesh
Node pair	Loop
Node voltage	Mesh current
Reference node	Loop around entire circuit part
Series paths	Parallel paths
Open circuit	Short circuit

A planar circuit, not containing mutually coupled inductances, with $N + 1$ nodes and M meshes will have a *dual* circuit with $M + 1$ nodes and N meshes. Each component in the first network is replaced by its dual component in the second network. The node-voltage equation for one circuit will have the same form as the mesh equations for the other, so the analysis of a circuit will also yield the performance of the dual network.* A simple example of dual circuits is shown in Fig. 5.31.

FIGURE 5.31
Duality of parallel and series circuits: (*a*) a parallel circuit; (*b*) a series circuit

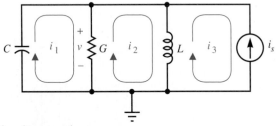

Node voltage equation

$$C\frac{d}{dt}v + Gv + \frac{1}{L}\int_{-\infty}^{t} v \, d\tau = i_s$$

Mesh current equations

$$
\begin{bmatrix}
\frac{1}{C}\int_{-\infty}^{t} d\tau + \frac{1}{G} & -\frac{1}{G} & 0 \\
-\frac{1}{G} & \frac{1}{G} + L\frac{d}{dt} & -L\frac{d}{dt} \\
0 & 0 & 1
\end{bmatrix}
\cdot
\begin{bmatrix}
i_1 \\
i_2 \\
i_3
\end{bmatrix}
=
\begin{bmatrix}
0 \\
0 \\
-i_s
\end{bmatrix}
$$

(a)

* For a more thorough discussion of duality, refer to Ernst A. Guilleman, *Introductory Circuit Theory*. New York: John Wiley & Sons, 1953.

FIGURE 5.31
Continued

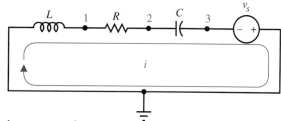

Mesh current equation

$$L\frac{d}{dt}i + Ri + \frac{1}{C}\int_{-\infty}^{t} i\ d\tau = v_s$$

Node voltage equations

$$\begin{bmatrix} \frac{1}{L}\int_{-\infty}^{t} d\tau + \frac{1}{R} & -\frac{1}{R} & 0 \\ -\frac{1}{R} & \frac{1}{R} + C\frac{d}{dt} & -C\frac{d}{dt} \\ 0 & 0 & 1 \end{bmatrix} \cdot \begin{bmatrix} v_1 \\ v_2 \\ v_3 \end{bmatrix} = \begin{bmatrix} 0 \\ 0 \\ -v_s \end{bmatrix}$$

(b)

As suggested in the preceding paragraph, the dual of a network can be constructed from a given circuit by inspection of the mesh-current and node-voltage equations. A convenient graphical method is to superimpose the dual network diagram on the original network diagram (refer to Fig. 5.32 as you read this paragraph). A node for the dual network is placed inside each mesh of the original network, and the reference node for the dual network is placed exterior to the original network. Each component that appears in a single mesh (or outer loop) of the original circuit becomes the dual component from the new node inside the mesh to the new reference node. Each new component is conveniently drawn across the original component. A component shared by two meshes in the original network yields a dual component shared by the two new nodes placed inside the meshes. This dual component is drawn across the original component. Source polarities are assigned so that if the source tends to force node k in one network positive, the dual source in the other network tends to force mesh current k positive.

To ensure that both mesh and node equations exhibit the dual relation, components connected to the dual reference node must be drawn so that the original

FIGURE 5.32
Graphical relationship between dual networks

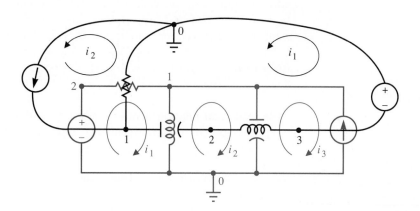

reference node is not enclosed. The method requires the counterclockwise assignment of mesh currents in the dual to maintain the signs for the mesh-current equations. We can correct this if desired by redrawing the dual as a mirror image, or with the top and bottom interchanged.

Observe that mutual inductance does not appear in Table 5.1 or Fig. 5.30. *Mutual inductance has no dual.* Therefore, a circuit with coupled coils has no dual.

REMEMBER

Dual pairs are listed in Table 5.1. Mutual inductance does not have a dual element. Once we have analyzed a network, we also have the analysis of the dual network.

EXERCISES

25. Use the graphical procedure to construct the dual of the parallel circuit of Fig. 5.31a. Compare your result with the circuit of Fig. 5.31b.

26. Use the graphical procedure to construct the dual of the series circuit shown in Fig. 5.31b. Compare your result with the circuit of Fig. 5.31a.

27. Use the graphical procedure to construct the dual of the network shown in Exercise 6.

5.9   Reciprocity*

Some electrical networks exhibit an interesting property called *reciprocity*. If reciprocity holds, the location of a current source and a voltage measurement can be interchanged, and the voltage measurement will not be changed. The same relationship also holds for the interchange of a voltage source and a current measurement. Reciprocity is of interest in theoretical developments, and also has implications for the type of components required to construct certain networks described in Chapter 17. We now formalize the definition of reciprocity.

DEFINITION

Reciprocal Network

For a **reciprocal network,** the voltage induced between any two terminals by a current source connected between any two terminals is unchanged if the location of the source and measurement are interchanged (refer to Fig. 5.33a, b). Similarly, the current induced in any branch by a voltage source in any branch is unchanged if the location of the source and measurement are interchanged (refer to Fig. 5.33c, d).

The reciprocal relation is always between voltage and current or current and voltage. Reciprocity does not generally apply between current source and current

---

* This section can be omitted without loss of continuity.

FIGURE 5.33
Properties of a reciprocal
network. (*a*) Interchange of
the position of current
source and voltage
measurement, $i_1 = i_b$
implies $v_2 = v_a$;
(*b*) interchange of the
position of voltage source
and current measurement,
$v_1 = v_b$ implies $i_2 = i_a$

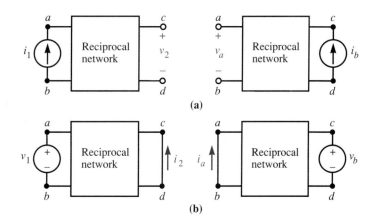

(a)

(b)

measurement or voltage source and voltage measurement. Typically other voltages and currents are altered when the source location is changed.

The terminals *a*, *b*, *c*, and *d* shown in Fig. 5.33a need not all be distinct. It is relatively easy to show that a single element—resistance, capacitance, or inductance—forms a reciprocal network. Coupled inductances also form a reciprocal network. Thus the elements—resistance, capacitance, and inductance (including coupled inductances)—exhibit reciprocity and are *reciprocal components*. A dependent source and an independent source do not form a reciprocal network, so a dependent source is a nonreciprocal component.

In general, a network of reciprocal devices forms a reciprocal network. A network that contains nonreciprocal components will not be a reciprocal network, but a *reciprocal relation* may exist between some terminals and branches.

As we observed in Chapter 4, networks that contain only resistance, capacitance, and inductance (including coupled inductances) can be described by a symmetrical system matrix (node or mesh transformation matrix) if KVL equations are eliminated from the node-voltage equations and KCL equations are eliminated from the mesh-current equations. This symmetry resulted when we followed a specific procedure to write the equations for a reciprocal network and was neither accidental nor inherent in the properties of the network. The proof of the following theorem can be found in more advanced texts.

THEOREM 6

A network can be described by a symmetrical system matrix if and only if it is a reciprocal network.

In many applications the entire network description may not be available. Only the relation between an input voltage or current and an output voltage or current may be available; for example,

$$\begin{bmatrix} T_{11} & T_{12} \\ T_{21} & T_{22} \end{bmatrix} \cdot \begin{bmatrix} v_1 \\ v_2 \end{bmatrix} = \begin{bmatrix} i_1 \\ i_2 \end{bmatrix} \tag{5.50}$$

If the square matrix is symmetrical, a reciprocal relation exists between the input and output variables.

EXAMPLE 5.14

Show that reciprocity holds for the interchange of a current source and a voltage measurement for the network of Fig. 5.34a.

FIGURE 5.34
An example of reciprocity where the location of a current source and voltage measurement are interchanged. $v_2 = v_a$

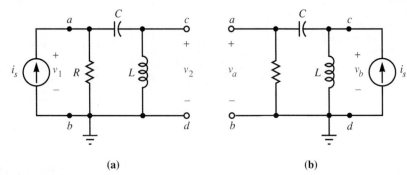

(a)   (b)

■ **SOLUTION**

Reciprocity can be established if you show that the equation relating response v_2 and input i_s for the network of Fig. 5.34a is the same as the equation relating response v_a and input i_s for the network of Fig. 5.34b.

The network of Fig. 5.34a is described by the following node-voltage equations:

$$\begin{bmatrix} C\dfrac{d}{dt} + \dfrac{1}{R} & -C\dfrac{d}{dt} \\ -C\dfrac{d}{dt} & C\dfrac{d}{dt} + \dfrac{1}{L}\int_{-\infty}^{t} d\tau \end{bmatrix} \cdot \begin{bmatrix} v_1 \\ v_2 \end{bmatrix} = \begin{bmatrix} i_s \\ 0 \end{bmatrix}$$

Perform the operation $C\,d/dt$ on the first equation and $C\,d/dt + (1/R)$ on the second equation. Addition of the first and second equations gives

$$\left(\frac{C}{R}\frac{d}{dt} + \frac{C}{L} + \frac{1}{RL}\int_{-\infty}^{t} d\tau \right) v_2 = C\frac{d}{dt} i_s$$

The network of Fig. 5.34b is described by the following node-voltage equations:

$$\begin{bmatrix} C\dfrac{d}{dt} + \dfrac{1}{R} & -C\dfrac{d}{dt} \\ -C\dfrac{d}{dt} & C\dfrac{d}{dt} + \dfrac{1}{L}\int_{-\infty}^{t} d\tau \end{bmatrix} \cdot \begin{bmatrix} v_a \\ v_b \end{bmatrix} = \begin{bmatrix} 0 \\ i_s \end{bmatrix}$$

Perform the operation $C\,d/dt + (1/L)\int_{-\infty}^{t} d\tau$ on the first equation and $C\,d/dt$ on the second. Addition of the first and second equations gives

$$\left(\frac{C}{R}\frac{d}{dt} + \frac{C}{L} + \frac{1}{RL}\int_{-\infty}^{t} d\tau \right) v_a = C\frac{d}{dt} i_s$$

Voltages v_2 and v_a are solutions to identical linear differential equations, so as predicted by the reciprocity theorem,

$$v_a = v_2 \quad \blacksquare$$

The reciprocity theorem has some practical application in design. An example of how a design can be adapted to a different application by the use of reciprocity is given in Problem 40. Reciprocity also gives some information about the type of components required to construct the two-port networks introduced in Chapter 17.

EXERCISES

Show that reciprocity holds for voltage v and current i in the networks of Exercises 28 and 29.

28. 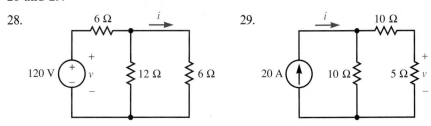 29.

30. Show that reciprocity holds for the interchange of a voltage source v_s and current measurement i in the network below.

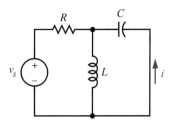

5.10 Summary

In this chapter, we presented a variety of concepts that help the engineer simplify the analysis of a circuit. We began with the definitions of linearity, time invariance, and superposition, and introduced source transformations, which provide an alternative method to solve for a specified response in some circuits. We next extended the idea of an equivalent source by the use of Thévenin's and Norton's theorems. We used these equivalent circuits to solve maximum power transfer problems. We concluded the chapter with optional sections on source mobility, duality, and reciprocity.

KEY FACTS

◆ Superposition holds for linear circuits.

◆ When you apply the superposition theorem, each independent source must be used once. The unused independent sources are set equal to zero. The dependent sources are retained.

- The superposition theorem states that the response due to all sources is equal to the sum of the responses due to each independent source acting alone, with all other independent sources set equal to zero.

- A 0-V voltage source is equivalent to a short circuit.

- A 0-A current source is equivalent to an open circuit.

- A voltage source of value v_s in series with a resistance of value R has the same terminal characteristics as a current source of value i_s in parallel with a resistance of value R, if the voltage, current, and resistance satisfy Ohm's law:

$$v_s = R i_s$$

- Source transformations can be used to eliminate nodes or meshes.

- Thévenin's and Norton's theorems apply to circuits with resistances, dependent sources, and independent sources.

- The Thévenin voltage is equal to the open-circuit voltage.

- The Thévenin resistance can be obtained if we set all independent sources in the circuit to zero (dependent sources are retained). Apply a source to two terminals. The ratio of terminal voltage to terminal current is the Thévenin resistance looking into these terminals.

- The Thévenin resistance is related to the open-circuit voltage and short-circuit current by Ohm's law:

$$R_{Th} = \frac{v_{oc}}{i_{sc}}$$

- The Norton current is equal to the short-circuit current.

- The Norton resistance is equal to the Thévenin resistance.

- If the open-circuit terminal voltage of a circuit is fixed and a fixed load resistance is connected to these terminals, the power delivered to the load is maximized by making the Thévenin resistance of the circuit as small as possible.

- If the open-circuit voltage and Thévenin resistance of a circuit are fixed, the power delivered to a load resistance is maximized when the load resistance is equal to the circuit Thévenin resistance.

- Source mobility lets us move the location of sources so that we can use source transformations or other network reduction techniques.

- The analysis of a network immediately gives us the analysis of the dual network.

- There is no dual for mutual inductance.

- The reciprocity relation holds between voltage and current, not between voltage and voltage or between current and current.

- Networks with dependent sources are not reciprocal networks.

Section 5.2

Use superposition to calculate current i_y or voltage v_x as indicated, for the networks shown in Problems 1 through 3. ⬛ 5.1, 5.2

1.

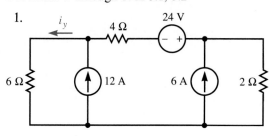

2.

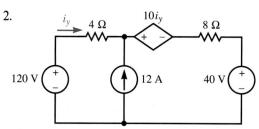

3.

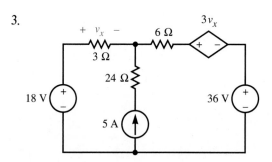

4. Use the principle of superposition to write three differential equations that can be solved for the component of current due to each independent voltage source in the following circuit. (Do not

solve the equations.) We will see in later chapters that this type of problem is one of the most important applications of superposition.

Section 5.3

5. Use source transformations to eliminate the two current sources from the network that follows. Solve for the two remaining mesh currents. Also solve for voltages v_1, v_2, and v_3. ⬛ 5.4

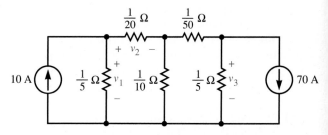

6. Use source transformations to eliminate the two voltage sources from the network that follows. Then solve for the two remaining node voltages. Also solve for currents i_1, i_2, and i_3.

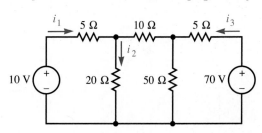

7. Use equivalent circuits to reduce the following circuit to a three-node circuit, and use node-voltage equations to determine voltage v from a single KCL equation.

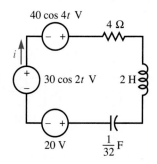

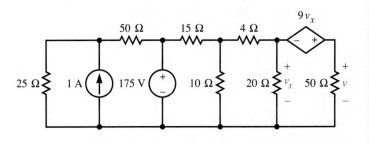

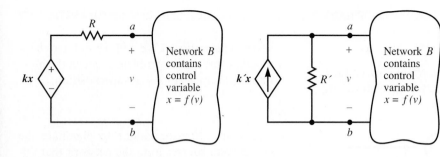

FIGURE 5.35

8. Analysis of electronic circuits is often facilitated by the performance of *source transformations on dependent sources*. Show that the dependent voltage source and series resistance shown in Fig. 5.35 is equivalent, with respect to its effect on network B, to the dependent current source and parallel resistance shown, if $R' = R$ and $k' = k/R$.

9. With the aid of the results obtained in Problem 8, convert the following network into a network that is equivalent with respect to terminals a and b, and consists of an independent voltage source, dependent voltage source, and series resistances.

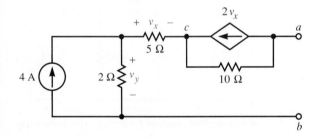

10. In the preceding network, replace the dependent current source with a dependent current source of value $2v_y$, reference arrow pointing toward the left, and repeat Problem 9.

Section 5.4

Determine the Thévenin and Norton equivalent circuits with respect to terminals a and b for each network shown in Problems 11 through 16. 5.1, 5.6

11.

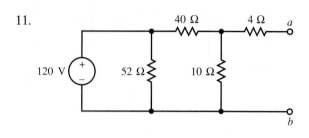

12.

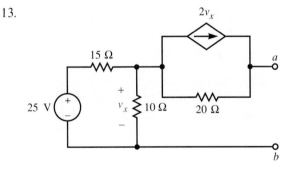

13.

14.

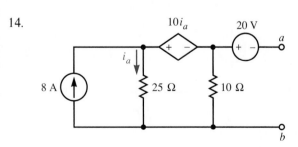

15.

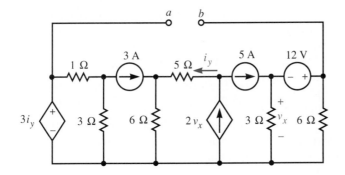

16.

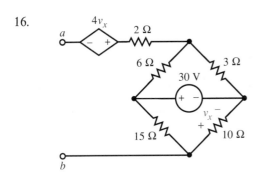

The Thévenin resistance found in (a) is the input resistance seen by the voltage source and 400-Ω resistance, the resistance found in (b) is the input resistance seen by the current source, and the resistance found in (c) is the output resistance seen by the 200-Ω load.

(d) Use the results from (a) and (c) to calculate the voltage gain v_{ef}/v_{ab} with the 200-Ω load connected and the 10-A current source not connected.

17. With the aid of the results of Problem 8, calculate the Thévenin equivalent circuit, with respect to terminals a and b, for the network of Problem 9.

18. Repeat Problem 17 for the network of Problem 10.

19. Determine the Thévenin and Norton equivalent circuits for the following network with respect to:
   (a) Terminals a and b when the voltage source and 400-Ω series resistance are not connected
   (b) Terminals c and d when the current source is not connected
   (c) Terminals e and f when the 200-Ω resistance is not connected

20. The terminal characteristics of a two-terminal device are experimentally measured and plotted on the following graph. Construct a Thévenin equivalent network as a model for the device. (Current i_{ab} is the current into terminal a of the device.)

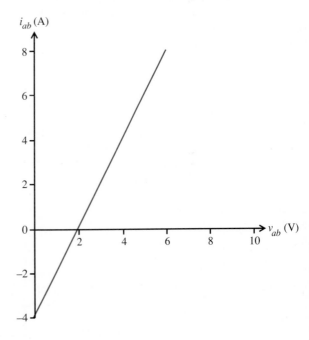

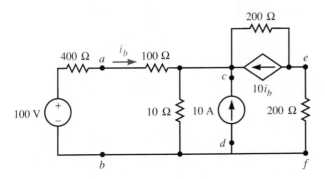

21. A 2-Ω speaker load is connected to the output of an amplifier driven by a signal generator. The output voltage v_{ab} is measured to be 4 cos 2000t V. The speaker load is changed to 8 Ω with the signal generator unchanged, and v_{ab} is measured to be 8 cos 2000t V. What is the Thévenin equivalent circuit model for the amplifier driven by the signal generator?

22. A resistive load is connected to a power supply, and the output voltage v_{ab} is measured to be 100 V when the load current $i_{ab} = 2$ A. The load is changed, and v_{ab} is measured to be 40 V when i_{ab} is 8 A. What is the Thévenin equivalent circuit model for the power supply?

23. Solve for the current through the nonlinear load shown below. The terminal equation for the load is $v = 2i^2$.

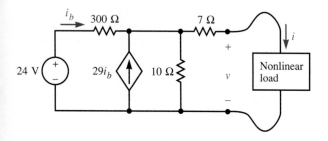

Section 5.6

For Problems 24 through 27, find the value of load resistance for each network that will absorb the

maximum power when connected to terminals a and b of the network shown in the circuit for the indicated problem, and calculate the power absorbed by the load.

24. Problem 11   25. Problem 12
26. Problem 13   27. Problem 14

For the networks in Problems 28 and 29, find the value of the unknown network parameter that will maximize the power delivered to the load circuit.

28.

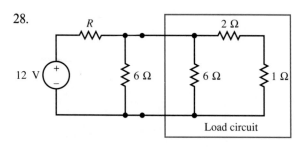

29. See Fig. 5.36.

30. A small-signal equivalent circuit for a field-effect transistor (FET) amplifier is shown in Fig. 5.37. Terminals G, S, and D are the gate, source, and drain terminals, respectively.
(a) Determine the value of R_2 that will absorb the maximum power.
(b) What power is absorbed by the value of R_2 found in (a)?

FIGURE 5.36

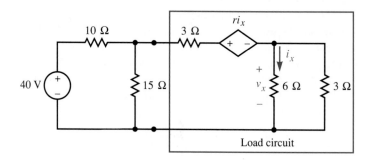

FIGURE 5.37

Section 5.7

Use source mobility to determine the voltage v_2 in the circuits of Problems 31 and 32.

31.

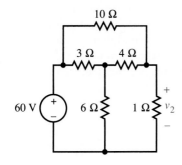

32.

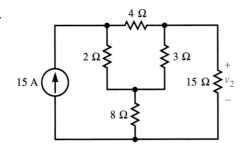

33. Use source mobility to reduce the following network to a simple series circuit that can be solved for voltage v_x.

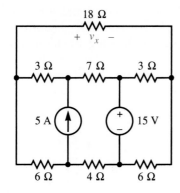

Section 5.8

34. Draw the dual of the network shown in Problem 1.
35. Draw the dual of the network shown in Problem 2.
36. Draw the dual of the network shown in Problem 3.

37. Construct the dual of the network shown in Fig. 4.7, and compare your result with the network of Fig. 4.18.

Section 5.9

Establish that a reciprocal relation does or does not exist between voltage v and current i in the networks in Problems 38 and 39.

38.

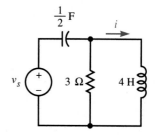

39.

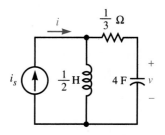

40. The relationship between an ideal voltage source voltage and resistance current is desired to be the same as the relationship between the voltage source voltage v_s and capacitance current i for the following network. Use the reciprocity theorem to design such a circuit.

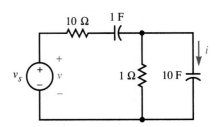

Comprehensive Problems

41. Use superposition to calculate currents i_a, i_b, and i_c for the following network. What effect does voltage v_a have on current i_b, and what effect does voltage v_b have on current i_a?

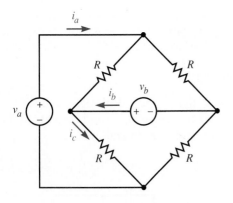

(c) The resistance R_c
if $R_a = R_b = 10\ \Omega$, $R_c = 13\ \Omega$, $R_e = 1\ \Omega$, and $\beta = 9$.

44. The small-signal model for a common collector transistor amplifier is shown in the following figure. It must have an output resistance R_o looking into terminals E and C of $10\ \Omega$ or less. What is the minimum value of β that will satisfy this requirement?

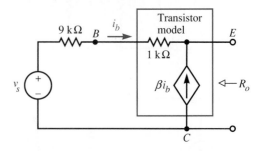

42. A simplified equivalent circuit for a grounded-base amplifier is shown below. Calculate the input resistance seen by the practical source, the voltage gain v_{cb}/v_{eb}, the current gain i_c/i_e, and the power gain $-v_{cb}i_c/v_{eb}i_e$, all with the 1000-Ω load connected.

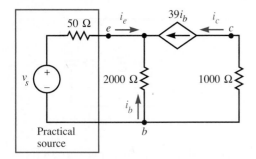

43. Transistors are often used in the *Darlington connection* (compound amplifier). A simplified small-signal equivalent circuit for this connection is shown in the following figure. Determine the Thévenin or Norton equivalent circuit for this network as seen by:
(a) The voltage source
(b) The resistance R_e

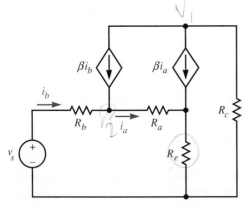

45. Semiconductor devices are linear only within certain voltage and current ranges and must be *biased* to operate within the proper range. You must design the *bias circuit* for a transistor amplifier with the large-signal equivalent circuit shown below. The collector current i_C must be 1 mA.
(a) Calculate the input resistance looking into terminals B and 0 with the connection between terminals A and B removed.
(b) Select resistances R_1 and R_2 so that the resistance looking into terminals A and 0 is four times the resistance looking into terminals B and 0, and so that collector cur-

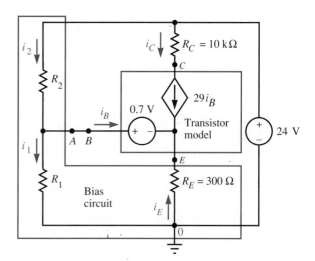

rent i_C is 1 mA when terminals A and B are connected. Refer to the first two steps of Problem 96 of Chapter 3 for help.

46. Solve for the voltage v across the nonlinear load shown in the circuit that follows. The terminal equation for the load is given by the graph.

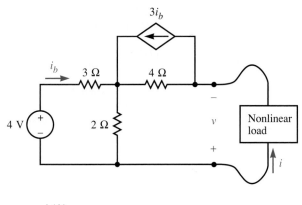

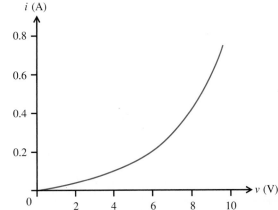

47. A small-signal equivalent circuit for a *phase splitter* utilizing a field-effect transistor (FET) follows.
    (a) Determine voltages v_1 and v_2.
    (b) For what value of R_D does $v_2 = -v_1$?

(c) When this equality is satisfied, what is the Thévenin resistance looking into terminals S and 0 and D and 0 with R_D and R_S connected?

(d) Calculate $\gamma_{1S} = \partial v_1/\partial R_S$, $\gamma_{1D} = \partial v_1/\partial R_D$, $\gamma_{2D} = \partial v_2/\partial R_D$, and $\gamma_{2S} = \partial v_2/\partial R_S$. These partial derivatives are a measure of the sensitivity of the two outputs due to loading by another network. When $R_D = R_S$, which partial derivative has the largest magnitude? Which has the smallest? Note that the voltage amplification factor μ is much greater than 1.

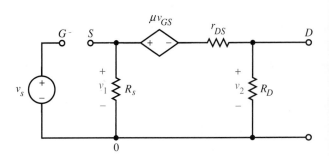

48. Construct a Thévenin equivalent circuit with respect to terminals a and b for the network in Fig. 5.38. Next, connect a 20-Ω feedback resistance between terminals a and c and construct the Thévenin equivalent with respect to terminals a and b. Compare your results. This is an example of the effect of feedback on an electronic amplifier. For the original network, an input resistance of 3 Ω is seen by the 12-V source with a 6-Ω series resistance. What is the input resistance with the 20-Ω feedback resistance connected? (Some of the output is fed back through this resistance and subtracted from the 12-V input.) We will investigate feedback circuits in Chapter 6.

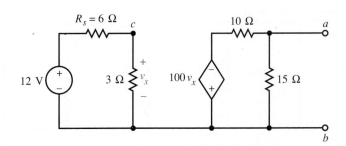

FIGURE 5.38

49. Calculate the voltage gain and output resistance looking into terminals S and D of the circuit in Fig. 5.39 if the resistance R is replaced by an open circuit.

50. The circuit in Fig. 5.39 is a simplified model of an FET amplifier. Both the voltage gain and the output resistance R_o (looking into terminals S and D) are a function of the feedback resistance R. Select R so that the output resistance is 1 kΩ. What is the voltage gain v_o/v_s for this value of R?

Although the component values are changed to simplify the arithmetic, the network used in Problems 51 and 52 is the small-signal equivalent circuit for a three-transistor amplifier with feedback.

(The 90-Ω resistance provides a feedback path from stage 3 to stage 1.)

51. For the circuit in Fig. 5.40, find the input resistance seen by the voltage source, the output resistance seen by the 10-Ω load resistance, the voltage gain v_o/v_s with the load disconnected, and the voltage gain v_o/v_s with the load connected.

52. Repeat Problem 51 with the switch closed.

53. Compare the voltage gains found in Problems 51 and 52. Why might feedback be desirable?

54. An amplifier with feedback is shown in Fig. 5.41. When the feedback path is indirectly opened by opening the switch, calculate the Thévenin

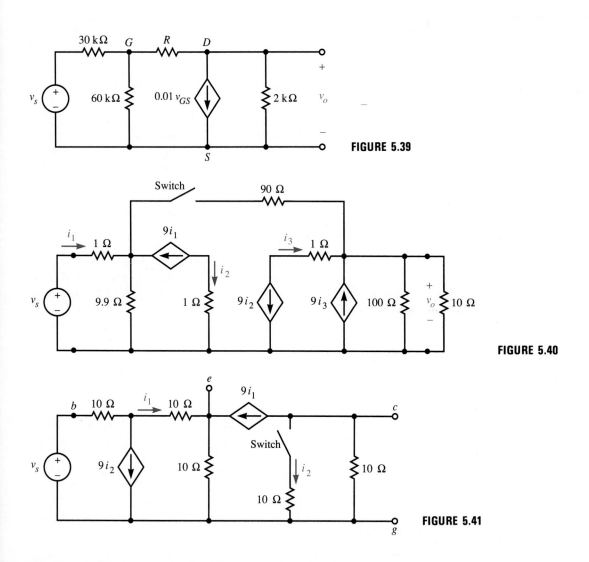

FIGURE 5.39

FIGURE 5.40

FIGURE 5.41

resistance looking into terminals b and g, c and g, and e and g. Also calculate the voltage gains v_{cg}/v_s and v_{eg}/v_s.

55. Repeat Problem 54 with the switch closed.

56. Calculate the input resistance seen by the voltage source, the output resistance looking into terminals c and d, and the voltage gain v_{cd}/v_{ab} for the three-transistor feedback amplifier model given in the circuit in Fig. 5.42, with the feedback path indirectly opened by opening the switch.

57. Repeat Problem 56 with the switch closed. Compare this result with that obtained in Problem 56.

58. Use PSpice to solve Problem 54 when v_s is 1 V.
59. Use PSpice to solve Problem 55 when v_s is 1 V.
60. Use PSpice to solve Problem 56 when v_s is 1 V.
61. Use PSpice to solve Problem 57 when v_s is 1 V.
62. Use PSpice to solve Problem 12. (Remember that no dangling nodes are allowed.)
63. Use PSpice to solve Problem 13.
64. Use PSpice to solve Problem 14. (Remember that no dangling nodes are allowed.)

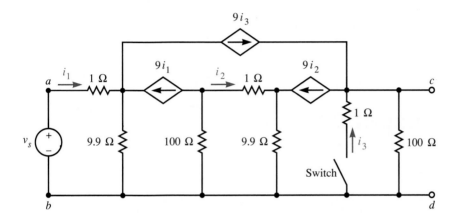

FIGURE 5.42

CHAPTER
6

Operational Amplifiers

We all know that the integrated circuit, or semiconductor chip, has revolutionized digital circuit design and given us the personal computer. Although less widely publicized, integrated circuits have also revolutionized linear circuit design by providing us with small building blocks containing many resistors and transistors. Perhaps the most versatile of these devices is the *operational amplifier*, or *op amp*.

Operational amplifiers are characterized by a high input resistance, low output resistance, and large voltage gain. Op amps are found in audio amplifier systems; they also appear in many instrumentation applications, such as digital multimeters, where they are used to amplify small voltages.

A detailed analysis of the voltages and currents internal to the op amp is seldom necessary. Usually we need consider only the terminal characteristics of the op amp. We often further simplify the analysis by introducing a four-terminal network component called the *ideal op amp.* As a tool for understanding op-amp circuits, we also introduce the concept of *negative feedback*.

6.1 An Op-Amp Model

An op amp, although once constructed as an interconnection of discrete transistors and resistors, is now built as an integrated circuit on a few square millimeters of silicon and costs less than one dollar. Op amps come in several types of packages, typically with eight or more terminals, some of which may have no internal connection. The dual-in-line package shown in Fig. 6.1 is quite popular.

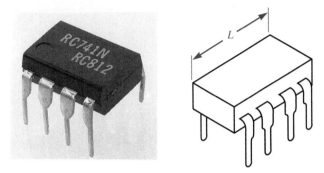

FIGURE 6.1
(*Left*) Op amp. Photograph by James Scherer. (*Right*) A dual-in-line package often used for operational amplifiers. The length L is less than $\frac{1}{2}$-inch

We shall consider only the external characteristics of the op amp. In the simplest case, only five connections are made to the op amp, as shown in Fig. 6.2.

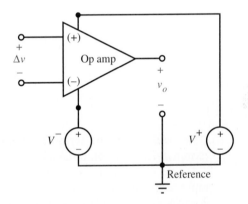

FIGURE 6.2
Op-amp symbol with power sources and reference shown

Constant voltages of value V^+ and V^-, shown in Fig. 6.2, supply the electric power necessary for operation of the amplifier. (Typical values for V^+ and V^- are $+15$ V and -15 V, respectively.) The reference symbol ($\rightleftharpoons$) shown in Fig. 6.2 represents the reference from which the constant voltages are measured. Physically, the reference is some common point, such as the power-supply case, to which the circuit is connected. This *chassis ground* is usually connected to the earth (*earth ground*). The output voltage v_o is a function of the *error voltage* Δv,* which is the voltage difference between the input voltage to the *noninverting* ($+$) input terminal and the input voltage to the *inverting* ($-$) input terminal.

Practically, the output voltage is constrained to fall between V^+ and V^-, as shown in Fig. 6.3. Throughout this chapter we will assume that our op amps are operating in the linear region indicated in Fig. 6.3, and so the output voltage v_o is proportional to the error voltage Δv.

* Do not be confused by the notation Δv. The delta (Δ) is used simply to denote "difference."

FIGURE 6.3
Op-amp output v_o as a
function of error voltage Δv

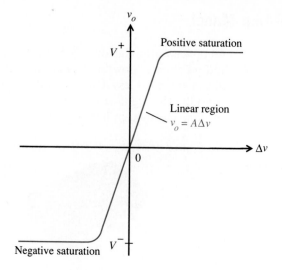

FIGURE 6.4
Op-amp model showing input
resistance R_i, output
resistance R_o, voltage gain
A, input current i_i, and
current to the reference i_g

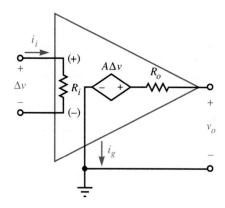

We can model the terminal characteristics of an op amp by standard two-terminal network components, as shown in Fig. 6.4.

In effect, an input resistance R_i is connected between the noninverting, or $(+)$, input terminal and the inverting, or $(-)$, input terminal. The output-circuit model consists of a dependent voltage source with a voltage gain of value A in series with an output resistance of value R_o. The connection to ground and current i_g account for the currents through the voltage sources shown in Fig. 6.2.

Typically the input resistance R_i is greater than 100 kΩ (greater than 10^{12} Ω for some op amps), the output resistance R_o is less than 100 Ω, and the voltage gain A is 10,000 or more. Although the voltage gain is large, its value is *not* precisely controlled. For this reason the op amp is used in circuits using negative feedback. We shall see that the assumption of very large gain and feedback leads to certain simplifying approximations.

6.2 Negative Feedback

The concept of negative feedback is profoundly important, not only in engineering, but also in many other areas of human activity and in biological systems. As an illustration, try to walk along a 100-m prescribed path with your eyes shut. You will surely veer to the left or the right as a result of variations in your steps. The same task is simple with your eyes open. You continually compare your position with your intended position along the path and make a correction that is proportional to the *negative of the position error*. This *negative feedback* causes your tracking error to be small.

Feedback is regularly used in electronic circuits, particularly in conjunction with op amps. The following example illustrates how a voltage v_{Fb}, which is a fraction of the output voltage v_o, can be fed back and subtracted from the input voltage v_s to obtain an error voltage Δv. We will show that the *closed-loop voltage gain*

$$G = \frac{v_o}{v_s} \tag{6.1}$$

is nearly independent of the *open-loop voltage gain A* for sufficiently large values of *A*.

EXAMPLE 6.1

⬇ *See Problem 6.3 of the PSpice manual.*

Calculate the output voltage of the noninverting op-amp circuit shown in Fig. 6.5.

FIGURE 6.5
A noninverting amplifier circuit that uses an op-amp with negative feedback

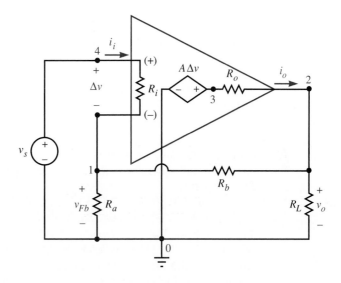

■ SOLUTION

We can easily analyze the circuit by the method of node voltages. Apply KCL to a surface enclosing node 1:

$$\frac{1}{R_a} v_{Fb} + \frac{1}{R_b}(v_{Fb} - v_o) + \frac{1}{R_i}(v_{Fb} - v_s) = 0 \tag{6.2}$$

Application of KCL to a surface enclosing node 2 gives

$$\frac{1}{R_L} v_o + \frac{1}{R_b}(v_o - v_{Fb}) + \frac{1}{R_o}[v_o - A(v_s - v_{Fb})] = 0 \tag{6.3}$$

The use of Cramer's rule and some complicated algebra, which we need not reproduce, gives us

$$v_o = \frac{(R_a + R_b)/R_a + (1/A) \cdot (R_o/R_i)}{1 + (1/A)\{[(R_L + R_o)/R_L][1 + (R_b/R_a) + (R_b/R_i)] + (R_o/R_a) + (R_o/R_i)\}} v_s$$

$$(6.4)$$

For any practical op-amp circuit,

$$A > 10{,}000 \qquad (6.5)$$

and

$$R_o \ll R_a \ll R_i \qquad (6.6)$$

so Eq. (6.4) is approximated by

$$v_o = \frac{\dfrac{R_a + R_b}{R_a}}{1 + \dfrac{1}{A}\left(\dfrac{R_L + R_o}{R_L}\right)\left(\dfrac{R_a + R_b}{R_a}\right)} v_s$$

$$= \frac{\dfrac{R_L}{R_L + R_o} A}{\left(\dfrac{R_L}{R_L + R_o} A\right)\left(\dfrac{R_a}{R_a + R_b}\right) + 1} v_s \quad\blacksquare \qquad (6.7)$$

In the above example, the inequality

$$R_a \ll R_i \qquad (6.8)$$

implies that the current through R_i can be neglected. An equivalent approximation is that

$$R_i = \infty \qquad (6.9)$$

If the op-amp gain A is large, the error voltage Δv must be small, and therefore i_i is very small. Therefore, Eq. (6.9) is a good approximation for most applications. The inequality

$$R_o \ll R_a \qquad (6.10)$$

implies that the feedback current through R_a and R_b does not cause a significant voltage across R_o. (The feedback circuit does not load the amplifier.) This approximation is valid for all practical circuits.

We will now infer some startling results from the preceding example.

Closed-Loop Voltage Gain

From Equation (6.7) we can see that the effect of the open-loop output resistance R_o is to reduce the open-loop gain A to an *effective open-loop gain A'* by the voltage divider relation for R_o and R_L. Typically R_o is much less than R_L and A' is approximately equal to A:

$$A' = \frac{R_L}{R_L + R_o} A \simeq A \qquad (6.11)$$

For this reason we will drop the prime and assume that A and A' are equal. We can now write Eq. (6.7) as

$$v_o = \frac{A}{A[R_a/(R_a + R_b)] + 1} v_s \tag{6.12}$$

We define the *feedback factor* as the fraction β of the output voltage v_o that is fed back and subtracted from a fraction α of the input voltage v_s to obtain the error voltage Δv. That is,

$$\Delta v = \alpha v_s - \beta v_o \tag{6.13}$$

We can use Eq. (6.13) to solve for v_o in any op-amp circuit:

$$v_o = A\,\Delta v = A(\alpha v_s - \beta v_o) \tag{6.14}$$

which yields the *general result*

$$v_o = \frac{\alpha A}{A\beta + 1} \cdot v_s \tag{6.15}$$

and the *closed-loop gain* G is defined by

$$G = \frac{v_o}{v_s} = \frac{\alpha A}{A\beta + 1} \tag{6.16}$$

We can easily determine α and β for the noninverting amplifier circuit of Fig. 6.5. With the assumption that R_i is infinity, the voltage divider relation and KVL give

$$\Delta v = v_s - \frac{R_a}{R_a + R_b} v_o \tag{6.17}$$

Comparison of Eqs. (6.17) and (6.13) shows that

$$\alpha = 1 \tag{6.18}$$

and

$$\beta = \frac{R_a}{R_a + R_b} \tag{6.19}$$

for the circuit of Fig. 6.5. Comparison of Eqs. (6.12) and (6.15) shows that they are of the same form, with α and β as given by Eqs. (6.18) and (6.19).

For values of *loop gain* $A\beta$ such that

$$A\beta \gg 1 \tag{6.20}$$

the closed-loop gain is approximately

$$G = \frac{v_o}{v_s} \simeq \frac{\alpha}{\beta} \tag{6.21}$$

For our example circuit,

$$G \simeq \frac{R_a + R_b}{R_a} \qquad (6.22)$$

The approximation becomes an equality as the loop gain becomes infinite. Thus *as long as the loop gain is large, the closed-loop gain is precisely determined by the feedback circuit* composed of resistances R_a and R_b and is independent of the open-loop gain A. This result is very important. The open-loop gain A varies substantially from one op amp to another, even for units with the same part number. In addition, the open-loop gain depends on temperature, supply voltage, and even the input signal. As we noted, the effective open-loop gain A' also depends on the load resistance R_L. However, as long as the effective loop gain is very large, the closed-loop gain is strictly determined by resistances R_a and R_b.

Closed-Loop Output Resistance

For the model of any practical op-amp circuit, the load current consists primarily of the op-amp output current. Therefore the short-circuit current* will be

$$i_{sc} = i_o\Big|_{v_o=0} = \frac{A\,\Delta v}{R_o}\Big|_{v_o=0}$$

$$= \frac{A(\alpha v_s - \beta v_o)}{R_o}\Big|_{v_o=0} = \frac{\alpha A v_s}{R_o} \qquad (6.23)$$

while the open-circuit voltage is given by Eq. (6.15). Thus the closed-loop output resistance R'_o is

$$R'_o = \frac{v_{oc}}{i_{sc}} = \frac{\alpha A}{A\beta + 1}\, v_s \cdot \frac{R_o}{\alpha A v_s}$$

$$= \frac{1}{A\beta + 1}\, R_o \qquad (6.24)$$

where v_{oc} is the open circuit value of v_o.

For large loop gains ($A\beta \gg 1$), the closed-loop output resistance is approximately

$$R'_o \simeq \frac{1}{\beta A}\, R_o \qquad (6.25)$$

Equation (6.25) is a general result for op-amp circuits. Thus the closed-loop output resistance R'_o is approximately equal to the already small open-loop output resistance divided by the loop gain. Because we always design for a loop gain much greater than 1, we can usually ignore the open-loop output resistance and simplify the op-amp model by assuming zero open-loop output resistance:

$$R_o = 0 \qquad (6.26)$$

---

* Short-circuiting a physical op-amp circuit will drive the op amp into the nonlinear region, but this is not a problem for our model.

Closed-Loop Error Voltage

Under the assumption of zero input current i_i and output resistance R_o, the error voltage Δv is

$$\Delta v = \frac{1}{A} v_o = \frac{1}{A} \cdot \frac{A}{A\beta + 1} v_s = \frac{1}{A\beta + 1} v_s \tag{6.27}$$

With a large loop gain ($A\beta \gg 1$) Eq. (6.27) gives us

$$\Delta v \simeq \frac{v_s}{A\beta} \tag{6.28}$$

Equation (6.28) implies that the error voltage is quite small relative to the input voltage for large loop gains. (This conclusion applies to all op-amp circuits operating in the linear range.) It follows that with Δv small and R_i large, the op-amp input current i_i must be very small.

$$i_i = \frac{\Delta v}{R_i} = \frac{v_s}{A\beta R_i} \tag{6.29}$$

Therefore the approximation that R_i is infinite introduces little error.

REMEMBER

Op amps are used in circuits with feedback. A large effective loop gain ($A\beta$) gives the following properties to a feedback amplifier:

1. The closed-loop gain G is primarily determined by the ratio α/β and is insensitive to changes in the open-loop gain A.
2. The closed-loop output resistance of the amplifier R_o' is very small.
3. The op-amp input current i_i is small, because R_i is large and Δv is small.

EXERCISES

See Problem 6.5 of the PSpice manual.

1. Use the method of node voltages to calculate the closed-loop gain of the circuit shown in Fig. 6.5. The open-loop input resistance is $R_i = 10$ kΩ, the open-loop output resistance is $R_o = 100$ Ω, and the open-loop gain is $A = 100$. The resistance values are $R_a = 1$ kΩ, $R_b = 9$ kΩ, and $R_L = 100$ Ω.

2. For the circuit of Fig. 6.5, assume that $R_o = 0$ and $R_i = \infty$. Use the voltage divider ratio to calculate β, and use Eq. (6.12) to calculate the closed-loop gain of the amplifier described in Exercise 1. Compare your result with that obtained in Exercise 1. (The loop gain is rather small, so your approximation may not be accurate.)

3. Repeat Exercise 2 with the use of Eq. (6.21).

4. Use the results of Exercise 1 to calculate the closed-loop output resistance of the amplifier as seen by R_L (neglect the current through R_b).

5. Use Eq. (6.24) to estimate the closed-loop output resistance of the amplifier described in Exercise 1. Compare your result with that obtained in Exercise 4.

6. Repeat Exercise 5 with the use of Eq. (6.25).

7. Use the results of Exercise 1 to calculate the closed-loop input resistance R_i of the amplifier.

8. Use Eq. (6.29) to estimate the closed-loop input resistance R_i of the amplifier described in Exercise 1. Compare your result with that obtained in Exercise 7.

6.3 Simplified Op-Amp Model

Analysis of circuits with feedback using the op-amp model of Fig. 6.4 becomes quite involved unless a computer program, such as PSpice, is used. We can reduce the labor involved and obtain useful results by simplifying our op-amp model.

In feedback circuits with large loop gain $A\beta$, the approximations

$$R_o = 0 \tag{6.30}$$

and

$$R_i = \infty \tag{6.31}$$

are reasonable, and the op-amp model of Fig. 6.5 reduces to a voltage-controlled voltage source. We will now repeat Example 6.1 using this simplified model.

EXAMPLE 6.2

See Problem 6.2 of the PSpice manual.

Determine the output voltage for the noninverting op-amp circuit shown in Fig. 6.6. The simplified op-amp model is used.

FIGURE 6.6
A noninverting amplifier circuit using an op-amp with negative feedback (simplified op-amp model shown)

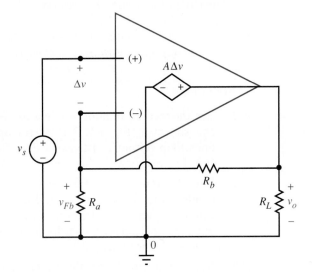

■ SOLUTION

The voltage divider relation gives

$$v_{Fb} = \frac{R_a}{R_a + R_b} v_o = \beta v_o \tag{6.32}$$

where
$$\beta = \frac{R_a}{R_a + R_b} \tag{6.33}$$

KVL easily gives
$$\Delta v = v_s - v_{Fb} = v_s - \beta v_o \tag{6.34}$$

[Equation (6.34) clearly shows that the feedback ratio β is the fraction of the output voltage v_o fed back and subtracted from the input voltage v_s to provide the error voltage Δv input to the op amp.] The control equation gives

$$\Delta v = \frac{1}{A} v_o \tag{6.35}$$

Substitution of Eq. (6.35) into Eq. (6.34) gives

$$\frac{1}{A} v_o = v_s - \beta v_o \tag{6.36}$$

which can be solved for
$$v_o = \frac{A}{A\beta + 1} v_s \tag{6.37}$$

This is, of course, the result given in Eq. (6.16). ■

As the loop gain becomes large, the last equation in the above example becomes

$$v_o = \frac{1}{\beta} v_s \tag{6.38}$$

which is equivalent to Eq. (6.21). Although we have greatly reduced the labor involved in analyzing op-amp circuits by the introduction of the simplified model, we will carry the simplification further in the next section.

REMEMBER

An approximate analysis of op-amp circuits can be performed if we assume that the op-amp input resistance R_i is infinite and that the op-amp output resistance R_o is zero. This gives a simplified op-amp model that is a voltage-controlled voltage source with gain A.

EXERCISES

⊞ See Examples 3.3 and 3.4, and Problems 6.1 and 6.5 of the PSpice manual.

An *inverting amplifier* circuit, which includes a simplified op-amp model, is shown in the following illustration.

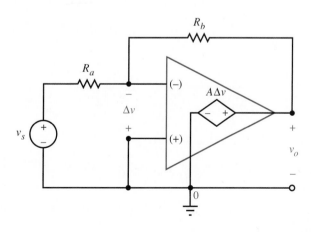

9. Determine the closed-loop voltage gain, $G = v_o/v_s$, for $R_a = 1\ \Omega$, $R_b = 10\ \Omega$, and $A = 100$.

10. Use the resistance values given in Exercise 9 and find the limit of G as the open-loop gain A becomes infinite. From this value, infer the feedback ratio β.

11. Calculate α and β as defined by Eq. (6.13). (Use superposition. Set $v_o = 0$ and calculate α. Then set $v_s = 0$ and calculate β.) Use these values of α and β to calculate the closed-loop gain, as given by Eq. (6.16).

6.4 The Ideal Op Amp

We have seen that for a sufficiently large open-loop gain A, the loop gain $A\beta$ will be large and the effect of the open-loop input resistance R_i and open-loop output resistance R_o is small. We will take advantage of this by introducing the *ideal op-amp* model. The parameters of the op-amp model of Fig. 6.4 are

---

DEFINITION

Ideal Op-Amp Parameters

$$R_o = 0 \qquad (6.39)$$

$$R_i = \infty \qquad (6.40)$$

$$A = \infty \qquad (6.41)$$

The ideal op-amp model is effectively a voltage-controlled voltage source with infinite voltage gain. We will use the ideal op-amp symbol shown in Fig. 6.7.

COMPONENT SYMBOL

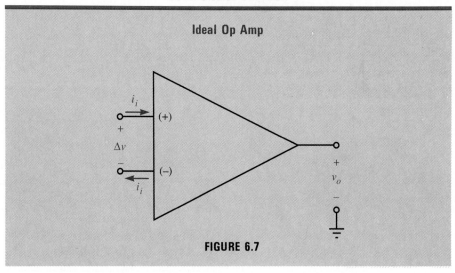

FIGURE 6.7

From Eqs. (6.28) and (6.29), the definition of the ideal op amp gives

Ideal Op-Amp Characteristics

$$\Delta v = 0 \qquad (6.42)$$

$$i_i = 0 \qquad (6.43)$$

A word of caution is in order:

CAUTION

We must exercise care when applying KCL with the ideal op-amp symbol, because the current to ground i_g in Fig. 6.4 is not shown in Fig. 6.7. It is safest to show the ground connection if KCL is applied to a surface enclosing the op-amp symbol.

We will now apply the ideal op-amp model to the analysis of two common op-amp circuits.

EXAMPLE 6.3

A noninverting voltage amplifier that uses feedback in conjunction with an operational amplifier is shown in Fig. 6.8. Use the ideal op-amp model to determine the closed-loop voltage gain v_o/v_s.

FIGURE 6.8
Noninverting voltage amplifier using an op amp with negative feedback (ideal op-amp model used)

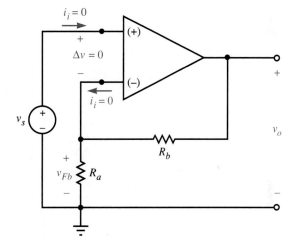

■ SOLUTION

Resistances R_a and R_b are effectively in series, because $i_i = 0$, so we can use the voltage divider relation to give

$$v_{Fb} = \frac{R_a}{R_a + R_b} v_o$$

(The subscript *Fb* designates "feedback." That is, v_{Fb} is the part of v_o fed back to the inverting input terminal.) The ideal op-amp assumption ($\Delta v = 0$) and KVL give

$$v_s = v_{Fb}$$

Because current $i_i = 0$, the voltage divider relation gives

$$v_s = \frac{R_a}{R_a + R_b} v_o$$

and the closed-loop voltage gain is

$$\frac{v_o}{v_s} = \frac{R_a + R_b}{R_a}$$

As shown in Section 6.2, this approximation is accurate as long as the loop gain is much greater than 1. (This is equivalent to the statement that the open-loop gain A is much greater than the closed-loop gain.) ■

From the preceding example we can see that an equivalent circuit for the non-inverting op-amp circuit of Fig. 6.8 is shown in Fig. 6.9.

FIGURE 6.9
An equivalent circuit for the noninverting amplifier of Fig. 6.8

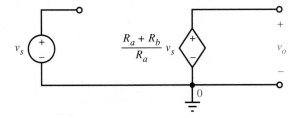

EXAMPLE 6.4

Use the ideal op-amp model to determine the closed-loop voltage gain of the *inverting voltage amplifier* of Fig. 6.10.

FIGURE 6.10
Inverting amplifier using an op amp with negative feedback

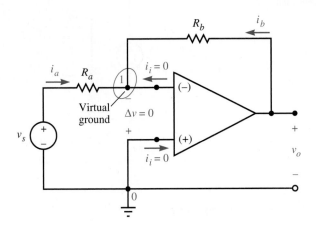

■ **SOLUTION**

The ideal op-amp model gives

$$v_{10} = -\Delta v = 0$$

Node 1 is said to be a *virtual ground*, because it is at the same voltage as the reference (ground). Thus

$$i_a = \frac{1}{R_a} v_s$$

and

$$i_b = \frac{1}{R_b} v_o$$

From KCL for the indicated closed surface,

$$-i_a - i_b = 0$$

Substitution from the two previous equations gives

$$-\frac{1}{R_a} v_s - \frac{1}{R_b} v_o = 0$$

Thus the closed-loop gain is

$$\frac{v_o}{v_s} = -\frac{R_b}{R_a}$$

Once again, the closed-loop gain is independent of the open-loop gain, if the open-loop gain is much greater than the closed-loop gain. ∎

From the preceding example, we can see that an equivalent circuit for the inverting op-amp circuit of Fig. 6.10 is given in Fig. 6.11.

FIGURE 6.11
An equivalent circuit for the inverting amplifier of Fig. 6.10

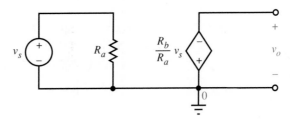

We can often simplify the analysis of networks containing two or more op amps by replacing portions of the network with equivalent circuits in this manner.

REMEMBER

The ideal op-amp model assumes zero output resistance ($R_o = 0$), infinite input resistance ($R_i = \infty$), and infinite open-loop gain ($A = \infty$). This implies that the error voltage is zero ($\Delta v = 0$) and the input current is zero ($i_i = 0$).

EXERCISES

12. Use the ideal op-amp model ($R_i = \infty$, $R_o = 0$, and $A = \infty$) to find an expression that relates v_o to v_1 and v_2 in the following circuit. What function does the circuit perform when $R_1 = R_2 = R_3$?

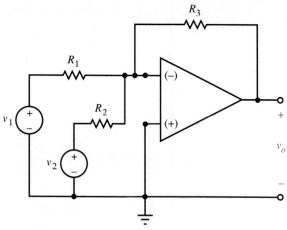

13. Construct an equivalent circuit for the preceding op-amp circuit, as was done in Figs. 6.9 and 6.11.

⬇ See Problem 6.6 of the PSpice manual.

14. Use the ideal op-amp model to find an expression for the voltage v_o for the following network. What operation does the circuit perform when $R_1 = R_a$ and $R_2 = R_b$?

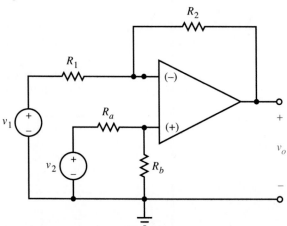

15. Construct an equivalent circuit for the preceding op-amp circuit, as was done for Figs. 6.9 and 6.11.

16. Use the ideal op-amp model to find an expression that relates v_o to v_s in the following circuit.

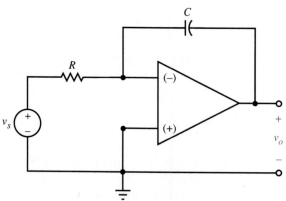

17. Interchange the location of R and C in the above circuit. What function does the circuit now perform? (This configuration is very sensitive to electrical noise and is seldom used.)

6.5 Node-Voltage Equations with Ideal Op Amps

We can easily apply the method of analysis by node voltages to networks containing op amps. If we use the op-amp model of Fig. 6.4, we simply replace the op amp by the chosen model. The network can then be analyzed by our standard procedures. If we use the ideal model of Fig. 6.7, we leave the op-amp symbol in the network diagram. We adapt our method of node-voltage analysis to the ideal op-amp model by using the following information:

1. The current into each input terminal of the ideal op amp is zero.
2. The voltage difference between the two input terminals of the ideal op amp is zero.
3. The output terminal of the ideal op-amp model is connected directly to the reference node by a voltage-controlled voltage source (with infinite gain).

The method is effectively demonstrated by the following example.

EXAMPLE 6.5

Calculate the node voltages of the network shown in Fig. 6.12.

FIGURE 6.12
A difference amplifier

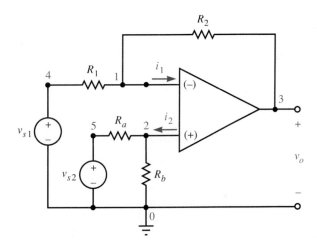

■ SOLUTION

First number the nodes as shown and indicate on the network diagram that the op-amp input currents and voltage difference are zero. The number of node-voltage equations is five, which is one less than the number of nodes. There will be two KCL equations, one for each node (except the reference node) that is not connected to a voltage source. (The output node, node 3, is connected to a voltage-controlled voltage source of infinite gain that is internal to the ideal op-amp symbol.) There is one KVL equation for each voltage source and one that includes the op-amp input voltage difference (Δv) of zero.

The KCL equations are

$$\frac{1}{R_1}(v_1 - v_4) + \frac{1}{R_2}(v_1 - v_3) = 0 \qquad \text{because } i_1 = 0 \tag{6.44}$$

$$\frac{1}{R_a}(v_2 - v_5) + \frac{1}{R_b}v_2 = 0 \qquad \text{because } i_2 = 0 \tag{6.45}$$

(Node 3, the output node, is connected to the reference node by a voltage-controlled source, so we do not write a KCL equation for a surface enclosing this node.) The KVL equations are

$$v_4 = v_{s1} \tag{6.46}$$

$$v_5 = v_{s2} \tag{6.47}$$

$$v_2 = v_1 \qquad \text{because } \Delta v = 0 \tag{6.48}$$

Actually, we would typically make use of the three KVL equations when writing the KCL equations (our shortcut procedure for analysis by node voltages), in which case we would write

$$\frac{1}{R_1}(v_1 - v_{s1}) + \frac{1}{R_2}(v_1 - v_3) = 0 \tag{6.44a}$$

$$\frac{1}{R_a}(v_1 - v_{s2}) + \frac{1}{R_b}v_1 = 0 \tag{6.45a}$$

These last two equations can easily be solved to give

$$v_3 = \frac{R_1 + R_2}{R_1} \cdot \frac{R_b}{R_a + R_b} v_{s2} - \frac{R_2}{R_1} v_{s1} \quad \blacksquare$$

REMEMBER

We can use node-voltage equations to analyze feedback circuits when the ideal op-amp model is assumed. The assumptions are

1. The op-amp input current is zero ($i_1 = i_2 = 0$).
2. The error voltage is zero ($\Delta v = 0$).
3. The output terminal is connected to the reference node by a voltage-controlled voltage source with infinite gain.

We can use the shortcut procedure (as outlined in Chapter 4) for writing node-voltage equations. This eliminates the KVL equations.

EXERCISES

See Problem 6.4 of the PSpice manual.

18. Use the method of node voltages to solve for the output voltage of the op-amp circuit at the top of page 263.

19. Construct an equivalent circuit for the following op-amp circuit.

See Problem 6.4 of the PSpice manual.

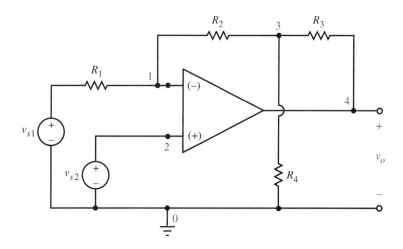

6.6 Summary

We introduced a network model for an op amp and used it in a feedback-amplifier circuit. We saw that as long as the loop gain $A\beta$ was large, the closed-loop gain G was primarily determined by the ratio α/β. We used this to simplify the op-amp model by assuming that the open-loop input resistance R_i was infinite and the open-loop output resistance R_o was zero. We further simplified our analysis by assuming that the open-loop gain A was infinite, which gave us the ideal op-amp model.

KEY FACTS

◆   The analysis presented assumes that the op amp operates in its linear region.

◆   Op amps are always used with feedback for linear operation.

◆   The closed-loop gain of an op-amp amplifier circuit is

$$G = \frac{\alpha A}{A\beta + 1}$$

where A is the effective open-loop gain, β is the feedback factor, and α is the fraction of the input signal applied to the input terminals with output voltage set equal to zero.

◆   The closed-loop output resistance is approximately

$$R'_o = \frac{1}{A\beta} R_o$$

where R_o is the open-loop output resistance and A is the open-loop gain.

◆   For large loop gain $A\beta$, we can simplify the op-amp model by assuming that the input resistance R_i of the op amp is infinite and the output resistance R_o is zero.

◆ If the loop gain is much greater than 1, we can use the ideal op-amp model; that is,

$$R_i = \infty$$

$$R_o = 0$$

$$A = \infty$$

This implies that the error voltage is

$$\Delta v = 0$$

and the op-amp input current is

$$i_i = 0$$

◆ The ideal op-amp model can be used for analysis by node voltages.

■ PROBLEMS ■

Section 6.1

1. We must measure the parameters of an op amp to determine the model as shown in Fig. 6.4. The op amp is connected to a power source as shown in Fig. 6.2. The output voltage and current are measured to be zero when both input terminals are connected to the reference. With an error voltage Δv of 1 mV, the open-circuit output voltage v_o is 5 V, and the input current is 0.1 mA. Connection of the output terminal to the reference through a resistance value of 100 Ω reduces the output voltage to 4 V. Determine the values R_i, R_o, and A for our op-amp model. Is this a practical method to measure the op-amp parameters when the open-loop gain is 100,000 or more?

Section 6.2

An inverting amplifier circuit is shown with the op-amp model in Fig. 6.13. The component values are $R_i = 10$ kΩ, $R_o = 100$ Ω, $A = 100$, $R_a = 1$ kΩ, $R_b = 10$ kΩ, and $R_L = 100$ Ω.

2. Determine the closed-loop voltage gain $G = v_o/v_s$, the closed-loop input resistance $R_i' = v_s/i_s$, and the output resistance seen by the load resistance R_L when the component values are as given above. Will the closed-loop input resistance R_i' change if the load resistance R_L is disconnected?

3. Repeat Problem 2 if the open-loop gain is $A = 1000$.

4. Use PSpice to solve Problem 2.

5. Use PSpice to solve Problem 3.

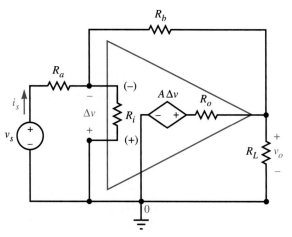

FIGURE 6.13

Section 6.3

The op-amp circuit shown below with the simplified op-amp model is used in Problems 6 through 9.

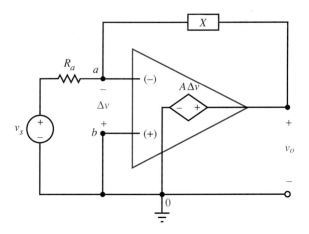

6. If the component X in the above circuit is a capacitance of value C, write a differential equation that relates the output voltage v_o to the input voltage v_s. What function does this circuit perform as the open-loop voltage gain A becomes infinite?

7. Repeat Problem 6 if component X is an inductance of value L. (This is not a practical circuit, because it is very sensitive to electrical noise.)

8. Box X in the preceding circuit represents a resistance of value R_b in parallel with a capacitance of value C. Write a differential equation that relates v_o to v_s. Determine the differential equation as the open-loop gain A becomes infinite.

9. Replace the voltage source shown in the preceding circuit with a short circuit, and insert the voltage source [(+) reference at the top] in place of the short circuit that connects terminal b to the reference. Then repeat Problem 6.

Problems 10 through 12 refer to the circuit in Fig. 6.14.

10. Develop an equivalent circuit for resistances R_1 and R_2 and the associated op amp. This should consist of a resistance R_{i1} and a voltage-controlled voltage source.

11. Develop an equivalent circuit for resistances R_3 and R_4 and the associated op amp. This should consist of a resistance R_{i2} and a controlled voltage source.

12. Combine the equivalent circuits obtained in Problems 10 and 11. Use this to determine voltage v_o as a function of the input voltage v_{s1}.

Section 6.4

13. Use the ideal op-amp model and repeat Problem 7.

14. Use the ideal op-amp model and repeat Problem 8.

15. Use the ideal op-amp model and repeat Problem 9.

16. Use the ideal op-amp model and repeat Problem 10.

17. Use the ideal op-amp model and repeat Problem 11.

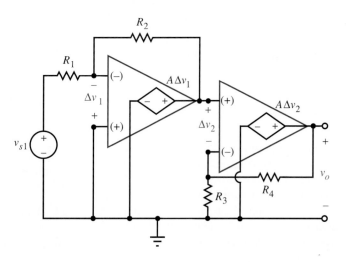

FIGURE 6.14

18. Use the ideal op-amp model and repeat Problem 12.

19. The inverting amplifier circuit shown below is used in an electronic instrument, called a digital multimeter, for measuring resistance. The constant voltage V_{ref} and the resistance R_{ref} are known. The voltage V_o is a measure of the unknown resistance R_x. Determine V_o as a function of V_{ref}, R_{ref}, and R_x.

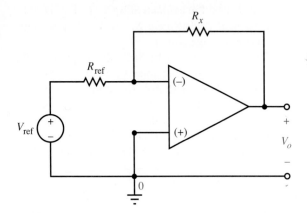

20. Use the ideal op-amp model to find the output current i for the following circuit. This is a *voltage-to-current converter*. Draw an equivalent circuit that consists of the independent source, a dependent source, a resistance, and the load device. Can you think of a practical limitation for the actual op-amp circuit?

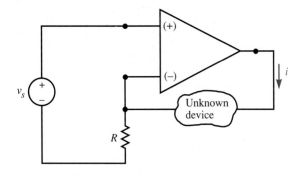

21. Use an ideal op-amp model and two resistances to design an inverting amplifier with an input resistance of 10 kΩ and a voltage gain of −5.

22. In the following figure, replace the circuit to the left of terminals a and b with its Thévenin equivalent circuit, and calculate the power absorbed by the 2-kΩ resistance.

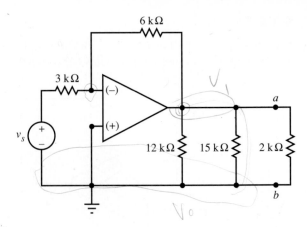

23. The op-amp circuit connected to the load in the following illustration is an example of a *negative converter*, or NC. Use the ideal op-amp model to determine the functional relationship between voltage v and current i,

$$v = f(i)$$

for this network, if the functional relationship between v_L and i_L is

$$v_L = g(i_L)$$

or $$i_L = g^{-1}(v_L)$$

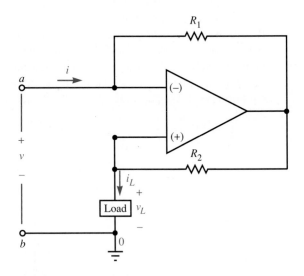

What is the equivalent circuit connected between terminals a and b in the preceding circuit if the load is

24. A resistance of value R?

25. An inductance of value L?

26. A capacitance of value C?

27. Use the ideal op-amp model and analysis by node voltages to determine the output of the *difference amplifier* shown in Fig. 6.15.

Write a set of node-voltage equations for the following two networks. (Do not solve the equations.)

28.

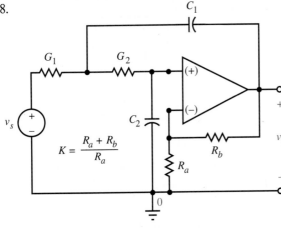

29.

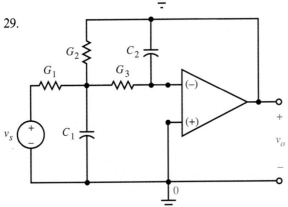

Problems 30 through 32 refer to the following circuit.

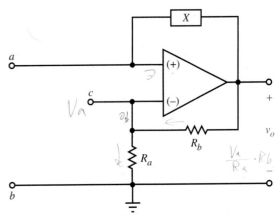

30. Determine the equivalent circuit looking into terminals a and b of the op-amp circuit shown above when component X is a resistance of value R. Also determine v_o as a function of v_{ab}.

31. Determine the equivalent circuit looking into terminals a and b of the preceding circuit if component X is a capacitance of value C.

32. Connect a resistance of value R between terminals a and b in the preceding op-amp circuit. What is the equivalent component connected between terminals c and b if component X is a resistance of value R_2? Also, determine v_o as a function of v_{cb}. What combination of resistance values will cause the resistance looking into terminals c and b to be negative? This can cause a problem, as we shall see in Chapter 8.

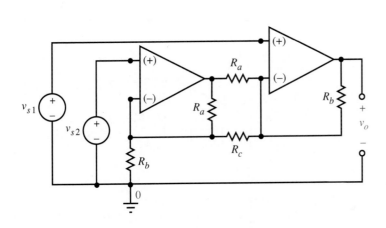

FIGURE 6.15

33. Calculate the output voltage v_o of the following difference amplifier.

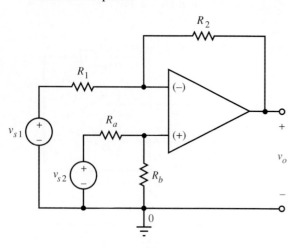

34. The difference amplifier shown above must have an output of

$$v_o = 2(v_{s2} - v_{s1})$$

and the resistance seen by each source must be 1 kΩ. Determine the values of R_1, R_2, R_a, and R_b.

35. Design a circuit with a single op amp that will have an output voltage that is directly proportional to an unknown conductance G. A reference voltage of 1 V is available, and the circuit must provide an output voltage with a magnitude of 1 V (± 1 V) when the conductance has a value of 1 mS.

36. Determine the output voltage V_o for the following op-amp circuit.

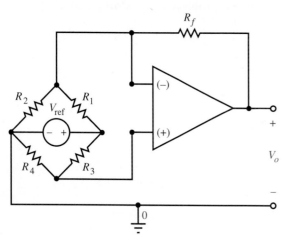

37. Apply the ideal op-amp model and determine the closed-loop gain $G = v_o/v_s$ of the following

circuit. (Observe the relative location of the inverting and noninverting terminals.) Although your analysis has yielded an answer, this result is wrong. The op amp will not function properly in this mode. To obtain an insight into the problem, work the following problem.

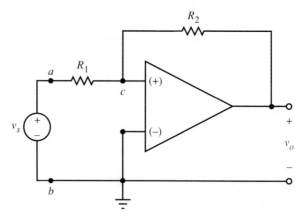

38. Use the simplified op-amp model of Section 6.3 to calculate the Thévenin resistance looking into terminals a and b (with the independent source disconnected) of the above circuit. Then calculate the Thévenin resistance looking into terminals c and b with the source connected. For what values of A is either of these resistances negative? We shall see in Chapter 8 that a negative resistance can lead to an unstable circuit (one that has an output that increases with time with zero input).

39. Repeat Problem 23 for the following circuit. (This circuit performs the function of a *gyrator*.)

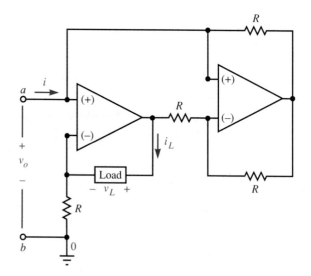

What is the equivalent circuit connected between terminals *a* and *b* in the preceding circuit if the load is

40. A resistance of value R?
41. An inductance of value L?
42. A capacitance of value C?
43. Determine the relationship between voltage v and current i in the circuit in Fig. 6.16. What is the relationship between voltage v and current i_L?

What is the equivalent circuit connected between terminals *a* and *b* in the circuit in Fig. 6.16 if the load is

44. A resistance of value R?
45. An inductance of value L?
46. A capacitance of value C?
47. Use superposition on the circuit of Problem 33. Calculate α and β as defined by Eq. (6.13). (There will be two values for α.) Then calculate the output voltage using Eq. (6.15).

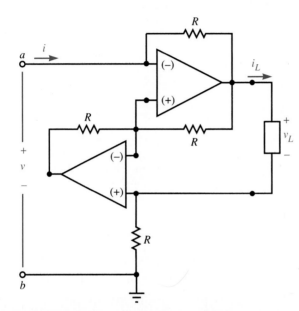

FIGURE 6.16

CHAPTER
7

Signal Models and
Signal Characterization

Electrical engineers often use the term *signal* or *waveform* to denote a voltage $v(t)$ or current $i(t)$ of interest in a physical circuit. Just as there is a mathematical model for the physical circuit the engineer is inspecting, so there are mathematical models for the signals (voltage and current) in the physical circuit. We find it convenient to characterize these mathematical signal models in various ways. This chapter describes several signal models and signal characterizations that are useful in electrical engineering. We will use these signal models extensively in the remainder of the text.

7.1 DC and Related Signals

A *dc* voltage or current waveform is one that is constant for all time, as illustrated in Fig. 7.1. (The term *dc* is applied to both voltage and current in spite of the fact that it is an abbreviation for *direct current*.)

FIGURE 7.1
A constant or dc voltage

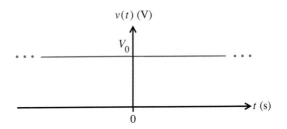

A dc voltage source of 9 V and a series resistance form an approximate model for a 9-V transistor radio battery.

The mathematical expression for the dc voltage of Fig. 7.1 is

---

DEFINITION

DC Signal

$$v(t) = V_0 \qquad -\infty < t < \infty \qquad (7.1)$$

A dc source is obviously not an exact model for any physical device. Nothing has been constant since the beginning of time. The unit step, discussed in the following paragraph, lets us include a starting time.

The Unit Step

See Problem 7.1 in the PSpice manual.

A waveform closely related to dc is one that is zero for $t < t_0$ and suddenly becomes some nonzero constant for $t > t_0$, as shown in Fig. 7.2.

FIGURE 7.2
A suddenly applied voltage

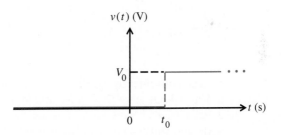

An abrupt shift in voltage or current between two possible values occurs frequently in the circuits of digital computers. We can mathematically describe the waveform shown in Fig. 7.2 in terms of the *unit step function*, written *u(t)*:

[The value of $u(0)$ is variously defined to be 0, $\frac{1}{2}$, or 1, or it is left undefined. The only restriction we need is that $|u(0)| < \infty$. Then the value of $u(0)$ will not influence any results obtained in this book or the analysis of any physical system.] You should sketch $u(t)$. Later in the book, when we use $u(t)$ in writing a voltage or current, we are referring to the unit step function. Observe that the transition or "step" occurs when the argument of the unit step changes from negative to positive. We can write the voltage depicted in Fig. 7.2,

$$v(t) = \begin{cases} 0 & t < t_0 \\ V_0 & t > t_0 \end{cases} \qquad (7.3)$$

in terms of the unit step function as

$$v(t) = V_0 u(t - t_0) \qquad (7.4)$$

Again, observe that the step occurs when the argument is zero.

The Unit Pulse

See Problems 7.2 and 7.3 in the PSpice manual.

Another common waveform, particularly in digital circuits, is a voltage or current that suddenly changes from zero to some value and back to zero after a short time. Consider the *unit pulse function*, written $\Pi(t)$:

DEFINITION

Unit Pulse Function

$$\Pi(t) = \begin{cases} 1 & |t| < \dfrac{1}{2} \\ 0 & |t| > \dfrac{1}{2} \end{cases} \qquad (7.5)$$

We can conveniently express this function as a sum of unit step functions, as shown in Fig. 7.3. Observe that these shifted unit step functions are 0 when the argument is negative and have value 1 when the argument is positive. We can easily time-scale and shift the unit pulse. For instance, in the equation

$$v(t) = \Pi\left(\frac{t - t_0}{T}\right) \qquad (7.6)$$

t_0 represents the center of the pulse (the argument is 0 for $t = t_0$) and T represents the pulse duration ($|t - t_0|/T = \frac{1}{2}$ for $t = t_0 \pm T/2$).

FIGURE 7.3
The unit pulse function
written as a sum of unit
step functions. (*a*) Unit
step advanced by $\frac{1}{2}$ s;
(*b*) the negative of a unit
step that is delayed by
$\frac{1}{2}$ s; (*c*) the sum of the
unit steps in *a* and *b*

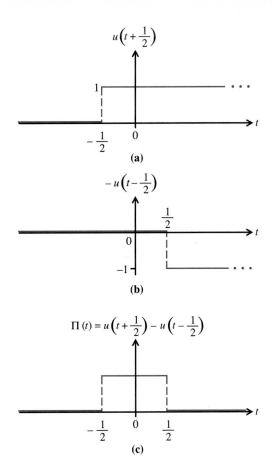

(a)

(b)

(c)

We can model many other waveforms occurring in physical systems in terms of appropriately scaled and time-shifted unit steps or pulses.

EXAMPLE 7.1

Write the following function with the use of the unit step function.

$$i(t) = \begin{cases} 0 & t < 0 \\ I_m \cos \omega_0 t \text{ A} & t > 0 \end{cases}$$

■ SOLUTION

$$i(t) = (I_m \cos \omega_0 t)u(t) \text{ A}$$

You should sketch $I_m \cos \omega_0 t$ and $u(t)$, and verify that the product is equal to $i(t)$ for any time t. (We will not concern ourselves with the value of $i(t)$ at $t = 0$, where the discontinuity occurs.) ■

EXAMPLE 7.2

The voltage across an inductance of L henries is given by

$$v = \begin{cases} 0 \text{ V} & t < 0 \text{ s} \\ V_0 \text{ V} & t \geq 0 \text{ s} \end{cases}$$

Write the voltage v and the inductance current i in terms of the unit step (assume the passive sign convention).

■ **SOLUTION**

We can easily write the voltage v in terms of the unit step:

$$v = V_0 u(t) \text{ V}$$

The inductance current is the integral of the voltage divided by the inductance:

$$i = \frac{1}{L} \int_{-\infty}^{t} v(\tau)\, d\tau = \frac{1}{L} V_0 \int_{-\infty}^{t} u(\tau)\, d\tau$$

$$i = \frac{1}{L} V_0 t u(t) \text{ A}$$

We can visualize the preceding integral by sketching the function $u(\tau)$. Select some time t on the τ axis. The area under the unit step to the left of time t is zero, if time t is less than zero. The area under the unit step to the left of time t is t, if time t is greater than zero. The integral of the unit step is often defined to be the *unit ramp function*:

$$r(t) = t u(t)$$

(You should sketch the unit ramp function.) The unit ramp is widely used in control system modeling, where, for example, a shaft angle can increase linearly with time as the shaft rotates. We will not need the unit ramp for our circuit analysis. ■

EXAMPLE 7.3

The voltage across an inductance of L henries is

$$v = V_0 u(t - t_0)$$

Determine the inductance current (assume the passive sign convention).

■ **SOLUTION**

We could determine the result by integration, but we recognize that this is simply the problem of Example 7.2 with the input delayed by t_0 s. Because the inductance is time-invariant, the current i is the same as the current obtained in Example 7.2, but delayed by t_0 s:

$$i = \frac{1}{L} V_0 (t - t_0) u(t - t_0)$$

$$= \frac{1}{L} V_0 r(t - t_0) \text{ A}$$

You should sketch this output current. ■

The dc waveform, the unit step, and the unit pulse are idealized mathematical *models* of some commonly occurring physical waveforms. The dc waveform is a model, because no physical waveform is constant forever. The unit step and unit pulse are models, because no physical signal can change its value by a finite amount instantaneously.

FIGURE 7.4
(a) Network with a switched source; (b) Thévenin equivalent for $t < t_0$; (c) Thévenin equivalent for $t > t_0$

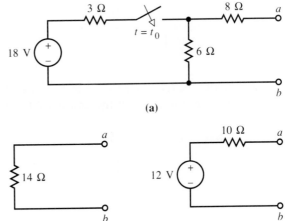

(a)

(b)

(c)

FIGURE 7.5
A unit step model for the switched source of Fig. 7.4a. (a) Unit step model; (b) Thévenin equivalent for $t < t_0$; (c) Thévenin equivalent for $t > t_0$; (d) Thévenin equivalent for all t

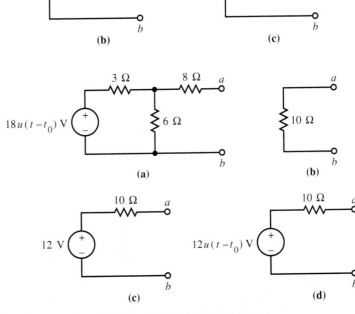

(a)

(b)

(c)

(d)

We often use the unit step to model a source that is switched on or off at some time t_0. Although this representation is useful, we should apply it with care. For example, suppose that an engineer proposed the network of Fig. 7.5a as a model for the network of Fig. 7.4a. Although this proposal may appear accurate at first glance, it contains a trap: The Thévenin equivalent for the two networks is the same for $t > t_0$, but different for $t < t_0$.

REMEMBER

A dc signal is constant for all time. A unit step is 0 when the argument is negative and has a value of 1 when the argument is positive. The unit pulse is centered about the point in time where the argument is zero. The unit pulse has a value of 1 when the argument has an absolute value less than $\frac{1}{2}$ and is 0 for all other values.

EXERCISES

Accurately sketch the functions given in Exercises 1 and 2 and write in terms of the unit step.

1.
$$i(t) = \begin{cases} 0 & t \leq 2 \\ 2 & 2 < t \leq 5 \\ 0 & t > 5 \end{cases}$$

2.
$$v(t) = \begin{cases} 0 & t \leq 0 \\ 1 & 0 < t \leq 1 \\ 2 & 1 < t \leq 3 \\ 5 & t > 3 \end{cases}$$

⬛ PWL and PULSE sources; see Sections A.3.3.1 and A.3.3.4, and Problems 7.1 and 7.2 of the PSpice manual.

Write the functions shown in Exercises 3 and 4 in terms of the unit step function, and then in terms of the unit pulse function.

3.

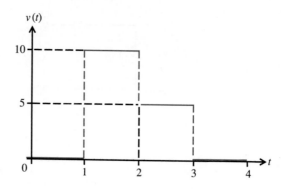

4.

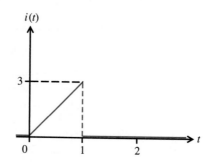

Draw the functions given in Exercises 5 through 10.

5. $u(t-1) + 2u(t-2) - u(t-3)$ 6. $2u(\sin \pi t)$

7. $3u(5-t)$ 8. $2tu(t-1)$

9. $2(t-1)u(t-1)$ 10. $2\Pi[(t-3/4)]$

The current in amperes given in Exercises 11 and 12 is applied to a 1-F capacitance. Calculate and sketch the resulting voltage.

11. $u(t-2)$ 12. $\Pi(t)$

7.2 Delta Functions*

⬛

In later chapters we will need to consider pulses that have a duration much shorter than the natural response time of our circuit (a mechanical analogy is a hammer blow that rings a bell). We will idealize a very short pulse with the *unit impulse* or *delta function*, $\delta(t)$.

---

* This material will not be needed until Chapter 15 and can be postponed.

Unit Impulse or Delta Function

The **unit impulse** or **delta function** $\delta(t)$ is defined by the integral relation

$$\int_{-\infty}^{\infty} f(t)\delta(t)\,dt = f(0) \tag{7.7}$$

for every function $f(t)$ that is defined and continuous[†] at $t = 0$.

The theory of the delta function is developed in a branch of mathematics known as the theory of distributions, but the properties of a delta function can be justified by treating $\delta(t)$ as a very short duration pulse of unit area. The idea of the delta function being a model for a very short duration pulse of unit area can be justified by considering the unit area pulse $(1/T)\Pi(t/T)$, as shown in Fig. 7.6a. It is easy to see that if T is small enough, Eq. (7.7) is approximated by

$$\int_{-\infty}^{\infty} \frac{1}{T} \Pi\left(\frac{1}{T}\right) f(t)\,dt = \int_{-T/2}^{T/2} \frac{1}{T} f(t)\,dt \tag{7.8a}$$

$$\simeq \int_{-T/2}^{T/2} \frac{1}{T} f(0)\,dt \tag{7.8b}$$

$$= \frac{1}{T} f(0)t \Big|_{-T/2}^{T/2} \tag{7.8c}$$

$$= f(0) \tag{7.8d}$$

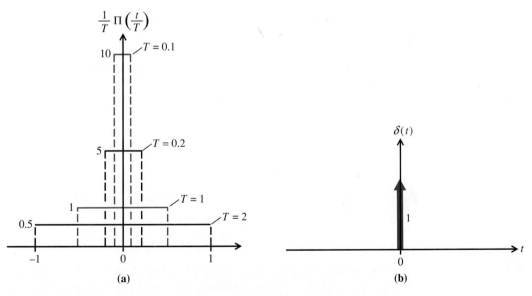

FIGURE 7.6 Approximation of the impulse or delta function by a rectangular pulse. (*a*) Pulses of unit area; (*b*) the impulse or delta function

[†] If $f(t)$ is discontinuous but bounded at $t = 0$, the integral in Eq. (7.7) is bounded, but we will leave the value undefined.

Notice that the step from Eq. (7.8a) to (7.8b) is valid whenever $f(t)$ is defined and continuous at $t = 0$, because under these conditions $f(t \pm (T/2)) \simeq f(t)$ for $T \simeq 0$. As long as the pulse $(1/T)\Pi(t/T)$ is so short that $f(t)$ is approximately constant for the pulse duration, a unit area pulse approximates an impulse. In the limit $T \to 0$, the step from Eq. (7.8a) to (7.8b) holds with equality. This limiting argument gives an *alternative definition* of the delta function:

$$\int_{-\infty}^{\infty} \delta(t)\,dt = 1 \tag{7.9}$$

and

$$\delta(t) = 0 \qquad t \neq 0 \tag{7.10}$$

The delta function is represented graphically as shown in Fig. 7.6b. We will use delta functions extensively, after we introduce Fourier transforms and Laplace transforms in Chapters 15 and 16, as a model for short-duration pulses.

Properties of the Delta Function

Several useful properties of the delta function are easily established. The delta function $\delta(t - t_0)$ is nonzero only for $t = t_0$. When we integrate the product of $\delta(t - t_0)$ and a function $f(t)$, we obtain only the value of $f(t)$ at time t_0. This *sifts out* the value of $f(t)$ at $t = t_0$ or *samples* $f(t)$ at time t_0. A simple change of variables $(\tau = t - t_0)$ lets us establish this *sifting* or *sampling property* of the unit impulse:

Sifting Property of the Unit Impulse

$$\int_{-\infty}^{\infty} f(t)\delta(t - t_0)\,dt = \int_{-\infty}^{\infty} f(\tau + t_0)\delta(\tau)\,d\tau \tag{7.11a}$$

$$= f(t_0) \tag{7.11b}$$

$$= \int_{-\infty}^{\infty} f(t_0)\delta(t - t_0)\,dt \tag{7.11c}$$

It can be shown (see Problem 89) that Eq. (7.7) implies that the impulse function is an even function:

$$\delta(t_0 - t) = \delta(t - t_0) \tag{7.12}$$

The delta function is often considered to be the derivative of the unit step,

$$\delta(t) = \frac{d}{dt}u(t) \tag{7.13}$$

If we view the unit impulse as a pulse of zero width and unit area, we can easily see that

$$\int_{-\infty}^{t} \delta(\tau)\,d\tau = \begin{cases} 0 & t < 0 \\ 1 & t > 0 \end{cases} = u(t) \tag{7.14}$$

We will use the delta function in Chapters 15 and 16 as an idealized model for physical signals. In some problems the use of the delta function will significantly simplify our work, compared with the use of a more exact model for a pulse.

REMEMBER

We can visualize a delta function as a very narrow pulse with an area of 1.

Evaluate the integrals given in Exercises 13 through 16.

13. $\displaystyle\int_{-\infty}^{\infty} (\cos 2\pi t)\delta(t)\, dt$

14. $\displaystyle\int_{-\infty}^{\infty} (\cos 2\pi t)\delta(t-1)\, dt$

15. $\displaystyle\int_{-\infty}^{\infty} (\cos 2\pi\alpha)\delta(t-\alpha)\, d\alpha$

16. $\displaystyle\int_{-\infty}^{\infty} t^3\delta(t-1)\, dt$

7.3 Exponential and Sinusoidal Signals*

In Chapter 8 we will see that a linear circuit naturally generates voltages and currents when a source changes value. The form of this natural response is determined by the value of the components and their interconnection, with all independent sources set equal to zero. We will also discover in Chapter 8 that the natural response includes exponential functions or the product of exponential functions and sines and cosines. For this reason we are very interested in exponential signals. In this section we will see that sines and cosines can also be written as exponential functions.

Exponential signals (voltages or currents) of the form

$$v(t) = V_0 e^{st} \tag{7.15}$$

or

$$i(t) = I_0 e^{st} \tag{7.16}$$

are fundamentally important in circuit theory. Three cases can occur:

1. $s = \sigma$; s is a real constant σ. $\tag{7.17}$
2. $s = j\omega$; s is an imaginary constant $j\omega$. $\tag{7.18}$
3. $s = \sigma + j\omega$; s is a complex constant having a real part σ and an imaginary part ω. $\tag{7.19}$

In Eqs. (7.18) and (7.19), the imaginary unit j is defined by

$$j = \sqrt{-1} \tag{7.20}$$

When s is real $(s = \sigma)$, the signal will have one of the forms shown in Fig. 7.7. A constant signal is just the special case of the exponential signal for $s = 0$. It may seem strange to consider complex signals, because they cannot occur in physical circuits. The reasons for our interest will be evident in Chapters 8 through 19.

FIGURE 7.7
Exponential signals for
real exponents

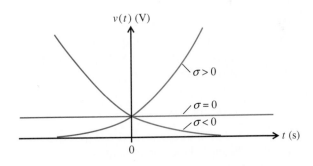

---

* This section uses concepts developed in Appendix B. It is a good idea to read Appendix B before you begin this section.

The Decaying Exponential

In Chapter 8 we will see that the natural response of a simple circuit containing a series resistance, capacitance, and voltage source is a *decaying exponential*. The decaying exponential ($\sigma < 0$) is often written

$$v(t) = V_0 e^{-t/\tau} \tag{7.21}$$

The parameter τ is called the *time constant*. This real exponential signal can be represented by its power series expansion about $t = 0$:

$$v(t) = V_0 \sum_{n=0}^{\infty} \frac{1}{n!} \left(\frac{-t}{\tau} \right)^n \tag{7.22}$$

In the neighborhood of $t = 0$, the signal can be approximated by the first two terms of its power series expansion:

$$v(t) \simeq V_0 \left(1 - \frac{t}{\tau} \right) \qquad |t| \ll \tau \tag{7.23}$$

This straight-line approximation is shown in Fig. 7.8. Notice that it intersects the vertical axis at V_0 and the horizontal axis at $t = \tau$. The value of the exponential at time $n\tau$ is

$$v(n\tau) = e^{-n}V_0 \qquad \text{for } n = 1, 2, \ldots \tag{7.24}$$

■ *EXP source; see Section A.3.3.3 and Problems 7.4 and 7.5 of the PSpice manual.*

Thus an exponentially decreasing signal decreases to 0.368 of its original value in an interval of time equal to one time constant and to 0.0067 of its initial value after five time constants. For this reason, we often say that an exponentially decreasing signal is negligible, compared with its original value, after five time constants.

FIGURE 7.8
The decaying exponential

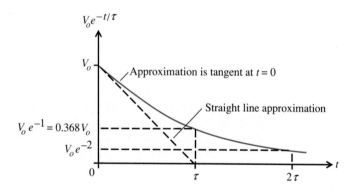

We shall see in Chapters 8 and 9 that the time constant of a circuit tells us how rapidly the transient voltage or current associated with an abrupt change in a source value dies out.

EXAMPLE 7.4

A 2-H inductance and a 3-F capacitance are connected in series. The current through these components is

$$i = 12e^{-2t} \text{ A} \qquad t \geq 0$$

Determine the voltage across the inductance and capacitance for $t > 0$, if the capacitance voltage at time zero is 0 V. What is the time constant for the voltages and currents?

■ **SOLUTION**

With the passive sign convention, the inductance voltage is

$$v_L = L\frac{d}{dt}i = 2\frac{d}{dt}(12e^{-2t})$$

$$= -48e^{-2t}\text{ V} \qquad t > 0$$

and the capacitance voltage is

$$v_C = v_C(0) + \frac{1}{C}\int_0^t i\,d\tau = 0 + \frac{1}{3}\int_0^t 12e^{-2\tau}\,d\tau$$

$$= 2 - 2e^{-2t}\text{ V} \qquad t > 0$$

The time constant for all voltages and currents is

$$\tau = \frac{1}{2}\text{ s} \quad ■$$

The Complex Exponential

Voltages and currents that have a sinusoidal variation with time (are sine or cosine functions of time) are very important in electrical engineering for several reasons:

1. The natural response of a circuit can contain sinusoidal terms.
2. Almost all electric power is generated with a sinusoidal voltage waveform.
3. Linear circuits with sinusoidal sources are easily analyzed.
4. Linear circuits with nonsinusoidal sources can be analyzed by representing each source as the sum of sinusoidal sources using Fourier series or transforms, as we will see in later chapters.

We will now show how to represent sinusoids (sines and cosines) in terms of complex exponential functions. The complex exponentials are more easily manipulated than sines and cosines, and we will use them throughout subsequent chapters.

The exponential with imaginary exponent ($s = j\omega$) is defined by a power series and is called a *complex exponential*.

DEFINITION

Complex Exponential

$$e^{j\omega t} = \sum_{n=0}^{\infty}\frac{1}{n!}(j\omega t)^n \tag{7.25}$$

Although complex signals cannot occur in physical circuits, we will show that complex exponentials are related to the trigonometric sines and cosines, and we will use them extensively in the remainder of the book. The series that represents the complex exponential can be separated into two terms, one containing even powers of j and one containing odd powers of j:

$$e^{j\omega t} = \left[\sum_{n=0}^{\infty} (-1)^n \frac{1}{(2n)!} (\omega t)^{2n} \right] + j\left[\sum_{n=0}^{\infty} (-1)^n \frac{1}{(2n+1)!} (\omega t)^{2n+1} \right] \quad (7.26)$$

Using the Taylor series expansions for the sine and cosine, we can write Eq. (7.26) as

Euler's Identity

$$e^{j\omega t} = \cos \omega t + j \sin \omega t \quad (7.27)$$

Euler's identity is of fundamental importance to linear circuit analysis and therefore should be memorized by all engineering students. The complex exponential is represented in the complex plane as shown in Fig. 7.9.

FIGURE 7.9
The complex exponential $e^{j\omega t}$

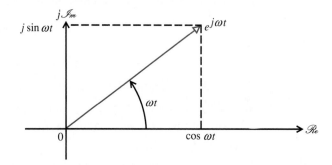

Obviously the real part* of the complex exponential is a cosine, and the imaginary part of the complex exponential is a sine.

$$\cos \omega t = \mathcal{R}e\left\{ e^{j\omega t} \right\} \quad (7.28)$$

$$\sin \omega t = \mathcal{I}m\left\{ e^{j\omega t} \right\} \quad (7.29)$$

$$\left| e^{j\omega t} \right| = 1 \quad (7.30)$$

From the result

$$e^{-j\omega t} = \cos(-\omega t) + j \sin(-\omega t)$$
$$= \cos \omega t - j \sin \omega t \quad (7.31)$$

addition or subtraction of $\frac{1}{2}e^{j\omega t}$ and $\frac{1}{2}e^{-j\omega t}$ gives

$$\cos \omega t = \frac{e^{j\omega t} + e^{-j\omega t}}{2} \quad (7.32)$$

$$\sin \omega t = \frac{e^{j\omega t} - e^{-j\omega t}}{2j} \quad (7.33)$$

---

* Refer to Appendix B for definitions of the real and imaginary operators used in Eqs. (7.28) and (7.29).

See Problem 7.6 in the PSpice manual.

A sinusoid* with arbitrary time origin and amplitude is shown in Fig. 7.10.

FIGURE 7.10
A sinusoid

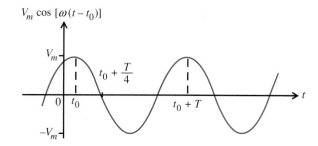

The cosine shown in Fig. 7.10 has a time delay of t_0 s and can be written as

$$V_m \cos \omega(t - t_0) = V_m \cos (\omega t - \omega t_0) \tag{7.34a}$$

$$= V_m \cos \left(\frac{2\pi}{T} t - \frac{2\pi}{T} t_0 \right) \tag{7.34b}$$

$$= V_m \cos (2\pi f t - 2\pi f t_0) \tag{7.34c}$$

$$= V_m \cos (2\pi f t + \theta) \tag{7.34d}$$

where V_m = amplitude
ω = angular frequency, rad/s
$f = \omega/2\pi$ = frequency, Hz $\qquad\qquad$ (7.35)
$T = 1/f$ = period, s $\qquad\qquad$ (7.36)
$\theta = -\omega t_0$ = phase shift, rad $\qquad\qquad$ (7.37)

CAUTION

The phase shift θ is often expressed in degrees and must be converted to radians before adding θ and ωt.

Damped Sinusoids

In Chapter 8, we will see that a signal closely related to a sinusoid, the *damped sinusoid*, occurs naturally in many circuits when they are disturbed by a suddenly changed source value. A *damped cosine* with phase shift can be written as

$$V_m e^{\sigma t} \cos (\omega t + \theta) = \mathscr{Re} \{ V_m e^{\sigma t + j(\omega t + \theta)} \}$$

$$= \mathscr{Re} \{ V_m e^{j\theta} e^{(\sigma + j\omega)t} \} \tag{7.38}$$

The undamped sinusoid is just a special case when $\sigma = 0$.

SIN source; see Section A.3.3.2 and Problems 7.6, 7.7, and 7.8 of the PSpice manual.

Refer to Fig. 7.11. We have all experienced a damped sinusoidal response in mechanical systems when a weight supported by a spring is disturbed—for example, when our car hits a bump and the shock absorbers (called dampers in England) are weak.

---

* The term *sinusoid* implies either the trigonometric sine or cosine function having arbitrary amplitude V_m, angular frequency ω, and phase θ.

FIGURE 7.11
The cosine with exponential
amplitude: (*a*) negatively
damped cosine, $\sigma > 0$;
(*b*) undamped cosine, $\sigma = 0$;
(*c*) damped cosine, $\sigma < 0$

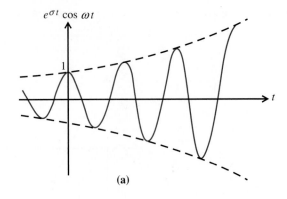

(a)

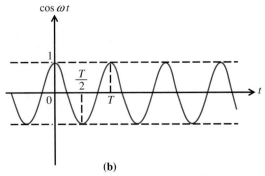

(b)

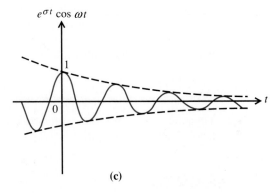

(c)

EXAMPLE 7.5

In Chapter 8 we will find that the natural response of a parallel *RLC* circuit with
$R = 2\ \Omega$, $L = 5$ H, and $C = 1$ F can be written as

$$v = (4 + j3)e^{(-1 + j2)t} + (4 - j3)e^{(-1 - j2)t} \text{ V}$$

if $v(0) = 8$ V and $d/dt(v)|_{t=0} = -20$ V/s. Use Euler's identity to write this as a func-
tion that includes the sine and cosine.

■ SOLUTION

$$v = e^{-t}[(4 + j3)(\cos 2t + j \sin 2t) + (4 - j3)(\cos 2t - j \sin 2t)]$$
$$= e^{-t}\{[(4 + j3) + (4 - j3)] \cos 2t + [(4 + j3) - (4 - j3)]j \sin 2t\}$$
$$= e^{-t}(8 \cos 2t - 6 \sin 2t) \ ■$$

A real exponential signal decays to e^{-1} (0.368) of its original value in one time constant, and to e^{-5} (0.0067) of its original value in five time constants. Euler's identity lets us write a complex exponential in terms of the trigonometric sine and cosine. The real part of a complex exponential is a cosine. The imaginary part of a complex exponential is a sine. We can write a sine or cosine in terms of complex exponentials.

EXERCISES

Evaluate the following functions for $t = 1$ s [read the cautionary note under Eq. (7.37)].

17. $e^{-0.2t}$ 18. $\cos 0.1t$

19. $\cos [0.1t + (\pi/6)]$ 20. $\cos (0.1t - 30°)$

21. $e^{j(0.1t)}$ 22. $e^{j[0.1t + (\pi/6)]}$

Find one solution for t in the following equations. (Be sure that both quantities in the argument in Problem 25 are in either radians or degrees before you add them.)

23. $e^{-2t} = 0.1$ 24. $5 \cos 2\pi60t = 4$

25. $5 \cos (2\pi60t + 36.87°) = 4$

7.4 Signal Characterization

A signal $x(t)$ (voltage or current) defined by a function such as

$$x(t) = 10 \cos 100t \qquad (7.39)$$

is *completely characterized* for all values of time t. This section describes two ways in which a signal can be *partially characterized*.

Continuity

The first partial characterization of a signal considered here involves continuity. The idea is that a continuous signal does not have any abrupt jumps (discontinuities). For example, the signals shown in Fig. 7.12b, d are continuous for the values of time shown. The remaining two signals shown in Fig. 7.12 are discontinuous for more than one value of time. For example, the signal of Fig. 7.12c has a discontinuity at time $t = T/2$. The concept of continuity will be of great importance in Chapter 9. To formally introduce the concept of continuity, we define the limit from the left:

$$x(t_0^-) = \lim_{\substack{t \to t_0 \\ t < t_0}} x(t) \qquad (7.40)$$

and the limit from the right:

$$x(t_0^+) = \lim_{\substack{t \to t_0 \\ t > t_0}} x(t) \qquad (7.41)$$

The quantities $x(t_0^-)$ and $x(t_0^+)$ are, respectively, the values of $x(t)$ just prior to and just after $t = t_0$. The concept of continuity, which is used extensively in Chapter 9, can now be stated as follows:

DEFINITION

Continuity

A signal $x(t)$ is said to be **continuous** at $t = t_0$ if and only if

$$x(t_0^+) = x(t_0) = x(t_0^-) \qquad (7.42)$$

A signal not continuous at $t = t_0$ is said to be discontinuous at $t = t_0$.

EXAMPLE 7.6

Given the signal

$$v(t) = u(t - t_0)$$

what are $v(t_0^-)$ and $v(t_0^+)$? Is the signal continuous at $t = t_0$?

■ SOLUTION

$$v(t_0^-) = 0$$
$$v(t_0^+) = 1$$

Since

$$v(t_0^-) \neq v(t_0^+)$$

the signal is discontinuous at $t = t_0$. ■

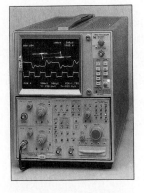

Voltage is measured with an oscilloscope. Courtesy of Tektronix, Inc.

Periodicity*

Just as the earth repeats its position around the sun with a period of one year, many signals repeat themselves with a period T. That is, if the signal has a value $v(t)$ at any time t, the signal will also have that same value at time $t + T$. For example, the sinusoidal voltage

$$v = 120\sqrt{2} \cos 2\pi 60t \text{ V} \qquad (7.43)$$

that supplies power for lights in our houses has a period of

$$T = \frac{1}{60} \text{ s} \qquad (7.44)$$

as given by Eq. (7.36). The formal definition of a periodic signal follows.

---

* The remainder of this chapter can be delayed until after Chapter 9 if desired.

Periodic Signal

A signal $x(t)$ is a **periodic signal** if and only if there exists a $T > 0$ such that for every t,

$$x(t + T) = x(t) \qquad (7.45)$$

The smallest T for which this is true is called the *period* of the signal.

The function $\cos(\omega t + \theta)$ is a periodic signal with period $T = 2\pi/\omega$. Other common periodic signal models are the square wave, triangular wave, and saw-tooth wave shown in Fig. 7.12.

FIGURE 7.12
Periodic signals

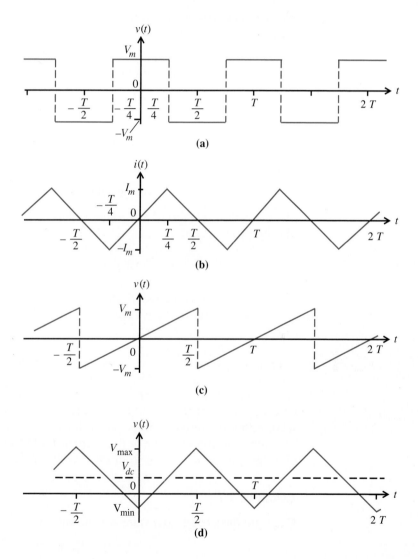

(a)

(b)

(c)

(d)

A continuous signal does not change instantaneously. A trigonometric sine or cosine is periodic and continuous.

EXERCISES

26. Which of the four functions shown in Fig. 7.12 are continuous at $t = T/4$?

Which of the following functions are periodic?

27. $\cos \pi t$

28. $\cos \pi t^2$

29. $u(\sin \pi t)$

7.5 Average Properties

We often find that certain average properties of signals are sufficient for performing an analysis. We will consider three such averages.

Average Value or DC Component

In later chapters, we will need to calculate the average or dc value of signals. The dc component of a signal is of great importance in electronics and electrochemistry.

DEFINITION

Average Value or DC Component

Over the interval t_0 to $t_0 + T$, the **dc component** of a signal is the **average value** of the signal:

$$X_{dc} = \frac{1}{T} \int_{t_0}^{t_0 + T} x(t)\, dt \qquad (7.46)$$

For a signal with period T, the average value over the infinite time interval is the same as the average value over any period.

From Eq. (7.46), if the area between $x(t)$ and the time axis is zero (area above the time axis is positive, and area below the time axis is negative), the dc component is zero. Thus the dc component of a sine or cosine waveform is zero, as are the dc components of the first three signals in Fig. 7.12. An example of a periodic wave with a nonzero dc component value is shown in Fig. 7.12d. The dotted line represents the average value or dc component of this last signal. The average value of a signal is quite different from the *average of the absolute value* of the signal.

EXAMPLE 7.7

Calculate the average (dc) value for the sinusoidal voltage

$$v(t) = V_m \sin \omega t \text{ V}$$

$$V_{dc} = \frac{1}{T} \int_0^T V_m \sin \omega t \, dt$$

$$= -\frac{1}{T} V_m \frac{1}{\omega} \cos \omega t \Big|_0^T$$

$$= -\frac{1}{T} V_m \frac{T}{2\pi} \cos \frac{2\pi}{T} t \Big|_0^T = 0 \text{ V } \blacksquare$$

Average Absolute Value

You will see in later courses in electronics that devices called *diodes* permit us to take the absolute value of a signal.

DEFINITION

Average Absolute Value

Over the time interval t_0 to $t_0 + T$, the **average absolute value** of a signal is

$$X_{aav} = \frac{1}{T} \int_{t_0}^{t_0 + T} |x(t)| \, dt \tag{7.47}$$

For a signal with period T, the average absolute value over the infinite time interval is the same as the average absolute value over one period.

The average absolute value is often of interest, particularly in electrochemical reactions.

EXAMPLE 7.8

For a sinusoid

$$v(t) = V_m \sin \omega t \text{ V}$$

calculate the average absolute value.

■ SOLUTION

$$V_{aav} = \frac{1}{T} \int_0^T \left| V_m \sin \frac{2\pi}{T} t \right| dt$$

$$= \frac{1}{T} \left[\int_0^{T/2} V_m \sin \frac{2\pi}{T} t \, dt + \int_{T/2}^T -V_m \sin \frac{2\pi}{T} t \, dt \right]$$

$$= \frac{2}{\pi} V_m = 0.637 V_m \text{ V } \blacksquare$$

Average Power for DC Signals

In Chapter 2 we showed that the *instantaneous power p* absorbed by a resistance is given by

$$p = Ri^2 \qquad (7.48)$$

or

$$p = \frac{1}{R} v^2 \qquad (7.49)$$

The *average power* absorbed over the interval t_0 to $t_0 + T$ is

$$P = \frac{1}{T} \int_{t_0}^{t_0 + T} p \, dt \qquad (7.50)$$

For a constant (dc) voltage or current, $i(t) = I$ and $v(t) = V$, Eq. (7.48) gives the average power as

$$P = RI^2 \qquad (7.51)$$

and

$$P = \frac{1}{R} V^2 \qquad (7.52)$$

Effective or RMS Value

In some equipment, such as an electric heater, the average power delivered to a resistance is the important quantity. Because of this, we often find it convenient to specify voltages and currents of arbitrary wave shapes by their *effective values* (effective heating value). When a sinusoidal voltage is specified, it is usually in terms of its effective value. For this reason, the 60-Hz, 120-V outlet in a house has a voltage waveform described by

$$v(t) = 120\sqrt{2} \cos (2\pi 60t + \theta) \text{ V} \qquad (7.53)$$

which has a maximum value of approximately 170 V. [We will soon see why the square root of 2 appears in Eq. (7.53).]

The effective value of a signal is determined from the average power delivered to a resistive load. Given an arbitrary current i through a resistance R, the average power absorbed by the resistance over the time interval t_0 to $t_0 + T$ is

$$P = \frac{1}{T} \int_{t_0}^{t_0 + T} p \, dt \qquad (7.54a)$$

$$= R \frac{1}{T} \int_{t_0}^{t_0 + T} i^2 \, dt \qquad (7.54b)$$

If we define the effective value I_{eff} of current i,

$$I_{\text{eff}}^2 = \frac{1}{T} \int_{t_0}^{t_0 + T} i^2 \, dt \qquad (7.55)$$

we see that Eq. (7.54b) is of the form

$$P = RI_{\text{eff}}^2 \qquad (7.56)$$

A similar development gives

$$V_{\text{eff}}^2 = \frac{1}{T} \int_{t_0}^{t_0 + T} v^2 \, dt \qquad (7.57)$$

and

$$P = \frac{1}{R} V_{\text{eff}}^2 \qquad (7.58)$$

From Eqs. (7.55) and (7.57), we see that the effective values are the square *root* of the average (*mean*) of the *square* of the voltage or current. We will take this as our definition:

DEFINITION

Root-Mean-Square (RMS) or Effective Value of a Signal

Over the time interval t_0 to $t_0 + T$,

$$X_{\text{rms}} = \sqrt{\frac{1}{T} \int_{t_0}^{t_0 + T} x^2 \, dt} \qquad (7.59)$$

r for root
m for mean
s for square

and

$$X_{\text{eff}} = \sqrt{\frac{1}{T} \int_{t_0}^{t_0 + T} x^2 \, dt} \qquad (7.60)$$

For a signal with period T, the **rms** or **effective value** over the infinite time interval is the same as the rms value over one period.

Thus the rms value is the same as the effective value:

$$I_{\text{eff}} = I_{\text{rms}} \qquad (7.61)$$

$$V_{\text{eff}} = V_{\text{rms}} \qquad (7.62)$$

EXAMPLE 7.9

A sinusoid

$$v(t) = V_m \cos \omega t \text{ V}$$

is applied to a resistance of R ohms. Calculate the effective or rms value of $v(t)$ and the power delivered to the resistance.

■ **SOLUTION**

$$
\begin{aligned}
V_{\text{rms}} &= \sqrt{\frac{1}{T} \int_{t_0}^{t_0 + T} V_m^2 \cos^2 \omega t \, dt} \\
&= \sqrt{\frac{1}{T} \int_{t_0}^{t_0 + T} \frac{V_m^2}{2} (1 - \cos 2\omega t) \, dt} \\
&= \frac{1}{\sqrt{2}} V_m \simeq 0.707 V_m \text{ V}
\end{aligned}
$$

and
$$P = \frac{V_{\text{rms}}^2}{R} = \frac{1}{2}\frac{V_m^2}{R} \blacksquare$$

From this last example we conclude that:

> The rms value of a sinusoid is the amplitude of the sinusoid divided by the square root of 2.

The square root of 2 relation *does not apply* to the relation between maximum and rms values for most other waveforms.

EXAMPLE 7.10

Calculate the rms value of a periodic signal $v(t)$ with period T, where the signal is defined to be

$$v(t) = \begin{cases} Kt \text{ V} & 0 \le t < \dfrac{T}{2} \\ 0 & \dfrac{T}{2} \le t < T \end{cases}$$

for a single period, where K is a positive constant.

■ SOLUTION

Sketch $v(t)$ for $-2T \le t \le 2T$. This will help visualize that

$$
\begin{aligned}
V_{\text{rms}} &= \sqrt{\frac{1}{T}\int_0^T v^2\,dt} \\
&= \sqrt{\frac{1}{T}\left[\int_0^{T/2}(Kt)^2\,dt + \int_{T/2}^T (0)^2\,dt\right]} \\
&= \sqrt{\frac{1}{T}\frac{K^2 T^3}{24}} = \frac{1}{2\sqrt{6}}KT = 0.2041 KT \text{ V} \blacksquare
\end{aligned}
$$

REMEMBER

We call the average value of a signal the dc component. The average absolute value is of particular interest in electrochemistry. The effective value and rms value of a signal are the same. The rms value is the square *root* of the *mean* square value. We can use the rms value of a signal to calculate the power delivered to a resistance. The rms value of a sinusoid is the amplitude (maximum value) of the sinusoid divided by the square root of 2. This square root of 2 relation does not hold for most other signals.

EXERCISES

Calculate the dc component of the periodic signal shown in the indicated figure.

30. Fig. 7.12b

31. Fig. 7.12d

Calculate the dc component of the periodic voltages or currents specified in Problems 32 through 35.

32. $i_a = 5$ A

33. $v_b = 4 \cos \omega t + 3 \cos \omega t$ V

34. $i_e = 4 + 3 \cos 2t$ A

35. $v_f = 8 \cos^2 \omega t$ V

Determine the average absolute value of the periodic signal shown in the indicated figure.

36. Fig. 7.12a

37. Fig. 7.12b

Determine the average absolute value of the periodic voltages or currents specified in Problems 38 and 39.

38. $i_b = 4 + 4 \cos 2t$ A

39. $v_d = 3 + 4 \cos 2t$ V

Calculate the rms (effective) value of the periodic signal shown in the indicated figure.

40. Fig. 7.12a

41. Fig. 7.12c

Calculate the rms (effective) value of the periodic signals specified in Problems 42 and 43.

42. $i_a = 5$ A

43. $v_d = 3 + 4 \cos 2t$ V

7.6 Energy and Power Signals

A simple voltage or current pulse, such as the unit pulse $\Pi(t)$, can supply only a finite amount of energy W to a resistance. In fact, for a 1-Ω resistance, the energy supplied by the unit pulse is 1 J. Finite energy signals are classified as *energy signals*.

DEFINITION

Energy Signal

A real signal for which the energy W is

$$W = \int_{-\infty}^{\infty} x^2(t)\, dt < \infty \qquad (7.63)$$

is called an **energy signal.**[†]

The quantity W given by Eq. (7.63) is called the *normalized energy* in the signal $x(t)$ and would be the energy delivered to a 1-Ω resistance by the voltage or current x. (The normalized energy will be used in Chapter 15.)

The average power (averaged over all time) delivered to a resistance by an energy signal is zero, because the energy is finite and the time interval is infinite. Some

---

[†] If complex signals are considered, $x^2(t)$ must be replaced by $x(t)x^*(t) = |x(t)|^2$, where $x^*(t)$ is the complex conjugate of $x(t)$.

signals, such as a sinusoid or other periodic signal, supply nonzero average power to a resistance and, therefore, supply infinite energy (over all time). These signals are not energy signals, but are classified as *power signals* (nonzero but finite average power).

DEFINITION

Power Signal

A real signal for which the average power is

$$0 < P = \lim_{T \to \infty} \frac{1}{T} \int_{-T/2}^{T/2} x^2(t)\, dt < \infty \tag{7.64}$$

is called a **power signal.**[†]

The quantity P given by Eq. (7.64) is the normalized power in the signal $x(t)$. P is the average power delivered to a 1-Ω resistance by $x(t)$. (The normalized power will be used in Chapter 14.) From the definitions, a signal cannot be both an energy signal and a power signal. A signal that has infinite average power is neither an energy signal nor a power signal. We will use these definitions in the following section as well as in later chapters.

EXAMPLE 7.11

Determine whether the following signals are energy or power signals.

(a) $v(t) = 2\Pi(t)$ V   (b) $v(t) = V_m \cos \omega t$ V

■ **SOLUTION**

(a)

$$W = \int_{-\infty}^{\infty} [2\Pi(t)]^2\, dt$$

$$= \int_{-1/2}^{1/2} 4\, dt = 4$$

The signal is an energy signal.

(b)

$$W = \int_{-\infty}^{\infty} (V_m \cos \omega t)^2\, dt$$

$$= \int_{-\infty}^{\infty} \frac{1}{2} V_m^2 (1 + \cos 2\omega t)\, dt = \infty$$

The signal is not an energy signal.

$$P = \lim_{T \to \infty} \frac{1}{T} \int_{-T/2}^{T/2} (V_m \cos \omega t)^2\, dt$$

[†] If complex signals are considered, $x^2(t)$ must be replaced by $x(t)x^*(t) = |x(t)|^2$.

$$= \lim_{T \to \infty} \frac{1}{T} \int_{-T/2}^{T/2} \frac{1}{2} V_m^2 (1 + \cos 2\omega t) \, dt$$

$$= \frac{1}{2} V_m^2 = V_{\text{rms}}^2$$

The signal is a power signal. ∎

Energy signals deliver a finite total energy to a resistance. We call the energy delivered by the signal to a 1-Ω resistance, the normalized energy in the signal. Power signals deliver a finite, nonzero average power to a resistance. We call the power delivered by the signal to a 1-Ω resistance, the normalized power in the signal.

EXERCISES

Determine whether the following are energy signals or power signals.

44. $v(t) = e^{-2t}u(t)$ V
45. $i(t) = 5\Pi(t) \cos \pi t$ A
46. $v(t) = 4 \cos^2 2t$ V

7.7 Orthogonality and Superposition of Power

In Chapter 5 we showed that superposition holds for voltages and currents. That is, the voltage and current response of a linear network to a number of independent sources is the *sum* of the responses obtained by applying each independent source once with other independent sources set equal to zero. We will now investigate whether superposition applies to mean square values of voltages and currents and to power. We need the definition of orthogonal* signals to answer this question.

DEFINITION

Orthogonal Signals

Two real signals, x_1 and x_2, are said to be **orthogonal** over an interval t_0 to $t_0 + T$ if and only if

$$\int_{t_0}^{t_0 + T} x_1(t)x_2(t) \, dt = 0^\dagger \qquad (7.65)$$

If both signals are power signals and the interval is from $-\infty$ to $+\infty$, the definition is modified as follows.

---

* Orthogonality of signals is a generalization of the idea of orthogonal vectors, which is, in turn, a generalization of the definition of perpendicular lines.

† If complex signals are considered, $x_2(t)$ must be replaced by its complex conjugate. If you are familiar with vector algebra, you will recognize that this integral is a generalized inner product.

Orthogonal Power Signals

Two real power signals, x_1 and x_2, are **orthogonal** over the infinite interval if and only if

$$\lim_{T \to \infty} \frac{1}{T} \int_{-T/2}^{T/2} x_1(t) x_2(t) \, dt = 0^* \tag{7.66}$$

We can establish the following results with only moderate difficulty:

$$\lim_{T \to \infty} \frac{1}{T} \int_{-T/2}^{T/2} \cos \omega_1 t \sin \omega_2 t \, dt = 0 \qquad \text{for all } \omega_1 \text{ and } \omega_2 \tag{7.67}$$

$$\lim_{T \to \infty} \frac{1}{T} \int_{-T/2}^{T/2} \cos \omega_1 t \cos \omega_2 t \, dt = \begin{cases} 0 & \omega_1 \neq \pm \omega_2 \\ 1 & \omega_1 = \omega_2 = 0 \\ \dfrac{1}{2} & \omega_1 = \pm \omega_2 \neq 0 \end{cases} \tag{7.68}$$

$$\lim_{T \to \infty} \frac{1}{T} \int_{-T/2}^{T/2} \sin \omega_1 t \sin \omega_2 t \, dt = \begin{cases} 0 & \omega_1 \neq \pm \omega_2 \\ 0 & \omega_1 = \omega_2 = 0 \\ \dfrac{1}{2} & \omega_1 = \omega_2 \neq 0 \\ -\dfrac{1}{2} & \omega_1 = -\omega_2 \neq 0 \end{cases} \tag{7.69}$$

From this we conclude the following:

1. Over the infinite interval, sines and cosines are orthogonal to each other whether they have the same or different frequencies.[†]
2. Cosines are orthogonal to cosines of different frequencies.
3. Sines are orthogonal to sines of different frequencies.

Orthogonality will allow us to answer the important question: Does superposition apply to the mean square values of voltages and currents? Consider the voltage $v(t)$ defined over all time:

$$v(t) = v_1(t) + v_2(t) + \cdots + v_N(t)$$

$$= \sum_{n=1}^{N} v_n(t) \tag{7.70}$$

The rms value squared of $v(t)$ is

---

[*] If complex signals are considered, $x_2(t)$ must be replaced by its complex conjugate. If you are familiar with vector algebra, you will recognize that this integral is a generalized inner product.

[†] If they have the same nonzero frequency, the sine and cosine must have the same phase shift. If they both have zero frequency, the phase shift for one must be such that this sinusoid has zero value.

$$V_{rms}^2 = \lim_{T \to \infty} \frac{1}{T} \int_{-T/2}^{T/2} v^2(t)\,dt = \lim_{T \to \infty} \frac{1}{T} \int_{-T/2}^{T/2} \left[\sum_{n=1}^{N} v_n(t) \right]^2 dt$$

$$= \lim_{T \to \infty} \frac{1}{T} \int_{-T/2}^{T/2} \left[\sum_{n=1}^{N} v_n(t) \right] \left[\sum_{m=1}^{N} v_m(t) \right] dt$$

$$= \sum_{n=1}^{N} \sum_{m=1}^{N} \lim_{T \to \infty} \frac{1}{T} \int_{-T/2}^{T/2} v_n(t) v_m(t)\,dt \qquad (7.71)$$

For the general case, no simplification is possible. For the *special case of orthogonal signals*, $v_1, v_2, \ldots, v_N$, the integrals in Eq. (7.71) are zero for $n \neq m$, so Eq. (7.71) becomes

$$V_{rms}^2 = \sum_{n=1}^{N} \lim_{T \to \infty} \frac{1}{T} \int_{-T/2}^{T/2} v_n^2(t)\,dt = \sum_{n=1}^{N} V_{rms_n}^2 \qquad (7.72)$$

This leads to the following result:

RMS Value of the Sum of Orthogonal Signals

For *orthogonal signals*, the squares of rms voltages add, and

$$V_{rms} = \sqrt{\sum_{n=1}^{N} V_{rms_n}^2} \qquad (7.73)$$

For sinusoids of different frequencies ($\omega_n \neq 0$) with amplitudes V_{m_n}, Eq. (7.73) becomes

$$V_{rms} = \sqrt{\sum_{n=1}^{N} \frac{1}{2} V_{m_n}^2} \qquad (7.74)$$

For *orthogonal* signals,

$$P_{av} = \frac{1}{R} V_{rms}^2 = \frac{1}{R} \sum_{n=1}^{N} V_{rms_n}^2 = \sum_{n=1}^{N} P_{av_n} \qquad (7.75)$$

where P_{av_n} is the average power absorbed from voltage component v_n. This proves that *for orthogonal signals, the average powers add* (superposition of power). *This is not true for signals that are not orthogonal.* Typically

$$P_{av} \neq \sum_{n=1}^{N} P_{av_n} \qquad (7.76)$$

EXAMPLE 7.12

For the network shown in Fig. 7.13, find the rms value of v and the average power absorbed by the resistance.

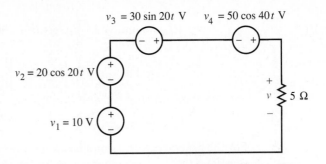

FIGURE 7.13

Voltage $v = \sum\limits_{n=1}^{4} v_n$ is the sum of orthogonal voltages

■ **SOLUTION**

The four signals are orthogonal [see the paragraph below Eq. (7.74)]; therefore

$$V_{\text{rms}} = \sqrt{10^2 + \left(\frac{20}{\sqrt{2}}\right)^2 + \left(\frac{30}{\sqrt{2}}\right)^2 + \left(\frac{50}{\sqrt{2}}\right)^2}$$

$$= \sqrt{2000} = 44.72 \text{ V}$$

and

$$P = \frac{1}{R} V_{\text{rms}}^2 = \frac{1}{5}(2000) = 400 \text{ W}$$

Note that the powers supplied by each source acting alone add to give

$$P = P_1 + P_2 + P_3 + P_4$$

$$= \frac{1}{5}(10)^2 + \frac{1}{5}\left(\frac{20}{\sqrt{2}}\right)^2 + \frac{1}{5}\left(\frac{30}{\sqrt{2}}\right)^2 + \frac{1}{5}\left(\frac{50}{\sqrt{2}}\right)^2$$

$$= 20 + 40 + 90 + 250 = 400 \text{ W}$$

so superposition of power holds, but only because the four voltages are orthogonal. ■

We will now work an example in which not all voltages are orthogonal.

EXAMPLE 7.13

Repeat Example 7.12 with v_4 changed to

$$v_4 = 50 \cos(20t - 53.13°) \text{ V}$$

■ **SOLUTION**

The rms values of v_1 through v_4 are the same, but v_4 is no longer orthogonal to v_2 and v_3. The problem could be worked by applying the definitions and integrating. A simpler way is to use the trigonometric identity

$$\cos(A + B) = \cos B \cos A - \sin B \sin A$$

This trigonometric identity lets us write voltage v_4 as a sum of a cosine and a sine component:

$$v_4 = 30 \cos 20t + 40 \sin 20t \text{ V}$$

Addition of v_1, v_2, v_3, and v_4 gives us

$$v = 10 + 50 \cos 20t + 70 \sin 20t \text{ V}$$

The three components are now orthogonal to each other, so

$$V_{\text{rms}} = \sqrt{10^2 + \left(\frac{50}{\sqrt{2}}\right)^2 + \left(\frac{70}{\sqrt{2}}\right)^2}$$

$$= \sqrt{3800} = 61.64 \text{ V}$$

$$P = \frac{1}{R} V_{\text{rms}}^2 = \frac{1}{5}(3800) = 760 \text{ W}$$

Note that the powers supplied by each source acting alone add to give

$$P_1 + P_2 + P_3 + P_4 = 400 \text{ W} \neq P$$

so superposition of power did not hold. ■

REMEMBER

Over the infinite interval, sines and cosines of different frequencies are orthogonal. Sines and cosines of the same nonzero frequency are orthogonal to each other if they have the same phase shift. Cosines are orthogonal to cosines of different frequencies, and sines are orthogonal to sines of different frequencies. Superposition of mean square values applies only to orthogonal signals. Superposition of power applies only to orthogonal signals.

EXERCISES

47. Use trigonometric identities and integration to show that $v_1(t) = 2 \sin 2t$ V and $v_2(t) = 4 \sin 4t$ V are orthogonal over the infinite interval.

Determine the rms value of the periodic voltages or currents given in Problems 48 through 52. (The components in Problem 52 are not orthogonal.)

48. $v_a = 4 \cos \omega t + 3 \cos \omega t$ V

49. $v_b = 4 \cos \omega t + 3 \sin \omega t$ V

50. $i_c = 4 \cos 5t + 4 \cos 10t$ A

51. $i_d = 4 + 3 \cos 2t$ A

52. $i_e = 2 \cos 2t + 2 \sin (2t + 45°)$ A

7.8 Summary

This chapter introduced the modeling of signals (voltages and currents) by mathematical functions. We first introduced the unit step function. The unit step function is 0 for negative time and takes on the value 1 for positive time. This provides a convenient notation to describe signals that are defined differently in separate time intervals. In an optional section we introduced the delta function, which is the idealized model for a very short duration pulse of unit area. (Delta functions will be used in Chapters 15 and 16 on Fourier and Laplace transform techniques.)

We then showed how the real exponential signal, which naturally occurs in linear systems, can be generalized to the complex exponential. This lets us write trigonometric sines and cosines as functions of complex exponentials. This complex exponential representation for sinusoids is fundamental to the analysis contained in the remainder of this text.

We followed the sections on signal modeling by a discussion of ways to partially characterize signals.

1. Continuous signals do not change instantaneously.
2. Periodic signals repeat themselves with a period T.
3. The dc component of a signal is its average value.
4. The average absolute value of a signal is of interest in chemical reactions.
5. The effective heating value of a signal is given by the root-mean-square (rms) value of the signal.
6. Two signals are orthogonal over a time interval if the integral of the product over this interval is zero.

We concluded the chapter by showing that superposition of power applies to orthogonal signals but not to signals in general.

KEY FACTS

◆ A dc signal is constant for all time.

◆ A unit step function $u(t)$ has a value of 0 when the argument is negative and a value of 1 when the argument is positive.

◆ We can visualize a delta function as a very short duration pulse of unit area.

◆ A real exponential signal decreases by a factor of $1/e$ in one time constant.

◆ A real exponential signal is usually considered negligible, compared with its original value, after five time constants.

◆ Euler's identity gives the complex exponential $e^{j\omega t}$ in terms of the trigonometric sine and cosine.

◆ The real part of a complex exponential $e^{j\omega t}$ is $\cos \omega t$.

◆ The imaginary part of a complex exponential $e^{j\omega t}$ is $\sin \omega t$.

◆ A continuous signal does not change instantaneously.

◆ A periodic signal with period T repeats itself every T seconds.

◆ The average value (dc component) of a signal over a time interval $t_0 \leq t \leq t_0 + T$ is

$$V_{\text{av}} = \frac{1}{T} \int_{t_0}^{t_0 + T} v(t) \, dt$$

◆ The average absolute value of a signal over a time interval $t_0 \leq t \leq t_0 + T$ is

$$V_{\text{aav}} = \frac{1}{T} \int_{t_0}^{t_0 + T} |v(t)| \, dt$$

- The rms (effective) value of a signal over a time interval $t_0 \leq t \leq t_0 + T$ is

$$V_{\text{rms}} = \sqrt{\frac{1}{T} \int_{t_0}^{t_0 + T} v^2(t)\, dt}$$

or

$$I_{\text{rms}} = \sqrt{\frac{1}{T} \int_{t_0}^{t_0 + T} i^2(t)\, dt}$$

- For periodic signals, an average value over one period is the same as the average value over the interval $-\infty < t < \infty$.

- We can obtain the rms value of a sinusoid by dividing the magnitude of the sinusoid (peak value) by the square root of 2. This relation is not valid for most other signals.

- The average power absorbed by a resistance is given by

$$P = \frac{1}{R} V_{\text{rms}}^2$$

$$= R I_{\text{rms}}^2$$

- The normalized energy in a signal is

$$W = \int_{-\infty}^{\infty} x^2(t)\, dt$$

- For an energy signal,

$$W < \infty$$

- The normalized power in a signal is

$$P = \lim_{T \to \infty} \frac{1}{T} \int_{-T/2}^{T/2} x^2(t)\, dt = 0$$

- For a power signal,

$$0 < P < \infty$$

- Two energy signals are orthogonal if and only if

$$\int_{-\infty}^{\infty} x_1(t) x_2(t)\, dt = 0$$

- Two power signals are orthogonal if and only if

$$\lim_{T \to \infty} \frac{1}{T} \int_{-T/2}^{T/2} x_1(t) x_2(t)\, dt = 0$$

- For orthogonal signals, superposition applies to mean square values and to power. Superposition of mean square values and superposition of power do *not* apply to signals that are not orthogonal.

- The rms value of the sum of N *orthogonal* signals is

$$V_{\text{rms}} = \sqrt{\sum_{n=1}^{N} V_{\text{rms}_n}^2}$$

■ PROBLEMS ■

Section 7.1

Sketch the functions listed in Problems 1 through 4, and write in terms of the unit step.

1.
$$v(t) = \begin{cases} 0 & t < 2 \\ 10e^{-(t-2)/4} & 2 < t < 6 \\ 0 & t > 6 \end{cases}$$

2.
$$v(t) = \begin{cases} 0 & t < 2 \\ 10e^{-t/4} & 2 < t < 6 \\ 0 & t > 6 \end{cases}$$

3. $i(t) = 10 \cos\left[\dfrac{2\pi}{T}(t - T)\right]\Pi\left(\dfrac{t - T}{T/2}\right)$

4. $i(t) = \sum\limits_{n=0}^{\infty} (n + 1)\Pi\left(t - \dfrac{1}{2} - n\right)$

Write the signals shown in Problems 5 and 6 in terms of the unit step.

5.

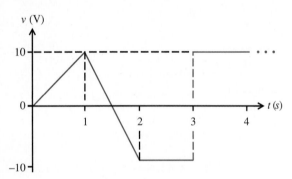

6.

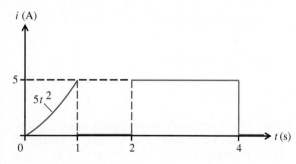

Problems 7 through 10, specify the voltage in volts applied to a 0.1-H inductance. Determine the inductance current as a function of time, $i(-\infty) = 0$, and sketch $i(t)$.

7. $e^{-2t}u(t)$

8. $e^{-2t}[u(t) - u(t - 2)]$

9. $e^{-2t}u(t)u(2 - t)$

10. $e^{-2t}\Pi\left(\dfrac{t - 1}{2}\right)$

11. Evaluate the integral

$$\int_{-\infty}^{t} \cdots \int_{-\infty}^{t_{n-1}} u(t_n)\,dt_n \cdots dt_1$$

12. Write the shifted unit ramp function $r(t - t_0)$ in terms of the unit step, and sketch the function.

Section 7.2

Sketch the functions given in Problems 13 through 17.

13. $v(t) = e^{-t}\delta(t - 1)$
14. $v(t) = (2 \cos \pi t)\delta(t)$
15. $i(t) = \sum\limits_{n=-\infty}^{\infty} \delta(t - nT)$
16. $i(t) = \sum\limits_{n=-\infty}^{\infty} (\cos \pi t)\delta(t - n)$
17. $i(t) = \sum\limits_{n=-\infty}^{\infty} (\sin \pi t)\delta(t - n)$

Problems 18 through 20 specify the current in amperes applied to a 0.5-F capacitance. Determine the capacitance voltage as a function of time, $v(-\infty) = 0$, and sketch $v(t)$.

18. $e^{-t}[\delta(t) - \delta(t - 2)]$
19. $(t^2 + 2t + 1)\delta(t - 2)$
20. $\sin 2\pi\tau\, u(\tau)\delta(t - \tau)$

Section 7.3

21. What is the smallest integer number of time constants required for an exponential signal to decay to less than 10 percent of its initial value? What percentage of its initial value is it at this time?

Sketch the signals given in Problems 22 through 26, and then write each as the real or imaginary part of a complex exponential.

22. $10 \cos 2\pi 6t$
23. $5 \sin (30t + 15°)$
24. $100 \cos (50t + \pi/4)$
25. $25e^{-3t} \cos (2\pi 5t + 30°)$
26. $e^{-3t} \cos 2\pi(t - 0.25)$

For Problems 27 through 30, write the signals given in the specified problem as the sum of complex exponentials.

27. Problem 22   28. Problem 23
29. Problem 24   30. Problem 25
31. Problem 26

For Problems 32 through 36, evaluate the signals given in the specified problem for $t = 0$.

32. Problem 22   33. Problem 23
34. Problem 24   35. Problem 25
36. Problem 26

Write the complex exponentials given in Problems 37 through 41 in terms of the trigonometric sine and cosine.

37. $5e^{j2t}$ 38. $10e^{(-3+j4)t}$
39. $15e^{-2+j3t}$ 40. $20e^{j(4t-3)}$
41. $30e^{-5t}(e^{j6t} + e^{-j6t})$

What phase shift in degrees and radians is introduced by the time delay in the sinusoids given in Problems 42 through 47? Write the function so that the phase shift is explicitly shown. Note the relation between frequency and phase shift for a fixed time delay. Also note the relation between time delay and phase shift for a fixed frequency.

42. $25 \cos 2\pi \, 20(t - 0.01)$
43. $100 \sin 2\pi 40(t - 0.01)$
44. $400 \cos 2\pi 60(t - 0.01)$
45. $40 \cos 2\pi 30(t - 0.005)$
46. $80 \sin 2\pi 30(t - 0.010)$
47. $120 \cos 2\pi 30(t - 0.020)$

What time delay will introduce the phase shift given in the sinusoids listed in Problems 48 through 55? Write the function so that the time delay is explicitly shown. Note the relation between time delay and phase shift for a fixed frequency. Also note

the relation between time delay and frequency for a fixed phase shift.

48. $240\sqrt{2} \cos (2\pi 60t - \pi/24)$
49. $240\sqrt{2} \cos (2\pi 60t - 15°)$
50. $240\sqrt{2} \cos (2\pi 60t - 30°)$
51. $240\sqrt{2} \cos (2\pi 60t - 60°)$
52. $100 \sin (2\pi 30t - 45°)$
53. $100 \sin (2\pi 60t - 45°)$
54. $100 \sin (2\pi 120t - 45°)$
55. $A \cos (\omega t + \theta) = 5 \cos 2t + 6 \sin 2t$

Section 7.4

Which of the functions listed in Problems 56 through 60 are continuous in the interval $-\infty < t < \infty$?

56. $\cos 2\pi t$ 57. $(\cos 2\pi t)u(t)$
58. $(\sin 2\pi t)u(t)$ 59. $e^{|t|}$
60. $u(\cos 2\pi t)$
61. Which of the functions listed in Problems 56 through 60 are periodic?

Section 7.5

62. Determine the dc component of $4 \cos^2 \omega t$.
63. Determine the average absolute value of the signal shown in Fig. 7.12d.
64. Determine the rms value of $4 \cos^3 \omega t$.

Section 7.6

Determine whether the following signals are energy signals or power signals.

65. $u(\sin \pi t)$ 66. $e^{-|t|}$
67. $e^{-t}(\cos t)u(t)$

Section 7.7

68. Show that

$$\lim_{T \to \infty} \frac{1}{T} \int_{-T/2}^{T/2} \cos \omega_1 t \cos \omega_2 t \, dt$$

$$= \begin{cases} 0 & \omega_1 \neq \pm\omega_2 \\ 1 & \omega_1 = \omega_2 = 0 \\ \dfrac{1}{2} & \omega_1 = \pm\omega_2 \neq 0 \end{cases}$$

This establishes the fact that cosines of different frequencies are orthogonal over the infinite interval.

69. Determine whether a constant voltage of value $\frac{1}{2}V_m$ is orthogonal over one period to the voltage given in Fig. 7.12c.

70. Determine whether the voltages given in Fig. 7.12a, c are orthogonal to each other over one period.

Determine the rms (effective) value of the signals listed in Problems 71 through 75.

71. $v(t) = 10 \cos 3t + 4 \cos 6t + 8 \sin 9t$ V
72. $i(t) = 6 \cos 2t + 8 \cos 2t$ A
73. $i(t) = 6 \cos 2t + 8 \sin 2t$ A
74. $v(t) = 6 + 8 \sin 2t$ V
75. $v(t) = 2 \cos 2\pi 60t - 10 \cos (2\pi 60t - 36.87°)$ V
76. Current i is periodic as shown below. Determine the rms value of this current.

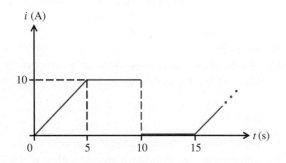

Comprehensive Problems

Determine whether the signals given in Problems 77 through 84 are energy or power signals and whether they are periodic or aperiodic. Determine the energy E for all energy signals, the power P for all power signals, and the period T for all periodic signals.

77. $10 \cos 2\pi t$
78. $10 \cos 2\pi t + 5 \cos 2t$
79. $e^{-2t}u(t)$ 80. $u(\cos 5t)$
81. $e^{\sin t}$ 82. $\delta(t)$

83. $2\Pi\left(\dfrac{t-1}{4}\right)$ 84. $\displaystyle\sum_{n=-\infty}^{\infty} \Pi\left(\dfrac{t-n}{1/4}\right)$

85. For the voltage $v(t)$ shown in Fig. 7.14, determine (a) the dc or average value, and (b) the rms or effective value.

86. Given a current

$$i(t) = A \sum_{n=-\infty}^{\infty} \Pi\left(\frac{t-nT}{\tau}\right) \qquad \tau \le T$$

(a) Sketch $i(t)$.
(b) Is $i(t)$ periodic? If so, what is the period?
(c) Find the dc component of $i(t)$.
(d) Find the average absolute value of $i(t)$.
(e) Find the rms or effective value of $i(t)$.

87. Repeat Problem 86 for

$$v(t) = \sum_{n=-\infty}^{\infty} h(t - 10n)$$

where $\quad h(t) = \begin{cases} 2 & 0 \le t < 4 \\ 8 & 4 \le t < 6 \\ 0 & \text{Otherwise} \end{cases}$

88. Voltmeters that read the average or dc value of a voltage are often adapted to read the rms value of a sinusoid. A *rectifier circuit* is added to take the absolute value of the voltage. The meter thus responds to V_{aav}.
(a) What constant must V_{aav} be multiplied by to indicate V_{rms} for a sinusoid?
(b) The meter is made to read V_{rms} directly for a sinusoid by recalibration of the scale with the constant found in (a). If a square wave is applied to the meter and a value of 10 V (rms) is indicated, what is the true rms value of the square wave?

89. Represent the unit impulse function as the sum of an even and an odd function, $\delta(t) = \delta_e(t) + \delta_o(t)$. Use this to prove that Eq. (7.7) implies that the unit impulse is an even function.

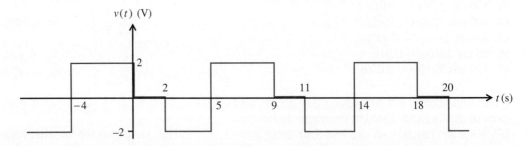

FIGURE 7.14

Source-Free Circuits

In earlier chapters we have seen that the response voltages and currents in a circuit containing energy storage elements (inductance and capacitance) are related to the input voltages and currents (due to independent sources) by differential equations. In this chapter we will introduce some definitions relating to differential equations and solve for the source-free response. That is, we solve for the response due to energy stored in inductances and capacitances after all independent sources have decreased to zero. To simplify our notation, we will introduce the operator notation, $p = d/dt$. This will lead to a very simple method to determine the response of source-free circuits. (In Chapter 9 we will extend the p operator technique to include the circuit response when sources are not zero.)

8.1 Differential Equations

We will begin by examining a parallel *RLC* circuit driven by a current source.

EXAMPLE 8.1

Obtain a differential equation that relates the response voltage $v(t)$ to the input current $i(t)$ in the network of Fig. 8.1.

FIGURE 8.1
A parallel *RLC* circuit

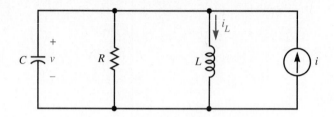

■ SOLUTION

Summation of the currents leaving the upper node gives

$$C \frac{d}{dt} v + \frac{1}{R} v + \frac{1}{L} \int_{-\infty}^{t} v \, d\tau - i = 0$$

This is an integro-differential equation, but we can reduce it to a second-order differential equation by differentiating once:

$$C \frac{d^2}{dt^2} v + \frac{1}{R} \frac{d}{dt} v + \frac{1}{L} v = \frac{d}{dt} i \quad ■$$

The differential equation we obtained in the preceding example is of the form

$$a_2 \frac{d^2}{dt^2} v + a_1 \frac{d}{dt} v + a_0 v = b_1 \frac{d}{dt} i \qquad (8.1a)$$

$$= g(t) \qquad (8.1b)$$

We refer to the known function $i = i(t)$ as the *input function* or *input signal*, and the unknown function $v = v(t)$ as the *response*, the *output signal*, or the *solution* to the differential equation. We can easily determine the *forcing function*, $g(t)$, by differentiation of $i(t)$.

Because the resistance, inductance, and capacitance are linear,* no product of the dependent variable v or its derivatives appears, and Eq. (8.1) is a *linear differential equation*. Because we assume that electromagnetic waves propagate instantaneously, the equation does not involve derivatives with respect to distance, but only with respect to a single independent variable t; therefore the equation is an *ordinary* or *lumped differential equation*. The coefficients a_k, where $k = 0, 1, 2$, are independent of t because the resistance, inductance, and capacitance are time-invariant,* so the equation has *constant coefficients*. The highest order of derivative appearing is 2, so this is a *second-order* equation. Thus Eq. (8.1) is a second-order, ordinary (lumped) linear differential equation with constant coefficients. We refer

* Refer to Section 5.1.

to such equations as LLTI differential equations, because they represent a *lumped*, *linear*, *time-invariant* network. Because a second-order equation describes the circuit of Fig. 8.1, we say this is a second-order circuit.

In Chapter 4 we saw that node-voltage or mesh-current equations for networks containing inductance and capacitance were simultaneous linear differential equations. In this chapter, we will see that a system of several equations in more than one unknown voltage (or unknown current) can be reduced to a higher-order linear differential equation in one unknown voltage (or unknown current). We will obtain an nth-order (the highest-order derivative of the output function is n) ordinary linear differential equation with constant coefficients, as defined below.

DEFINITION

nth-Order Ordinary (Lumped) Linear Differential Equation with Constant Coefficients (nth-Order LLTI Differential Equation)

$$a_n \frac{d^n}{dt^n} v + a_{n-1} \frac{d^{n-1}}{dt^{n-1}} v + \cdots + a_0 v$$

$$= b_m \frac{d^m}{dt^m} i + b_{m-1} \frac{d^{m-1}}{dt^{m-1}} i + \cdots + b_0 i \tag{8.2a}$$

$$= g(t) \tag{8.2b}$$

The input is the known function $i = i(t)$, and the output is the function $v = v(t)$. The forcing function is $g(t)$.

To simplify our discussion, we introduce another definition.

DEFINITION

nth-Order Circuit

A circuit described by an nth-order LLTI differential equation is an **nth-order LLTI circuit**.

For notational efficiency, we define the *derivative operator*

$$p^n = \frac{d^n}{dt^n} \tag{8.3}$$

EXAMPLE 8.2

Write the differential equation

$$C \frac{d^2}{dt^2} v + \frac{1}{R} \frac{d}{dt} v + \frac{1}{L} v = \frac{d}{dt} i$$

obtained in Example 8.1 in terms of the p operator.

■ **SOLUTION**

The differential equation written in operator notation is

$$Cp^2v + \frac{1}{R}pv + \frac{1}{L}v = pi$$

Because differentiation is a linear transformation, we can write the preceding equation more compactly as

$$\left(Cp^2 + \frac{1}{R}p + \frac{1}{L}\right)v = pi$$

This equation is of the form

$$\mathscr{A}(p)v = \mathscr{B}(p)i$$

(this is read, "$\mathscr{A}(p)$ operating on v is equal to $\mathscr{B}(p)$ operating on i") where

$$\mathscr{A}(p) = Cp^2 + \frac{1}{R}p + \frac{1}{L}$$

and

$$\mathscr{B}(p) = p \quad ■$$

As in the preceding example, operator notation lets us write the general nth-order LLTI differential equation of Eq. (8.2) as

$$\mathscr{A}(p)v = \mathscr{B}(p)i \tag{8.4a}$$

$$= g(t) \tag{8.4b}$$

where

$$\mathscr{A}(p) = a_np^n + a_{n-1}p^{n-1} + \cdots + a_0 \tag{8.5}$$

and

$$\mathscr{B}(p) = b_mp^m + b_{m-1}p^{m-1} + \cdots + b_0 \tag{8.6}$$

The mth-degree polynomial in the operator p, $\mathscr{B}(p)$, simply defines the calculations required to determine $g(t)$ from i.

EXAMPLE 8.3

If the input current of Examples 8.1 and 8.2 is

$$i = I_me^{st}$$

determine the forcing function $g(t)$.

■ **SOLUTION**

$$g(t) = \mathscr{B}(p)i$$

$$= pI_me^{st} = \frac{d}{dt}I_me^{st}$$

$$= sI_me^{st} \quad ■$$

The operator $\mathscr{A}(p)$ characterizes the response of the differential equation and therefore the response of the related circuit. If we replace the operator p in $\mathscr{A}(p)$

with the complex variable s, we have an nth-degree polynomial $\mathcal{A}(s)$ that we call the *characteristic polynomial* of the differential equation. The equation

$$\mathcal{A}(s) = 0 \qquad (8.7)$$

is the *characteristic equation*. This characteristic equation plays an important role in the solution of the corresponding differential equation.

EXAMPLE 8.4

Write the characteristic equation for the circuit of Example 8.1.

■ SOLUTION

In Example 8.2 we found

$$\mathcal{A}(p) = Cp^2 + \frac{1}{R}p + \frac{1}{L}$$

so the characteristic equation is

$$\mathcal{A}(s) = Cs^2 + \frac{1}{R}s + \frac{1}{L} = 0 \quad ■$$

REMEMBER

Ordinary (lumped) linear time-invariant (LLTI) differential equations describe circuits that have values of resistance, inductance, and capacitance and control parameters that are constant. The operator p^n implies the nth derivative with respect to time. We can write an LLTI differential equation in the form $\mathcal{A}(p)v = \mathcal{B}(p)i$, where i is the input and v is the output. The characteristic equation $\mathcal{A}(s) = 0$ plays an important role in the solution of the corresponding LLTI differential equation.

EXERCISES

Write the following differential equations in operator notation, and write the characteristic equation. Eliminate any integrals by differentiation. (The known input function is on the right.)

1. $\dfrac{d}{dt}i + \dfrac{1}{4}i = 5e^{-2t}$
2. $\dfrac{d^2}{dt^2}i + 6\dfrac{d}{dt}i + 8i = 0$

3. $\dfrac{d}{dt}v + 5v + 6\displaystyle\int_{-\infty}^{t} v\,d\tau = 24e^{-3t}$
4. $\dfrac{d^2}{dt^2}i + 9\dfrac{d}{dt}i = 20e^{-t}$

5. $\dfrac{d^2}{dt^2}v + 6\dfrac{d}{dt}v + 9v = 0$
6. $\dfrac{d^2}{dt^2}i + 9i = 36$

8.2 Distinct Roots of the Characteristic Equation

In this chapter we consider only the response of circuits to energy initially stored in the inductance and capacitance elements. That is, we consider circuits that are *source free* for the time interval of investigation. A *source-free* or *homogeneous differential equation* describes such a circuit.

Homogeneous Differential Equation

An LLTI differential equation of the form

$$\mathscr{A}(p)v_c = 0 \tag{8.8}$$

is a **homogeneous differential equation**.

Engineers frequently call the solution to a homogeneous equation the *source-free response* or *natural response*. In Chapter 9 we will see that the solution to the homogeneous equation is also a part of the solution to the equation where the forcing function is not zero. We will call the natural response the *complementary response* because it completes the response when added to the component of the response that is unique to a particular input.

Complementary Response

The solution to the homogeneous differential equation [Eq. (8.8)] is the **complementary response** (complementary solution).

Solution of Eq. (8.8) requires n integrations for an nth-order LLTI differential equation. As a result, the complementary response contains n constants of integration.

Several techniques are available to solve LLTI differential equations. The p operator technique is chosen, because it leads naturally to steady-state sinusoidal analysis, which will be introduced in Chapter 10.

We will begin by deriving the s-shift theorem, which we will use to solve homogeneous linear differential equations. First, recognize that the differentiation rule for a product of two functions gives

$$\frac{d}{dt}(e^{-s_1 t}v) = e^{-s_1 t}\frac{d}{dt}v - s_1 e^{-s_1 t}v \tag{8.9}$$

Equation (8.9) can be rearranged and written in operator notation to give us the following result.

THEOREM 1

S-Shift Theorem

$$p(e^{-s_1 t}v) = e^{-s_1 t}(p - s_1)v \tag{8.10}$$

We will now use the s-shift theorem to solve a first-order homogeneous differential equation. Multiplication of the first-order homogeneous equation

$$(p - s_1)v_c = 0 \tag{8.11}$$

by e^{-s_1t} gives us

$$e^{-s_1t}(p - s_1)v_c = 0 \qquad (8.12)$$

From Eq. (8.10), Eq. (8.12) is equivalent to

$$p(e^{-s_1t}v_c) = 0 \qquad (8.13)$$

Therefore the solution (v_c) to Eq. (8.13) is the same as the solution (v_c) to Eq. (8.11). We can write Eq. (8.13) as

$$\frac{d}{dt}(e^{-s_1t}v_c) = 0 \qquad (8.14)$$

We can easily integrate Eq. (8.14) with respect to time to obtain the solution (v_c):

$$\int \frac{d}{dt}(e^{-s_1t}v_c)\,dt = e^{-s_1t}v_c - A_1 = 0 \qquad (8.15)$$

where A_1 is a constant of integration. Multiplication of Eq. (8.15) by e^{s_1t} gives us the solution to Eq. (8.11):

$$v_c = A_1 e^{s_1t} \qquad (8.16)$$

Observe that *the coefficient s_1 in the exponent* of Eq. (8.16) *is the root to the characteristic equation:*

$$s - s_1 = 0 \qquad (8.17)$$

Therefore the *root of the characteristic equation determines the complementary response, except for the value of the constant of integration.*
 We will now analyze a simple source-free circuit.

EXAMPLE 8.5

Energy has been stored in the capacitance of Fig. 8.2, and the switch is opened at $t = 0$. The capacitance voltage that exists immediately after the switch is opened is $v(0^+) = 4$ V. Determine and sketch v for $t > 0$.

FIGURE 8.2
A circuit that is source free
for $t > 0$

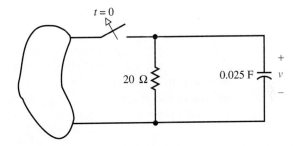

■ **SOLUTION**

Summation of the currents leaving the top node for $t > 0$ gives

$$0.025\,\frac{d}{dt}v + \frac{1}{20}v = 0$$

which can be written as

$$0.025(p + 2)v = 0$$

The characteristic equation

$$0.025(s + 2) = 0$$

has a single root

$$s_1 = -2$$

which gives the solution

$$v_c = A_1 e^{-2t}$$

(The subscript c has been added to remind us that it is the solution to a homogeneous equation.) Because the circuit is source free, the complementary response v_c is the complete response v.

We use the *initial value* of $v[v(0^+) = 4 \text{ V}]$ to evaluate the constant of integration:

$$v(0^+) = 4 = A_1 e^{-2(0)}$$

which gives

$$A_1 = 4$$

and

$$v = 4e^{-2t} \text{ V} \qquad t > 0$$

This is sketched in Fig. 8.3. ∎

FIGURE 8.3
Complementary response of a first-order circuit

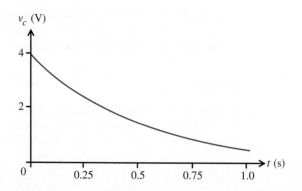

We can apply the p operator to the solution of nth-order differential equations that arise in the analysis of circuits with n energy storage elements. The operator $\mathcal{A}(p)$ can be manipulated in much the same way as an algebraic polynomial, because of the properties of a derivative. Thus we can factor the operator as if it were a polynomial. This permits us to write

$$\mathcal{A}(p)v_c = 0 \tag{8.18}$$

in factored form:

$$a_n(p - s_1)(p - s_2) \cdots (p - s_n)v_c = 0 \tag{8.19}$$

where $s_1, s_2, \ldots, s_n$ are the n roots of the corresponding characteristic equation

$$\mathcal{A}(s) = a_n(s - s_1)(s - s_2) \cdots (s - s_n) = 0 \tag{8.20}$$

If the roots are all distinct ($s_i \neq s_j$, $i \neq j$), we can apply the s-shift theorem and integration n times, as in Eq. (8.15), to prove the following theorem.

Complementary Response for Distinct Roots

For an nth-order homogeneous linear differential equation with constant coefficients,

$$\mathcal{A}(p)v_c = 0 \tag{8.21}$$

the characteristic equation can be written in the factored form as

$$a_n(s - s_1)(s - s_2) \cdots (s - s_n) = 0 \tag{8.22}$$

If the roots* $s_1, s_2, \ldots, s_n$ of the characteristic equation are all distinct, Eq. (8.21) has the solution

$$v_c = A_1 e^{s_1 t} + A_2 e^{s_2 t} + \cdots + A_n e^{s_n t} \tag{8.23}$$

called the *complementary response*, where

$$A_j \qquad j = 1, 2, \ldots, n \tag{8.24}$$

are the n constants of integration.

EXAMPLE 8.6

For the circuit used in Example 8.1, $C = 2$ F, $R = \frac{1}{22}$ Ω, and $L = \frac{1}{20}$ H. Current i is zero for $t \geq 0$, so the circuit is described by the homogeneous equation

$$(2p^2 + 22p + 20)v = 0 \qquad t > 0$$

Determine and sketch the response for this equation.

■ **SOLUTION**

The characteristic equation is

$$2s^2 + 22s + 20 = 0$$

which we can write in factored form as

$$2(s + 1)(s + 10) = 0$$

The roots of this equation are

$$s_1 = -1$$

and

$$s_2 = -10$$

Theorem 2 gives the solution to the homogeneous differential equation as (a subscript c is placed on v to remind us that it is the solution to a homogeneous equation)

$$v_c = A_1 e^{-t} + A_2 e^{-10t} \text{ V} \qquad t > 0$$

The forcing function is zero, so the complementary response v_c is also the complete response v. Additional information, such as the value of v at time $t = 0$ and the

---

* In later chapters we will call the roots of the characteristic equation the *poles* of the circuit.

inductance current at time $t = 0$, is required to evaluate the constants of integration, A_1 and A_2. In Fig. 8.4, we sketch the response for $A_1 = A_2 = 1$. ∎

FIGURE 8.4
Complementary response of
a second-order circuit with
distinct (first-order) roots

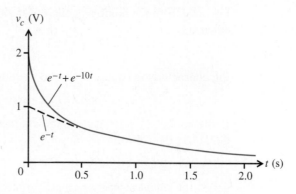

REMEMBER

A homogeneous differential equation, $\mathscr{A}(p)v = 0$, describes a source-free circuit. We call the solution to a homogeneous differential equation the complementary response, the source-free response, or the natural response. The form of the complementary response is determined by the characteristic equation, $\mathscr{A}(s) = 0$. If the roots of the characteristic equation, $s_1, s_2, \ldots, s_n$, are all distinct, the complementary response is of the form

$$v(t) = A_1 e^{s_1 t} + A_2 e^{s_2 t} + \cdots + A_n e^{s_n t} \tag{8.25}$$

where $A_1, A_2, \ldots, A_n$ are constants of integration.

EXERCISES

Determine the solution of the homogeneous differential equations given in Exercises 7 through 11.

7. $\dfrac{d}{dt} i + \dfrac{1}{4} i = 0$

8. $\dfrac{d}{dt} v + 5v + 6 \displaystyle\int_{-\infty}^{t} v \, d\tau = 0$

9. $\dfrac{d^2}{dt^2} v + 4 \dfrac{d}{dt} v + 5v = 0$

10. $\dfrac{d^2}{dt^2} i + 9 \dfrac{d}{dt} i = 0$

11. Determine and sketch the current i through a series connection of a resistance of $2\ \Omega$ and an inductance of $\frac{1}{10}$ H for $t > 0$, if $i(0^+) = 10$ A.

Find a homogeneous LLTI differential equation of the lowest order that is satisfied by the solutions in Exercises 12 through 14.

12. $5e^{-2t}$

13. $72e^{-2t} + 9e^{-3t}$

14. $5 + 13e^{-2t}$

8.3 Repeated Roots of the Characteristic Equation

Occasionally a circuit is described by a characteristic equation with repeated roots; that is, not all roots are distinct. We can apply the s-shift theorem and integration, as in Eq. (8.15), to reduce a second-order differential equation to a first-order differential equation. A second application of the procedure yields the following theorem.

THEOREM 3

Complementary Response for Second-Order Roots

A second-order homogeneous linear differential equation that can be written as

$$(p - s_1)^2 v_c = 0 \qquad (8.26)$$

has a characteristic equation

$$(s - s_1)^2 = 0 \qquad (8.27)$$

with *second-order* root

$$s = s_1 \qquad (8.28)$$

Equation (8.26) has a complementary response of the form

$$v_c = (A_0 + A_1 t)e^{s_1 t} \qquad (8.29)$$

where A_0 and A_1 are the two constants of integration.

We will now revisit our second-order circuit of Example 8.1.

EXAMPLE 8.7

For the circuit used in Example 8.1, $C = 2$ F, $R = \frac{1}{12}$ Ω, and $L = \frac{1}{18}$ H. Current i is zero for $t > 0$, so the circuit is described by the KCL equation

$$2 \frac{d}{dt} v + 12v + 18 \int_{-\infty}^{t} v \, d\tau = 0$$

which can be differentiated once to give

$$(2p^2 + 12p + 18)v = 0 \qquad t > 0$$

Determine and sketch the response for this circuit, given the initial conditions $v(0^+) = 2$ V and $i_L(0^+) = -38$ A.

■ SOLUTION

The characteristic equation is

$$2s^2 + 12s + 18 = 0$$

which can be written in factored form as

$$2(s + 3)^2 = 0$$

We have a second-order root

$$s_1 = -3$$

Theorem 3 gives the solution to the homogeneous equation as

$$v_c = (A_0 + A_1 t)e^{-3t} \qquad t > 0$$

which is also the complete response v.

We will now use our initial conditions to solve for the constants of integration (A_0 and A_1). We will first use the initial value of the voltage:

$$v(0^+) = 2 = [A_0 + A_1(0^+)]e^{-3(0^+)}$$

which gives us

$$A_0 = 2 \tag{8.30}$$

We need the value of the derivative of v at $t = 0^+$ to obtain a second equation in terms of A_0 and A_1. We can easily obtain this by evaluating the KCL equation for the top node at $t = 0^+$. We differentiate the exponential expression for v and use the numerical values of $v(0^+)$ where the derivative is not needed. The original KCL equation evaluated at $t = 0^+$ is

$$2\frac{d}{dt}[(A_0 + A_1 t)e^{-3t}]\Big|_{t=0^+} + 12v(0^+) + i_L(0^+) = 0$$

which gives

$$2(-3A_0 + A_1) + 12(2) + (-38) = 0$$

This gives us

$$-3A_0 + A_1 = 7 \tag{8.31}$$

Equations (8.30) and (8.31) give

$$A_0 = 2$$

and

$$A_1 = 13$$

Therefore

$$v = (2 + 13t)e^{-3t} \text{ V} \qquad t > 0$$

This is sketched in Fig. 8.5. ∎

FIGURE 8.5
Complementary response of a second-order circuit with second-order roots

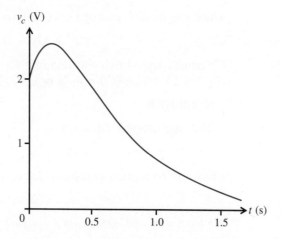

Although electric circuits are frequently described by high-order linear differential equations, the characteristic equation rarely has roots of higher order than 2. However, roots of higher order than 2 cause no mathematical problems. Application of the s-shift theorem and integration n times, as in Eq. (8.15), yields the following extension of Theorems 2 and 3.

THEOREM 4

Complementary Response for Repeated Roots

For an nth-order homogeneous linear differential equation with constant coefficients,

$$\mathscr{A}(p)v_c = 0 \tag{8.32}$$

the characteristic equation can always be written in the factored form as

$$a_n(s - s_1)^{m_1}(s - s_2)^{m_2} \cdots (s - s_l)^{m_l} = 0 \tag{8.33}$$

where

$$m_1 + m_2 + \cdots + m_l = n \tag{8.34}$$

and $s_1, s_2, \ldots, s_l$ are distinct.

The complementary response is of the form

$$v_c = P_1(t)e^{s_1 t} + P_2(t)e^{s_2 t} + \cdots + P_l(t)e^{s_l t} \tag{8.35}$$

where

$$P_k(t) = A_{k,0} + A_{k,1}t + \cdots + A_{k,m_k - 1}t^{m_k - 1} \tag{8.36}$$

and

$$A_{k,j} \text{ for } k = 1, \ldots, l \quad \text{and} \quad j = 0, \ldots, m_k - 1 \tag{8.37}$$

are the n constants of integration.

REMEMBER

Mth-order factors of the form $(s - s_1)^m$ in the characteristic polynomial introduce terms of the form $(A_0 + A_1 t + \cdots + A_{m-1}t^{m-1})e^{s_1 t}$ into the complementary response, where $A_0, A_1, \ldots, A_{m-1}$ are constants of integration. In particular, when $m = 2$, the corresponding term is $(A_0 + A_1 t)e^{s_1 t}$. For a second-order circuit, we require two equations to evaluate the constants of integration. The first equation is obtained from the initial value of the variable. The second must include the initial value of the derivative of the variable. This second equation is a KVL or KCL equation.

EXERCISES

Determine the solution to the homogeneous differential equations in Exercises 15 through 17. (Constants of integration are not to be evaluated.)

15. $\dfrac{d^2}{dt^2}i + 6\dfrac{d}{dt}i + 9i = 0$ 　　　　　　16. $p^2(p + 2)v = 0$

17. $(p^3 + 3p^2 + 3p + 1)v = 0$

An inductance $L = 2$ H, a resistance $R = 8\ \Omega$, and a capacitance $C = \frac{1}{8}$ F are connected in series.

18. Determine the second-order differential equation in terms of current i, that describes this circuit.

19. Solve the differential equation developed in Exercise 18.

20. Determine the constants of integration that appear in the solution of Exercise 19 if the initial current is $i(0^+) = 12$ A and the initial capacitance voltage is (passive sign convention) $v(0^+) = 36$ V.

Determine a homogeneous differential equation of lowest order that is satisfied by the solutions given in Exercises 21 through 24.

21. $(2 + 3t)e^{-6t}$

22. $4e^{-5t} + (7 + 9t)e^{-8t}$

23. $5t + e^{-2t}$

24. $(5 + 6t + 7t^2)e^{-t}$

8.4 Stability

Examination of the responses calculated in Examples 8.5, 8.6, and 8.7 reveals that in each case the complementary response asymptotically approaches zero as t becomes infinite. This is because the roots of the characteristic equation are negative, which causes the exponential response to decay to zero with time. These are examples of stable circuits.

DEFINITION

Stable Network

A network described by an LLTI differential equation whose characteristic equation has only roots with negative real parts is called a **stable network**, because the complementary response will asymptotically approach zero as t approaches infinity. A network that is not stable is either *marginally stable* or *unstable*.

We now consider a circuit that is unstable.

EXAMPLE 8.8

The value of current i in the circuit of Fig. 8.6 is 10 A at time zero. Determine $i(t)$ for $t \geq 0$.

FIGURE 8.6
An unstable first-order circuit

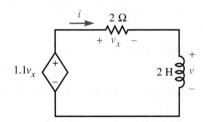

■ **SOLUTION**

Writing a KVL equation around the loop gives us

$$-1.1(2i) + 2i + 2\frac{d}{dt}i = 0$$

After dividing this equation by 2, we can write it as

$$(p - 0.1)i = 0$$

The characteristic equation

$$s - 0.1 = 0$$

has the root

$$s_1 = 0.1$$

The root is real and positive, so the circuit is unstable with natural response

$$i_c = A_1 e^{0.1t} \qquad t \geq 0$$

that is also the complete response i.

We can easily evaluate the constant of integration

$$i(0^+) = 10 = A_1 e^{0.1(0)}$$

which gives

$$A_1 = 10$$

and

$$i = 10e^{0.1t} \text{ A} \qquad t \geq 0 \blacksquare$$

The response found in the preceding example is sketched in Fig. 8.7. We can see that the positive root ($s_1 = 0.1$) causes the response to naturally increase with time. Physically it is impossible for current to increase without bound. Practically the current will become so large that some component will become nonlinear or be destroyed and thus alter the circuit.

FIGURE 8.7
Complementary response of an unstable first-order circuit

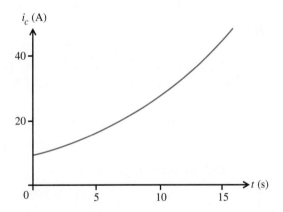

We will next consider a *marginally stable* circuit, which is often considered to be a special case of an unstable circuit.

EXAMPLE 8.9

Energy has been stored in the capacitances of Fig. 8.8, and the switch is opened at $t = 0$. The initial capacitance voltages just after the switch is opened are $v_1(0^+) = 12$ V and $v_2(0^+) = 0$. Calculate and sketch the voltage v_1 for $t > 0$.

FIGURE 8.8
A marginally stable circuit

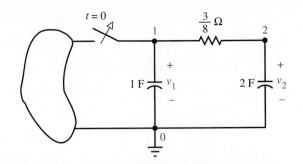

■ **SOLUTION**

We begin by writing two node-voltage equations (KCL equations):

$$\frac{d}{dt}v_1 + \frac{8}{3}(v_1 - v_2) = 0 \qquad t > 0 \tag{8.38}$$

and

$$\frac{8}{3}(v_2 - v_1) + 2\frac{d}{dt}v_2 = 0 \qquad t > 0 \tag{8.39}$$

We can write these two simultaneous differential equations in matrix form as

$$\begin{bmatrix} p + \dfrac{8}{3} & -\dfrac{8}{3} \\ -\dfrac{8}{3} & 2p + \dfrac{8}{3} \end{bmatrix} \cdot \begin{bmatrix} v_1 \\ v_2 \end{bmatrix} = \begin{bmatrix} 0 \\ 0 \end{bmatrix}$$

We can develop a single differential equation in terms of v_1 by using Cramer's rule as we would for algebraic equations:

$$\begin{vmatrix} p + \dfrac{8}{3} & -\dfrac{8}{3} \\ -\dfrac{8}{3} & 2p + \dfrac{8}{3} \end{vmatrix} v_1 = \begin{vmatrix} 0 & -\dfrac{8}{3} \\ 0 & 2p + \dfrac{8}{3} \end{vmatrix}$$

Evaluation of the determinants gives us the differential equation

$$(2p^2 + 8p)v_1 = 0$$

The characteristic equation

$$\mathscr{A}(s) = 2s^2 + 8s = 2s(s + 4) = 0$$

has roots

$$s_1 = 0$$

and

$$s_2 = -4$$

The real part of s_1 is zero and not negative, so the complementary response contains a constant term. The characteristic equation represents a marginally stable circuit. (Since the real part of this first-order root is zero, rather than positive, the complementary response does not grow with time, but it does not asymptotically approach zero.) These roots give a complementary response of

$$v_{1c} = A_1 e^{0t} + A_2 e^{-4t}$$
$$= A_1 + A_2 e^{-4t} \qquad t > 0 \qquad (8.40)$$

that is also the complete response v_1.

We will solve for the constants of integration (A_1 and A_2). We first use the value of v_1 at $t = 0^+$ and Eq. (8.40):

$$v_1(0^+) = 12 = A_1 + A_2 e^{-4(0)}$$

which gives

$$A_1 + A_2 = 12$$

We need the value of the derivative of v_1 at $t = 0^+$ to obtain a second equation in terms of A_1 and A_2. We can easily obtain this equation by evaluating our very first KCL equation [Eq. (8.38)] at $t = 0^+$. We differentiate the exponential expression for v_1 and use numerical values of v_1 and v_2 where the derivative is not needed. Equation (8.38) becomes

$$\frac{d}{dt}(A_1 + A_2 t e^{-4t})\Big|_{t=0^+} + \frac{8}{3}(12 - 0) = 0$$

which gives

$$-4A_2 e^{-4t}\Big|_{t=0^+} + 32 = 0$$

Substitution of 0^+ for t gives

$$A_2 = 8$$

Thus

$$A_1 + A_2 = 12$$

yields

$$A_1 = 4$$

and

$$v_{1c} = 4 + 8e^{-4t} \text{ V} \qquad t > 0$$

The response is sketched in Fig. 8.9. Observe that for this marginally stable circuit, the complementary response does not go to zero as time goes to infinity, but it does not go to infinity. ∎

FIGURE 8.9
Complementary response of a marginally stable circuit

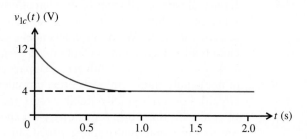

We next consider a third-order equation.

EXAMPLE 8.10

Determine the complementary response for a network described by the source-free equation

$$\frac{d^3}{dt^3}i + 4\frac{d^2}{dt^2}i + 5\frac{d}{dt}i + 2i = 0 \qquad t > 0$$

■ SOLUTION

The differential equation can be written in operator notation as

$$(p^3 + 4p^2 + 5p + 2)i_c = 0 \qquad t > 0$$

Solve for the roots of the characteristic equation:

$$\mathscr{A}(s) = s^3 + 4s^2 + 5s + 2 = (s + 1)^2(s + 2) = 0$$

A simple root is $s = -2$, and a second-order root is $s = -1$. The roots are real and negative, so the circuit is stable. The complementary response [see Eq. (8.35)] is

$$i_c = A_1e^{-2t} + (A_2 + A_3t)e^{-t} \qquad t > 0$$

Additional information, such as the value of i and its first two derivatives at $t = 0^+$, would be required to evaluate the constants of integration A_1, A_2, and A_3. The response is sketched in Fig. 8.10 for $A_1 = -1$ and $A_2 = A_3 = 1$. Observe that this is a stable circuit, and so the complementary response goes to zero as time goes to infinity. ■

FIGURE 8.10
Complementary response for a stable third-order circuit

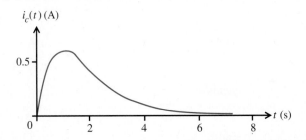

In the next two sections we will examine two first-order circuits of practical importance.

REMEMBER

If the real parts of all roots of the characteristic equation are negative, the equation represents a stable circuit. For a stable circuit, the complementary response goes to zero with time. When all independent sources are set equal to zero in a stable circuit, all voltages and currents must eventually go to zero.

Any first-order roots with zero real part cause the circuit to be marginally stable (a marginally stable circuit is often considered to be a special case of an unstable circuit). For a marginally stable circuit, the complementary response does not go to zero with time, but does not grow exponentially with time. When all sources are zero in a marginally stable circuit, all voltages and currents either asymptotically become constant or, as we shall later see in Section 8.7, oscillate with fixed amplitude.

Any roots of order greater than 1 with zero real part cause the circuit to be unstable. Roots with a positive real part also cause the circuit to be unstable. For an unstable circuit, the complementary response grows without bound as time increases. Thus, even with all independent sources set equal to zero, the voltages and currents in an unstable circuit will continue to increase with time. In practice, the voltages or currents will eventually be limited by nonlinearities in the physical devices or by self-destruction of the circuit.

EXERCISES

25. Determine and sketch the voltage $v(t)$ for $t > 0$ for the following source-free circuit, if $v(0^+) = 10$ V (source free implies no independent sources). Is the circuit stable?

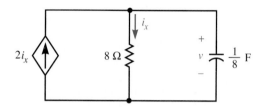

26. Refer to the circuit of Example 8.9, which is shown in Fig. 8.8. Use Cramer's rule and the node-voltage equations to obtain a second-order differential equation that can be solved for v_2. Solve this equation, and evaluate the constants of integration. Is the response for v_2 of the same form (except for the constants of integration) as the response for v_1?

8.5 The Source-Free *RC* Circuit

Consider the circuit shown in Fig. 8.11. Energy is stored in the capacitance by network X, so when the switch is opened at $t = 0$, a voltage $v(0^+) = V_0$ appears across the capacitance. Application of KCL to a surface enclosing the top node gives

$$C\frac{d}{dt}v + \frac{1}{R}v = 0 \qquad t > 0 \tag{8.41}$$

which is a source-free or homogeneous first-order linear differential equation. We can write Eq. (8.41) as

$$\left(p + \frac{1}{RC}\right)v_c = 0, \qquad t > 0 \tag{8.42}$$

FIGURE 8.11
An *RC* circuit that is source free for $t > 0$

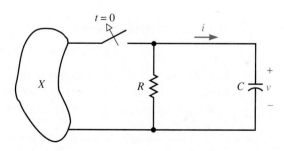

(A subscript c is added to v as a reminder that it is the solution to a homogeneous equation.) The characteristic equation

$$s + \frac{1}{RC} = 0 \qquad (8.43)$$

has a root

$$s = -\frac{1}{RC} \qquad (8.44)$$

This gives a complementary response of

$$v_c = Ae^{-(1/RC)t} \qquad t > 0 \qquad (8.45)$$

⬛ *See Problem 8.1 of the PSpice manual.*

Because the circuit is source free, the complementary response v_c is the complete response v, so that $v(t) = v_c(t)$ for $t > 0$. We can evaluate the constant of integration from the value of v at $t = 0^+$:

$$v(0^+) = V_0 = Ae^0 = A \qquad (8.46)$$

Thus the capacitance voltage is

$$v = V_0 e^{-t/RC} \qquad t > 0 \qquad (8.47)$$

This is shown in Fig. 8.12.

FIGURE 8.12
Source-free response of a simple *RC* circuit

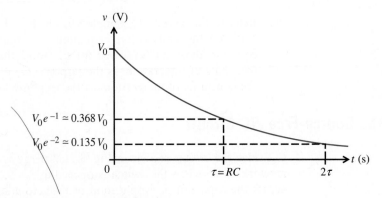

With the *time constant* defined as

$$\tau = RC \qquad (8.48)$$

the source-free response is $\quad v = V_0 e^{-t/\tau} \qquad t > 0 \qquad (8.49)$

Note that after one time constant,

$$v(\tau) = V_0 e^{-1} \simeq 0.368 V_0 \qquad (8.50)$$

after two time constants,

$$v(2\tau) = V_0 e^{-2} \simeq 0.135 V_0 \qquad (8.51)$$

and after five time constants,

$$v(5\tau) = V_0 e^{-5} \simeq 0.0067 V_0 \qquad (8.52)$$

For this reason we often say that in a source-free circuit the voltage across a capacitance is negligible, compared with its original value, after five time constants. We can find capacitance current i from the terminal equation

$$i = C \frac{d}{dt} v = C \frac{d}{dt} V_0 e^{-t/RC} = -\frac{1}{R} V_0 e^{-t/RC} \qquad t > 0 \qquad (8.53)$$

or by applying KCL to the original circuit:

$$i = -\frac{1}{R}v \qquad (8.54)$$

The source-free response (voltage or current) of an RC circuit is of the form $Ae^{-t/\tau}$. The time constant is $\tau = RC$.

EXERCISES

27. Obtain a differential equation in terms of voltage v, and solve for the response of the source-free circuit shown below if $v(0^+) = 10$ V.

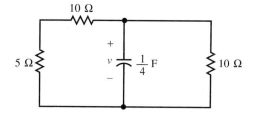

28. For the following circuit, write a differential equation in terms of i, and solve for the source-free response if $v(5) = 10$ V. (Source free implies no independent sources.)

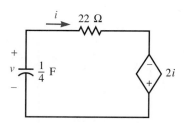

8.6   The Source-Free *RL* Circuit

Consider the circuit shown in Fig. 8.13. Energy is stored in the inductance by network X. The switch is opened at $t = 0$, giving a current $i(0^+) = I_0$ through the inductance.

FIGURE 8.13
An *RL* circuit that is
source free for $t > 0$

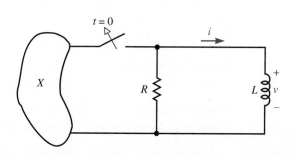

Application of KVL quickly gives

$$L\frac{d}{dt}i + Ri = 0 \qquad t > 0 \tag{8.55}$$

or

$$\left(p + \frac{R}{L}\right)i_c = 0 \tag{8.56}$$

The characteristic equation

$$s + \frac{R}{L} = 0 \tag{8.57}$$

has a root

$$s = -\frac{R}{L} \tag{8.58}$$

This gives a complementary response of

$$i_c = Ae^{-(R/L)t} \qquad t > 0 \tag{8.59}$$

that is also the complete response i.

■ See Problem 8.2 of the PSpice manual.
The constant of integration can be evaluated from the value of i at time 0^+:

$$i(0^+) = I_0 = Ae^0 = A \tag{8.60}$$

Thus the inductance current is

$$i = I_0 e^{-(R/L)t} \qquad t > 0 \tag{8.61}$$

as shown in Fig. 8.14.

FIGURE 8.14
Source-free response of a simple *RL* circuit

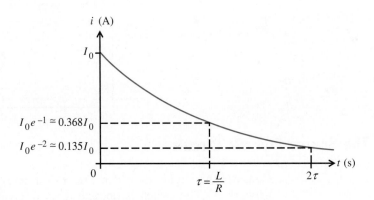

With the *time constant* defined as

$$\tau = \frac{L}{R} \tag{8.62}$$

the source-free response is

$$i = I_0 e^{-t/\tau} \qquad t > 0 \tag{8.63}$$

As with capacitance voltage, we often say that the current through an inductance is negligible, compared with its original value, after five time constants.

We can calculate inductance voltage v from the terminal equation

$$v = L\frac{d}{dt}i = L\frac{d}{dt}I_0 e^{-(R/L)t} = -RI_0 e^{-(R/L)t} \qquad t > 0 \qquad (8.64)$$

or by applying KVL to the series RL circuit:

$$v = -Ri \qquad (8.65)$$

We should note that the results obtained in this section could have been obtained from those of the preceding section by the use of duality.

REMEMBER

The source-free response (current or voltage) of an RL circuit is of the form $Ae^{-t/\tau}$. The time constant is $\tau = L/R$.

EXERCISES

29. Obtain a differential equation in terms of current i, and solve for the response in the following source-free circuit if $i(0^+) = 10$ A.

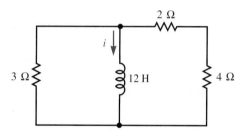

30. For the following circuit, write a differential equation in terms of voltage v, and solve for the source-free response if $i(5) = 10$ A. (Source free implies no independent sources.)

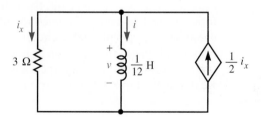

8.7 Complex Roots of the Characteristic Equation

The technique that is contained in Theorems 2, 3, and 4 is all that is required for us to find the complementary response, with the exception of evaluation of the constants of integration, for any LLTI differential equation. If some of the roots of the characteristic equation are complex, which is often the case in networks containing both inductance and capacitance, we can write the solution in a more convenient form.

EXAMPLE 8.11

The KCL equation that describes the voltage across a parallel combination of a 2-F capacitance and an $\frac{1}{8}$-H inductance is

$$2\frac{d}{dt}v + 8\int_{-\infty}^{t} v\,d\tau = 0 \qquad t > 0$$

Determine the solution.

■ SOLUTION

The equation can be differentiated once, divided by 2, and written in operator notation as

$$(p^2 + 4)v_c = 0$$

The roots of the characteristic equation

$$\mathscr{A}(s) = s^2 + 4 = 0$$

are

$$s = \pm\sqrt{-4} = \pm j2$$

From Theorem 2, the resulting complementary response is

$$v_c = A_1 e^{j2t} + A_2 e^{-j2t}$$

Euler's identity can be used to write v_c as

$$v_c = A_1(\cos 2t + j\sin 2t) + A_2(\cos 2t - j\sin 2t)$$
$$= (A_1 + A_2)\cos 2t + j(A_1 - A_2)\sin 2t$$

This is of the form

$$v_c = A\cos 2t + B\sin 2t \qquad t > 0$$

We will see in Theorem 5 that we can write this final equation *directly* once the roots of the characteristic equation are known. *We need not write an equation containing the complex exponential.* ■

Observe that the roots of the characteristic equation in the preceding example have zero real part. As a result, the voltage v across the elements does not asymptotically tend to zero as t goes to infinity. It oscillates forever. This equation describes a marginally stable circuit. The complementary response is the complete response, because the equation is homogeneous. The constants of integration are typically evaluated by using the value of v and its derivative at some time t_0, as we did in Example 8.9. The response is sketched in Fig. 8.15 for $A = B = 1$.

FIGURE 8.15
Complementary response of a marginally stable second-order circuit

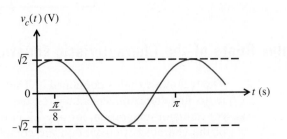

With Example 8.11 for guidance, we can prove the following theorem.

THEOREM 5

Complex Roots of Order 1

First-order roots of the characteristic equation of the form[†]

$$s_1 = \sigma + j\omega \qquad \text{of order 1} \tag{8.66}$$

and

$$s_2 = s_1^* = \sigma - j\omega \qquad \text{of order 1} \tag{8.67}$$

introduce terms into the complementary response that can be written as

$$e^{\sigma t}(A \cos \omega t + B \sin \omega t) \tag{8.68}$$

where A and B are constants of integration.

PROOF

Theorem 2 states that first-order roots of the form given by Eqs. (8.66) and (8.67) introduce terms in the complementary response that can be written by the use of Euler's identity as

$$
\begin{aligned}
A_1 e^{(\sigma + j\omega)t} + A_2 e^{(\sigma - j\omega)t} &= e^{\sigma t}(A_1 e^{j\omega t} + A_2 e^{-j\omega t}) \\
&= e^{\sigma t}[(A_1 \cos \omega t + jA_1 \sin \omega t) + (A_2 \cos \omega t - jA_2 \sin \omega t)] \\
&= e^{\sigma t}[(A_1 + A_2) \cos \omega t + j(A_1 - A_2) \sin \omega t] \\
&= e^{\sigma t}(A \cos \omega t + B \sin \omega t) \tag{8.69}
\end{aligned}
$$

where

$$A = A_1 + A_2 \tag{8.70}$$

and

$$B = j(A_1 - A_2) \tag{8.71}$$

We now apply Theorem 5 to a stable second-order circuit.

EXAMPLE 8.12

Determine the complementary response for the following KCL equation:

$$\frac{1}{52} \frac{d}{dt} v + \frac{1}{13} v + \frac{1}{4} \int_{-\infty}^{t} v \, d\tau = 0 \qquad t > 0$$

which describes the voltage across a parallel RLC circuit with $C = \frac{1}{52}$ F, $R = 13\ \Omega$, and $L = 4$ H.

■ SOLUTION

Multiplication by 52 and differentiation lets us write the equation in operator notation as

$$(p^2 + 4p + 13)v_c = 0$$

[†] Any complex roots of a polynomial with real coefficients must appear as complex conjugate pairs. This will be established in Chapter 13.

The roots of the characteristic equation

$$s^2 + 4s + 13 = 0$$

can be obtained from the standard formula for the solution of a quadratic equation:

$$s = \frac{-4 \pm \sqrt{16 - 52}}{2}$$

$$= -2 \pm j3$$

The complementary response is given by Theorem 5 as

$$v_c = e^{-2t}(A \cos 3t + B \sin 3t)$$

The real parts of the roots of the characteristic equation are negative and equal to -2, so this differential equation represents a stable system, and the complementary response, which is sketched in Fig. 8.16 for $A = B = 1$, will decay to zero as time t goes to infinity. ■

FIGURE 8.16
Complementary response of
a stable second-order circuit

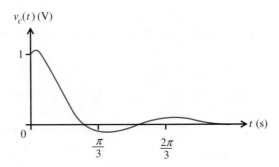

On rare occasions, complex roots are of order greater than 1. We can use Theorem 4 and Euler's identity to establish the following result.

THEOREM 6

Complex Roots of Order Greater than 1

Mth-order roots of the characteristic equation of the form

$$s_1 = \sigma + j\omega \qquad \text{of order } m \tag{8.72}$$

and

$$s_2 = s_1^* = \sigma - j\omega \qquad \text{of order } m \tag{8.73}$$

introduce terms into the complementary response that we can write as

$$e^{\sigma t}[(A_0 + A_1 t + \cdots + A_{m-1}t^{m-1}) \cos \omega t$$
$$+ (B_0 + B_1 t + \cdots + B_{m-1}t^{m-1})] \sin \omega t \tag{8.74}$$

where A_i and B_i, $i = 0, \ldots, m-1$, are the constants of integration.

In the following section, we will investigate how the relationship between R, L, and C in a series circuit determines whether the roots of the characteristic equation are real and distinct, second-order, or complex.

Complex roots of the characteristic equation always appear in complex conjugate pairs. A pair of roots of the form $s_1 = \sigma_1 + j\omega_1$ and $s_2 = s_1^*$ introduce a term into the complementary response of the form

$$e^{\sigma_1 t}(A \cos \omega_1 t + B \sin \omega_1 t)$$

EXERCISES

Find the solution to the homogeneous differential equations in Exercises 31 and 32.

31. $(p^2 + 4p + 5)v = 0$ 32. $(p^2 + 9)i = 0$

Find a homogeneous LLTI differential equation of the lowest order that is satisfied by the solutions given in Exercises 33 through 38.

33. $3 \cos 3t + 4 \sin 3t$

34. $5 \cos 3t$

35. $e^{-2t}(3 \cos 5t + 4 \sin 5t)$

36. $e^{-t}[(2 + t) \cos 6t + (3 + 9t) \sin 6t]$

37. $e^{-t} + e^{-2t} \sin 6t$

38. $t^2 + \cos 4t$

8.8   Source-Free Series *RLC* Circuits

Electrical networks containing resistance, inductance, and capacitance have an interesting property: The roots of the characteristic equation may be real, imaginary, or complex. The roots may also be distinct (first order) or of higher order. This phenomenon is demonstrated by the series *RLC* circuit shown in Fig. 8.17.

FIGURE 8.17
A series *RLC* circuit

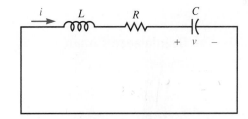

We can find current i by writing and solving the KVL equation

$$L\frac{d}{dt}i + Ri + \frac{1}{C}\int_{-\infty}^{t} i\,d\tau = 0 \tag{8.75}$$

  See Example 3.6 and Problem 8.3 in the PSpice manual.

After dividing by L, we can differentiate the equation once and write it in operator notation as

$$\left(p^2 + \frac{R}{L}p + \frac{1}{LC}\right)i = 0 \tag{8.76}$$

We can write the characteristic equation

$$s^2 + \frac{R}{L}s + \frac{1}{LC} = 0 \tag{8.77}$$

in the standard form

$$s^2 + 2\zeta\omega_0 s + \omega_0^2 = 0 \tag{8.78}$$

The *undamped natural frequency* ω_0 for the series *RLC* circuit is found by equating the coefficients $1/LC$ and ω_0^2 in Eqs. (8.77) and (8.78):

$$\omega_0 = \frac{1}{\sqrt{LC}} \tag{8.79}$$

Equating the coefficients of s in Eqs. (8.77) and (8.78) yields the *damping ratio* ζ for the series *RLC* circuit:

$$\zeta = \frac{1}{2} R \sqrt{\frac{C}{L}} \tag{8.80}$$

The roots of the characteristic equation are

$$s = -\zeta\omega_0 \pm \omega_0 \sqrt{\zeta^2 - 1} \tag{8.81a}$$

$$= -\alpha \pm \omega_0 \sqrt{\zeta^2 - 1} \tag{8.81b}$$

with the *damping coefficient* α for the series circuit given by

$$\alpha = \zeta\omega_0 \tag{8.82a}$$

$$= \frac{R}{2L} \tag{8.82b}$$

There are three cases, each of which is of interest:

1. Overdamped: $\zeta > 1$. The two roots are real, negative, and distinct.
2. Critically damped: $\zeta = 1$. The two roots are real and negative, but identical.
3. Underdamped: $\zeta < 1$. The two roots are imaginary or complex with a negative real part.

The Overdamped Circuit

If

$$\alpha > \omega_0 \tag{8.83}$$

then, from Eq. (8.82a),

$$\zeta = \frac{\alpha}{\omega_0} > 1 \tag{8.84}$$

and the two roots are real, negative, and distinct:

$$s = -\alpha \pm \sqrt{\alpha^2 - \omega_0^2} \tag{8.85a}$$

$$= \sigma_1, \sigma_2 \tag{8.85b}$$

From Theorem 2, the complementary response is of the form

$$i_c = A_1 e^{\sigma_1 t} + A_2 e^{\sigma_2 t} \tag{8.86}$$

The complementary response of an overdamped circuit is not oscillatory (a mechanical analogy is the ride of an automobile that has very stiff shock absorbers).

EXAMPLE 8.13

For the series RLC circuit shown in Fig. 8.17, let $R = \frac{25}{8}$ Ω, $L = \frac{1}{64}$ H, and $C = \frac{1}{100}$ F. Find the source-free response if the initial conditions are $i(0^+) = 8$ A and $v(0^+) = 16$ V.

■ SOLUTION

The KVL equation written around the loop is

$$\frac{1}{64}\frac{d}{dt}i + \frac{25}{8}i + 100 \int_{-\infty}^{t} i\,d\tau = 0 \qquad t > 0$$

or, after differentiation and multiplication by 64,

$$(p^2 + 200p + 6400)i = 0 \qquad t > 0$$

The characteristic equation

$$s^2 + 200s + 6400 = 0$$

has roots

$$s = -40, -160$$

The roots are real, negative, and distinct, so the network is overdamped. The complementary response is thus [see Eq. (8.23)]

$$i_c = A_1 e^{-40t} + A_2 e^{-160t}$$

Because the circuit is source-free, the complementary response i_c is the complete response i. Now evaluate the constants of integration by use of the initial conditions:

$$i(0^+) = A_1 e^{-40(0)} + A_2 e^{-160(0)} = 8$$

or

$$A_1 + A_2 = 8 \tag{8.87}$$

Knowledge of the first derivative of i is needed to obtain a second equation containing A_1 and A_2. This is easily found from the original KVL equation evaluated at $t = 0^+$. The KVL equation

$$L\frac{d}{dt}i + Ri + \frac{1}{C}\int_{-\infty}^{t} i\,d\tau = 0$$

becomes

$$L\frac{d}{dt}i\bigg|_{t=0^+} + Ri(0^+) + v(0^+) = 0$$

The derivative is

$$\frac{d}{dt}i\bigg|_{t=0^+} = (-40A_1 e^{-40t} - 160A_2 e^{-160t})\bigg|_{t=0^+}$$

$$= -40A_1 - 160A_2$$

so

$$L(-40A_1 - 160A_2) + Ri(0^+) + v(0^+) = 0$$

Substitution for R, L, $i(0^+)$, and $v(0^+)$ gives

$$\frac{5}{8} A_1 + \frac{5}{2} A_2 = 41 \qquad (8.88)$$

Equations (8.87) and (8.88) can be solved for $A_1 = -\frac{56}{5}$ and $A_2 = \frac{96}{5}$ to give

$$i = -\frac{56}{5} e^{-40t} + \frac{96}{5} e^{-160t} \qquad t > 0$$

It follows from the element values that

$$\alpha = \frac{R}{2L} = 100$$

$$\omega_0 = \frac{1}{\sqrt{LC}} = 80 \text{ rad/s}$$

and the damping ratio is greater than 1, as expected for an overdamped circuit:

$$\zeta = \frac{\alpha}{\omega_0} = \frac{5}{4} > 1 \quad \blacksquare$$

The Critically Damped Circuit

If the series resistance is decreased, the power absorbed (Ri^2) will decrease, and the natural response or complementary response will change.

If

$$\alpha = \omega_0 \qquad (8.89)$$

then

$$\zeta = \frac{\alpha}{\omega_0} = 1 \qquad (8.90)$$

and the two roots are real and negative, but not distinct. Thus

$$s = -\alpha \pm \omega_0 \sqrt{1 - 1}$$
$$= -\alpha, -\alpha$$
$$= \sigma \qquad (8.91)$$

From Theorem 3, the second-order root gives a complementary response of the form [see Eq. (8.29)]

$$i_c = (A_0 + A_1 t)e^{\sigma t}$$
$$= (A_0 + A_1 t)e^{-(R/2L)t} \qquad (8.92)$$

A critically damped circuit has the smallest damping ratio possible without causing an oscillatory response.

EXAMPLE 8.14

For the series RLC circuit shown in Fig. 8.17, change the resistance to $R = \frac{5}{2}\,\Omega$, and leave the inductance, capacitance, and initial conditions the same values as in Example 8.13. Find the source-free response.

SOLUTION

The KVL equation written around the loop is

$$\frac{1}{64}\frac{d}{dt}i + \frac{5}{2}i + 100\int_{-\infty}^{t}i\,d\tau = 0 \qquad t > 0$$

or, after differentiation and multiplication by 64,

$$(p^2 + 160p + 6400)i = 0 \qquad t > 0$$

The characteristic equation

$$s^2 + 160s + 6400 = 0$$

has roots

$$s = -80, -80$$

Thus the equation has a second-order root, and the network is critically damped. The complementary response is of the form

$$i_c = (A_0 + A_1t)e^{-80t}$$

Current i_c is also the complete response i.

The constants of integration are evaluated as in the previous example and give the complete response

$$i = (8 - 1664t)e^{-80t} \text{ A} \qquad t < 0$$

It follows from the element values that

$$\alpha = \frac{R}{2L} = 80$$

$$\omega_0 = \frac{1}{\sqrt{LC}} = 80$$

and the damping ratio is 1, as expected for a critically damped circuit:

$$\zeta = \frac{\alpha}{\omega_0} = 1 \quad \blacksquare$$

The Underdamped Circuit

If the series resistance is decreased below the value that gives critical damping, the Ri^2 losses are reduced to a level that causes the natural response or complementary response to become oscillatory. If

$$\alpha < \omega_0 \tag{8.93}$$

then

$$\zeta = \frac{\alpha}{\omega_0} < 1 \tag{8.94}$$

and the two roots form a complex conjugate pair with negative real part:

$$s = -\alpha \pm \omega_0\sqrt{\zeta^2 - 1} \tag{8.95a}$$

$$= -\alpha \pm j\omega_0\sqrt{1 - \zeta^2} \tag{8.95b}$$

$$s = -\alpha \pm j\omega_d \qquad (8.96)$$

with the *damped natural frequency* ω_d defined by

$$\omega_d = \omega_0\sqrt{1 - \zeta^2} \qquad (8.97)$$

From Theorem 5, the complex roots give a complementary response of the form

$$i_c = e^{-\alpha t}(A\cos\omega_d t + B\sin\omega_d t) \qquad (8.98)$$

The name *damped natural frequency* is applied to ω_d because this is the radian frequency of the sinusoidal factor. The undamped natural frequency ω_0 is the radian frequency of the sinusoidal factor when the damping ratio is zero (the series resistance is zero). From Eqs. (8.94) and (8.97), $\omega_d < \omega_0$.

EXAMPLE 8.15

For the series *RLC* circuit shown in Fig. 8.17, change the resistance to $R = \frac{3}{2}\,\Omega$, and leave the inductance, capacitance, and initial conditions the same. Find the source-free response.

■ SOLUTION

The KVL equation written around the loop is

$$\frac{1}{64}\frac{d}{dt}i + \frac{3}{2}i + 100\int_{-\infty}^{t} i\,d\tau = 0 \qquad t > 0$$

or, after differentiation and multiplication by 64,

$$(p^2 + 96p + 6400)i = 0 \qquad t > 0$$

The characteristic equation

$$s^2 + 96s + 6400 = 0$$

has roots

$$s = -48 \pm j64$$

Thus the equation has complex roots, and the network is underdamped.
The complementary response is therefore

$$i_c = e^{-48t}(A\cos 64t + B\sin 64t)$$

The constants of integration are evaluated as previously and give the complete response

$$i = e^{-48t}(8\cos 64t - 22\sin 64t) \qquad t > 0$$

It follows from the element values that

$$\alpha = \frac{R}{2L} = 48$$

$$\omega_0 = \frac{1}{\sqrt{LC}} = 80$$

and the damping ratio is less than 1, as expected for an underdamped circuit:

$$\zeta = \frac{\alpha}{\omega_0} = \frac{3}{5} < 1 \quad ■$$

The response of the three series circuits investigated is shown in Fig. 8.18. Note that for the overdamped and critically damped cases, the natural response may change signs once or not at all, depending on the initial conditions. For the underdamped case, the natural response will repeatedly change signs. The smaller the damping coefficient, the more slowly the amplitude of the oscillations will decrease (an automobile with worn-out shock absorbers has a small damping ratio). We should make two observations at this point:

1. In practical circuits the numbers do not usually work out to be integers, although they were in our examples.
2. The critically damped case is rather unusual in practice.

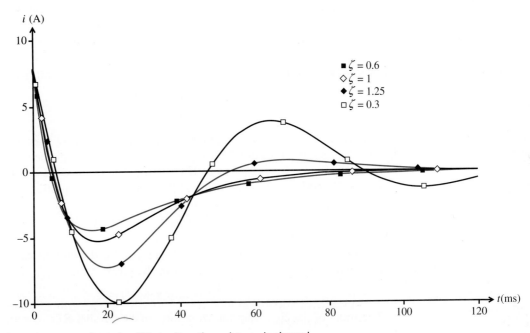

FIGURE 8.18   Natural response of a series *RLC* circuit as the resistance is changed

REMEMBER

A series *RLC* circuit can be overdamped, critically damped, or underdamped. An increase in the resistance increases the damping.

EXERCISES

A series *RLC* circuit is source free for $t > 0$ and has $L = 5$ H and $C = \frac{1}{125}$ F. The inductance current at $t = 0^+$ is $i(0^+) = 4$ A, and the capacitance voltage is $v(0^+) = 200$ V. (The passive sign convention is used.) Write a differential equation in i, and solve for $i(t)$ for $t > 0$ if:

39. $R = 130\ \Omega$ 40. $R = 50\ \Omega$

41. $R = 30\ \Omega$

8.9 Source-Free Parallel *RLC* Circuits

Like series *RLC* circuits, the parallel *RLC* circuit shown in Fig. 8.19 can also exhibit an overdamped, critically damped, or underdamped response.

FIGURE 8.19
A parallel *RLC* circuit

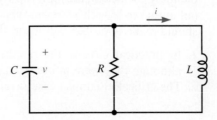

⊞ *See Problem 8.4 of the PSpice manual.*

Voltage v can be found by writing and solving the KCL equation

$$C \frac{d}{dt} v + \frac{1}{R} v + \frac{1}{L} \int_{-\infty}^{t} v \, d\tau = 0 \tag{8.99}$$

This equation can be differentiated once and written in operator notation as

$$\left(p^2 + \frac{1}{RC} p + \frac{1}{LC} \right) v = 0 \tag{8.100}$$

(This result could have been anticipated from the analysis of the series *RLC* circuit and the duality concept.) This equation yields the complementary response or natural response for the parallel *RLC* circuit. The form of the complementary response is determined by the roots of the characteristic equation

$$\left(s^2 + \frac{1}{RC} s + \frac{1}{LC} \right) = 0 \tag{8.101}$$

The unity of the natural response of all second-order circuits can be recognized if we write the characteristic equation in the form

$$(s^2 + 2\zeta\omega_0 s + \omega_0^2) = 0 \tag{8.102}$$

The roots of the characteristic equation for the parallel *RLC* circuit are

$$s = -\zeta\omega_0 \pm \omega_0\sqrt{\zeta^2 - 1} \tag{8.103a}$$

$$= -\alpha \pm \omega_0\sqrt{\zeta^2 - 1} \tag{8.103b}$$

The only difference between the form of the complementary response for the series *RLC* circuit and that for the parallel *RLC* circuit is the dependence of ζ on R, L, and C. As for the series circuit, there are three cases: overdamped, $\zeta > 1$; critically damped, $\zeta = 1$; and underdamped, $\zeta < 1$. For the *parallel RLC* circuit,

$$\zeta = \frac{1}{2R} \sqrt{\frac{L}{C}} = \frac{1}{2RC} \sqrt{LC} \tag{8.104}$$

$$\omega_0 = \frac{1}{\sqrt{LC}} \tag{8.105}$$

and

$$\alpha = \frac{1}{2RC} \tag{8.106}$$

As always, the damping ratio is

$$\zeta = \frac{\alpha}{\omega_0} \qquad (8.107)$$

Because of the dual relationship that exists between the series and parallel circuits, only one example will be given here. (Refer to Examples 8.6 and 8.7 for two additional examples.)

EXAMPLE 8.16

For the parallel *RLC* circuit shown in Fig. 8.19, let $R = \frac{2}{3}\,\Omega$, $L = \frac{1}{100}$ H, and $C = \frac{1}{64}$ F. Find the complementary response if $i(0^+) = 16$ A and $v(0^+) = 8$ V.

■ **SOLUTION**

The KCL equation for a surface enclosing the upper node is

$$\frac{1}{64}\frac{d}{dt}v + \frac{3}{2}v + 100\int_{-\infty}^{t} v\,d\tau = 0 \qquad t > 0$$

or, after differentiation and multiplication by 64,

$$(p^2 + 96p + 6400)v = 0$$

The characteristic equation

$$s^2 + 96s + 6400 = 0$$

has roots

$$s = -48 \pm j64$$

The roots are complex, so the network is underdamped. The complementary response is therefore

$$v_c = e^{-48t}(A\cos 64t + B\sin 64t) \qquad t > 0$$

Because the circuit is source free, the complementary response is the complete response. The constants of integration are easily evaluated from the initial conditions:

$$v(0^+) = e^{-48(0)}\{A\cos[64(0)] + B\sin[64(0)]\} = 8$$

or

$$A = 8 \qquad (8.108)$$

Knowledge of the derivative of v is needed to obtain a second equation containing A and B. This is easily found from the original KCL equation evaluated at $t = 0^+$. The KVL equation

$$C\frac{d}{dt}v + \frac{1}{R}v + \frac{1}{L}\int_{-\infty}^{t} v\,d\tau = 0$$

when evaluated at $t = 0^+$, becomes

$$C\frac{d}{dt}v\bigg|_{t=0^+} + \frac{1}{R}v(0^+) + \frac{1}{L}\int_{-\infty}^{0^+} v\,d\tau = 0$$

The derivative is

$$\frac{d}{dt} v\Big|_{t=0^+} = e^{-48t}(-64A \sin 64t + 64B \cos 64t) - 48e^{-48t}(A \cos 64t + B \sin 64t)\Big|_{t=0^+}$$

$$= 64B - 48A$$

$$= 64B - 384$$

so

$$C(64B - 384) + \frac{1}{R} v(0^+) + i(0^+) = 0$$

Substitution for R, C, $i(0^+)$, and $v(0^+)$ gives

$$\frac{1}{64}(64B - 384) + \frac{3}{2}(8) + 16 = 0 \qquad (8.109)$$

and

$$B = -22$$

The resulting solution is then

$$v = e^{-48t}(8 \cos 64t - 22 \sin 64t) \qquad t > 0$$

It follows from the element values that

$$\alpha = \frac{1}{2RC} = 48$$

$$\omega_0 = \frac{1}{\sqrt{LC}} = 80 \text{ rad/s}$$

and the damping ratio is less than 1, as expected for an underdamped circuit:

$$\zeta = \frac{\alpha}{\omega_0}$$

$$= \frac{3}{5} < 1 \quad \blacksquare$$

If the resistance in the previous example were decreased to $\frac{2}{5}\,\Omega$ with all the elements and initial conditions the same, the circuit would be critically damped, with a response of

$$v = (8 - 1664t)e^{-80t} \qquad t > 0 \qquad (8.110)$$

A further decrease in R to $\frac{8}{25}\,\Omega$ would result in an overdamped circuit, with response

$$v = -\frac{56}{5} e^{-40t} + \frac{96}{5} e^{-160t} \qquad t > 0 \qquad (8.111)$$

Although the definitions of damping coefficient, damping ratio, undamped resonant frequency, and damped resonant frequency are important, *memorization of the equations* for application to *RLC* circuits *is not recommended*. It is usually better to write the appropriate differential equations with the use of KVL and KCL and then solve them.

A parallel RLC circuit is the dual of a series RLC circuit, and can be overdamped, critically damped, or underdamped. An increase in the resistance of a parallel RLC circuit decreases the damping. An increase in the resistance of a series RLC circuit increases the damping.

EXERCISES

A parallel RLC circuit with nodes a and b is source free for $t > 0$ and has $L = 4$ H and $C = \frac{1}{64}$ F. The inductance current at $t = 0^+$ is $i_{ab}(0^+) = 15$ A, and the capacitance voltage is $v_{ab}(0^+) = 64$ V. Write a differential equation, and solve for $v_{ab}(t)$ if:

42. $R = \frac{32}{5}\,\Omega$ 43. $R = 8\,\Omega$

44. $R = \frac{32}{3}\,\Omega$

8.10   Source-Free Response by PSpice

In Chapter 4, we used SPICE (the version PSpice) to calculate the dc response of circuits. We can use the same program to analyze source-free circuits with initial values of capacitance voltage and inductance current specified. We only need to change the program control statement from .OP or .DC to .TRAN, and request a printer plot with .PLOT. We discuss the .TRAN statement after the following example. The .PLOT statement is discussed on page 176 of Chapter 4.

EXAMPLE 8.17

A parallel RLC circuit is as shown in Fig. 8.20. The initial conditions are $i(0) = 16$ A and $v(0) = 8$ V. Plot the capacitance voltage and inductance current for $0 \le t < 0.1$ s at 0.004-s intervals.

FIGURE 8.20
A source-free parallel RLC circuit

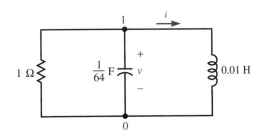

■ SOLUTION

We will use the node numbers as indicated in Fig. 8.20. Our resistance statement is the same as it is for dc analysis, but we must specify the initial values of inductance current and capacitance voltage (see Chapter 4, page 171). The solution control statement is .TRAN, and the .PLOT statement gives us a printer plot. The circuit file is as follows.

```
INPUT FILE FOR EXAMPLE 8.17

*THIS IS A SOURCE-FREE PARALLEL RLC CIRCUIT.

*THE RESISTANCE VALUE IS CHOSEN TO MAKE THIS

*AN UNDERDAMPED CIRCUIT.

*

*INITIAL CONDITIONS: I = 16 A, V = 8 V

R    1    0    1

L    1    0    0.010         IC = 16

C    1    0    0.015625      IC = 8

.TRAN    .004    0.100       UIC

.PLOT    TRAN    V(1)        I(L)

.END
```

Our output file contains, along with a reproduction of the circuit file and some unneeded information, the printer plot shown in Fig. 8.21.

FIGURE 8.21
Printer plot of the natural response of a second-order circuit

```
LEGEND:

*: V(1)
+: I(L)

   TIME          V(1)
(*)----------  -1.0000E+01  -5.0000E+00   0.0000E+00   5.0000E+00   1.0000E+01
(+)----------  -1.0000E+01   0.0000E+00   1.0000E+01   2.0000E+01   3.0000E+01

  0.000E+00  8.000E+00 .           .           .           +         .   *    .
  4.000E-03  2.328E+00 .           .           .         *   + .          .
  8.000E-03 -2.287E+00 .           .       *   .         +   .           .
  1.200E-02 -5.662E+00 .           .    *  .           +   .           .
  1.600E-02 -7.773E+00 .      *     .           +   .           .
  2.000E-02 -8.716E+00 .  *         .           +   .           .
  2.400E-02 -8.669E+00 .  *         .       +   .           .
  2.800E-02 -7.858E+00 .      *     .   +   .           .           .
  3.200E-02 -6.528E+00 .        *   .+   .           .           .
  3.600E-02 -4.914E+00 .         + *.           .           .
  4.000E-02 -3.223E+00 .         +  .   *   .           .
  4.400E-02 -1.624E+00 .         +  .       *   .           .
  4.800E-02 -2.391E-01 .         +  .           *.           .
  5.200E-02  8.562E-01 .         +  .           .  *        .
  5.600E-02  1.629E+00 .         +  .           .     *     .
  6.000E-02  2.083E+00 .          + .           .      *    .
  6.400E-02  2.249E+00 .          + .           .       *   .
  6.800E-02  2.177E+00 .         +  .           .       *   .
  7.200E-02  1.927E+00 .          +.            .        *  .
  7.600E-02  1.560E+00 .          + .           .         * .
  8.000E-02  1.137E+00 .          .+            .           .
  8.400E-02  7.074E-01 .          .+            .  *        .
  8.800E-02  3.110E-01 .          .+            . *         .
  9.200E-02 -2.424E-02 .          . +           *           .
  9.600E-02 -2.820E-01 .          .+           *.           .
  1.000E-01 -4.587E-01 .          .+           *.           .
```

We see that the left-hand column of numbers in Fig. 8.21 gives the time at 0.004-s intervals between 0 and 0.1 s. The second column gives the value of the node voltage v_1 at the indicated time. ■

We now examine the .TRAN solution control statement.

The .TRAN Statement

.TRAN statement; See Section B.2.1.3 of the PSpice manual.

The .TRAN statement we used in Example 8.17 is used for analysis of both source-free circuits and circuits with sources. (We will analyze driven circuits in Chapter 9.) The .TRAN statement with its parameters is of the form

```
.TRAN     TSTEP     TFINAL     [TNOPRINT]     [UIC]
```

The parameter TSTEP is the time step value (in seconds) for printing or plotting (this is not the time increment used for the numerical calculations). Parameter TFINAL is the final value (in seconds) of time considered. Parameter TNOPRINT suppresses all output prior to this time, but the calculations are always performed beginning at time zero and ending at time TFINAL. The default value of TNOPRINT is zero. The parameter UIC specifies that the initial conditions given in the capacitance and inductance statements should be used. (In Chapter 9 we will see that when UIC is omitted, the initial conditions are calculated with the use of .OP.)

A graphics package, which is called with .PROBE, is included in PSpice and permits a high-quality graphical output, if a suitable printer is available. We will discuss .PROBE in Chapter 9.

REMEMBER

The .TRAN solution control statement causes the source-free response to be calculated. The UIC parameter specifies the use of the initial conditions given in the capacitance and inductance component specification statements.

EXERCISES

See Problems 8.1 through 8.7 in the PSpice manual.

45. Use PSpice to calculate voltage v for the RC circuit described in Example 8.5. Plot voltage v for $0 \le t \le 1.5$ s with a step size of 0.05 s.

46. Use PSpice to calculate current i for the series RLC circuit described in Example 8.13. Plot current i for $0 \le t \le 0.048$ s with a step size of 0.002 s.

47. Repeat Exercise 46 with resistance R changed to 1 Ω. Plot current i for $0 \le t \le 0.096$ s with a step size of 0.002 s.

8.11   Summary

We began this chapter with a brief introduction to linear differential equations. We introduced the operator $p = d/dt$ for notational convenience and as an aid to solve the differential equations. We analyzed source-free circuits and the resulting homogeneous linear differential equations. The solution to homogeneous differential equations is called the source-free response, the natural response, or the complementary response.

The form of the complementary response is determined by the roots of the characteristic equation.

1. If the roots of the characteristic equation are real, the response is the sum of real exponentials.
2. If the roots of the characteristic equation are complex, the response contains an oscillatory factor.
3. Repeated roots introduce polynomial factors.

In addition, the stability of a circuit is determined from the roots of the corresponding characteristic equation.

1. If the real part of all roots is negative, the circuit is stable.
2. If all roots with a zero real part are first-order and no roots have a positive real part, the circuit is marginally stable.
3. If one or more roots have a positive real part, the circuit is unstable. Higher-order roots with a zero real part also cause the circuit to be unstable.

KEY FACTS

- The differential operator p is defined by

$$p^n = \frac{d^n}{dt^n}$$

- We can write an LLTI differential equation in the form

$$\mathscr{A}(p)v = \mathscr{B}(p)i$$

where the input is $i = i(t)$ and the response is $v = v(t)$.

- A source-free circuit is described by a homogeneous differential equation

$$\mathscr{A}(p)v = 0$$

- The solution to a homogeneous differential equation is the complementary response (natural response or source-free response).

- The roots of the characteristic equation

$$\mathscr{A}(s) = a_n(s - s_1)^{m_1}(s - s_2)^{m_2} \cdots (s - s_l)^{m_l} = 0$$

determine the complementary response.

- First-order factors $(s - s_i)$, $(m_i = 1)$, give terms in the response of the form

$$A_i e^{s_i t}$$

where A_i is a constant of integration.

- Second-order factors $(s - s_i)^2$, $(m_i = 2)$, give terms in the response of the form

$$(A_0 + A_1 t)e^{s_i t}$$

- First-order complex conjugate roots

$$s_1 = \sigma_1 + j\omega_1$$

$$s_2 = s_1^* = \sigma_1 - j\omega_1$$

give terms in the response of the form

$$e^{\sigma_1 t}(A \cos \omega_1 t + B \sin \omega_1 t)$$

- Initial conditions are used to calculate the constants of integration.

- If the real part of all roots of the characteristic equation is negative, the corresponding circuit is stable.

- A circuit with first-order roots with a zero real part and no roots with a positive real part is often said to be marginally stable.

- If one or more roots have a positive real part, the circuit is unstable.

- The time constant of an RC circuit is

$$\tau = RC$$

- The time constant of an RL circuit is

$$\tau = L/R$$

- RLC circuits can be overdamped, underdamped, or critically damped.

PROBLEMS

Section 8.1

Write the differential equations given in Problems 1 through 5 in operator notation. Write the characteristic equation, and determine the roots of the equation.

1. $\dfrac{d}{dt} i + 7i = 0$

2. $2\dfrac{d}{dt} i + 48i + 286 \int i\, dt = 0$

3. $\dfrac{d^4}{dt^4} v + 24 \dfrac{d^3}{dt^3} v + 144 \dfrac{d^2}{dt^2} v = 0$

4. $\dfrac{d^3}{dt^3} v + 12 \dfrac{d^2}{dt^2} v + 47 \dfrac{d}{dt} v + 60v = 0$

5. $\dfrac{d^2}{dt^2} i - 3 \dfrac{d}{dt} i + 2i = 0$

Section 8.2

Determine the complementary response for the differential equations in Problems 6 through 11. (Do not evaluate the constants of integration.)

6. $(p + 7)i = 0$
7. $(5p - 15)v = 0$
8. $(2p^2 + 30p + 100)i = 0$

9. $2\dfrac{d}{dt} i + 48i + 286 \int i\, dt = 0$

10. $5(p^2 + p)v = 0$
11. $(p + 1)(p + 2)(p + 3)i = 0$

12. For the following circuit, write a KVL equation, and solve for current $i = i(t)$, $t > 0$, if $i(0^+) = 8$ A.

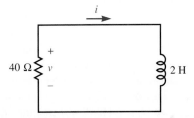

13. For the preceding circuit, $i(0^+) = 8$ A. Use this and Ohm's law to calculate $v(0^+)$. Write a KCL equation (in terms of voltage v) for the top node. Differentiate this integral equation once to obtain a first-order differential equation. Solve the equation for $v = v(t)$, $t > 0$, and use the value of $v(0^+)$ to calculate the constant of integration.

14. For the following circuit, $v(0^+) = 8$ V. Use this and Ohm's law to calculate $i(0^+)$. Write a KVL equation (in terms of current i) clockwise around the loop. Differentiate this integral equation once to obtain a first-order differen-

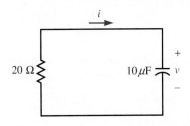

tial equation. Solve for current $i = i(t)$, $t > 0$, and use the value of $i(0^+)$ to calculate the constant of integration.

15. For the following circuit, write a KVL equation that can be solved for current i. Differentiate this integro-differential equation once to obtain a second-order differential equation.

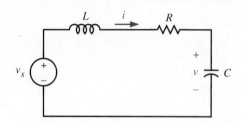

16. In the preceding circuit, $L = 2$ H, $R = 6\ \Omega$, and $C = \frac{1}{4}$ F. If $v_s = 0$ for $t > 0$, determine the response $i = i(t)$ for $t > 0$. (Insufficient information is given to calculate the constants of integration.)

Section 8.3

17. For the circuit given in Problem 15, $L = 2$ H, $R = 12\ \Omega$, and $C = \frac{1}{18}$ F. Voltage v_s is zero for $t > 0$. The initial conditions are $i(0^+) = 2$ A and $v(0^+) = -38$ V. Determine and sketch the response $i = i(t)$ for $t > 0$.

18. A parallel RLC circuit is source free for all time greater than or equal to zero. The component values are $L = \frac{1}{8}$ H, $R = \frac{1}{8}\ \Omega$, and $C = 2$ F. The initial value of the inductance current is $i(0^+) = 4$ A, and the initial value of the capacitance voltage (passive sign convention) is $v(0^+) = 2$ V. Determine and sketch the response $v = v(t)$ for $t > 0$.

19. Use the initial conditions and Ohm's law to calculate the resistance voltage (passive sign convention) $v_R(0^+)$ in Problem 17. Now use KVL to calculate the inductance voltage (passive sign convention) at $t = 0^+$.

20. Use the initial conditions (passive sign convention) to calculate the resistance current and capacitance current at $t = 0^+$ for the circuit of Problem 18.

Section 8.4

Which of the differential equations given in Problems 21 through 28 represent stable circuits?

21. $(p^2 + 2p + 1)v = 0$ 22. $(p^2 - 1)i = 0$

23. $(p^2 - p + 2)v = 0$ 24. $(p^2 + 4p + 5)i = 0$
25. $(p^2 + p)v = 0$ 26. $(p^3 + p^2)v = 0$
27. $(p^2 - 4p + 4)i = 0$ 28. $(p^2 + 7p + 10)v = 0$
29. Determine the values of A for which the following circuit is stable.

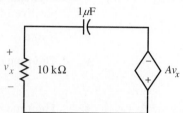

30. The initial currents in the following source-free circuit are $i_1(0) = 12$ A and $i_2(0) = 0$ A. Write two mesh equations and use Cramer's rule to reduce these to one second-order equation in terms of current i_1. (This procedure is demonstrated in Example 8.9.) Solve for current i_1, including evaluation of the constants of integration. Is the circuit stable? Calculate the energy stored in the inductances as a function of time.

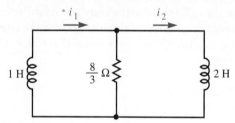

Section 8.5

31. Refer to the circuit of Problem 29. Determine the time constant as a function of the dependent voltage source gain A. What is the effective capacitance seen by the 10-kΩ resistance? (This is a function of the gain A.)

Use equivalent circuits to reduce the networks in Problems 32 and 33 to a simple series RC circuit, then calculate voltage v_{12} for $t > 0$, if $v_{12}(0^+) = 10$ V.

32.

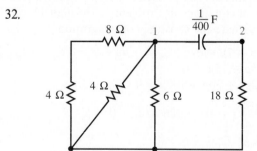

33.

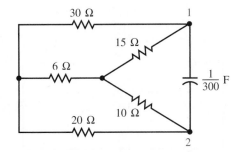

34. In the following source-free circuit, write a differential equation, and determine the parameter β that will result in a time constant of 0.02 s. Use this value of β and calculate $v(t)$, $t > 0$, if $v(0^+) = 40$ V.

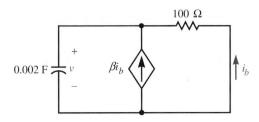

Section 8.6

35. For the amplifier circuit that follows, $v_s = 0$ for $t \geq -1$ s and $i(0) = 10$ A. Use Thévenin's theorem to simplify the circuit and calculate the time constant. What is the effective resistance seen by the inductance? Determine and sketch current i for $0 \leq t \leq 1$ s.

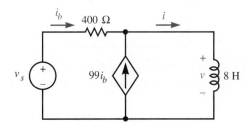

36. Use the Thévenin equivalent circuit calculated in the previous problem and the initial value $i(0^+) = 10$ A to calculate $v(0)$ by the use of Ohm's law and KVL.

37. For the following circuit, the initial current is $i(0) = 24$ A. Calculate $i(t)$ for $t > 0$. What is the time constant?

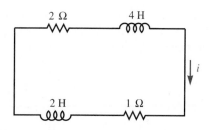

38. Write a differential equation, and determine the time constant of the following source-free network. Find $i(t)$, $t > 0$, if $i(0^+) = 80$ A.

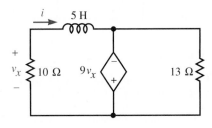

Section 8.7

For Problems 39 through 44, determine the response for the homogeneous equation (the response should be written without complex exponentials), and state whether the equation represents a stable, marginally stable, or unstable circuit.

39. $(p^2 + 2p + 10)v = 0$

40. $5\dfrac{d}{dt}i + 40i + 100\displaystyle\int i\,dt = 0$

41. $(p^2 + 4)v = 0$

42. $(p + 1)(p^2 + 6p + 10)i = 0$

43. $(p^2 + 2p + 5)^2 v = 0$

44. $(p^4 + 6p^2 + 9)i = 0$

45. Given the following source-free circuit, write a differential equation, and solve for $v(t)$, $t > 0$, if $v(0^+) = 20$ V and $i(0^+) = 1$ A.

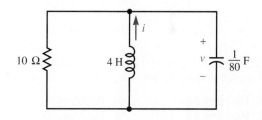

46. The initial current for a source-free series RLC circuit is $i(0^+) = 8$ A, and the capacitance voltage is $v(0^+) = 48$ V (passive sign convention). If $L = 2$ H, $R = 8\ \Omega$, and $C = \frac{1}{40}$ F, determine current $i = i(t)$ for $t > 0$.

47. In the following circuit, voltage v_s is zero for $t > -1$ s, $v(0) = 10$ V, $i(0) = 20$ A, and $\beta = 49$. Determine voltage $v = v(t)$ for $t \geq 0$.

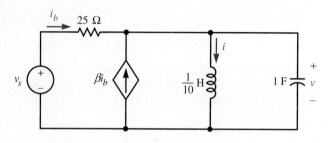

Section 8.8

For a source-free series RLC circuit with $L = 2$ H and $C = 4$ F, write a KVL equation.

48. For what range of resistance values is the circuit overdamped?
49. For what resistance value is the circuit critically damped?
50. For what range of resistance values is the circuit underdamped?

Section 8.9

Problems 51 through 53 refer to a parallel RLC circuit with $L = 4$ H and $C = \frac{1}{80}$ F. Write a KCL equation.

51. For what range of resistance values is the circuit overdamped?
52. For what resistance value is the circuit critically damped?
53. For what range of resistance values is the circuit underdamped?
54. What value of current gain β causes the circuit of Problem 47 to be critically damped?

Section 8.10

Use PSpice to solve the problems indicated in Problems 55 through 60 (look carefully at Problem 60 before you try to solve it).

55. Problem 34   56. Problem 35
57. Problem 38   58. Problem 45
59. Problem 47   60. Problem 30
61. Use PSpice to calculate and plot voltage v_1 and current i_L in the following circuit for the time interval $0 < t \leq 1$ s if $v_1(0^+) = 36$ V and $i_L(0^+) = 0$ A.

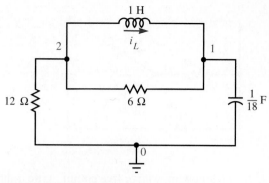

Comprehensive Problems

62. Write a differential equation, and determine the source-free response $i(t)$, $t > 0$, of the following network if $v_1(0^+) = 20$ V and $i(0^+) = 2$ A.

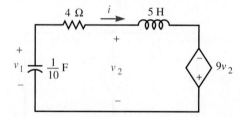

63. Given an LC network as shown below with $v(0^+) = 40$ V and $i(0^+) = 10$ A, find $v(t)$ and $i(t)$ for $t > 0$. Calculate the energy stored in the capacitance and inductance as a function of time, and then calculate the total energy stored in the network as a function of t for $t > 0$.

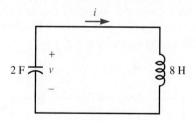

64. Determine the output voltage v_o for $t > 0$ in the following op-amp circuit if $v(0^+) = 2$ V.

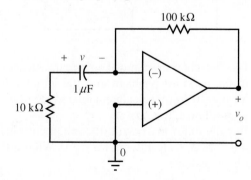

65. Replace the capacitance in the preceding circuit with a 1-mH inductance. The initial inductance current (reference arrow pointing to the right) is $i(0^+) = 0.1$ mA. Determine v_o for $t > 0$.

66. Determine voltage v_o for $t > 0$ in the following op-amp circuit if $v(0^+) = 10$ V. What is the time constant of this circuit?

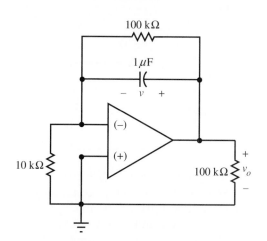

67. Replace the capacitance in the preceding circuit with a 1-mH inductance having an initial current (reference arrow pointing to the left) of $i(0^+) = 0.1$ mA. Determine $v_o(t)$ for $t > 0$.

68. For the following op-amp circuit, the initial conditions are $v_a(0^+) = 0$ and $v_b(0^+) = 10$ V. Determine $v_o(t)$ for $t > 0$.

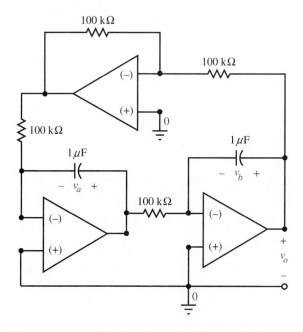

69. For the source-free circuit that follows, $i_1(0) = 6$ A and $i_2(0^+) = 1$ A. Write two mesh-current equations for this circuit. Use Cramer's rule to reduce the mesh equations to a second-order differential equation in terms of current i_1. Solve this equation for current i_1 (evaluate the constants of integration).

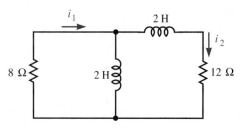

70. Solve for mesh currents i_1 and i_2 for $t > 0$ in the following circuit, if $v(0^+) = 4$ V and $i_1(0^+) = 8$ A. To simplify the work, use the operator $1/p$ to represent integration.

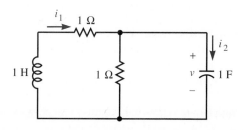

71. Solve Problem 61 without the use of PSpice.

72. Determine current $i(t)$ for $t > 0$ in the following network (use mesh equations) if $v_1(0^+) = 10$ V and $v_2(0^+) = 0$ V.

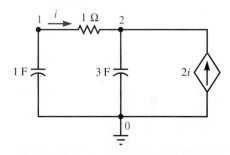

73. Use node-voltage analysis to solve for voltages v_1 and v_2 in the preceding circuit. Then use Cramer's rule to reduce the two simultaneous equations to an equation in terms of one node voltage.

74. A simplified equivalent circuit of a tuned-drain oscillator that uses a field-effect transistor is shown in the following figure. The oscillator

will oscillate at an angular frequency ω. This will occur when two roots of the characteristic equation are $s = \pm j\omega$. Show that this occurs when

$$\omega = \frac{1}{\sqrt{LC}} \sqrt{1 + \frac{R}{r}}$$

and

$$M = \frac{CrR + L}{gr}$$

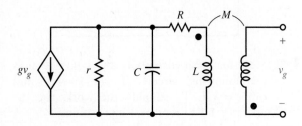

75. For the following source-free circuit:
   (a) Write a KVL equation and determine the time constant.
   (b) Find $i(t)$, $t > 0$, if $i(0^+) = 48$ A.

(c) Find the energy stored in the circuit at $t = 0^+$. Is the answer uniquely determined?
(d) Determine the energy stored in the circuit at $t = \infty$. Is the answer uniquely determined?
(e) What is the change in stored energy between $t = 0^+$ and $t = \infty$? Is the answer unique?

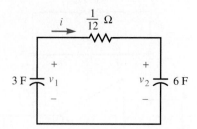

76. For the preceding source-free circuit:
   (a) Write two node-voltage equations.
   (b) Determine $v_1(t)$, $t > 0$, and $v_2(t)$, $t > 0$, if $v_1(0^+) = 4$ V and $v_2(0^+) = 0$ V.
   (c) Determine the energy stored in the circuit at $t = 0^+$.
   (d) Determine the energy stored in the circuit at $t = \infty$.

Driven Circuits

In Chapter 8, we analyzed source-free circuits and learned to solve the homogeneous differential equations that describe such circuits. From an engineering point of view, we solved for what happens when the electric power in a circuit is turned off.

In this chapter we will analyze driven circuits, that is, circuits that include independent sources (drivers), and learn to solve the differential equations that describe these circuits. We are effectively solving for the response of a circuit when the electric power is turned on.

The complete response of a driven circuit will be seen to consist of two parts: the complementary response (as found in Chapter 8) and the particular response. We will show that, in some very important cases, the circuit reaches a steady state some time after the power is turned on, and only the particular response remains. This is the condition that is reached soon after you turn on an electrical appliance.

9.1 The Particular Response

In Chapter 8, we analyzed source-free circuits. These circuits contained resistance, energy storage elements (inductance and capacitance), and dependent sources, but no independent sources. We will now analyze circuits that also include independent sources.

The response of the simple first-order circuit shown in Fig. 9.1 is obtained from the KCL equation for a surface enclosing the top right-hand node:

$$C \frac{d}{dt} v + \frac{1}{R}(v - v_s) = 0 \tag{9.1}$$

FIGURE 9.1
A first-order circuit

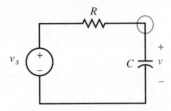

We can replace d/dt by p and write this equation as

$$C\left(p + \frac{1}{RC}\right)v = \frac{1}{R}v_s \tag{9.2}$$

This first-order differential equation is of the form

$$a_1(p - s_1)v = g(t) \tag{9.3}$$

The s-shift theorem

$$p(e^{-s_1 t}v) = e^{-s_1 t}(p - s_1)v \tag{9.4}$$

from Chapter 8 can be used to solve Eq. (9.3). Multiplication of Eq. (9.3) by $e^{-s_1 t}$ gives

$$e^{-s_1 t}a_1(p - s_1)v = e^{-s_1 t}g(t) \tag{9.5}$$

From Eq. (9.4), we can see that the solution to Eq. (9.5) is the same as the solution to the following equation:

$$a_1 p(e^{-s_1 t}v) = e^{-s_1 t}g(t) \tag{9.6}$$

Multiplication of Eq. (9.6) by $1/a_1$ gives

$$p(e^{-s_1 t}v) = \frac{1}{a_1} e^{-s_1 t}g(t) \tag{9.7}$$

Multiplication of both sides of this equation by dt puts the left-hand side in the form of a differential that we can easily integrate:

$$\int d(e^{-s_1 t}v) = \frac{1}{a_1} \int e^{-s_1 t}g(t)\, dt \tag{9.8}$$

This gives

$$e^{-s_1 t}v = \frac{1}{a_1} \int e^{-s_1 t}g(t)\, dt \tag{9.9}$$

which can be rearranged to give

$$v = \frac{1}{a_1} e^{s_1 t} \int e^{-s_1 t} g(t)\, dt \tag{9.10}$$

The indefinite integral implicitly includes an additive constant of integration. If the constant of integration is shown explicitly, Eq. (9.10) becomes

$$v = A e^{s_1 t} + \frac{1}{a_1} e^{s_1 t} \int e^{-s_1 t} g(t)\, dt \tag{9.11}$$

We recognize the first term, containing the constant of integration A, as the *complementary response* v_c that we found in Chapter 8. The second term is the *particular response* v_p and depends on the particular forcing function $g(t)$. We can then write the *complete response* given by Eq. (9.11) in the form

$$v = v_c + v_p \tag{9.12}$$

The complementary response v_c, given by Theorem 2 of Chapter 8 is

$$v_c = A e^{s_1 t} \tag{9.13}$$

and the particular response v_p is given by the integral

$$v_p = \frac{1}{a_1} e^{s_1 t} \int e^{-s_1 t} g(t)\, dt \tag{9.14}$$

with the *constant of integration set equal to zero*.

We learned how to solve for the complementary response in Chapter 8. In Sections 9.1 through 9.4, we will concentrate on solving for the particular response component of the complete solution.

EXAMPLE 9.1

Determine the particular response v_p for $t \geq 0$ in the RC circuit shown in Fig. 9.1, if $v_s = 40 e^{-3t}$ V for $t \geq 0$, $R = 20\ \Omega$, and $C = 0.025$ F.

■ **SOLUTION**

Summation of the currents leaving the top right-hand node gives us

$$0.025 \frac{d}{dt} v + \frac{1}{20}(v - v_s) = 0$$

which we can write as

$$0.025(p + 2)v = \frac{1}{20} v_s$$

Equation (9.14) gives the particular response:

$$v_p = \frac{1}{0.025} e^{-2t} \int e^{2t} \left(\frac{1}{20} v_s\right) dt = 40 e^{-2t} \int e^{2t} \left(\frac{1}{20} 40 e^{-3t}\right) dt$$

$$= 40 e^{-2t}(-2 e^{-t}) = -80 e^{-3t}\ \text{V} \qquad t \geq 0 \quad ■$$

We can easily extend the procedure used in Example 9.1 to the more general nth-order LLTI differential equation

$$\mathscr{A}(p)v = \mathscr{B}(p)i \tag{9.15}$$

The operator $\mathscr{A}(p)$ in Eq. (9.15) can be written in factored form to give

$$a_n(p - s_1)(p - s_2) \cdots (p - s_n)v = g(t) \tag{9.16}$$

where

$$g(t) = \mathscr{B}(p)i \tag{9.17}$$

and the roots $s_1, s_2, \ldots, s_n$ of the characteristic equation need not be distinct. The s-shift theorem and integration can be used to reduce Eq. (9.16) to a differential equation of order $n - 1$. Repeated application of the s-shift theorem and integration gives the solution to Eq. (9.16) as

$$v = v_c + v_p \tag{9.18}$$

where v_c is the complementary response as determined in Chapter 8, and v_p is the particular response given by the following theorem.

THEOREM 1

The Particular Response by Integration

The particular response for the nth order LLTI differential equation

$$a_n(p - s_1)(p - s_2) \cdots (p - s_n)v = g(t) \tag{9.19}$$

is given by the n-fold integral

$$v_p = \frac{1}{a_n} e^{s_n t} \int e^{(s_{n-1} - s_n)t} \left\{ \int \cdots \left[\int e^{(s_1 - s_2)t} \left(\int e^{-s_1 t} g(t)\, dt \right) dt \right] \cdots dt \right\} dt \tag{9.20}$$

by setting all constants of integration equal to zero. The roots $s_1, s_2, \ldots, s_n$ are not required to be distinct.

Although Theorem 1 appears formidable, the following example of a second-order circuit should clarify the procedure.

EXAMPLE 9.2

Determine the particular response for $t > 0$ in the circuit of Fig. 9.2 if $i_s = 40e^{-3t}$ A for $t \geq 0$.

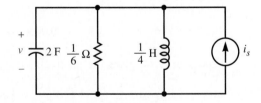

■ **SOLUTION**

Summation of the currents leaving the top node gives

$$2\frac{d}{dt}v + 6v + 4\int_{-\infty}^{t} v\, d\tau - i_s = 0$$

We can differentiate this once and write the equation in operator notation:

$$2(p + 1)(p + 2)v = pi_s$$

Equation (9.20) gives the particular solution:

$$v_p = \frac{1}{2} e^{-2t} \int e^{(-1+2)t} \left(\int e^t(-120e^{-3t}) \, dt \right) dt$$

$$= \frac{1}{2} e^{-2t} \int e^{(-1+2)t}(60e^{-2t}) \, dt$$

$$= \frac{1}{2} e^{-2t}(-60e^{-t})$$

$$= -30e^{-3t} \text{ V} \qquad t > 0 \quad \blacksquare$$

In principle, we can use Theorem 1 to determine the particular response for any forcing function $g(t)$. In practice, the integrals can be difficult to evaluate. In the following section we will see how to obtain the particular response for an important class of input functions without integration. This reduces the calculus problem to one of simple algebra.

REMEMBER

We can obtain the complete response of an nth-order LLTI differential equation by an n-fold integration. This complete response consists of two parts. One part is the complementary response, which contains n constants of integration. The complementary response is easily found by the methods presented in Chapter 8. We call the second part, which does not include constants of integration, the particular response. If we set the constants of integration equal to zero, only the particular response remains.

EXERCISES

Use Theorem 1 to calculate the particular response for the following LLTI differential equations.

1. $(p + 4)v = 24e^{-6t}$
2. $(p + 4)v = 24$
3. $(2p + 4)v = 12e^{-6t}$
4. $(2p + 4)v = 12e^{-2t}$
5. $(2p^2 + 14p + 20)i = 24e^{-3t}$
6. $(2p^2 + 14p + 20)v = 24e^{-2t}$
7. $(2p^2 + 20p + 50)i = 100e^{-5t}$

9.2 Exponential Inputs

Although a signal (voltage or current) can be generated as almost any function of time, some signals are encountered much more frequently than others. Batteries used in automobiles and portable appliances supply constant voltages, and virtually all electric power is supplied to industry by sinusoidal voltages. In Chapter 7 we saw that a constant was a special case of an exponential time function and that the complex exponential is related to the trigonometric sine and cosine by Euler's identity. Therefore the response of a linear circuit to an exponential source is of great practical importance in engineering. The following development will show how to obtain the particular response for an exponential input function without integration.

The derivative of an exponential input function

$$i = Ke^{st} \tag{9.21}$$

is

$$pi = pKe^{st} = sKe^{st} \tag{9.22}$$

and for an nth-order ordinary linear differential equation with exponential input,

$$\begin{aligned} \mathscr{A}(p)v = \mathscr{B}(p)Ke^{st} \\ = \mathscr{B}(s)Ke^{st} \\ = g(t) \end{aligned} \tag{9.23}$$

For a first-order equation,

$$\begin{aligned} a_1(p + s_1)v = \mathscr{B}(p)Ke^{st} \\ = \mathscr{B}(s)Ke^{st} \end{aligned} \tag{9.24}$$

If $s \neq s_1$, Theorem 1 [Eq. (9.20)] gives

$$\begin{aligned} v_p = \frac{1}{a_1} e^{s_1 t} \int e^{-s_1 t} \mathscr{B}(s)Ke^{st}\, dt \\ \\ = \frac{1}{a_1(s - s_1)} \mathscr{B}(s)Ke^{st} \\ \\ = \frac{\mathscr{B}(s)}{\mathscr{A}(s)} Ke^{st} \qquad \mathscr{A}(s) \neq 0 \end{aligned} \tag{9.25}$$

We obtain the particular solution by simply replacing the operator p in the differential equation by the coefficient s in the exponent. This is a simple algebra problem.

In a similar manner, we can apply Theorem 1 to higher-order LLTI differential equations to prove the following theorem.

THEOREM 2

Particular Response for Exponential Inputs with $\mathscr{A}(s) \neq 0$

An LLTI differential equation with an exponential input function

$$\mathscr{A}(p)v = \mathscr{B}(p)Ke^{st} \tag{9.26}$$

has a particular response

$$v_p = \frac{\mathscr{B}(s)}{\mathscr{A}(s)} Ke^{st} \qquad \text{if } \mathscr{A}(s) \neq 0 \tag{9.27}$$

Once obtained, Theorem 2 is easily justified without integration. The differentiation implied by $\mathscr{B}(p)Ke^{st}$, on the right-hand side of Eq. (9.26), obviously results in $\mathscr{B}(s)Ke^{st}$, which is an exponential. Equality for all time implies that the left-hand side of Eq. (9.26) must also be an exponential. This implies that v_p is an exponential, with the result that $\mathscr{A}(p)v_p$ is equal to $\mathscr{A}(s)v_p$. Therefore we can replace p by s on both sides of Eq. (9.26) and divide by $\mathscr{A}(s)$ to obtain Eq. (9.27).

For the unusual case where $s = s_1$ in Eq. (9.24), $\mathscr{A}(s) = 0$, and Theorem 2 does not apply. We can still use Theorem 1 to solve such problems. For a first-order equation,

$$v_p = \frac{1}{a_1} e^{s_1 t} \int e^{-s_1 t} \mathscr{B}(s_1) K e^{s_1 t} \, dt$$

$$= \frac{1}{a_1} \mathscr{B}(s) t e^{st}, \qquad \mathscr{A}(s) = 0 \qquad (9.28)$$

We can use Theorem 1 to extend the result given by Eq. (9.28) to higher-order equations and obtain Theorem 3.

THEOREM 3

Particular Response for Exponential Inputs with $\mathscr{A}(s) = 0$

For an LLTI differential equation with an exponential input

$$\mathscr{A}(p)v = \mathscr{B}(p)K e^{st} \qquad (9.29)$$

with $\mathscr{A}(s) = 0$, $\mathscr{A}(p)$ can be written as

$$\mathscr{A}(p) = (p - s)^m \mathscr{A}_1(p) \qquad (9.30)$$

where m is a positive integer and

$$\mathscr{A}_1(s) \neq 0 \qquad (9.31)$$

For this case,

$$v_p = \frac{\mathscr{B}(s)}{\mathscr{A}_1(s)} \frac{t^m}{m!} K e^{st} \qquad (9.32)$$

We now apply Theorems 2 and 3 to some simple examples.

EXAMPLE 9.3

The circuit of Fig. 9.1 used in Example 9.1 is described by the equation

$$(0.025p + 0.05)v = \frac{1}{20}(40e^{-3t}) \qquad t > 0$$

Use Theorem 2 to find the particular response.

■ **SOLUTION**

From Theorem 2 [Eq. (9.27)],

$$v_p = \frac{\mathscr{B}(s)}{\mathscr{A}(s)} K e^{st}$$

$$= \frac{1/20}{0.025(-3) + 0.05} 40e^{-3t}$$

$$v_p = -80e^{-3t} \text{ V} \qquad t > 0$$

This is the same result obtained by integration in Example 9.1. ∎

The procedure of Theorem 2 is just as easily applied to second-order circuits.

EXAMPLE 9.4

The circuit of Fig. 9.2 in Example 9.2 is described by the equation

$$(2p^2 + 6p + 4)v = p40e^{-3t} \qquad t \geq 0$$

Determine the particular response for $t > 0$.

■ **SOLUTION**

$$v_p = \frac{-3}{2(-3)^2 + 6(-3) + 4} 40e^{-3t}$$

$$= -30e^{-3t} \text{ V} \qquad t > 0$$

This is, of course, the same result obtained in Example 9.2. ∎

We now work two examples for which Theorem 3 applies.

EXAMPLE 9.5

The source in Example 9.1 is changed so that the circuit is described by the equation

$$(0.025p + 0.05)v = \frac{1}{20}(40e^{-2t}) \qquad t > 0$$

Determine the particular response

■ **SOLUTION**

The coefficient in the exponent is

$$s = -2$$

and

$$\mathscr{A}(s) = \mathscr{A}(-2) = 0.025(-2) + 0.05 = 0$$

We can write $\mathscr{A}(p)$ as

$$\mathscr{A}(p) = 0.025(p + 2)$$
$$= (p + 2)\mathscr{A}_1(p)$$

where

$$\mathscr{A}_1(p) = 0.025$$

From Theorem 3 [Eq. (9.32)],

$$v_p = \frac{\mathscr{B}(s)}{\mathscr{A}_1(s)} tKe^{st} = \frac{1/20}{0.025} t40e^{-2t}$$

$$= 80te^{-2t} \text{ V} \qquad t > 0 \blacksquare$$

Application of Theorem 3 to a second-order equation is also straightforward.

EXAMPLE 9.6

⊞ *See Problem 9.1 in the PSpice manual.*

The source in Example 9.2 is changed so that the circuit is described by

$$(2p^2 + 6p + 4)v = p(40e^{-2t}) \qquad t \geq 0$$

Determine the particular solution.

■ SOLUTION

The coefficient of the exponent is

$$s = -2$$

and

$$\mathscr{A}(s) = \mathscr{A}(-2) = 2(-2)^2 + 6(-2) + 4 = 0$$

so we must use Theorem 3. We can write

$$\mathscr{A}(p) = (p + 2)2(p + 1)$$
$$= (p + 2)\mathscr{A}_1(p)$$

where

$$\mathscr{A}_1(p) = 2(p + 1)$$

From Theorem 3,

$$v_p = \frac{\mathscr{B}(s)}{\mathscr{A}_1(s)} tKe^{st} = \frac{(-2)}{2(-2 + 1)} t40e^{-2t}$$
$$= 40te^{-2t} \text{ V} \qquad t \geq 0 \quad ■$$

The following example demonstrates that we can easily determine the particular response to exponential inputs for equations of higher order than 2.

EXAMPLE 9.7

Find the particular response for the network equation

$$(p^3 + 4p^2 + 5p + 2)i = (2p^2 + 4p + 1)v$$

if (a) $v = 8e^{-3t}$ V and (b) $v = 8e^{-t}$ V.

■ SOLUTION

(a) Because

$$\mathscr{A}(s) = \mathscr{A}(-3) \neq 0$$

Theorem 2 applies, and

$$i_p = \frac{\mathscr{B}(-3)}{\mathscr{A}(-3)} 8e^{-3t} = \frac{2(-3)^2 + 4(-3) + 1}{(-3)^3 + 4(-3)^2 + 5(-3) + 2} 8e^{-3t}$$
$$= -14e^{-3t} \text{ A}$$

(b) Because

$$\mathscr{A}(s) = \mathscr{A}(-1) = 0$$

we must write the operator $\mathscr{A}(p)$ as

$$\mathscr{A}(p) = (p + 1)^2(p + 2) = (p + 1)^2 \mathscr{A}_1(p)$$

where

$$\mathscr{A}_1(p) = p + 2$$

as in Eq. (9.30) of Theorem 3. Therefore the particular response is obtained from Eq. (9.32):

$$i_p = \frac{\mathscr{B}(-1)}{\mathscr{A}_1(-1)} \frac{t^2}{2!} 8e^{-t} = \frac{2(-1)^2 + 4(-1) + 1}{-1 + 2} \frac{t^2}{2!} 8e^{-t}$$

$$= -4t^2 e^{-t} \text{ A} \ \blacksquare$$

In Chapter 7 we observed that we can write a constant K as

$$K = Ke^{0t} \tag{9.33}$$

This lets us adapt Theorems 2 and 3 to constant inputs by recognizing that $s = 0$ for a dc input.

EXAMPLE 9.8

Find the particular response for the differential equation

$$(p^2 + 4p + 4)v = 12$$

■ **SOLUTION**

Observe that the forcing function

$$g(t) = 12$$
$$= 12e^{0t}$$

is an exponential with $s = 0$. Thus $\mathscr{A}(s) \neq 0$ and the particular solution is given by Eq. (9.27) of Theorem 2:

$$v_p = \frac{1}{\mathscr{A}(0)} 12e^{0t}$$

$$= \frac{1}{0 + 0 + 4} 12$$

$$= 3 \text{ V} \ \blacksquare$$

REMEMBER

For an exponential input, if $\mathscr{A}(s) \neq 0$, the particular response is obtained by replacing p by s in the LLTI differential equation

$$\mathscr{A}(p)v = \mathscr{B}(p)Ke^{st} \tag{9.34}$$

because pe^{st} is equal to se^{st}. For a dc input,

$$s = 0 \tag{9.35}$$

Calculate the particular response component of current i in the following circuit for the input voltages specified in Exercises 8 through 12.

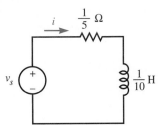

8. $v_s = 100e^{-5t}$ V

9. $v_s = 100e^{-2t}$ V

10. $v_s = 100$ V

11. $v_s = 100e^{5t}$ V

12. $v_s = 100e^{2t}$ V

Calculate the particular response for the LLTI differential equations in Exercises 13 through 16.

13. $(2p^2 + 14p + 20)v = 24e^{-3t}$

14. $(2p^2 + 14p + 20)v = 24e^{-2t}$

15. $(2p^2 + 20p + 50)i = 100e^{-5t}$

16. $(p^2 + 2p + 5)i = 20$

9.3 Superposition

In Chapter 5 we saw how the principle of superposition can be used to analyze networks of resistances and dependent sources that are driven by more than one independent source. A more important application of superposition is to networks that also contain inductance and capacitance and two or more independent sources. The superposition theorem is restated in terms of linear differential equations in the following theorem.

THEOREM 4

Superposition

The particular response for the LLTI differential equation with constant coefficients

$$\mathscr{A}(p)v_p = \mathscr{B}_1(p)i_1 + \mathscr{B}_2(p)i_2 \tag{9.36}$$

is unique, and the particular response is

$$v_p = v_{1p} + v_{2p} \tag{9.37}$$

where v_{1p} is the particular response for

$$\mathscr{A}(p)v_{1p} = \mathscr{B}_1(p)i_1 \tag{9.38}$$

and v_{2p} is the particular response for

$$\mathscr{A}(p)v_{2p} = \mathscr{B}_2(p)i_2 \tag{9.39}$$

PROOF

⬇ *See Problem 9.6 in the PSpice manual.*

This theorem follows because the particular response can be obtained by repeated integration (Theorem 1) with all constants of integration set equal to zero. If all constants of integration are set equal to zero, the indefinite integral is a linear transformation (the integral of the sum of two functions is the sum of the integrals of each of the two functions).

EXAMPLE 9.9

Find the particular response for the equation

$$(p^3 + 4p^2 + 5p + 2)v = (2p^2 + 4p + 1)(8e^{-3t} + 8e^{-t})$$

■ **SOLUTION**

This equation is of the form

$$\mathscr{A}(p)v = \mathscr{B}(p)(i_1 + i_2)$$

where

$$i_1 = 8e^{-3t}$$

and

$$i_2 = 8e^{-t}$$

The particular solution for input i_1 was determined in Example 9.7a:

$$v_{1p} = -14e^{-3t} \text{ V}$$

and the particular solution for input i_2 was calculated in Example 9.7b:

$$v_{2p} = -4t^2 e^{-t} \text{ V}$$

Thus Theorem 4 (superposition) gives the particular solution to the original equation as

$$v_p = v_{1p} + v_{2p} = -14e^{-3t} - 4t^2 e^{-t} \text{ V} \quad ■$$

REMEMBER

The particular response for two or more sources acting simultaneously is the sum of the particular response for each source acting independently.

EXERCISES

Find the particular response to the following differential equations.

17. $(p + 3)i = 12(1 + 2e^{-t} + 3e^{-3t})$ 18. $(p^2 + 3p + 1)v = 24 + 12e^{-2t}$

19. $(p^2 + 6p + 5)v = 10 + 110e^{2t}$

9.4 Sinusoidal Inputs

In engineering the sinusoid is the most widely encountered input function. For example, sinusoidal voltages supply power to our homes. We will use Euler's identity and superposition to adapt Theorems 2 and 3 to sinusoidal inputs. Remember, the coefficient s of the exponential e^{st} in the input function can be real, imaginary, or complex.* Therefore we can use Theorem 2 or 3 to solve an equation of the form

$$\mathscr{A}(p)v_p = \mathscr{B}(p)Ke^{j\phi}e^{j\omega t}$$
$$= \mathscr{B}(p)Ke^{j(\omega t + \phi)} \tag{9.40}$$

Euler's identity,

$$e^{j\omega t} = \cos \omega t + j \sin \omega t \tag{9.41}$$

let us write Eq. (9.40) as

$$\mathscr{A}(p)v_p = \mathscr{B}(p)[K \cos (\omega t + \phi) + jK \sin (\omega t + \phi)] \tag{9.42}$$

The particular solution v_p is complex and can be written as

$$v_p = v_{pR} + jv_{pI} \tag{9.43}$$

where v_{pR} is the real part of v_p:

$$v_{pR} = \mathscr{R}e\ \{v_p\} \tag{9.44}$$

and v_{pI} is the imaginary part of v_p:

$$v_{pI} = \mathscr{I}m\ \{v_p\} \tag{9.45}$$

This lets us write Eq. (9.40) as

$$\mathscr{A}(p)(v_{pR} + jv_{pI}) = \mathscr{B}(p)[K \cos (\omega t + \phi) + jK \sin (\omega t + \phi)] \tag{9.46}$$

The real parts of the two sides of Eq. (9.46) must be equal, as must be the imaginary parts. Since the coefficients in $\mathscr{A}(p)$ and $\mathscr{B}(p)$ are real for an equation describing an actual system, Eq. (9.46) can be written as two equations:

$$\mathscr{A}(p)v_{pR} = \mathscr{B}(p)K \cos (\omega t + \phi) \tag{9.47}$$

$$\mathscr{A}(p)v_{pI} = \mathscr{B}(p)K \sin (\omega t + \phi) \tag{9.48}$$

Equations (9.40) through (9.45) provide an easy method to find the solution to an equation of the form of Eq. (9.47) or (9.48):

1. Replace the sinusoidal input function, $K \cos (\omega t + \phi)$ or $K \sin (\omega t + \phi)$, with the complex exponential input function $Ke^{j(\omega t + \phi)}$.
2. Find the particular response for this complex exponential input.
3. If the original input function is a *cosine*, the particular response for this input is given by the *real part* of the particular response for the complex exponential input.
4. If the original input function is a *sine*, the particular response for this input is given by the *imaginary part* of the particular response for the complex exponential input.

Therefore Theorem 2 immediately yields the following theorem.

---

* It is a good idea to become familiar with Appendix B at this point.

Particular Response for Cosine Inputs with $\mathscr{A}(j\omega) \neq 0$

To find the particular response for the LLTI differential equation

$$\mathscr{A}(p)v = \mathscr{B}(p)K \cos (\omega t + \phi) \tag{9.49}$$

replace $\cos (\omega t + \phi)$ by $e^{j(\omega t + \phi)}$ and find the particular response for this input function by the method of Theorem 2. The real part of this particular response is the particular response for Eq. (9.49). The result is

$$v_p = \mathscr{R}e \left\{ \frac{\mathscr{B}(j\omega)}{\mathscr{A}(j\omega)} K e^{j(\omega t + \phi)} \right\} \tag{9.50a}$$

$$= \left| \frac{\mathscr{B}(j\omega)}{\mathscr{A}(j\omega)} \right| K \cos \left(\omega t + \phi + \left/ \frac{\mathscr{B}(j\omega)}{\mathscr{A}(j\omega)} \right. \right) \quad \text{if } \mathscr{A}(j\omega) \neq 0 \tag{9.50b}$$

where

$$\left/ \frac{\mathscr{B}(j\omega)}{\mathscr{A}(j\omega)} \right. = \arctan \frac{\mathscr{I}m \left\{ \mathscr{B}(j\omega) \right\}}{\mathscr{R}e \left\{ \mathscr{B}(j\omega) \right\}} - \arctan \frac{\mathscr{I}m \left\{ \mathscr{A}(j\omega) \right\}}{\mathscr{R}e \left\{ \mathscr{A}(j\omega) \right\}} \tag{9.51}$$

(This theorem is the basis for steady-state sinusoidal analysis as presented in Chapters 10 and 11.)

We can apply Theorem 5 directly for a sine input function by using the trigonometric identity

$$\sin (\omega t + \phi) = \cos \left(\omega t + \phi - \frac{\pi}{2} \right) \tag{9.52}$$

Alternatively;

Theorem 5 is adapted to a sine input by replacing cos by sin in Eqs. (9.49) and (9.50b) and $\mathscr{R}e \left\{ \cdot \right\}$ by $\mathscr{I}m \left\{ \cdot \right\}$ in Eq. (9.50a).

For a cosine excitation with $\mathscr{A}(j\omega) \neq 0$, the particular response is of the form given by Theorem 5, and is periodic with time. Such a response is said to be *cyclostationary* or *stationary*[*] and is called the *steady-state sinusoidal response*. Note that the angular frequency ω of the steady-state response is the same as that of the input function. The steady-state sinusoidal response is considered in great detail in Chapters 10 and 11.

If $\mathscr{A}(j\omega) = 0$ (an unusual occurrence[†]), we can use Theorem 3 to obtain the particular response for a cosine input. This gives Theorem 6.

---

[*] The term *stationary* has a somewhat different meaning when random signals are discussed.

[†] The case where $\mathscr{A}(j\omega) = 0$ can occur only if the circuit is marginally stable or unstable.

THEOREM 6

Particular Response for Cosine Inputs with $\mathscr{A}(j\omega) = 0$

To find the particular response for the LLTI differential equation

$$\mathscr{A}(p)v = \mathscr{B}(p)K \cos (\omega t + \phi) \qquad (9.53)$$

replace $\cos (\omega t + \phi)$ by $e^{j(\omega t + \phi)}$ and find the particular response for this input function by the method of Theorem 3. The real part of this particular response is the particular response for Eq. (9.53). The result, if $\mathscr{A}(j\omega) = 0$, is

$$v_p = \mathscr{R}e \left\{ \frac{\mathscr{B}(j\omega)}{\mathscr{A}_1(j\omega)} K \frac{t^m}{m!} e^{j(\omega t + \phi)} \right\}$$

$$= \left| \frac{\mathscr{B}(j\omega)}{\mathscr{A}_1(j\omega)} \right| K \frac{t^m}{m!} \cos \left(\omega t + \phi + \left/ \frac{\mathscr{B}(j\omega)}{\mathscr{A}_1(j\omega)} \right. \right) \qquad (9.55)$$

where

$$\left/ \frac{\mathscr{B}(j\omega)}{\mathscr{A}_1(j\omega)} \right. = \text{arc} \tan \frac{\mathscr{I}m \left\{ \mathscr{B}(j\omega) \right\}}{\mathscr{R}e \left\{ \mathscr{B}(j\omega) \right\}} - \text{arc} \tan \frac{\mathscr{I}m \left\{ \mathscr{A}_1(j\omega) \right\}}{\mathscr{R}e \left\{ \mathscr{A}_1(j\omega) \right\}} \qquad (9.56)$$

and

$$\mathscr{A}(p) = (p - j\omega)^m \mathscr{A}_1(p) \qquad \mathscr{A}_1(j\omega) \neq 0 \qquad (9.57)$$

We can adapt Theorem 6 to a sine input function in the same way we adapted Theorem 5 to a sine input.

If $\mathscr{A}(j\omega) = 0$, the response is oscillatory and tends to increase with time due to the factor t^m, and the response is said to be *nonstationary*.

EXAMPLE 9.10

See Example 3.7 and Problem 9.3 in the PSpice manual.

Write a differential equation that can be solved for the output voltage v shown on the network diagram of Fig. 9.3. Solve for the particular response v_p of v for $t > 0$, if $v_s = 15/\sqrt{2} \cos 3t$ V.

FIGURE 9.3
Network for Example 9.10

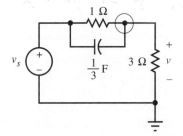

■ SOLUTION

A KCL equation for the indicated node is

$$\frac{1}{3} \frac{d}{dt} (v - v_s) + (v - v_s) + \frac{1}{3} v = 0$$

which we can rearrange and write in operator notation as

$$(p + 4)v = (p + 3)v_s$$

$$= (p + 3)\frac{15}{\sqrt{2}}\cos 3t \qquad t > 0$$

Replace $\cos 3t$ by e^{j3t} and find the particular solution to this equation:

$$(p + 4)v = (p + 3)\frac{15}{\sqrt{2}}e^{j3t}$$

The particular response is given by Theorem 2 as

$$v_p = \frac{\mathscr{B}(j3)}{\mathscr{A}(j3)}\frac{15}{\sqrt{2}}e^{j3t}$$

The particular solution for the cosine forcing function is given by the real part of the above equation, as stated in Theorem 5. Thus

$$v_p = \mathscr{R}e\left\{\frac{\mathscr{B}(j3)}{\mathscr{A}(j3)}\frac{15}{\sqrt{2}}e^{j3t}\right\}$$

$$= \mathscr{R}e\left\{\frac{3 + j3}{4 + j3}\frac{15}{\sqrt{2}}e^{j3t}\right\}$$

$$= \mathscr{R}e\left\{\frac{\sqrt{3^2 + 3^2}\,e^{j\,\text{arc tan}\,(3/3)}}{\sqrt{4^2 + 3^2}\,e^{j\,\text{arc tan}\,(3/4)}}\cdot\frac{15}{\sqrt{2}}e^{j3t}\right\}$$

$$= \mathscr{R}e\left\{\frac{45}{5}e^{j[3t + \text{arc tan}\,(3/3) - \text{arc tan}(3/4)]}\right\}$$

$$= \mathscr{R}e\left\{\frac{45}{5}e^{j[3t + (\pi/4) - 0.644]}\right\}$$

$$= 9\cos\left(3t + \frac{\pi}{4} - 0.644\right)\text{V}$$

$$= 9\cos(3t + 8.13°)\text{V} \blacksquare$$

REMEMBER

The particular response for a sinusoidal input [with $\mathscr{A}(j\omega) \neq 0$] is a sinusoid of the same frequency. Only the amplitude and phase of the response differ from those of the input. We obtain the particular response to a sinusoid by replacing the cosine with a complex exponential. We then determine the particular response for this complex exponential. For a cosine input, the particular response is the real part of the complex exponential response. For a sine input, the particular response is the imaginary part of the complex exponential response.

EXERCISES

Find the particular response for the differential equations given in Exercises 20 through 24.

20. $(p + 4)i = 100 \cos (3t + 45°)$

21. $(p^2 + 2p + 5)v = 20 \cos 2t$

22. $(p + 5)v = 120 \sin 5t$

23. $(p^2 + 4)i = 50 \cos 3t$

24. $(p^2 + 4)i = 50 \cos 2t$

9.5 The Complete Response

In Section 9.1 we observed that the *complete response* v for the differential equation

$$\mathscr{A}(p)v = \mathscr{B}(p)i \qquad (9.58)$$

consists of two parts, the *complementary response* v_c (as found in Chapter 8), which includes the constants of integration and satisfies the differential equation

$$\mathscr{A}(p)v_c = 0 \qquad (9.59)$$

and the *particular response* v_p (as found in Sections 9.1 through 9.4), which satisfies the equation

$$\mathscr{A}(p)v_p = \mathscr{B}(p)i \qquad (9.60)$$

and is obtained by setting all constants of integration equal to zero.

We will now work two examples in which we determine the complete response to differential equations and evaluate the constants of integration for specified initial conditions.

CAUTION

The *complete response*

$$v = v_c + v_p \qquad (9.61)$$

must be used when determining the constants of integration.

EXAMPLE 9.11

Find the complete response v for $t > 0$, given $v(0^+) = 6$ V, for the network of Fig. 9.4.

FIGURE 9.4
Network for Example 9.11

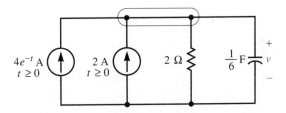

■ **SOLUTION**

See Problem 9.6 in the PSpice manual.

Application of KCL to a surface enclosing the top node yields

$$-4e^{-t} - 2 + \frac{1}{2}v + \frac{1}{6}\frac{d}{dt}v = 0 \qquad t > 0$$

which we can write as

$$(p + 3)v = 12 + 24e^{-t} \qquad t > 0$$

We calculate the complementary response by the method introduced in Chapter 8. The characteristic equation

$$s + 3 = 0$$

has a root

$$s = -3$$

so the complementary response is

$$v_c = Ae^{-3t}$$

Use superposition to find the particular response:

$$v_p = v_{1p} + v_{2p} = \frac{1}{\mathscr{A}(0)}\, 12 + \frac{1}{\mathscr{A}(-1)}\, 24e^{-t}$$

$$= \frac{1}{0 + 3}\, 12 + \frac{1}{-1 + 3}\, 24e^{-t} = 4 + 12e^{-t}$$

The complete response is

$$v = v_p + v_c$$
$$= 4 + 12e^{-t} + Ae^{-3t} \qquad t > 0$$

We can now use the initial condition $v(0^+) = 6$ V to evaluate the constant of integration (note that we are using the *complete response* at $t = 0^+$):

$$v(0^+) = 4 + 12e^{-0} + Ae^{-0} = 6$$

Thus

$$A = -10$$

and

$$v = 4 + 12e^{-t} - 10e^{-3t} \text{ V} \qquad t > 0 \quad \blacksquare$$

EXAMPLE 9.12

Find the complete response v for the network shown in Fig. 9.5, given that $v(0^+) = 13$ V and $i(0^+) = \frac{14}{3}$ A.

FIGURE 9.5
Network for Example 9.12

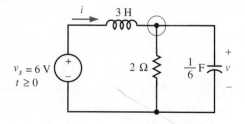

■ SOLUTION

The KCL equation for a surface enclosing the top node is

$$\frac{1}{3}\int_{-\infty}^{t} (v - v_s)\, d\tau + \frac{1}{2}v + \frac{1}{6}\frac{d}{dt}v = 0$$

We can differentiate this equation once and multiply it by 6 to give

$$(p^2 + 3p + 2)v = 12 \qquad t > 0$$

We proceed as in Chapter 8. The roots of the characteristic equation

$$s^2 + 3s + 2 = 0$$

are

$$s_1 = -1$$
$$s_2 = -2$$

From these roots, the complementary response is

$$v_c = A_1 e^{-t} + A_2 e^{-2t}$$

Equation (9.27) gives the particular response as

$$v_p = \frac{1}{\mathscr{A}(0)}\, 12$$
$$= \frac{1}{0 + 0 + 2}\, 12$$
$$= 6$$

and the complete response is

$$v = v_p + v_c$$
$$= 6 + A_1 e^{-t} + A_2 e^{-2t} \qquad t > 0$$

Knowledge of both $v(0^+)$ and $i(0^+)$ is required to evaluate the constants of integration. The first equation is obtained by evaluation of the *complete response* $v(t)$ at $t = 0^+$. This gives

$$v(0^+) = 6 + A_1 e^{-0} + A_2 e^{-2(0)} = 13$$

which reduces to

$$A_1 + A_2 = 7 \tag{9.62}$$

The value of dv/dt at $t = 0^+$ is required to obtain the second equation. This is easily obtained from the original KCL equation evaluated at $t = 0^+$:

$$-i(0^+) + \frac{1}{2} v(0^+) + \frac{1}{6} \frac{d}{dt} (6 + A_1 e^{-t} + A_2 e^{-2t})\Big|_{t = 0^+} = 0$$

Substitution of numerical values into this equation gives

$$-\frac{14}{3} + \frac{13}{2} + \frac{1}{6}(-A_1 - 2A_2) = 0$$

which reduces to

$$A_1 + 2A_2 = 11 \tag{9.63}$$

We can solve Eqs. (9.62) and (9.63) for A_1 and A_2 to give

$$v = 6 + 3e^{-t} + 4e^{-2t} \text{ V} \qquad t > 0$$

The input voltage v_s and the output voltage v are sketched in Fig. 9.6. ∎

FIGURE 9.6
The input and response of
the network for Fig. 9.5

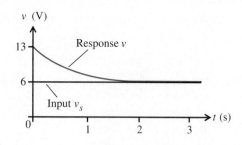

REMEMBER

The complete response is the sum of the complementary response and the particular response. When evaluating the constants of integration, we must use the complete response.

EXERCISES

⬛ *See Example 3.7 and
Problems 9.1 through 9.6 in
the PSpice manual.*

Find the complete response for the differential equations in Exercises 25 through 28. Do not evaluate the constants of integration.

25. $(p^2 + p)v = 5 + e^{-t}$

26. $(p^2 + 10p + 9)i = 10 \cos 2t$

27. $(p^2 + 2p + 37)v = 120 \cos 6t$

28. $(p^3 + 3p^2 + 3p + 1)i = 40$

Use the complete response found for Exercise 25 and evaluate the constants of integration for the initial conditions specified in Exercises 29 and 30.

29. $v(0^+) = 5$ V and dv/dt evaluated at $t = 0^+$ is 17 V/s.

30. $v(2^+) = 5$ V and dv/dt evaluated at $t = 2^+$ is 17 V/s.

31. Find voltage v for $t > 0$ in the network below if $v(0^+) = 300$ V and $i(0^+) = 6$ A.

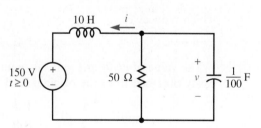

9.6 Initial Conditions with Sources Constant Prior to t_0

In Section 9.5 we have seen that the last step in determining a network's complete response is evaluation of the constants of integration that appear in the sum

$$v = v_p + v_c \tag{9.64}$$

We observed that this step required knowledge of *initial conditions* such as the value of a voltage or current at some initial time t_0. This section explains how to find initial conditions. The key to finding initial conditions lies in the following two theorems.

Continuity of Capacitance Voltage

Capacitance voltage v_C must be continuous if the current is not infinite.* Therefore, for any physical system,

$$v_C(t_0^+) = v_C(t_0) = v_C(t_0^-) \qquad (9.65)$$

THEOREM 8

Continuity of Inductance Current

Inductance current i_L must be continuous if the voltage is not infinite.† Therefore, for any physical system,

$$i_L(t_0^+) = i_L(t_0) = i_L(t_0^-) \qquad (9.66)$$

Theorem 7 follows because the terminal equation for capacitance is

$$i_C = C \frac{d}{dt} v_C \qquad (9.67)$$

This implies that an instantaneous change in voltage v_C results in an infinite current i_C. An instantaneous change in capacitance voltage would correspond to an instantaneous change in stored energy, which is physically impossible.

Similarly, because the terminal equation for inductance is

$$v_L = L \frac{d}{dt} i_L \qquad (9.68)$$

inductance current cannot change instantaneously unless the voltage is infinite. An instantaneous change in inductance current would also correspond to an instantaneous change in stored energy, which is again physically impossible. Therefore, inductance current is continuous (Theorem 8).

In the following example, we use Theorem 7, and the fact that the complementary response of a stable circuit asymptotically approaches zero with time, to evaluate the initial conditions.

EXAMPLE 9.13

Refer to the network of Fig. 9.7. The initial value of v_o at time t_x is known:

$$v_o(t_x) = 30 \text{ V}$$

and
$$v_s(t) = 50 \text{ V} \qquad t > t_x$$

(a) Determine $v_o(t)$ for $t > t_x$.
(b) Describe what happens to the solution for $t > t_x + 5(\frac{1}{4})$.
(c) Describe what happens in the limit as $t_x \to -\infty$.

---

* An impulse is infinite in magnitude, so an impulse of capacitance current will cause capacitance voltage to be discontinuous at the point in time at which the impulse occurs.
† An impulse in inductance voltage will cause a discontinuity in inductance current.

FIGURE 9.7
Determining initial conditions
for an *RC* circuit

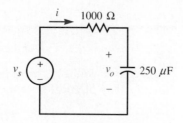

■ **SOLUTION**

(a) Continuity of capacitance voltage (Theorem 7) assures us that

$$v_o(t_x^+) = v_o(t_x) = v_o(t_x^-) = 30 \text{ V}$$

Application of KCL to a surface enclosing the upper right-hand node yields

$$\frac{1}{1000}(v_o - v_s) + 0.00025 \frac{d}{dt} v_o = 0$$

After multiplication by 4000, we can write this equation in operator notation as

$$(p + 4)v_o = 4v_s$$

The characteristic equation

$$s + 4 = 0$$

gives the complementary response

$$v_c = Ae^{-4t}$$

The particular response is

$$v_p = \frac{4}{0 + 4} 50 = 50 \text{ V} \qquad t > t_x$$

The complete solution is

$$v_o = v_p + v_c$$
$$= 50 + Ae^{-4t} \qquad t > t_x$$

We can now use the initial condition to evaluate the constant of integration:

$$v_o(t_x^+) = 30 = 50 + Ae^{-4t_x}$$

which gives

$$A = -20e^{4t_x}$$

Therefore

$$v_o(t) = 50 - 20e^{-4(t - t_x)} \text{ V} \qquad t > t_x$$

This is plotted in Fig. 9.8a for $t_x = 0$.

(b) Examination of the solution reveals that if $t > t_x + 5(\frac{1}{4})$—that is, if t is more than five time constants after t_x—then the complementary response term in $v_o(t)$ is small compared with its original value, and $v_o(t)$ is approximated by $v_p(t)$.

$$v_o(t) \simeq v_p(t)$$

$$= 50 \text{ V} \qquad t > t_x + 5\left(\frac{1}{4}\right)$$

This case is plotted in Fig. 9.8b.

(c) The condition $t_x = -\infty$ means that the change in source voltage from 30 to 50 V occurred an infinitely long time in the past. Use of this condition in $v_o(t)$ as found in (a) yields

$$v_o(t) = \lim_{t_x \to -\infty} (50 - 20e^{-4(t-t_x)})$$

$$= 50 \text{ V}$$

$$= v_p(t) \qquad t > t_0 > -\infty$$

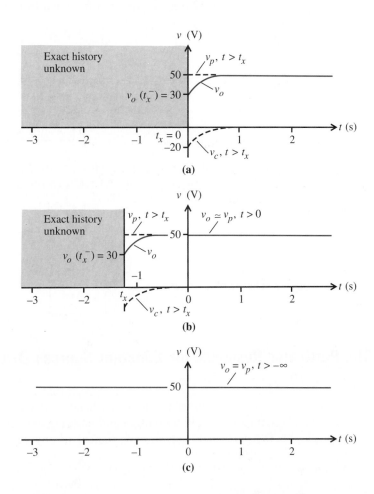

FIGURE 9.8
The output voltage of the circuit of Fig. 9.7 when $v(t_x^-) = 30$ V and $v_s(t) = 50$ V for $t > t_x$. (a) v_o for $t_x = 0$; (b) v_o for $t_x = -5(1/4)$; (c) v_o for $t_x = -\infty$

Notice that in the limit as $t_x \to -\infty$, the complementary response goes to zero for all $t > -\infty$. This means that

$$v(t_0) = v_p(t_0) \qquad t_0 > -\infty$$

This case is plotted in Fig. 9.8c. ∎

In Example 9.13, we saw that when a constant source is applied to a stable circuit, the circuit eventually reaches *steady state* (equilibrium), and the response becomes constant. We can use the particular response to determine initial conditions.

Capacitance voltage and inductance current are continuous unless impulses of current or voltage are present. If a source has been constant for a very long time, the complementary response of a stable circuit decays to zero (has negligible value), and only the particular response remains. The particular response is constant, and the circuit has reached steady state (equilibrium).

EXERCISES

32. Use a differential equation obtained from KVL to solve for current i in the following circuit.

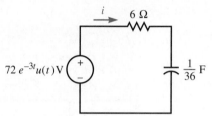

33. Find $i(0^-)$, $v(0^-)$, $i(0^+)$, $v(0^+)$, and $v(t)$, $t > 0$, for the following circuit if $i_s = 2 + 4u(t)$ A. Next, use KCL to obtain an algebraic equation that gives current $i(t)$, $t > 0$.

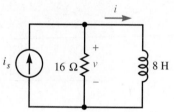

9.7 The Particular Response for Constant Sources Directly from the Circuit

In Sections 9.2 and 9.6 we have seen that the particular response is constant for a constant source, if $\mathscr{A}(0) \neq 0$. Therefore the derivatives of the particular response components of all voltages and currents are zero.* Specifically, this implies that the particular response component of any capacitance current i_{Cp} is zero for a constant source:

$$i_{Cp} = C \frac{d}{dt} v_{Cp} = 0 \qquad (9.69)$$

The particular response component of any inductance voltage v_{Lp} is also zero for a constant source:

$$v_{Lp} = L \frac{d}{dt} i_{Lp} = 0 \qquad (9.70)$$

The implication of the above two equations is important.

---

* This is implied when p is replaced by 0 in the operator $\mathscr{B}(p)/\mathscr{A}(p)$ to find the particular solution to a constant input for $\mathscr{A}(0) \neq 0$.

This provides a simple method to find particular responses for constant sources, and thus initial conditions for stable circuits, *directly from the network*.

EXAMPLE 9.14

(a) Find the initial value of v_o at $t = 0^-$, in the single-loop *RC* circuit of Fig. 9.9, given that

$$v_s = \begin{cases} 50 \text{ V} & t < 0 \\ 20 \text{ V} & t \geq 0 \end{cases}$$

(b) Determine $v_o(t)$ for $t > 0$.

FIGURE 9.9
Circuit for Example 9.9

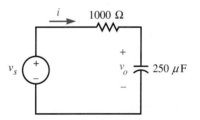

■ SOLUTION

(a) Replace the capacitance by an open circuit, as shown in Fig. 9.10, which is valid for $t < 0$.

FIGURE 9.10
Equivalent circuit for determining the particular response for the circuit in Fig. 9.9 when $t < 0$

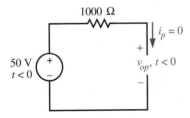

From KVL,

$$-50 + 1000(0) + v_{op} = 0$$

or

$$v_{op} = 50 \text{ V} \qquad t < 0$$

The initial condition is then

$$v_o(0^-) = v_{op}(0^-) = 50 \text{ V}$$

(This is the particular response for $t < 0$ that is found in Example 9.13.)

(b) To obtain the particular response, the capacitance is replaced by an open circuit, as shown in Fig. 9.11. The source now has a value of 20 V for $t > 0$.

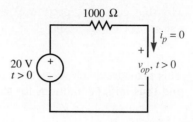

This circuit yields

$$v_p = v_{op} = 20 \text{ V} \qquad t > 0$$

As found in Example 9.13,

$$v_c = Ae^{-4t}$$

so

$$v_o = v_p + v_c$$
$$= 20 + Ae^{-4t} \qquad t > 0$$

Because capacitance voltage is continuous,

$$v_o(0^+) = v_o(0^-) = 50 \text{ V}$$

Therefore

$$v_o(0^+) = 50 = 20 + Ae^{-4(0)}$$

and

$$v_o = 20 + 30e^{-4t} \text{ V} \qquad t > 0$$

The input v_s and response v_o are sketched in Fig. 9.12. ∎

FIGURE 9.12
The input voltage v_s and response v_o for the network of Example 9.14

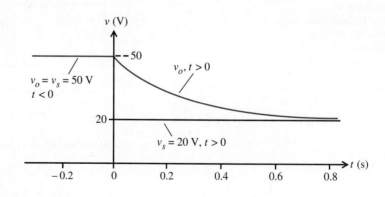

EXAMPLE 9.15

Determine the voltage $v(t)$, $t > 0$, for the network shown in Fig. 9.13.

FIGURE 9.13
A second-order circuit

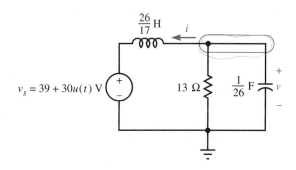

FIGURE 9.14
An equivalent circuit for
determining the particular
response of the circuit in
Fig. 9.13 when $t < 0$

SOLUTION

Recognize that for a source that has been constant for a very long time, inductance appears to be a short circuit and capacitance an open circuit. This gives the equivalent circuit, good for $t < 0$, shown in Fig. 9.14.

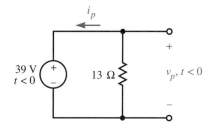

This obviously yields

$$v(0^-) = v_p(0^-) = 39 \text{ V} \qquad t < 0$$

Continuity of capacitance voltage assures us that

$$v(0^+) = v(0^-) = 39 \text{ V}$$

Inductance current is easily found from Fig. 9.14:

$$i(0^-) = i_p(0^-) = -3 \text{ A} \qquad t < 0$$

Continuity of inductance current assures us that

$$i(0^+) = i(0^-) = -3 \text{ A}$$

The particular solution for $t > 0$ can be found from an equivalent circuit as in Fig. 9.14 where the voltage source now has a value of 69 V. This gives

$$v_p = 69 \text{ V} \qquad t > 0$$

The KCL equation in terms of v for the surface enclosing the top junction of Fig. 9.13 is*

$$\frac{1}{26} \frac{d}{dt} v + \frac{1}{13} v + \frac{17}{26} \int (v - v_s) \, dt = 0$$

Multiplication of this equation by 26 and differentiation yields

$$(p^2 + 2p + 17)v = 17v_s$$

* We can write the integro-differential equation with either a definite or indefinite integral.

The characteristic equation

$$s^2 + 2s + 17 = 0$$

has roots

$$s = -1 \pm j4$$

This gives a complementary response of

$$v_c = e^{-t}(A \cos 4t + B \sin 4t)$$

The complete solution is

$$v = v_p + v_c$$
$$= 69 + e^{-t}(A \cos 4t + B \sin 4t) \text{ V} \qquad t > 0$$

We must now evaluate the constants of integration. We obtain the first equation for evaluating the constants of integration from

$$v_c(0^+) = 69 + e^0(A \cos 0^+ + B \sin 0^+) = 39$$

which gives

$$A = -30$$

The second equation required for evaluation of the constants of integration requires knowledge of dv/dt. We obtain this from the original KCL equation evaluated at $t = 0^+$:

$$\frac{1}{26}\frac{d}{dt}\left[69 + e^{-t}(A \cos 4t + B \sin 4t)\right]\Big|_{t=0^+} + \frac{1}{13}v(0^+) + i(0^+) = 0$$

Use of the numerical values for A, $v(0^+)$, and $i(0^+)$ gives

$$\frac{1}{26}\left[e^0(-120 \sin 0^+ + 4B \cos 0^+) - e^0(-30 \cos 0^+ + B \sin 0^+)\right] + \frac{39}{13} + (-3) = 0$$

or

$$B = -7.5$$

The response of the network is then

$$v = 69 + e^{-t}\left[-30 \cos 4t - 7.5 \sin 4t\right] \text{ V} \qquad t > 0$$

The input voltage and response are sketched in Fig. 9.15. ∎

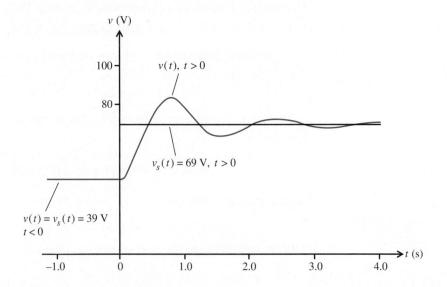

FIGURE 9.15
Input v_s and response v for the network of Fig. 9.13

For a stable network, we can find the particular response for constant sources directly from the network diagram. To do this, replace all capacitances by open circuits and all inductances by short circuits. If the sources have been constant for a very long time, only the particular response remains in a stable circuit. The particular response can be used to find initial conditions.

EXERCISES

34. Use the method presented in this section to rework Exercise 33.

35. For the following circuit, find voltage v, and plot v for $0.1 < t < 0.5$ s, if $v_s = 48$ V for $t < 0$ and $v_s = 24e^{-8t}$ V for $t \geq 0$.

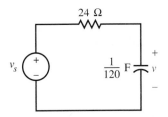

36. Find current i for $t > 0$ in the following circuit, if $v_s = 200[1 + u(t)]$ V. Is the network overdamped, critically damped, or underdamped?

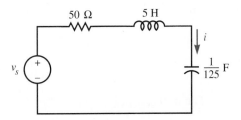

9.8 Initial Conditions with Time-Varying Sources Prior to t_0

We now address the problem of determining initial conditions in circuits whose sources are time-varying prior to t_0. In a stable system (a system for which the real parts of all roots of the characteristic equation are negative), the complementary response always decays asymptotically to zero. Therefore, *after a sufficiently long time* the output of a stable circuit will be (approximately) just the particular response.* We observed this for a constant input in Example 9.13. A common case in engineering is when a source has been sinusoidal for a very long time, so that the system has reached the *sinusoidal steady state*. If at time t_0 the sinusoidal source suddenly changes, the initial conditions are still found from the particular response prior to t_0.

---

* This assumes that the complementary response decays to zero faster than the particular response, which is always the case for stable systems with constant or sinusoidal inputs.

EXAMPLE 9.16

For $t > 0$, find the capacitance voltage v_o in Fig. 9.16, when v_s has been

$$v_s = 50 \cos 3t \text{ V}$$

for a long time, but changes to

$$v_s = 20 \text{ V} \qquad t > 0$$

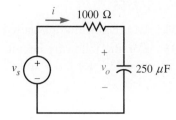

■ SOLUTION

As found in Example 9.13, v_o is described by the differential equation

$$(p + 4)v_o = 4v_s$$

Thus

$$(p + 4)v_o = (4)50 \cos 3t \qquad t < 0$$

The particular response is

$$v_{op} = \mathscr{R}e \left\{ \frac{4}{\mathscr{A}(j3)} 50e^{j3t} \right\}$$

$$= \frac{4}{|\mathscr{A}(j3)|} 50 \cos \left[3t - \arctan \frac{\mathscr{I}m \{\mathscr{A}(j3)\}}{\mathscr{R}e \{\mathscr{A}(j3)\}} \right]$$

$$= 40 \cos (3t - 36.87°) \text{ V} \qquad t < 0$$

The initial condition needed at $t = 0^+$ is

$$v_o(0^+) = v_o(0^-) = v_{op}(0^-) = 40 \cos (0 - 36.87°)$$
$$= 32 \text{ V}$$

(Observe that this value *could not* be obtained by replacing the capacitance by an open circuit, because the time-varying source for $t < 0$ prevents the derivative of the voltage from being zero at $t = 0^-$.) For $t > 0$, $v_s = 20$ V, and the differential equation becomes

$$(p + 4)v_o = (4)20 \qquad t > 0$$

The particular response is

$$v_p = \frac{4}{\mathscr{A}(0)} 20$$

$$= \frac{4}{0 + 4} 20 = 20 \text{ V} \qquad t > 0$$

(This particular response could have been found by replacing the capacitance with an open circuit, because the source is constant for $t > 0$.) The complementary response, obtained from the root $s = -4$ of the characteristic equation, is

$$v_{oc} = Ae^{-4t} \qquad t > 0$$

The complete response is

$$v_o = v_{oc} + v_{op}$$
$$= Ae^{-4t} + 20 \qquad t > 0$$

Now use the value of v_o at time 0^+ to evaluate the constant of integration A (remember that the complete response must be used):

$$v_o(0^+) = Ae^0 + 20 = 32 \text{ V}$$

which gives

$$A = 12$$

The result is

$$v_o = 12e^{-4t} + 20 \text{ V} \qquad t > 0$$

which is sketched in Fig. 9.17. ■

FIGURE 9.17
Input v_s and response v_o for the network of Example 9.16

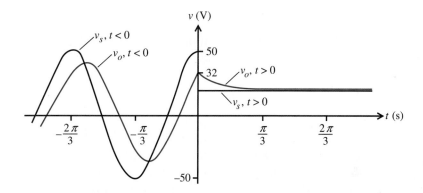

REMEMBER

If a source is time-varying, we cannot find the particular response by replacing capacitances by open circuits and inductances by short circuits. If a sinusoidal source has been applied to a stable circuit for a very long time, only the particular response remains. We use this particular response to determine initial conditions.

EXERCISES

37. Repeat Exercise 33 if $i_s = \{(2 \cos 2t)[1 - u(t)] + (2 \cos 1.5t)u(t)\}$ A.
38. Repeat Exercise 36 if $v_s = [(200 \cos 4t)u(-t) + 200u(t)]$ V.

9.9 Stable Circuits with Switches

A common occurrence in practical circuits is that a throw of a switch changes the circuit configuration. This introduces no unusual difficulties. Just use the circuit model appropriate for the time interval under consideration. We illustrate the procedure in the following two examples.

EXAMPLE 9.17

See Problem 9.8 in the PSpice manual.

The switch shown in Fig. 9.18 has been closed for a very long time and is opened at $t = 0$. Solve for $i(t)$, $t > 0$.

FIGURE 9.18
A switched circuit

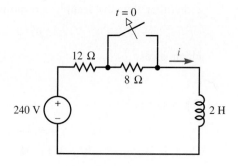

■ SOLUTION

The KVL equation with the switch closed is

$$-240 + 12i + 2\frac{d}{dt}i = 0$$

or

$$(p + 6)i = 120$$

Since the switch has been closed for a very long time, only the particular response remains:

$$i_p(t) = \frac{120}{0 + 6} = 20 \text{ A} \qquad t < 0$$

(This result can also be found if we recognize that the inductance appears as a short circuit to a source that has been constant for a very long time.) Continuity of inductance current gives

$$i(0^+) = i(0^-) = i_p(0^-) = 20 \text{ A}$$

For $t > 0$, the KVL equation would be

$$-240 + 20i + 2\frac{d}{dt}i = 0$$

or

$$(p + 10)i = 120$$

The characteristic equation

$$s + 10 = 0$$

gives a complementary response of

$$i_c = Ae^{-10t} \qquad t > 0$$

The particular response is

$$i_p = \frac{120}{0 + 10} = 12 \text{ A} \qquad t > 0$$

(This can also be found if we replace the inductance by a short circuit, because the source is constant for $t > 0$.) Addition of i_p and i_c gives

$$i = 12 + Ae^{-10t} \qquad t > 0$$

Use of the initial condition gives

$$i(0^+) = 20 = 12 + Ae^{-0}$$

or

$$A = 8$$

Thus the complete response is

$$i = 12 + 8e^{-10t} \text{ A} \qquad t > 0$$

You should sketch the current for $-0.1 < t < 0.3$ s. ■

EXAMPLE 9.18

⬛ *See Problem 9.7 in the PSpice manual.*

Find the $v(t)$ for the network shown in Fig. 9.19. The switch has been closed for a very long time and is opened at $t = 0$ and then reclosed at $t = 0.5$ s.

FIGURE 9.19
A switched *RC* network

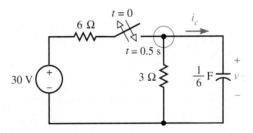

■ SOLUTION

Steady state has been reached at $t = 0^-$, because the switch was closed for a very long time. This implies that the capacitance appears to be an open circuit. Application of KCL to a surface enclosing the top node gives

$$\frac{1}{6}(v - 30) + \frac{1}{3}v + 0 = 0 \qquad t = 0^-$$

which reduces to

$$v(0^-) = 10 \text{ V}$$

(This is the particular response for $t < 0$.) After the switch is opened, KCL yields

$$\frac{1}{3}v + \frac{1}{6}\frac{d}{dt}v = 0 \qquad 0 < t < 0.5 \text{ s}$$

which can be rearranged to give

$$(p + 2)v = 0 \qquad 0 < t < 0.5 \text{ s}$$

The characteristic equation

$$s + 2 = 0$$

gives the complementary response, which is also the complete response:

$$v_c = Ae^{-2t}$$

The initial condition at $t = 0^+$ yields the value of A:

$$v(0^+) = v(0^-) = 10 = Ae^0 = A$$

Thus

$$v = 10e^{-2t} \qquad 0 < t < 0.5 \text{ s}$$

The value of v at $t = 0.5^-$ s,

$$v(0.5^-) = 10e^{-2(0.5)} = 10e^{-1} \simeq 3.68 \text{ V}$$

can be used as the *initial condition for* $t > 0.5$ s. After the switch is closed, the voltage is described by the KCL equation

$$\frac{1}{6}(v - 30) + \frac{1}{3}v + \frac{1}{6}\frac{d}{dt}v = 0 \qquad t > 0.5 \text{ s}$$

which can be written as

$$(p + 3)v = 30 \qquad t > 0.5 \text{ s}$$

The characteristic equation

$$s + 3 = 0$$

gives the complementary response

$$v_c = Be^{-3t}$$

(Note that when the switch was closed the circuit changed and thus the time constant changed.) The particular response is

$$v_p = \frac{\mathscr{B}(0)}{\mathscr{A}(0)}30 = \frac{1}{0 + 3}30 = 10$$

which gives the complete response

$$v = v_p + v_c = 10 + Be^{3t} \qquad t > 0.5 \text{ s}$$

We can evaluate constant B from the initial condition:

$$v(0.5^-) = 10e^{-1} = v(0.5^+) = 10 + Be^{-3(0.5)}$$

which gives

$$B = 10(e^{-1} - 1)e^{3(0.5)} \simeq 10(-0.632)e^{-3(0.5)}$$

Thus

$$v = 10(1 - 0.632e^{-3(t - 0.5)}) \text{ V} \qquad t > 0.5 \text{ s}$$

The results are combined to give

$$v = \begin{cases} 10 \text{ V} & t = 0 \\ 10e^{-2t} \text{ V} & 0 < t < 0.5 \text{ s} \\ 10(1 - 0.632e^{-3(t - 0.5)}) \text{ V} & t > 0.5 \text{ s} \end{cases}$$

This is plotted in Fig. 9.20. ∎

FIGURE 9.20
Response of a switched *RC*
network

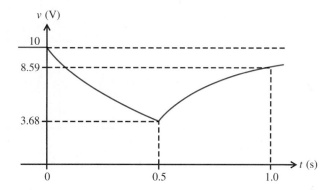

REMEMBER

When you evaluate initial conditions, remember that capacitance voltage and
inductance current are continuous. When you write the differential equations that
describe the circuit, use the circuit that corresponds to the switch position for the
time interval considered.

EXERCISES

39. The switch shown in the following circuit has been closed for a very long
time and is opened at $t = 0$. Find current i for $t > 0$.

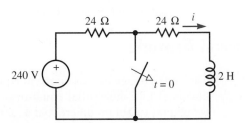

40. At $t = 0$ the switch shown in the following circuit is closed after having been
open for a very long time. Find current i for $t > 0$ if $v_a = 16 \cos 2t$ V and
$v_b = 32$ V for all time.

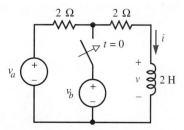

41. Find current $i(t)$ for $t > 0$ if the switch shown in the following circuit has
been closed for a very long time and is opened at $t = 0$. Is the network overdamped, critically damped, or underdamped with the switch closed? With
the switch open?

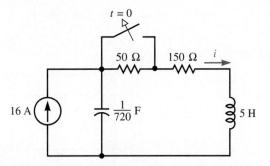

The switch in the following circuit has been closed for a very long time and is opened at $t = 0$ for Exercises 42 and 43.

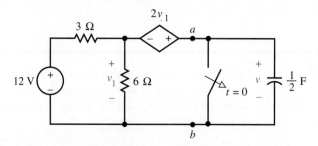

42. Make a Norton equivalent for the network to the left of terminals a and b, and use this to solve for voltage v for $t > 0$.

43. Use node-voltage equations to obtain a single differential equation that can be solved for voltage v for $t > 0$. This should be the same as the differential equation obtained in Exercise 42.

9.10 More General Networks*

The node-voltage or mesh-current equations describing many networks are simultaneous LLTI differential equations. We can use Cramer's rule or other matrix techniques to obtain differential equations that contain a single unknown voltage or current. We can then solve these by our standard techniques. We illustrate this by the following example.

EXAMPLE 9.19

Given the network shown in Fig. 9.21, determine node voltage v_1 for $t > 0$ when $v_s = 36u(t)$ V.

FIGURE 9.21
An *RLC* network that is neither a series nor a parallel circuit

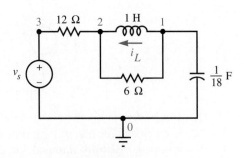

* This section can be omitted without loss of continuity.

■ **SOLUTION**

First determine the initial conditions. Because the source is zero for $t < 0$,

$$v_1(0^+) = v_1(0^-) = 0$$

$$i_L(0^+) = i_L(0^-) = 0$$

A KCL equation for a surface enclosing node 2 at $t = 0^+$ gives

$$\frac{1}{12}[v_2(0^+) - 36] - i_L(0^+) + \frac{1}{6}[v_2(0^+) - v_1(0^+)] = 0$$

or

$$v_2(0^+) = 12 \text{ V}$$

which will be used later.

The node-voltage equations (KCL equations) for nodes 1 and 2 are

$$\frac{1}{18}\frac{d}{dt}v_1 + \frac{1}{6}(v_1 - v_2) + \frac{1}{1}\int(v_1 - v_2)\,dt = 0$$

and

$$\frac{1}{12}(v_2 - v_s) + \frac{1}{6}(v_2 - v_1) + \frac{1}{1}\int(v_2 - v_1)\,dt = 0$$

To simplify notation we will use the inverse operator,

$$\frac{1}{p} = \int(\cdot)\,dt$$

to write the node-voltage equations in matrix form:

$$\begin{bmatrix} \frac{1}{18}p + \frac{1}{6} + \frac{1}{p} & -\frac{1}{6} - \frac{1}{p} \\ -\frac{1}{6} - \frac{1}{p} & \frac{1}{4} + \frac{1}{p} \end{bmatrix} \cdot \begin{bmatrix} v_1 \\ v_2 \end{bmatrix} = \begin{bmatrix} 0 \\ \frac{1}{12}v_s \end{bmatrix}$$

Cramer's rule gives the differential equation to be solved for v_1:

$$\begin{vmatrix} \frac{1}{18}p + \frac{1}{6} + \frac{1}{p} & -\frac{1}{6} - \frac{1}{p} \\ -\frac{1}{6} - \frac{1}{p} & \frac{1}{4} + \frac{1}{p} \end{vmatrix} \cdot v_1 = \begin{vmatrix} 0 & -\frac{1}{6} - \frac{1}{p} \\ \frac{1}{12}v_s & \frac{1}{4} + \frac{1}{p} \end{vmatrix}$$

Evaluation of the determinants gives

$$\frac{1}{p}(p^2 + 5p + 6)v_1 = \left(1 + \frac{6}{p}\right)v_s$$

Multiplication of the equation by p (this is equivalent to differentiation) yields

$$(p^2 + 5p + 6)v_1 = (p + 6)v_s$$

We can easily show that the solution is

$$v_1 = 36 + A_1 e^{-2t} + A_2 e^{-3t} \qquad t > 0$$

We obtain the first equation required for finding the initial conditions by evaluating the preceding equation for v_1 at $t = 0^+$:

$$v_1(0^+) = 36 + A_1 e^{-2(0)} + A_2 e^{-3(0)} = 0$$

which gives

$$A_1 + A_2 = -36 \tag{9.71}$$

To obtain the second equation, which includes the derivatives of v_1, we evaluate the original KCL equation for a surface enclosing node 1 at $t = 0^+$:

$$\frac{1}{18}(-2A_1 - 3A_2) + \frac{1}{6}[v_1(0^+) - v_2(0^+)] + i_L(0^+) = 0$$

This gives

$$2A_1 + 3A_2 = -36 \tag{9.72}$$

Solving Eqs. (9.71) and (9.72) for A_1 and A_2 yields

$$\begin{bmatrix} A_1 \\ A_2 \end{bmatrix} = \begin{bmatrix} -72 \\ 36 \end{bmatrix}$$

Substitution of these values into the equation for v_1 completes the example. ∎

The order of the differential equation obtained by this procedure will be equal to or less than the number of energy storage elements. Exceptions in which equality will not hold include networks with series or parallel capacitances or inductances.

REMEMBER

We can use Cramer's rule to reduce a system of first- or second-order simultaneous LLTI differential equations to a higher-order LLTI equation containing only one unknown voltage or current.

EXERCISES

44. Given the network shown below, if $i(0^+) = 1$ A and $v(0^+) = 3$ V, write two KCL equations, and solve for $v(t)$, $t > 0$, if $v_s = 12$ V for $t > 0$.

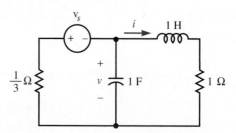

45. For the network of Exercise 44, write two mesh-current equations, and solve for current $i(t)$ for $t > 0$.

46. The source in the network of Exercise 44 has a value $v_s = [36 - 12u(t)]$ V (you must calculate the initial conditions). Determine voltages $v_1(t)$ and $v_2(t)$ for $t > 0$.

9.11   The Complete Response with PSpice

In Chapter 4 we used the computer program SPICE (the version PSpice) to calculate the dc response of circuits, and in Chapter 8 we used this program to calculate the response of source-free circuits with specified initial conditions. The inclusion of sources does not significantly complicate our situation. We only need to add data statements that describe the sources. This will be demonstrated with a simple example in which the initial condition is specified and the source is constant for $t > 0$.

EXAMPLE 9.20

The source v_s in Fig. 9.22 is a constant 4 V for $t > 0$. If $v_1(0)$ is 2 V, use PSpice to plot voltage v_1 for $0 \le t \le 4$ s in 0.2-s steps.

FIGURE 9.22
Example circuit for transient analysis by PSpice

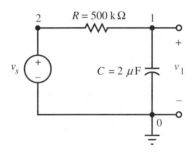

■ SOLUTION

The following circuit file is all that is required to obtain a printer plot of v_1 for $0 \le t \le 4$ s (the constant source is specified by line 5 of the circuit file):

```
INPUT FILE FOR EXAMPLE 9.20

*CALCULATE AND PLOT THE COMPLETE RESPONSE.

R       2    1    500K

C       1    0    2U      IC=2

VS      2    0    DC    4

.TRAN    0.2    4      UIC

.PLOT    TRAN    V(1)

.END
```

Save this file under the name EX9-20.CIR. We can run PSpice using this as the input file in exactly the same way we did for dc circuits in Chapter 4 and source-free circuits in Chapter 8. The output file contains the printer plot of Fig. 9.23.

FIGURE 9.23
Response of the circuit for Example 9.20

```
    TIME        V(1)
(*)----------      2.0000E+00    2.5000E+00    3.0000E+00    3.5000E+00    4.0000E+00
                              - - - - - - - - - - - - - - - - - - - - - - - - - - - -
  0.000E+00  2.000E+00 *             .             .             .             .
  2.000E-01  2.362E+00 .         *   .             .             .             .
  4.000E-01  2.659E+00 .             .   *         .             .             .
  6.000E-01  2.902E+00 .             .       *     .             .             .
  8.000E-01  3.101E+00 .             .             . *           .             .
  1.000E+00  3.264E+00 .             .             .     *       .             .
  1.200E+00  3.398E+00 .             .             .         *   .             .
  1.400E+00  3.507E+00 .             .             .             *             .
  1.600E+00  3.596E+00 .             .             .             .  *          .
  1.800E+00  3.670E+00 .             .             .             .    *        .
  2.000E+00  3.730E+00 .             .             .             .      *      .
  2.200E+00  3.779E+00 .             .             .             .       *     .
  2.400E+00  3.819E+00 .             .             .             .        *    .
  2.600E+00  3.852E+00 .             .             .             .         *   .
  2.800E+00  3.879E+00 .             .             .             .          *  .
  3.000E+00  3.901E+00 .             .             .             .          *  .
  3.200E+00  3.919E+00 .             .             .             .           * .
  3.400E+00  3.933E+00 .             .             .             .           * .
  3.600E+00  3.945E+00 .             .             .             .            *.
  3.800E+00  3.955E+00 .             .             .             .            *.
  4.000E+00  3.963E+00 .             .            ..             .            *.
                              - - - - - - - - - - - - - - - - - - - - - - - - - - - -
```

An analysis of the circuit gives an output

$$v_1 = 4 - 2e^{-t}\,\text{V} \qquad t > 0$$

You should verify that this matches the printout at $t = 2$ s. ■

Sources that are constant for all time greater than zero are specified by using DC for the type in the component statement for the source. We can obtain a pulse by replacing DC by PULSE. This source type specification must be followed by numbers that specify pulse amplitude, position, rise and fall times, and duration. We use a pulse source in Example 9.21, but we precede it by a general description of the PULSE type of source.

Pulse Sources

PULSE source; see Sections 3.2.2 and A.3.3.4 of the PSpice manual.

A pulse voltage (or current) source is specified by a component statement of this type:

```
VNAME     N+     N-     PULSE(VS,VP,[TD],[TR],[TF],[PW],[PER])
```

where VNAME is the source name, N+ and N− are the node connections, and the PULSE parameters, shown on the pulse of Fig. 9.24, are as summarized in Table 9.1. (The unit of time is seconds, and the unit of voltage is volts.)

TABLE 9.1
Pulse parameters

	PARAMETER	DEFAULT VALUE		UNITS
VS	Initial voltage	None		Volts
VP	Pulsed voltage	None		Volts
[TD]	Delay time	0		Seconds
[TR]	Rise time	TSTEP	As given in the .TRAN statement	Seconds
[TF]	Fall time	TSTEP	As given in the .TRAN statement	Seconds
[PW]	Pulse width	TFINAL	As given in the .TRAN statement	Seconds
[PER]	Period	TFINAL	As given in the .TRAN statement	Seconds

FIGURE 9.24
The pulse source

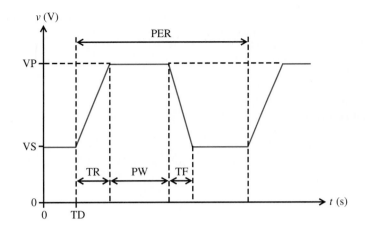

Thus the default value TFINAL for PER gives a single pulse for the analysis. We demonstrate the pulse source in Example 9.21. ∎

EXAMPLE 9.21

Replace the voltage source in the circuit of Example 9.20 (Fig. 9.22) by a pulse source that is defined by

$$v_s = \begin{cases} 0 & t < 1 \text{ s} \\ 4 \text{ V} & 1 < t < 2 \text{ s} \\ 0 & t > 2 \text{ s} \end{cases}$$

Let PSpice calculate the initial conditions, and plot voltages v_1 and v_2 for $0 \le t \le 4$ s in 0.2-s steps.

∎ SOLUTION

The problem statement specifies zero rise and fall times for the pulse. If a rise time (TR) or fall time (TF) of zero is used in the PULSE source, the program defaults to the step size (in our case, 0.2 s). We will specify zero for TR and TF, but a smaller rise or fall time than TSTEP can be used. Our circuit file is as follows (the pulse source is specified by line 5 of the circuit file):

```
INPUT FILE FOR EXAMPLE 9.21

*FOR EXAMPLE 9.21 WE USE A PULSE INPUT.

R      2    1      500K

C      1    0      2U

VS     2    0      PULSE(0,4,1,0,0,1)

.TRAN     0.2     4

.PLOT     TRAN     V(1),     V(2)

.END
```

We save this circuit file under a convenient name, and use it as the input file for PSpice. After Pspice is run, the output file gives us the printer plot shown in Fig. 9.25.

FIGURE 9.25
Pulse input and response for the circuit of Example 9.21

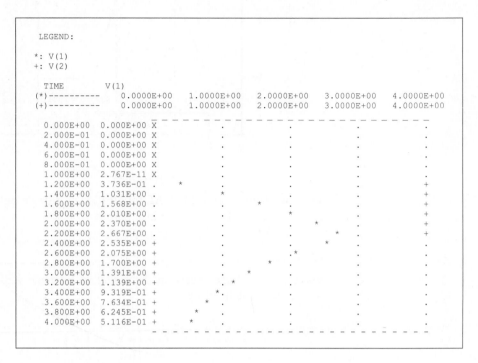

The calculated response for this circuit is

$$v_1(t) = \begin{cases} 0 & t < 1 \text{ s} \\ 4 - 4e^{-(t-1)} \text{ V} & 1 < t < 2 \text{ s} \\ (4 - 4e^{-1})e^{-(t-2)} \text{ V} & t > 2 \text{ s} \end{cases}$$

If we compare the calculated values at $t = 1.4$ s and 2.4 s with the printed values at these times, we will find a considerable difference. This is caused by the default rise and fall times (TSTEP = 0.2 s) for the pulse. As an exercise, change the rise and fall times to 0.01 s in the pulse specification, and repeat this example. Compare the results to those of this example. ■

Other sources are available for transient analysis. In particular, we can use a sinusoidal source.

Sinusoidal Sources

SIN source; see Sections 3.2.2 and A.3.3.2 of the PSpice manual.

For transient analysis, a damped sinusoidal source is specified by a component data statement as follows:

```
VNAME     N+     N-     SIN(VO,VA,[FREQ],[TD],[ALPHA],[PHASE])
```

where VNAME is the source name, N+ and N− are the node connections, and SIN specifies a sinusoidal source. The units on the sin parameters are volts, hertz, seconds, seconds^{-1}, and degrees for the appropriate parameters. We summarize the parameters in Table 9.2.

TABLE 9.2
SIN parameters

	PARAMETER	DEFAULT VALUE	UNITS
VO	Offset voltage	None	Volts
VA	Peak amplitude	None	Volts
[FREQ]	Frequency	1/TSTOP	Hertz
[TD]	Delay	0	Seconds
[DF]	Damping factor	0	Seconds^{-1}
[PHASE]	Phase	0	Degrees

The voltage starts at VO and remains at this value until time TD. The voltage then becomes an exponentially damped sinusoid described by

$$\text{VNAME} = \text{VO} + \text{VA} \cdot \text{SIN}\left[2\pi \cdot \left\{\text{FREQ} \cdot (t - \text{TD}) + \frac{\text{PHASE}}{360}\right\}\right]$$
$$\cdot \, e^{-(t-\text{TD})\cdot\text{ALPHA}} \qquad t \geq 0 \tag{9.73}$$

EXAMPLE 9.22

See Example 3.7 in the PSpice manual.

Replace the voltage source in Fig. 9.22 by a sinusoidal input that begins at $t = 0$:

$$v_s = [4 \sin 2\pi(0.5)t]u(t) \text{ V}$$

Repeat Example 9.21.

■ **SOLUTION**

The circuit file is as follows (the sinusoidal source is specified by line 5 of the circuit file):

```
INPUT FILE FOR EXAMPLE 9.22

*FOR EXAMPLE 9.22 WE USE a SINE INPUT.

R      2     1     500K

C      1     0     2U

VS     2     0     SIN(0,4,0.5,0,0,0)

.TRAN    0.2    4

.PLOT    TRAN    V(1),    V(2)

.END
```

This circuit file is saved under a convenient name, such as EX9-22.CIR, and used as the input file for PSpice. The output file contains the printer plot shown in Fig. 9.26. (We have sketched connecting lines to make the plot more readable.) ■

FIGURE 9.26
The sinusoidal input voltage
and the output voltage for
the circuit of Fig. 9.22

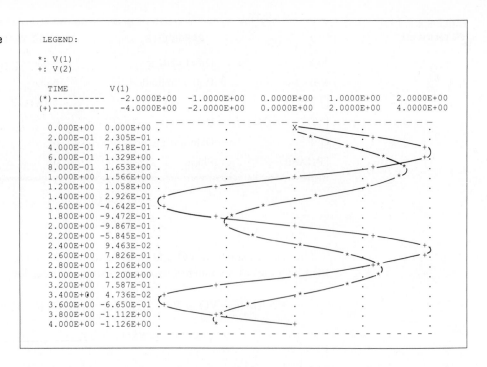

```
LEGEND:

*: V(1)
+: V(2)

  TIME          V(1)
(*)----------   -2.0000E+00  -1.0000E+00   0.0000E+00   1.0000E+00   2.0000E+00
(+)----------   -4.0000E+00  -2.0000E+00   0.0000E+00   2.0000E+00   4.0000E+00

 0.000E+00   0.000E+00 .                               X
 2.000E-01   2.305E-01 .                          . * .        . +
 4.000E-01   7.618E-01 .                          .      *      .
 6.000E-01   1.329E+00 .                          .           *  .
 8.000E-01   1.653E+00 .                          .        . +
 1.000E+00   1.566E+00 .                     +    .
 1.200E+00   1.058E+00 .            +         .
 1.400E+00   2.926E-01 (+                          .     *
 1.600E+00  -4.642E-01 .+               .      *
 1.800E+00  -9.472E-01 .         +  *
 2.000E+00  -9.867E-01 .         *         .      +
 2.200E+00  -5.845E-01 .             *  .                  . +
 2.400E+00   9.463E-02 .                  . *              .
 2.600E+00   7.826E-01 .                          .      *
 2.800E+00   1.206E+00 .                               +*
 3.000E+00   1.200E+00 .                     +         *
 3.200E+00   7.587E-01 .            +              *
 3.400E+00   4.736E-02 (+                    *
 3.600E+00  -6.650E-01 .+               *
 3.800E+00  -1.112E+00 .         (+*
 4.000E+00  -1.126E+00 .         *                +
```

Observe that PSpice gives a printer plot when the .PLOT statement is used, whereas some versions of SPICE generate graphics. This is not a problem, because PSpice supports a graphics postprocessor.

The .PROBE Command

.PROBE; see Section 3.2.3 in the PSpice manual.

With PSpice we can obtain high-quality graphics by replacing the .PLOT statement with the .PROBE statement:

```
.PROBE[OUTVAR]
```

OUTVAR is a list of output variables (from .DC, .TRAN, or .AC analysis) that is written to a data file named PROBE.DAT for use by the probe graphics postprocessor. Note that no analysis type precedes the output variables. If a list of output variables is omitted, all node voltages and device currents are stored in the file.

PROBE is menu-driven and guides us through the steps required to generate a plot of the desired variables. In addition to permitting us to plot any of the variables stored in PROBE.DAT, we can also plot functions of these variables. For example, if we request a plot of $v * i$, the product of v and i (which is instantaneous power) is plotted.

EXAMPLE 9.23

See Example 3.6 in the PSpice manual.

Modify the circuit file for Example 9.22 by replacing the .PLOT statement on line 8 with a .PROBE statement. Run PSpice for this input file, and use PROBE to plot node voltages v_1 and v_2 and the instantaneous power that is absorbed by the capacitance.

The circuit file is as follows:

```
INPUT FILE FOR EXAMPLE 9.23

*FOR EXAMPLE 9.23 WE ALSO USE A SINE INPUT.

*FOR EXAMPLE 9.23 WE USE .PROBE GRAPHICS.

R       2    1      500K

C       1    0      2U

VS      2    0      SIN(0,4,0.5,0,0,0)

.TRAN     0.2      4

.PROBE

.END
```

We must save this circuit file and use it as the input file for PSpice. After the program runs, a rectangular grid appears on the screen with a request for information. We simply hit the return for the default (add variables). The program requests the variables, so we type V(1), V(2), and enter. A good-quality graph then fills our screen. We will now add the curve for the instantaneous power absorbed by the capacitance to the existing plot. We will scale this by plotting 100,000 times this power, so that the curves will be of similar magnitude. Request "add a trace," hit the return, and type 100000*V(1)*I(C). A second return gives a plot similar to that shown in Fig. 9.27. The quality of the printed output depends on the particular

FIGURE 9.27
The plot for Example 9.23 using the .PROBE statement

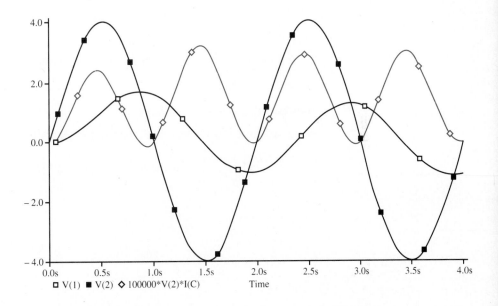

system configuration used. Our particular system gives the plot shown in Fig. 9.27 when a hard copy is requested. Some printers will require nonstandard page length specifications to fit the hard copy plot on one 8.5 × 11″ sheet of paper. ∎

Switches

Switches are modeled in PSpice by the use of voltage-controlled switches or current-controlled switches, which are special cases of controlled resistance. We will first use PSpice to rework Example 9.18, and then we will describe the switch models in more detail. A current-controlled switch is used in the following example. The model chosen will specify a switch resistance of 0.001 Ω when the control current is less than or equal to 0 A, and a switch resistance of 1000 Ω when the control current is greater than or equal to 2 A. The current source i_{sw} provides a control current through the 0-V voltage source v_s.

EXAMPLE 9.24

See Problems 9.7 and 9.8 in the PSpice manual.

Use the PSpice model shown in Fig. 9.28 to solve Example 9.18, in which the switch has been closed for a very long time and is opened at $t = 0$ and reclosed at $t = 0.5$ s.

FIGURE 9.28
The PSpice model for the switched circuit of Fig. 9.19

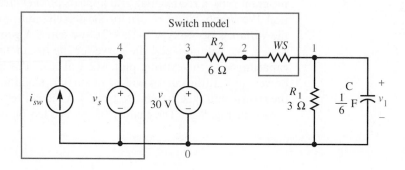

■ **SOLUTION**

The following circuit file describes the circuit model shown in Fig. 9.28.

```
CIRCUIT FILE FOR EXAMPLE 9.24

*THIS EXAMPLE USES A CURRENT CONTROLLED RESISTANCE

*TO SIMULATE A SWITCH.

R1    1    0    3

R2    2    3    6

C     1    0    .16667

V     3    0    DC    30
```

```
*THE NEXT BLOCK SIMULATES THE SWITCH

ISW      0      4      PULSE(0,2,0,0,0,.5)

VS       4      0      DC      0

WS       1      2      VS      WMOD

.MODEL     WMOD     ISWITCH     (RON=.1E3 ROFF=.001 ION=2 IOFF=0)

*SWITCH MODEL COMPLETE

.TRAN     .001     1

.PROBE

.END
```

After running PSpice, we can use PROBE to plot v_1. The result is shown in Fig. 9.29. ■

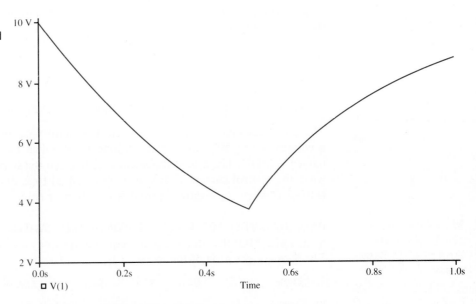

FIGURE 9.29
The response of the switched *RC* circuit of Fig. 9.19 as obtained by PSpice

We will now discuss our switch models in more detail.

🔲 *Current-controlled switch; see Sections 3.2.2 and A.5.2 in the PSpice manual.*

DATA STATEMENT FOR A CURRENT-CONTROLLED SWITCH   A data statement and a .MODEL statement are required to specify a current-controlled switch. The data statement is of the form

```
WNAME     N+     N-     VNAME     MODNAME
```

TABLE 9.3
MODEL parameters for a current-controlled switch

NAME	DESCRIPTION	DEFAULT VALUE	UNITS
RON	On-state resistance	1	Ohms
ROFF	Off-state resistance	$1E+6$	Ohms
ION	Minimum control current for the on state	0.001	Amperes
IOFF	Maximum control current for the off state	0	Amperes

WNAME is a unique switch name beginning with W. N+ and N− represent the switch node connections (passive sign convention), and VNAME is the name of a voltage source through which the controlling current flows (passive sign convention). MODNAME is the name of the .MODEL statement that specifies the switch parameters. The .MODEL statement for a current-controlled switch is of the form

.MODEL; see Section B.2.5 of the PSpice manual.

```
.MODEL    MODNAME    ISWITCH (RON=XX ROFF=XX ION=XX IOFF=XX)
```

where MODNAME is the model name specified in the data statement WNAME, and ISWITCH specifies that the model is for a current-controlled switch. The model parameters are given in Table 9.3.

Both RON and ROFF must have values greater than zero and less than 10^{12} Ω. Resistance RON can be *greater than or less than* ROFF. We will use the current-controlled switch of Example 9.24 as an example. The data statement and model statement are

```
WS    1    2    VS    WMOD

.MODEL    WMOD    ISWITCH (RON=1E3 ROFF=.001 ION=2 IOFF=0)
```

The data statement specifies a switch WS controlled by the current through a voltage source VS with switch parameters specified by a .MODEL statement named WMOD. The model statement specifies that the switch resistance is 1000 Ω when the control current is greater than or equal to 2, and the switch resistance is 0.001 Ω when the control current is less than or equal to 0 A.

DATA STATEMENT FOR A VOLTAGE-CONTROLLED SWITCH   Both a data statement and a .MODEL statement are required to specify a voltage-controlled switch. The data statement is of the form

Voltage-controlled switch; see Sections 3.2.2 and A.5.1 of the PSpice manual.

```
SNAME    N+    N-    NC+    NC-    MODNAME
```

SNAME is a unique switch name beginning with S. N+ and N− represent the switch node connections, NC+ and NC− specify the node connections for the controlling voltage (passive sign convention), and MODNAME is the name of the .MODEL statement that specifies the switch parameters. The model statement for a voltage-controlled switch is of the form

```
.MODEL    MODNAME    VSWITCH (RON=XX ROFF=XX VON=XX VOFF=XX)
```

where MODNAME is the model name given in the data statement, and VSWITCH specifies that the model is for a voltage-controlled switch. The model parameters are given in Table 9.4.

TABLE 9.4
MODEL parameters for a
voltage-controlled switch

NAME	DESCRIPTION	DEFAULT VALUE	UNITS
RON	On-state resistance	1	Ohms
ROFF	Off-state resistance	1E + 6	Ohms
VON	Minimum control voltage for the on state	1	Volts
VOFF	Maximum control voltage for the off state	0	Volts

The permissible values for the switch resistance are the same for voltage-controlled and current-controlled switches. In both switch models, the switch resistance changes continuously as the control variable traverses the range from the on value to the off value. To avoid numerical problems, the ratio of ROFF to RON should be no greater than 10^{12}. It is also advisable to select the on and off values for the control parameters so that the control variable does not cause the switching to occur any faster than is necessary to accurately model the system.

An example of the data and model statements for a voltage-controlled switch is

```
SW1     1     2     3     4     SMOD

.MODEL     SMOD     (VON=10     VOFF=1)
```

This models a voltage-controlled switch connected between nodes 1 and 2. The switch is on when v_{34} is greater than or equal to 10 V and off when v_{34} is less than or equal to 1 V. The default values of switch resistance are used: RON = 1 Ω and ROFF = 1E6 Ω.

REMEMBER

We can use PSpice to determine the response of a driven circuit. We can specify initial conditions or use the program to calculate initial conditions.

EXERCISES

*See Problems 9.1
through 9.8 in the PSpice
manual.*

47. Use PSpice to solve for $v(t)$ and $i(t)$ in the circuit of Exercise 44. Use the .PLOT statement to get a printer plot of $v(t)$ and $i(t)$. Use a time interval and step size that give a meaningful graph.

48. Repeat Exercise 47, but use .PROBE to obtain a better-quality graphical output.

49. Use PSpice to solve Exercise 46, and use the .PLOT statement to obtain a printer plot of v_1 and v_2. Select a reasonable time interval and step size.

50. Repeat Exercise 49, but use .PROBE to obtain a better-quality graphical output.

51. Use PSpice and a voltage-controlled switch to calculate and plot current i for the circuit of Example 9.17. (Select ROFF = 10^4 Ω and RON = 10^{-3} Ω.)

9.12 Summary

In this chapter we solved differential equations that describe networks with sources. We used the s-shift theorem and integration to obtain a solution. We observed that this complete solution has two parts; one part includes the constants of integration and is the complementary response as found in Chapter 8, and the second part is the particular response due to the particular input. We then showed how we can obtain the particular response for exponential inputs Ke^{st} by simply replacing the operator p by s [if $\mathscr{A}(s) \neq 0$]. We used the complex exponential input to extend this simple procedure to sinusoidal inputs. This is the basis for steady-state sinusoidal analysis, presented in Chapters 10 and 11. We then used the complete solution to obtain initial conditions and evaluate the constants of integration. We concluded by using PSpice to calculate the complete response.

KEY FACTS

◆ Ordinary (lumped) linear time-invariant (LLTI) differential equations describe lumped linear time-invariant (LLTI) networks.

◆ The response for LLTI networks consists of two parts: the complementary response and the particular response.

◆ The complementary response is determined from the characteristic equation by the method presented in Chapter 8.

◆ The complementary response contains the constants of integration.

◆ The particular response for an nth-order LLTI differential equation is obtained from an n-fold integration with all constants of integration set equal to zero.

◆ For exponential inputs, the particular response is obtained by replacing the operator p by the coefficient s of the exponent, if $\mathscr{A}(s) \neq 0$.

◆ For cosine inputs, the particular response is the real part of the particular response for the complex exponential input.

◆ For sine inputs, the particular response is the imaginary part of the particular response for the complex exponential input.

◆ Capacitance voltage is continuous if capacitance current does not include an impulse.

◆ Inductance current is continuous if inductance voltage does not include an impulse.

◆ If a dc source or sinusoidal source has been applied to a stable circuit for a very long time, only the particular response is of significant amplitude.

◆ The previous three facts are used to determine initial conditions.

◆ The complete response must be used to evaluate constants of integration.

◆ One equation for evaluation of constants of integration is obtained by evaluation of the solution at $t = 0^+$ (or t_0^+).

- A second equation that includes the derivative can always be obtained from the original KCL or KVL equation evaluated at $t = 0^+$ (or t_0^+).

- Cramer's rule can be used to reduce a system of simultaneous differential equations to a differential equation in terms of a single voltage or current.

- PSpice can be used to obtain the complete response for *RLC* circuits with both independent and dependent sources.

- For PSpice, .PLOT gives a printer plot.

- For PSpice, .PROBE gives a good-quality graphical display.

■ PROBLEMS ■

Section 9.1

By the use of Theorem 1, determine the particular response for the differential equations listed in Problems 1 through 12.

1. $(p + 4)i = (p + 1)12e^{-2t}$
2. $2(p + 4)v = (p + 2)12e^{-2t}$
3. $2(p + 2)v = (p + 3)12e^{-2t}$
4. $4(p^2 + 7p + 6)i = (p + 1)64e^{-3t}$
5. $(p^2 + 7p + 6)v = (p + 1)64e^{-t}$
6. $2(p^2 + p)i = 12e^{-3t}$
7. $(p^2 + 10p + 25)v = 50e^{-5t}$
8. $2(p^2 + p)v = 12e^{-t}$
9. $(p + 4)v = (p + 1)12 + 24e^{-4t}$
10. $(p^2 + 6p + 8)i = (p + 4)72$
11. $2(p^2 + p)i = 36$
12. $(p^2 + 2p + 5)v = \delta(t)$
13. Determine the particular response component of current i in the following circuit if $v_s = 24tu(t)$ V.

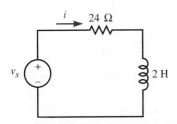

14. Determine the particular response component of voltage v in the following network if $v_s = 36u(t)$ V.

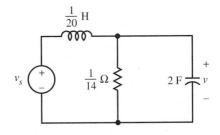

15. Solve for the particular response component of voltage v in the preceding circuit if $v_s = 36e^{-5t}u(t)$ V.

Section 9.2

By the use of Theorem 2 or 3, determine the particular response for the differential equations listed in Problems 16 through 19.

16. $(p + 6)i = (p + 36)4, t > 0$
17. $(p^2 + 3p + 2)v = e^{-100t}, t > 0$
18. $(p^2 + 9p + 14)i = 84, t > 0$
19. $(p^2 + 6p + 9)v = 24e^{-3t}, t > 0$
20. Determine the particular response component of voltage v in the following circuit if $i_s = 20u(t)$ A.

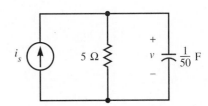

21. Repeat Problem 20 if $i_s = 10e^{-2t}u(t)$ A.
22. Repeat Problem 20 if $i_s = 10e^{-10t}u(t)$ A.
23. Determine the particular response component of current i in the following circuit if $i_s = 10u(t)$ A.

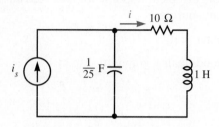

24. Repeat Problem 23 if $i_s = 36e^{-2t}$ A.
25. Repeat Problem 23 if $i_s = 36e^{-5t}$ A.

Section 9.3

Solve for the particular response of the differential equations in Problems 26 through 28.

26. $(p + 9)v = 36 + 72e^{-3t}$
27. $(p^2 + 12p + 2)i = 12 + 50e^{-3t}$
28. $(p^2 + 12p + 6)i = 18 + 28e^{-2t} + 90e^{2t}$
29. Use superposition to solve for the particular response component of voltage v in the following circuit if $v_s = 36e^{-2t}$ V for $t > 0$ and $i_s = 2$ A for $t > 0$.

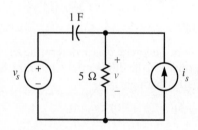

Section 9.4

Find the particular response for each of the following differential equations. Which particular solutions can be considered steady-state solutions?

30. $(p + 48)v = (2p + 1) \cos 14t$
31. $(p - 12)v = 39 \sin 3t$
32. $(p + 24)i = 52 \cos (2t + 15°)$
33. $(p^2 + 9)v = 36 \cos 3t$
34. $(p^2 + 3p + 2)i = 4 \cos t + 8 \cos 2t$
35. $(p + 4)i = 48e^{-t} \cos 4t$
36. $(p + 8)v = 64e^{-t} \cos 4t$
37. $(p + 6)i = 40e^{-3t} \cos [4t + (\pi/6)]$
38. Determine the particular response component of voltage v in the circuit in Fig. 9.30.

Section 9.5

Find the complete response of each of the following differential equations for $t > 0$, and evaluate the constants of integration.

39. $(p + 4)v = 100$, $v(0^+) = 20$ V
40. $(p + 4)v = 100 + 48e^{-t}$, $v(0^+) = 20$ V
41. $(d/dt)v + 6v + 9 \int_{-\infty}^{t} v \, d\tau = 36e^{-4t}$, $v(0^+) = 12$ V, and $9 \int_{-\infty}^{0} v \, d\tau = 4$ A
42. $(d/dt)v + 10v + 9 \int_{-\infty}^{t} v \, d\tau = 6 - \cos 3t$, $v(0^+) = 18$ V, and $9 \int_{-\infty}^{0} v \, d\tau = 12$ A
43. $(p^2 + 3p + 2)i = 4 \cos t + 8 \cos 2t$, $i(0^+) = 0$ A, and $(di/dt)|_{t=0^+} = 0$ A/s
44. $(p + 6)v = 40e^{-3t} \cos [4t + (\pi/6)]$, $v(0^+) = 120$ V
45. For the following network, find voltage v for $t > 0$ if $v(0^+) = 0$.

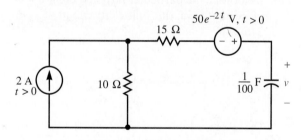

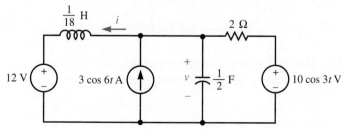

FIGURE 9.30

46. For the following network, find voltage v for $t > 0$ if $v(0^+) = 40$ V.

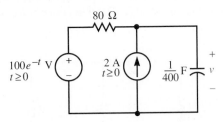

47. The initial conditions for the following network are $i_1(0^+) = -1$ A and $i_2(0^+) = -2$ A. Find $v(t)$ for $t > 0$.

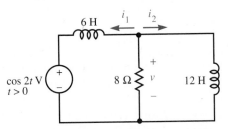

48. For the following network, find voltage $v(t)$ for $t > 0$, when $v(0^+) = 100$ V and $i(0^+) = 2$ A are given and $v_s = 50$ V for $t > 0$.

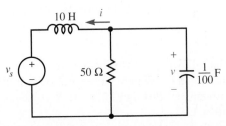

49. For the following network, find $i(t)$ for $t > 0$ if $v(0^+) = 500$ V, $i(0^+) = 15$ A, and $i_s = 12e^{-3t}$ A for $t > 0$.

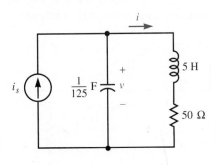

Section 9.6

50. Determine voltage v for $t > 0$ in the following circuit if $v_s = 50[2 + u(t)]$ V.

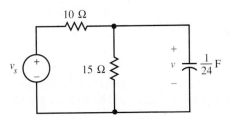

51. Determine current i for $t > 0$ in the following circuit if $i_s = 30[2 - u(t)]$ A.

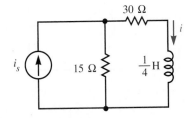

52. Determine current i in the following circuit.

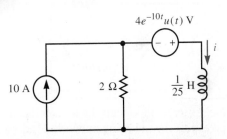

53. Determine voltage v for $t > 0$ in the following circuit if $i_s = 5 + (20 \cos 3t)u(t)$ A.

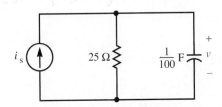

54. For the following circuit, find voltage $v(t)$ for $t > 0$ if $R = 5\ \Omega$, $L = 1$ H, and $C = \frac{1}{10}$ F.

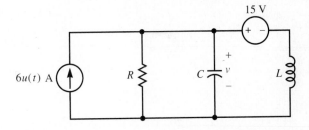

55. Repeat Problem 54 if $R = 400\ \Omega$, $L = \frac{20}{3}$ H, and $C = 10\ \mu$F.
56. Repeat Problem 54 if $R = 25\ \Omega$, $L = 10$ H, and $C = 4000\ \mu$F.
57. Repeat Problem 54 if $R = 25\ \Omega$, $L = 125$ mH, and $C = 1250\ \mu$F.

Section 9.7

58. Use the technique presented in Section 9.7 to solve for voltage v in the following circuit.

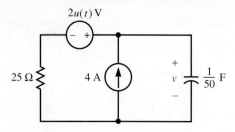

59. Solve for current i in the following circuit.

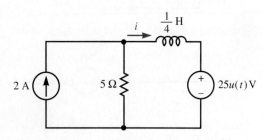

60. No other components have ever been connected to the following network. Find voltage v if $v_s = 200 - 100u(t)$ V. Is the network overdamped, underdamped, or critically damped?

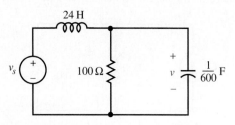

61. Repeat Problem 60 if v_s is changed to $v_s = 100 + 100u(t - 2)$ V.

Section 9.8

Calculate voltage $v(t)$ for the following network with voltage v_s as specified in Problems 62 through 65.

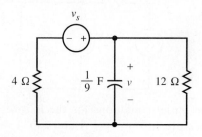

62. $4e^{6t}u(-t) + 2u(t)$ V
63. $8e^{-3|t|}$ V
64. $(20 \cos 4t)[1 - u(t)]$ V
65. $24e^{4t} + 12u(t)$ V
66. Repeat Problem 60 if voltage v_s is changed to $v_s = 100[(\cos 5t)u(-t) + (\cos 3t)u(t)]$ V.
67. In a series RLC circuit, voltage v_s is $24 \cos 2t$ V for $t < 0$ and 24 V for $t > 0$. The element values are $R = 8\ \Omega$, $L = 1$ H, and $C = \frac{1}{12}$ F. Determine current $i(t)$ for $t > 0$. (Assume the passive sign convention.)
68. The sources all go to zero for $t > 0$ in the circuit of Problem 38. Determine $v(0^+)$ and $i(0^+)$.

Section 9.9

For the following network, calculate the requested information if the switches are actuated as indicated in Problems 69 through 72.
(a) $v(0^-)$, $i(0^-)$
(b) $v(0^+)$, $i(0^+)$
(c) The complementary response for v, $t > 0$
(d) The particular response for v, $t > 0$
(e) The complete response for v, $t > 0$, including evaluation of the constants of integration

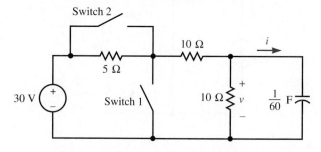

69. Both switches have been open for a very long time. At $t = 0$, switch 1 is closed.

70. Switch 2 has been open for a very long time, switch 1 has been closed for a very long time, and switch 1 is opened at $t = 0$.

71. Both switches have been open for a very long time, and switch 2 is closed at $t = 0$.

72. Switch 1 has been open for a very long time, switch 2 has been closed for a very long time, and switch 2 is opened at $t = 0$.

73. What would happen if switches 1 and 2 in the preceding circuit were both closed at the same time?

For the network in Fig. 9.31, calculate the requested information if the switches are actuated as indicated in Problems 74 through 77.

(a) $i(0^-)$, $v(0^-)$
(b) $i(0^+)$, $v(0^+)$
(c) The complementary response for i, $t > 0$
(d) The particular response for i, $t > 0$
(e) The complete response for i, $t > 0$, including evaluation of the constants of integration

74. Both switches have been open for a very long time. Switch 1 is closed at $t = 0$.

75. Switch 2 has been open for a very long time, switch 1 has been closed for a very long time, and switch 1 is opened at $t = 0$.

76. Both switches have been open for a very long time, and switch 2 is closed at $t = 0$.

77. Switch 1 has been open for a very long time, switch 2 has been closed for a very long time, and switch 2 is opened at $t = 0$.

78. The switch in the network in Fig. 9.32 has been open for a very long time and is closed at $t = 0$. Find voltage $v(t)$ for $t > 0$.

79. For the following network, the switch has been open for a very long time and is closed at $t = 0$. Find $i(t)$ for $t > 0$.

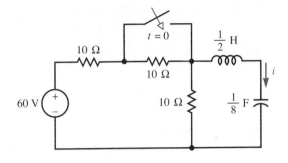

80. The switch in the following network has been open for a very long time and is closed at $t = 0$. Find $i(t)$ for $t > 0$.

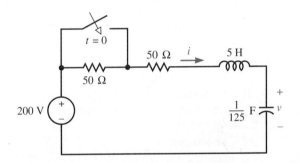

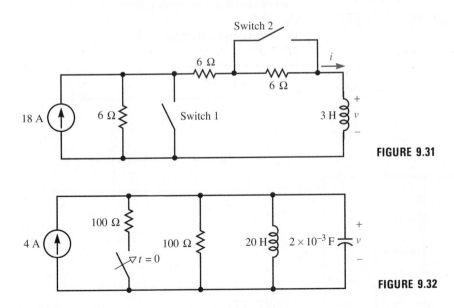

FIGURE 9.31

FIGURE 9.32

81. For the following network, the switch has been open since $t = -\infty$ and is closed at $t = 0$. Find $v(t)$ for $t > 0$.

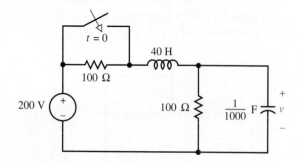

82. Repeat Problem 81 if the switch has been closed for a very long time and is opened at $t = 0$.

83. In the network in Fig. 9.33, switch 1 has been open for a very long time, switch 2 has been closed for a very long time, and switch 1 is closed at $t = 0$. Find:
    (a) $v_1(0^-)$, $i_1(0^-)$, $v_2(0^-)$, $i_2(0^-)$.
    (b) $v_1(0^+)$, $i_1(0^+)$, $v_2(0^+)$, $i_2(0^+)$.
    (c) $v_1(\infty)$, $i_1(\infty)$, $v_2(\infty)$, $i_2(\infty)$. (These are the particular responses for $t > 0$, because the source is constant for $t > 0$.)
    (d) The natural response for v_1, $t > 0$. (The form of the natural response will be the same for all voltages and currents.)
    (e) The complete response for v_1 when $t > 0$, including evaluation of the constants of integration.

84. Repeat Problem 83 if switches 1 and 2 have been closed for a very long time, and switch 1 is opened at $t = 0$.

85. Repeat Problem 83 if switches 1 and 2 have been open for a very long time, and switch 2 is closed at $t = 0$.

86. Repeat Problem 83 if switch 1 has been open for a very long time, switch 2 has been closed for a very long time, and switch 2 is opened at $t = 0$.

87. The switch shown in the following network has been closed for a very long time and is opened at $t = 0$. Find:
    (a) $v_C(0^-)$, $v_L(0^-)$, $i_C(0^-)$, and $i_L(0^-)$
    (b) $v_C(0^+)$, $v_L(0^+)$, $i_C(0^+)$, and $i_L(0^+)$
    (c) $v_C(\infty)$, $v_L(\infty)$, $i_C(\infty)$, and $i_L(\infty)$

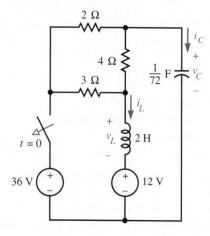

88. Repeat Problem 87 if the switch has been open for a very long time and is closed at $t = 0$.

Section 9.10

89. Solve for the mesh currents for $t > 0$ in the following network if $i_1(0^+) = 1$ A, $i_2(0^+) = 2$ A, and $v_s = 120e^{-3t}$ V for $t \geq 0$.

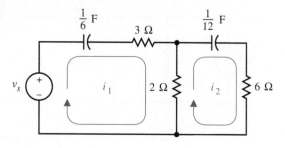

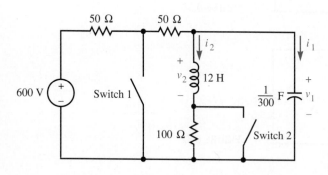

FIGURE 9.33

90. Determine the mesh currents for $t > 0$ in the circuit in Fig. 9.34.

91. Solve for the mesh currents in the following network if $v_s = 10[1 + u(t)]$ V.

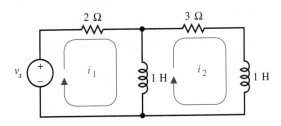

92. Solve for node voltages v_1 and v_2 in the network in Fig. 9.35.

93. The initial conditions for the following network are $v_1(0^-) = 6$ V and $v_2(0^-) = 5$ V. The current source is $i_s = 20$ A, $t \geq 0$. Determine voltages v_1 and v_2 for $t > 0$.

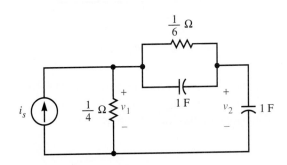

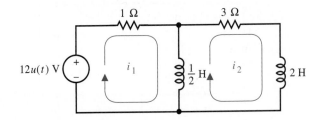

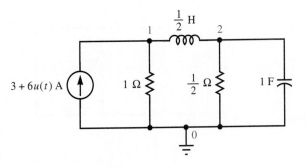

94. Solve for voltages v_1 and v_2 in the following network if $i_s = 30 + 60u(t)$ A.

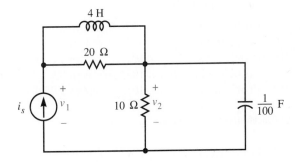

95. For the following network, solve for the mesh currents, and find i_L, i_C, v_L, and v_C for $t > 0$ if $i_s = 4[1 + u(t)]$ A.

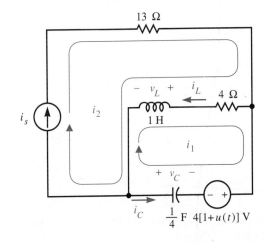

FIGURE 9.34

FIGURE 9.35

Section 9.11

In Problems 96 through 105, use PSpice to solve the indicated problem.

96. Problem 48	97. Problem 50
98. Problem 51	99. Problem 54
100. Problem 58	101. Problem 79
102. Problem 90	103. Problem 93
104. Problem 94	105. Problem 95

Comprehensive Problems

106. For the following network, find $v(t)$ if $C = \frac{1}{24}$ F, $v_{s1} = 5e^{-4t}u(t)$ V, and $v_{s2} = 10$ V for $t > -\infty$.

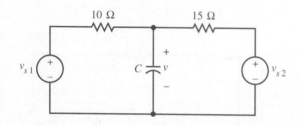

107. For the preceding circuit, $v_{s2} = 0$ V, $v_{s1} = 15\Pi(t - 0.5)$ V [$\Pi(t)$ is the unit pulse function], and (a) $C = \frac{1}{60}$ F; (b) $C = \frac{1}{12}$ F; (c) $C = 1/6$ F; (d) $C = 1/3$ F; (e) $C = 5/3$ F. On the same graph, plot $v(t)$ for all values of capacitance in (a) through (e) for $0 \le t \le 4$ s.

For the networks shown in Problems 108 through 110, find current i and voltage v_x for $t > 0$.

108.

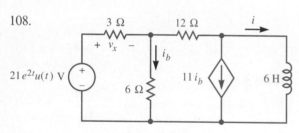

109.

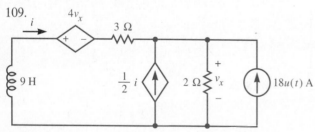

110.

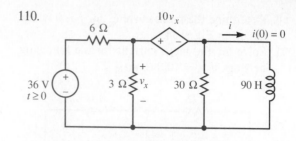

111. The switch in the following network has been closed for a very long time and is opened at $t = 0$. Find voltage v for $t > 0$.

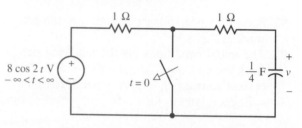

112. Repeat Problem 111 if the switch has been open for a very long time and is closed at $t = 0$.

For the networks shown in Problems 113 through 115, find voltage v_x for $t > 0$. What is the equivalent capacitance seen by the network to the left of terminals a and b?

113.

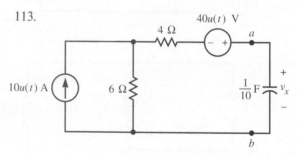

114.

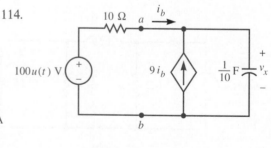

115.

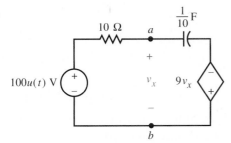

116. Analyze the network shown below if $v_s = 12e^{-t}u(t)$ V.
   (a) Write one KVL equation around the left-hand mesh, and solve for i_3. Write one KVL equation around the outer loop, and solve for i_2. Find $i_1 = i_2 + i_3$.
   (b) Write two mesh equations, and solve for i_1 and i_2. Find $i_3 = i_1 - i_2$. Compare your results with those obtained in (a).

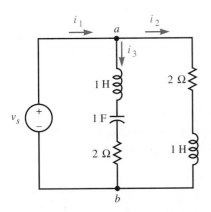

117. Solve for the voltage v_{ab} in the following network if $i_s = 10u(t)$ A.

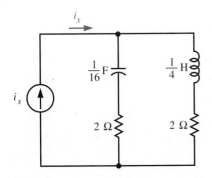

118. Replace the current source in the preceding network with a voltage source of value $20u(t)$ V, with the (+) terminal at the top of the page, and solve for current i_x.

For the networks in Problems 119 through 121, write a differential equation that relates the output voltage v to the input voltage v_s. What is the complementary response? If $v_s = V_s \cos \omega t$ V, what is the particular response? (These are peculiar circuits.)

119.

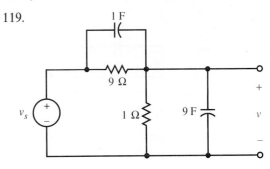

120.

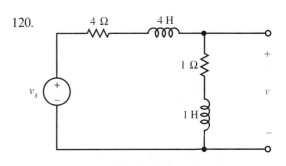

121.

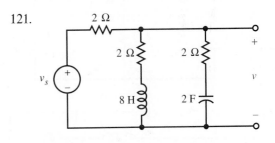

122. Determine the value of capacitance C in the following network so that the phase angle on the particular response component of current i_2 is in phase with voltage v_s when the input voltage is $v_s = 100 \cos 10t$ V. Determine the

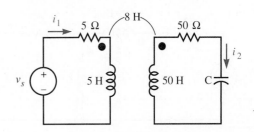

particular response component of current i_1 for this value of C and v_s.

123. Determine the output voltage of the following op-amp circuit when:
   (a) $v_s = 10u(t)$ V
   (b) $v_s = (10 \cos 100t)u(t)$ V.

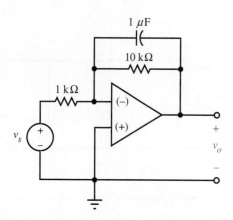

Use node-voltage equations to solve for the node voltages in Problems 124 through 126 for $t > 0$, if $i_a(0) = i_b(0) = 0$.

124.

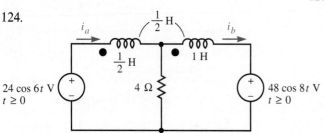

125.

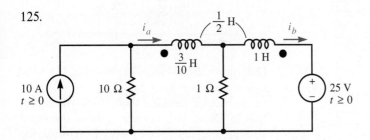

126.

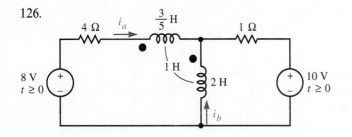

127. The circuit shown below contains an ideal op amp. Write a single differential equation that can be solved for v_2. If $v_s = \cos \omega_0 t$ V, the particular solution for v_2 will be of the form

$$v_{2p} = A(\omega_0) \cos [\omega_0 t + \phi(t)]$$

Plot $20 \log_{10} A(\omega_0)$ for $0.1 \le \omega_0 \le 100$ rad/s on semilogarithmic paper.

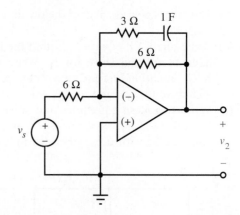

128. For the preceding op-amp circuit, replace the 3-Ω resistance with a short circuit and place a 1-F capacitance in series with the voltage source. Calculate v_2 if $v_s = 10u(t)$ V.

129. The following network is a model for an oscillator using an electromechanical relay. The switch S2 is manually closed at $t = 0$. When current $i \geq 6(1 - e^{-1})$ A, the relay, modeled by 2 H in series with 4 Ω, causes switch S2 to be opened. When $i < 6e^{-1}(1 - e^{-1})$ A, the relay causes switch S2 to be closed. The time required for the mechanical switching, as well as the change in inductance as the switching occurs, is ignored for simplicity. Calculate i, and draw a graph showing i as a function of time from $t = 0$ until the third opening of S2. Is i periodic for $t > t_0$? If so, for what value of t_0, and what is the period T?

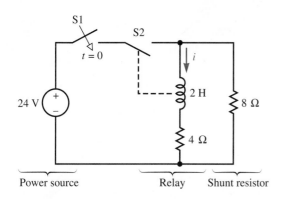

Power source   Relay   Shunt resistor

130. The following problem occurs in digital logic circuits. When circuits are connected together by a twisted pair (two wires twisted together), the capacitive coupling between the wires can falsely trigger a logic gate. The capacitance per foot of a 30 AWG wire size twisted pair is approximately 15 pF. If the voltage v introduced on the logic gate exceeds 1.4 V in peak value, a false trigger may result. The equivalent circuit is shown below, where

$$v_s = 1 \times 10^9 [r(t) - r(t - 3 \times 10^{-9})]$$

with $r(t)$ the unit ramp $[r(t) = tu(t)]$. Will false triggering occur if the twisted pair has a length of 2 ft? What minimum length of twisted pair will cause v to exceed 1.4 V in peak value?

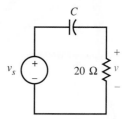

131. Given that $i_s = 2e^{-2t}u(t)$ A, determine node voltages v_1 and v_2 for $t > 0$ in the following circuit.

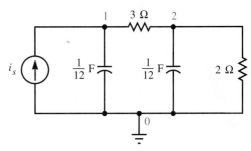

132. (a) Find a differential equation that relates the output voltage $v = v_{ab}$ to the input voltage v_s in the following network.
    (b) With $R = 1$ Ω, $C = 1$ F, and $v_s = V_s u(t)$ V, find the complete response v_{ab}.

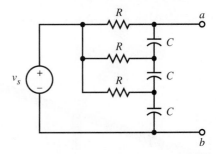

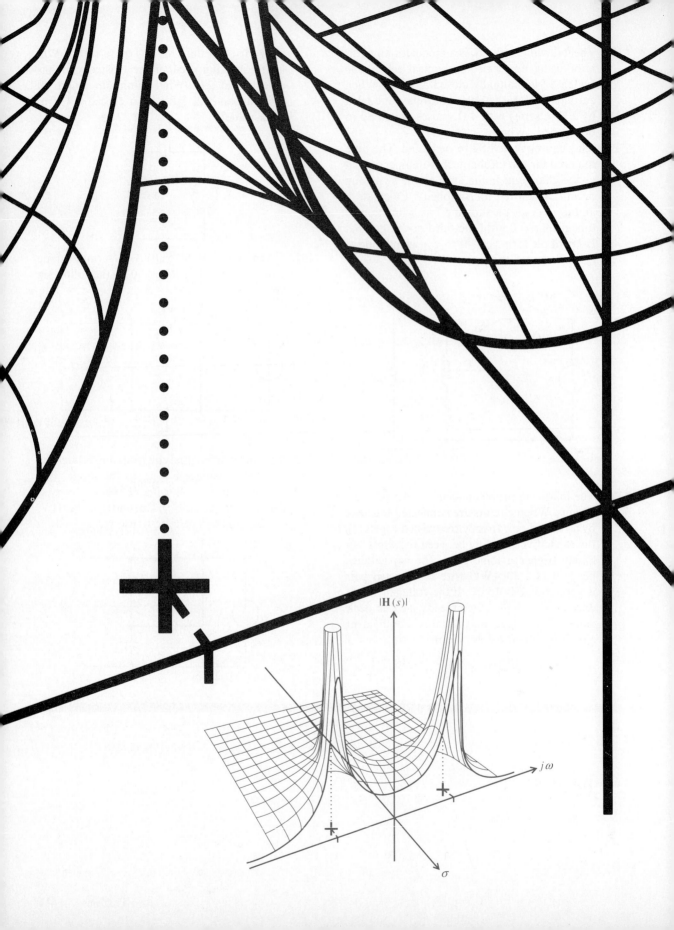

FREQUENCY-DOMAIN CIRCUIT ANALYSIS

When a stable LLTI circuit is excited by a sinusoidal input, the steady-state response is sinusoidal with the same frequency as the input. In Part Three we find the amplitude and phase of the response by writing and solving an algebraic equation—bypassing calculus completely. The algebraic method constitutes a body of LLTI circuit analysis known as *frequency-domain* circuit analysis. Frequency-domain circuit analysis provides us with a profound understanding of circuit response.

CHAPTER
10

The Sinusoidal Steady State

In Chapter 9 we saw that when a sinusoidal input is applied
to a stable LLTI circuit, the steady-state response is always
sinusoidal. This is one of the most important phenomena in
electrical engineering. Of all possible periodic inputs, the
sinusoidal signal is the *only* real one that reproduces itself
throughout an arbitrary stable LLTI circuit without changing its
functional form or "shape."

We shall see in Chapters 14 and 15, that physical signals
can be represented as sums or integrals of sinusoids. By
studying the sinusoidal steady state in detail, therefore, we
prepare a foundation for a new and deep understanding of
how circuits respond to all physical signals.

Our objective in this chapter is to describe the sinusoidal
steady state with the use of the new concepts of *phasor,
impedance, admittance,* and *transfer function*. These concepts
are central to much of electrical engineering and to all that
follows in this book. They are described using complex
arithmetic. We urge you, therefore, to master Appendix B before
proceeding. We begin this chapter by reviewing pertinent
results from Chapter 9.

10.1   Pertinent Results from Chapter 9

⊟ *See Example 3.7 in the PSpice manual.*

We have said that the steady-state response of any stable LLTI circuit to a sinusoidal input is always a sinusoid with the same frequency as the input. The response has this property regardless of whether the input and output are voltages or currents. A typical response is shown in Fig. 10.1. From Chapter 9 we know that the complete response is given by the sum of the particular response $v_p(t)$ and the complementary response $v_c(t)$. The steady-state response is sinusoidal for the following reasons.

1. The particular response $v_p(t)$ of every stable LLTI circuit to a sinusoidal input is a sinusoid with the same frequency. This result was derived in Theorem 5 of Chapter 9, where it was shown that the particular solution of the LLTI equation

$$\mathscr{A}(p)v = \mathscr{B}(p)i \tag{10.1}$$

for the input

$$i(t) = K \cos (\omega t + \phi) \tag{10.2}$$

is, for a stable circuit,

$$v_p(t) = \left| \frac{\mathscr{B}(j\omega)}{\mathscr{A}(j\omega)} \right| K \cos \left(\omega t + \phi + \left| \frac{\mathscr{B}(j\omega)}{\mathscr{A}(j\omega)} \right. \right) \tag{10.3}$$

FIGURE 10.1
Typical response of a stable, linear, time-invariant network to a sinusoidal input applied at $t = 0$. (a) Network with input $i(t)$ and output $v(t)$, (b) input $i(t)$, (c) output $v(t)$

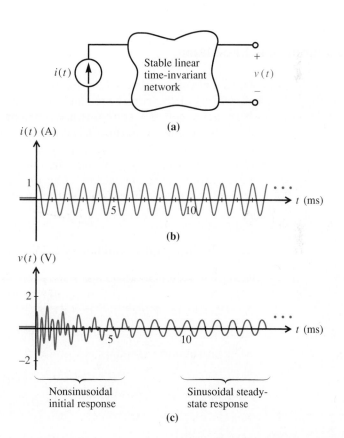

2. By definition, the complementary response $v_c(t)$ of every stable LLTI circuit approaches zero asymptotically as t increases:

$$v_c(t) \to 0 \qquad \text{as} \qquad t \to \infty \tag{10.4}$$

These two statements imply that the complete response

$$v(t) = v_p(t) + v_c(t) \tag{10.5}$$

of every stable LLTI circuit to the sinusoidal input, Eq. (10.2), asymptotically approaches the sinusoid of Eq. (10.3). When $v_c(t)$ becomes negligibly small compared with $v_p(t)$, the circuit is said to be operating in the sinusoidal steady state, and the output is

$$v(t) = \left| \frac{\mathscr{B}(j\omega)}{\mathscr{A}(j\omega)} \right| K \cos \left(\omega t + \phi + \left/ \frac{\mathscr{B}(j\omega)}{\mathscr{A}(j\omega)} \right. \right) \tag{10.6}$$

In the following section we shall introduce the new concepts of phasors and impedance to focus on the relation between the sinusoidal input $i(t)$ of Eq. (10.2) and the steady-state response $v(t)$ of Eq. (10.6).

REMEMBER

The steady-state response of a stable LLTI network to a sinusoidal input is a sinusoid with the same frequency as the input.

10.2 Phasors and Impedance

When a circuit has reached the sinusoidal steady state, all the current and voltage waveforms in the circuit are sinusoidal with the same frequency as the source. *Phasors* are complex constants used to represent the amplitudes and phases of these sinusoidal waveforms. *Impedances* are other complex constants used to relate voltage phasors to current phasors. In Chapter 11 we shall see that by using phasors and impedance, we can find the amplitudes and phases of sinusoidal waveforms by using only algebra. Our present goal is to explain in detail the concepts of phasor and impedance and to show how they are related.

Phasors

Let us begin with the definition of a phasor.

DEFINITION

Phasor

The **phasor** corresponding to a sinusoid

$$i(t) = I_m \cos (\omega t + \phi_{\mathbf{I}}) \tag{10.7}$$

is defined as
$$\mathbf{I} = I_m \underline{/\phi_{\mathbf{I}}} \tag{10.8}$$

FIGURE 10.2
The phasor $\mathbf{I} = I_m\underline{/\theta_I}$

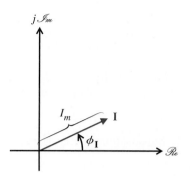

The phasor $\mathbf{I}$ represents the amplitude and phase of $i(t)$, as illustrated in Fig. 10.2. The subscript m in I_m signifies that I_m is the *maximum* value or amplitude of $i(t)$ and that I_m is the *magnitude* of the phasor $\mathbf{I}$. The subscript I in $\phi_\mathbf{I}$ denotes that $\phi_\mathbf{I}$ is the phase angle of $i(t)$ and the angle of $\mathbf{I}$. As an example, if $i(t) = 3\cos(100t + 25°)$, then $\mathbf{I} = 3\underline{/25°}$. The units of a phasor are always the same as those of the associated sinusoidal function. In Eq. (10.7) we have assumed that the sinusoidal function is a current, $i(t)$. Therefore, like $i(t)$, the phasor $\mathbf{I}$ has the unit ampere.

Recall that in Chapter 9 we used complex exponentials to determine the particular response of LLTI circuits when the input was a sinusoid. We replaced $i(t)$ with a complex exponential

$$\mathbf{i}(t) = \mathbf{I}e^{j\omega t} = I_m e^{j(\omega t + \phi_\mathbf{I})} \tag{10.9}$$

because we knew from Euler's identity that

$$i(t) = I_m \cos(\omega t + \phi_\mathbf{I})$$
$$= \mathscr{R}e\,\{\mathbf{i}(t)\}$$
$$= \mathscr{R}e\,\{\mathbf{I}e^{j\omega t}\} \tag{10.10}$$

As shown in Fig. 10.3a, we can visualize the complex exponential $\mathbf{i}(t)$ as a *rotating phasor*. The real input, $i(t)$, is the projection of the rotating phasor $\mathbf{i}(t)$ onto the real axis. A plot of the real input $i(t)$ versus t is shown in Fig. 10.3b.

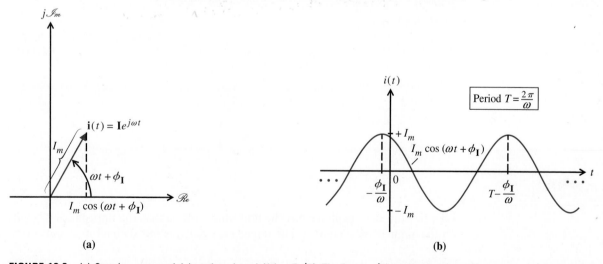

FIGURE 10.3 (a) Complex exponential (rotating phasor) $\mathbf{i}(t) = \mathbf{I}e^{j\omega t}$. The factor $e^{j\omega t}$ causes the phasor $\mathbf{I}$ to rotate counterclockwise with angular velocity ω. The real current is obtained by taking the real part of $\mathbf{i}(t)$. (b) Corresponding real current $i(t) = I_m \cos(\omega t + \phi_\mathbf{I})$

Impedance

If the input to the LLTI equation, Eq. (10.1), is the complex exponential $\mathbf{i}(t)$ of Eq. (10.9), then the particular response $\mathbf{v}_p(t)$ is also a complex exponential. This result follows directly from Theorem 2 of Chapter 9 with $s = j\omega$.[†] Thus, *we can write the particular solution for the rotating phasor input* $\mathbf{i}(t)$ *as another rotating phasor:*

$$\mathbf{v}_p(t) = \mathbf{V}e^{j\omega t} \tag{10.11}$$

Substitution of Eqs. (10.11) and (10.9) into Eq. (10.1) yields

$$\mathscr{A}(p)\mathbf{V}e^{j\omega t} = \mathscr{B}(p)\mathbf{I}e^{j\omega t} \tag{10.12}$$

After differentiations, this becomes

$$\mathscr{A}(j\omega)\mathbf{V}e^{j\omega t} = \mathscr{B}(j\omega)\mathbf{I}e^{j\omega t} \tag{10.13}$$

and we see that we have simply replaced the derivative operator p in Eq. (10.12) by $j\omega$. If both sides are multiplied by $e^{-j\omega t}$, this becomes

$$\mathscr{A}(j\omega)\mathbf{V} = \mathscr{B}(j\omega)\mathbf{I} \tag{10.14}$$

which is a complex algebraic equation that does not depend on time. We can solve this equation for the voltage phasor:

$$\mathbf{V} = \mathbf{Z}\mathbf{I} \tag{10.15}$$

where we define the *impedance*, $\mathbf{Z}$, as follows.

DEFINITION

Impedance

The **impedance** of a stable circuit whose LLTI equation is

$$\mathscr{A}(p)v = \mathscr{B}(p)i \tag{10.16}$$

is defined as

$$\mathbf{Z} = \frac{\mathscr{B}(j\omega)}{\mathscr{A}(j\omega)} \tag{10.17a}$$

Equivalently, the impedance is the ratio of phasor voltage $\mathbf{V}$ to phasor current $\mathbf{I}$:

$$\mathbf{Z} = \frac{\mathbf{V}}{\mathbf{I}} \tag{10.17b}$$

Like resistance, impedance has the unit ohm. Notice that the impedance depends on the angular frequency ω. The impedance $\mathbf{Z}(j\omega)$ is not defined if $\mathscr{A}(j\omega) = 0$.

[†] We can assume that $\mathscr{A}(j\omega) \neq 0$ because if $\mathscr{A}(j\omega) = 0$ then the circuit would not be stable (see page 318).

It is convenient to represent the relation between **V** and **I** in circuit form, as illustrated in Fig. 10.4. The circuit is characterized by its impedance **Z**, which is the ratio of **V** to **I** as given by Eq. (10.17b). Since **V**, **I**, and **Z** are complex quantities, they do not exist physically. Yet the analogy between Eq. (10.15) and Ohm's law, $V = RI$, is so strong that Eq. (10.15) is often called the *ac version of Ohm's law*. It follows from Eq. (10.15) that the impedance **Z** determines the magnitude, $V_m = |\mathbf{V}|$, and the angle, $\phi_{\mathbf{V}} = \underline{/\mathbf{V}}$, of the phasor **V**:

$$V_m = |\mathbf{Z}|I_m \qquad (10.18a)$$

$$\phi_{\mathbf{V}} = \underline{/\mathbf{Z}} + \phi_{\mathbf{I}} \qquad (10.18b)$$

where $|\mathbf{Z}|$ and $\underline{/\mathbf{Z}}$ are, respectively, the magnitude and the angle of **Z**. We see from Eq. (10.18a) that the magnitude of the voltage phasor equals the *product* of the magnitude of the impedance and the magnitude of the current phasor. We see from Eq. (10.18b) that the angle of the voltage phasor equals the *sum* of the angle of the impedance and the angle of the current phasor.

Equation (10.14) is called the *frequency-domain* version of the time-domain LLTI Eq. (10.1). Figure 10.4b is called the *frequency-domain* representation of the *time-domain* LLTI circuit of Fig. 10.4a. In much of our future work, we shall be concerned with frequency-domain equations and representations, because they describe sinusoidal steady-state voltages and currents in a mathematically compact and convenient way.

FIGURE 10.4
Representations of LLTI circuit: (*a*) time domain, $\mathscr{A}(p)v(t) = \mathscr{B}(p)i(t)$; (*b*) frequency domain, $\mathbf{V} = \mathbf{ZI}$

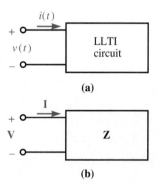

If the input to a stable LLTI circuit is a rotating phasor $\mathbf{i}(t) = \mathbf{I}e^{j\omega t}$ then the particular response is the rotating phasor $\mathbf{v}_p(t) = \mathbf{V}e^{j\omega t}$ where $\mathbf{V} = \mathbf{ZI}$. The impedance is given by $\mathbf{Z} = \mathscr{B}(j\omega)/\mathscr{A}(j\omega)$ where $\mathscr{B}(p)$ and $\mathscr{A}(p)$ are the operators found in the circuit's LLTI equation $\mathscr{A}(p)v(t) = \mathscr{B}(p)i(t)$.

EXERCISES

1. Define in your own words the following terms: (a) phasor, (b) rotating phasor, and (c) impedance.

2. Discuss the equation $\mathbf{V} = \mathbf{ZI}$. Specifically, explain (a) how this equation was obtained, and (b) how it relates the magnitude and angle of **V** to the magnitude and angle of **I**.

10.3 Relation of Phasors and Impedance to Real Input and Output

We have seen that the particular response to a rotating phasor input $\mathbf{i}(t)$ is a rotating phasor, $\mathbf{v}_p(t)$. $\mathbf{v}_p(t)$ can be broken down into its real and imaginary components. According to Section 9.4, the real particular solution $v_p(t)$ for the real input $i(t)$ of Eq. (10.7) is the real part of $\mathbf{v}_p(t)$. Thus, the particular response to the sinusoidal input, Eq. (10.7), is

$$v_p(t) = \mathscr{R}e\,\{\mathbf{V}e^{j\omega t}\}$$
$$= V_m \cos{(\omega t + \phi_{\mathbf{V}})} \qquad (10.19)$$

where V_m and $\phi_{\mathbf{V}}$ are given by Eq. (10.18). Equation (10.19) follows from the fact that $\mathscr{A}(p)$ and $\mathscr{B}(p)$ are real linear operators.

It is helpful to visualize the relationship between $\mathbf{I}$ and $\mathbf{V}$ and $i(t)$ and $v_p(t)$ as shown in Figs. 10.5 and 10.6. The relationship between the phasors $\mathbf{V}$ and $\mathbf{I}$ is illustrated in Fig. 10.5. The rotating phasors $\mathbf{i}(t) = \mathbf{I}e^{j\omega t}$ and $\mathbf{v}_p(t) = \mathbf{V}e^{j\omega t}$ are illustrated in Fig. 10.6a.

FIGURE 10.5
The phasors **V** and **I**, where **V** = **ZI**. In the figure, $V_m = |\mathbf{Z}|I_m$ and $\phi_{\mathbf{V}} = \underline{/\mathbf{Z}} + \phi_{\mathbf{I}}$

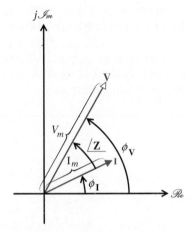

We see from Eqs. (10.10) and (10.19) that the real input waveform

$$i(t) = I_m \cos{(\omega t + \phi_{\mathbf{I}})} \qquad (10.20)$$

is the projection of the rotating current phasor $\mathbf{i}(t)$ onto the real axis, and the associated real steady-state output waveform

$$v_p(t) = \mathscr{R}e\,\{\mathbf{V}e^{j\omega t}\}$$
$$= V_m \cos{(\omega t + \phi_{\mathbf{V}})} \qquad (10.21)$$

is the projection of the rotating voltage phasor $\mathbf{v}_p(t)$ onto the real axis.

Plots of $i(t)$ and $v_p(t)$ versus t are given in Fig. 10.6b.

If instead of the cosine input of Eq. (10.7), we had considered a sine input,

$$i(t) = I_m \sin{(\omega t + \phi_{\mathbf{I}})}$$
$$= \mathscr{I}m\,\{\mathbf{I}e^{j\omega t}\} \qquad (10.22)$$

then the particular response would have been (as was shown on page 364)

$$v_p(t) = \mathscr{I}m\left\{\mathbf{V}e^{j\omega t}\right\}$$

$$= V_m \sin(\omega t + \phi_\mathbf{V}) \qquad (10.23)$$

where V_m and $\phi_\mathbf{V}$ are again determined by Eq. (10.18). These sine functions are the projections of $\mathbf{i}(t)$ and $\mathbf{v}_p(t)$, respectively, onto the imaginary axis $\mathscr{I}m$ in Fig. 10.6.

FIGURE 10.6
(*a*) Rotating current and voltage phasors. Both phasors rotate counterclockwise with angular velocity ω. (*b*) Corresponding real current and voltage waveforms. Both waveforms are sinusoidal with angular frequency ω

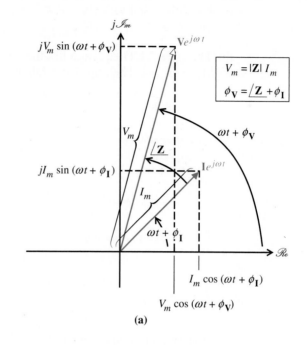

$$V_m = |\mathbf{Z}| \, I_m$$
$$\phi_\mathbf{V} = \underline{/\mathbf{Z}} + \phi_\mathbf{I}$$

(a)

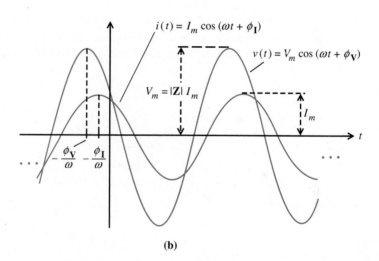

(b)

EXAMPLE 10.1

The relation between input current i and output voltage v of a certain circuit is given by the LLTI equation

$$\frac{d}{dt}v + 2v = 4\frac{d}{dt}i + 4i$$

Assume that $i(t) = 5 \cos(\omega t + 45°)$ A, where $\omega = 3$ rad/s. Find (a) the phasor $\mathbf{I}$, (b) the impedance $\mathbf{Z}$, (c) the phasor $\mathbf{V}$, and (d) the particular response $v_p(t)$. Interpret your results graphically.

■ **SOLUTION**

(a) It follows from Eqs. (10.7) and (10.8) that

$$\mathbf{I} = I_m \underline{/\phi_\mathbf{I}} = 5\underline{/45°} \text{ A}$$

(b) The given LLTI equation has the form $\mathscr{A}(p)v = \mathscr{B}(p)i$, where $\mathscr{A}(p) = p + 2$ and $\mathscr{B}(p) = 4p + 4$. The impedance follows from this and Eq. (10.17a):

$$\mathbf{Z} = \frac{\mathscr{B}(j\omega)}{\mathscr{A}(j\omega)} = \frac{4j\omega + 4}{j\omega + 2}$$

This result can also be obtained by substituting $\mathbf{i}(t) = \mathbf{I}e^{j\omega t}$ and $\mathbf{v}_p(t) = \mathbf{V}e^{j\omega t}$ into the LLTI equation and solving for $\mathbf{V}/\mathbf{I}$. Since the input angular frequency is $\omega = 3$ rad/s, this becomes

$$\mathbf{Z} = \frac{4 + j12}{2 + j3} = \frac{12.65\underline{/71.6°}}{3.61\underline{/56.3°}} = 3.5\underline{/15.3°} \ \Omega$$

(c) The voltage phasor is given by Eq. (10.15):

$$\mathbf{V} = \mathbf{ZI} = (3.5\underline{/15.3°})(5\underline{/45°}) = 17.5\underline{/60.3°} \text{ V}$$

(d) The result of (c) tells us that the amplitude of the voltage is 17.5 V and the phase is 60.3°. Thus

$$v_p(t) = 17.5 \cos(\omega t + 60.3°) \text{ V}$$

where the angular frequency ω is the same as that of the input, $\omega = 3$ rad/s. The phasors $\mathbf{I}$ and $\mathbf{V}$ are illustrated in Fig. 10.7. Figure 10.8a depicts the rotating current and voltage phasors and their projections onto the real axis. These projections are, respectively, the sinusoidal input current $i(t)$ and the sinusoidal steady-state output voltage $v(t)$. Plots of $i(t)$ and $v(t)$ versus t are given in Fig. 10.8b. ■

FIGURE 10.7
Phasors for Example
10.1. The impedance
$\mathbf{Z} = 3.5\underline{/15.3°} \ \Omega$

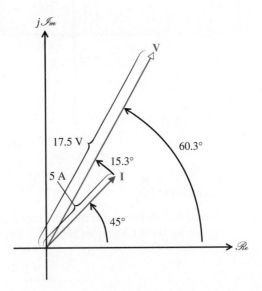

FIGURE 10.8
(*a*) Rotating current and voltage phasors for Example 10.1. The current and voltage phasors both rotate counterclockwise with angular velocity 3 rad/s. (*b*) Corresponding real waveforms. The sinusoidal current and voltage waveforms have the same angular frequency, 3 rad/s

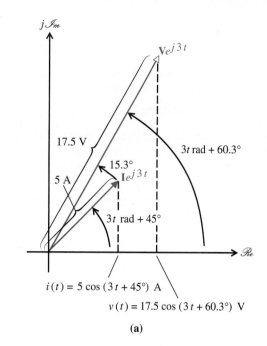

$i(t) = 5 \cos (3t + 45°)$ A

$v(t) = 17.5 \cos (3t + 60.3°)$ V

(a)

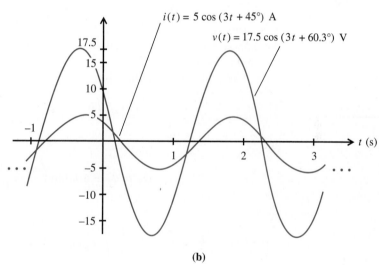

(b)

EXAMPLE 10.2

(a) Determine the impedance of the series *RLC* circuit of Fig. 10.9.

FIGURE 10.9
Series *RLC* circuit

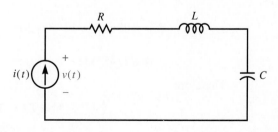

(b) Find the steady-state voltage response to an input

$$i(t) = I_m \cos(\omega t + \phi_I)$$

(c) Evaluate your answers to (a) and (b) for $R = 1\ \text{k}\Omega$, $L = 2\ \text{mH}$, $C = 0.001\ \mu\text{F}$, and $i(t) = 3 \cos(10^6 t + 15°)$ mA.

■ SOLUTION

⊞ See Problems 10.1 through 10.3 in the PSpice manual.

(a) The terminal equation of the series RLC circuit is, by KVL,

$$v(t) = L\frac{d}{dt}i + Ri + \frac{1}{C}\int_{-\infty}^{t} i(\alpha)\,d\alpha$$

Differentiated with respect to t, this becomes

$$\frac{d}{dt}v = L\frac{d^2}{dt^2}i + R\frac{d}{dt}i + \frac{1}{C}i$$

Thus $\mathscr{A}(p) = p$ and $\mathscr{B}(p) = Lp^2 + Rp + 1/C$. It follows from Eq. (10.17a) that

$$\mathbf{Z} = \frac{L(j\omega)^2 + Rj\omega + (1/C)}{j\omega} = j\omega L + R + \frac{1}{j\omega C}$$

$$= \sqrt{R^2 + \left(\omega L - \frac{1}{\omega C}\right)^2}\ \bigg/\!\arctan\left(\frac{\omega L - (1/\omega C)}{R}\right)$$

This result may also be obtained by substituting $i(t) = \mathbf{I}e^{j\omega t}$ and $v_p(t) = \mathbf{V}e^{j\omega t}$ into the differential equation and solving for $\mathbf{V}/\mathbf{I}$. Notice that both the magnitude and the angle of $\mathbf{Z}$ depend on ω.

(b) The particular response is given by Eq. (10.19), where

$$V_m = \sqrt{R^2 + \left(\omega L - \frac{1}{\omega C}\right)^2}\ I_m$$

and

$$\phi_\mathbf{V} = \phi_\mathbf{I} + \arctan\left(\frac{\omega L - (1/\omega C)}{R}\right)$$

Thus

$$v_p(t) = \sqrt{R^2 + \left(\omega L - \frac{1}{\omega C}\right)^2}\ I_m \cos\left[\omega t + \phi_\mathbf{I} + \arctan\left(\frac{\omega L - (1/\omega C)}{R}\right)\right]$$

This is also the sinusoidal steady-state response after the complementary response becomes negligible.

(c) For the given numerical values, we have

$$\mathbf{I} = 3\underline{/15°}\ \text{mA}$$

and

$$\mathbf{Z} = j10^6 \times 2 \times 10^{-3} + 10^3 + \frac{1}{j10^6 \times 10^{-9}} = 10^3 + j10^3\ \Omega$$

$$= \sqrt{2}\underline{/45°}\ \text{k}\Omega$$

Therefore

$$\mathbf{V} = \mathbf{Z}\mathbf{I} = 3\sqrt{2}\underline{/60°}\ \text{V}$$

and the steady-state response is

$$v(t) = 3\sqrt{2} \cos (10^6 t + 60°) \text{ V} \quad \blacksquare$$

REMEMBER

The phasors $\mathbf{I} = I_m / \underline{\phi_I}$ and $\mathbf{V} = V_m / \underline{\phi_V}$ contain the amplitude and phase information of the sinusoidal waveforms $i(t) = I_m \cos (\omega t + \phi_I)$ and $v_p(t) = V_m \cos (\omega t + \phi_V)$, respectively. According to the ac version of Ohm's law, $\mathbf{V} = \mathbf{ZI}$, where $\mathbf{Z}$ is the impedance defined by Eq. (10.17).

EXERCISES

3. Explain in your own words the *practical* significance of the equation $\mathbf{V} = \mathbf{ZI}$.
4. A certain LLTI circuit has input $i = 10 \cos (100t + 20°)$ mA and steady-state output $v = 25 \cos (100t - 65°)$ V.
   (a) Write down the phasors $\mathbf{I}$ and $\mathbf{V}$.
   (b) Determine $\mathbf{Z}$. Express your answer in polar form.
5. Repeat Example 10.1 for $\omega = 2$ rad/s.
6. The impedance in Example 10.1 was derived with the use of Eq. (10.17a). Show that the same result is obtained by substituting $\mathbf{i}(t) = \mathbf{I}e^{j\omega t}$ and $\mathbf{v}_p(t) = \mathbf{V}e^{j\omega t}$ into the given differential equation and solving for $\mathbf{V}/\mathbf{I}$. Are the two methods really different?

10.4 Impedances of *R*, *L*, and *C*

Until now the network components R, L, and C have been defined by differential equations relating terminal current $i(t)$ and voltage $v(t)$. In the sinusoidal steady state, these terminal equations can, in effect, be replaced by algebraic equations of the form $\mathbf{V} = \mathbf{ZI}$, where $\mathbf{Z}$ is the component's impedance, and phasors $\mathbf{I}$ and $\mathbf{V}$ represent terminal current and voltage. As a prerequisite step in the development of the totally algebraic network analysis of Chapter 11, we now derive the impedances of R, L, and C.

The terminal equation of a resistance is

$$v(t) = Ri(t) \tag{10.24}$$

By substituting Eqs. (10.9) and (10.11) into this, we obtain

$$\mathbf{V}e^{j\omega t} = R\mathbf{I}e^{j\omega t} \tag{10.25}$$

or

$$\mathbf{V} = R\mathbf{I} \tag{10.26}$$

Therefore the impedance of a resistance is

$$\mathbf{Z} = R \tag{10.27}$$

The same result can be obtained directly from Eq. (10.17a) by use of $\mathscr{A}(p) = 1$ and $\mathscr{B}(p) = R$. (The two methods are equivalent.)

For an inductance, we have

$$v(t) = L\frac{d}{dt}i \qquad (10.28)$$

which by substitution of Eqs. (10.9) and (10.11) becomes

$$\mathbf{V}e^{j\omega t} = j\omega L\mathbf{I}e^{j\omega t} \qquad (10.29)$$

or

$$\mathbf{V} = j\omega L\mathbf{I} \qquad (10.30)$$

Therefore, for an inductance,

$$\mathbf{Z} = j\omega L = \omega L\underline{/90^\circ} \qquad (10.31)$$

Again, this result could have been obtained from Eq. (10.17a) by the use of $\mathscr{A}(p) = 1$ and $\mathscr{B}(p) = Lp$.

Finally, for a capacitance,

$$C\frac{d}{dt}v = i \qquad (10.32)$$

Here we will use Eq. (10.17a) for variety. Since $\mathscr{A}(p) = Cp$ and $\mathscr{B}(p) = 1$ in Eq. (10.32), Eq. (10.17a) yields

$$\mathbf{Z} = \frac{1}{j\omega C} = \frac{1}{\omega C}\underline{/-90^\circ} \qquad (10.33)$$

These results are summarized in Fig. 10.10. Figure 10.11 illustrates the corresponding phasor diagrams.

FIGURE 10.10
Frequency-domain representation of
(*a*) resistance, $\mathbf{V} = R\mathbf{I}$,
(*b*) inductance, $\mathbf{V} = j\omega L\mathbf{I}$,
and (*c*) capacitance,
$\mathbf{V} = (1/j\omega C)\mathbf{I}$

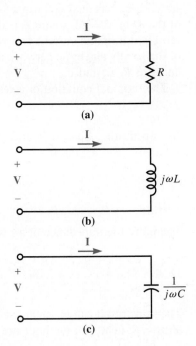

FIGURE 10.11
Voltage and current
phasors for (*a*) resistance,
$\mathbf{V} = R\mathbf{I}$, (*b*) inductance,
$\mathbf{V} = j\omega L\mathbf{I}$, and (*c*)
capacitance, $\mathbf{V} = (1/j\omega C)\mathbf{I}$

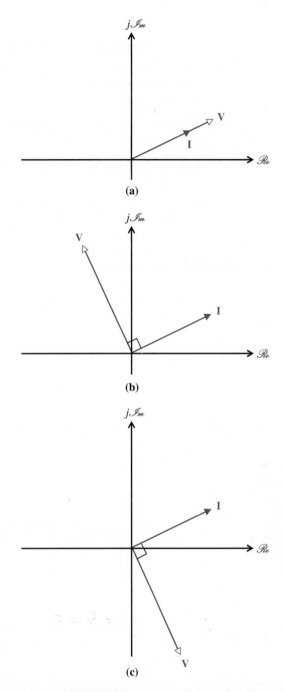

(a)

(b)

(c)

Recall that in Example 10.2 we determined that the impedance of a series *RLC* network was given by $\mathbf{Z} = j\omega L + R + 1/j\omega C$. We can now see that this impedance equals the sum of the impedances of *L*, *R*, and *C*. This result suggests that impedances in series may always be added, just as resistances in series may always be added. In Chapter 11 we shall prove that this is indeed the case.

REMEMBER

The impedances of the elements *R*, *L*, and *C* are *R*, $j\omega L$, and $1/j\omega C$, respectively.

7. Derive the impedance of a resistance R using Eq. (10.17a).

8. Derive the impedance of an inductance L using Eq. (10.17a).

9. Derive the impedance of a capacitance C from the LLTI terminal equation, $i(t) = \mathbf{I}e^{j\omega t}$, and $v_p(t) = \mathbf{V}e^{j\omega t}$.

In Exercises 10 and 11, assume that $R = 1\ \text{k}\Omega$, $C = 1\ \mu\text{F}$, $L = 1\ \text{H}$, and $i(t) = 1\cos(1000t - 45°)\ \text{mA}$.

10. Redraw the frequency-domain components of Fig. 10.10, and indicate the appropriate numerical values for the phasors and impedances.

11. Draw the numerical-valued *rotating* current and voltage phasors, and indicate their projections on the real axis for each component, R, L, and C. Give the formulas for the real terminal voltages.

10.5 Concepts Related to Impedance

The concepts of phasor and impedance give rise to several important related concepts. For easy reference, we place their definitions together in this section.

Driving-Point and Transfer Impedance

When we move beyond the two-terminal elements R, L, and C to networks, two situations can arise, as illustrated in Fig. 10.12. If the voltage phasor response $\mathbf{V}$, appears across the terminals of the current phasor source $\mathbf{I}$, as shown in Fig. 10.12a, then $\mathbf{Z}$ is called a *driving-point impedance*. If the voltage response appears across *another pair* of terminals, as shown in Fig. 10.12b, then $\mathbf{Z}$ is called a *transfer impedance*.

FIGURE 10.12
(*a*) Driving-point impedance
and (*b*) transfer impedance

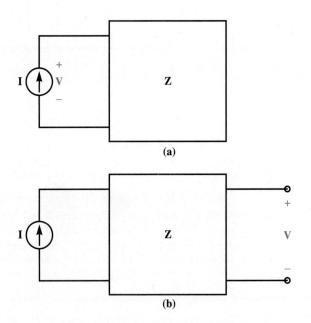

(a)

(b)

Driving-Point and Transfer Admittance

Assume that a sinusoidal voltage $v(t)$ is applied to a stable LLTI circuit. We know that the steady-state current response $i(t)$ is sinusoidal with the same frequency. Thus we can represent $v(t)$ and $i(t)$ with the use of phasors $\mathbf{V}$ and $\mathbf{I}$. The ratio of current phasor response to voltage phasor input $\mathbf{Y} = \mathbf{I}/\mathbf{V}$ is called an *admittance* and has the unit of siemens. If the current response is in the same branch as the voltage source as shown in Fig. 10.13a, then $\mathbf{Y}$ is called a *driving-point admittance*. The driving-point admittance of any LLTI circuit equals the reciprocal of the corresponding driving-point impedance

$$\mathbf{Y} = \mathbf{Z}^{-1} \tag{10.34}$$

If the current response is in a branch other than that of the voltage source as shown in Fig. 10.13b, then $\mathbf{Y}$ is called a *transfer admittance*. Transfer admittances and impedances do not satisfy Eq. (10.34).

FIGURE 10.13
(*a*) Driving-point admittance and (*b*) transfer admittance

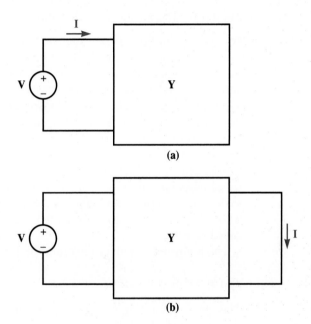

(a)

(b)

Components of Impedance and Admittance

Every impedance can be expressed in the rectangular form $\mathbf{Z}(j\omega) = R(\omega) + jX(\omega)$. $R(\omega)$ and $X(\omega)$ are the *resistive* and the *reactive components* of the impedance, respectively, or more simply, the *resistance* and the *reactance*. Every admittance can be expressed in the rectangular form $\mathbf{Y}(j\omega) = G(\omega) + jB(\omega)$. $G(\omega)$ and $B(\omega)$ are the *conductive* and the *susceptive components* of the admittance, respectively, or more simply, the *conductance* and the *susceptance*. The units of $\mathbf{Z}(\omega)$, $R(\omega)$, and $X(\omega)$ are ohms, whereas those of $\mathbf{Y}(\omega)$, $G(\omega)$, and $B(\omega)$ are siemens.

To determine the relationships between driving-point resistance, reactance, conductance, and susceptance, we simply substitute $\mathbf{Y} = G(\omega) + jB(\omega)$ and $\mathbf{Z} = R(\omega) + jX(\omega)$ into $\mathbf{Y} = \mathbf{Z}^{-1}$, and equate the real and the imaginary parts of both sides. The results are tabulated in Table 10.1. Notice from the table that con-

TABLE 10.1

The relationship between resistance $R(\omega)$, reactance $X(\omega)$, conductance $G(\omega)$, and susceptance $B(\omega)$. $\mathbf{Z}(j\omega)$ is any driving-point impedance, and $\mathbf{Y}(j\omega)$ is the corresponding driving-point admittance: $\mathbf{Z}(j\omega) = \mathbf{Y}^{-1}(j\omega)$. This table does not apply to transfer impedance or transfer admittance.

$$\mathbf{Y}(j\omega) = G(\omega) + jB(\omega) \qquad\qquad \mathbf{Z}(j\omega) = R(\omega) + jX(\omega)$$

$$G(\omega) = \frac{R(\omega)}{R^2(\omega) + X^2(\omega)} \qquad\qquad R(\omega) = \frac{G(\omega)}{G^2(\omega) + B^2(\omega)}$$

$$B(\omega) = \frac{-X(\omega)}{R^2(\omega) + X^2(\omega)} \qquad\qquad X(\omega) = \frac{-B(\omega)}{G^2(\omega) + B^2(\omega)}$$

ductance is *not* the reciprocal of resistance, but depends on both resistance and reactance. Table 10.1 does not apply to transfer impedance or transfer admittance.

Inductive, Resistive, and Capacitive Networks

We know from Eq. (10.18b) that the phase ϕ_V of a sinusoidal voltage waveform equals the sum of the angle of the impedance, $\underline{/\mathbf{Z}}$, and the phase of the current, ϕ_I. If the impedance angle $\underline{/\mathbf{Z}}$ is positive, then the voltage is said to *lead* the current. An inductance is the most elementary example of an impedance that has a positive angle ($\underline{/j\omega L} = 90°$). Consequently, any network that has a positive driving-point impedance angle is called *inductive*. If an impedance angle is negative, then we say that the voltage *lags* the current. A capacitance is the most elementary example of an impedance that has a negative angle ($\underline{/1/j\omega C} = -90°$). Consequently, any network that has a negative driving-point impedance angle is called *capacitive*. If an impedance angle is zero, then the current and voltage are said to be *in phase*. A resistance is the most elementary example of an impedance whose angle is zero. Any network whose driving-point impedance angle is zero is called *resistive*.

The following examples illustrate the preceding concepts.

EXAMPLE 10.3

The driving-point impedance of a certain circuit operating at 100 rad/s is $\mathbf{Z}(j100) = 5\underline{/36.87°}\ \Omega$.
  (a) Find the driving-point resistance and reactance.
  (b) Find the driving-point admittance, conductance, and susceptance.
  (c) Is the network inductive, resistive, or capacitive?

■ **SOLUTION**

(a) Writing $\mathbf{Z}$ in rectangular form, we find that $\mathbf{Z}(j100) = 4 + j3\ \Omega$. The resistance is therefore $R(100) = 4\ \Omega$, and the reactance is $X(100) = 3\ \Omega$.

(b) The driving-point admittance is

$$\mathbf{Y}(j100) = \frac{1}{\mathbf{Z}(j100)} = \frac{1}{5}\underline{/-36.87°}\ \text{S}$$

which in rectangular form equals $\mathbf{Y}(j100) = 0.16 - j0.12$ S. The conductance is therefore $G(100) = 0.16$ S, and the susceptance is $B(100) = -0.12$ S.

(c) Since the impedance angle is $36.87°$, the voltage leads the current by $36.87°$. Therefore the circuit is inductive. ■

EXAMPLE 10.4

(a) Determine the driving-point admittance of the parallel RLC circuit of Fig. 10.14. Comment on your result.

(b) Assume that $R = 10\ \Omega$, $L = 1$ mH, $C = 30\ \mu\text{F}$, and $v(t) = 25 \sin(10^4 t + 70°)$ mV. Find the steady-state current $i(t)$.

FIGURE 10.14
Parallel RLC circuit

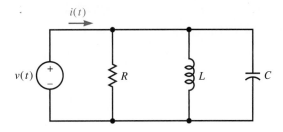

■ SOLUTION

(a) The driving-point admittance can be found from $Y = 1/Z$, where Z is the driving-point impedance. Since Z is unknown, however, it is easier to obtain Y from $Y = I/V$. To obtain the ratio I/V, we can start with the circuit's terminal equation, which, by KCL, is

$$i(t) = C\frac{d}{dt}v + \frac{1}{R}v + \frac{1}{L}\int_{-\infty}^{t} v(\alpha)\,d\alpha$$

where we have taken care to write the input variable v on the right-hand side. When differentiated, this becomes

$$\frac{d}{dt}i = C\frac{d^2}{dt^2}v + \frac{1}{R}\frac{d}{dt}v + \frac{1}{L}v$$

which has the form

$$\mathscr{A}(p)i = \mathscr{B}(p)v$$

where $\mathscr{A}(p) = p$, $\mathscr{B}(p) = Cp^2 + (1/R)p + 1/L$. Notice that confusion regarding which operator is $\mathscr{A}(p)$ and which is $\mathscr{B}(p)$ can be avoided if we always write the input variable (here v) on the right-hand side of the differential equation. It follows that the driving-point admittance is

$$Y = \frac{I}{V} = \frac{C(j\omega)^2 + (1/R)j\omega + (1/L)}{j\omega}$$

or

$$Y = j\omega C + \frac{1}{R} + \frac{1}{j\omega L}$$

$$= \sqrt{\left(\frac{1}{R}\right)^2 + \left(\omega C - \frac{1}{\omega L}\right)^2} \, \bigg/ \arctan\left[R\left(\omega C - \frac{1}{\omega L}\right)\right]$$

This result can also be obtained by substituting $\mathbf{v}(t) = \mathbf{V}e^{j\omega t}$ and $\mathbf{i}(t) = \mathbf{I}e^{j\omega t}$ into the differential equation and solving for I/V. Notice that the admittance of the parallel RLC network is given by the sum of the admittances of R, L, and C.

(b) For the given numerical values, we have

$$\mathbf{V} = 25\underline{/70°} \text{ mV}$$

and

$$\mathbf{Y} = j10^4 \times 30 \times 10^{-6} + \frac{1}{10} + \frac{1}{j10^4 \times 10^{-3}} = 0.1 + j0.2 \text{ S}$$

$$= 0.224\underline{/63.4°} \text{ S}$$

Therefore

$$\mathbf{I} = \mathbf{VY} = 5.6\underline{/133.4°} \text{ mA}$$

and

$$i(t) = \mathscr{I}m\ \{(5.6\underline{/133.4°})e^{j10^4 t}\}$$

$$= 5.6 \sin(10^4 t + 133.4°) \text{ mA} \quad \blacksquare$$

EXERCISES

12. Define the following terms: (a) driving-point impedance, (b) transfer impedance, (c) driving-point admittance, and (d) transfer admittance.

13. Define the following terms: (a) resistance, (b) reactance, (c) conductance, and (d) susceptance.

14. The driving-point admittance of a certain circuit is $\mathbf{Y} = 3 + j4$ S. Find the conductance, susceptance, resistance, and reactance. State whether the circuit is inductive, resistive, or capacitive.

15. Find the expressions for the driving-point resistance, reactance, conductance, susceptance, and admittance of a series RLC circuit.

16. Devise a circuit composed of two resistances R_1 and R_2 to illustrate that Eq. (10.34) applies for driving-point admittance and does *not* apply for transfer admittance.

10.6 Transfer Functions

The impedance $\mathbf{Z}(j\omega)$ of Eq. (10.17a) is an example of what is more generally referred to as a *transfer function*. Consider any stable LLTI circuit that is driven by exactly one sinusoidal input. The steady-state response of this circuit is sinusoidal with the same frequency as the input. Therefore we can represent both the input and the output with the use of phasors. The transfer function gives us the ratio of the output phasor to the input phasor. The transfer function itself is given by a ratio of polynomials in $j\omega$ of the form $\mathscr{B}(j\omega)/\mathscr{A}(j\omega)$, where $\mathscr{B}(p)$ and $\mathscr{A}(p)$ are, respectively, the differential operators operating on the network's time-domain input and output. An arbitrary transfer function is usually denoted by $\mathbf{H}(j\omega)$, or more simply $\mathbf{H}$. Thus,

$$\mathbf{H} = \frac{\mathbf{B}}{\mathbf{A}} = \frac{\mathscr{B}(j\omega)}{\mathscr{A}(j\omega)} \tag{10.35}$$

where **B** denotes the output phasor (**V** or **I**), and **A** denotes the input phasor (**V** or **I**).[†] The term *transfer function* can be applied to both impedance and admittance, but it is most often used when the input and the output have the same dimensions. Any transfer function can be determined either directly from Eq. (10.35) or, equivalently, by substitution of the appropriate complex exponentials into the circuit's LLTI equation. In Chapter 11 we will see that any transfer function can be obtained directly from the circuit without the use of differential equations. The following example illustrates the basic definition given by Eq. (10.35).

EXAMPLE 10.5

The circuit of Fig. 10.15 depicts a voltage divider consisting of a resistance and an inductance. Voltages $v_s(t)$ and $v_o(t)$ are, respectively, the input and the output.

FIGURE 10.15

RL voltage divider

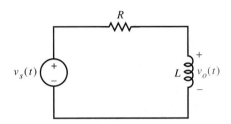

(a) Determine the voltage transfer function.
(b) Determine the steady-state output for an input $v_s(t) = V_{sm} \cos(\omega t + \phi_s)$.
(c) Evaluate your answers to (a) and (b) for $R = 2\,\Omega$, $L = 0.1$ H, and $v_s(t) = 10 \cos 8t$ V.

■ SOLUTION

(a) By KVL and Ohm's law, we have

$$v_o(t) + \frac{R}{L} \int_{-\infty}^{t} v_o(\alpha) \, d\alpha = v_s(t)$$

Differentiated, this becomes

$$\frac{d}{dt} v_o(t) + \frac{R}{L} v_o(t) = \frac{d}{dt} v_s(t)$$

This equation has the form $\mathscr{A}(p)v_o(t) = \mathscr{B}(p)v_s(t)$ where $\mathscr{A}(p) = p + R/L$ and $\mathscr{B}(p) = p$. The voltage transfer function is given by Eq. (10.35) using $\mathscr{A}(j\omega) = j\omega + R/L$ and $\mathscr{B}(j\omega) = j\omega$. The result is

$$\mathbf{H} = \frac{\mathbf{V}_o}{\mathbf{V}_s} = \frac{j\omega}{j\omega + (R/L)} = \frac{j\omega L}{R + j\omega L}$$

$$= \frac{\omega L}{\sqrt{R^2 + (\omega L)^2}} \left| \underline{90° - \arctan\left(\frac{\omega L}{R}\right)} \right.$$

This result can also be obtained by substituting $v_s(t) = \mathbf{V}_s e^{j\omega t}$ and $v_o(t) = \mathbf{V}_o e^{j\omega t}$ into the differential equation and solving for $\mathbf{V}_o/\mathbf{V}_s$.

[†] The phasors **A** and **B** should not be confused with the polynomials $\mathscr{A}(j\omega)$ and $\mathscr{B}(j\omega)$. They are completely different quantities.

(b) The particular response to $v_s(t) = V_{sm} \cos(\omega t + \phi_s)$ is

$$v_{op}(t) = V_{om} \cos(\omega t + \phi_o)$$

where $V_{om} = |\mathbf{H}| V_{sm}$ and $\phi_o = \underline{/\mathbf{H}} + \phi_s$. This result is obtained by projecting the rotating phasor $\mathbf{v}_o(t) = \mathbf{V}_o e^{j\omega t}$ onto the real axis, where $\mathbf{V}_o = \mathbf{H}\mathbf{V}_s$. The steady-state response equals the particular response after the complementary response is negligible:

$$v_o(t) = v_{op}(t) = \frac{\omega L}{\sqrt{R^2 + (\omega L)^2}} V_{sm} \cos\left[\omega t + \phi_s + 90° - \arctan\left(\frac{\omega L}{R}\right)\right]$$

(c) For the given numerical values, we find

$$\mathbf{V}_s = 10\underline{/0°} \text{ V}$$

$$\mathbf{H} = \frac{j0.8}{2 + j0.8} = 0.37\underline{/68.2°}$$

and
$$\mathbf{V}_o = 3.7\underline{/68.2°} \text{ V}$$

Therefore the particular response is

$$v_{op}(t) = 3.7 \cos(8t + 68.2°) \text{ V}$$

This is the steady-state response after the complementary response is negligible. ■

In Example 10.2 we saw that the driving-point impedance of a series RLC circuit is $\mathbf{Z} = R + j\omega L + (1/j\omega C)$, the sum of the impedances of R, L, and C. We saw in Example 10.4 that the driving-point admittance of a parallel RLC circuit is $\mathbf{Y} = (1/R) + (1/j\omega L) + j\omega C$, the sum of the admittances of R, L, and C. Finally, in Example 10.5 we found that the transfer function of an RL voltage divider is $\mathbf{H} = j\omega L/(R + j\omega L)$, which may be recognized as being analogous to a voltage divider relation with resistances replaced by impedances. All three results suggest that circuits involving element impedances and phasor voltages and currents may be analyzed with the use of only algebra—just as circuits containing only resistances and dc sources may be analyzed with the use of only algebra. This idea, which is one of the truly great ideas in circuit theory, is the topic of Chapter 11.

REMEMBER

A circuit's *transfer function* equals the ratio of the output phasor to the input phasor.

EXERCISES

17. Define transfer function. Is a driving-point impedance a transfer function?

The following circuit is a current divider consisting of a resistance and an inductance. Currents $i_s(t)$ and $i_o(t)$ are, respectively, the input and the output.

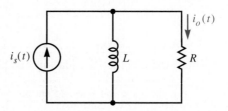

18. Determine the current transfer function $\mathbf{H} = \mathbf{I}_o/\mathbf{I}_s$.

19. Determine the steady-state output for an input $i_s(t) = 5 \cos 1000t$ mA if $R = 2.2 \ \Omega$ and $L = 1$ mH.

10.7 Summary

This is a basic chapter that provides the foundation for all that follows in this book. We began by stating that for *every* stable LLTI circuit, a sinusoidal input always produces a sinusoidal particular response. Since the complementary response of a stable circuit always dies out as time increases, the complete response to a sinusoidal input will eventually be a sinusoid with the same frequency as the input. When this happens, the circuit has reached the sinusoidal steady state. The major topic of this chapter involved the use of phasors and transfer functions to determine the amplitude and the phase of the sinusoidal particular response. A phasor (for example, $\mathbf{I}$) is the complex amplitude of a complex exponential, as in $\mathbf{I}e^{j\omega t}$. If the complex exponential (or rotating phasor) $\mathbf{I}e^{j\omega t}$ is mathematically considered to be the input to a stable LLTI circuit with input-output equation $\mathscr{A}(p)v(t) = \mathscr{B}(p)i(t)$, then the particular response is the rotating phasor $\mathbf{V}e^{j\omega t}$, where $\mathbf{V} = \mathbf{Z}\mathbf{I}$ and $\mathbf{Z} = |\mathbf{Z}|\underline{/\theta} = \mathscr{B}(j\omega)/\mathscr{A}(j\omega)$ is the impedance. In practice, we are interested in a circuit's response to a real sinusoidal input, such as $I_m \cos(\omega t + \phi_I)$. By regarding this real input as the real part of $\mathbf{I}e^{j\omega t}$, we can obtain the real particular response by taking the real part of $\mathbf{V}e^{j\omega t}$, where $\mathbf{V} = \mathbf{Z}\mathbf{I}$; that is, $\mathscr{R}e\{\mathbf{V}e^{j\omega t}\} = |\mathbf{Z}|I_m \cos(\omega t + \phi_I + \theta)$. Thus, the sinusoidal input and output are projections of the rotating phasors onto the real axis. Figure 10.16 summarizes these key ideas. If you understand this figure and can apply the concepts involved to simple examples, then you have learned a great deal.

FIGURE 10.16
Evolution of the impedance concept. (*a*) Time-domain: The network is described by a linear differential equation with constant coefficients.
(*b*) Particular response to input $\mathbf{I}e^{j\omega t}$: If $i(t) = \mathbf{I}e^{j\omega t}$ then $v_p(t) = \mathbf{V}e^{j\omega t}$ where $\mathbf{V} = \mathbf{Z}\mathbf{I}$ and $\mathbf{Z} = \mathscr{B}(j\omega)/\mathscr{A}(j\omega)$. This solution requires that $\mathscr{A}(j\omega) \neq 0$, which is the case for every stable network. $\mathbf{Z}$ is the impedance of the network.
(*c*) Frequency-domain: The relationship of $\mathbf{V}$ to $\mathbf{I}$ is indicated without the $e^{j\omega t}$'s.

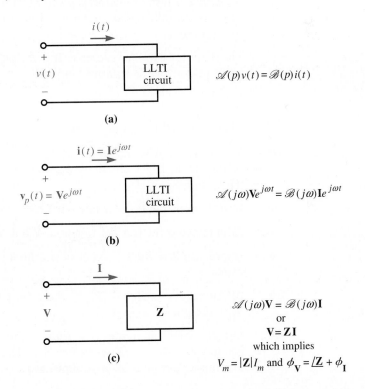

$i(t)$

$v(t)$ — LLTI circuit — $\mathscr{A}(p)v(t) = \mathscr{B}(p)i(t)$

(a)

$i(t) = \mathbf{I}e^{j\omega t}$

$v_p(t) = \mathbf{V}e^{j\omega t}$ — LLTI circuit — $\mathscr{A}(j\omega)\mathbf{V}e^{j\omega t} = \mathscr{B}(j\omega)\mathbf{I}e^{j\omega t}$

(b)

$\mathbf{I}$

$\mathbf{V}$ — $\mathbf{Z}$ — $\mathscr{A}(j\omega)\mathbf{V} = \mathscr{B}(j\omega)\mathbf{I}$
or
$\mathbf{V} = \mathbf{Z}\mathbf{I}$
which implies
$V_m = |\mathbf{Z}|I_m$ and $\phi_{\mathbf{V}} = \underline{/\mathbf{Z}} + \phi_{\mathbf{I}}$

(c)

FIGURE 10.16
Continued
(*d*) Particular
response to sinusoidal
inputs: The real part
operator $\mathcal{R}e \{\ \}$ is applied
to $\mathbf{I}e^{j\omega t}$ and $\mathbf{V}e^{j\omega t}$.
(*e*) The imaginary part
operator $\mathcal{I}m \{\ \}$ is
applied to $\mathbf{I}e^{j\omega t}$ and $\mathbf{V}e^{j\omega t}$.

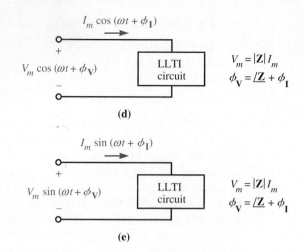

KEY FACTS

♦ The steady-state response of a stable LLTI circuit to a sinusoidal input is a sinusoid with the same frequency as the input.

♦ A phasor, for example, $\mathbf{A}$, represents both the amplitude and the phase of a sinusoidal voltage or current. The magnitude of $\mathbf{A}$, $|\mathbf{A}| = A_m$, is the amplitude, and the angle of $\mathbf{A}$, $\underline{/\mathbf{A}} = \phi_\mathbf{A}$, is the phase.

♦ A cosine input waveform is the projection of the rotating phasor $\mathbf{A}e^{j\omega t}$ onto the real axis of the complex plane: $A_m \cos (\omega t + \phi_\mathbf{A}) = \mathcal{R}e \{\mathbf{A}e^{j\omega t}\}$.

♦ A circuit's transfer function is given by $\mathbf{H} = \mathcal{B}(j\omega)/\mathcal{A}(j\omega)$, where $\mathcal{A}(p)$ and $\mathcal{B}(p)$ are the differential operators of the LLTI equation relating excitation to response.

♦ The output phasor, for example, $\mathbf{B} = B_m \underline{/\phi_\mathbf{B}}$, is obtained by multiplying the input phasor by the transfer function:

$$\mathbf{B} = \mathbf{HA}$$

Thus $B_m = |\mathbf{H}|A_m$ and $\phi_\mathbf{B} = \underline{/\mathbf{H}} + \phi_\mathbf{A}$.

♦ When the input is $A_m \cos (\omega t + \phi_\mathbf{A}) = \mathcal{R}e \{\mathbf{A}e^{j\omega t}\}$, the steady-state output of the stable LLTI circuit is $B_m \cos (\omega t + \phi_\mathbf{B}) = \mathcal{R}e \{\mathbf{B}e^{j\omega t}\}$, where $B_m = |\mathbf{H}|A_m$ and $\phi_\mathbf{B} = \underline{/\mathbf{H}} + \phi_\mathbf{A}$.

♦ Impedance is the transfer function: $\mathbf{Z} = \mathbf{V}/\mathbf{I}$

♦ Admittance is the transfer function: $\mathbf{Y} = \mathbf{I}/\mathbf{V}$

♦ Impedance $\mathbf{Z} = R(\omega) + jX(\omega)$, where $R(\omega)$ is the resistance and $X(\omega)$ is the reactance.

♦ Admittance $\mathbf{Y} = G(\omega) + jB(\omega)$, where $G(\omega)$ is the conductance and $B(\omega)$ is the susceptance.

♦ Driving-point admittance $\mathbf{Y} = G(\omega) + jB(\omega)$ is related to driving-point impedance $\mathbf{Z} = R(\omega) + jX(\omega)$ by $\mathbf{Y} = \mathbf{Z}^{-1}$. The relation $\mathbf{Y} = \mathbf{Z}^{-1}$ does not hold for transfer admittances and impedances.

Section 10.1

1. Explain in your own words the meaning of the term *sinusoidal steady state*.

The complementary response of a circuit is the solution to the homogeneous equation $\mathscr{A}(p)v_c(t) = 0$, for which the input is set to zero. Reasoning from physical grounds, we might expect that *every* physical circuit made up of Rs, Ls, and Cs would be stable. (With the input set to zero, any energy initially present in the circuit should gradually be dissipated by the resistances; thus $v_c(t) \to 0$ as $t \to \infty$.) For the circuits shown in Problems 2 through 4, find $v_c(t)$ and discuss this idea. In Problem 4, consider three possibilities: $\alpha < 1$, $\alpha = 1$, and $\alpha > 1$.

2.

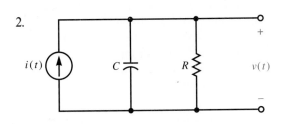

3.

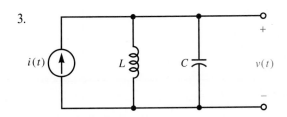

4.

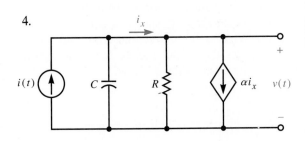

Section 10.2

5. Let $\mathscr{A}(p) = a_n p^n + a_{n-1} p^{n-1} + \cdots + a_1 p + a_0$, where p denotes d/dt. An engineer claims that "since $\mathscr{A}(p)e^{j\omega t} = \mathscr{A}(j\omega)e^{j\omega t}$, by canceling the $e^{j\omega t}$'s it follows that $\mathscr{A}(p) = \mathscr{A}(j\omega)$." The engineer cannot possibly be correct; a differential

operator is not a complex constant. What is wrong with the reasoning?

6. Give a detailed derivation of Eq. (10.13).

Section 10.3

7. (a) Find the expression for the impedance of a parallel RL circuit. Put your result in rectangular form.
   (b) Assume that $R = 1$ kΩ and $L = 1$ mH. At what value of ω does $\underline{/Z} = 45°$?

Problems 8 through 13 refer to the following circuit.

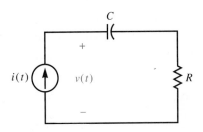

8. Find the differential equation relating input $i(t)$ and output $v(t)$.
9. Use rotating phasors to derive the frequency-domain version of the differential equation. Your result should have the form $\mathscr{A}(j\omega)\mathbf{V} = \mathscr{B}(j\omega)\mathbf{I}$.
10. Use the equation derived in Problem 9 to obtain the driving-point impedance of the circuit.
11. Find the particular response for an input $i(t) = \mathbf{I}e^{j\omega t}$.
12. Sketch $\mathbf{V}$ and $\mathbf{I}$ for $\mathbf{I} = 100\underline{/20°}$ mA, $R = 1$ kΩ, $C = 1$ μF, and $\omega = 1$ krad/s.
13. Use the results of Problems 11 and 12 to determine the particular response $v_p(t)$ when $i(t) = I_m \cos(\omega t + \phi_1)$. Use the numerical values given in Problem 12.
14. A series circuit consisting of a 1-Ω resistance, a 1-μF capacitance, and a 1-μH inductance is driven by a current source $i(t) = 1 \cos \omega t$ mA. Use phasors to determine the steady-state voltage $v(t)$ across the input terminals assuming that $\omega = 100$ krad/s. Plot $i(t)$ and $v(t)$ versus t.
15. Repeat Problem 14 if $\omega = 1$ Mrad/s.
16. Repeat Problem 14 if $\omega = 10$ Mrad/s.
17. (a) Show that $\mathscr{R}e\{\mathscr{A}(p)\mathbf{v}(t)\} = \mathscr{A}(p)\mathscr{R}e\{\mathbf{v}(t)\}$ if $\mathscr{A}(p) = 3p + 2$, $p = d/dt$, and $\mathbf{v}(t)$ is *any*

complex function. [*Hint*: Write $v(t)$ in the rectangular form: $\mathbf{v}(t) = v_R(t) + jv_I(t)$.]

(b) Generalize your analysis in (a) by assuming that $\mathscr{A}(p) = a_n p^n + a_{n-1}p^{n-1} + \cdots + a_0$, where the coefficients are arbitrary *real* numbers.

(c) Show that $\mathscr{R}e\{\mathscr{B}(p)\mathbf{i}(t)\} = \mathscr{B}(p)\,\mathscr{R}e\{\mathbf{i}(t)\}$ if $\mathbf{i}(t)$ is *any* complex function, $\mathscr{B}(p) = b_m p^m + b_{m-1}p^{m-1} + \cdots + b_0$ and the coefficients are arbitrary *real* numbers.

(d) What do the results of (b) and (c) have to do with Eq. (10.21)?

18. A linear time-invariant network has a current input $3\cos(\omega t + 50°)$ A and a voltage output $6\cos(\omega t + 70°)$ V. Find the associated impedance.

19. If the voltage output of the network in Problem 18 is $-3\cos\omega t$ V, what is the input? Put your answer in the form $A\cos(\omega t + \phi)$.

Section 10.4

20. Use phasors to determine the steady-state voltage across each component of a series RLC circuit in which $R = 1\ \text{k}\Omega, L = 1\ \mu\text{H}, C = 1\ \mu\text{F}$, and $i(t) = 3\cos(2 \times 10^6 t + 30°)$ mA. Plot $i(t)$ and each voltage.

21. Check your answers to Problem 20 with the use of the time-domain terminal equations for $R, L,$ and C.

Section 10.5

Refer to Table 10.1 for Problems 22 through 25.

22. Use the equation $\mathbf{Y} = \mathbf{Z}^{-1}$ to derive the expressions shown for $G(\omega)$ and $B(\omega)$.

23. Use the equation $\mathbf{Z} = \mathbf{Y}^{-1}$ to derive the expressions shown for $R(\omega)$ and $X(\omega)$.

24. Show that $R(\omega)G(\omega) = 1$ if and only if $X(\omega) = 0$.

25. Show that $X(\omega) = 0$ if and only if $B(\omega) = 0$.

26. Plot the impedance angle of a series RLC circuit versus ω, for $R = 1\ \Omega, L = 1\ \text{mH}$, and $C = 1000\ \mu\text{F}$. Indicate the frequencies for which the circuit is (a) capacitive, (b) resistive, and (c) inductive. ▪ 10.6

Section 10.6

27. Derive the voltage transfer function $\mathbf{V}_o/\mathbf{V}_s$ of the following circuit from the appropriate dif-

ferential equation. Does your result have the form of a voltage divider relation with resistances replaced by impedances?

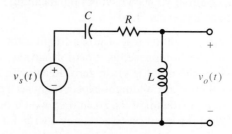

28. Use your results from Problem 27 to determine the steady-state response $v_o(t)$ in the preceding circuit for $v_s(t) = V_m\cos(\omega t + \phi_V)$.

29. Use your results from Problem 27 to determine the steady-state response $v_o(t)$ in the preceding circuit for $v_s(t) = V_m\sin(\omega t + \phi_V)$.

30. Find the current transfer function $\mathbf{I}_o/\mathbf{I}_s$ of the circuit shown below from the appropriate differential equation. Comment on any interesting feature of your result.

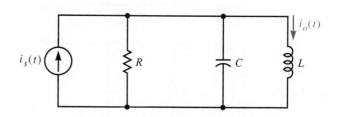

31. Use your answer to Problem 30 to determine the steady-state response $i_o(t)$ in the preceding circuit for $i_s(t) = I_m\sin(\omega t + \phi_I)$.

32. (a) What is the differential equation that relates the output $v_o(t)$ to the input $v_s(t)$ for the circuit shown below?

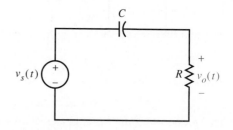

(b) Use your answer to (a) to obtain the transfer function

$$\mathbf{H}(j\omega) = \frac{\mathbf{V}_o}{\mathbf{V}_s}$$

(c) Give the magnitude and the angle of $\mathbf{H}(j\omega)$.

33. In the circuit for Problem 32, if $v_s(t) = \mathbf{V}_s e^{j\omega t}$, what is $v_o(t)$?

34. (a) In the circuit for Problem 32, if $v_s(t) = V_{sm}\cos(\omega t + \phi_s)$, what is $v_o(t)$?
    (b) Give the numerical form of your answer in (a) if $v_s(t) = 100\cos(10^3 t + 20°)$, $R = 1\ \text{k}\Omega$, and $C = 1\ \mu\text{F}$.

35. (a) Find the differential equation relating input $v_s(t)$ and output $v_o(t)$ for the circuit shown below.

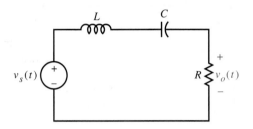

    (b) Find the voltage transfer function.
    (c) Find $v_{op}(t)$ if $L = 2\ \text{H}$, $C = 1\ \text{F}$, $R = 1\ \Omega$, and $v_s(t) = 100\cos(t + 15°)\ \text{mV}$.

In the circuits in Problems 36 through 38:
    (a) Find the differential equations relating excitation and response.
    (b) Find the transfer functions.
    (c) Find the steady-state responses for real sinusoidal excitation with arbitrary amplitude, frequency, and phase.

36.

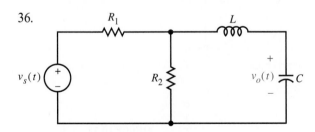

37.

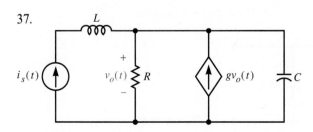

38.

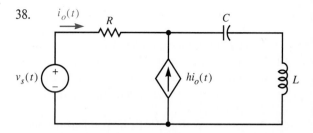

39. Start with the appropriate differential equation and derive the voltage transfer function $\mathbf{H} = \mathbf{V}_o/\mathbf{V}_s$ for the op-amp circuit shown below. Use an ideal op-amp model. ☐ 10.4

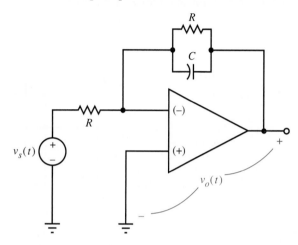

40. Repeat Problem 39 for the circuit shown below.

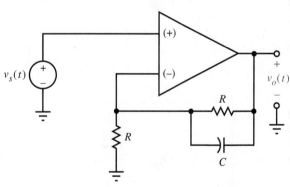

41. (a) Start from the appropriate differential equation to determine the voltage transfer function of the ideal op-amp circuit shown in Fig. 10.17. ☐ 10.5
    (b) Specialize your result of (a) for $R_1 = 1\ \text{k}\Omega$, $C = 1000\ \mu\text{F}$, and $R_2 = R_3 = 2.2\ \text{k}\Omega$.
    (c) Let $v_s(t) = 50\cos(t + 30°)u(t)\ \text{mV}$. Compute the particular response for the numerical values of (b). (Assume that the capacitance is initially uncharged.)

(d) Is the circuit stable? What would have been the answer to (a) if the capacitance was initially charged? Discuss.

42. Determine the complete response of the ideal op-amp circuit in Fig. 10.18. Assume that the capacitances are uncharged the instant before the switch closes.

43. The circuit in Fig. 10.19 is an ideal model of a tuned common-emitter transistor amplifier. Determine the voltage transfer function value for $\omega = 10^4$ rad/s. Assume that $\beta = 100$.

44. Repeat Problem 43 for $\omega = 10^6$ rad/s.

45. Repeat Problem 43 for $\omega = 10^8$ rad/s.

46. The switch closes at $t_0 = 0$ in the circuit shown below. A circuit designer wants the sinusoidal steady state to begin the instant the switch closes. Show how this is possible if the capacitance is given the right initial voltage $v_C(0^-)$, and state the value of $v_C(0^-)$.

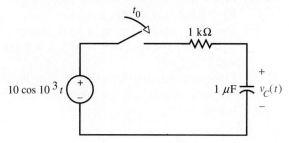

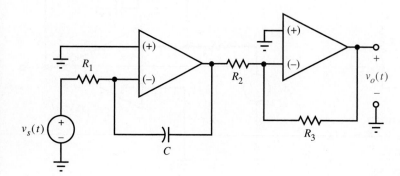

FIGURE 10.17

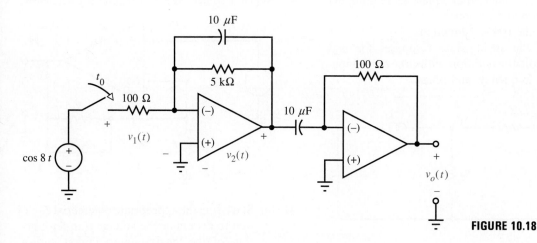

FIGURE 10.18

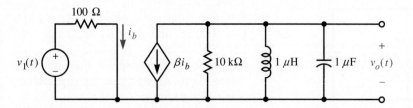

FIGURE 10.19

47. Assume that the circuit of Problem 46 initially contains no stored energy. A circuit designer wants the sinusoidal steady state to begin the instant the switch closes. At what instant t_0 should the switch close?

48. Repeat Problem 47 for the circuit in Fig. 10.20.

49. The circuit in Fig. 10.21 consists of two circuits joined by a buffer amplifier.
   (a) Show that $v_3(t)$ and $v_1(t)$ are related by the LLTI equation $\mathscr{A}(p)v_3 = \mathscr{B}(p)v_1$, where $\mathscr{A}(p) = \mathscr{A}_2(p)\mathscr{A}_1(p)$ and $\mathscr{B}(p) = \mathscr{B}_2(p)\mathscr{B}_1(p)$.
   (b) Show that the transfer function is $\mathbf{H} = \mathbf{V}_3/\mathbf{V}_1 = \mathbf{H}_1\mathbf{H}_2$, where $\mathbf{H}_1 = \mathscr{B}_1(j\omega)/\mathscr{A}_1(j\omega)$ and $\mathbf{H}_2 = \mathscr{B}_2(j\omega)/\mathscr{A}_2(j\omega)$
   (c) Generalize your result from (a) to N circuits joined by $N-1$ buffer amplifiers.

50. An LLTI circuit consists of two circuits in series as shown in Fig. 10.22.
   (a) Show that $v(t)$ and $i(t)$ are related by $\mathscr{A}(p)v = \mathscr{B}(p)i$, where $\mathscr{A}(p) = \mathscr{A}_1(p)\mathscr{A}_2(p)$ and $\mathscr{B}(p) = \mathscr{A}_2(p)\mathscr{B}_1(p) + \mathscr{A}_1(p)\mathscr{B}_2(p)$.
   (b) Use your result from (a) to show that impedances in series add. This result will be derived another way in Chapter 11.

51. Consider any two LLTI circuits in parallel as shown in Fig. 10.23.
   (a) Show that $v(t)$ and $i(t)$ are related by $\mathscr{A}(p)i = \mathscr{B}(p)v$, where $\mathscr{A}(p) = \mathscr{A}_1(p)\mathscr{A}_2(p)$ and $\mathscr{B}(p) = \mathscr{A}_2(p)\mathscr{B}_1(p) + \mathscr{A}_1(p)\mathscr{B}_2(p)$.
   (b) Use your result from (a) to show that admittances in parallel add. This result will be derived another way in Chapter 11.

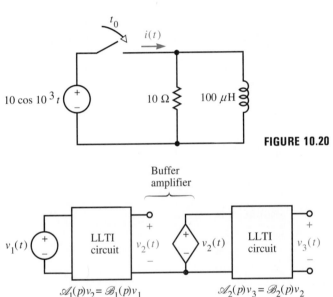

FIGURE 10.20

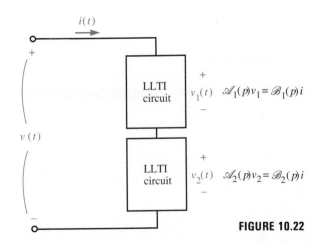

FIGURE 10.21

FIGURE 10.22

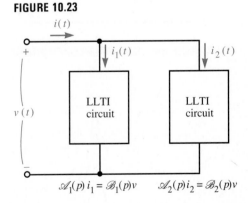

FIGURE 10.23

Comprehensive Problems

For Problems 52 through 60, determine an equivalent circuit that contains no more than two components for the circuit to the right of terminals a and b in the following illustration. What is the driving-point impedance seen by the voltage source? Determine the steady-state current $i(t)$ or explain why it cannot be found. Does the particular response exist?

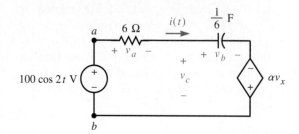

52. $v_x = v_a$, $\alpha = \frac{1}{2}$ 53. $v_x = v_a$, $\alpha = 1$
54. $v_x = v_a$, $\alpha = \frac{3}{2}$ 55. $v_x = v_b$, $\alpha = \frac{1}{2}$
56. $v_x = v_b$, $\alpha = 1$ 57. $v_x = v_b$, $\alpha = \frac{3}{2}$

58. $v_x = v_c$, $\alpha = \frac{1}{2}$ 59. $v_x = v_c$, $\alpha = 1$
60. $v_x = v_c$, $\alpha = \frac{3}{2}$
61. Determine currents $i_1(t)$, $i_2(t)$ and $i_3(t)$ and the driving-point impedance $\mathbf{Z}_{ab}$ of the following circuit if the input is $v_{ab}(t) = 150 \cos 2t$ V.

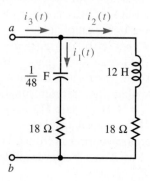

62. Use superposition to determine the steady-state voltage $v(t)$ for the circuit in Fig. 10.24 if $i_a(t) = 90 \cos t$ A and $i_b(t) = 18 \cos 2t$ A.
63. Use superposition to determine the steady-state current $i(t)$ in the circuit in Fig. 10.25.

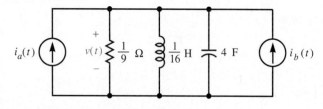

FIGURE 10.24

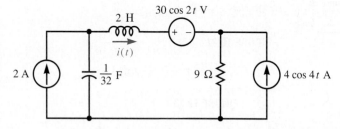

FIGURE 10.25

AC Circuit Analysis

When a stable, linear, time-invariant circuit is driven by a
sinusoidal source, the particular response of every voltage and
every current in the circuit is sinusoidal with the same frequency
as the source. This fact implies that there is no need to use
differential equations or calculus to find the voltage and current
waveforms in a linear, time-invariant circuit operating in the
sinusoidal steady state. Since all the waveforms are known to be
sinusoidal with the same frequency as the source, all that
remains is to determine the amplitudes and phases of the
sinusoids. The amplitudes and phases are constants that can be
calculated with the use of algebra.

The purpose of this chapter is to describe the principal
algebraic techniques that are used to find the amplitudes and
phases of sinusoidal steady-state response waveforms. We
present the foundations for these techniques in Section 11.1.
The circuit analysis techniques themselves are described in
Sections 11.2 through 11.6. We shall see that all the circuit
analysis techniques are simply reformulations of methods
encountered earlier.

Sinusoidal steady-state circuit analysis is often referred to as *ac circuit analysis*. The letters "ac" are an abbreviation for "alternating current," a term that originated in the nineteenth century. Today the term *ac waveform* is used to describe either a current or a voltage waveform that varies sinusoidally with time. The term *ac circuit* is used to describe a linear, time-invariant circuit that is operating in the sinusoidal steady state.*

11.1 Foundations of AC Circuit Analysis

One of the most ingenious and useful ideas in circuit theory is to determine the amplitudes and phases of the sinusoidal waveforms in an ac circuit with the use of algebra—not calculus. The development of this algebra for ac circuit analysis involves the steps described in this section.

Frequency-Domain Independent Sources

The first step in the development of an algebraic system for ac circuit analysis was taken in Chapter 10, where we represented independent time-domain sources by corresponding phasors. The phasor sources are called *frequency-domain* sources and contain information equivalent to that contained in the corresponding time-domain sources. However, there is an ambiguity that we need to resolve now. Theoretically, there are two possible time-domain waveforms corresponding to each phasor. The reason is that the time-domain waveform is obtained by multiplying the phasor by $e^{j\omega t}$ and operating on the product with *either* the real part operator, $\mathcal{R}e\{\cdot\}$, *or* the imaginary part operator, $\mathcal{I}m\{\cdot\}$. For example, the frequency-domain voltage source phasor $\mathbf{V}_s = V_{sm}\underline{/\phi_\mathbf{V}}$ can represent either the time-domain source

$$
\begin{aligned}
v_s(t) &= \mathcal{R}e\left\{\mathbf{V}_s e^{j\omega t}\right\} \\
&= V_{sm}\cos\left(\omega t + \phi_\mathbf{V}\right)
\end{aligned}
\tag{11.1}
$$

or the time-domain source

$$
\begin{aligned}
v_s(t) &= \mathcal{I}m\left\{\mathbf{V}_s e^{j\omega t}\right\} \\
&= V_{sm}\sin\left(\omega t + \phi_\mathbf{V}\right)
\end{aligned}
\tag{11.2}
$$

Because the same phasor can represent either a cosine function or a sine function, depending on which operator is used, it is necessary that the choice of operator, $\mathcal{R}e\{\cdot\}$ or $\mathcal{I}m\{\cdot\}$, be specified at the onset of every ac circuit analysis. The choice between $\mathcal{R}e\{\cdot\}$ and $\mathcal{I}m\{\cdot\}$ is ultimately a matter of convenience or personal preference. Nevertheless, the $\mathcal{R}e\{\cdot\}$ operator has become the almost universal choice of modern textbooks and practicing electrical engineers. Therefore, the real part operator will be used in this book unless stated otherwise.

As illustrated in Example 11.1, when the $\mathcal{R}e\{\cdot\}$ operator is used, time-domain sources represented by a sine function must be converted to the cosine function with the use of the trigonometric identity $\sin x = \cos\left(x - 90°\right)$.

---

* There is another meaning of the term *ac* in which the time variation is not restricted to the sinusoidal. In this usage, "ac" means simply "time-varying," as opposed to "dc," which means "constant." In this book, the letters "ac" will be used *only* to denote *sinusoidal* time variation.

EXAMPLE 11.1

Figure 11.1a illustrates a linear network driven by three sinusoidal sources of the same frequency. Construct the frequency-domain equivalent.

FIGURE 11.1
Time domain–frequency domain source transformations: (*a*) time domain; (*b*) frequency domain

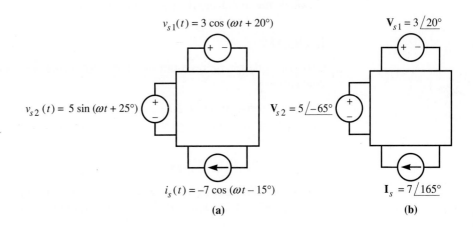

$v_{s1}(t) = 3 \cos(\omega t + 20°)$

$\mathbf{V}_{s1} = 3\underline{/20°}$

$v_{s2}(t) = 5 \sin(\omega t + 25°)$

$\mathbf{V}_{s2} = 5\underline{/-65°}$

$i_s(t) = -7 \cos(\omega t - 15°)$

$\mathbf{I}_s = 7\underline{/165°}$

(a) **(b)**

■ SOLUTION

By the convention assumed in this book, the time-domain voltage source $v_{s1}(t)$ of Fig. 11.1a is transformed into the frequency domain by noting that

$$v_{s1}(t) = 3 \cos(\omega t + 20°)$$
$$= \mathscr{R}e\left\{3e^{j20°}e^{j\omega t}\right\}$$

[as in Eq. (11.1)]. Therefore

$$\mathbf{V}_{s1} = 3\underline{/20°}$$

Similarly, for

$$i_s(t) = -7 \cos(\omega t - 15°)$$

we obtain

$$\mathbf{I}_s = -(7\underline{/-15°}) = 7\underline{/165°}$$

The phasor representation of the third source is obtained by first converting the sine function into a cosine function by means of the identity

$$\sin(\omega t + \phi) = \cos(\omega t + \phi - 90°)$$

Therefore

$$v_{s2}(t) = 5 \sin(\omega t + 25°)$$
$$= 5 \cos(\omega t - 65°)$$

and accordingly,

$$\mathbf{V}_{s2} = 5\underline{/-65°}$$

The frequency-domain representation of Fig. 11.1a is shown in Fig. 11.1b. ■

Notice that in Fig. 11.1a, the three independent sinusoidal sources all have the same frequency. If this were not the case, then Fig. 11.1b would have no meaning.

In a frequency-domain representation like that of Fig. 11.1b, it is always necessary that the phasors refer to sinusoidal functions of time that have the same frequency. It is also necessary that the phasors be related to their time-domain counterparts via the same operator, $\mathscr{R}e\{\cdot\}$ or $\mathscr{I}m\{\cdot\}$.

EXERCISES

Give the phasor representation of the following sources. (Assume the $\mathscr{R}e\{\cdot\}$ convention.)

1. $i(t) = 6\cos(1000t + 30°)$
2. $v(t) = -5\cos(1000t + 30°)$
3. $i(t) = 4\sin(1000t + 30°)$
4. $v(t) = -3\sin(1000t + 30°)$

Frequency-Domain Dependent Sources

Let us now extend the phasor source representation to include dependent sources. Consider Fig. 11.2a, which depicts a current-controlled voltage source.

FIGURE 11.2
(*a*) Time-domain and (*b*) frequency-domain forms of a current-controlled voltage source

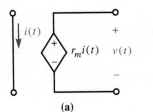 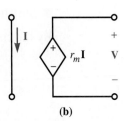

(a)   (b)

In the sinusoidal steady state, $i(t)$ and $v(t)$ will be sinusoidal functions of t. To obtain the phasor representation of Fig. 11.2a, we replace $i(t)$ and $v(t)$ by rotating phasors:

$$\mathbf{i}(t) = \mathbf{I}e^{j\omega t} \tag{11.3}$$

and

$$\begin{aligned}\mathbf{v}(t) &= r_m\mathbf{i}(t)\\ &= r_m\mathbf{I}e^{j\omega t}\end{aligned} \tag{11.4}$$

Now Eq. (11.4) can be rewritten as

$$\mathbf{v}(t) = \mathbf{V}e^{j\omega t} \tag{11.5}$$

where

$$\mathbf{V} = r_m\mathbf{I} \tag{11.6}$$

We obtain the phasor or *frequency-domain* dependent source from this equation. The corresponding circuit symbol is shown in Fig. 11.2b. Similar results for the remaining types of dependent sources are obtained the same way. Figure 11.3 summarizes the results. Notice that dependent sources work the same way in the frequency domain as they do in the time domain.

FIGURE 11.3
Frequency-domain versions
of dependent sources:
(a) voltage-controlled
voltage source;
(b) current-controlled
voltage source;
(c) voltage-controlled
current source;
(d) current-controlled
current source

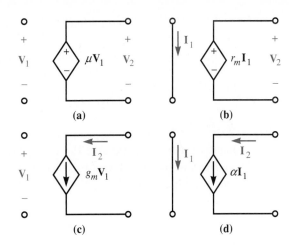

REMEMBER

Dependent sources work the same way in the frequency domain as they do in the time domain.

EXERCISE

5. Derive the entries in Fig. 11.3.

Frequency-Domain Terminal Equations for *R, L, C,* and Coupled Inductances

The next step in our development is to use the rotating phasors $\mathbf{I}e^{j\omega t}$ and $\mathbf{V}e^{j\omega t}$ to transform the time-domain equations of the LLTI network elements into frequency-domain equations. Part of this step was taken in Chapter 10, where we obtained the frequency-domain equations for *R, L,* and *C* shown in Fig. 11.4. Each

FIGURE 11.4
Time-domain and frequency-domain terminal equations for *R, L,* and *C*

Time domain	$i(t)$ R, $v(t) = Ri(t)$	$i(t)$ L, $v(t) = L\dfrac{d}{dt}i$	$i(t)$ C, $v(t) = \dfrac{1}{C}\displaystyle\int_{-\infty}^{t} i(\tau)d\tau$
Frequency domain	$\mathbf{I}$ R, $\mathbf{V} = R\mathbf{I}$	$\mathbf{I}$ $j\omega L$, $\mathbf{V} = j\omega L\mathbf{I}$	$\mathbf{I}$ $\dfrac{1}{j\omega C}$, $\mathbf{V} = \dfrac{1}{j\omega C}\mathbf{I}$

of the frequency-domain equations is an expression of Ohm's law for alternating current:

$$\mathbf{V} = \mathbf{Z}(j\omega)\mathbf{I} \tag{11.7}$$

or

$$\mathbf{I} = \mathbf{Y}(j\omega)\mathbf{V} \tag{11.8}$$

$\mathbf{Z}(j\omega)$ and $\mathbf{Y}(j\omega)$ are, respectively, the impedance and the admittance of the element, and $\mathbf{V} = V_m\underline{/\phi_\mathbf{V}}$ and $\mathbf{I} = I_m\underline{/\phi_\mathbf{I}}$ are voltage and current phasors. The real waveforms associated with Eqs. (11.7) and (11.8) are given by the projections of the rotating phasors $\mathbf{I}e^{j\omega t}$ and $\mathbf{V}e^{j\omega t}$ onto the real axis. It follows from Eq. (11.7) that if

$$i(t) = \mathcal{R}e\,\{\mathbf{I}e^{j\omega t}\} = I_m\cos(\omega t + \phi_\mathbf{I}) \tag{11.9}$$

then

$$
\begin{aligned}
v(t) &= \mathcal{R}e\,\{\mathbf{V}e^{j\omega t}\} \\
&= \mathcal{R}e\,\{\mathbf{Z}(j\omega)\mathbf{I}e^{j\omega t}\} \\
&= \underbrace{|\mathbf{Z}(j\omega)|I_m}_{V_m}\cos(\omega t + \underbrace{\underline{/\mathbf{Z}(j\omega)} + \phi_\mathbf{I}}_{\phi_\mathbf{V}})
\end{aligned} \tag{11.10}
$$

Similar results follow from Eq. (11.8).

Algebraic relationships can also be derived for the phasor voltages and currents of coupled inductances. Consider, for example, the terminal equations for the coupled inductances in Fig. 11.5a:

$$
\begin{aligned}
v_1(t) &= L_1\frac{d}{dt}i_1 + M\frac{d}{dt}i_2 \\
v_2(t) &= M\frac{d}{dt}i_1 + L_2\frac{d}{dt}i_2
\end{aligned} \tag{11.11a}
$$

By substituting $\mathbf{i}_1(t) = \mathbf{I}_1e^{j\omega t}$, $\mathbf{i}_2(t) = \mathbf{I}_2e^{j\omega t}$, $\mathbf{v}_1(t) = \mathbf{V}_1e^{j\omega t}$, and $\mathbf{v}_2(t) = \mathbf{V}_2e^{j\omega t}$ and differentiating as indicated, the above equations become

$$
\begin{aligned}
\mathbf{V}_1e^{j\omega t} &= j\omega L_1\mathbf{I}_1e^{j\omega t} + j\omega M\mathbf{I}_2e^{j\omega t} \\
\mathbf{V}_2e^{j\omega t} &= j\omega M\mathbf{I}_1e^{j\omega t} + j\omega L_2\mathbf{I}_2e^{j\omega t}
\end{aligned} \tag{11.11b}
$$

By multiplying both sides of each equation by $e^{-j\omega t}$, we obtain

$$
\begin{aligned}
\mathbf{V}_1 &= j\omega L_1\mathbf{I}_1 + j\omega M\mathbf{I}_2 \\
\mathbf{V}_2 &= j\omega M\mathbf{I}_1 + j\omega L_2\mathbf{I}_2
\end{aligned} \tag{11.11c}
$$

These are the frequency-domain equations for coupled inductances. The frequency-domain circuit model is shown in Fig. 11.5b. Equations (11.11c) can also be "solved" so that $\mathbf{I}_1$ and $\mathbf{I}_2$ are expressed as functions of $\mathbf{V}_1$ and $\mathbf{V}_2$. The result is shown in Fig. 11.5d. The corresponding integral equations for $i_1(t)$ and $i_2(t)$ are given in Fig. 11.5c. Algebraic expressions for mutually coupled inductances with three or more windings can be obtained in the same way. A term accounting for the mutual inductance between each pair of windings is simply included.

FIGURE 11.5
(a), (c) Time-domain and
(b), (d) frequency-domain
forms for coupled
inductances

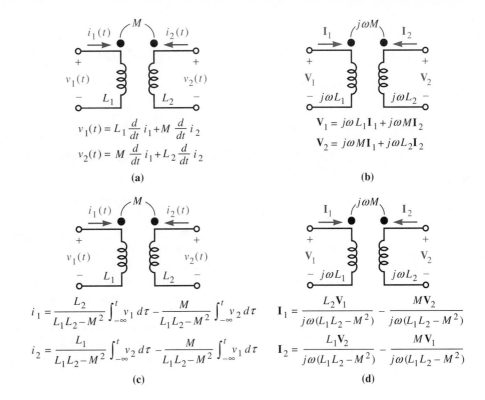

$$v_1(t) = L_1 \frac{d}{dt} i_1 + M \frac{d}{dt} i_2$$

$$v_2(t) = M \frac{d}{dt} i_1 + L_2 \frac{d}{dt} i_2$$

(a)

$$\mathbf{V}_1 = j\omega L_1 \mathbf{I}_1 + j\omega M \mathbf{I}_2$$

$$\mathbf{V}_2 = j\omega M \mathbf{I}_1 + j\omega L_2 \mathbf{I}_2$$

(b)

$$i_1 = \frac{L_2}{L_1 L_2 - M^2} \int_{-\infty}^{t} v_1 \, d\tau - \frac{M}{L_1 L_2 - M^2} \int_{-\infty}^{t} v_2 \, d\tau$$

$$i_2 = \frac{L_1}{L_1 L_2 - M^2} \int_{-\infty}^{t} v_2 \, d\tau - \frac{M}{L_1 L_2 - M^2} \int_{-\infty}^{t} v_1 \, d\tau$$

(c)

$$\mathbf{I}_1 = \frac{L_2 \mathbf{V}_1}{j\omega(L_1 L_2 - M^2)} - \frac{M \mathbf{V}_2}{j\omega(L_1 L_2 - M^2)}$$

$$\mathbf{I}_2 = \frac{L_1 \mathbf{V}_2}{j\omega(L_1 L_2 - M^2)} - \frac{M \mathbf{V}_1}{j\omega(L_1 L_2 - M^2)}$$

(d)

The Frequency-Domain Forms of KCL and KVL

We are now ready to take the final step in the development of the algebraic system for ac circuit analysis. *This step is to express Kirchhoff's current and voltage laws in terms of current and voltage phasors.* The results of this step are called *frequency-domain* or *phasor* versions of KCL and KVL. The reexpressed laws have exactly the same form as KCL and KVL for time-domain currents and voltages, with the exception that the currents and voltages are specified by phasors.

Recall that Kirchhoff's current law states that the algebraic sum of all the currents leaving any closed surface in a network is zero. That is,

$$\sum_{n=1}^{N} i_n(t) = 0 \tag{11.12}$$

where $i_n(t)$ is the nth current of the N currents leaving the closed surface at time t. In the sinusoidal steady state, we know that each of the currents $i_n(t)$, $1 \le n \le N$, is sinusoidal and has the same frequency. Therefore

$$i_n(t) = I_{mn} \cos(\omega t + \phi_n) \tag{11.13}$$

for $n = 1, 2, \ldots, N$, where the I_{mn}'s and ϕ_n's are the amplitudes and phases of the currents. Therefore, in the sinusoidal steady state, KCL, Eq. (11.12), becomes

$$\sum_{n=1}^{N} I_{mn} \cos(\omega t + \phi_n) = 0 \tag{11.14}$$

which is

$$\sum_{n=1}^{N} \mathscr{R}e\left\{\mathbf{I}_n e^{j\omega t}\right\} = 0 \tag{11.15}$$

or, equivalently,

$$\mathscr{R}e\left\{\sum_{n=1}^{N} \mathbf{I}_n e^{j\omega t}\right\} = 0 \tag{11.16}$$

where the phasor $\mathbf{I}_n$ has magnitude I_{mn} and angle ϕ_n. To satisfy Eq. (11.16), it is sufficient that

$$\sum_{n=1}^{N} \mathbf{I}_n e^{j\omega t} = 0 \tag{11.17}$$

Equation (11.17) is not a physical law, but *contains* the physical law described by Eq. (11.14). That is, if Eq. (11.17) is true, then Eq. (11.14) is true* because of the properties of the real part operator $\mathscr{R}e\left\{\cdot\right\}$.

A phasor equation can be obtained from Eq. (11.17) if we multiply both sides by $e^{-j\omega t}$. The result is the *frequency-domain form of KCL*.

Frequency-Domain Form of Kirchhoff's Current Law

$$\sum_{n=1}^{N} \mathbf{I}_n = 0 \tag{11.18}$$

where $\mathbf{I}_n$ is the nth current phasor of the N current phasors leaving the closed surface.

The frequency-domain form of KCL states that the sum of all the phasor currents exiting any closed surface is zero. Equation (11.18) is also referred to as the *phasor form of KCL*, or simply as KCL.

EXAMPLE 11.2

See Problem 11.5 in the PSpice manual.

Figure 11.6a illustrates an $N = 3$-branch node with known currents exiting via branches 1 and 2. The known currents are

$$i_1(t) = 5\cos(\omega t + 30°)$$

and

$$i_2(t) = 5\cos(\omega t + 150°)$$

Use the phasor form of KCL to determine $i_3(t)$.

---

* Observe that if Eq. (11.17) is true, then so is

$$\sum_{n=1}^{N} I_{mn}\sin(\omega t + \phi_n) = 0$$

This is the result of taking the imaginary part of Eq. (11.17). It follows that regardless of how one chooses to describe the sinusoidal branch currents [i.e., using cosines throughout as in Eq. (11.14) or using sines throughout as above], Eq. (11.17), and thus Eq. (11.18), will ensure that the physical KCL law is correctly accounted for.

FIGURE 11.6
(a) Time-domain and
(b) frequency-domain
currents

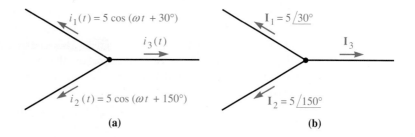

(a)   (b)

SOLUTION

We begin by representing the currents $i_1(t)$, $i_2(t)$, and $i_3(t)$, respectively, as the real parts of the complex exponentials $\mathbf{I}_1 e^{j\omega t}$, $\mathbf{I}_2 e^{j\omega t}$, and $\mathbf{I}_3 e^{j\omega t}$, where $\mathbf{I}_1 = 5\underline{/30°}$ and $\mathbf{I}_2 = 5\underline{/150°}$. The phasor form of KCL, Eq. (11.14), then indicates that, in Fig. 11.6b,

$$5\underline{/30°} + 5\underline{/150°} + \mathbf{I}_3 = 0$$

Therefore

$$\begin{aligned}
\mathbf{I}_3 &= -5\underline{/30°} - 5\underline{/150°} \\
&= -(4.33 + j2.5) - (-4.33 + j2.5) \\
&= -j5 \\
&= 5\underline{/-90°}
\end{aligned}$$

The phasors are shown in Fig. 11.7. The time-domain current represented by $\mathbf{I}_3$ is

$$\begin{aligned}
i_3(t) &= \mathcal{R}e\ \{\mathbf{I}_3 e^{j\omega t}\} \\
&= 5\cos(\omega t - 90°) \quad \blacksquare
\end{aligned}$$

FIGURE 11.7
The current phasors sum to zero (KCL)

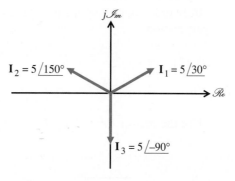

We emphasize that KCL indicates that the sum of all current *phasors* leaving any closed surface is zero. It does *not* follow that the peak or rms values of current sum to zero. For instance, in the above example, phasors $5\underline{/30°}$, $5\underline{/150°}$, and $5\underline{/-90°}$ sum to zero, but the peak currents 5, 5, and 5 do not.

Let us now consider Kirchhoff's voltage law. Kirchhoff's voltage law states that the algebraic sum of all voltage drops taken around any closed path in a network is zero. That is,

$$\sum_{n=1}^{N} v_n(t) = 0 \qquad\qquad (11.19)$$

where $v_n(t)$ is the voltage drop, taken in the direction of the path along the nth segment of the N segments in the closed path. In the sinusoidal steady state, each of the voltages $v_n(t)$, $1 \leq n \leq N$, will be sinusoidal. Following the same reasoning that led to Eq. (11.16), we can represent the sinusoidal voltages as the real parts of complex exponential functions. In terms of these complex exponentials, KVL becomes

$$\sum_{n=1}^{N} \mathbf{V}_n e^{j\omega t} = 0 \qquad (11.20)$$

where the phasor $\mathbf{V}_n$ represents the amplitude and the phase of the sinusoidal voltage $v_n(t)$. Equation (11.20) can be multiplied by $e^{-j\omega t}$, yielding the *frequency-domain form of KVL*.

Frequency-Domain Form of Kirchhoff's Voltage Law

$$\sum_{n=1}^{N} \mathbf{V}_n = 0 \qquad (11.21)$$

where $\mathbf{V}_n$ is the phasor voltage drop, taken in the direction of the path, along the nth segment of the N segments in the closed path.

The frequency-domain form of KVL states that the sum of all the phasor voltage drops around a closed path is zero. Equation (11.21) is also referred to as the *phasor form of KVL*, or simply as KVL.

EXAMPLE 11.3

Figure 11.8a illustrates an $N = 4$-branch mesh with known voltage drops across three of its four components. These known voltages are given by the time-domain expressions

$$v_1(t) = \cos \omega t$$

$$v_2(t) = -2 \cos (\omega t + 45°)$$

$$v_3(t) = -3 \sin \omega t$$

Use the phasor form of KVL to determine $v_4(t)$.

FIGURE 11.8
(*a*) Time-domain and (*b*) frequency-domain voltage drops

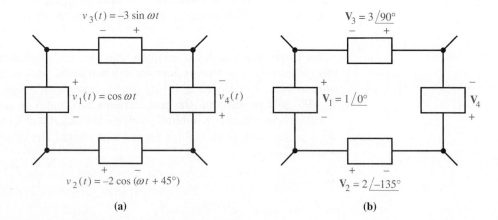

(a)   (b)

■ **SOLUTION**

We represent the voltages $v_1(t)$, $v_2(t)$, $v_3(t)$, and $v_4(t)$, respectively, as the real parts of the complex sinusoids $\mathbf{V}_1 e^{j\omega t}$, $\mathbf{V}_2 e^{j\omega t}$, $\mathbf{V}_3 e^{j\omega t}$, and $\mathbf{V}_4 e^{j\omega t}$, where $\mathbf{V}_1 = 1\underline{/0°}$, $\mathbf{V}_2 = 2\underline{/-135°}$, and $\mathbf{V}_3 = 3\underline{/+90°}$. The phasor form of KVL, Eq. (11.21), tells us that

$$1\underline{/0°} + 2\underline{/-135°} + 3\underline{/+90°} + \mathbf{V}_4 = 0$$

or

$$\mathbf{V}_4 = -1\underline{/0°} - 2\underline{/-135°} - 3\underline{/+90°}$$

and therefore

$$\mathbf{V}_4 = -(1 + j0) - (-\sqrt{2} - j\sqrt{2}) - (0 + j3)$$
$$= 0.4142 - j1.5858$$
$$= 1.6389\underline{/-75.36°}$$

The phasors are shown in Fig. 11.9. The time-domain current represented by $\mathbf{V}_4$ is

$$v_4(t) = \mathcal{R}e\ \{\mathbf{V}_4 e^{j\omega t}\}$$
$$= 1.6389 \cos(\omega t - 75.36°) \quad ■$$

FIGURE 11.9
The voltage phasors sum to zero (KVL)

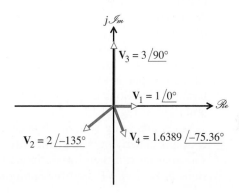

Notice that KVL indicates that the sum of the *phasor* voltage drops around a closed path is zero. It does *not* indicate that the peak or rms values of voltage sum to zero. After all, peak and rms values are always nonnegative, so the *only* way they *could* sum to zero would be for each and every peak and rms value to be zero.

REMEMBER

KCL and KVL work the same way in the frequency domain as they do in the time domain.

EXERCISE

6. Three phasor currents satisfy $\mathbf{I}_1 = \mathbf{I}_2 + \mathbf{I}_3$. Express I_{m1} in terms of I_{m2}, I_{m3}, $\underline{/I_2}$, and $\underline{/I_3}$. (*Hint:* Draw the phasors $\mathbf{I}_1$, $\mathbf{I}_2$, and $\mathbf{I}_3$ as vectors in the complex plane and use trigonometry.)

Relationship Between Time- and Frequency-Domain Circuits

We have noticed that the ac version of Ohm's law, $\mathbf{V} = \mathbf{ZI}$, has the same mathematical form as Ohm's law, $V = RI$. We also know that the frequency-domain versions of KCL and KVL have the same mathematical form as the time-domain versions of KCL and KVL. The profound implication of these facts is that we can use current and voltage phasors to analyze ac networks in exactly the same way that we use dc current and voltage variables to analyze networks composed of resistances and dc sources. *This equivalence of ac and dc circuit analysis applies to the sinusoidal steady-state response of all LLTI networks.*

A detailed summary of the concepts underlying frequency-domain circuit analysis is given in Fig. 11.10. Figure 11.10a illustrates the time-domain representation of a simple circuit that has two sinusoidal sources with the same frequency ω that were connected a long time ago (mathematically, at $t = -\infty$). Because this is a stable LLTI network, the complementary responses have decayed to zero, and all the voltages and currents are now sinusoidal with the same frequency as the sources. Thus, the circuit is operating in the sinusoidal steady state with frequency ω. The sinusoidal functions are indicated in Fig. 11.10b, in which the amplitudes and phases are constants to be determined. Note that every sinusoid has been written in the same functional form (which we choose to be a cosine). Figure 11.10c shows that the cosine functions are the real parts of the complex exponential functions. Figure 11.10d shows the same circuit with the voltages and currents described by complex exponential functions. There is no physical meaning to these complex voltages and currents. We can progress from Fig. 11.10c to 11.10d because of *mathematical* properties of the operators $\mathscr{R}e\{\cdot\}$, $\mathscr{A}(p)$, and $\mathscr{B}(p)$ described earlier. Figure 11.10e is identical to Fig. 11.10d except that the $e^{j\omega t}$'s have been omitted. This corresponds to the cancellation of the $e^{j\omega t}$'s in all KCL, KVL, and element terminal equations in the network. Figure 11.10e is the frequency-domain representation of Fig. 11.10a.

When analyzing a circuit like that in Fig. 11.10a, you should understand the concepts underlying Fig. 11.10b through d, but draw only the result in Fig. 11.10e. The analysis can then be continued by application of the frequency-domain versions of KCL, KVL, and the component terminal equations.

EXAMPLE 11.4

See Problem 11.1 in the PSpice manual.

Use the frequency-domain versions of KCL, KVL, and the element terminal equations to determine $i_L(t)$ in the circuit of Fig. 11.10a.

■ SOLUTION

Application of KCL at the top two nodes in the network of Fig. 11.10e yields

$$\mathbf{I}_L = \mathbf{I}_1 \qquad \mathbf{I}_R + \mathbf{I}_C = \mathbf{I}_1 + \mathbf{I}_s$$

Application of KVL around the three meshes yields

$$\mathbf{V}_R + \mathbf{V}_L = \mathbf{V}_s \qquad \mathbf{V}_R = \mathbf{V}_C \qquad \mathbf{V}_C = \mathbf{V}_o$$

FIGURE 11.10 Summary of time-frequency transformations: (*a*) circuit with two sinusoidal sources with the same frequency ω; (*b*) sinusoidal functions in which the amplitudes and phases are constants to be determined; (*c*) cosine functions are the real parts of the complex exponential functions; (*d*) voltages and currents described by complex exponential functions; (*e*) frequency-domain representation of (*a*)

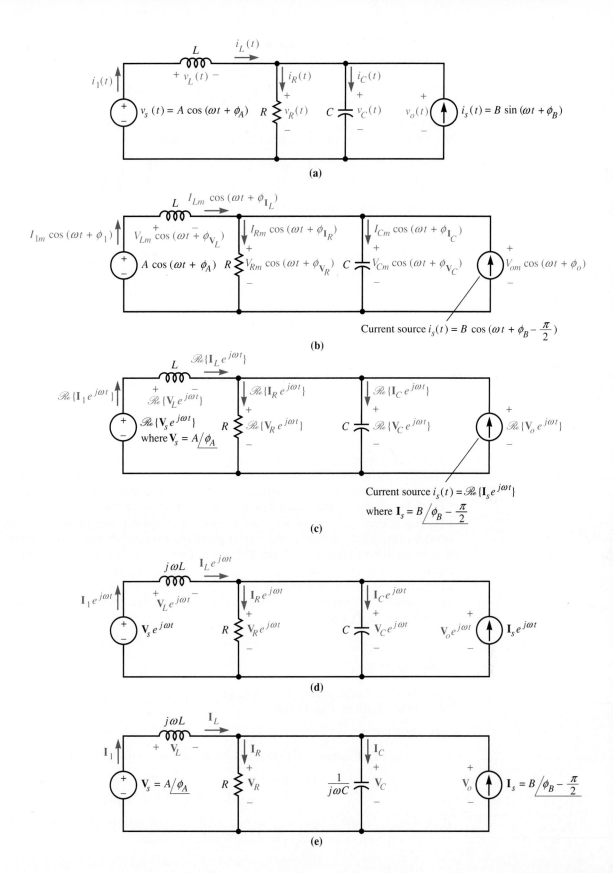

(a)

(b)

Current source $i_s(t) = B \cos(\omega t + \phi_B - \frac{\pi}{2})$

(c)

Current source $i_s(t) = \mathcal{R}e\{\mathbf{I}_s e^{j\omega t}\}$

where $\mathbf{I}_s = B \big/ \underline{\phi_B - \frac{\pi}{2}}$

(d)

(e)

where $\mathbf{V}_o$ is the (phasor) voltage across the current source. The element terminal equations yield three more equations,

$$\mathbf{V}_L = j\omega L \mathbf{I}_L \qquad \mathbf{V}_C = \frac{1}{j\omega C}\mathbf{I}_C \qquad \mathbf{V}_R = \mathbf{I}_R R$$

for a total of eight equations in the eight unknowns $\mathbf{I}_R$, $\mathbf{I}_C$, $\mathbf{I}_L$, $\mathbf{I}_1$, $\mathbf{V}_R$, $\mathbf{V}_C$, $\mathbf{V}_L$, and $\mathbf{V}_o$. These equations may be solved for $\mathbf{I}_L$. The result is

$$\mathbf{I}_L = \frac{(1 + j\omega RC)\mathbf{V}_s}{R - \omega^2 RLC + j\omega L} - \frac{R\mathbf{I}_s}{R - \omega^2 RLC + j\omega L}$$

The corresponding time-domain solution is

$$i_L(t) = \mathscr{R}e\left\{\mathbf{I}_L e^{j\omega t}\right\} = I_{Lm}\cos(\omega t + \phi_{IL})$$

As a numerical example, assume that $\omega = 10^3$ rad/s, $R = 1\ \text{k}\Omega$, $C = 1\ \mu\text{F}$, $L = 0.5\ \text{H}$, $A = 30\ \text{V}$, $\phi_A = -25°$, $B = 10\ \text{mA}$, and $\phi_B = 15°$. Then

$$\mathbf{I}_L = \frac{(1 + j)(30\underline{/-25°})}{500 + j500} - \frac{10\underline{/15°} - 90°}{500 + j500}$$

$$= \frac{37.279 + j24.1699}{500 + j500}$$

$$= 0.06283\underline{/-12.04°}\ \text{A}$$

and
$$i_L(t) = 62.83\cos(1000t - 12.04°)\ \text{mA} \quad\blacksquare$$

Example 11.4 was solved by direct application of KCL, KVL, and the element terminal equations. As in the time domain, this method is fundamental but often unnecessarily long. We will describe much faster methods later.

Before conclusion of the present topic, it is worthwhile to notice that the transformation between time and frequency domains works both ways. Notice from Eq. (10.1) and (10.14) that the differential operator $p = d/dt$ enters a time-domain equation in exactly the same way that $j\omega$ enters a frequency-domain equation. In other words, differentiation in the time domain *always* corresponds to multiplication by $j\omega$ in the frequency domain. Therefore, if we start with a frequency-domain equation that has the form of Eq. (10.14), the corresponding time-domain equation can be derived simply by replacing $j\omega$ with $p = d/dt$ and phasors $\mathbf{V}$ or $\mathbf{I}$ with their time-domain counterparts $v(t)$ and $i(t)$. This process often provides an easy way to derive an LLTI equation.

EXAMPLE 11.5

Find the differential equation relating $i_L(t)$ to the independent sources, $v_s(t)$ and $i_s(t)$, in the circuit of Fig. 11.10a.

■ SOLUTION

As we saw in Example 11.4, phasor $\mathbf{I}_L$ is related to phasors $\mathbf{V}_s$ and $\mathbf{I}_s$ by

$$\mathbf{I}_L = \frac{(1 + j\omega RC)\mathbf{V}_s}{R - \omega^2 RLC + j\omega L} - \frac{R\mathbf{I}_s}{R - \omega^2 RLC + j\omega L}$$

If we substitute d/dt for $j\omega$, we obtain a mathematical jumble. However, by first multiplying both sides by $R - \omega^2 RLC + j\omega L$, we obtain

$$[R + (j\omega)^2 RLC + j\omega L]\mathbf{I}_L = (1 + j\omega RC)\mathbf{V}_s - R\mathbf{I}_s$$

[We used the identity $-\omega^2 = (j\omega)^2$.] The time-domain equation corresponding to this frequency-domain equation is

$$(R + RLCp^2 + Lp)i_L(t) = (1 + RCp)v_s(t) - Ri_s(t)$$

or equivalently,

$$RLC \frac{d^2}{dt^2} i_L + L \frac{d}{dt} i_L + Ri_L = RC \frac{d}{dt} v_s + v_s - Ri_s$$

The validity of this equation can be checked if we substitute in the complex exponentials $i_L = \mathbf{I}_L e^{j\omega t}$, $v_s = \mathbf{V}_s e^{j\omega t}$, and $i_s = \mathbf{I}_s e^{j\omega t}$ and simplify. The result will be the original frequency-domain equation relating the phasors $\mathbf{I}_L$, $\mathbf{V}_s$, and $\mathbf{I}_s$. Of course, the validity can also be checked if we use the time-domain methods described in earlier chapters. ■

> The axioms of network theory, KCL and KVL, have been shown to hold for phasors. The introduction of impedance has extended Ohm's law, $v(t) = Ri(t)$, to the phasor form $\mathbf{V} = \mathbf{ZI}$. It follows that any relationship in the time domain that is justified by KCL, KVL, and Ohm's law has an equivalent relationship in the frequency domain. *Therefore all the time-domain LLTI network analysis techniques developed in earlier chapters can be reformulated in the frequency domain.*

In the following sections, we will accomplish this reformulation. This should serve as a review of many of the circuit analysis techniques we developed earlier.

REMEMBER

Every frequency-domain concept and procedure we will encounter in this chapter is associated with a corresponding time-domain concept or procedure encountered earlier in this book.

EXERCISES

7. Explain in your own words the relationship between Fig. 11.10e and 11.10a.
8. The voltage transfer function of a certain circuit is given by

$$\mathbf{H}(j\omega) = \frac{1}{1 - \omega^2 + j\omega}$$

What is the differential equation relating the output voltage $v_o(t)$ to the input voltage $v_s(t)$?

11.2 Simple Circuits

Let us now apply the method of ac circuit analysis to some elementary circuits. In this section we consider only circuits that do not contain dependent sources. Ac circuits that do not contain dependent sources can often be analyzed by inspection. An ability to do this depends on knowledge of the simple circuits shown in

Figs. 11.11 through 11.14. These circuits have already been drawn in the frequency domain, directing attention to the sinusoidal steady-state response of the corresponding time-domain circuits. The **Z**s and **Y**s represent arbitrary impedances and admittances.

FIGURE 11.11
Admittances in parallel

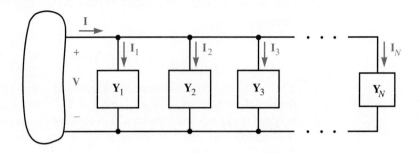

Parallel Circuit

Consider the parallel circuit of Fig. 11.11. The voltage drop from the top to the bottom of each element of Fig. 11.11 is $+\mathbf{V}$ (by the phasor form of KVL). The current phasors $\mathbf{I}_n$ are related to $\mathbf{V}$ by $\mathbf{I}_n = \mathbf{Y}_n\mathbf{V}$, where $\mathbf{Y}_n$ is the admittance of the nth element. Application of the phasor form of KCL to the top node gives

$$\mathbf{I} = \mathbf{Y}_1\mathbf{V} + \mathbf{Y}_2\mathbf{V} + \cdots + \mathbf{Y}_N\mathbf{V} \tag{11.22}$$

or

$$\mathbf{I} = \mathbf{Y}_p\mathbf{V} \tag{11.23}$$

where $\mathbf{Y}_p$ is the equivalent admittance:

$$\mathbf{Y}_p = \mathbf{Y}_1 + \mathbf{Y}_2 + \cdots + \mathbf{Y}_N \tag{11.24}$$

This shows that *admittances that are in parallel add*. Expressed in terms of impedances $\mathbf{Z}_p = 1/\mathbf{Y}_p$ and $\mathbf{Z}_n = 1/\mathbf{Y}_n$, where $n = 1, 2, \ldots, N$, Eq. (11.24) becomes

$$\frac{1}{\mathbf{Z}_p} = \frac{1}{\mathbf{Z}_1} + \frac{1}{\mathbf{Z}_2} + \cdots + \frac{1}{\mathbf{Z}_N} \tag{11.25}$$

or

$$\mathbf{Z}_p = \cfrac{1}{\cfrac{1}{\mathbf{Z}_1} + \cfrac{1}{\mathbf{Z}_2} + \cdots + \cfrac{1}{\mathbf{Z}_N}} \tag{11.26}$$

For the special case $N = 2$, Eq. (11.26) becomes

$$\mathbf{Z}_p = \cfrac{1}{\cfrac{1}{\mathbf{Z}_1} + \cfrac{1}{\mathbf{Z}_2}} = \mathbf{Z}_1 \,\|\, \mathbf{Z}_2 \tag{11.27}$$

where the symbol "$\|$" should be read "in parallel with." This symbol provides a convenient notation for the mathematical operation given in Eq. (11.27). It is easy to verify that $\|$ is both commutative

$$\mathbf{Z}_1 \,\|\, \mathbf{Z}_2 = \mathbf{Z}_2 \,\|\, \mathbf{Z}_1 \tag{11.28}$$

and associative,

$$(\mathbf{Z}_1 \| \mathbf{Z}_2) \| \mathbf{Z}_3 = \mathbf{Z}_1 \| (\mathbf{Z}_2 \| \mathbf{Z}_3) \tag{11.29}$$

Since $\|$ is associative there is no need to use parentheses for the parallel combination of three or more impedances. For instance, the parallel combination in Eq. (11.29) can be written without ambiguity as $\mathbf{Z}_1 \| \mathbf{Z}_2 \| \mathbf{Z}_3$.

Series Circuit

Figure 11.12 shows a series circuit that is the dual of the network of Fig. 11.11.

FIGURE 11.12
Impedances in series

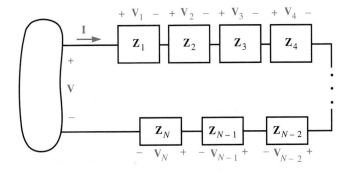

The current through each impedance is $\mathbf{I}$ (by the phasor form of KCL). The voltage phasors $\mathbf{V}_n$ are related to $\mathbf{I}$ by $\mathbf{V}_n = \mathbf{Z}_n\mathbf{I}$, where $\mathbf{Z}_n$ is the impedance of the nth element. Application of KVL around the path gives

$$\mathbf{V} = \mathbf{Z}_1\mathbf{I} + \mathbf{Z}_2\mathbf{I} + \cdots + \mathbf{Z}_N\mathbf{I} \tag{11.30}$$

or
$$\mathbf{V} = \mathbf{Z}_s\mathbf{I} \tag{11.31}$$

where $\mathbf{Z}_s$ is the equivalent impedance:

$$\mathbf{Z}_s = \mathbf{Z}_1 + \mathbf{Z}_2 + \cdots + \mathbf{Z}_N \tag{11.32}$$

Therefore *impedances that are in series add.*

Current and Voltage Dividers

The network of Fig. 11.13 is an ac current divider. The network of Fig. 11.14 is an ac voltage divider.

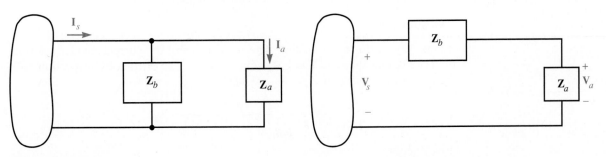

FIGURE 11.13   Current divider   **FIGURE 11.14**   Voltage divider

It is left as an exercise (Exercise 10) for you to show that for the current divider of Fig. 11.13,

$$\mathbf{I}_a = \frac{\mathbf{Z}_b}{\mathbf{Z}_a + \mathbf{Z}_b}\,\mathbf{I}_s = \frac{\mathbf{Y}_a}{\mathbf{Y}_a + \mathbf{Y}_b}\,\mathbf{I}_s \qquad (11.33)$$

where
$$\mathbf{Y}_a = 1/\mathbf{Z}_a$$

and
$$\mathbf{Y}_b = 1/\mathbf{Z}_b$$

Similarly, for the voltage divider of Fig. 11.14,

$$\mathbf{V}_a = \frac{\mathbf{Z}_a}{\mathbf{Z}_a + \mathbf{Z}_b}\,\mathbf{V}_s = \frac{\mathbf{Y}_b}{\mathbf{Y}_a + \mathbf{Y}_b}\,\mathbf{V}_s \qquad (11.34)$$

Circuit Analysis by Inspection

See Problem 11.2 in the PSpice manual.

The preceding results for simple circuits can often be combined to provide immediate answers to seemingly complicated network problems. The method is called *circuit analysis by inspection*.

EXAMPLE 11.6

Find **V** by inspection of the network of Fig. 11.15.

FIGURE 11.15
Circuit of Example 11.6

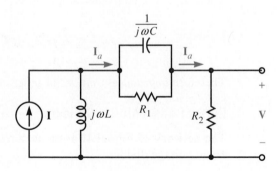

■ SOLUTION

One way to do this is to first determine $\mathbf{I}_a$ using the current-divider relation, Eq. (11.33), and then set $\mathbf{V} = \mathbf{I}_a R_2$. The result can be written by inspection with the use of $\mathbf{Z}_b = j\omega L$ and $\mathbf{Z}_a = R_1 \parallel (1/j\omega C) + R_2$:

$$\mathbf{V} = \frac{j\omega L}{j\omega L + R_1 \parallel \dfrac{1}{j\omega C} + R_2}\,R_2\mathbf{I}$$

which becomes, after some routine algebraic manipulations,

$$V = \frac{-\omega^2 + j\dfrac{\omega}{R_1 C}}{-\omega^2 + \left(\dfrac{R_1 + R_2}{L}\right)\left(\dfrac{1}{R_1 C}\right) + j\omega\left(\dfrac{1}{R_1 C} + \dfrac{R_2}{L}\right)} R_2 I \quad \blacksquare$$

Many ac circuits cannot be readily analyzed by inspection. The systematic circuit analysis techniques described in the following two sections provide a means to analyze them.

Admittances in parallel add as in Eq. (11.24). Impedances in series add as in Eq. (11.32). Currents and voltages divide as in Eqs. (11.33) and (11.34), respectively.

EXERCISES

9. Verify Eqs. (11.28) and (11.29).

10. Derive Eqs. (11.33) and (11.34).

11. Inspect the illustration below to find the quantities requested: (a) V/I, (b) V_o/V, (c) I_1/V, (d) I_2/V, (e) I_1/I, and (f) I_2/I.

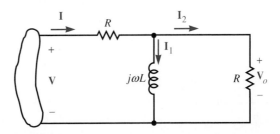

12. Evaluate the transfer function V/I of Example 11.6 in the limit $\omega \to \infty$. Explain why your result is correct by inspection of the circuit.

11.3   AC Node-Voltage Analysis

The technique of node-voltage analysis provides a way to systematically analyze any ac circuit.

Setting Up the Node-Voltage Equations

The procedure to set up the node-voltage equations in the frequency domain is basically the same as that in the time domain. Time-domain node-voltage analysis was described in Section 4.2. It is a good idea to review this material—particularly the shortcut method on page 135—before continuing with the following frequency-domain examples.

EXAMPLE 11.7

Write a set of node-voltage equations for the network shown in Fig. 11.16.

FIGURE 11.16
Node-voltage example

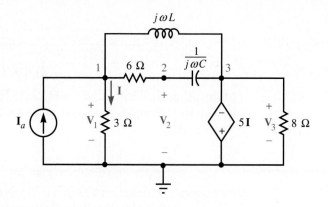

■ SOLUTION

For a surface enclosing node 1, KCL gives

$$-\mathbf{I}_a + \frac{1}{3}\mathbf{V}_1 + \frac{1}{6}(\mathbf{V}_1 - \mathbf{V}_2) + \frac{1}{j\omega L}(\mathbf{V}_1 - \mathbf{V}_3) = 0$$

or

$$\left(\frac{1}{2} + \frac{1}{j\omega L}\right)\mathbf{V}_1 - \frac{1}{6}\mathbf{V}_2 - \frac{1}{j\omega L}\mathbf{V}_3 = \mathbf{I}_a$$

For a surface enclosing node 2, KCL yields

$$\frac{1}{6}(\mathbf{V}_2 - \mathbf{V}_1) + j\omega C(\mathbf{V}_2 - \mathbf{V}_3) = 0$$

or

$$-\frac{1}{6}\mathbf{V}_1 + \left(j\omega C + \frac{1}{6}\right)\mathbf{V}_2 - j\omega C\mathbf{V}_3 = 0$$

KVL gives the third node-voltage equation:

$$\mathbf{V}_3 = -5\mathbf{I} = -5\left(\frac{1}{3}\mathbf{V}_1\right) = -\frac{5}{3}\mathbf{V}_1$$

The three node-voltage equations can be written in the matrix form

$$\begin{bmatrix} \frac{1}{2} + \frac{1}{j\omega L} & -\frac{1}{6} & -\frac{1}{j\omega L} \\ -\frac{1}{6} & j\omega C + \frac{1}{6} & -j\omega C \\ \frac{5}{3} & 0 & 1 \end{bmatrix}\begin{bmatrix} \mathbf{V}_1 \\ \mathbf{V}_2 \\ \mathbf{V}_3 \end{bmatrix} = \begin{bmatrix} \mathbf{I}_a \\ 0 \\ 0 \end{bmatrix}$$

where the left-hand matrix is the *node transformation matrix*. ■

EXAMPLE 11.8

Write the node-voltage equations for the network shown in Fig. 11.17.

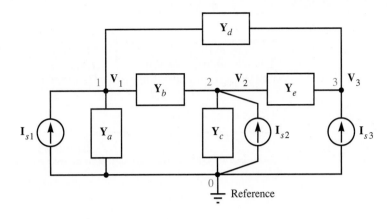

FIGURE 11.17
All node voltages are measured with respect to the reference

■ SOLUTION

Because this network contains only current sources, all the node equations will be KCL equations. These equations are

Node 1 $\mathbf{Y}_a\mathbf{V}_1 + \mathbf{Y}_b(\mathbf{V}_1 - \mathbf{V}_2) + \mathbf{Y}_d(\mathbf{V}_1 - \mathbf{V}_3) - \mathbf{I}_{s1} = 0$

Node 2 $\mathbf{Y}_b(\mathbf{V}_2 - \mathbf{V}_1) + \mathbf{Y}_c\mathbf{V}_2 + \mathbf{Y}_e(\mathbf{V}_2 - \mathbf{V}_3) - \mathbf{I}_{s2} = 0$

Node 3 $\mathbf{Y}_e(\mathbf{V}_3 - \mathbf{V}_2) + \mathbf{Y}_d(\mathbf{V}_3 - \mathbf{V}_1) - \mathbf{I}_{s3} = 0$

These can be written in matrix form as

$$\begin{bmatrix} \mathbf{Y}_a + \mathbf{Y}_b + \mathbf{Y}_d & -\mathbf{Y}_b & -\mathbf{Y}_d \\ -\mathbf{Y}_b & \mathbf{Y}_b + \mathbf{Y}_c + \mathbf{Y}_e & -\mathbf{Y}_e \\ -\mathbf{Y}_d & -\mathbf{Y}_e & \mathbf{Y}_e + \mathbf{Y}_d \end{bmatrix} \begin{bmatrix} \mathbf{V}_1 \\ \mathbf{V}_2 \\ \mathbf{V}_3 \end{bmatrix} = \begin{bmatrix} \mathbf{I}_{s1} \\ \mathbf{I}_{s2} \\ \mathbf{I}_{s3} \end{bmatrix}$$

Observe that the node transformation matrix on the left-hand side is symmetric. The node transformation matrix of a network containing no dependent sources can always be arranged to be symmetric. ■

Now we will describe a convenient interpretation of the KCL equations for networks composed of only admittances and independent current sources, as illustrated by Example 11.8. Consider the KCL equation for node 1. Observe that the coefficient of $\mathbf{V}_1$ is the sum of the admittances directly connecting node 1 to the other nodes in the network, the coefficient of $\mathbf{V}_2$ is the negative of the admittance directly connecting node 1 to node 2, and the coefficient of $\mathbf{V}_3$ is the negative of the admittance directly connecting node 1 to node 3. Now look at the KCL equation for node 2. The coefficient of $\mathbf{V}_1$ is the negative of the admittance directly connecting node 2 to node 1, the coefficient of $\mathbf{V}_2$ is the sum of the admittances directly connecting node 2 to other nodes in the network, and the coefficient of $\mathbf{V}_3$ is the negative of the admittance directly connecting node 3 to node 1. Finally, consider the KCL equation for node 3. The coefficients of $\mathbf{V}_1$, $\mathbf{V}_2$, and $\mathbf{V}_3$ in this equation have a similar interpretation. (What is it?) Observe further that the right-hand sides of the equations are simply the independent phasor current source inputs to the nodes for which KCL is written.

This interpretation can be generalized to an arbitrary $(N + 1)$-node network composed of only admittances and independent current sources. For this purpose we make the following definitions.

The **self-admittance** at node k, $\mathbf{Y}_{kk}$, is the sum of the admittances of the branches that connect node k directly to the other nodes of the network.

The **mutual admittance** between nodes k and l, $\mathbf{Y}_{kl}$, is the negative of the sum of the admittances of the branches that connect node k directly to node l.

The node-voltage equations for such a network are all KCL equations and have the matrix form

$$\begin{bmatrix} \mathbf{Y}_{11} & \mathbf{Y}_{12} & \cdots & \mathbf{Y}_{1N} \\ \mathbf{Y}_{21} & \mathbf{Y}_{22} & \cdots & \mathbf{Y}_{2N} \\ \vdots & & & \\ \mathbf{Y}_{N1} & \mathbf{Y}_{N2} & \cdots & \mathbf{Y}_{NN} \end{bmatrix} \begin{bmatrix} \mathbf{V}_1 \\ \mathbf{V}_2 \\ \vdots \\ \mathbf{V}_N \end{bmatrix} = \begin{bmatrix} \mathbf{I}_{s1} \\ \mathbf{I}_{s2} \\ \vdots \\ \mathbf{I}_{sN} \end{bmatrix} \qquad (11.35)$$

where $\mathbf{I}_{sk}$ is the sum of all source currents entering node k. The square matrix on the left-hand side is the network's *admittance matrix*. By definition, $\mathbf{Y}_{kl} = \mathbf{Y}_{lk}$, so that an admittance matrix is symmetric. An example of an admittance matrix was given in Example 11.8. Observe that the admittance matrix in this equation could have been written directly by inspection of the circuit in Fig. 11.17.

Let us now return to node-voltage analysis of more general circuits.

EXAMPLE 11.9

Set up the node-voltage equations for the ideal op-amp circuit of Fig. 11.18, where $\mathbf{Z}_1$, $\mathbf{Z}_2$, $\mathbf{Z}_a$, and $\mathbf{Z}_b$ are arbitrary impedances. Solve for $\mathbf{V}_o$.

FIGURE 11.18
Op-amp circuit. All node voltages are measured with respect to the reference

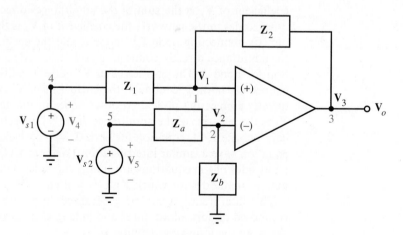

By inspection, $\mathbf{V}_4 = \mathbf{V}_{s1}$ and $\mathbf{V}_5 = \mathbf{V}_{s2}$. Using these results, we can apply KCL to a surface enclosing node 1:

$$\frac{1}{\mathbf{Z}_1}(\mathbf{V}_1 - \mathbf{V}_{s1}) + \frac{1}{\mathbf{Z}_2}(\mathbf{V}_1 - \mathbf{V}_3) = 0$$

Application of KCL to a surface enclosing node 2 yields

$$\frac{1}{\mathbf{Z}_a}(\mathbf{V}_1 - \mathbf{V}_{s2}) + \frac{1}{\mathbf{Z}_b}\mathbf{V}_1 = 0$$

where we have set $\mathbf{V}_2 = \mathbf{V}_1$ because the op amp is ideal. Solving for $\mathbf{V}_3 = \mathbf{V}_o$ yields

$$\mathbf{V}_o = -\frac{\mathbf{Z}_2}{\mathbf{Z}_1}\mathbf{V}_{s1} + \left(\frac{\mathbf{Z}_b}{\mathbf{Z}_a + \mathbf{Z}_b}\right)\left(\frac{\mathbf{Z}_1 + \mathbf{Z}_2}{\mathbf{Z}_1}\right)\mathbf{V}_{s2}$$

This example should be compared with Example 6.5. ■

We have shown how to set up the node-voltage equations. The next step is to solve them. There are several techniques to obtain numerical solutions, one of which is Cramer's rule. Other techniques (used in computers) are computationally more efficient. In the next section, we will use Cramer's rule to obtain the *theoretical* solution. This derivation will provide the start for our later investigation of Thévenin's theorem.

REMEMBER

Node-voltage analysis works the same way in the frequency domain as it does in the time domain.

EXERCISES

In Exercises 13 and 14, all elements are specified in terms of their admittances. The units are amperes, volts, and siemens.

13. Set up the node-voltage equations for the circuit shown.

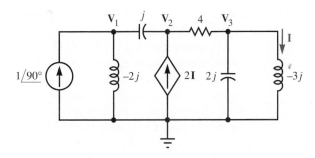

14. Set up and solve the node-voltage equations for the following circuit. Label the nodes as shown.

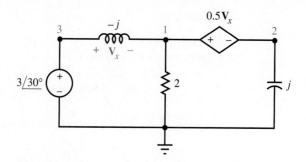

General Form of the Node-Voltage Equations and Solution by Cramer's Rule

Node-voltage analysis of an $(N + 1)$-node network with use of the procedure on page 135 yields a system of N equations:

$$
\left.
\begin{array}{c}
\text{KCL} \\
\text{equations}
\end{array}
\left\{
\begin{array}{c}
\\ \\
\end{array}
\right.
\begin{array}{c}
\text{KVL} \\
\text{equations}
\end{array}
\left\{
\begin{array}{c}
\\ \\
\end{array}
\right.
\right.
\begin{bmatrix}
T_{11} & T_{12} & \cdots & T_{1N} \\
\vdots & \vdots & & \vdots \\
T_{M1} & T_{M2} & & T_{MN} \\
\vdots & \vdots & & \vdots \\
T_{N1} & T_{N2} & \cdots & T_{NN}
\end{bmatrix}
\begin{bmatrix}
V_1 \\
\vdots \\
V_M \\
\vdots \\
V_N
\end{bmatrix}
=
\begin{bmatrix}
I_{s1} \\
\vdots \\
V_{sM} \\
\vdots \\
V_{sN}
\end{bmatrix}
\left.
\begin{array}{c}
\\ \\
\end{array}
\right\}
\begin{array}{c}
\text{Due to} \\
\text{current sources}
\end{array}
\qquad (11.36)
$$

The square matrix is the node transformation matrix. The letters T_{ij} are used to denote its elements, which, of course, depend on the circuit.

The form of the node-voltage equations is simplified considerably when the circuit is made up of passive elements and independent current sources only. Such a circuit was shown in Example 11.8, where it was noted that the node-voltage equations could be set up by inspection with the use of the concepts of self- and mutual admittance.

The analytical expression for the phasor node voltages is given by Cramer's rule:

$$
V_n = \frac{\Delta_{1n}}{\Delta} I_{s1} + \frac{\Delta_{2n}}{\Delta} I_{s2} + \cdots + \frac{\Delta_{(M-1)n}}{\Delta} I_{s(M-1)} + \frac{\Delta_{Mn}}{\Delta} V_{sM} + \cdots + \frac{\Delta_{Nn}}{\Delta} V_{sN}
$$

$$(11.37)$$

for $n = 1, 2, \ldots, N$, where Δ is the determinant of the node transformation matrix, and Δ_{ij} is the cofactor for row i and column j (see Appendix A).

We can see from Eq. (11.37) that the superposition principle of Chapter 5 applies to frequency-domain circuit analysis. Equation (11.37) indicates that the nth node-voltage phasor is given by a weighted sum or superposition of voltages caused by the independent sources. A simple interpretation of the ratios Δ_{mn}/Δ can be obtained if we observe that if the independent source phasors were zero for all m except for $m = k$, then Eq. (11.37) would become

$$
V_n =
\begin{cases}
\dfrac{\Delta_{kn}}{\Delta} I_{sk} & \text{for } 1 \leq k \leq M - 1 \\[2mm]
\dfrac{\Delta_{kn}}{\Delta} V_{sk} & \text{for } M \leq k \leq N
\end{cases}
\qquad (11.38)
$$

Therefore Δ_{kn}/Δ is simply the transfer function from source k to the voltage at node n. We see from Eq. (11.38) that the ratio Δ_{kn}/Δ is

1. A driving-point impedance for $n = k = 1, 2, \dots, M - 1$
2. A transfer impedance for $n \neq k$, $k = 1, 2, \dots, M - 1$
3. A voltage transfer function for $k = M, M + 1, \dots, N$

The *physical* interpretation of the Cramer's rule solution, Eq. (11.38), for transfer impedance is illustrated in Fig. 11.19. The transfer impedance, $\mathbf{V}_n/\mathbf{I}_{sk} = \Delta_{kn}/\Delta$, can be determined experimentally by the following procedure:

1. Set all the independent sources in the network to zero. Inject a current $i_{sk}(t) = |\mathbf{I}_{sk}| \cos(\omega t + \phi_{\mathbf{I}_{sk}})$ into node k, as shown in Fig. 11.19. Set the amplitude $|\mathbf{I}_{sk}|$ and phase $\phi_{\mathbf{I}_{sk}}$ to any convenient value ($|\mathbf{I}_{sk}| \neq 0$).
2. Measure the resulting steady-state sinusoidal voltage drop $v_n(t) = |\mathbf{V}_n| \cos(\omega t + \phi_{\mathbf{V}_n})$ from node n to the reference node.
3. Calculate Δ_{kn}/Δ with the use of Eq. (11.38), where $\mathbf{V}_n = |\mathbf{V}_n|\underline{/\phi_{\mathbf{V}_n}}$ and $\mathbf{I}_{sk} = |\mathbf{I}_{sk}|\underline{/\phi_{\mathbf{I}_{sk}}}$.

FIGURE 11.19
Interpretation of the
Cramer's rule solution

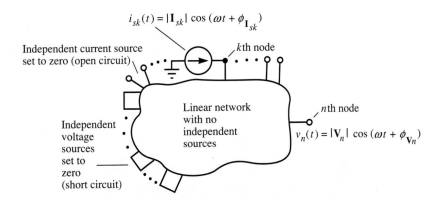

A similar interpretation applies for driving-point impedance ($n = k$). The voltage transfer functions can be determined if we use Eq. (11.38) as follows:

1. Set all the independent sources to zero except $v_{sk}(t) = |\mathbf{V}_{sk}| \cos(\omega t + \phi_{\mathbf{V}_{sk}})$. This is the independent voltage source represented by the phasor $\mathbf{V}_{sk}$ for $M \leq k \leq N$.
2. Measure the resulting steady-state sinusoidal voltage drop $v_n(t) = |\mathbf{V}_n| \cos(\omega t + \phi_{\mathbf{V}_n})$ from node n to the reference node.
3. Calculate Δ_{kn}/Δ with the use of Eq. (11.38), where $\mathbf{V}_n = |\mathbf{V}_n|\underline{/\phi_{\mathbf{V}_n}}$ and $\mathbf{V}_{sk} = |\mathbf{V}_{sk}|\underline{/\phi_{\mathbf{V}_{sk}}}$.

Example 11.10 provides additional insight into the meaning of the Cramer's rule solution [Eq. (11.37)].

EXAMPLE 11.10

⬛ *See Problem 11.4 in the PSpice manual.*

(a) Determine the node-voltage equations for the circuit of Fig. 11.20a.
(b) Use Cramer's rule to determine the impedances $\mathbf{V}_n/\mathbf{I}_3$, $n = 1, 2, 3$, where $\mathbf{I}_3$ is a current injected into node 3 with return at the reference node.

FIGURE 11.20
Determination of transfer impedance Δ_{kn}/Δ. (a) Original circuit; (b) configuration for measuring transfer impedances $V_1/I_{s3} = \Delta_{31}/\Delta$ and $V_2/I_{s3} = \Delta_{32}/\Delta$ and driving-point impedance $V_3/I_{s3} = \Delta_{33}/\Delta$. The units are volts, amperes, and siemens

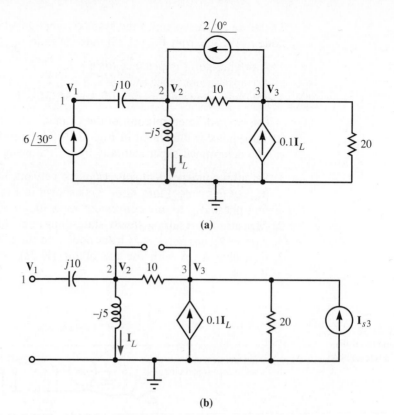

(a)

(b)

■ SOLUTION

(a) All the node-voltage equations for this circuit are KCL equations.

Node 1 $j10(V_1 - V_2) - 6\underline{/30°} = 0$

Node 2 $j10(V_2 - V_1) - j5V_2 + 10(V_2 - V_3) - 2\underline{/0°} = 0$

Node 3 $10(V_3 - V_2) + 20V_3 - 0.1I_L + 2\underline{/0°} = 0$

where $I_L = -j5V_2$. After eliminating I_L, we obtain the following matrix equation:

$$
\begin{bmatrix}
j10 & -j10 & 0 \\
-j10 & 10 + 5j & -10 \\
0 & -10 + 0.5j & 30
\end{bmatrix}
\begin{bmatrix}
V_1 \\
V_2 \\
V_3
\end{bmatrix}
=
\begin{bmatrix}
6\underline{/30°} \\
2 \\
-2
\end{bmatrix}
$$

(b) According to the Cramer's rule solution, Eq. (11.37), each node voltage V_1, V_2, and V_3 can be interpreted as a superposition of responses due to current sources I_{s1}, I_{s2}, and I_{s3}. That is,

$$
V_n = \frac{\Delta_{1n}}{\Delta} I_{s1} + \frac{\Delta_{2n}}{\Delta} I_{s2} + \frac{\Delta_{3n}}{\Delta} I_{s3} \qquad n = 1, 2, 3
$$

where

$$
\begin{bmatrix}
I_{s1} \\
I_{s2} \\
I_{s3}
\end{bmatrix}
=
\begin{bmatrix}
6\underline{/30°} \\
2 \\
-2
\end{bmatrix}
$$

In this interpretation, current $\mathbf{I}_{si}$ enters node i and has its return at the reference node.

The significance of the ratio Δ_{3n}/Δ for $n = 1, 2, 3$ is illustrated in Fig. 11.20b, which depicts the original circuit with independent sources set to zero and current $\mathbf{I}_{s3}$ injected into node 3. From Eq. (11.38) and the matrix equation of (a) it follows that the transfer impedance from node 3 to node 1 is

$$\frac{\mathbf{V}_1}{\mathbf{I}_{s3}} = \frac{\Delta_{31}}{\Delta} = \frac{\begin{vmatrix} -j10 & 0 \\ 10 + 5j & -10 \end{vmatrix}}{\begin{vmatrix} j10 & -j10 & 0 \\ -j10 & 10 + 5j & -10 \\ 0 & -10 + 0.5j & 30 \end{vmatrix}} = \frac{j100}{1450 + j2000} \simeq 0.04\underline{/36°}\ \Omega$$

the transfer impedance from node 3 to node 2 is

$$\frac{\mathbf{V}_2}{\mathbf{I}_{s3}} = \frac{\Delta_{32}}{\Delta} = \frac{-\begin{vmatrix} j10 & 0 \\ -j10 & -10 \end{vmatrix}}{\Delta} = \frac{j100}{1450 + j2000} \simeq 0.04\underline{/36°}\ \Omega$$

and the driving-point impedance at node 3 is

$$\frac{\mathbf{V}_3}{\mathbf{I}_{s3}} = \frac{\Delta_{33}}{\Delta} = \frac{\begin{vmatrix} j10 & -j10 \\ -j10 & 10 + 5j \end{vmatrix}}{\Delta} = \frac{50 + j100}{1450 + j2000} \simeq 0.045\underline{/9.4°}\ \Omega\ \blacksquare$$

11.4   AC Mesh-Current Analysis

As we stated in Section 4.5, mesh-current circuit analysis applies only to planar circuits. The procedure summarized on page 155 applies to ac mesh-current analysis in the frequency domain as well as in the time domain. The following examples illustrate the method.

EXAMPLE 11.11

Write a set of mesh-current equations for the network shown in Fig. 11.21.

FIGURE 11.21
Circuit for Example 11.11

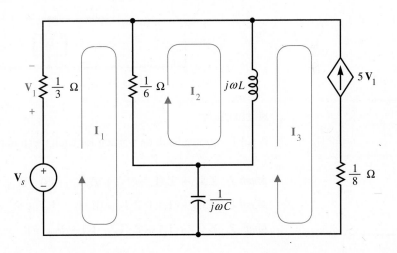

■ SOLUTION

Use the mesh currents assigned in Fig. 11.21. KVL for a closed path around mesh 1 gives the first mesh-current equation:

$$-\mathbf{V}_s + \frac{1}{3}\mathbf{I}_1 + \frac{1}{6}(\mathbf{I}_1 - \mathbf{I}_2) + \frac{1}{j\omega C}(\mathbf{I}_1 - \mathbf{I}_3) = 0$$

The second mesh-current equation is obtained by application of KVL to a closed path around mesh 2:

$$\frac{1}{6}(\mathbf{I}_2 - \mathbf{I}_1) + j\omega L(\mathbf{I}_2 - \mathbf{I}_3) = 0$$

KCL gives the third mesh-current equation:

$$\mathbf{I}_3 = -5\mathbf{V}_1 = -5\left(\frac{1}{3}\mathbf{I}_1\right) = -\frac{5}{3}\mathbf{I}_1$$

The three mesh-current equations can be written as

$$\begin{bmatrix} \frac{1}{2} + \frac{1}{j\omega C} & -\frac{1}{6} & -\frac{1}{j\omega C} \\ -\frac{1}{6} & j\omega L + \frac{1}{6} & -j\omega L \\ \frac{5}{3} & 0 & 1 \end{bmatrix} \begin{bmatrix} \mathbf{I}_1 \\ \mathbf{I}_2 \\ \mathbf{I}_3 \end{bmatrix} = \begin{bmatrix} \mathbf{V}_s \\ 0 \\ 0 \end{bmatrix}$$

where the left-hand matrix is the mesh transformation matrix. ■

EXAMPLE 11.12

Write the mesh-current equations for the network of Fig. 11.22.

FIGURE 11.22
Circuit for Example 11.12

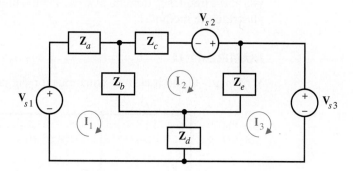

■ SOLUTION

A KVL equation can be written for each mesh in this circuit. The KVL equations are

Mesh 1 $\mathbf{Z}_a\mathbf{I}_1 + \mathbf{Z}_b(\mathbf{I}_1 - \mathbf{I}_2) + \mathbf{Z}_d(\mathbf{I}_1 - \mathbf{I}_3) - \mathbf{V}_{s1} = 0$

Mesh 2 $\mathbf{Z}_b(\mathbf{I}_2 - \mathbf{I}_1) + \mathbf{Z}_c\mathbf{I}_2 + \mathbf{Z}_e(\mathbf{I}_2 - \mathbf{I}_3) - \mathbf{V}_{s2} = 0$

Mesh 3 $\mathbf{Z}_d(\mathbf{I}_3 - \mathbf{I}_1) + \mathbf{Z}_e(\mathbf{I}_3 - \mathbf{I}_2) + \mathbf{V}_{s3} = 0$

The matrix form is

$$
\begin{bmatrix}
\mathbf{Z}_a + \mathbf{Z}_b + \mathbf{Z}_d & -\mathbf{Z}_b & -\mathbf{Z}_d \\
-\mathbf{Z}_b & \mathbf{Z}_b + \mathbf{Z}_c + \mathbf{Z}_e & -\mathbf{Z}_e \\
-\mathbf{Z}_d & -\mathbf{Z}_e & \mathbf{Z}_d + \mathbf{Z}_e
\end{bmatrix}
\begin{bmatrix}
\mathbf{I}_1 \\
\mathbf{I}_2 \\
\mathbf{I}_3
\end{bmatrix}
=
\begin{bmatrix}
\mathbf{V}_{s1} \\
\mathbf{V}_{s2} \\
-\mathbf{V}_{s3}
\end{bmatrix}
$$

Notice that the mesh transformation matrix on the left-hand side is symmetric. The mesh transformation matrix of a network containing no dependent sources can always be arranged to be symmetric. ■

The mesh-current equations of networks composed of only impedances and independent voltage sources, such as that in Example 11.12, have an interpretation that is the dual of that occurring for Example 11.8. Consider the KVL equation for mesh 1. The coefficient of $\mathbf{I}_1$ is the sum of the impedances making up mesh 1, the coefficient of $\mathbf{I}_2$ is the negative of the impedance that is common to meshes 1 and 2, and the coefficient of $\mathbf{I}_3$ is the negative of the impedance that is common to meshes 1 and 3. Similarly, for the KVL equation of mesh 2, the coefficient of $\mathbf{I}_1$ is the negative of the impedance that is common to meshes 2 and 1, the coefficient of $\mathbf{I}_2$ is the sum of the impedances making up mesh 2, and the coefficient of $\mathbf{I}_3$ is the negative of the impedance that is common to meshes 3 and 1. The coefficients of $\mathbf{I}_1$, $\mathbf{I}_2$, and $\mathbf{I}_3$ for the third mesh equation have a similar interpretation. Observe further that the right-hand sides of the equations are simply the independent source phasor voltages in the meshes for which KVL is written. The above interpretations can be generalized to an arbitrary N-mesh network composed of only impedances and independent voltage sources. For this purpose we make the following definitions.

DEFINITION

The **self-impedance** of mesh k, $\mathbf{Z}_{kk}$, is the sum of the impedances composing mesh k.

DEFINITION

The **mutual impedance** of meshes k and l, $\mathbf{Z}_{kl}$, is the negative of the sum of all impedances in common with meshes k and l.

The mesh equations for such a network are all KVL equations and have the matrix form:

$$
\begin{bmatrix}
\mathbf{Z}_{11} & \mathbf{Z}_{12} & \cdots & \mathbf{Z}_{1N} \\
\mathbf{Z}_{21} & \mathbf{Z}_{22} & \cdots & \mathbf{Z}_{2N} \\
\vdots & & & \\
\mathbf{Z}_{N1} & \mathbf{Z}_{N2} & & \mathbf{Z}_{NN}
\end{bmatrix}
\begin{bmatrix}
\mathbf{I}_1 \\
\mathbf{I}_2 \\
\vdots \\
\mathbf{I}_N
\end{bmatrix}
=
\begin{bmatrix}
\mathbf{V}_{s1} \\
\mathbf{V}_{s2} \\
\vdots \\
\mathbf{V}_{sN}
\end{bmatrix}
\tag{11.39}
$$

where $\mathbf{V}_{sk}$ is the sum of all independent phasor voltage source drops taken counterclockwise in mesh k (if we assume clockwise mesh currents). The square matrix on the left-hand side is the network's *impedance matrix*. By definition, $\mathbf{Z}_{kl} = \mathbf{Z}_{lk}$.

Therefore the impedance matrix is symmetric. An example of an impedance matrix was given in Example 11.12. Observe that the impedance matrix in Example 11.12 could have been written directly by inspection of the circuit in Fig. 11.22.

Let us now return to mesh-current analysis of general planar circuits.

EXAMPLE 11.13

Write the mesh-current equations for the circuit of Fig. 11.23.

FIGURE 11.23
Circuit for Example 11.13

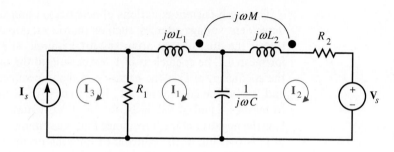

■ SOLUTION

Application of KVL around mesh 1 yields

$$R_1(\mathbf{I}_1 - \mathbf{I}_3) + j\omega L_1\mathbf{I}_1 + j\omega M\mathbf{I}_2 + \frac{1}{j\omega C}(\mathbf{I}_1 - \mathbf{I}_2) = 0$$

Similarly, for mesh 2,

$$\frac{1}{j\omega C}(\mathbf{I}_2 - \mathbf{I}_1) + j\omega L_2\mathbf{I}_2 + j\omega M\mathbf{I}_1 + R_2\mathbf{I}_2 + \mathbf{V}_s = 0$$

By substituting $\mathbf{I}_3 = \mathbf{I}_s$ into the above equations and rearranging, we obtain

$$\begin{bmatrix} R_1 + j\omega L_1 + \dfrac{1}{j\omega C} & -\dfrac{1}{j\omega C} + j\omega M \\[2mm] -\dfrac{1}{j\omega C} + j\omega M & \dfrac{1}{j\omega C} + j\omega L_2 + R_2 \end{bmatrix} \begin{bmatrix} \mathbf{I}_1 \\ \mathbf{I}_2 \end{bmatrix} = \begin{bmatrix} R_1\mathbf{I}_s \\ -\mathbf{V}_s \end{bmatrix} \quad ■$$

Once the mesh-current equations have been set up, the next step is to solve them. In our future work, we will not need a general theoretical expression for the solution. Therefore we can proceed to our next topic.

REMEMBER

Mesh-current analysis works the same way in the frequency domain as it does in the time domain.

EXERCISES

15. Set up the mesh-current equations for $\mathbf{I}_1$, $\mathbf{I}_2$, and $\mathbf{I}_3$ for the following circuit. All impedance values are in ohms.

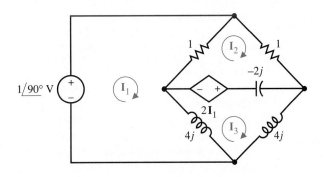

16. Set up the mesh-current equations for the following circuit.

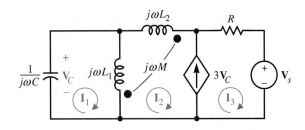

11.5 Superposition

In Section 5.2 we showed that the response of a linear network to a number of independent sources is equal to the sum of the responses to each independent source acting on the network alone. Theorem 4 of Chapter 9 states that this *superposition* property also applies to the particular response. Indeed, the principle of superposition is the fundamental principle in the theory of linear circuits. Therefore we use it often in this text. In this section, we use superposition to determine a network's steady-state response to two or more sinusoidal sources whose frequencies are different.

See Problem 11.8 in the PSpice manual.

EXAMPLE 11.14

Two sinusoidal sources that have *different* frequencies have been driving the *RL* network of Fig. 11.24 for a long time (that is, since $t = -\infty$). Determine $v(t)$.

FIGURE 11.24
Circuit for Example 11.14

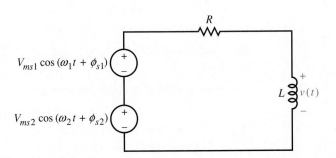

■ **SOLUTION**

The complementary response $v_c(t)$ is zero by now (that is, for all finite t), because the network is stable and the sources were applied at $t = -\infty$. Therefore the complete response $v(t)$ is just the particular response. Because there are two sinusoidal inputs, the particular response of this network contains *two* sinusoidal terms (called *components*); one component has frequency ω_1, and the other has frequency ω_2. Frequency-domain concepts apply only to the *individual* components. The total response $v(t)$ is found by adding the individual sinusoidal responses caused by each source acting alone. The time- and frequency-domain networks for each source acting alone are shown in Fig. 11.25. The complex exponential factor $e^{j\omega_1 t}$ is implicit in Fig. 11.25b, and the factor $e^{j\omega_2 t}$ is implicit in Fig. 11.25d. The output-voltage phasors in Fig. 11.25b, d can be determined with the use of the voltage divider relation [Eq. (11.34)].

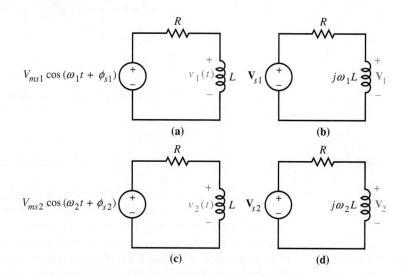

FIGURE 11.25
Pertinent to superposition

$$\mathbf{V}_1 = \frac{j\omega_1 L}{R + j\omega_1 L}\,\mathbf{V}_{s1} = \mathbf{H}(j\omega_1)\mathbf{V}_{s1}$$

$$\mathbf{V}_2 = \frac{j\omega_2 L}{R + j\omega_2 L}\,\mathbf{V}_{s2} = \mathbf{H}(j\omega_2)\mathbf{V}_{s2}$$

where we have defined

$$\mathbf{H}(j\omega) = \frac{j\omega L}{R + j\omega L}$$

$$= \frac{\omega L}{\sqrt{R^2 + \omega^2 L^2}}\;\left|\frac{\pi}{2} - \tan^{-1}\left(\frac{\omega L}{R}\right)\right.$$

The corresponding rotating phasors are

$$\mathbf{V}_1 e^{j\omega_1 t} = \mathbf{H}(j\omega_1)\mathbf{V}_{s1} e^{j\omega_1 t}$$

and

$$\mathbf{V}_2 e^{j\omega_2 t} = \mathbf{H}(j\omega_2)\mathbf{V}_{s2} e^{j\omega_2 t}$$

The real parts give the particular response in Fig. 11.25a, c:

$$v_1(t) = |\mathbf{H}(j\omega_1)|V_{ms1} \cos\left[\omega_1 t + \phi_{s1} + \underline{/\mathbf{H}(j\omega_1)}\right]$$

$$= \frac{\omega_1 L}{\sqrt{R^2 + \omega_1^2 L^2}} V_{ms1} \cos\left[\omega_1 t + \phi_{s1} + \frac{\pi}{2} - \tan^{-1}\left(\frac{\omega_1 L}{R}\right)\right]$$

and

$$v_2(t) = |\mathbf{H}(j\omega_2)|V_{ms2} \cos\left[\omega_2 t + \phi_{s2} + \underline{/\mathbf{H}(j\omega_2)}\right]$$

$$= \frac{\omega_2 L}{\sqrt{R^2 + \omega_2^2 L^2}} V_{ms2} \cos\left[\omega_2 t + \phi_{s2} + \frac{\pi}{2} - \tan^{-1}\left(\frac{\omega_2 L}{R}\right)\right]$$

These are the individual components of the total particular response of the circuit of Fig. 11.24. By superposition, the total response is

$$v(t) = v_1(t) + v_2(t)$$

$$= \frac{\omega_1 L}{\sqrt{R^2 + \omega_1^2 L^2}} V_{ms1} \cos\left[\omega_1 t + \phi_{s1} + \frac{\pi}{2} - \tan^{-1}\left(\frac{\omega_1 L}{R}\right)\right]$$

$$+ \frac{\omega_2 L}{\sqrt{R^2 + \omega_2^2 L^2}} V_{ms2} \cos\left[\omega_2 t + \phi_{s2} + \frac{\pi}{2} - \tan^{-1}\left(\frac{\omega_2 L}{R}\right)\right] \blacksquare$$

EXAMPLE 11.15

In the circuit of Fig. 11.26, a sinusoidal current source with frequency ω_3 has been added to the RL network of the previous example. Determine $v(t)$.

FIGURE 11.26
Independent sources with different frequencies must be considered one at a time

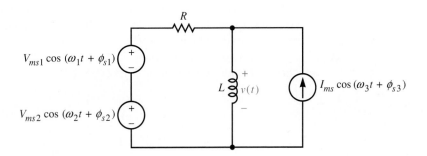

■ **SOLUTION**

Consider the three independent sources individually. The sinusoidal output components resulting from the voltage sources have already been determined for this network. (When the current source is set to zero, it is in effect an open circuit, and the network becomes that of Fig. 11.24.) The effect of the current source is found by setting the voltage sources to zero, as illustrated in Fig. 11.27.

FIGURE 11.27
Voltage sources set to zero

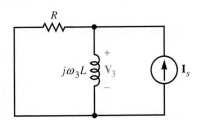

The output-voltage phasor produced by the current source is, by inspection of Fig. 11.27,

$$\mathbf{V}_3 = (j\omega_3 L \parallel R)\mathbf{I}_s = \frac{j\omega_3 LR}{R + j\omega_3 L}\mathbf{I}_s$$

The complex exponential response is consequently

$$\mathbf{V}_3 e^{j\omega_3 t} = \frac{j\omega_3 LR}{R + j\omega_3 L}\mathbf{I}_s e^{j\omega_3 t}$$

The real part gives the sinusoidal steady-state response due to the current source:

$$v_3(t) = \frac{\omega_3 LR}{\sqrt{R^2 + \omega_3^2 L^2}}I_{ms}\cos\left[\omega_3 t + \phi_{s3} + \frac{\pi}{2} - \tan^{-1}\left(\frac{\omega_3 L}{R}\right)\right]$$

By superposition, the total response $v(t)$ in Fig. 11.26 is

$$v(t) = v_1(t) + v_2(t) + v_3(t)$$

$$= \frac{\omega_1 L}{\sqrt{R^2 + \omega_1^2 L^2}}V_{ms2}\cos\left[\omega_1 t + \phi_{s1} + \frac{\pi}{2} - \tan^{-1}\left(\frac{\omega_1 L}{R}\right)\right]$$

$$+ \frac{\omega_2 L}{\sqrt{R^2 + \omega_2^2 L^2}}V_{ms2}\cos\left[\omega_2 t + \phi_{s2} + \frac{\pi}{2} - \tan^{-1}\left(\frac{\omega_2 L}{R}\right)\right]$$

$$+ \frac{\omega_3 LR}{\sqrt{R^2 + \omega_3^2 L^2}}I_{ms}\cos\left[\omega_3 t + \phi_{s3} + \frac{\pi}{2} - \tan^{-1}\left(\frac{\omega_3 L}{R}\right)\right] \blacksquare$$

Up to this point we have shown how to use superposition to find a network's response to sinusoidal sources with different frequencies. Superposition can also be used to advantage when the sources have the same frequency. Consider once again the two-source network in Fig. 11.10e. With the use of superposition, the current $\mathbf{I}_L$ can be written by *inspection* as

$$\mathbf{I}_L = \frac{\mathbf{V}_s}{j\omega L + R \parallel \dfrac{1}{j\omega C}} - \frac{\dfrac{1}{j\omega L}}{\dfrac{1}{j\omega L} + \dfrac{1}{R} + j\omega C}\mathbf{I}_s \tag{11.40}$$

We obtain the terms in Eq. (11.40) by assuming that each source acts alone. The first term is the input (phasor) voltage divided by the driving-point impedance seen by the voltage source. The second term follows from the current divider relation [Eq. (11.33)]. For the numerical values used in Example 11.4 ($\omega = 10^3$ rad/s, $R = 1$ kΩ, $C = 1$ μF, $L = 0.5$ H, $\mathbf{V}_s = 30\underline{/-25°}$ V, and $\mathbf{I}_s = 10\underline{/-75°}$ mA), the driving-point impedance is

$$j\omega L + R \parallel \frac{1}{j\omega C} = 0.5 \text{ k}\Omega \tag{11.41}$$

and the current transfer function is

$$-\frac{\dfrac{1}{j\omega L}}{\dfrac{1}{j\omega L} + \dfrac{1}{R} + j\omega C} = -(\sqrt{2}\underline{/-45°}) \tag{11.42}$$

Therefore
$$\mathbf{I}_L = \frac{30\underline{/-25°}}{0.5} - (\sqrt{2}\underline{/-45°})(10\underline{/-75°}) \text{ mA}$$

$$= 60\underline{/-25°} - 10\sqrt{2}\underline{/-120°} \text{ mA} \qquad (11.43)$$

and

$$i_L(t) = \mathscr{R}e\left\{\mathbf{I}_L e^{j1000t}\right\}$$

$$= 60\cos(1000t - 25°) - 10\sqrt{2}\cos(1000t - 120°) \text{ mA} \qquad (11.44)$$

This result agrees with that given in Example 11.4, but it displays the contributions from the two sources individually.

REMEMBER

When sources have different frequencies, superposition is applied to the *time-domain waveforms* arising from the different sources. *It is never correct to add non-rotating phasors that represent sinusoids with different frequencies.* For instance, in Example 11.14, it would be a *fundamental mistake* to define a total response phasor

$$\mathbf{V}_1 + \mathbf{V}_2 = \cancel{\frac{j\omega_1 L}{R + j\omega_1 L}\mathbf{V}_{s1} + \frac{j\omega_2 L}{R + j\omega_2 L}\mathbf{V}_{s2}} \qquad \cancel{(11.45)}$$

Equation (11.37), however, shows us that we *can*, add phasors that represent sinusoids having the *same* frequencies. An example of this is given by Eq. (11.40).

EXERCISES

17. Find $i(t)$ for the following circuit. Express your answer in the form $i(t) = A\cos(\omega_1 t + \phi_A) + B\sin(\omega_2 t + \phi_B)$.

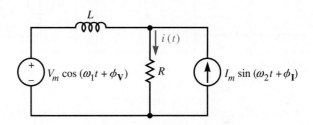

18. Find $v_o(t)$ in the following circuit. Express your answer in the form $v_o(t) = B_1\cos(\omega_1 t + \phi_1) + B_2\cos(\omega_2 t + \phi_2) + B_3\cos(\omega_3 t + \phi_3)$.

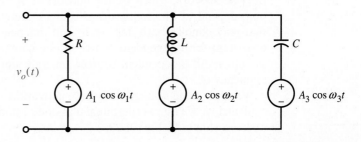

11.6 Thévenin and Norton Equivalent Circuits of AC Circuits

Recall from Section 5.4 that the terminal equation of any network composed of resistances and sources has the form

$$v(t) = v_{oc}(t) - i(t)R_{Eq} \qquad (11.46)$$

This result is illustrated in Fig. 11.28, where network A is the network composed of resistances and sources, $v_{oc}(t)$ is the voltage appearing across terminals a and b when the terminals are open-circuited, and R_{Eq} is the Thévenin equivalent resistance of network A. The two circuits enclosed by red in Fig. 11.28b, c have the same terminal equation as network A of Fig. 11.28a. These circuits are called, respectively, the Thévenin equivalent circuit and the Norton equivalent circuit of network A. Thévenin and Norton equivalent circuits are important because we can use them in place of network A whenever we want to determine the effect of network A on another circuit.

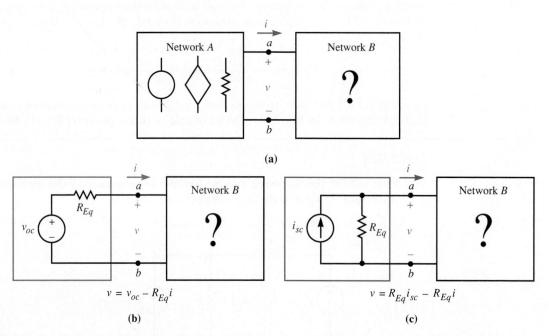

FIGURE 11.28 Replacing a network of resistances and independent and dependent sources by an equivalent network. (*a*) Networks coupled at only two terminals; (*b*) network *A* replaced by Thévenin equivalent (enclosed by red); (*c*) network *A* replaced by Norton equivalent (enclosed by red)

In this section we will derive the Thévenin and Norton equivalents of an arbitrary ac network, which we call network A in Fig. 11.29. Network A may contain both dependent and independent sources. Network B is an arbitrary ac network that may contain both dependent and independent sources. No control voltage or current for a dependent source in one network (A or B), however, is found in the other. All independent sources generate sinusoidal waveforms with the same frequency ω.

We apply node-voltage analysis to network A to derive its Thévenin and Norton equivalents. We number terminal a as node 1 and select terminal b as the reference node, as shown in Fig. 11.29.

FIGURE 11.29
Ac network coupled at only
two terminals

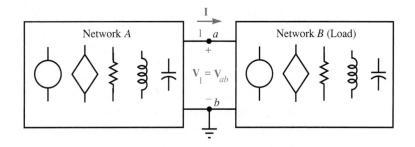

The node-voltage equations are given by Eq. (11.36).

$$
\begin{array}{c}
\text{KCL}\\
\text{equations}
\end{array}
\left\{
\begin{array}{c}
\text{KVL}\\
\text{equations}
\end{array}
\right.
\left[
\begin{array}{cccc}
T_{11} & T_{12} & \cdots & T_{1N}\\
\vdots & \vdots & & \vdots\\
T_{M1} & T_{M2} & \cdots & T_{MN}\\
\vdots & \vdots & & \vdots\\
T_{N1} & T_{N2} & \cdots & T_{NN}
\end{array}
\right]
\left[
\begin{array}{c}
V_1\\
\vdots\\
V_M\\
\vdots\\
V_N
\end{array}
\right]
=
\left[
\begin{array}{c}
I_{s1}-I\\
\vdots\\
V_{sM}\\
\vdots\\
V_{sN}
\end{array}
\right]
$$

- Load current
- Due to current sources
- Due to voltage sources

(11.47)

Node transformation matrix of network A

In Eq. (11.47), the phasor I_{sj} is the sum of the independent source currents of network A into node 1. Since the load draws current I, the total current entering node 1 is $I_{s1} - I$. The phasor V_{sk}, $(M \le k \le N)$, is the sum of the voltage-source phasors of network A in the loop for which KVL is written. The elements T_{mn}, $(1 \le m \le N, 1 \le n \le N)$, are functions of $j\omega$, where ω is the common frequency of the sinusoidal waveforms present throughout the network.

By Cramer's rule,

$$V_{ab} = V_1$$

$$= \frac{\Delta_{11}}{\Delta} I_{s1} + \frac{\Delta_{21}}{\Delta} I_{s2} + \cdots + \frac{\Delta_{(M-1)1}}{\Delta} I_{s(M-1)} + \frac{\Delta_{M1}}{\Delta} V_{sM} + \cdots$$

$$+ \frac{\Delta_{N1}}{\Delta} V_{sN} - \frac{\Delta_{11}}{\Delta} I \tag{11.48}$$

The quantities Δ_{nm} are the nmth cofactors of the node transformation matrix, and Δ is its determinant. Consequently, the Δ_{nm}'s and Δ are functions of $j\omega$. We notice that if network B is an open circuit, then $I = 0$ and V_{ab} is, by definition, the open-circuit voltage phasor, V_{oc}:

$$V_{oc} = V_{ab}\big|_{I=0}$$

$$= \frac{\Delta_{11}}{\Delta} I_{s1} + \frac{\Delta_{21}}{\Delta} I_{s2} + \cdots + \frac{\Delta_{(M-1)1}}{\Delta} I_{s(M-1)} + \frac{\Delta_{M1}}{\Delta} V_{sM} + \cdots + \frac{\Delta_{N1}}{\Delta} V_{sN} \tag{11.49}$$

Substitution of Eq. (11.49) into Eq. (11.48) yields

$$V_{ab} = V_{oc} - Z_{ab}(j\omega)I \tag{11.50}$$

where we have defined

$$\mathbf{Z}_{ab}(j\omega) = \frac{\Delta_{11}(j\omega)}{\Delta(j\omega)} \tag{11.51}$$

$\mathbf{Z}_{ab}(j\omega)$ is called the *equivalent impedance* of network A looking into terminals a and b. We notice next that if network B is a short circuit, then $\mathbf{V}_{ab} = 0$ and $\mathbf{I}$ is, by definition, the *short-circuit phasor current* $\mathbf{I}_{sc}$. We can obtain an expression for $\mathbf{I}_{sc}$ by setting $\mathbf{V}_{ab} = 0$ in Eq. (11.50) and solving for $\mathbf{I}$. The result is

$$\mathbf{I}_{sc} = \frac{\mathbf{V}_{oc}}{\mathbf{Z}_{ab}(j\omega)} \tag{11.52}$$

Equation (11.52) can also be used to determine $\mathbf{Z}_{ab}(j\omega)$ if $\mathbf{I}_{sc} \neq 0$.

$$\mathbf{Z}_{ab}(j\omega) = \frac{\mathbf{V}_{oc}}{\mathbf{I}_{sc}} \tag{11.53}$$

Equations (11.50) and (11.53) are the fundamental definitions of the Thévenin and Norton equivalents of an ac circuit. They indicate that the terminal equation relating the phasors $\mathbf{V}_{ab}$ and $\mathbf{I}$ in the network of Fig. 11.29 is identical to the terminal equations in the networks of Fig. 11.30a, b. The networks of Fig. 11.30a, b are, respectively, the Thévenin and Norton equivalents of the network of Fig. 11.29. The equivalency of Figs. 11.29 and 11.30a, b means that the sinusoidal steady-state terminal voltage $v_{ab}(t)$ and current $i(t)$ will be identical for all three circuits. This is a very important result that applies to *any* ac network A.

FIGURE 11.30
Ac network A replaced by
(*a*) Thévenin equivalent
(enclosed by red);
(*b*) Norton equivalent
(enclosed by red)

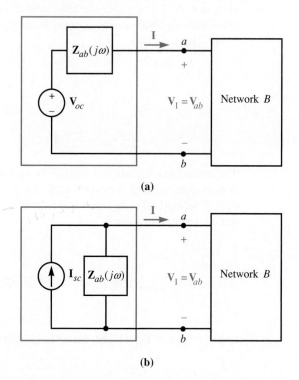

(a)

(b)

As a matter of notation, $\mathbf{Z}_{ab}$ is sometimes denoted $\mathbf{Z}_{Th}$ for Thévenin equivalent impedance (Fig. 11.30a), $\mathbf{Z}_N$ for Norton equivalent impedance (Fig. 11.30b), or $\mathbf{Z}_{Eq}$ for either.

Procedure to Determine $\mathbf{Z}_{ab}$

See Problem 11.6 in the PSpice manual.

Let us next derive a procedure to determine $\mathbf{Z}_{ab}$. Refer to Eq. (11.49) to observe that the phasor $\mathbf{V}_{oc}$ is a weighted summation of all the independent source phasors. If all the independent sources are set to zero, then $\mathbf{V}_{oc}$ becomes zero and Eq. (11.50) becomes $\mathbf{V}_{ab} = -\mathbf{Z}_{ab}(j\omega)\mathbf{I}$. This suggests the following procedure to determine $\mathbf{Z}_{ab}(j\omega)$.

> Disconnect the load from terminals a and b and set all independent sources of network A equal to zero, as shown in Fig. 11.31a. If no dependent sources are present, series and parallel combinations of impedance can often be used to find the equivalent impedance $\mathbf{Z}_{ab}(j\omega)$ looking into terminals a and b. If dependent sources are present, connect an independent source between terminals a and b, and define $\mathbf{I}_{ab}$ as the current entering network A at terminal a, as shown in Fig. 11.31b. If a current source is used, calculate or measure $\mathbf{V}_{ab}$. If a voltage source is used, calculate or measure $\mathbf{I}_{ab}$. The equivalent impedance of network A is $\mathbf{Z}_{ab}(j\omega) = \mathbf{V}_{ab}/\mathbf{I}_{ab}$.

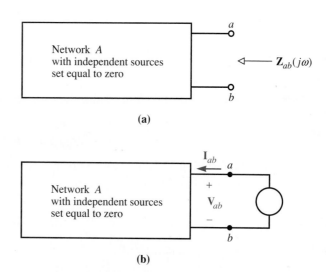

FIGURE 11.31 Network A with independent sources set to zero. (a) Meaning of $\mathbf{Z}_{ab}$; (b) calculation of $\mathbf{Z}_{ab}$ as ratio of driving-point voltage phasor to current phasor: $\mathbf{Z}_{ab} = \mathbf{V}_{ab}/\mathbf{I}_{ab}$. The external source can be either a voltage source or a current source

The following examples illustrate the procedure.

EXAMPLE 11.16

Find the Thévenin and Norton equivalents of the circuit of Fig. 11.32 (looking into terminals a and b).

FIGURE 11.32
Circuit for Example 11.16

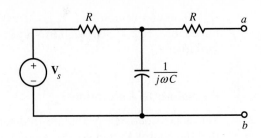

■ **SOLUTION**

We will first determine the open circuit voltage $\mathbf{V}_{oc}$ by the voltage-divider equation:

$$\mathbf{V}_{oc} = \frac{\dfrac{1}{j\omega C}}{\dfrac{1}{j\omega C} + R} \, \mathbf{V}_s$$

$$= \frac{1}{1 + j\omega RC} \, \mathbf{V}_s$$

Next, we find the equivalent impedance $\mathbf{Z}_{ab}$ by setting $\mathbf{V}_s = 0$. By replacing the voltage source with a short circuit, we find by inspection that

$$\mathbf{Z}_{ab}(j\omega) = R + R \, \| \, \frac{1}{j\omega C}$$

$$= \frac{R(2 + j\omega RC)}{1 + j\omega RC}$$

We use Eq. (11.52) to obtain the short-circuit current:

$$\mathbf{I}_{sc} = \frac{\mathbf{V}_{oc}}{\mathbf{Z}_{ab}(j\omega)} = \frac{\mathbf{V}_s}{R(2 + j\omega RC)}$$

The resulting Thévenin and Norton equivalents of the network of Fig. 11.32 are shown in Fig. 11.33. ■

FIGURE 11.33
(a) Thévenin and
(b) Norton equivalents
for Example 11.16

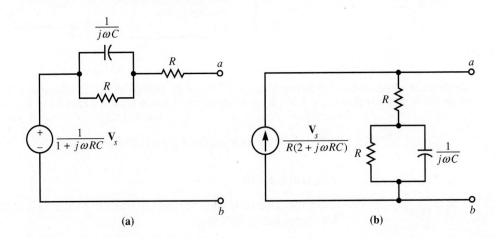

EXAMPLE 11.17

Find the Thévenin and Norton equivalents of the network of Fig. 11.34.

FIGURE 11.34
Circuit for Example 11.17

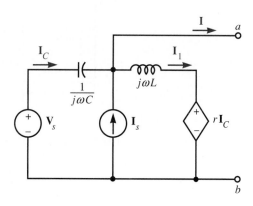

■ SOLUTION

First find $\mathbf{V}_{oc}$ (open terminals a and b). Because $\mathbf{I} = 0$, the inductance current phasor is $\mathbf{I}_1 = \mathbf{I}_s + \mathbf{I}_C$. KVL then yields

$$-\mathbf{V}_s + \frac{1}{j\omega C}\,\mathbf{I}_C + j\omega L(\mathbf{I}_C + \mathbf{I}_s) + r\mathbf{I}_C = 0$$

so that

$$\mathbf{I}_C = \frac{\mathbf{V}_s - j\omega L\mathbf{I}_s}{\dfrac{1}{j\omega C} + j\omega L + r}$$

The open-circuit voltage phasor is

$$\mathbf{V}_{oc} = \mathbf{V}_s - \frac{1}{j\omega C}\,\mathbf{I}_C$$

$$= \mathbf{V}_s - \left(\frac{\mathbf{V}_s - j\omega L\mathbf{I}_s}{1 - \omega^2 LC + j\omega rC}\right)$$

$$= \frac{(-\omega^2 LC + j\omega rC)\mathbf{V}_s + j\omega L\mathbf{I}_s}{1 - \omega^2 LC + j\omega rC}$$

We will find the Thévenin equivalent impedance $\mathbf{Z}_{ab}(j\omega)$ by setting the independent sources in the network to zero and applying a source to terminals a and b, as shown in Fig. 11.35. The driving-point impedance $\mathbf{Z}_{ab}(j\omega)$ is the ratio of the voltage phasor $\mathbf{V}_{ab}$ to the current phasor $\mathbf{I}_{ab} = -\mathbf{I}$. We can write a KCL equation at node a:

$$j\omega C\mathbf{V}_{ab} + (\mathbf{V}_{ab} - r\mathbf{I}_C)\frac{1}{j\omega L} = \mathbf{I}_{ab}$$

where

$$\mathbf{I}_C = -j\omega C\mathbf{V}_{ab}$$

FIGURE 11.35
Method for finding $\mathbf{Z}_{ab}(j\omega)$

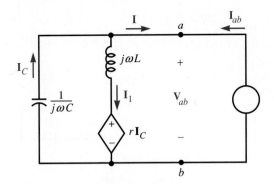

By eliminating $\mathbf{I}_C$ from these equations, we find that

$$\mathbf{V}_{ab} = \frac{j\omega L}{1 - \omega^2 LC + j\omega rC}\,\mathbf{I}_{ab}$$

Therefore

$$\mathbf{Z}_{ab}(j\omega) = \frac{j\omega L}{1 - \omega^2 LC + j\omega rC}$$

Finally, to obtain $\mathbf{I}_{sc}$, we use Eq. (11.52), with $\mathbf{V}_{oc}$ and $\mathbf{Z}_{ab}(j\omega)$ as given above. The result is

$$\mathbf{I}_{sc} = \mathbf{I}_s + \left(j\omega + \frac{r}{L}\right)C\mathbf{V}_s$$

The Thévenin and Norton equivalents of the network of Fig. 11.34 are shown in Fig. 11.36. ∎

FIGURE 11.36
(a) Thévenin and
(b) Norton equivalents
for Example 11.17

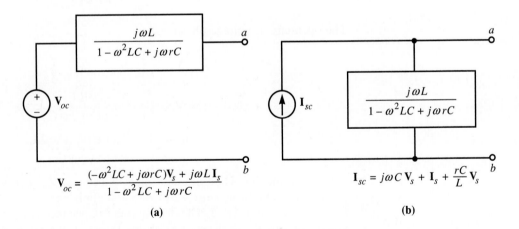

$$\mathbf{V}_{oc} = \frac{(-\omega^2 LC + j\omega rC)\mathbf{V}_s + j\omega L\,\mathbf{I}_s}{1 - \omega^2 LC + j\omega rC}$$

(a)

$$\mathbf{I}_{sc} = j\omega C\,\mathbf{V}_s + \mathbf{I}_s + \frac{rC}{L}\mathbf{V}_s$$

(b)

We can see from the two examples above that the Thévenin and Norton equivalent impedance $\mathbf{Z}_{ab}(j\omega)$ of a network can be a complicated function of ω. However, for a fixed value of ω, denoted by $\omega = \omega_0$, the function $\mathbf{Z}_{ab}(j\omega)$ becomes simply a complex number, and in this case the Thévenin and Norton equivalent circuits can be simplified. To show this, we denote

$$\mathbf{Z}_{ab}(j\omega_0) = \mathbf{Z}_0 \qquad\qquad (11.54)$$

We then write $\mathbf{Z}_0$ in rectangular form as

$$\mathbf{Z}_0 = R_0 + jX_0 \tag{11.55}$$

where

$$R_0 = \mathscr{R}e\,\{\mathbf{Z}_0\} \tag{11.56}$$

and

$$X_0 = \mathscr{I}m\,\{\mathbf{Z}_0\} \tag{11.57}$$

If X_0 is positive, then we can write Eq. (11.55) as

$$\mathbf{Z}_0 = R_0 + j\omega_0 L_0 \tag{11.58}$$

where L_0 is defined by

$$L_0 = \frac{X_0}{\omega_0} \tag{11.59}$$

and has the unit of henry. If X_0 is negative, then we can write Eq. (11.55) as

$$\mathbf{Z}_0 = R_0 + \frac{1}{j\omega_0 C_0} \tag{11.60}$$

where C_0 is defined by

$$C_0 = \frac{-1}{\omega_0 X_0} \tag{11.61}$$

and has the unit of farad. The implication of Eqs. (11.58) and (11.60) is that, *for a fixed frequency* $\omega = \omega_0$, the Thévenin and Norton equivalent impedance $\mathbf{Z}_{ab}(j\omega_0)$ can always be represented very simply as either a resistance R_0 in series with an inductance L_0, or a resistance R_0 in series with a capacitance C_0. (R_0 can be negative for a network containing dependent sources.)

EXAMPLE 11.18

Suppose that in Fig. 11.34, $C = 2\ \mu\text{F}$, $L = 1\ \text{mH}$, $r = 30\ \Omega$, $\mathbf{V}_s = 10\underline{/0°}$ V, $\mathbf{I}_s = 5\underline{/90°}$ A, and $\omega_0 = 10^4$ rad/s. Find the Thévenin and Norton equivalent circuits.

■ SOLUTION

For the given numerical values, the equivalent impedance (Fig. 11.36)

$$\mathbf{Z}_{ab}(j\omega) = \frac{j\omega L}{1 - \omega^2 LC + j\omega rC}$$

becomes the complex number

$$\mathbf{Z}_{ab}(j\omega_0) = 6 + j8\ \Omega$$

Since the reactance is positive, Eqs. (11.58) and (11.59) apply, where

$$R_0 = 6\ \Omega$$

and

$$L_0 = \frac{8}{10^4} = 0.8\ \text{mH}$$

Similarly, from Fig. 11.36, we have

$$\mathbf{V}_{oc} = \frac{(-\omega^2 LC + j\omega rC)\mathbf{V}_s + j\omega L\mathbf{I}_s}{1 - \omega^2 LC + j\omega rC}$$

$$\mathbf{I}_{sc} = j\omega C\mathbf{V}_s + \mathbf{I}_s + \frac{rC}{L}\mathbf{V}_s$$

which become

$$\mathbf{V}_{oc} = 52.34\underline{/136.5°} \text{ V}$$

and

$$\mathbf{I}_{sc} = 5.234\underline{/83.4°} \text{ A}$$

The resulting Thévenin and Norton equivalents are shown in Fig. 11.37. ∎

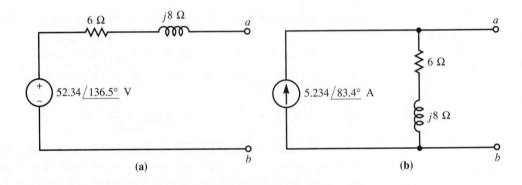

(a)   **(b)**

FIGURE 11.37 (a) Thévenin and (b) Norton equivalents for Example 11.18. The angular frequency, $\omega = 10^4$ rad/s

---

REMEMBER

The Thévenin and Norton equivalent impedance of a network is, in general, a complicated function of angular frequency ω. For a *fixed* angular frequency, this equivalent impedance can always be represented either as a resistance in series with an inductance or as a resistance in series with a capacitance.

EXERCISES

Find the Thévenin and Norton ac equivalent circuits of each circuit shown.

19.   20.

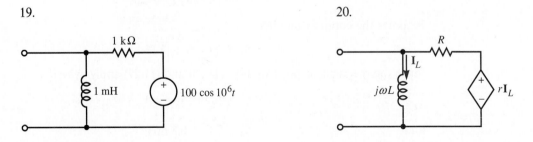

21. Determine the Thévenin and Norton equivalent circuits of the network shown.

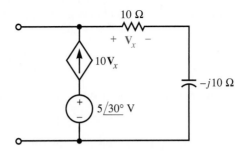

11.7   Thévenin and Norton Equivalents of Arbitrary LLTI Circuits

We can now derive the Thévenin and Norton equivalents of an arbitrary network of sources and resistances, inductances, and capacitances, shown as network *A* in Fig. 11.38. Network *A* is an arbitrary network of *R*s, *L*s (including mutually coupled inductances), *C*s, and independent and dependent sources. The independent source waveforms are arbitrary—they can be nonsinusoidal. Arbitrary initial conditions exist for capacitance voltage and inductance current. Network *B* is also arbitrary. It can be active, nonlinear and time-varying. The only restriction is that no control voltage or current for a dependent source in one network (*A* or *B*) is found in the other.

FIGURE 11.38
Network *A* coupled at only two terminals to arbitrary network *B*

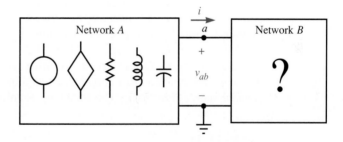

What we actually need here is network *A*'s time-domain terminal equation, expressed in the simplest possible form. We can obtain this equation directly from Eq. (11.50). Equation (11.50) is the frequency-domain version of a corresponding differential equation. This differential equation can be derived by the method of Section 11.1. The impedance $\mathbf{Z}_{ab}(j\omega)$ of Eq. (11.51) is the ratio of two polynomials in $j\omega$. It is helpful to write this ratio in the more familiar notation

$$\mathbf{Z}_{ab}(j\omega) = \frac{\mathscr{B}(j\omega)}{\mathscr{A}(j\omega)} \tag{11.62}$$

where $\mathscr{B}(j\omega) = \Delta_{11}(j\omega)$ and $\mathscr{A}(j\omega) = \Delta(j\omega)$. The polynomials $\mathscr{B}(j\omega)$ and $\mathscr{A}(j\omega)$ have the general form

$$\mathscr{B}(j\omega) = (j\omega)^m b_m + (j\omega)^{m-1} b_{m-1} + \cdots + j\omega b_1 + b_0 \tag{11.63}$$

$$\mathscr{A}(j\omega) = (j\omega)^n a_n + (j\omega)^{n-1} a_{n-1} + \cdots + j\omega a_1 + a_0 \tag{11.64}$$

where, of course, the coefficients $a_0, a_1, \ldots, a_n, b_0, b_1, \ldots, b_m$ depend on the network and can be found by any convenient method. For examples of the polynomials $\mathscr{B}(j\omega)$ and $\mathscr{A}(j\omega)$, see $\mathbf{Z}_{ab}(j\omega)$ in the preceding three examples.

Substitution of Eq. (11.62) into Eq. (11.50) yields

$$\mathbf{V}_{ab} = \mathbf{V}_{oc} - \frac{\mathscr{B}(j\omega)}{\mathscr{A}(j\omega)} \mathbf{I} \tag{11.65}$$

By multiplying both sides by $\mathscr{A}(j\omega)$, we obtain

$$\mathscr{A}(j\omega)\mathbf{V}_{ab} = \mathscr{A}(j\omega)\mathbf{V}_{oc} - \mathscr{B}(j\omega)\mathbf{I} \tag{11.66}$$

According to Section 11.1, the time-domain equation corresponding to Eq. (11.66) can be obtained by replacing phasors by functions of t and $j\omega$ by d/dt. The result can be written compactly in the familiar operator notation

$$\mathscr{A}(p)v_{ab}(t) = \mathscr{A}(p)v_{oc}(t) - \mathscr{B}(p)i(t) \tag{11.67}$$

where $p = d/dt$. An expression for the short-circuit current, $i_{sc}(t)$, can be obtained by setting $v_{ab}(t)$ equal to zero in Eq. (11.67) and rearranging the result. This yields the following relation between $i_{sc}(t)$ and $v_{oc}(t)$:

$$\mathscr{A}(p)v_{oc}(t) = \mathscr{B}(p)i_{sc}(t) \tag{11.68}$$

Equation (11.67) is the time-domain version of Eq. (11.66). As with any other time-domain network equation, its application is *not* restricted to sinusoidal waveforms. *It follows that Eq. (11.67) is the terminal equation of network A of Fig. 11.38 (including independent sources). This equation applies regardless of the external constraints imposed on $i(t)$ and $v_{ab}(t)$ by network B.* Because Eq. (11.65) is the frequency-domain terminal equation of the Thévenin and Norton equivalent circuits of Fig. 11.30 as well as of network A in Fig. 11.29, Eq. (11.67) is also the time-domain terminal equation of these circuits. In other words, *the time-domain versions of the Thévenin and Norton circuits of Fig. 11.30 apply for arbitrary network B and independent source waveforms.* The following example illustrates this.

EXAMPLE 11.19

In the circuit of Fig. 11.39, the source voltage is $v_s(t) = 10u(t)$ V, and the capacitance has an initial voltage $v_C(0^-) = -5$ V.

FIGURE 11.39
Circuit for Example 11.19

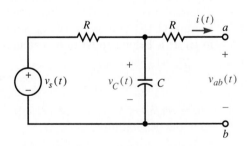

(a) Find the Thévenin and Norton equivalent circuits for $t \geq 0$.
(b) Find the terminal equation of the circuit.

FIGURE 11.40
(a) Thévenin and (b) Norton equivalents for Example 11.19

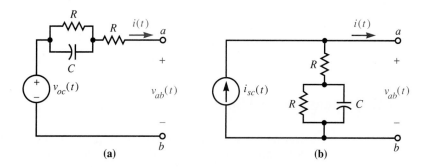

(a) (b)

■ **SOLUTION**

(a) The frequency-domain Thévenin and Norton equivalents were derived in Example 11.16 and shown in Fig. 11.33. To find the Thévenin and Norton equivalents for the problem at hand, we need only redraw Fig. 11.33 in the time domain. The result is shown in Fig. 11.40, where $v_{oc}(t)$ and $i_{sc}(t)$ are determined from Fig. 11.39. Let us first determine $v_{oc}(t)$. The open-circuit voltage $v_{oc}(t)$ equals $v_C(t)$. The capacitance C is in series with the resistance R connected to the source, and therefore the form of the complete response is simply $v_{oc}(t) = v_C(t) = A + Be^{-t/RC}$. We are given $v_C(0^-) = -5$. Therefore, $A + B = -5$. We are also given $v_s(t) = 10u(t)$, so $v_C(\infty) = 10$. Therefore, $A = 10$. Consequently, $B = -15$, and the Thévenin equivalent voltage in Fig. 11.40a is

$$v_{oc}(t) = 10 - 15e^{-t/RC} \text{ V} \qquad \text{for } t \geq 0$$

Now we will determine $i_{sc}(t)$. The short-circuit current $i_{sc}(t)$ equals $i(t)$ when terminal a is shorted to terminal b [$v_{ab}(t) = 0$]. With terminal a shorted to terminal b, the capacitance sees an equivalent resistance $R \parallel R = R/2$. Therefore, the time constant is $RC/2$, and the form of the complete response is $i_{sc}(t) = a + be^{-2t/RC}$. We are given $v_C(0^-) = -5$ V, so $i_{sc}(0^-) = -5/R$ A. We are also given $v_s(t) = 10u(t)$, so $i_{sc}(\infty) = 10/2R = 5/R$ A. This implies that the Norton equivalent current in Fig. 11.40b is

$$i_{sc}(t) = \frac{5}{R} - \frac{10}{R}e^{-2t/RC} \text{ A} \qquad \text{for } t \geq 0$$

(b) The terminal equation of the network can be found from Eqs. (11.62) and (11.67) with the use of

$$\mathbf{Z}_{ab}(j\omega) = \frac{R(2 + j\omega RC)}{1 + j\omega RC}$$

from which we have $\mathcal{A}(p) = 1 + RCp$, $\mathcal{B}(p) = R(2 + RCp)$, and

$$\left(1 + RC\frac{d}{dt}\right)v_{ab}(t) = \left(1 + RC\frac{d}{dt}\right)v_{oc}(t) - R\left(2 + RC\frac{d}{dt}\right)i(t)$$

By substituting the formula for $v_{oc}(t)$ into this equation, we obtain the result

$$RC\frac{d}{dt}v_{ab} + v_{ab}(t) = 10 - R^2C\frac{d}{dt}i(t) - 2Ri(t) \qquad t \geq 0$$

The initial energy stored in the capacitance of the Thévenin or Norton equivalent circuits is zero. ■

22. Repeat Example 11.19 with the capacitance replaced with an inductance. Assume that the circuit is at rest (that is, has no stored energy) for $t < 0$.

11.8 AC Response with PSpice

In Chapter 4 we calculated the dc response of circuits by use of PSpice. In Chapter 8 we also used PSpice to determine the response of source-free circuits, and in Chapter 9 we used it to solve for the complete response of circuits that included independent sources. In this section we will show how to determine the sinusoidal steady-state response of circuits with the use of PSpice.

The input file (circuit file) required to perform ac circuit analysis with PSpice is very similar to that required for dc analysis. Three changes are required.

1. The source type must be specified as AC rather than DC. The source statement now requires specification of both the magnitude and the angle of the source phasor.
2. The solution control statement must be .AC rather than .OP or .DC. This statement must include the source frequency in hertz (or range of frequencies, as will be shown in Chapter 13).
3. The .PRINT statement must have AC specified for the type. This statement will cause the real part, the imaginary part, the magnitude, and the angle of any requested node voltage or component current to be stored in the output file for printing.

These three statements are discussed in more detail after the following example.

EXAMPLE 11.20

Use PSpice to calculate the phasor voltage $\mathbf{V}_o$ for the circuit shown in Fig. 11.41 if $v_s = 10 \cos(t + 45°)$ V. Print the real part, the imaginary part, the magnitude, and the angle of $\mathbf{V}_o$. Also print the amplitude and phase of current i_R.

FIGURE 11.41
Example circuit for ac analysis by PSpice

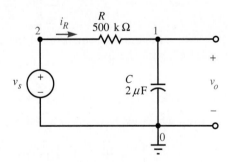

■ SOLUTION

The phasor input voltage is $\mathbf{V}_s = 10\underline{/45°}$ V and the frequency is

$$ f = \frac{\omega}{2\pi} = \frac{1}{2\pi} = 0.159155 \text{ Hz} $$

We construct the following circuit file and save it under the name EX11-20.CIR.

```
CIRCUIT FILE FOR EXAMPLE 11.20

R       2     1     500K

C       1     0     2U

VS      2     0     AC        10      45

.AC     LIN   1     .159155         .159155

.PRINT    AC      VR(1)     VI(1)     VM(1)

.PRINT    AC      VP(1)     IM(R)     IP(R)

.END
```

We use this as the input file for PSpice. The output file is stored in EX11-20.OUT if the default is used. In addition to the original circuit description, the output file includes the source frequency and the real part, the imaginary part, the magnitude, and the angle of V_1, as requested in the .PRINT statement:

```
FREQ            VR(1)          VI(1)          VM(1)

1.592E-01       7.071E+00      -1.264E-06     7.071E+00


FREQ            VP(1)          IM(R)          IP(R)

1.592E-01       -1.024E-05     1.414E-05      9.000E+01
```

■

Data Statement for AC Sources

⏎ *AC sources; see Sections A.3.2, A.3.2.1, and A.3.2.2 of the PSpice manual.*

Specification of ac sources must include the name, node connections, amplitude and phase in degrees:

```
SNAME     N+     N-     [TYPE]     VALUE     PHASE
```

SNAME is a unique source name. The first letter must be V for an independent voltage source and I for an independent current source. N+ and N− identify the node connections (passive sign convention). TYPE is AC for ac analysis. VALUE is the amplitude in volts or amperes. PHASE is the phase in degrees.

■ VOLTAGE SOURCE EXAMPLE

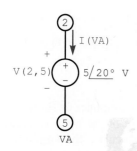

A voltage source name VA is connected with the (+) reference mark at node 2 and the (−) reference mark at node 5, and has a phasor value of $12\underline{/20°}$ V. This is specified by the statement

```
VA     2     5     AC     12     20
```
■

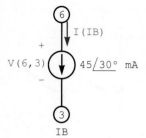

V(6,3) ↓ I(IB) 45/30° mA

IB

A current source named IB is connected from node 6 to node 3 with the reference arrow pointing toward node 3. The source has a value of $45\underline{/30°}$ mA. This is specified by the statement

```
IB      6      3      AC      45M      30  ■
```

Solution Control Statement for AC Analysis

The .AC solution control statement is used to calculate the steady-state ac response of a circuit. This statement permits calculation of a number of frequencies in one run of the program. The frequency values can be spaced linearly or logarithmically. We will postpone the use of this swept frequency capability until Chapter 13. For now we will restrict our calculations to a single frequency by specifying the linear sweep option with one frequency point and setting the beginning frequency and ending frequency equal. The statement *for a single-frequency calculation* is of the form

.AC command; see Section B.2.1 of the PSpice manual.

```
.AC     LIN     I     FSTART     FEND
```

where the starting frequency (FSTART) is the source frequency in hertz, the ending frequency (FEND) is equal to FSTART, and the frequency increment I is greater than zero. In Example 11.20, the frequency is 0.159155 Hz.

The .Print Statement

When used with the .AC solution control statement, the .PRINT statement has the form

.PRINT command; see Section B.2.9 of the PSpice manual.

```
.PRINT     AC     [OUTPUT VALUE LIST]     ...
```

The output value list can contain any node voltage or component current. The real part, imaginary part, magnitude, and angle of the requested phasor can be plotted if we add the proper suffix to the variable name.

Suffix	Output Value
None	Magnitude
M	Magnitude
DB	$20 \log_{10}$ (magnitude)
P	Phase
R	Real part
I	Imaginary part
G	Group delay (group delay will not be considered)

For examples, refer to the .PRINT statements in Examples 11.20 and 11.21.

EXAMPLE 11.21

The small-signal equivalent circuit for a common-emitter amplifier is shown in Fig. 11.42. Calculate the steady-state amplitude and phase of the node voltages when the input voltage is $v_s(t) = 4 \cos 2\pi 1000t$ V. Also calculate the amplitude and phase of the current $i_{R_2}(t)$.

FIGURE 11.42
Small-signal equivalent
circuit for Example 11.21.
A dummy source has been
inserted to force calculation
of current i_b

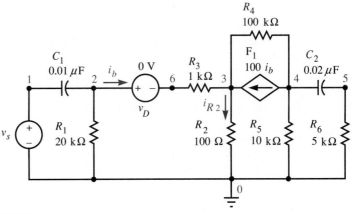

■ SOLUTION

Observe that a dummy source of 0 V has been inserted in series with R_3 to force calculation of the control current i_b. (*Caution*: Insertion of this source introduced an extra node, so R_3 and C_1 are no longer connected to the same node.) A suitable circuit file for ac analysis of this circuit is

```
CIRCUIT FILE FOR EXAMPLE 11.21

VS        1        0      AC       4      0

C1        1        2      .01U

*INSERT DUMMY VOLTAGE SOURCE

VD        2        6      AC       0      0

R1        2        0      20K

R2        3        0      100

R3        6        3      1K

R4        3        4      100K

F1        4        3      VD       100

R5        4        0      10K

C2        4        5      .02U

R6        5        0      5K

.AC       LIN      1      1000     1000

*THREE PRINT STATEMENTS USED SO

*ONLY FOUR OUTPUT VALUES PER LINE.

.PRINT AC      VM(1)      VP(1)      VM(2)      VP(2)

.PRINT AC      VM(3)      VP(3)      VM(4)      VP(4)

.PRINT AC      VM(5)      VP(5)      IM(R2)     IP(R2)

.END
```

After editing out some unwanted output, the output file contains the following values for the magnitudes and angles of the node-voltage phasors and the magnitude and angle of the requested current phasor.

FREQ	VM(1)	VP(1)	VM(2)	VP(2)
1.000E+03	4.000E+00	0.000E+00	1.608E+00	6.718E+01

FREQ	VM(3)	VP(3)	VM(4)	VP(4)
1.000E+03	1.457E+00	6.732E+01	7.980E+01	-1.426E+02

FREQ	VM(5)	VP(5)	IM(R2)	IP(R2)
1.000E+03	4.245E+01	-8.472E+01	1.457E-02	6.732E+01

■

REMEMBER

For steady-state ac analysis with PSpice:

1. The source type must be specified as AC. The magnitude and angle of the source phasor must be listed. The angle is in degrees.
2. The .AC solution control statement must be used. The frequency is in hertz.
3. The .PRINT statement must have AC specified for the type. The suffix attached to the voltage or current name determines whether the real part, the imaginary part, the magnitude, or the angle of the voltage or current phasor is printed.

EXERCISES

Use PSpice to calculate the magnitude and angle of the phasor node voltages and current I_1 in each of the following circuits, if $i_s(t) = 5 \cos 4t$ A and $v_s(t) = 100 \cos 10t$ V.

23.

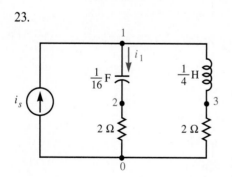

24.

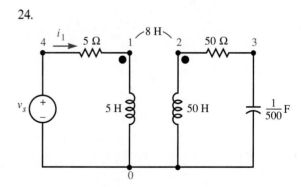

11.9 Summary

In this chapter, we described the methods of ac circuit analysis. These methods apply to circuits operating in the sinusoidal steady state, where, as was described in Chapter 10, every voltage and current waveform is sinusoidal with frequency

ω. In ac circuit analysis, the amplitude and the phase of a sinusoid, for example, $v(t) = V_m \cos(\omega t + \phi)$, are compactly represented by a phasor, $\mathbf{V} = V_m \underline{/\phi}$. The ac version of Ohm's law, $\mathbf{V} = \mathbf{Z}\mathbf{I}$, is used to relate the terminal voltage phasor $\mathbf{V}$ and current phasor $\mathbf{I}$ for R, L, and C. Similarly, algebraic terminal equations involving phasor voltage and current are used for coupled inductances and controlled sources. The ac version of KCL,

$$\sum_{\substack{\text{Closed} \\ \text{surface}}} \mathbf{I}_n = 0 \tag{11.69}$$

and KVL,

$$\sum_{\substack{\text{Closed} \\ \text{path}}} \mathbf{V}_n = 0 \tag{11.70}$$

apply. The implication of all this is that ac circuits can be analyzed with the use of methods similar to those used for circuits composed of resistances and dc sources only. We showed, for example, that impedances in series add and admittances in parallel add. We developed voltage and current divider relations for impedances similar to those of Chapter 3. We formulated the systematic circuit analysis techniques of node-voltage analysis and mesh-current analysis of Chapter 4 for ac circuits. We then applied the principle of superposition to circuits containing more than one sinusoidal source. At this point we showed how to apply ac circuit analysis techniques to circuits containing sinusoidal sources with possibly *different* frequencies by analyzing the ac circuit for each frequency separately. We then extended the concept of Thévenin and Norton equivalent circuits to ac circuits. In an optional section, we developed the Thévenin and Norton equivalents of LLTI circuits with nonsinusoidal sources.

KEY FACTS

♦ Ac circuit analysis relies on the ac version of Ohm's law, $\mathbf{V} = \mathbf{Z}\mathbf{I}$, and the phasor versions of KCL,

$$\sum_{\substack{\text{Closed} \\ \text{surface}}} \mathbf{I}_n = 0$$

and KVL,

$$\sum_{\substack{\text{Closed} \\ \text{path}}} \mathbf{V}_n = 0$$

♦ Impedances in series add.

♦ Admittances in parallel add.

♦ Algebraic voltage and current divider relations apply for phasor voltages and currents.

♦ Algebraic node-voltage and mesh-current circuit analysis applies for phasor voltages and currents.

♦ When we apply superposition to ac circuit analysis, independent sources with different frequencies *must* be considered separately.

♦ It is *never* correct to add nonrotating phasors that represent sinusoids that have different frequencies.

- The Thévenin equivalent of an ac circuit consists of the open-circuit voltage phasor, $\mathbf{V}_{oc}$, in series with the Thévenin equivalent impedance.

- The Norton equivalent of an ac circuit consists of the short-circuit current phasor, $\mathbf{I}_{sc}$, in parallel with the Norton equivalent impedance.

- The Thévenin and Norton equivalent impedances are equal and are given by $\mathbf{Z}_{Eq} = \mathbf{V}_{oc}/\mathbf{I}_{sc}$.

■ PROBLEMS ■

Section 11.1

Give the formula for the time-domain waveform associated with each of the following phasors. The units are milliamperes and volts. Use the $\mathscr{R}e\,\{\cdot\}$ convention.

1. $\mathbf{I} = 6\underline{/30^\circ}$
2. $\mathbf{I} = 1 + j$
3. $\mathbf{V} = 7e^{-j45^\circ}$
4. $\mathbf{V} = 10j$

5. (a) Find the numerical values of the phasor currents $\mathbf{I}_1$, $\mathbf{I}_2$, and $\mathbf{I}_3$ corresponding to the time-domain currents of Fig. 11.43. Put your answers in polar form.
   (b) Use your answer from (a) to find $i_1(t)$. Express your answer in the form $i_1(t) = A\cos(100t + \phi)$.
   (c) Carefully plot $i_1(t)$, $i_2(t)$, and $i_3(t)$. Check your results by noting from the plots that $i_1(t) = i_2(t) + i_3(t)$.
   (d) Refer to a phasor diagram of the equation

$\mathbf{I}_1 = \mathbf{I}_2 + \mathbf{I}_3$ to explain why the amplitude of $i_1(t)$ is *not* 50.
   (e) Explain why the amplitude of $i_1(t)$ is *not* 50 by referring to your plots from (c).

6. A voltage is given by $v(t) = 3\cos 100t - 3\sin 100t$.
   (a) Find the phasor representation, $\mathbf{V}$, of $v(t)$ with the use of $\mathscr{R}e\,\{\cdot\}$ convention.
   (b) Sketch $\mathbf{V}$ in the complex plane.
   (c) Use your answer from (a) and the formula $v(t) = \mathscr{R}e\,\{\mathbf{V}e^{j100t}\}$ to write $v(t) = A\cos(100t + \phi)$.
   (d) Check your answer to (c) using trigonometric identities.

7. Use phasors to find $v_x(t)$ in the equation $v_x(t) + \sum_{n=1}^{7}\cos(\omega t + n45^\circ) = 0$. *Hint:* Analytically add the phasor representations of $\cos(\omega t + n45^\circ)$ for $1 \le n \le 7$.

8. For the circuit in Fig. 11.44, find $\mathbf{I}_1$ if $\mathbf{I}_4 = 4\underline{/90^\circ}$, $\mathbf{I}_5 = 5\underline{/45^\circ}$, and $\mathbf{I}_6 = 6\underline{/135^\circ}$ A.

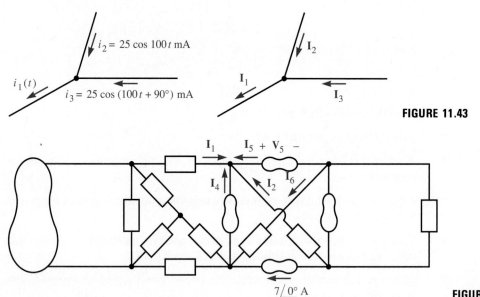

$i_2 = 25\cos 100t$ mA

$i_1(t)$

$i_3 = 25\cos(100t + 90^\circ)$ mA

I_2

I_1

I_3

FIGURE 11.43

I_1 $I_5 + V_5 -$

I_4 I_2 I_6

$7\underline{/0^\circ}$ A

FIGURE 11.44

9. Find the differential equation relating $v(t)$ to $i(t)$ for a circuit with transfer impedance

$$Z = \frac{1}{1 + j\omega 3} \ \Omega$$

10. Find the differential equation relating $v(t)$ to $i(t)$ for a circuit with transfer admittance

$$Y = \frac{j\omega}{2 + 4j\omega} \ S$$

11. Find the differential equation relating $v(t)$ to $i(t)$ if

$$\frac{I}{V} = j\omega C + \frac{1}{R} + \frac{1}{j\omega L}$$

12. An ac circuit operates at $\omega = 100$ rad/s. We are given the system of equations

$$\begin{bmatrix} 3 & 2j \\ 1 & -2 \end{bmatrix} \begin{bmatrix} I_1 \\ I_2 \end{bmatrix} = \begin{bmatrix} 12\underline{/60°} \\ 4 \end{bmatrix}$$

Find $i_1(t)$ and $i_2(t)$.

Section 11.2

Find the driving-point impedance of each circuit by inspection.

13.

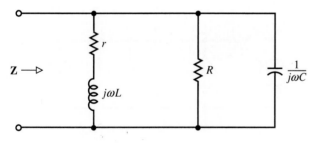

14.

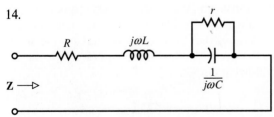

15. Use your answer to Problem 13 to find the differential equation relating driving-point current input to driving-point voltage output for the circuit in Problem 13.

16. Repeat Problem 15 for the circuit in Problem 14.

17. For the following circuit, find V_a and I_b by inspection.

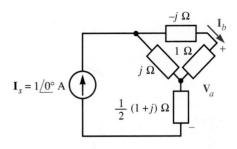

18. For the following circuit, find V_a, V_b, and I_c by a Δ-to-Y transformation (see page 100).

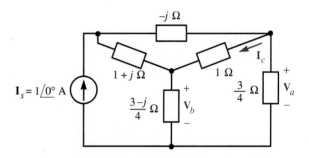

19. (a) Use the voltage divider equation to determine V_R, V_L, and V_C in the series $RL\bar{C}$ circuit shown below.
    (b) Plot $v_R(t)$, $v_L(t)$, and $v_C(t)$.
    (c) Show that $v_L(t) + v_C(t) = 0$. Discuss.

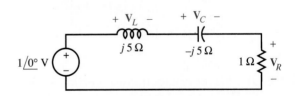

20. (a) For what value(s) of Z does the driving-point impedance Z_{in} of the circuit shown below equal Z?

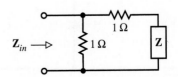

(b) Let $\mathbf{Z}_0$ denote the answer to (a). Show that the driving-point impedance of the circuit shown below also equals $\mathbf{Z}_0$.

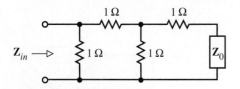

21. Find the driving-point impedance of the infinite ladder network shown below.

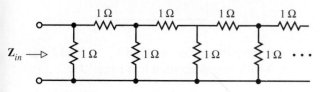

22. Compare the driving-point impedances of the circuits shown below.

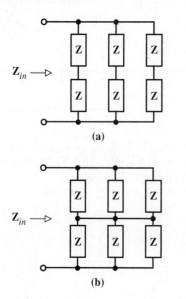

(a)

(b)

23. (a) Find the voltage transfer function $\mathbf{V}_2/\mathbf{V}_1 = \mathbf{H}_{21}(j\omega)$ by inspection of the circuit shown below.

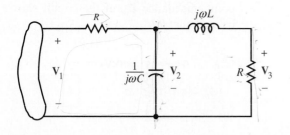

(b) Now find $\mathbf{V}_3/\mathbf{V}_2$ by inspection.
(c) Combine your results from (a) and (b) to determine $\mathbf{V}_3/\mathbf{V}_1 = \mathbf{H}_{31}(j\omega)$.
(d) Use inspection to check your answer to (c).

Find the differential equations relating the following for the circuit of Problem 23.

24. $v_2(t)$ to $v_1(t)$
25. $v_3(t)$ to $v_2(t)$
26. $v_3(t)$ to $v_1(t)$
27. By inspection, find the formulas for $\mathbf{I}_2$, $\mathbf{I}_3$, $\mathbf{V}_1$, and $\mathbf{V}_4$ in the circuit below.

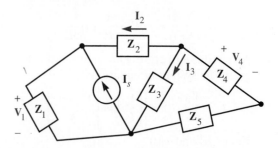

Section 11.3

28. Set up the node-voltage equations for the circuit shown below.

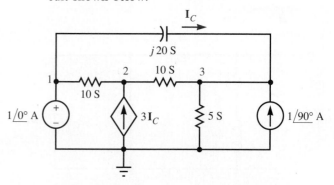

29. Transform the node-voltage equations obtained in Problem 28 to the time domain. Assume that the capacitance has a value of 2 F.
30. Set up the node-voltage equations for the following circuit.

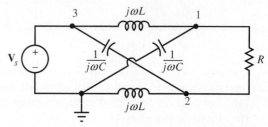

31. The circuit in Fig. 11.45 is an ac model for a single-stage transistor amplifier. Set up the node-voltage equations.

32. Set up and solve the node-voltage equations for the circuit in Fig. 11.46.

33. (a) Set up, in matrix form, the node-voltage equations for the Δ (top) and Y (bottom)

circuits driven by current sources shown in Fig. 11.47.

(b) Give the admittance matrix for each circuit.

(c) Find the relationship between the Δ-circuit admittances and the Y-circuit admittances that would make the Δ and Y circuits electrically equivalent. Specifically, solve for Y_a, Y_b, and Y_c in terms of Y_1, Y_2, and Y_{12}.

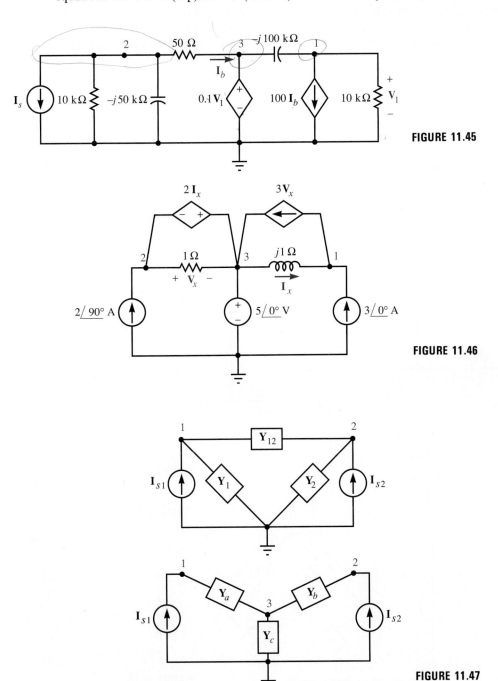

FIGURE 11.45

FIGURE 11.46

FIGURE 11.47

34. (a) Set up, in matrix form, the node-voltage equations for the circuit in Fig. 11.48.
    (b) What is the network's admittance matrix?
    (c) Solve for the node voltages.
35. (a) Set up the node-voltage equations for the finite-gain op-amp circuit in Fig. 11.49.
    (b) Solve for $\mathbf{V}_o$.
    (c) Let $A \to \infty$ in your answer to (b). Does your result agree with that in Example 11.9?
36. Figure 11.50 depicts a basic four-arm ac bridge used to measure impedance. The bridge is said

to be balanced when $\mathbf{V}_{ac} = 0$. Show that the bridge is balanced if $\mathbf{Z}_1\mathbf{Z}_4 = \mathbf{Z}_2\mathbf{Z}_3$.

37. The ac bridge of Problem 36 can be used to measure a capacitor modeled by a capacitance C_x in series with a resistance R_x by setting $\mathbf{Z}_1 = R_1$, $\mathbf{Z}_2 = R_2$, $\mathbf{Z}_3 = R_3 + (1/j\omega C_3)$, and $\mathbf{Z}_4 = \mathbf{Z}_x = R_x + (1/j\omega C_x)$.
    (a) Show that when the bridge is balanced, C_x and R_x are given by $C_x = C_3 R_1/R_2$ and $R_x = R_2 R_3/R_1$, independent of ω.
    (b) How many of the bridge elements must be made adjustable to obtain a balance?

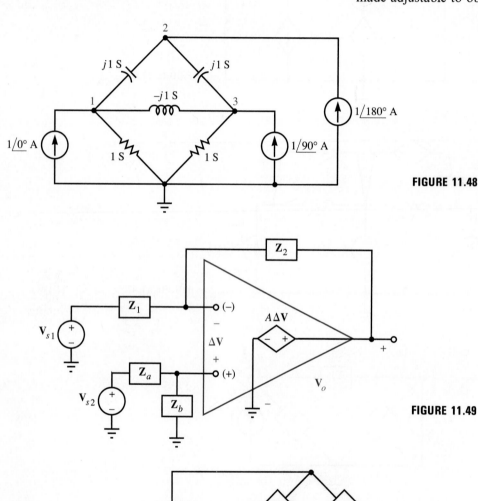

FIGURE 11.48

FIGURE 11.49

FIGURE 11.50

38. The ac bridge of Problem 36 can be used to measure an inductor (modeled by an inductance L_x in series with a resistance R_x) by setting $\mathbf{Z}_1 = R_1 \parallel (1/j\omega C_1)$, $\mathbf{Z}_2 = R_2$, $\mathbf{Z}_3 = R_3$, and $\mathbf{Z}_4 = \mathbf{Z}_x$. This is called a Maxwell bridge.
   (a) Show that when the bridge is balanced, L_x and R_x are given by $L_x = R_2 R_3 C_1$ and $R_x = R_2 R_3 / R_1$.
   (b) How many bridge elements must be made adjustable to obtain a balance?

39. (a) Set up, in matrix form, the node-voltage equations for $\mathbf{V}_1$ and $\mathbf{V}_2$ for the circuit in Fig. 11.51. 🔛 11.7
   (b) Transform the equations of (a) into the time domain.

(c) Solve the equations of (a) for $\mathbf{V}_2$. (Eliminate $\mathbf{V}_1$.)

Section 11.4

40. Write a set of mesh-current equations for the circuit in Fig. 11.52.
41. Assign mesh currents to the circuit of Problem 32. Then set up and solve the mesh-current equations.
42. Write the mesh-current equations for the circuit in Fig. 11.53.
43. Set up and solve the mesh-current equation for the circuit in Fig. 11.54.
44. Can *any* ac circuit be solved using mesh-current analysis?

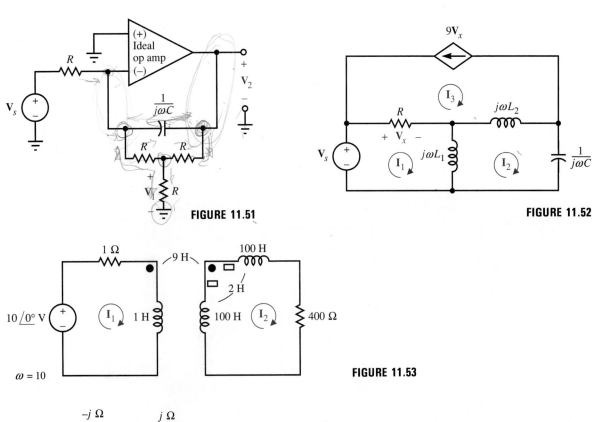

FIGURE 11.51

FIGURE 11.52

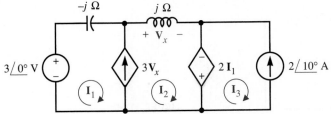

FIGURE 11.53

FIGURE 11.54

45. (a) Set up, in matrix form, the mesh-current equations for the following circuit.
    (b) What is the circuit's impedance matrix?
    (c) Solve for the mesh currents.

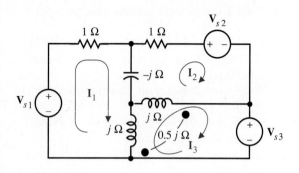

46. Set up the mesh-current equations for the following network.

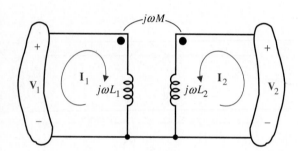

47. Set up the mesh-current equations for the following network.

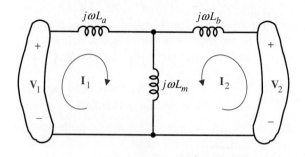

48. How must L_a, L_b, and L_m be related to L_1, L_2, and M in order for the Y circuit of Problem 47 to be electrically equivalent to the coupled inductances of Problem 46?

49. By assuming that currents I_1 and I_2 in Problem 46 are created by current sources, determine the impedance matrix for coupled coils.

Section 11.5

50. Find $i(t)$ in the circuit in Fig. 11.55.

51. Find $v(t)$ in the circuit in Fig. 11.56.

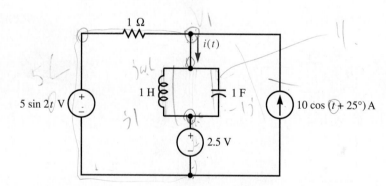

FIGURE 11.55

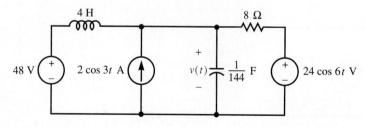

FIGURE 11.56

52. Find $v_o(t)$ in the circuit in Fig. 11.57.

53. Determine $i_o(t)$ in the circuit in Fig. 11.58 without the use of mesh-current or node-voltage equations.

54. Find $v_o(t)$ in the circuit in Fig. 11.59.

55. Find $v_o(t)$ for the ideal op-amp circuit in Fig. 11.60.

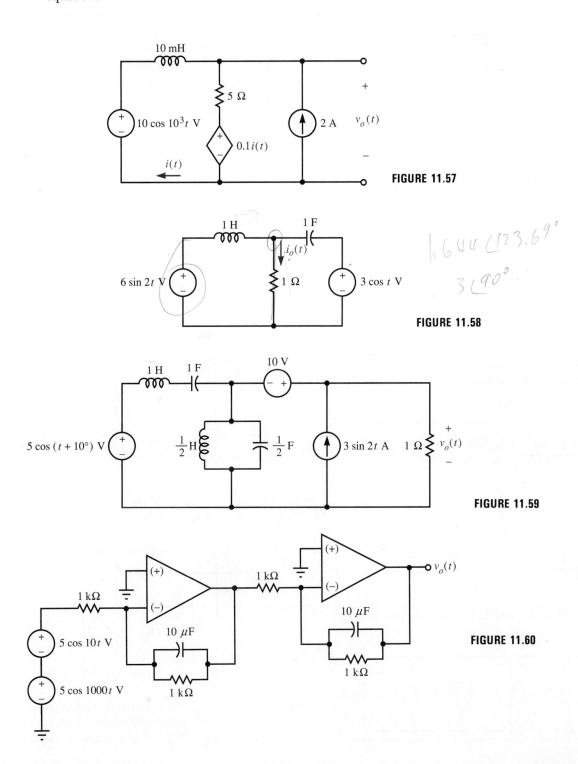

FIGURE 11.57

FIGURE 11.58

FIGURE 11.59

FIGURE 11.60

Section 11.6

56. Find the Thévenin and Norton equivalents of the following circuit.

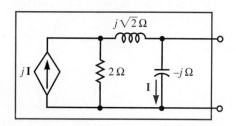

Find the Thévenin and Norton equivalents of each of the following circuits.

57.

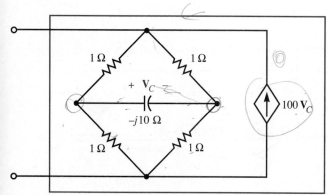

58.

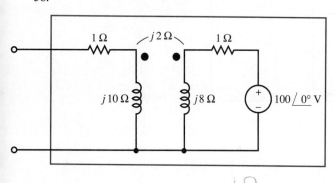

59. (a) Find $\mathbf{V}_{oc}$ and $\mathbf{I}_{sc}$ for the following circuit.

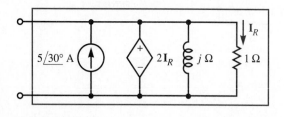

(b) Find $\mathbf{Z}_{Eq}$ with use of the formula $\mathbf{Z}_{Eq} = \mathbf{V}_{oc}/\mathbf{I}_{sc}$.

(c) Derive $\mathbf{Z}_{Eq}$ with another method, and check your result from (b).

60. We are given the Thévenin equivalents of each of two separate circuits. The two circuits are then connected in parallel. Find the Thévenin equivalent of the parallel circuit.

61. Find the Thévenin and Norton equivalents of the following circuit.

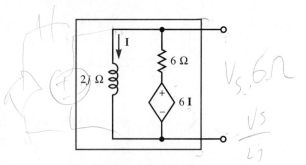

62. The following circuit does not have a Thévenin equivalent. Why not?

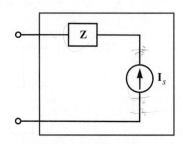

63. The following circuit does not have a Norton equivalent. Why not?

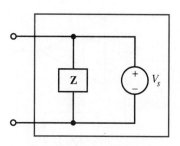

64. A linear time-invariant electrical network, network A, is terminated (separately) with a 1-Ω resistance and a 1000-μF capacitance as shown

below. The respective currents (steady-state responses) are $i_1(t) = 5 \cos(1000t - 45°)$ A and $i_2(t) = 10 \cos(1000t - 45°)$ A. Find the Thévenin and Norton equivalents of network A.

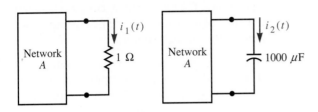

65. Assume that the capacitance or resistance shown in the circuit of Problem 64 is replaced with a 1-mH inductance. What current $i_3(t)$ will flow through the inductance?

66. When two ac networks, A and B, are connected together as shown in the first illustration below, $\mathbf{I}_{ab} = 1\underline{/0°}$ A and $\mathbf{V}_{aa'} = \sqrt{2}\underline{/-45°}$ V. When the same networks are connected as shown in the second illustration, $\mathbf{I}_{ab'} = 3\underline{/0°}$ A and

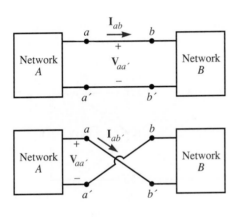

$\mathbf{V}_{aa'} = \sqrt{2}\underline{/45°}$ V. Find the Thévenin and Norton equivalents of each ac network A and B.

67. (a) Find the Thévenin equivalent circuit looking back into the output of the finite-gain op-amp circuit in Fig. 11.61.
    (b) What happens to your answer from (a) in the limit $A \to \infty$?

Section 11.7

68. Find the Thévenin and Norton equivalents of the coupled-coil circuit shown below. Assume that the circuit is at rest for $t < 0$.

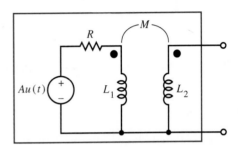

69. Find the Thévenin and Norton equivalents of the RL voltage divider shown below. Assume that the circuit is at rest for $t < 0$.

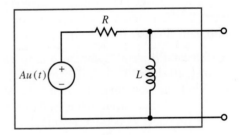

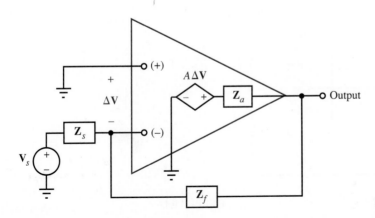

FIGURE 11.61

70. Find the Thévenin and Norton equivalents of the circuit shown below, assuming that $\omega_1 = \omega_2 = 10^3$ rad/s.

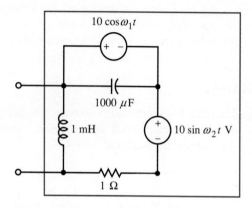

71. Repeat Problem 61 for $\omega_1 = 10^3$ rad/s and $\omega_2 = 2 \times 10^3$ rad/s.

Section 11.8

Use PSpice to find the phasors in Problems 72 and 73.

72. Find the node voltages in the op-amp circuit of Problem 35. Assume that $\mathbf{Z}_a$, $\mathbf{Z}_b$, and $\mathbf{Z}_1$ are 1-kΩ resistances and that $\mathbf{Z}_2$ is the impedance of a 1-μF capacitance, where $f = 1$ kHz. Let $\mathbf{V}_{s1} = 2$ mV$\underline{/0°}$ and $\mathbf{V}_{s2} = 0$, and let the open-loop gain A equal 10^5.

73. (a) Repeat Problem 72 for open-loop gain A of 1, 10, 100, 10^3, 10^4, 10^6, and 10^7.
   (b) What can you conclude about the dependence of the output-voltage phasor $\mathbf{V}_o$ on the amplifier's open-loop gain?

Problems 74 through 77 refer to the circuit in Fig. 11.62. The input voltage is $v_s = \cos 2\pi 1000t$ V. Use PSpice to solve the problems.

74. Replace capacitance C_1 by a short circuit and capacitance C_2 by an open circuit. Calculate all node voltages, and determine the impedance seen by the voltage source.

75. Calculate a value of C_1 that has an impedance of magnitude one-tenth the input resistance calculated in Problem 74. Use this capacitance value, calculate all node voltages, and determine the impedance seen by the voltage source ($C_2 = 0$).

76. Use the value of C_1 found in Problem 75. Let $C_2 = 50$ μF. Then repeat Problem 75. Compare the results with those obtained in Problem 75. Has the input impedance changed?

77. Repeat Problem 75, but use a capacitance C_1 of ten times the value determined in Problem 75 and let $C_2 = 50$ μF.

Comprehensive Problems

Use any method to find the driving-point impedance of the networks in Problems 78 and 79.

78.

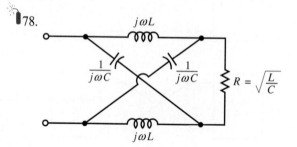

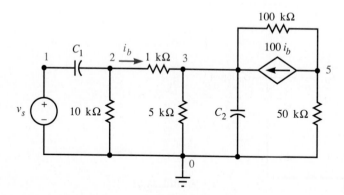

FIGURE 11.62

79. See Fig. 11.63.

80. (a) The feedback ratio of the voltage amplifier circuit model in Fig. 11.64 is $V_{Fb}/V_0 = \beta = Z_a/(Z_a + Z_b)$. Show that the closed-loop gain $G = V_o/V_s$ is given by

$$G = \frac{A\beta}{A\beta + 1}\frac{1}{\beta}$$

where $A = A(j\omega)$ is the open-loop gain of the op amp.

(b) Under what condition does $G \simeq 1/\beta$?

(c) What is the relative error $[G - (1/\beta)]/G$ in the approximation of (b)?

81. Consider the circuit in Fig. 11.65.
(a) Show that

$$G = \frac{V_o}{V_s} = -\frac{Z_b}{Z_a}$$

for $|A\beta| \gg 1$, where β is the feedback factor, $\beta = Z_a/(Z_a + Z_b)$.

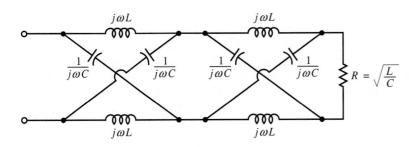

FIGURE 11.63

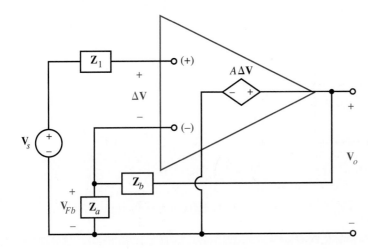

FIGURE 11.64

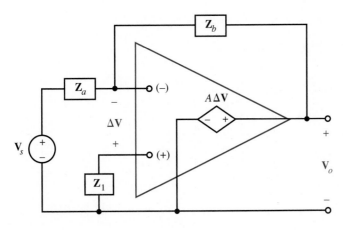

FIGURE 11.65

(b) Assume that $\mathbf{Z}_a = 1/j\omega C$ and $\mathbf{Z}_b = R$. Use the approximation in (a) to obtain a differential equation relating $v_o(t)$ to $v_s(t)$.

(c) Repeat (b) for $\mathbf{Z}_a = R$ and $\mathbf{Z}_b = 1/j\omega C$.

The electrical energy required to keep an automobile battery charged is supplied by an alternating current generator, called an *alternator* that is driven by the engine. The ac voltage from the alternator is converted to dc for the battery. For our purposes, we can model the alternator as a voltage source $v_s(t) = \alpha n \cos nt$ V in series with an inductance of value $L = 0.4$ mH, and the load as a variable resistance of value R_L as shown in the following circuit. The variable n is the engine speed, which has a minimum value of $n = n_1 = 1000$ rpm and a maximum value of $n = n_2 = 10,000$ rpm. The parameter α is adjusted by a *voltage regulator* that attempts to maintain a constant amplitude for voltage v_L. However, α has a maximum value of α_m. For $n = n_1$ this limits the maximum amplitude of the load current i_L to 40 A, if the amplitude of v_L is maintained at 12 V.

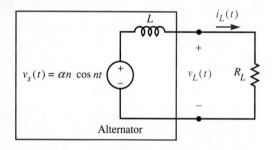

$v_s(t) = \alpha n \cos nt$

Alternator

82. What is the maximum amplitude αn of the alternator *internal voltage* $v_s(t)$ for $n = n_1$?

83. What is the amplitude of the short-circuit current $(R_L = 0)$ for $n = n_1$?

84. The engine speed is increased to $n = n_2$. What is the amplitude of the internal voltage $v_s(t)$?

85. What is the amplitude of the short-circuit current $(R_L = 0)$ for $n = n_2$?

86. What is the maximum amplitude that current i_L can have if the amplitude of the load voltage $v_L(t)$ is maintained at 12 V?

87. Explain why an automobile alternator is not equipped with a *current regulator* to limit the maximum available current.

PSpice gives the particular response for a sinusoidal input to an unstable circuit. The particular response, of course, is not necessarily the steady-state response. Use PSpice to perform a steady-state analysis on the following circuit for the values in Problems 88 through 90 if $v_s(t) = 100 \cos 2t$ V.

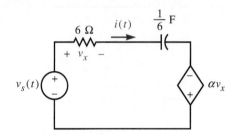

88. $\alpha = \frac{1}{2}$

89. $\alpha = 1$

90. $\alpha = \frac{3}{2}$

91. Do the circuits in Problems 88 through 90 really have a steady-state solution?

CHAPTER 12

Power in AC Circuits

In this chapter, we investigate the flow of energy in ac circuits. In any ac circuit, the instantaneous power $p(t)$ delivered to any network part is given by the product of a sinusoidal voltage and a sinusoidal current. By plotting $p(t)$ and by using various trigonometric identities, we discover properties of energy flow in ac circuits that are both fascinating and practically useful. We find, for example, that ac networks not only consume energy at an average rate but also borrow and return energy to their sources. These and other related facts are important in the design of virtually every ac circuit and source, whether on the microelectronic scale or on the scale of large equipment used in factories.

We begin with a basic description of energy flow between an ac network and a source. This leads to the basic definitions of average power P, reactive power Q, and complex power S. We then state and prove the complex power balance theorem, which is akin to a conservation of energy law for an ac network. Finally, we derive the conditions under which maximum average power is exchanged between an ac source and an electrical load.

12.1 Instantaneous Power and Average Power

In this section we investigate the flow of energy between a sinusoidal source and a passive LLTI network. The source can be a sinusoidal voltage source such as the ac outlet in your home, a sinusoidal current source, or, more generally, a network containing sinusoidal sources operating at a common frequency ω. The voltage and the current at the terminals of the network are shown in Fig. 12.1, where

$$v(t) = V_m \cos{(\omega t + \phi_\mathbf{V})} \qquad (12.1)$$

and

$$i(t) = I_m \cos{(\omega t + \phi_\mathbf{I})} \qquad (12.2)$$

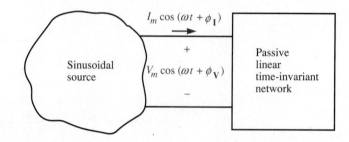

The instantaneous power entering the network is the product of voltage $v(t)$ and current $i(t)$:

$$p(t) = v(t)i(t) = V_m \cos{(\omega t + \phi_\mathbf{V})} I_m \cos{(\omega t + \phi_\mathbf{I})} \qquad (12.3)$$

The unit of $p(t)$ is the watt. It is informative to write Eq. (12.3) in another form. Using the trigonometric identity

$$\cos A \cos B = \frac{1}{2} \cos{(A - B)} + \frac{1}{2} \cos{(A + B)} \qquad (12.4)$$

we find that

$$p(t) = \frac{1}{2} V_m I_m \cos{(\phi_\mathbf{V} - \phi_\mathbf{I})} + \frac{1}{2} V_m I_m \cos{(2\omega t + \phi_\mathbf{V} + \phi_\mathbf{I})} \qquad (12.5)$$

or

$$p(t) = \frac{1}{2} V_m I_m \cos\theta + \frac{1}{2} V_m I_m \cos{(2\omega t + \phi_\mathbf{V} + \phi_\mathbf{I})} \qquad (12.6)$$

where

$$\theta = \phi_\mathbf{V} - \phi_\mathbf{I} \qquad (12.7)$$

is the angle of the network's driving-point impedance. The factor $\cos\theta$ in Eq. (12.6) is called the *power factor* of the network and is symbolized as PF:

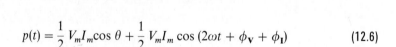

Power Factor
$\mathrm{PF} = \cos\theta \qquad (12.8)$

The angle θ is called the *power-factor angle*. If $\theta > 0$, then the current lags the voltage, and if $\theta < 0$, then the current leads the voltage. Because $\cos(-\theta) = \cos(+\theta)$, it is not possible to determine whether the current is leading or lagging the voltage from the value of PF alone. To convey this information, we refer to a *lagging power factor* for $\theta > 0$ (the current lags the voltage) and a *leading power factor* for $\theta < 0$ (the current leads the voltage). Substitution of Eq. (12.8) into Eq. (12.6) yields

$$p(t) = \frac{1}{2} V_m I_m \cdot \text{PF} + \frac{1}{2} V_m I_m \cos(2\omega t + \phi_{\mathbf{V}} + \phi_{\mathbf{I}}) \tag{12.9}$$

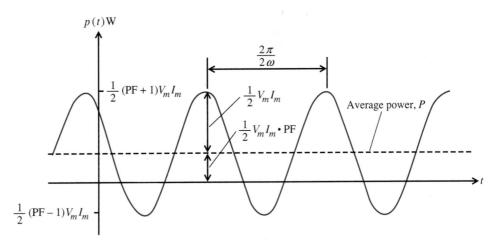

FIGURE 12.2 The instantaneous power $p(t)$ entering a passive LLTI network [Plot of Eq. (12.9)]

Equation (12.9) is plotted in Fig. 12.2. Observe that the instantaneous power entering the network is periodic with angular frequency 2ω. There are two components in $p(t)$: a constant component $\frac{1}{2} V_m I_m \cdot$ PF and a sinusoidal component with amplitude $\frac{1}{2} V_m I_m$ and angular frequency 2ω. According to Eq. (12.8), the largest possible value of a power factor is unity; PF ≤ 1. Therefore $p(t)$ will generally assume both positive and negative values, as is illustrated in Fig. 12.2. When $p(t)$ is positive, energy flows into the network from the source. When $p(t)$ is negative, energy flows back out of the network into the source. Thus the source is not only delivering energy that is being consumed by the network, it is also lending energy that is returned by the network. The *average power* delivered to the network is the constant (first) term in Eq. (12.9), because the average value of a sinusoid is zero. The average power delivered to the network is also called the *real power* or the *active power*, and is denoted by P:

$$P = \frac{1}{2} V_m I_m \cdot \text{PF} = \frac{1}{2} V_m I_m \cos\theta \tag{12.10}$$

The unit of average power is, of course, the watt.

Equation (12.10) has an important theoretical implication. Remember that P must be nonnegative for a passive load. Therefore $\cos\theta$ must be nonnegative, and

Digital wattmeter. Courtesy
of Valhalla Scientific, Inc.

so θ must be in the range $-\pi/2 \le \theta \le +\pi/2$. This means that *the resistive component of the driving-point impedance of every passive network is nonnegative*;

$$R(\omega) = \mathcal{R}e\left\{\mathbf{Z}(j\omega)\right\} \ge 0$$

EXAMPLE 12.1

A factory draws 120-A rms current from a 12,470-V rms 60-Hz source at a power factor of 0.75 lagging. What is the average power consumed?

■ SOLUTION

From Example 7.9, the peak voltage and current V_m and I_m in Eq. (12.10) are related to the rms (or effective) voltage V_{rms} and current I_{rms} by

$$V_m = \sqrt{2}\, V_{\text{rms}}$$

and

$$I_m = \sqrt{2}\, I_{\text{rms}}$$

Accordingly, Eq. (12.10) becomes

$$P = V_{\text{rms}} I_{\text{rms}} \cdot \text{PF}$$

which is, in this example,

$$P = 12{,}470 \times 120 \times 0.75 = 1.12 \text{ MW} \quad ■$$

EXERCISES

1. Define and give the units of (a) instantaneous power, (b) average power, and (c) the power factor.

■ *See Problem 12.1 in the PSpice manual.*

2. A sinusoidal current with amplitude 10 A is applied to an impedance $\mathbf{Z} = 10\underline{/45^\circ}\ \Omega$.
   (a) Determine the power factor. (Be sure to indicate whether it is leading or lagging.)
   (b) Determine the average power supplied.

3. Show why driving-point resistance must be nonnegative for every passive network.

12.2 Apparent Power

Notice that the formula [Eq. (12.10)] for the average power supplied by a source depends not only on the amplitudes of voltage and current but also on the power factor. The product

$$\frac{1}{2} V_m I_m = \frac{P}{\text{PF}} \tag{12.11}$$

is called the *apparent power*. The unit of apparent power is called the *volt-ampere* (VA) to emphasize that apparent power is simply one-half the product of the voltage and current amplitudes. This product is clearly *not* the average power P. As we see from Eq. (12.11), apparent power is computed by *dividing* the average power by the power factor.

Apparent power has practical significance for an electric utility company, because a utility company must supply both average and apparent power to a customer. To illustrate, suppose that a factory requires an average power P at a given voltage amplitude V_m. We can see from Eq. (12.11) that the *apparent* power will be larger than P if the power factor is less than 1. Thus, the current amplitude I_m that must be supplied will be larger for PF < 1 than it would be for PF $= 1$, even though the average power P supplied is the same in either case. A larger current amplitude cannot be supplied without additional cost to the utility company. The additional cost arises because greater current in the transmission wires connecting the factory to the utility company's generators will be accompanied by more power dissipation (through heat) in the transmission lines. For this reason, a utility company will often raise its rates for industrial customers who operate at low power factors.

EXAMPLE 12.2

A fully loaded 50-hp induction motor operates at 75 percent efficiency and a power factor of 0.8 lagging. Find the apparent power.

■ SOLUTION

Efficiency, η, is the ratio of output power to input power. The conversion between watts and horsepower is 1 hp $= 745.7$ W. Consequently, the input power P is

$$P = \frac{P_{\text{out}}}{\eta} = \frac{50 \times 745.7}{0.75} \text{ W}$$

$$= 49.7 \text{ kW}$$

Therefore the apparent power is, from Eq. (12.11),

$$\frac{1}{2} V_m I_m = \frac{49.7}{0.8} \text{ kVA} = 62.1 \text{ kVA} \quad ■$$

4. Define and give the unit of apparent power.
5. A sinusoidal current with amplitude 10 A is delivered to an impedance $\mathbf{Z} = 10\underline{/45^\circ}\ \Omega$. Determine the apparent power supplied.
6. Explain the practical significance of apparent power.

12.3   Reactive Power

Significant insight into the flow of energy between a source and a network can be obtained from certain trigonometric identities. These trigonometric identities will show us that the instantaneous power supplied to a passive network can be regarded as consisting of two parts. The first part describes energy that always flows into the network and is dissipated in it. The second part describes energy that is only borrowed and returned by the network.

The first step is to use the identity $\cos{(A + B)} = \cos A \cos B - \sin A \sin B$ to rewrite $v(t)$ as

$$
\begin{aligned}
v(t) &= V_m \cos{(\omega t + \phi_\mathbf{V})} \\
&= V_m \cos{(\omega t + \phi_\mathbf{I} + \theta)} \\
&= \underbrace{V_m \cos \theta \cos{(\omega t + \phi_\mathbf{I})}}_{\text{In phase with } i(t)} - \underbrace{V_m \sin \theta \sin{(\omega t + \phi_\mathbf{I})}}_{\substack{\text{In phase quadrature} \\ \text{with } i(t)}}
\end{aligned} \tag{12.12}
$$

The first term on the right-hand side of Eq. (12.12) is *in phase* with $i(t) = I_m \cos{(\omega t + \phi_\mathbf{I})}$ because it has the same phase angle $\phi_\mathbf{I}$ as $i(t)$. The second term is *in phase quadrature* with $i(t)$ because it is 90° out of phase with $i(t)$ [$\cos{(\omega t + \phi_\mathbf{I})}$ leads $\sin{(\omega t + \phi_\mathbf{I})}$ by 90°]. By using Eq. (12.12), we can express the instantaneous power as

$$
\begin{aligned}
p(t) &= v(t)i(t) \\
&= V_m I_m \cos \theta \cos^2{(\omega t + \phi_\mathbf{I})} - V_m I_m \sin \theta \cos{(\omega t + \phi_\mathbf{I})} \sin{(\omega t + \phi_\mathbf{I})} \quad (12.13)
\end{aligned}
$$

Next, use $\cos^2 A = \frac{1}{2}(1 + \cos 2A)$ and $\cos A \sin A = \frac{1}{2}\sin 2A = -\frac{1}{2}\cos{[2A + (\pi/2)]}$ to obtain

$$
p(t) = \frac{1}{2} V_m I_m \cos \theta [1 + \cos{(2\omega t + 2\phi_\mathbf{I})}] + \frac{1}{2} V_m I_m \sin \theta \cos{\left(2\omega t + 2\phi_\mathbf{I} + \frac{\pi}{2}\right)}
\tag{12.14}
$$

which is

$$
p(t) = \underbrace{P[1 + \cos{(2\omega t + 2\phi_\mathbf{I})}]}_{\substack{\text{Energy flow into} \\ \text{the network}}} \overset{\text{W\,Jts}}{} + \underbrace{Q \cos{\left(2\omega t + 2\phi_\mathbf{I} + \frac{\pi}{2}\right)}}_{\substack{\text{Energy borrowed and} \\ \text{returned by} \\ \text{the network}}} \overset{\text{VAR}}{}
\tag{12.15}
$$

where

$$P = \frac{1}{2} V_m I_m \cos \theta \qquad (12.16)$$

and

$$Q = \frac{1}{2} V_m I_m \sin \theta \qquad (12.17)$$

The letter Q serves as a reminder that the second term in Eq. (12.15) comes from that part of $v(t)$ that is *in phase quadrature* with $i(t)$. The terms in Eq. (12.15) are plotted in Fig. 12.3 for $\theta = 45°$. The in-phase and quadrature-phase components of $v(t)$ are responsible for the first and second terms, respectively, on the right-hand side of Eq. (12.15). The first term is nonnegative if $\cos \theta \geq 0$, which must be the case for a passive network. Therefore the in-phase component of $v(t)$ produces a periodically fluctuating but continuous flow of energy into the network. The average value of the first term is P, as defined in Eq. (12.16). Therefore *the in-phase component of voltage is responsible for the delivery of the power consumed.* The second term in Eq. (12.15) is sinusoidal and therefore has an average value of zero. Therefore *the quadrature-phase component of voltage is responsible for the periodic two-way exchange of energy between the source and the network.* As far as the second term of Eq. (12.15) is concerned, the network is only borrowing energy and returning it to the source. The amount of energy borrowed depends on the amplitude, Q, of the second term in Eq. (12.15), as defined by Eq. (12.17). Q is called the *reactive* power and has the unit of *volt-ampere reactive* (VAR). As indicated earlier, the active power P is the average rate at which energy is consumed by the network. Equation (12.15) shows us that the reactive power Q is the maximum rate at which energy is borrowed by the network. Notice from Eqs. (12.16) and (12.17) and $\theta = \underline{/Z}$ that $Q = 0$ if $\mathbf{Z}$ is purely real, whereas $P = 0$ if $\mathbf{Z}$ is purely imaginary. Thus a resistance does not borrow and return energy, and an inductance and a capacitance do not consume energy.

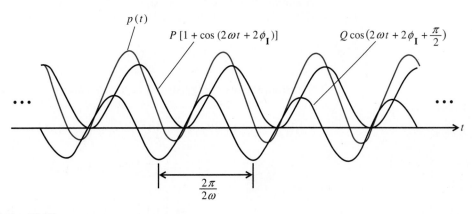

FIGURE 12.3 The terms in Eq. (12.15)

Relation Between Reactive Power and Stored Energy for L and C

There is an important relationship between the reactive power entering an inductance and the average energy that is stored in it. From the relation $\mathbf{V} = j\omega L\mathbf{I}$, it follows that $V_m = \omega L I_m$ and $\theta = 90°$. Use of these results in Eq. (12.17) leads to the following expression for the reactive power entering an inductance:

$$Q = \frac{1}{2} V_m I_m \sin \theta$$

$$= \frac{1}{2} I_m^2 \omega L \qquad (12.18)$$

The *instantaneous energy* stored in the inductance is, according to Eq. (2.43),

$$w_L(t) = \frac{1}{2} L i^2(t)$$

$$= \frac{1}{2} L I_m^2 \cos^2 (\omega t + \phi_\mathbf{I})$$

$$= \frac{1}{4} L I_m^2 + \frac{1}{4} L I_m^2 \cos (2\omega t + 2\phi_\mathbf{I}) \qquad (12.19)$$

The *average energy* stored in the inductance is the first term on the right-hand side of Eq. (12.19),

$$W_{L,\text{ave}} = \frac{1}{4} L I_m^2 \qquad (12.20)$$

By comparing Eqs. (12.20) and (12.18), we can see that the reactive power entering the inductance can be expressed in terms of the average stored energy as

$$Q = 2\omega W_{L,\text{ave}} \qquad (12.21)$$

That is, the maximum rate of energy exchange between a source and an inductance is equal to the product of the average energy stored in the inductance and twice the angular frequency of the source.

A similar result applies for capacitance. From Eq. (12.17) and the relation $\mathbf{I} = j\omega C\mathbf{V}$, it follows that the reactive power entering a capacitance is

$$Q = -\frac{1}{2} V_m^2 \omega C \qquad (12.22)$$

The instantaneous energy stored in a capacitance is, by Eq. (2.30),

$$w_C(t) = \frac{1}{2} C v^2(t)$$

$$= \frac{1}{2} C V_m^2 \cos^2 (\omega t + \phi_\mathbf{V})$$

$$= \frac{1}{4} C V_m^2 + \frac{1}{4} C V_m^2 \cos (2\omega t + 2\phi_\mathbf{V}) \qquad (12.23)$$

The average energy stored is

$$W_{C,\text{ave}} = \frac{1}{4} C V_m^2 \tag{12.24}$$

Therefore the reactive power entering a capacitance is related to the average stored energy by

$$Q = -2\omega W_{C,\text{ave}} \tag{12.25}$$

Notice that the reactive power entering a capacitance is negative, and that the reactive power entering an inductance is positive. For this reason, capacitors are said to be "sources of reactive power,"[†] whereas inductors are said to "consume reactive power." More generally, a passive network with a positive reactance will have a positive impedance angle, and will therefore be associated with a positive reactive power input Q. Such a network is inductive, and is said to "consume reactive power." A passive network with a negative reactance is associated with a negative reactive power input Q. Such a network is capacitive, and is said to be a "source of reactive power." If a network is inductive, some or all of its reactive power requirements can be supplied from a capacitive network rather than from the source. This procedure is illustrated in Example 12.4.

EXERCISES

7. Define and give the unit of reactive power.

8. Define phase quadrature, and explain why the symbol Q is appropriate for reactive power.

📁 *See Problem 12.2 in the PSpice manual.*

9. A sinusoidal current with amplitude 10 A is input to an impedance $\mathbf{Z} = 10\underline{/45°}\ \Omega$. Determine the reactive power supplied.

12.4 Complex Power

Convenient expressions for the average and the reactive power entering a network can be derived directly from the frequency-domain representation of Fig. 12.4, in which $\mathbf{V} = V_m e^{j\phi_V}$ and $\mathbf{I} = I_m e^{j\phi_I}$ are the phasors associated with $v(t)$ and $i(t)$, respectively. Here

$$\mathbf{V} = \mathbf{ZI} \quad \text{and} \quad \mathbf{I} = \mathbf{YV} \tag{12.26}$$

$$V_m = |\mathbf{Z}| I_m \quad \text{and} \quad I_m = |\mathbf{Y}| V_m \tag{12.27}$$

and

$$\phi_V = \underline{/\mathbf{Z}} + \phi_I \quad \text{and} \quad \phi_I = \underline{/\mathbf{Y}} + \phi_V \tag{12.28}$$

where $\mathbf{Z}$ and $\mathbf{Y}$ are the driving-point impedance and the driving-point admittance of the network, respectively.

[†] Some people do not like this terminology because, after all, the difference in sign merely indicates that the borrowing and returning of energy occur at different times for the inductance and capacitance. Although this objection makes a valid point, the terminology is widely accepted.

FIGURE 12.4
Frequency-domain
representation of Fig. 12.1

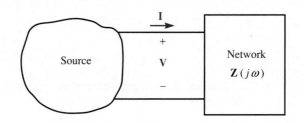

In connection with Fig. 12.4, we define the *complex power* entering the network as follows.

DEFINITION

The **complex power** entering the network of Fig. 12.4 is defined as

$$S = \frac{1}{2} VI^*$$
(12.29)

The unit of $\mathbf{S}$ is the volt-ampere (VA). Like other complex quantities used in electrical engineering, complex power has no physical significance in itself, but provides a convenient way to derive physically meaningful quantities. For example, because $\mathbf{V} = V_m e^{j\phi_V}$ and $\mathbf{I} = I_m e^{j\phi_I}$, then

$$\mathbf{S} = \frac{1}{2} V_m I_m e^{j(\phi_V - \phi_I)}$$
(12.30)

or

$$\mathbf{S} = \frac{1}{2} V_m I_m e^{j\theta} = \frac{1}{2} V_m I_m \underline{/\theta}$$
(12.31)

where $\theta = \phi_V - \phi_I$. Therefore the magnitude of $\mathbf{S}$, $|\mathbf{S}|$, is the apparent power $\frac{1}{2} V_m I_m$, and the angle of $\mathbf{S}$, θ, is the angle of the network's impedance $\mathbf{Z}$. Furthermore, Euler's identity enables Eq. (12.31) to be rewritten as

$$\mathbf{S} = \frac{1}{2} V_m I_m (\cos \theta + j \sin \theta)$$

$$= \frac{1}{2} V_m I_m \cos \theta + j \frac{1}{2} V_m I_m \sin \theta$$
(12.32)

so that from the definitions in Eqs. (12.16) and (12.17),

$$\mathbf{S} = P + jQ$$
(12.33)

Therefore the real part of $\mathbf{S}$ is the average power P, and the imaginary part of $\mathbf{S}$ is the reactive power Q. Equations (12.31) and (12.33) are illustrated in Fig. 12.5. Figure 12.6, which is derived directly from Fig. 12.5, is called the *power triangle*.

FIGURE 12.5
Complex power S

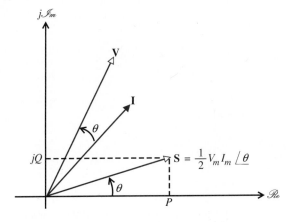

FIGURE 12.6
The power triangle

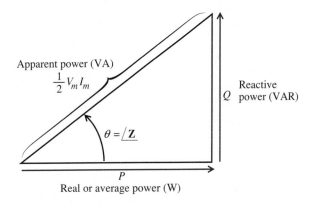

Some interesting results can be derived by an inspection of the power triangle; for example, by the Pythagorean theorem,

$$\frac{1}{2}\,V_m I_m = \sqrt{P^2 + Q^2} \tag{12.34}$$

and by the trigonometric definition of $\cos\theta$,

$$PF = \frac{P}{\sqrt{P^2 + Q^2}} \tag{12.35}$$

Equations (12.34) and (12.35) are useful because they express apparent power and power factor as functions of average power and reactive power.

Additional results follow from Eq. (12.29) and the relation $\mathbf{V} = \mathbf{Z}(j\omega)\mathbf{I} = [R(\omega) + jX(\omega)]\mathbf{I}$:

$$\mathbf{S} = \frac{1}{2}\,\mathbf{V}\mathbf{I}^*$$

$$= \frac{1}{2}\,\mathbf{Z}I_m^2$$

$$= \frac{1}{2}\,I_m^2 R(\omega) + j\,\frac{1}{2}\,I_m^2 X(\omega) \tag{12.36}$$

Therefore, by comparison with Eq. (12.33), we have

$$P = \frac{1}{2} I_m^2 R(\omega)$$ (12.37)

and

$$Q = \frac{1}{2} I_m^2 X(\omega)$$ (12.38)

According to Eqs. (12.37) and (12.38), for a given current amplitude I_m, the average power P entering a network depends on the network's resistance $R(\omega) = \mathscr{R}e\,\{\mathbf{Z}(\omega)\}$ and the reactive power Q depends on the network's reactance $X(\omega) = \mathscr{I}m\,\{\mathbf{Z}(j\omega)\}$. Similar results can be obtained with the use of Eq. (12.29) and the formula $\mathbf{I} = \mathbf{Y}(j\omega)\mathbf{V} = [G(\omega) + jB(\omega)]\mathbf{V}$:

$$\mathbf{S} = \frac{1}{2}\,\mathbf{VI}^* = \frac{1}{2}\,\mathbf{Y}^*V_m^2$$

$$= \frac{1}{2}\,V_m^2 G(\omega) - j\frac{1}{2}\,V_m^2 B(\omega)$$ (12.39)

from which

$$P = \frac{1}{2}\,V_m^2 G(\omega)$$ (12.40)

and

$$Q = -\frac{1}{2}\,V_m^2 B(\omega)$$ (12.41)

FIGURE 12.7
Decomposition of the phasor **V**

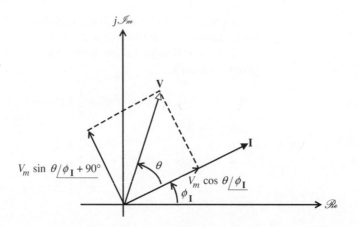

A final insight is contained in Fig. 12.7, which shows the driving-point voltage and current phasors **V** and **I** as vectors in the complex plane. The vector **V** is viewed as having two components, one parallel to the vector **I** and the other perpendicular to **I**. These two components are $V_m \cos \theta \underline{/\phi_{\mathbf{I}}}$ and $V_m \sin \theta \underline{/\phi_{\mathbf{I}} + 90°}$, respectively, and they correspond to the in-phase and phase-quadrature components of $v(t)$ shown in Eq. (12.12). The substitution of

$$\mathbf{V} = \underbrace{V_m \cos \theta e^{j\phi_{\mathbf{I}}}}_{\text{Parallel to }\mathbf{I}} + \underbrace{V_m \sin \theta e^{j[\phi_{\mathbf{I}} + (\pi/2)]}}_{\text{Orthogonal to }\mathbf{I}} \tag{12.42}$$

into $\mathbf{S} = \frac{1}{2}\mathbf{V}\mathbf{I}^*$ yields

$$\begin{aligned}
\mathbf{S} &= \frac{1}{2} \left[V_m \cos \theta e^{j\phi_{\mathbf{I}}} + V_m \sin \theta e^{j[\phi_{\mathbf{I}} + (\pi/2)]} \right] \mathbf{I}^* \\
&= \frac{1}{2} (V_m \cos \theta e^{j\phi_{\mathbf{I}}} + jV_m \sin \theta e^{j\phi_{\mathbf{I}}}) I_m e^{-j\phi_{\mathbf{I}}} \\
&= \frac{1}{2} V_m I_m \cos \theta + j \frac{1}{2} V_m I_m \sin \theta \\
&= P + jQ \tag{12.43}
\end{aligned}$$

Therefore the component of **V** parallel to **I** is responsible for the average power $P = \mathscr{R}e\{\mathbf{S}\}$ supplied to the network. The component of **V** orthogonal to **I** is responsible for the reactive power $Q = \mathscr{I}m\{\mathbf{S}\}$ supplied to the network.[†] The component of **V** parallel to **I** corresponds to the component of $v(t)$ that is in phase with $i(t)$. The component of **V** that is orthogonal to **I** corresponds to the component of $v(t)$ that is in phase quadrature with $i(t)$.

EXAMPLE 12.3

A coil draws 1-A peak current at a 0.6 lagging power factor from a 120-V rms 60-Hz source. Assume that the coil is modeled by a series RL circuit (where R represents the winding resistance of the coil), and find (a) the complex power entering the coil and (b) the values of R and L.

■ **SOLUTION**

(a) The complex power entering the coil is $\mathbf{S} = P + jQ$, where from Eq. (12.16),

$$P = \frac{1}{2}(1)(\sqrt{2} \cdot 120)(0.6)$$

$$= 50.9 \text{ W}$$

Because PF $= \cos \theta = 0.6$ lagging, $\sin \theta = \sin(\cos^{-1} 0.6) = 0.8$. The reactive power is, from Eq. (12.17),

$$Q = \frac{1}{2}(1)(\sqrt{2} \cdot 120)(0.8)$$

$$= 67.9 \text{ VAR}$$

[†] The term *orthogonal* is used here in the sense of *perpendicular*. It is worth noting that Eq. (12.42) is the frequency-domain form of Eq. (12.12). In Eq. (12.12), the component of $v(t)$ that is in phase quadrature with $i(t)$ is orthogonal to $i(t)$ in the sense discussed in Section 7.7.

FIGURE 12.8
Power triangle for Example
12.3

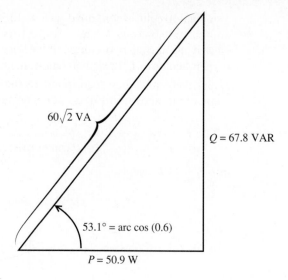

$60\sqrt{2}$ VA

$Q = 67.8$ VAR

$53.1° = \arccos(0.6)$

$P = 50.9$ W

Therefore

$$\mathbf{S} = 50.9 + j67.8 \text{ VA}$$

The power triangle is shown in Fig. 12.8.

(b) The driving-point impedance of a series RL circuit is

$$\mathbf{Z}(j\omega) = R + j\omega L$$

Therefore, from Eq. (12.37) with $R(\omega) = R$,

$$R = \frac{2P}{I_m^2}$$

$$= \frac{(2)(50.9)}{1^2}$$

$$= 101.8 \ \Omega$$

and from Eq. (12.38) with $X(\omega) = \omega L$,

$$L = \frac{2Q}{\omega I_m^2}$$

$$= \frac{(2)(67.8)}{(2\pi 60)(1^2)}$$

$$= 0.36 \text{ H} \quad \blacksquare$$

REMEMBER

Average power, reactive power, apparent power, and power factor can be obtained from complex power $\mathbf{S}$.

EXERCISES

10. Define and give the unit of complex power.

11. Sketch the power triangle from memory.

12. A sinusoidal current with an amplitude of 10 A is applied to an impedance $\mathbf{Z} = 10\underline{/45°}\ \Omega$.
    (a) Determine the complex power supplied.
    (b) Sketch the power triangle.
    (c) Give the average power, the reactive power, and the apparent power.

13. A sinusoidal voltage with an effective amplitude of 120 V rms and frequency of 60 Hz is applied to a series RL circuit with impedance $\mathbf{Z} = 100 + j377\ \Omega$. Sketch the power triangle, and indicate the numerical values and engineering significance of its sides and angles.

12.5 The Complex Power Balance Theorem

This section describes an important property of energy flow in ac circuits. This property is an extension of the power balance equation that was described in Section 1.5.

FIGURE 12.9
Parallel loads on an ac line

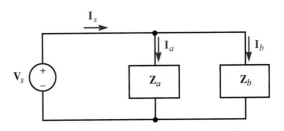

We introduce the basic idea by considering two load impedances $\mathbf{Z}_a$ and $\mathbf{Z}_b$ connected in parallel across a sinusoidal voltage source (or "ac line"), as illustrated in Fig. 12.9. The line current is

$$\mathbf{I}_s = \mathbf{I}_a + \mathbf{I}_b \tag{12.44}$$

where $\mathbf{I}_a$ and $\mathbf{I}_b$ are the phasor currents in the individual loads. The complex power *supplied* by the source is

$$\mathbf{S}_s = \frac{1}{2}\mathbf{V}_s\mathbf{I}_s^*$$

$$= \frac{1}{2}\mathbf{V}_s\mathbf{I}_a^* + \frac{1}{2}\mathbf{V}_s\mathbf{I}_b^*$$

$$= \mathbf{S}_a + \mathbf{S}_b \tag{12.45}$$

which is the *sum* of the complex powers *delivered* to the loads. This neat result can also be written as

$$-\mathbf{S}_s + \mathbf{S}_a + \mathbf{S}_b = 0 \tag{12.46a}$$

or

$$\sum \mathbf{S}_i = 0 \tag{12.46b}$$

where the sum is over the complex powers entering all components in the network, including the source.

Equations (12.45) and (12.46) are examples of the following theorem.

Complex Power Balance Theorem

In an arbitrary ac network, the sum of the complex powers entering all network components (including the sources) equals zero.

This theorem is expressed mathematically as

$$\sum_{\substack{\text{All} \\ \text{components}}} \frac{1}{2} \mathbf{VI}^* = 0 \tag{12.47a}$$

or

$$\sum_{\substack{\text{All} \\ \text{components}}} \mathbf{S}_i = 0 \tag{12.47b}$$

In Eqs. (12.47a) and (12.47b), the passive sign convention is used for each component. $\mathbf{S}_i$ is the complex power delivered to the ith component, and the summation is over all components, including the sources. If the active sign convention is used for the sources, then we have

$$\underbrace{\sum \mathbf{S}_i}_{\substack{\text{Supplied} \\ \text{by sources}}} = \underbrace{\sum \mathbf{S}_i}_{\substack{\text{Delivered to} \\ \text{passive} \\ \text{elements}}} \tag{12.47c}$$

An example of this statement is given in Eq. (12.45), where the voltage source of Fig. 12.9 was the supplier of complex power.

The proof of the complex power balance equation is on page 525. The basic ingredients of the proof can be understood if we consider the four-node network of Fig. 12.10. The components represented by rectangles are arbitrary impedances (including open and short circuits), independent sources, and dependent sources. Subscripts pq have been used to identify the currents. The order of subscripts in

FIGURE 12.10
Pertinent to Eqs. (12.48) through (12.54)

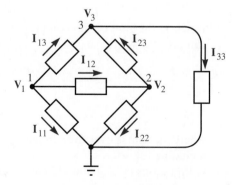

$\mathbf{I}_{pq}$ signifies that the arrow points from node p to node q for $p \neq q$. For $p = q$, the arrow points from node p to the reference node (see Fig. 12.10). If we apply KCL at the (nonreference) nodes, we obtain the following three equations:

Node 1 $\qquad \mathbf{I}_{11} + \mathbf{I}_{12} + \mathbf{I}_{13} = 0$ $\hfill$ (12.48)

Node 2 $\qquad -\mathbf{I}_{12} + \mathbf{I}_{22} + \mathbf{I}_{23} = 0$ $\hfill$ (12.49)

Node 3 $\qquad -\mathbf{I}_{13} - \mathbf{I}_{23} + \mathbf{I}_{33} = 0$ $\hfill$ (12.50)

If we multiply the conjugates of Eqs. (12.48), (12.49), and (12.50) by $\frac{1}{2}\mathbf{V}_1, \frac{1}{2}\mathbf{V}_2$, and $\frac{1}{2}\mathbf{V}_3$, respectively, we obtain

$$\frac{1}{2}\mathbf{V}_1\mathbf{I}_{11}^* + \frac{1}{2}\mathbf{V}_1\mathbf{I}_{12}^* + \frac{1}{2}\mathbf{V}_1\mathbf{I}_{13}^* = 0 \tag{12.51}$$

$$-\frac{1}{2}\mathbf{V}_2\mathbf{I}_{12}^* + \frac{1}{2}\mathbf{V}_2\mathbf{I}_{22}^* + \frac{1}{2}\mathbf{V}_2\mathbf{I}_{23}^* = 0 \tag{12.52}$$

$$-\frac{1}{2}\mathbf{V}_3\mathbf{I}_{13}^* - \frac{1}{2}\mathbf{V}_3\mathbf{I}_{23}^* + \frac{1}{2}\mathbf{V}_3\mathbf{I}_{33}^* = 0 \tag{12.53}$$

Finally, by adding Eqs. (12.51), (12.52), and (12.53), we obtain

$$\frac{1}{2}\mathbf{V}_1\mathbf{I}_{11}^* + \frac{1}{2}(\mathbf{V}_1 - \mathbf{V}_2)\mathbf{I}_{12}^* + \frac{1}{2}(\mathbf{V}_1 - \mathbf{V}_3)\mathbf{I}_{13}^* + \frac{1}{2}\mathbf{I}_{22}^*\mathbf{V}_2$$

$$+ \frac{1}{2}(\mathbf{V}_2 - \mathbf{V}_3)\mathbf{I}_{23}^* + \frac{1}{2}\mathbf{I}_{33}^*\mathbf{V}_3 = 0 \tag{12.54}$$

The terms on the left-hand side of Eq. (12.54) represent the complex powers entering the individual components of the network of Fig. 12.10. [For example, $\frac{1}{2}(\mathbf{V}_1 - \mathbf{V}_2)\mathbf{I}_{12}^*$ is the complex power entering the component between nodes 1 and 2.] Equation (12.54) confirms the validity of the complex power balance theorem for the network of Fig. 12.10.

General Proof of the Complex Power Balance Theorem

By using the notation and approach leading to Eq. (12.54), we can prove the complex power balance theorem for the arbitrary $(N + 1)$-node network depicted in Fig. 12.11. We begin by arbitrarily selecting a reference node, numbering the remaining nodes, and assigning node voltages to the nonreference nodes. This step is an implicit statement of KVL. The next step is to apply KCL. KCL requires that the phasor currents leaving a surface enclosing any node p sum to zero:

$$\sum_{q=1}^{N} \mathbf{I}_{pq} = 0 \tag{12.55}$$

We conjugate Eq. (12.55) and multiply it by $\frac{1}{2}\mathbf{V}_p$ to get

$$\sum_{q=1}^{N} \frac{1}{2}\mathbf{V}_p\mathbf{I}_{pq}^* = 0 \tag{12.56}$$

Next, we sum Eq. (12.56) over $p = 1, 2, \ldots, N$ to obtain

$$\sum_{p=1}^{N} \sum_{q=1}^{N} \frac{1}{2}\mathbf{V}_p\mathbf{I}_{pq}^* = 0 \tag{12.57}$$

FIGURE 12.11
Notation for arbitrary ac
network

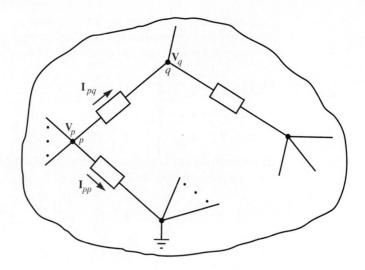

But

$$\sum_{p=1}^{N} \sum_{q=1}^{N} \frac{1}{2} \mathbf{V}_p \mathbf{I}_{pq}^* = \sum_{p=1}^{N} \frac{1}{2} \mathbf{V}_p \mathbf{I}_{pp}^* + \sum_{\substack{p=1 \\ p \neq q}}^{N} \sum_{q=1}^{N} \frac{1}{2} \mathbf{V}_p \mathbf{I}_{pq}^* \qquad (12.58)$$

The second summation on the right-hand side can be written as

$$\sum_{\substack{p=1 \\ p \neq q}}^{N} \sum_{q=1}^{N} \frac{1}{2} \mathbf{V}_p \mathbf{I}_{pq}^* = \sum_{p=1}^{N-1} \sum_{\substack{q=2 \\ q > p}}^{N} \frac{1}{2} \mathbf{V}_p \mathbf{I}_{pq}^* + \sum_{p=2}^{N} \sum_{\substack{q=1 \\ q < p}}^{N-1} \frac{1}{2} \mathbf{V}_p \mathbf{I}_{pq}^* \qquad (12.59)$$

We can simplify the right-hand side of Eq. (12.59) by changing summation indices. Change p to i and q to k in the double sum where $q > p$. Change p to k and q to i in the double sum where $q < p$. This yields

$$\sum_{\substack{p=1 \\ p \neq q}}^{N} \sum_{q=1}^{N} \frac{1}{2} \mathbf{V}_p \mathbf{I}_{pq}^* = \sum_{i=1}^{N-1} \sum_{\substack{k=2 \\ k > i}}^{N} \frac{1}{2} \mathbf{V}_i \mathbf{I}_{ik}^* + \sum_{k=2}^{N} \sum_{\substack{i=1 \\ i < k}}^{N-1} \frac{1}{2} \mathbf{V}_k \mathbf{I}_{ki}^*$$

$$= \sum_{i=1}^{N-1} \sum_{\substack{k=2 \\ k > i}}^{N} \frac{1}{2} (\mathbf{V}_i \mathbf{I}_{ik}^* + \mathbf{V}_k \mathbf{I}_{ki}^*)$$

$$= \sum_{i=1}^{N-1} \sum_{\substack{k=2 \\ k > i}}^{N} \frac{1}{2} (\mathbf{V}_i - \mathbf{V}_k) \mathbf{I}_{ik}^* \qquad (12.60)$$

where the last step follows from $\mathbf{I}_{ki} = -\mathbf{I}_{ik}$, $i \neq k$. By substituting Eq. (12.60) into Eq. (12.58), and equating the result to zero as in Eq. (12.57), we obtain

$$\sum_{p=1}^{N} \frac{1}{2} \mathbf{V}_p \mathbf{I}_{pp}^* + \sum_{i=1}^{N-1} \sum_{\substack{k=2 \\ k > i}}^{N} \frac{1}{2} (\mathbf{V}_i - \mathbf{V}_k) \mathbf{I}_{ik}^* = 0 \qquad (12.61)$$

Notice that $\frac{1}{2} \mathbf{V}_p \mathbf{I}_{pp}^*$ is the complex power delivered to the element connected between node p and the reference node. Notice also that $\frac{1}{2} (\mathbf{V}_i - \mathbf{V}_k) \mathbf{I}_{ik}^*$ is the complex power delivered to the element connected between nodes i and k. We see, therefore, that the individual terms in Eq. (12.61) are the complex powers absorbed by the elements in the network and that these complex powers sum to zero. This proves the complex power balance theorem for an arbitrary ac network.

EXAMPLE 12.4

⚠ *See Problem 12.3 in the PSpice manual.*

An inductive load draws 1 kW at PF = 0.9 lagging from a 120-V rms source. In an effort to raise the power factor seen by the source, a capacitive load is placed in parallel with the inductive load, as shown in Fig. 12.12. The capacitive load draws 10 W at PF = 0.02 leading.

FIGURE 12.12
Circuit for Example 12.4

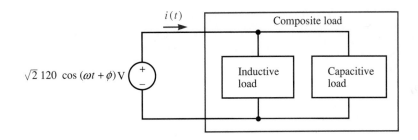

(a) Find the complex power supplied by the source.
(b) Find the rms (effective) current supplied by the source.
(c) Find the rms current supplied to the inductive load.
(d) Find the impedance of the composite load (enclosed in red).
(e) Find the power factor of the composite load.
(f) Explain why the power factor of the composite load is greater than that of the inductive load.

■ SOLUTION

(a) By the complex power balance theorem, the complex power supplied by the source, $\mathbf{S}_s$, equals the sum of the complex powers entering the loads. Thus

$$\mathbf{S}_s = \mathbf{S}_1 + \mathbf{S}_2$$

where $\mathbf{S}_1$ and $\mathbf{S}_2$ denote the complex powers supplied to the inductive and capacitive loads, respectively. The complex power supplied to the inductive load is

$$\mathbf{S}_1 = 1 \text{ kW} + j1 \tan(\cos^{-1} 0.9) \text{ kVAR}$$
$$= 1000 + j484.3 \text{ VA}$$

The complex power supplied to the capacitive load is

$$\mathbf{S}_2 = 10 \text{ W} - j10 \tan(\cos^{-1} 0.02) \text{ VAR}$$
$$= 10 - j499.9 \text{ VA}$$

Therefore the source supplies

$$\mathbf{S}_s = 1010 - j15.6 \text{ VA}$$

The power triangles associated with $\mathbf{S}_1$, $\mathbf{S}_2$, and $\mathbf{S}_s$ are shown in Fig. 12.13.

(b) The effective value of $i(t)$ can be determined from

$$|\mathbf{S}_s| = \frac{1}{2} V_m I_m$$

Recall that $I_m = \sqrt{2} I_{\text{rms}}$ and $V_m = \sqrt{2} V_{\text{rms}}$. Therefore

$$|\mathbf{S}_s| = V_{\text{rms}} I_{\text{rms}}$$

FIGURE 12.13
Power triangles for Example
12.4: (*a*) inductive load;
(*b*) capacitive load;
(*c*) composite load. (The
power triangles are
not drawn to scale.)

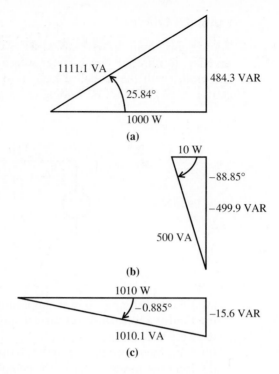

1111.1 VA

484.3 VAR

25.84°

1000 W

(a)

10 W

−88.85°

−499.9 VAR

500 VA

(b)

1010 W

−0.885°

−15.6 VAR

1010.1 VA

(c)

Solving for I_{rms}, we have

$$I_{\text{rms}} = \frac{|\mathbf{S}_s|}{V_{\text{rms}}}$$

$$= \frac{|1010 - j15.6|}{120} \text{ A rms}$$

$$= \frac{1010.1}{120} = 8.417 \text{ A rms}$$

(c) The rms current supplied to the inductive load follows from

$$I_{1\,\text{rms}} = \frac{|\mathbf{S}_1|}{V_{\text{rms}}} = \frac{1111.1}{120} = 9.259 \text{ A rms}$$

where we obtained $|\mathbf{S}_1|$ from the power triangle (Fig. 12.13a).

(d) Because $\mathbf{S}_s = \frac{1}{2}\mathbf{VI}^*$ and $\mathbf{Z} = \mathbf{V}/\mathbf{I}$,

$$\mathbf{Z} = \frac{2\mathbf{S}_s}{|\mathbf{I}|^2} = \frac{2\mathbf{S}_s}{I_m^2} = \frac{\mathbf{S}_s}{I_{\text{rms}}^2} = 14.25 - j0.22 \ \Omega$$

(e)

$$\text{PF} = \cos{(\underline{/\mathbf{Z}})} = \cos\left(\tan^{-1}\frac{-0.22}{14.25}\right) = 0.9999 \text{ (leading)} \simeq 1$$

(f) The inductive load has a power factor of only 0.9, because it requires 484.3
VAR reactive power as well as 1 kW average power. When the capacitive load is
placed in the circuit, its reactive power requirement (-499.9 VAR) almost cancels
that of the inductive load (484.3 VAR). Thus the reactive power required by the

composite load is only -15.6 VAR. In this sense, the capacitive load *supplies* the reactive power needed by the inductive load. The result is an increase in the power factor seen by the source from 0.9 (inductive load alone) to almost unity (composite load). The use of one load to supply reactive power needed by another load in an effort to raise power factor is called *power factor correction.* ■

EXAMPLE 12.5

Analyze the distribution of complex power in the parallel *RLC* network of Fig. 12.14.

FIGURE 12.14
Parallel *RLC* circuit

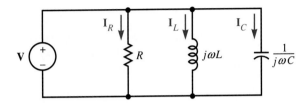

■ **SOLUTION**

The complex powers entering the resistance, inductance, and capacitance are, respectively,

$$\mathbf{S}_R = P_R + jQ_R$$

$$= \frac{1}{2}\frac{V_m^2}{R} + j0$$

$$\mathbf{S}_L = P_L + jQ_L$$

$$= 0 + j\left(\frac{1}{2}V_m^2 \frac{1}{\omega L}\right)$$

and

$$\mathbf{S}_C = P_C + jQ_C$$

$$= 0 + j\left(-\frac{1}{2}V_m^2 \omega C\right)$$

By the complex power balance theorem, the sum of these complex powers equals the complex power supplied by the source $\mathbf{S}_s$:

$$\mathbf{S}_s = \mathbf{S}_R + \mathbf{S}_L + \mathbf{S}_C$$

$$= P_R + j(Q_L + Q_C)$$

$$= \frac{1}{2}\frac{V_m^2}{R} + j\left(\frac{1}{2}V_m^2 \frac{1}{\omega L} - \frac{1}{2}V_m^2 \omega C\right)$$

The power triangle is shown in Fig. 12.15. The real power supplied by the source is the term $\frac{1}{2}V_m^2/R$, which is the power dissipated by the resistance. The reactive power supplied by the source is the sum of the reactive powers consumed by the inductance and the capacitance. In view of the negative sign of Q_C, the capacitance can be thought of as a supplier of reactive power. By this interpretation, the reactive power supplied by the source is the difference between the reactive power consumed

FIGURE 12.15
Power triangle for parallel
RLC circuit

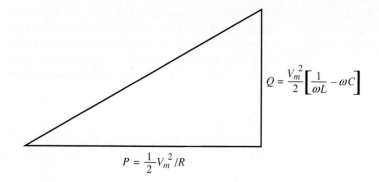

$$Q = \frac{V_m^2}{2}\left[\frac{1}{\omega L} - \omega C\right]$$

$$P = \frac{1}{2}V_m^2 / R$$

by the inductance and that supplied by the capacitance. Recall from Eqs. (12.21) and (12.25) that Q_L and Q_C can be expressed in terms of the average stored energies $W_{L,\text{ave}}$ and $W_{C,\text{ave}}$ in the inductance and the capacitance, respectively. This leads to

$$\mathbf{S}_s = \frac{1}{2}\frac{V_m^2}{R} + j2\omega(W_{L,\text{ave}} - W_{C,\text{ave}})$$

The reactive power supplied by the source is equal to 2ω times the *excess* of the average stored magnetic-field energy compared with the average stored electric-field energy. ∎

The analysis contained in Example 12.5 can be generalized. Consider an arbitrary network consisting of resistances, inductances, capacitances, and independent sources operating in the sinusoidal steady state at frequency ω. Because the passive elements are Rs, Ls, and Cs, Eq. (12.47c) can be written as

$$\sum_{\substack{\text{All}\\\text{sources}}} \mathbf{S}_s = \sum_{\substack{\text{All}\\\text{resistances}}} \mathbf{S}_R + \sum_{\substack{\text{All}\\\text{inductances}}} \mathbf{S}_L + \sum_{\substack{\text{All}\\\text{capacitances}}} \mathbf{S}_C$$

$$= \sum_{\substack{\text{All}\\\text{resistances}}} \left(\frac{1}{2}\right)\frac{V_m^2}{R} + \sum_{\substack{\text{All}\\\text{inductances}}} j2\omega W_{L,\text{ave}} - \sum_{\substack{\text{All}\\\text{capacitances}}} j2\omega W_{C,\text{ave}}$$

$$= \sum_{\substack{\text{All}\\\text{resistances}}} \left(\frac{1}{2}\right)\frac{V_m^2}{R} + j2\omega\left(\sum_{\substack{\text{All}\\\text{inductances}}} W_{L,\text{ave}} - \sum_{\substack{\text{All}\\\text{capacitances}}} W_{C,\text{ave}}\right) \qquad (12.62\text{a})$$

The first term on the right-hand side of Eq. (12.62a) is the total average power dissipated in the resistances. The second term is $j2\omega$ times the difference between the total average energy stored in the inductances and the total average energy stored in the capacitances. With the obvious definitions, Eq. (12.62a) can be rewritten as

$$\mathbf{S} = P + j2\omega\{W_{\text{tot ave magnetic}} - W_{\text{tot ave electric}}\} \qquad (12.62\text{b})$$

where $\mathbf{S}$ is the total complex power supplied by the sources, P is the total average power dissipated in the resistances, $W_{\text{tot ave magnetic}}$ is the total average energy stored in the inductances, and $W_{\text{tot ave electric}}$ is the total average energy stored in the capacitances.

FIGURE 12.16
Power triangle for any
circuit composed of
Rs, Ls, and Cs

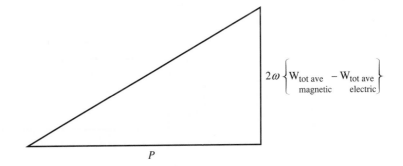

The power triangle is shown in Fig. 12.16. Equation (12.62b) represents a theoretical result that has remarkable generality, since it applies to *any* network made up of Rs, Ls, and Cs. We can see from it that any such network will appear *resistive* to a source at a particular angular frequency ω if and only if the total average energy stored in the network's magnetic fields equals the total average energy stored in the network's electric fields.

REMEMBER

The sum of the complex powers entering all network components equals zero.

EXERCISE

14. Find the total complex power supplied by the sources in the circuit shown below.

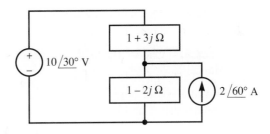

12.6 Maximum Power Transfer

In Section 5.6 we solved the problem of maximizing the power delivered to a load resistance R_L from a source with internal resistance R_s. In this section, we extend the problem to ac circuits. We consider the circuit of Fig. 12.17. In this figure the source is represented by its Thévenin equivalent circuit inside the red rectangle. $\mathbf{V}_s$ is the open-circuit source (phasor) voltage, and $\mathbf{Z}_s$ is the equivalent source impedance. $\mathbf{Z}_L$ is an arbitrary load impedance. The load current and voltage are

$$\mathbf{I}_L = \frac{\mathbf{V}_s}{\mathbf{Z}_s + \mathbf{Z}_L} \qquad (12.63)$$

and

$$\mathbf{V}_L = \frac{\mathbf{Z}_L}{\mathbf{Z}_s + \mathbf{Z}_L} \mathbf{V}_s \qquad (12.64)$$

FIGURE 12.17
Ac source and load
impedance

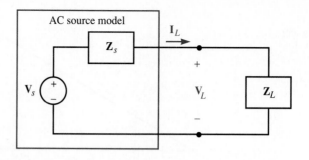

The complex power delivered to the load is

$$S_L = \frac{1}{2} V_L I_L^*$$

$$= \frac{1}{2} \frac{|V_s|^2 Z_L}{|Z_s + Z_L|^2} \qquad (12.65)$$

Notice that the complex power delivered depends on the amplitude of the voltage source $V_{sm} = |V_s|$ but does not depend on its phase $\phi_s = \underline{/V_s}$. The impedances Z_L and Z_s in the right-hand side of Eq. (12.65) are both functions of the source frequency ω. Therefore, if the source frequency is varied, $S_L = P_L + jQ_L$ will vary. One way to maximize the average power delivered to the load is to leave the circuit as it is but choose the source frequency to maximize P_L. This approach is related to the topic of *frequency response*, described in Chapter 13. In the analysis that follows, we will assume that the frequency of the source is fixed.

For a fixed source frequency ω and open-circuit voltage amplitude V_{sm}, there are two cases of practical interest.

Case 1 Choose the source impedance Z_s to maximize $P_L = \mathscr{R}e\{S_L\}$ for a fixed load impedance Z_L.

Case 2 Choose the load impedance Z_L to maximize P_L for a fixed source impedance Z_s.

An analysis of both problems is facilitated by expressing Z_L and Z_s in terms of their resistive and reactive parts:

$$Z_L = R_L + jX_L \qquad (12.66)$$

$$Z_s = R_s + jX_s \qquad (12.67)$$

By substituting Eqs. (12.66) and (12.67) into (12.65), we find that

$$P_L = \frac{1}{2} \frac{V_{sm}^2 R_L}{(R_s + R_L)^2 + (X_s + X_L)^2} \qquad (12.68)$$

and

$$Q_L = \frac{1}{2} \frac{V_{sm}^2 X_L}{(R_s + R_L)^2 + (X_s + X_L)^2} \qquad (12.69)$$

Consider case 1, in which Z_L is fixed and Z_s is to be chosen for maximum power transfer. Because Z_L is fixed, it suffices to choose R_s and X_s to minimize the denominator in the preceding expression for P_L, Eq. (12.68). We can do this by choosing R_s to minimize $(R_s + R_L)^2$ and X_s to minimize $(X_s + X_L)^2$. Unless we wish to construct a negative resistance, the quantity $(R_s + R_L)^2$ is minimized by choosing

$$R_s = 0 \qquad (12.70)$$

Negative reactance is easy to obtain. Therefore we minimize the quantity $(X_s + X_L)^2$ by choosing

$$X_s = -X_L \qquad (12.71)$$

The resulting maximum average power delivered is obtained by substituting Eqs. (12.70) and (12.71) into Eq. (12.68). The result is

$$\max_{\mathbf{Z}_s} \{P_L\} = \frac{1}{2}\frac{V_{sm}^2}{R_L} \quad \text{W} \qquad (12.72)$$
$$\text{(Case 1)}$$

which depends only on the open-circuit voltage amplitude and the load resistance $R_L = \mathscr{R}e\{\mathbf{Z}_L\}$. Representative plots of Eq. (12.68) are given in Fig. 12.18. (The figure assumes that $-X_L$ and R_L are fixed positive numbers.) Notice in Fig. 12.18b that regardless of the value of R_s, a peak occurs when $X_s = -X_L$. Notice in Fig. 12.18a that regardless of the value of X_s, a peak occurs at $R_s = 0$. Therefore P_L *is always increased by a decrease in source resistance or by a decrease in the absolute value of the sum of source and load reactance.*

FIGURE 12.18
P_L versus R_s and X_s for a fixed load impedance

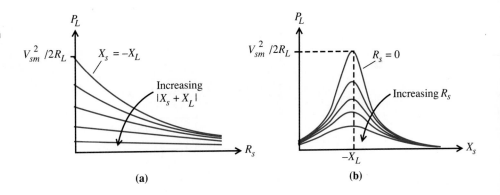

Refer now to case 2, in which $\mathbf{Z}_s$ is fixed and $\mathbf{Z}_L$ is to be chosen for maximum power transfer. It is easy to see that if we choose

$$X_L = -X_s \qquad (12.73)$$

we minimize the term $(X_s + X_L)^2$ in Eq. (12.68). This leaves

$$P_L = \frac{1}{2}\frac{V_{sm}^2 R_L}{(R_s + R_L)^2} \qquad (12.74)$$

which (by elementary calculus) has a maximum when

$$R_L = R_s \qquad (12.75)$$

The maximum delivered power is

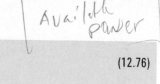
Available power

$$\max_{\mathbf{Z}_L} \{P_L\} = \frac{V_{sm}^2}{8R_s} \quad \text{W} \qquad (12.76)$$
(Case 2)

The above value of P_L is called the *available power* of the source. It depends solely on the open-circuit voltage amplitude and the source resistance $R_s = \mathscr{R}e\,\{\mathbf{Z}_s\}$. The conditions for maximum power transfer stated in Eqs. (12.73) and (12.75) can be written more compactly as

$$\mathbf{Z}_L = R_s - jX_s \qquad (12.77a)$$

or equivalently,

$$\mathbf{Z}_L = \mathbf{Z}_s^* \qquad (12.77b)$$

See Problem 12.4 in the PSpice manual.

A load impedance satisfying Eq. (12.77) is said to be *matched* to the source. Representative plots of Eq. (12.68) are given in Fig. 12.19. (The figure assumes that X_s is a negative number.) Notice in Fig. 12.19b that regardless of the value of R_L, a peak occurs when $X_L = -X_s$. Notice in Fig. 12.19a that a peak occurs at $R_L = R_s$ only if $X_L = -X_s$.

FIGURE 12.19
P_L versus R_L and X_L for a fixed source impedance

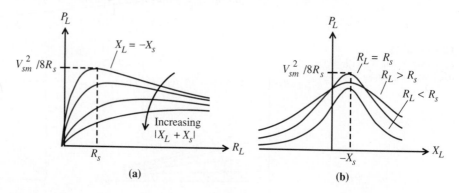

(a)

(b)

EXAMPLE 12.6

(a) Find the load impedance $\mathbf{Z}_L$ that maximizes the average power drawn from the network shown in Fig. 12.20, where $\omega = 10$ Mrad/s, $R = 2\ \Omega$, $C = 0.1\ \mu\text{F}$, $L = 0.2\ \mu\text{H}$, and $\mathbf{V}_s = 10\underline{/37.6°}$ mV.

(b) Find the average power delivered to the load impedance from (a).

FIGURE 12.20
Circuit for Example 12.6

if $\mathbf{Z}_L = R_L$, (totally resistive)
to maximize P_L, you get
$R_L = |\mathbf{Z}_s|$.

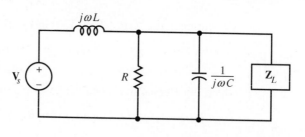

■ SOLUTION

(a) Maximum power transfer occurs when $\mathbf{Z}_L = \mathbf{Z}_s^*$, where $\mathbf{Z}_s$ is the equivalent impedance of the source. The equivalent impedance $\mathbf{Z}_s$ can be found by inspection if we set the independent source in the source network to zero. Thus

$$\mathbf{Z}_s = j\omega L \,\|\, R \,\|\, \frac{1}{j\omega C}$$

$$= \frac{G}{G^2 + (\omega C - 1/\omega L)^2} + j\frac{-(\omega C - 1/\omega L)}{G^2 + (\omega C - 1/\omega L)^2}$$

where $G = 1/R$. For the given numerical values, this simplifies to

$$\mathbf{Z}_s = 1 - j\,\Omega$$

The matched load impedance is $\mathbf{Z}_L = \mathbf{Z}_s^*$; that is,

$$\mathbf{Z}_L = 1 + j\,\Omega$$

Note that $\mathbf{Z}_L$ can be synthesized by a 1-Ω resistance in series with a 0.1-μH inductance.

(b) The power delivered under matched conditions can be found from Eq. (12.76) with the use of $R_s = 1$ and $V_{sm} = |\mathbf{V}_{oc}|$, where $\mathbf{V}_{oc}$ is the Thévenin equivalent voltage of the source network. Disconnecting $\mathbf{Z}_L$, we find

$$\mathbf{V}_{oc} = \frac{R \,\|\, (1/j\omega C)}{R \,\|\, (1/j\omega C) + j\omega L}\,\mathbf{V}_s$$

For the given numerical values, this simplifies to

$$\mathbf{V}_{oc} = \left[\frac{-2j/(2 - j)}{-2j/(2 - j) + 2j}\right] 10\underline{/37.6^\circ}$$

$$= \left(\frac{-2j}{2 + 2j}\right) 10\underline{/37.6^\circ}$$

$$= 5\sqrt{2}\underline{/-97.4^\circ}\,\text{mV}$$

Therefore

$$V_{sm} = 5\sqrt{2}\,\text{mV}$$

and

$$P_L = \frac{(5\sqrt{2})^2}{8}\,\mu\text{W}$$

$$= 6.25\,\mu\text{W} \quad ■$$

REMEMBER

We maximize the average power dissipated in a fixed load impedance $\mathbf{Z}_L = R_L + jX_L$ by choosing the source impedance as $\mathbf{Z}_s = 0 - jX_L$. We maximize the average power supplied by a source having fixed source impedance $\mathbf{Z}_s = R_s + jX_s$ by choosing the load impedance as $\mathbf{Z}_L = R_s - jX_s$.

For Exercises 15 through 18, use the circuit shown below.

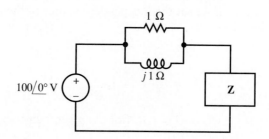

15. What is the value of the maximum average power that can be delivered to impedance **Z**?

16. What value of impedance **Z** will consume the maximum average power?

17. What is the value of the maximum average power that can be delivered to the 1-Ω resistance?

18. What value of impedance **Z** is required to maximize the average power delivered to the 1-Ω resistance?

19. The short-circuit current $\mathbf{I}_{sc}$ from an ac network is $2\underline{/20°}$ A, and the open-circuit voltage $\mathbf{V}_{oc}$ is $10\underline{/65°}$ V. What is the available power?

12.7   Summary

The topic of this chapter was energy flow in ac circuits. We began by describing instantaneous power, $p(t) = v(t)i(t)$, supplied to a circuit operating in the sinusoidal steady state. This led to the conclusion that the average power entering the circuit is $P = \frac{1}{2}V_m I_m \cos\theta$. We defined and discussed the significance of apparent power, $\frac{1}{2}V_m I_m$, and reactive power, $Q = \frac{1}{2}V_m I_m \sin\theta$. We defined complex power as $\mathbf{S} = \frac{1}{2}\mathbf{VI}^*$. Complex power has no physical meaning in itself. However, since $\mathbf{S} = P + jQ = \frac{1}{2}V_m I_m\underline{/\theta}$, complex power does provide a convenient analytical tool to calculate average, reactive, and apparent power. We then stated and proved the complex power balance theorem. According to this theorem, the total complex power delivered to all elements in an ac circuit (including sources) equals zero; that is, the total complex power supplied by the sources equals the total complex power delivered to the remaining elements of the circuit. Using this theorem, we showed that the complex power supplied to any network composed of Rs, Ls, and Cs is given by

$$\mathbf{S} = P + j2\omega\{W_{\text{tot ave magnetic}} - W_{\text{tot ave electric}}\}$$

Finally, we extended the maximum power transfer problem treated in Chapter 5 to ac circuits. We considered two cases. In the first case, the load impedance was fixed, but the source impedance could be chosen. Here, maximum average power was delivered to the fixed load impedance $\mathbf{Z}_L$ when the source impedance $\mathbf{Z}_s$ was chosen so that $\mathcal{R}e\{\mathbf{Z}_s\} = 0$, $\mathcal{I}m\{\mathbf{Z}_s\} = -\mathcal{I}m\{\mathbf{Z}_L\}$. In the second case, the source impedance was fixed, but the load impedance could be chosen. Here, the maximum average power was supplied by the fixed source when $\mathbf{Z}_L = \mathbf{Z}_s^*$.

KEY FACTS

- Instantaneous power: $\quad p(t) = v(t)i(t)$ W

- Average power: $\qquad P = \dfrac{1}{2} V_m I_m \cdot \text{PF}$ W

  where PF is the power factor $\text{PF} = \cos\theta$ with $\theta = \underline{/\mathbf{Z}}$.

- P is the average rate at which energy is *dissipated* by the network.

- Reactive power: $\quad Q = \dfrac{1}{2} V_m I_m \sin\theta$ VAR

- Q is the peak rate at which energy is *borrowed* and *returned* by the network.

- Apparent power is $\frac{1}{2} V_m I_m$ VA, and so is simply one-half the product of the voltage and current amplitudes.

- Complex power: $\qquad \mathbf{S} = \dfrac{1}{2} \mathbf{V}\mathbf{I}^*$ VA

- Complex power: $\quad \mathbf{S} = P + jQ = \dfrac{1}{2} V_m I_m \underline{/\theta}$ VA

- Complex power balance theorem:

$$\underbrace{\sum \mathbf{S}_i}_{\substack{\text{Supplied} \\ \text{by sources}}} = \underbrace{\sum \mathbf{S}_i}_{\substack{\text{Delivered} \\ \text{to passive} \\ \text{elements}}}$$

- Complex power:

$$\mathbf{S} = P + j2\omega\{W_{\text{tot ave magnetic}} - W_{\text{tot ave electric}}\}$$

- Maximum average power is delivered to a fixed impedance $\mathbf{Z}_L = R_L + jX_L$ when the source impedance $\mathbf{Z}_s$ is chosen as $\mathbf{Z}_s = -jX_L$.

- Maximum average power is supplied by a source with internal impedance $\mathbf{Z}_s = R_s + jX_s$ when the load impedance $\mathbf{Z}_L$ is chosen as $\mathbf{Z}_L = R_s - jX_s = \mathbf{Z}_s^*$. When $\mathbf{Z}_L = \mathbf{Z}_s^*$, we say that $\mathbf{Z}_L$ is *matched* to $\mathbf{Z}_s$.

■ PROBLEMS ■

Sections 12.1 to 12.3

1. A 60-Hz source delivers 480 VA at PF $= 0.707$ lagging. Assume that the current phase angle $\phi_\mathbf{I}$ equals zero, and plot the instantaneous power delivered $p(t)$.

2. What is the peak instantaneous power delivered by the source in Problem 1?

For Problems 3 through 7, the instantaneous power supplied by a voltage source is given by $p(t) = 10 + 10\cos(200t - 20°) + 5\sin(200t - 20°)$ mW.

3. Find P.
4. Find Q.
5. Find the apparent power.
6. Find the peak instantaneous power.
7. Find the frequency of the voltage source in hertz.
8. A series RLC circuit is driven by a sinusoidal source with angular frequency ω. Find the relation between R, L, C, and ω such that the reactive power delivered equals zero.
9. Repeat Problem 8 for a parallel RLC circuit.

10. A sinusoidal source delivers 14.14 VA at $f =$ 10 Hz to a parallel RC circuit. Find the values of R and C if PF = 0.707 leading and the source voltage is $4\sqrt{2}$ V rms.

11. A voltage source is supplying 2 VAR to a series RL network at a frequency of 60 Hz. The circuit element values are $R = 3.77\ \Omega$ and $L = 10$ mH. Find:
    (a) The power factor of the network
    (b) The average power supplied to the network
    (c) The apparent power supplied to the network
    (d) The effective voltage of the source
    (e) The effective current supplied to the network.

Section 12.4

12. A 120-V rms sinusoidal voltage is applied to a load impedance $\mathbf{Z}$. Measurements reveal that the apparent power entering the load is 60 VA and that the power factor is PF = 0.866 lagging.
    (a) What is the peak current supplied?
    (b) State the value of $\mathbf{Z}$ in polar form.
    (c) State the complex power $\mathbf{S}$ in polar form.
    (d) Sketch the power triangle, indicating the values of $|\mathbf{S}|$, $\underline{/\mathbf{S}}$, P, and Q.
    (e) Assume that the load is a series RL circuit, so that $\mathbf{Z} = R + j\omega L$. Assume further that the frequency is 60 Hz. Determine the values of R and L.

13. A load draws 10,000 VA from a sinusoidal source at a power factor of 0.707 leading.
    (a) What is the average power delivered to the load?
    (b) What is the reactive power delivered to the load?
    (c) Find the peak current if the effective source voltage is 120 V rms.
    (d) The load consists of a resistance R in parallel with a capacitance C. What is the expression for the complex power $\mathbf{S} = \frac{1}{2}\mathbf{VI}^*$ entering the load? State your answer in terms of V_{eff}, R, and $j\omega C$.
    (e) Find R and C if $V_{\text{eff}} = 120$ V rms and $f = 60$ Hz.

14. A certain capacitor is modeled by a capacitance C in parallel with a resistance R.
    (a) Find the admittance of the parallel RC model.

(b) The capacitor consumes $\mathbf{S} = 0.05 - j5$ VA when driven by a 7.07-V rms 60-Hz sinusoidal source. Determine the values of R and C.
(c) What is the average stored energy?

15. Derive the expression for the complex power entering a series RLC circuit as a function of R, L, C, ω, and the rms terminal current I_{rms}. State the resulting expressions for P and Q.

16. Derive the expression for the complex power entering a series RLC circuit as a function of R, L, C, ω, and the rms terminal voltage V_{rms}. State the resulting expressions for P and Q.

17. Repeat Problem 15 for a parallel RLC circuit.

18. Repeat Problem 16 for a parallel RLC circuit.

19. A current $i(t) = 30 \cos(5t + 20°)$ A is delivered to a circuit. The resulting driving-point terminal voltage is $v(t) = 10 \cos(5t + 65°)$ V.
    (a) What is the driving-point impedance of the circuit? (Put your answer in polar form.)
    (b) What is the complex power entering the circuit? (Put your answer in polar form.)
    (c) What is the apparent power entering the circuit?

20. A sinusoidal current $i(t) = I_m \cos(\omega t + \phi_{\mathbf{I}})$ is delivered to a series RLC circuit.
    (a) Determine the complex power entering each element.
    (b) Sketch the power triangle of the series circuit and indicate its parameters.
    (c) At what frequency does the source supply only real power?
    (d) Under what condition does the source supply only reactive power?

21. A 1-V rms sinusoidal source with angular frequency ω is connected to a series RLC circuit in which $R = 0.1\ \Omega$, $L = 1\ \mu H$, and $C = 1\ \mu F$. Make smooth plots of the following quantities versus ω for $0 \leq \omega \leq 2$ Mrad/s: (a) power factor, (b) average power delivered, (c) reactive power delivered, and (d) apparent power. Comment on your results. (*Hint*: The curves vary rapidly in the vicinity of $\omega = 1$ Mrad/s.) ◘ 12.5

Section 12.5

22. All waveforms in the following circuit are sinusoidal with frequency 60 Hz. System 1 consumes 1 W at PF = 0.8 lagging. System 2 consumes 1 W at PF = 0.9 lagging. The effective value of $i(t)$ is $I_{\text{eff}} = 1$ A rms. Find

(a) The effective value of $v(t)$
(b) The reactive power entering the red region
(c) The driving-point impedance $\mathbf{Z}_{in}$ of the enclosed system
(d) The apparent power entering the red region
(e) The power factor of the enclosed system

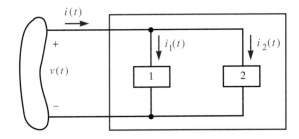

23. A 120-V rms sinusoidal voltage source provides power to two loads connected in series. Load 1 consumes 10 W at PF = 0.8 lagging. Load 2 consumes 14 W at PF = 0.9 leading. Find:
    (a) The peak current
    (b) The rms voltage across *each* load
    (c) The power factor of the series combination of loads 1 and 2
    (d) The complex power supplied by the source
24. Two loads are connected in parallel across a sinusoidal source. The effective current supplied is 10 A rms. Load 1 consumes 12 kW at PF = 0.707 leading. Load 2 consumes 16 kW at PF = 0.866 lagging. Find:
    (a) The rms current through each load
    (b) The peak voltage
    (c) The power factor of the parallel combination
    (d) The complex power supplied by the source
25. A 120-V rms 60-Hz sinusoidal voltage source drives a 50-kW induction motor that operates with a lagging power factor of 0.8. What is the peak current supplied by the source?
26. A 120-V rms 60-Hz sinusoidal voltage source drives a load composed of a motor and a capacitance connected in parallel. The motor consumes 50 kW at a lagging power factor of 0.8, and the capacitance has value C.
    (a) What value of C is required to make the power factor of the parallel load equal unity?

(b) Assuming that the capacitance is as computed in (a), what is the peak current supplied by the source? Compare your answer with the answer to Problem 25.
27. Two loads are in parallel across a 70.7-V rms line. Load 1 draws 100 VA at a lagging power factor of 0.342. Load 2 draws -33.55 VAR at a power factor of 0.819.
    (a) What is the peak line voltage?
    (b) What is the complex power entering load 1?
    (c) What is the complex power entering load 2?
    (d) What is the rms current supplied by the source?
    (e) A third load is placed in parallel across the 70.7-V rms line. What should the reactance of the third load be in order to make the power factor of the combined loads equal unity? (Assume that the third load is purely reactive.)
28. Two loads are connected in series with a 1-A rms sinusoidal current source. Load 1 is an inductance with value L. Load 2 draws $\sqrt{2}$ VA at PF = 0.707 leading.
    (a) Find the value of L that makes the circuit look purely resistive to the source at a frequency $\omega = 0.5$ rad/s.
    (b) Find the peak instantaneous power supplied by the source for $\omega = 0.5$ rad/s and the value of L determined in (a).
29. A sinusoidal source is applied to a network made up of Rs, Ls, and Cs. Use Eq. (12.62b) and the relations $\mathbf{S}_s = \frac{1}{2}\mathbf{V}_s\mathbf{I}_s^*$ and $\mathbf{Z}(j\omega) = \mathbf{V}_s/\mathbf{I}_s$ to show that the driving-point impedance of the network can be written in the form

$$\mathbf{Z}(j\omega) = \frac{P + j2\omega(W_{\text{tot ave magnetic}} - W_{\text{tot ave electric}})}{\frac{1}{2}I_{sm}^2}$$

30. Analyze the distribution of complex power in the coupled inductance circuit shown below.

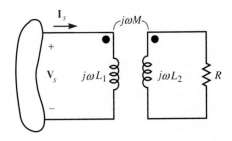

31. (a) Analyze the distribution of complex power in an ac series RLC circuit driven by a current source $\mathbf{I}_s$.
    (b) Draw the associated power triangle.

Section 12.6

32. A practical ac source has open-circuit voltage $\mathbf{V}_s = 1\underline{/0^\circ}$ V and Thévenin equivalent source impedance $\mathbf{Z}_s$. The available power from the source is 0.125 W. If an impedance $\mathbf{Z}_L = j1\ \Omega$ is attached to the terminals of the source, the voltage across that impedance is $\mathbf{V}_L = 1\underline{/90^\circ}$ V. Find $\mathbf{Z}_s$.

33. When a LLTI network is terminated in a 1-Ω resistance, the steady-state current in the resistance is $i_R(t) = 5\cos(1000t - 45^\circ)$ A. When the same network is terminated in a 1000-μF capacitance, the steady-state current is $i_C(t) = 10\cos(1000t - 45^\circ)$ A. Suppose that the network is now terminated in an arbitrary load impedance $\mathbf{Z}_L$.
    (a) Give the expression for the average power supplied to $\mathbf{Z}_L$. Use numerical values wherever possible.
    (b) Find the value of $\mathbf{Z}_L$ that consumes the maximum average power.

34. When two ac networks A and B are joined together as connection 1 shown below, $\mathbf{I}_{ab} = 1\underline{/0^\circ}$ A and $\mathbf{V}_{aa'} = \sqrt{2}\underline{/-45^\circ}$ V. When the same networks are joined as connection 2, $\mathbf{I}_{ab'} = 3\underline{/0^\circ}$ A and $\mathbf{V}_{aa'} = \sqrt{2}\underline{/45^\circ}$ V.

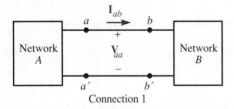

Connection 1

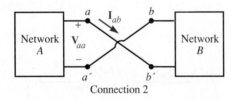

Connection 2

    (a) Find the maximum average power that can be dissipated in a load impedance that is connected from terminal a to terminal a' in connection 1 (with networks A and B in place).

(b) Find the load impedance of (a) that consumes maximum average power.

35. Repeat Problem 34 if the load impedance is connected from terminal a to terminal a' in connection 2 (with networks A and B in place).

36. Consider the active network shown below.

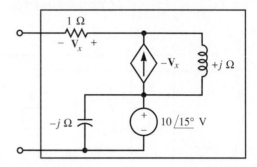

    (a) What load impedance should be connected to the network to consume the maximum average power?
    (b) What is the maximum average power that can be obtained from the network?

37. What is the value of $\mathbf{Z}$ that will absorb the maximum average power for the circuit shown below?

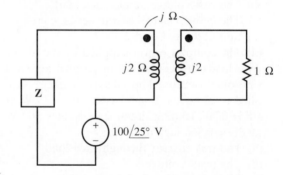

38. (a) Determine the values of M and R such that maximum average power is supplied by the source network shown.
    (b) With M and R as determined in (a), what is the average power supplied by the source?

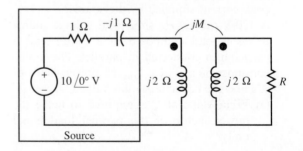

Source

Comprehensive Problems

39. Two loads are connected in parallel to a 240 V rms source. The complex power absorbed by load 1 is $S_1 = 10/\underline{36.8°}$ kVA. If load 2 is purely reactive, find the complex power S_2 so that the combined load has a lagging power factor of 0.9. What is the impedance of load 2?

A factory has three major loads. A motor delivers 10 hp at an overall efficiency of 0.85 and a lagging power factor of 0.8 (1 hp = 0.7457 kW). A resistance furnace requires 15 kW, and an ozone generator requires 8 kVA at a leading power factor of 0.6. The line voltage is 240 V rms. Determine the quantities requested in Problems 40 through 43 for the combined loads.

40. Complex power
41. Real power
42. Rms line current
43. Power factor

You rented a low-budget, 90-year-old apartment. The apartment is wired with 15-A rms circuits. The voltage is 120 V rms, 60 Hz. Your television requires 3 A rms and is a resistive load. You find a bargain on a 120-V rms air conditioner, which requires 15 A rms at a 0.8 lagging power factor, and you wish to use both units at once. Neglect turn on transients. Solve Problems 44 through 48.

44. What total complex power is required by the combined load?

45. What real power is required by the combined load?

46. What is the current required by the combined load?

47. You inadvertently install a 20-A rms fuse. What is the percentage increase in current and resistive heating in the apartment wires above the value at 15 A rms ? Is this safe?

48. You find a bargain in capacitors at the surplus store. Can you use these to reduce the current to 15 A rms for the combined loads? If so, what value of C is required?

49. The following circuit is operating in the steady state.
    (a) Determine the average power delivered to the 9-Ω resistance.
    (b) Does superposition apply to instantaneous voltages and currents?
    (c) Does superposition apply to phasor voltages and currents?
    (d) Does superposition apply to instantaneous power?
    (e) Does superposition apply to average power?
    (f) Is some special property satisfied by the sources in this circuit so that one of the above is true for this circuit that is not true in general?

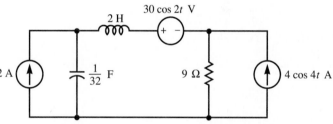

CHAPTER
13

Frequency Response

We find a circuit's *frequency response* by plotting the circuit's transfer function $\mathbf{H}(j\omega)$ versus ω. This plot reveals that LLTI circuits can be used to separate or *filter* sinusoidal signals that have different frequencies. Such filters are used in radio and television tuners to separate one broadcast channel from another.

The topic of frequency response is important not only in radio and television engineering but also in virtually every physical system or phenomenon that is linear and time-invariant. Because this topic is so important, we study it in considerable depth in this chapter.

We begin by giving some simple examples of frequency response. In these examples, we show how a circuit can separate sinusoidal signals whose frequencies are different. Then we embark on our major development, which is to discover the underlying reason any given circuit has the frequency response it has. During this development, we encounter several basically new concepts, including *resonance, complex frequency,* and *poles and zeros.* We then describe Bode plots, which are a practical and commonly used format to plot frequency response. We conclude with an introduction to using PSpice to plot frequency response.

13.1 A New Viewpoint: Response Is a Function of Frequency

The frequency-domain description of networks is much more than a labor-saving device for us to compute a network's sinusoidal steady-state output. It also provides us with insight into how we might use a circuit. This statement is illustrated in the following discussion.

High-Pass Filter

See Problem 13.1 in the PSpice manual.

Figure 13.1 depicts an ordinary RL voltage divider. By examining the transfer function of this circuit, we will find that we can use the circuit to *separate* or *filter* sinusoidal waveforms.

FIGURE 13.1
A high-pass filter

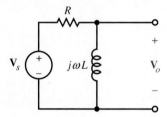

We see from an inspection of Fig. 13.1 that the circuit's voltage transfer function is

$$\mathbf{H}(j\omega) = \frac{\mathbf{V}_0}{\mathbf{V}_s} = \frac{j\omega L}{R + j\omega L} \tag{13.1a}$$

which has magnitude

$$|\mathbf{H}(j\omega)| = \frac{\omega L}{\sqrt{R^2 + \omega^2 L^2}} = \frac{\dfrac{\omega L}{R}}{\sqrt{1 + \left(\dfrac{\omega L}{R}\right)^2}} \tag{13.1b}$$

and angle

$$\underline{/\mathbf{H}(j\omega)} = \frac{\pi}{2} - \arctan \frac{\omega L}{R} \tag{13.1c}$$

Let us plot the magnitude and angle of the voltage transfer function versus angular frequency ω. The results are shown in Fig. 13.2. When viewed as a function of frequency* this way, $|\mathbf{H}(j\omega)|$ and $\underline{/\mathbf{H}(j\omega)}$ are called the *amplitude* and *phase characteristics* of the circuit, respectively, and $\mathbf{H}(j\omega) = |\mathbf{H}(j\omega)|\underline{/\mathbf{H}(j\omega)}$ is called the *frequency response* of the circuit. The plots in Fig. 13.2 show us the relative amplitude and phase of the sinusoidal steady-state output:

$$v_o(t) = |\mathbf{H}(j\omega)| V_{sm} \cos\left(\omega t + \phi_{\mathbf{V}_s} + \underline{/\mathbf{H}(j\omega)}\right) \tag{13.2}$$

with respect to the circuit input

$$v_s(t) = V_{sm} \cos\left(\omega t + \phi_{\mathbf{V}_s}\right) \tag{13.3}$$

at any frequency ω.

---

* The term *frequency* is used in this book to denote either angular frequency ω (in radians per second) or "frequency" f (in hertz). Since $\omega = 2\pi f$, ω and f differ only in the choice of units: One hertz is equivalent to 2π radians per second.

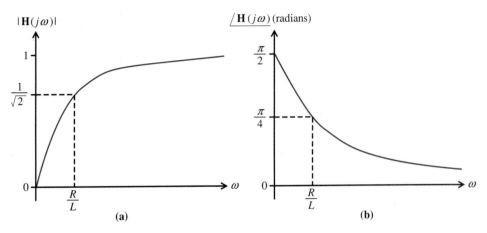

FIGURE 13.2 Frequency response characteristics of the circuit of Fig. 13.1. (*a*) Amplitude characteristic; (*b*) phase characteristic

The point R/L on the ω axis provides a convenient (but basically arbitrary) demarcation separating the frequency axis into two parts. We see by an inspection of Fig. 13.2a that the magnitude of the voltage transfer function $|\mathbf{H}(j\omega)|$ is small, $|\mathbf{H}(j\omega)| \ll 1$, for $\omega \ll R/L$. This tells us that the amplitude of the output sinusoid is small compared with the amplitude of the input sinusoid when the frequency of the input, ω, is small compared with R/L. We also see from Fig. 13.2 that $|\mathbf{H}(j\omega)|$ is nearly unity for $\omega \gg R/L$. This tells us that the amplitudes of the sinusoidal output and input are nearly the same when the frequency of the input is large compared with R/L.

It is easy to explain the shape of the amplitude characteristic of Fig. 13.2a in terms of circuit. As ω approaches zero, the magnitude of the impedance of the inductance $|j\omega L|$ approaches zero. For $\omega = 0$, this magnitude is zero and the inductance is equivalent to a short circuit, which, of course, leads to zero voltage output. As ω approaches infinity, $|j\omega L|$ also approaches infinity. In the limit $\omega \to \infty$, the inductance is equivalent to an open circuit, and this gives us unity voltage gain, because there is then no voltage drop across R.

Now let us consider how the behavior depicted by Fig. 13.2a can be used to separate one signal from another. Suppose that the input to the circuit is the sum of two sinusoidal waveforms as illustrated in Fig. 13.3. For the numerical circuit values given and excluding round-off error, the value of R/L equals 3.14×10^5 rad/s. The angular frequencies of the input sinusoids are $\omega = 3.14 \times 10^4$ rad/s and $\omega = 3.14 \times 10^6$ rad/s. Notice that the amplitudes of the input sinusoids each equal 3 V. It follows directly from Eqs. (13.1b) and (13.1c) that the circuit's transfer function is $\mathbf{H}(j3.14 \times 10^4) = 0.0995\underline{/84.3°}$ for $\omega = 3.14 \times 10^4$ and $\mathbf{H}(j3.14 \times 10^6) = 0.995\underline{/5.7°}$ for $\omega = 3.14 \times 10^6$. The steady-state output is, by superposition, the sum:

$$v_o(t) = (0.0995)(3.0) \cos (3.14 \times 10^4 t + 10° + 84.3°)$$
$$+ (0.995)(3.0) \cos (3.14 \times 10^6 + 27° + 5.7°)$$
$$= 0.299 \cos (3.14 \times 10^4 t + 94.3°) + 2.99 \cos (3.14 \times 10^6 t + 32.7°) \text{ V} \quad (13.4)$$

The remarkable feature of this expression is that the amplitudes of the two output sinusoidal signals differ by a factor of 10. We see that the circuit has passed the higher-frequency input component with relatively little change in amplitude and phase, but has attenuated or *rejected* the lower-frequency input component by a

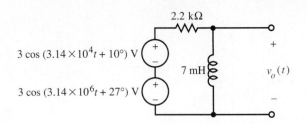

FIGURE 13.3
High-pass filter with
$R/L = 3.14 \times 10^5$ rad/s

$3\cos(3.14 \times 10^4 t + 10°)$ V

$3\cos(3.14 \times 10^6 t + 27°)$ V

2.2 kΩ

7 mH

$v_o(t)$

factor of approximately 10. In effect, the circuit does indeed *filter* its input signals, *passing* the higher-frequency component and (approximately) removing or *rejecting* the lower-frequency component. For this reason we call the circuit a *high-pass filter*.

Notice that the amplitude characteristic of Fig. 13.2 is down by a factor of $1/\sqrt{2}$ at $\omega = R/L$ compared with its maximum of unity. A factor of $1/\sqrt{2}$ in amplitude corresponds to a factor of $(1/\sqrt{2})^2 = 1/2$ in power. For this reason we call R/L the *half-power* frequency of the filter. We call the frequency interval $0 \leq \omega \leq R/L$ the *stopband* of the filter: Signals whose frequencies are in this range are (approximately) stopped or rejected by the filter. The frequency interval $R/L < \omega < \infty$ is called the *passband* of the filter: Signals whose frequencies are in this range are (approximately) passed without change by the filter.

EXERCISES

1. Use Eq. (13.1a) to show that $\mathbf{H}(j\omega) = (1/\sqrt{2})\underline{/45°}$ for $\omega = R/L$.

2. Find $v_o(t)$ in Fig. 13.3 if the angular frequencies of the inputs to the circuit are 3.14×10^3 rad/s and 3.14×10^7 rad/s rather than 3.14×10^4 rad/s and 3.14×10^6 rad/s.

3. A voltage $A_1 \cos(\omega_1 t + \phi_1)$ is applied across the terminals of a resistance of R. The average power that is delivered is 1 W. The voltage is then changed to $A_2 \cos(\omega_2 t + \phi_2)$, where $A_2 = A_1(1/\sqrt{2})$, $\omega_2 = 16\omega_1$, and $\phi_2 = \phi_1 + 27°$. What is the average power delivered?

4. For the circuit shown below:
   (a) Determine and sketch the frequency response characteristics.
   (b) Find the half-power frequency.
   (c) Identify the stopband and the passband.

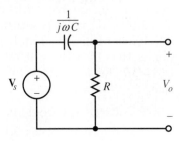

$\dfrac{1}{j\omega C}$

V_s

R

V_o

Low-Pass Filter

A second example of frequency response is provided by the *RC* voltage divider of Fig. 13.4. Its frequency response

FIGURE 13.4
A low-pass filter

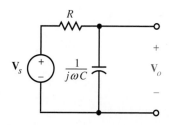

$$\mathbf{H}(j\omega) = \frac{1}{1 + j\omega RC} = \frac{1}{\sqrt{1 + (\omega RC)^2}} \underline{/-\arctan(\omega RC)} \qquad (13.5)$$

is plotted in Fig. 13.5. We can see from the plots that the circuit passes low-frequency sinusoidal inputs with little change in their amplitude and phase, but rejects high-frequency inputs. The demarcation between "low frequency" and "high frequency" occurs in the vicinity of the half-power frequency $\omega = 1/RC$. This circuit is called a *low-pass filter*. The frequency interval $0 \le \omega \le 1/RC$ is the *passband* of this filter, and $1/RC$ is its *half-power bandwidth* in radians per second. The *stopband* is the interval $1/RC < \omega < \infty$.

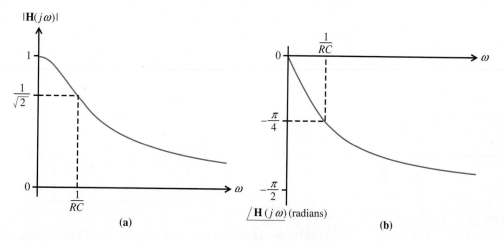

FIGURE 13.5 Frequency response characteristics of the circuit of Fig. 13.4. (*a*) Amplitude characteristic; (*b*) phase characteristic

The *RL* and *RC* circuits of Figs. 13.1 and 13.4 are elementary examples of high-pass and low-pass filters. Much better filters can be obtained with the use of sophisticated circuit design techniques.

EXERCISES

5. Repeat Exercise 4 for the circuit shown below.

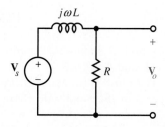

6. Assume that the input to the circuit shown below is $v_s(t) = 10 + 7 \cos 0.1t + 5 \cos t + 3 \cos 10t$. Find $v_o(t)$.

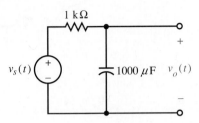

13.2 Resonance

In the preceding section we showed that circuits can be used to filter sinusoidal signals that have different frequencies. We saw that the *kind* of filtering done by a circuit depends on the shape of the circuit's amplitude response characteristic. This observation leads us to ask *why* a given circuit has an amplitude characteristic with a particular shape. This is an important question because, if we know the reasons for the shape of an amplitude characteristic, we will be able to design better filters. The answer is intimately connected to the phenomenon of resonance, which is the topic we consider now.

⚡ *See Problem 13.2 in the PSpice manual.*

FIGURE 13.6
RLC circuit where $R = 0.1\ \Omega$, $L = 1\ \mu H$, and $C = 1\ \mu F$

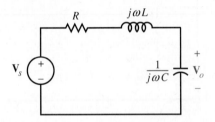

Consider the series *RLC* circuit of Fig. 13.6. The voltage transfer function of this circuit is (by the voltage divider relation)

$$\mathbf{H}(j\omega) = \frac{\mathbf{V}_o}{\mathbf{V}_s} = \frac{1/j\omega C}{R + j\omega L + (1/j\omega C)} = \frac{1}{1 - \omega^2 LC + j\omega RC} \qquad (13.6)$$

for which

$$|\mathbf{H}(j\omega)| = \frac{1}{\sqrt{(1 - \omega^2 LC)^2 + (\omega RC)^2}} \qquad (13.7)$$

and

$$\underline{/\mathbf{H}(j\omega)} = -\tan^{-1} \frac{\omega RC}{1 - \omega^2 LC} \qquad (13.8)$$

Figure 13.7 shows the amplitude and phase characteristics for $R = 0.1\ \Omega$, $L = 1\ \mu H$, and $C = 1\ \mu F$.

The most prominent feature of the frequency response is the sharp peak that occurs in the amplitude characteristic. We call this peak a *resonance* peak. We

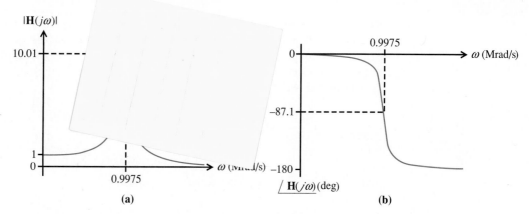

FIGURE 13.7 Frequency response characteristics of the circuit of Fig. 13.6 for $R = 0.1\ \Omega$, $L = 1\ \mu H$, and $C = 1\ \mu F$. (*a*) Amplitude characteristic; (*b*) phase characteristic

call the value of ω at which the peak occurs the *maximum-response resonance frequency*, ω_{mr}. A formula for ω_{mr} can be found if we set the derivative of $|\mathbf{H}(j\omega)|$ with respect to ω to zero and solve for ω. The result is

$$\omega_{mr} = \sqrt{\frac{1}{LC} - 2\left(\frac{R}{2L}\right)^2}$$

$$= 997.5\ \text{krad/s} = 0.9975\ \text{Mrad/s} \tag{13.9}$$

at which $|\mathbf{H}| = 10.01$ and $\underline{/\mathbf{H}} = -87.1°$.

Since the amplitude characteristic for frequencies near 0.9975 Mrad/s is relatively large compared with that for other frequencies, a practical use of the circuit would be to increase the amplitudes of sinusoidal voltages whose frequencies are near 0.9975 Mrad/s relative to those with other frequencies. We will examine this application in detail in Section 13.5. In the next subsection, we will examine the more basic issue of *why* the peak occurs.

Why Resonance Occurs

Our objective in this subsection is to understand why there is a peak in the series *RLC* circuit's amplitude characteristic (Fig. 13.7a). How is it possible for the output amplitude to be a factor of 10.01 *greater* than the input amplitude? Let us first try to discover an underlying mathematical explanation. We can do this by writing the denominator of $\mathbf{H}(j\omega)$ [Eq. (13.6)],

$$\mathscr{A}(j\omega) = 1 - \omega^2 LC + j\omega RC \tag{13.10}$$

in factored form. To factor $\mathscr{A}(j\omega)$, we first find the roots to the characteristic equation

$$\mathscr{A}(s) = 1 + s^2 LC + sRC = 0 \tag{13.11}$$

For the circuit values given, the roots are complex. The roots are

$$s_1 = -\alpha + j\omega_d \tag{13.12}$$

and

$$s_2 = -\alpha - j\omega_d \tag{13.13}$$

where $\alpha = R/2L = 4 \times 10^4$ is the *damping coefficient* and $\omega_d = \sqrt{(1/LC)-(R/2L)^2} = 0.9987$ Mrad/s is the *damped natural frequency*. Recall that the quantities s_1, s_2, α, and ω_d were encountered previously in connection with the complementary or natural response of an underdamped series RLC circuit (Section 8.8). With s_1 and s_2 determined, we can now write $\mathscr{A}(s)$ in its factored form:*

$$\mathscr{A}(s) = LC(s - s_1)(s - s_2) \tag{13.14}$$

Therefore

$$\mathscr{A}(j\omega) = LC(j\omega - s_1)(j\omega - s_2) \tag{13.15}$$

and we can write the transfer function as follows:

$$\mathbf{H}(j\omega) = \frac{1}{1 - \omega^2 LC + j\omega RC} = \frac{1}{LC(j\omega - s_1)(j\omega - s_2)} \tag{13.16}$$

The advantage of writing the transfer function in factored form is that we can draw the factors $j\omega - s_1$ and $j\omega - s_2$ as vectors in the complex plane, as shown in Fig. 13.8. The factor

$$j\omega - s_1 = j\omega - (-\alpha + j\omega_d) = \alpha + j(\omega - \omega_d) \tag{13.17}$$

in Eq. (13.16) is represented as a vector starting at the root s_1 and ending at the point $j\omega$. Similarly, the factor

$$j\omega - s_2 = j\omega - (-\alpha - j\omega_d) = \alpha + j(\omega + \omega_d) \tag{13.18}$$

is represented as a vector starting at the root s_2 and ending at $j\omega$. The amplitude characteristic is determined by the inverse of the product of the lengths of these two vectors; that is,

$$|\mathbf{H}(j\omega)| = \frac{1/LC}{|j\omega - s_1||j\omega - s_2|} \tag{13.19}$$

FIGURE 13.8
Characteristic equation roots and input frequency as points in the complex plane

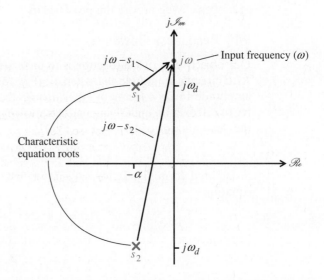

* The equality $s^2LC + sRC + 1 = LC(s - s_1)(s - s_2)$ is established by the fact that both sides are polynomials with the same degree, 2, have the same roots, s_1 and s_2, and have the same coefficient, LC, multiplying the highest-degree variable, s^2.

The phase characteristic is determined by the negative of the sum of the angles of the two vectors

$$\underline{/\mathbf{H}(j\omega)} = -\left(\underline{/j\omega - s_1} + \underline{/j\omega - s_2}\right) \tag{13.20}$$

Figure 13.8 provides a *geometric* interpretation of the combined influence of the input frequency ω and the characteristic equation roots s_1 and s_2 in the determination of $\mathbf{H}(j\omega)$ of Eq. (13.16). The mathematical reason for the peak in $|\mathbf{H}(j\omega)|$ and for the abrupt phase change in $\underline{/\mathbf{H}(j\omega)}$ in Fig. 13.7 is related to the behavior of the vector $j\omega - s_1$.

We can see from Fig. 13.8 that if ω is varied in the range $\omega_d - \alpha$ to $\omega_d + \alpha$, the vector $j\omega - s_1$ undergoes a relatively pronounced change in both its length $|j\omega - s_1|$ and its angle $\underline{/j\omega - s_1}$. The magnitude $|j\omega - s_1|$ is minimum when $\omega = \omega_d$, and therefore the maximum of $|\mathbf{H}(j\omega)|$ occurs at approximately this frequency. The conclusion is as follows.

It is the "closeness" of the input frequency (represented as the point $j\omega$ in Fig. 13.8) to the circuit's characteristic equation root s_1 (shown as the point s_1) that causes the resonance peak in the amplitude characteristic.

This is the *mathematical* explanation for the peak in the amplitude characteristic of Fig. 13.7a.

Now that we have found the mathematical explanation for the resonance peak of Fig 13.7a, we can also understand the physical reason for this peak. Recall from Section 8.2 that the *natural* response of a circuit is determined from the roots of the circuit's characteristic equation. For the series *RLC* circuit of Fig. 13.6, the roots are complex and the natural response has the form given by Eq. (8.98), namely,

$$v_{oc}(t) = e^{-\alpha t}(A \cos \omega_d t + B \sin \omega_d t) \tag{13.21}$$

where the constants A and B depend on initial conditions. What is important here is the fact that the natural response of the series *RLC* circuit of Fig. 13.6 is a *damped sinusoidal oscillation* with angular frequency ω_d. We have seen that the peak in $|\mathbf{H}(j\omega)|$ occurs when the circuit is driven by a sinusoidal source whose frequency ω is approximately equal to ω_d. We conclude that the resonance peak of Fig. 13.7a is caused by the periodic *reinforcement* that occurs when the circuit is driven at a frequency that is approximately equal to the circuit's damped natural frequency.

Examples of similar interactions between an input function and a system's natural response abound in nature. Consider a child on a swing. To go higher and higher, the child pumps the swing at the times that best reinforce the swing's natural oscillations. A series of small but correctly timed periodic inputs by the child results in a large oscillatory response. Another example of resonance occurs when a soprano sings a certain note that makes a glass goblet shatter. Another occurs when an earthquake tremor is amplified in a poorly designed structure. These examples serve to illustrate that resonance can have harmful as well as beneficial consequences.

In this section we have gained insight into the frequency response of a series *RLC* circuit by plotting the input frequency and the characteristic equation roots in the complex plane as in Fig. 13.8. In the next section, we will obtain even greater insight by combining the graphical approach of Fig. 13.8 with a new concept called *complex frequency*.

A characteristic equation root close to the imaginary axis of the complex plane causes a resonance peak in the amplitude characteristic.

EXERCISES

7. For the circuit shown below,
   (a) Plot the amplitude and phase characteristics,
   (b) Determine the values of the damping coefficient and the damped natural frequency.

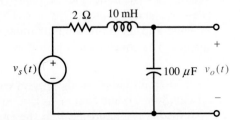

8. Assume that the input to the circuit in Exercise 7 is $v_s(t) = 10 + 7 \cos 50t + 5 \cos 1000t + 3 \cos 2000t$. Find $v_o(t)$.

9. Verify that

$$\omega_{mr} = \sqrt{\frac{1}{LC} - 2\left(\frac{R}{2L}\right)^2}$$

is the maximum-response resonance frequency associated with the transfer function of Eq. (13.6). (*Hint*: You can simplify your derivation by noting that the maximum of $|\mathbf{H}(j\omega)|$ must occur at the same frequency as the *minimum* of $|\mathbf{H}(j\omega)|^{-2}$.)

10. Show the root of the characteristic equation of the following circuit as a point in the complex plane. Use a ruler and protractor to help plot the amplitude and phase characteristics. Assume that $R = 1 \text{ k}\Omega$ and $C = 1000 \ \mu\text{F}$. (*Hint*: To use a ruler and protractor, you need to represent $\mathbf{H}(j\omega)$ by a drawing analogous to Fig. 13.8.)

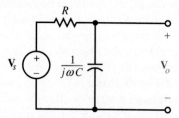

13.3   Complex Frequency

In the previous section, we interpreted the shape of an *RLC* circuit's amplitude and phase characteristics geometrically by plotting the input frequency and the characteristic equation roots as points in the complex plane. The geometric interpretation was somewhat limited because the input frequency was plotted as the point $j\omega$ and was therefore constrained to lie along the imaginary axis. In this section we will show that additional insight into the shape of the amplitude characteristic can be obtained by generalizing the notion of frequency to include any point in the complex plane. This is the basic idea behind complex frequency.

Complex frequency is defined as $s = \sigma + j\omega$, where $\sigma = \mathcal{R}e\{s\}$ is the *neper** frequency and $\omega = \mathcal{I}m\{s\}$ is the *angular* frequency. The domain of all possible values of s is called the *complex-frequency plane* or the *s plane*. A particular value of s is a point in the s plane, as depicted in Fig. 13.9. In the illustration, the neper frequency is -0.02 Mnepers per second (MNp/s) and the angular frequency is 1.1 Mrad/s.

FIGURE 13.9
Depiction of a particular value of s as a point in the s plane. Here $s = -0.02$ MNp/s $+$ $j1.1$ Mrad/s

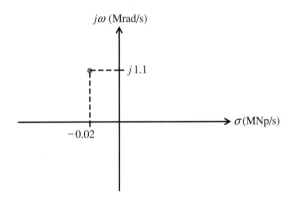

Let us investigate the results of replacing the argument, $j\omega$, in the transfer function of Eq. (13.6) by $s = \sigma + j\omega$. The substitution of s for $j\omega$ in Eq. (13.6) yields

$$\mathbf{H}(s) = \frac{\mathbf{V}_o}{\mathbf{V}_s} = \frac{1/sC}{R + sL + (1/sC)} = \frac{1}{s^2 LC + sRC + 1} = \frac{1/LC}{(s - s_1)(s - s_2)} \quad (13.22)$$

where s_1 and s_2 are the roots of the characteristic Eq. (13.11) given by Eqs. (13.12) and (13.13). Since we have called s the complex frequency, we will now call s_1 and s_2 *complex natural frequencies* of the circuit. It follows from Eq. (13.22) that

$$|\mathbf{H}(s)| = \frac{1/LC}{|s - s_1||s - s_2|} \quad (13.23)$$

and

$$\phi_H = -\underline{/s - s_1} - \underline{/s - s_2} \quad (13.24)$$

Equations (13.22) through (13.24) can be interpreted geometrically in terms of the vectors $s - s_1$ and $s - s_2$ shown in Fig. 13.10. The factors $|s - s_1|$ and $|s - s_2|$ in Eq. (13.23) are the lengths of the vectors $s - s_1$ and $s - s_2$, respectively. We see, therefore, that $|\mathbf{H}(s)|$ becomes arbitrarily large as $s \to s_1$ or $s \to s_2$. The function $|\mathbf{H}(s)|$ is plotted as a function of s in Fig. 13.11. Notice that $|\mathbf{H}(s)|$ looks like a thin rubber sheet that is supported over the s plane by poles located on the complex natural frequencies s_1 and s_2. Observe that the intersection of $|\mathbf{H}(s)|$ with the plane $\sigma = 0$ is the amplitude characteristic $|\mathbf{H}(j\omega)|$ previously shown in Fig. 13.7a for $\omega > 0$.

---

* The term *neper* is derived from the Naperian (natural) logarithm ln.

FIGURE 13.10
Characteristic equation roots
and complex input frequency
as points in the complex
plane

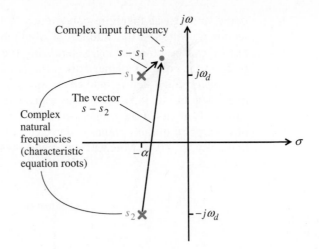

Figure 13.11 illustrates one advantage of using the complex frequency s in place of $j\omega$ in a transfer function $\mathbf{H}(j\omega)$. The substitution of s for $j\omega$ has provided us with the *rubber sheet analogy*. The rubber sheet analogy provides the means to make a quick, approximate sketch of a circuit's amplitude characteristic by inspection of the factored algebraic expression for $\mathbf{H}(s)$. The amplitude characteristic $|\mathbf{H}(j\omega)|$ is always given by the intersection of $|\mathbf{H}(s)|$ (represented by the rubber sheet) and the plane defined by $\sigma = 0$. We will describe the rubber sheet analogy in greater detail in Section 13.4.

In this subsection we have defined the concept of complex frequency s and have illustrated one way in which it can be used. In the following subsection we show that complex frequency is also a useful tool to analyze the particular response of an LLTI circuit to a *damped* sinusoidal input.

FIGURE 13.11
Frequency response
magnitude as a function of
complex frequency

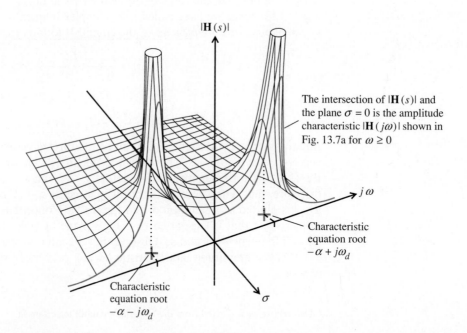

11. For the *RC* low-pass filter shown:
    (a) Make a rough three-dimensional sketch (analogous to that in Fig. 13.11) of the voltage transfer function magnitude, $|\mathbf{V}_o/\mathbf{V}_s| = |\mathbf{H}(s)|$.
    (b) Explain in your own words how your sketch is related to the amplitude characteristic $|\mathbf{H}(j\omega)|$ shown in Fig. 13.5a.

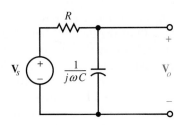

Spiraling Phasors and Damped Sinusoids

We know from our work in Chapter 10 that the complex exponential $\mathbf{V}e^{st}$ can be interpreted as a *rotating phasor* when $s = j\omega$. We now ask, "How can we interpret $\mathbf{V}e^{st}$ when $s = \sigma + j\omega$ is an *arbitrary* complex number? To answer this question, we write

$$\mathbf{V}e^{st} = (V_m e^{j\phi_\mathbf{V}})e^{(\sigma + j\omega)t} = V_m e^{\sigma t}e^{j(\omega t + \phi_\mathbf{V})} = V_m e^{\sigma t}\underline{/\omega t + \phi_\mathbf{V}} \qquad (13.25)$$

According to Eq. (13.25), multiplication of phasor $\mathbf{V}$ by e^{st} causes the phasor to rotate at an angular rate ω and vary in magnitude at an exponential rate σ, as illustrated in Fig. 13.12. Accordingly, $\mathbf{V}e^{st}$ may be interpreted as a *spiraling phasor*.

FIGURE 13.12
The spiraling phasor
$\mathbf{V}e^{st} = V_m e^{\sigma t}\,\underline{/\omega t + \phi_V}$.
The projection on the $\mathcal{R}e$ axis is $V_m e^{\sigma t}\cos{(\omega t + \phi_\mathbf{V})}$.
(a) $\sigma < 0$; (b) $\sigma = 0$;
(c) $\sigma > 0$

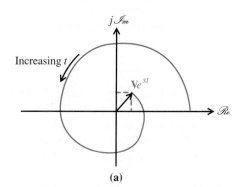

(a)

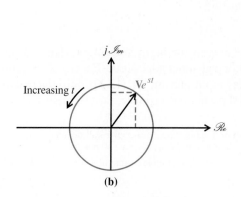

(b)

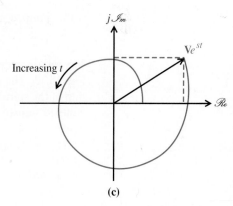

(c)

The tip of the phasor spirals outward for $\sigma > 0$ and inward for $\sigma < 0$. Rotation without spiraling occurs for $\sigma = 0$. The projection of the spiraling phasor onto the real axis is the damped sinusoid.

$$\mathcal{R}e\left\{\mathbf{V}e^{st}\right\} = V_m e^{\sigma t} \cos\left(\omega t + \phi_{\mathbf{V}}\right) \tag{13.26}$$

Similarly, the projection onto the imaginary axis is the damped sinusoid

$$\mathcal{I}m\left\{\mathbf{V}e^{st}\right\} = V_m e^{\sigma t} \sin\left(\omega t + \phi_{\mathbf{V}}\right) \tag{13.27}$$

The neper frequency σ determines the rate of exponential growth ($\sigma > 0$) or decay ($\sigma < 0$) of the damped sinusoid. For $\sigma = 0$, the amplitude of the sinusoid is constant, because the phasor rotates without spiraling. The $\sigma = 0$ case is the one we studied in Chapters 10 through 12.

Recall that in Chapter 10 we used rotating phasors to find the particular response of an LLTI circuit to a sinusoidal input. We can similarly use spiraling phasors to find the particular response of a circuit to a damped sinusoidal input. The method is based on two facts.

The first fact is that the particular response of an LLTI equation

$$\mathcal{A}(p)v_{op}(t) = \mathcal{B}(p)v_s(t) \tag{13.28}$$

to a spiraling phasor input*

$$\mathbf{v}_s(t) = \mathbf{V}_s e^{st} \tag{13.29}$$

is also a spiraling phasor:

$$\mathbf{v}_{op}(t) = \mathbf{V}_o e^{st} \tag{13.30}$$

The input and output phasors are related by

$$\mathbf{V}_o = \mathbf{H}(s)\mathbf{V}_s \tag{13.31}$$

$\mathbf{H}(s)$ is the transfer function

$$\mathbf{H}(s) = \frac{\mathbf{V}_o}{\mathbf{V}_s} = \frac{\mathcal{B}(s)}{\mathcal{A}(s)} \tag{13.32}$$

This follows from Theorem 2 of Chapter 9.[†] Alternatively, we can establish it by simply substituting Eqs. (13.29) and (13.30) into Eq. (13.28), performing the differentiations, and solving for $\mathbf{V}_o$.

The second fact is that the particular response of the LLTI equation, Eq. (13.28), to a damped sinusoidal input

$$v_s(t) = V_{sm} e^{\sigma t} \cos\left(\omega t + \phi_{\mathbf{V}_s}\right) = \mathcal{R}e\left\{\mathbf{V}_s e^{st}\right\} \tag{13.33}$$

is the damped sinusoid

$$v_{op}(t) = V_{om} e^{\sigma t} \cos\left(\omega t + \phi_{\mathbf{V}_o}\right) = \mathcal{R}e\left\{\mathbf{V}_o e^{st}\right\} \tag{13.34}$$

The relationship between the input and output damped sinusoids is illustrated in Fig. 13.13. The damped sinusoidal input, Eq. (13.33) is represented by the projection of the spiraling input phasor, Eq. (13.29), onto the real axis. The corresponding damped sinusoidal particular response, Eq. (13.34), is represented by the projection

---

* Do not let the notation confuse you. Subscript s still denotes *source* as it did in previous chapters. The s in the exponent is the complex frequency $s = \sigma + j\omega$.

† We assume that $\mathcal{A}(s) \neq 0$ for the complex value of $s = \sigma + j\omega$ of the input.

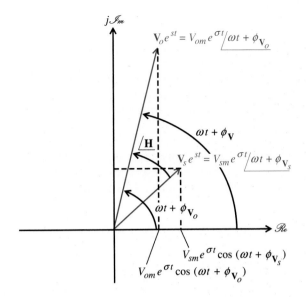

FIGURE 13.13
Input and output spiraling phasors and their projections. The real input and output are the projections of the spiraling phasors onto the real axis.
$V_{om} = |\mathbf{H}| V_{sm}$;
$\phi_{\mathbf{V}_0} = \underline{/\mathbf{H}} + \phi_{\mathbf{V}s}$

of the spiraling output phasor, Eq. (13.30), onto the real axis. Notice from Eqs. (13.33) and (13.34) that the output damped sinusoid has the same neper frequency σ and angular frequency ω as the input.

Equation (13.34) can be proved by reasoning similar to that which led to Theorem 5 of Chapter 9. The essential difference is that $j\omega$ is replaced by s and sinusoids are replaced by damped sinusoids. The practical result is that phasors can be used to calculate LLTI circuit response to a damped sinusoidal input.

In the preceding discussion we assumed that the input and output were both voltages. Of course, similar considerations apply for currents.

EXAMPLE 13.1

(a) Start with the appropriate differential equation and determine the impedance $\mathbf{Z}(s)$ of a series RLC circuit.

(b) Find the particular response voltage at the input terminals to a damped sinusoidal input $i(t) = 3e^{\sigma t} \cos (\omega t + 10°)$ A, where $\sigma = -0.02$ MNp/s and $\omega = 1.1$ Mrad/s. Assume that $R = 0.1\ \Omega$, $L = 1\ \mu$H, and $C = 1\ \mu$F.

■ SOLUTION

(a) The differential equation relating driving-point current to voltage is

$$v = Ri + L\frac{d}{dt}i + \frac{1}{C}\int_{-\infty}^{t} i\, d\tau$$

or

$$\frac{d}{dt}v = L\frac{d^2}{dt^2}i + R\frac{d}{dt}i + \frac{1}{C}i$$

which is

$$\mathscr{A}(p)v = \mathscr{B}(p)i$$

where $\mathscr{A}(p) = p$ and $\mathscr{B}(p) = Lp^2 + Rp + (1/C)$. The impedance is given by

$$Z(s) = \frac{V}{I} = \frac{\mathcal{B}(s)}{\mathcal{A}(s)} = sL + R + \frac{1}{sC}$$

We can obtain the same result by substituting $i(t) = Ie^{st}$ and $v_p(t) = Ve^{st}$ into the differential equation and solving for V/I. $Z(s)$ exists provided that $s \neq 0$.

(b) The input is a damped sinusoid: $i(t) = 3e^{\sigma t} \cos(\omega t + 10°)$ A, where $\sigma = -0.02$ MNp/s and $\omega = 1.1$ Mrad/s. The corresponding phasor current is

$$\mathbf{I} = 3\underline{/10°} \text{ A}$$

and the associated complex frequency is $s = -0.02$ MNp/s $+ j1.1$ Mrad/s. To find the particular response voltage, we evaluate the impedance $Z(s)$ of (a) for $s = -2 \times 10^4 + j1.1 \times 10^6$. This yields

$$Z(-2 \times 10^4 + j1.1 \times 10^6) = \underbrace{-0.02 + j1.1}_{sL} + \underbrace{0.1}_{R} + \underbrace{\frac{1}{-0.02 + j1.1}}_{\frac{1}{sC}}$$

$$= 0.202\underline{/71.6°} \ \Omega$$

The output phasor is

$$\mathbf{V} = \mathbf{ZI} = (0.202\underline{/71.6°})(3\underline{/10°}) = 0.606\underline{/81.6°} \text{ V}$$

Therefore the particular response to the damped sinusoidal input $i(t)$ is the damped sinusoid

$$v_p(t) = 0.606e^{\sigma t} \cos(\omega t + 81.6°) \text{ V}$$

The values of σ and ω are always the same as those of the input: $\sigma = -0.02$ MNp/s and $\omega = 1.1$ Mrad/s. ∎

REMEMBER

The particular response of an LLTI equation to a spiraling phasor input $\mathbf{V}_s e^{st}$ is a spiraling phasor $\mathbf{H}(s)\mathbf{V}_s e^{st}$ where $\mathbf{H}(s)$ is the transfer function. The projections of the spiraling phasors onto the real axis of the complex plane are the corresponding damped sinusoidal input and output.

EXERCISES

12. Derive Eq. (13.31) by substituting Eqs. (13.29) and (13.30) into Eq. (13.28).

13. A 1-Ω resistance is connected in parallel with a 1-mH inductance.
    (a) Starting from the appropriate differential equation, show that the driving-point impedance $Z(s)$ is given by

    $$Z(s) = \frac{sLR}{sL + R}$$

    (b) A current $i(t) = 10e^{-3000t} \cos(2000t + 20°)$ mA is applied to the parallel circuit. Determine the particular response voltage $v_p(t)$ appearing across the input terminals.

14. Repeat Exercise 13 for $i(t) = 10 \cos(1000t + 20°)$ mA.

15. Repeat Exercise 13 for $i(t) = 10e^{-3000t}$ mA.

13.4 s-Domain Circuit Analysis

In Chapter 10 we used rotating phasors $\mathbf{i}(t) = \mathbf{I}e^{j\omega t}$ and $\mathbf{v}(t) = \mathbf{V}e^{j\omega t}$ to transform the differential time-domain terminal equations of R, L, and C into algebraic frequency-domain equations. We can similarly use spiraling phasors $\mathbf{i}(t) = \mathbf{I}e^{st}$ and $\mathbf{v}(t) = \mathbf{V}e^{st}$ to transform the differential terminal equations of R, L, and C into algebraic equations of the form

$$\mathbf{V} = \mathbf{Z}(s)\mathbf{I} \tag{13.35}$$

where $\mathbf{Z}(s)$ is the impedance of the element. The derivations are identical to those on pages 425 to 426 with the exception that $j\omega$ is replaced by s. Similar derivations apply to coupled inductances. The results are summarized in Table 13.1.

TABLE 13.1
Element terminal equations in the s domain

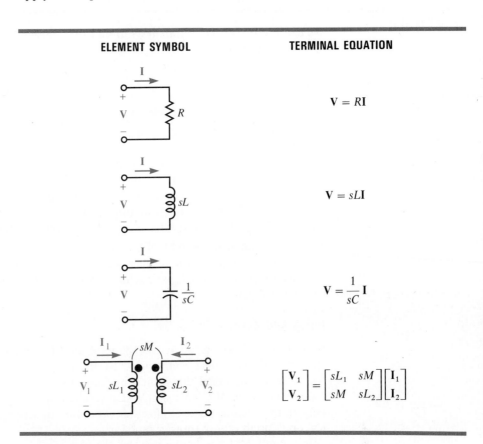

ELEMENT SYMBOL	TERMINAL EQUATION
R	$\mathbf{V} = R\mathbf{I}$
sL	$\mathbf{V} = sL\mathbf{I}$
$\dfrac{1}{sC}$	$\mathbf{V} = \dfrac{1}{sC}\mathbf{I}$
sM, sL_1, sL_2	$\begin{bmatrix} \mathbf{V}_1 \\ \mathbf{V}_2 \end{bmatrix} = \begin{bmatrix} sL_1 & sM \\ sM & sL_2 \end{bmatrix}\begin{bmatrix} \mathbf{I}_1 \\ \mathbf{I}_2 \end{bmatrix}$

Spiraling phasors also transform Kirchhoff's laws and the terminal equations of dependent sources into algebraic equations involving phasors. The results are identical to Eqs. (11.18) and (11.21) and Fig. 11.3, with the understanding that $j\omega$ is replaced by s. The practical result is that the s-domain transfer function of any LLTI circuit can be obtained by the phasor circuit analysis methods discussed in Chapter 11.

A simple example is provided by our problem of finding the driving-point impedance of a series RLC circuit. The complex-frequency (or s-domain) circuit is shown in Fig. 13.14. By using the fact that impedances in series add, we can

FIGURE 13.14
s-domain circuit

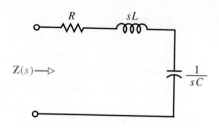

write $\mathbf{Z}(s) = R + sL + (1/sC)$ by inspection. Notice that this result agrees with that obtained in Example 13.1(a) as it must.

REMEMBER

We can use phasor circuit analysis to find the transfer function $\mathbf{H}(s) = \mathscr{B}(s)/\mathscr{A}(s)$ of any LLTI circuit.

EXERCISES

16. Derive the entries in Table 13.1.

17. A current $i(t)$ is applied to a parallel *RL* circuit, where $R = 1\ \mathrm{k\Omega}$ and $L = 1\ \mathrm{mH}$.
    (a) Draw the time-domain circuit.
    (b) Draw the *s*-domain circuit.
    (c) Write the driving-point impedance $\mathbf{Z}(s)$ by inspection.
    (d) Find the particular response voltage for $i(t) = 5 \cos 10^6 t$ mA. *Hint*: To obtain $\mathbf{Z}(j\omega)$, just set $s = j\omega$ in your answer to (c).

18. Show that when Kirchhoff's current law is transformed into the complex-frequency domain, the result is Eq. (11.18). *Hint*: Start with Eq. (11.12) and specialize to $i_n(t) = I_{mn}e^{\sigma t} \cos (\omega t + \phi_n) = \mathscr{R}e \left\{ \mathbf{I}_n e^{st} \right\}$ where $\mathbf{I}_n = I_{mn}/\underline{\phi_n}$ and $s = \sigma + j\omega$.

Poles and Zeros

In Section 13.2 we saw that the shape of an *RLC* circuit's frequency response characteristic is critically influenced by the roots of the circuit's characteristic equation. We obtained this insight by factoring the denominator of the circuit's transfer function $\mathbf{H}(s)$. This simple step resulted in a graphical interpretation for $\mathbf{H}(s)$. In this section we generalize this procedure to arrive at a completely graphical description of any transfer function.

In general, both the numerator and the denominator of a transfer function are polynomials in *s*:

$$\mathbf{H}(s) = \frac{\mathscr{B}(s)}{\mathscr{A}(s)} \tag{13.36}$$

Let us now write both the numerator and the denominator of Eq. (13.36) in factored form. To factor the denominator, we first find the roots to the characteristic equation

$$\mathscr{A}(s) = a_n s^n + a_{n-1}s^{n-1} + \cdots + a_0 = 0 \tag{13.37}$$

From now on, we will denote the distinct roots of the characteristic equation by

$s_{p1}, s_{p2}, \ldots, s_{pl}$. We let m_i denote the order (multiplicity) of root s_{pi}, $i = 1, 2, \ldots, l$. The factored form of $\mathscr{A}(s)$ is given by

$$\mathscr{A}(s) = a_n(s - s_{p1})^{m_1}(s - s_{p2})^{m_2} \cdots (s - s_{pl})^{m_l} \tag{13.38}$$

Similarly, to factor the numerator, we first find the roots of the equation

$$\mathscr{B}(s) = b_m s^m + b_{m-1} s^{m-1} + \cdots + b_0 = 0 \tag{13.39}$$

We let $s_{z1}, s_{z2}, \ldots, s_{zr}$ denote the distinct roots and let q_i denote the order of root s_{zi}, $i = 1, 2, \ldots, r$. The factored form of $\mathscr{B}(s)$ is

$$\mathscr{B}(s) = b_m(s - s_{z1})^{q_1}(s - s_{z2})^{q_2} \cdots (s - s_{zr})^{q_r} \tag{13.40}$$

By substituting Eqs. (13.38) and (13.40) into Eq. (13.36), we obtain

$$\mathbf{H}(s) = K \frac{(s - s_{z1})^{q_1}(s - s_{z2})^{q_2} \cdots (s - s_{zr})^{q_r}}{(s - s_{p1})^{m_1}(s - s_{p2})^{m_2} \cdots (s - s_{pl})^{m_l}} \tag{13.41}$$

where $K = b_m/a_n$. The following definitions apply to Eq. (13.41).

1. The transfer function is said to have *poles* at the complex natural frequencies $s = s_{p1}, s_{p2}, \ldots, s_{pl}$. These frequencies are the roots of the characteristic equation $\mathscr{A}(s) = 0$. Since the poles are the roots of the characteristic equation, they determine the form of the circuit's natural (complementary) response (Theorems 2 through 4 of Chapter 8).
2. m_i is the *order* of the pole s_{pi}. The pole is called *simple* if $m_i = 1$.
3. The transfer function is said to have *zeros* at the complex frequencies s_{z1}, $s_{z2}, \ldots, s_{zr}$. These frequencies are the roots of the equation $\mathscr{B}(s) = 0$.
4. q_i is called the *order* of the zero s_{zi}. The zero is called *simple* if $q_i = 1$.
5. The poles and the zeros are called the *critical frequencies* of the circuit.
6. A plot of the critical frequencies of $\mathbf{H}(s)$ as points in the plane is called a *pole-zero plot* or *pole-zero constellation*. The poles are depicted by $\times$s, and the zeros are depicted by $\bigcirc$s. All poles and zeros are assumed to be simple unless indicated otherwise, and the value of the constant K is shown in a box. A typical pole-zero plot is shown in Fig. 13.15.

FIGURE 13.15
A typical pole-zero plot

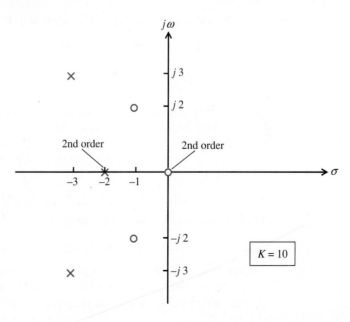

We can see from Eq. (13.41) that the transfer function of a circuit is completely determined by the pole-zero plot and the constant K. Indeed, the differential equation relating the output and input of a circuit can be derived from the pole-zero plot. Therefore, the pole-zero plot provides a *complete* description of the circuit's input-output properties.

EXAMPLE 13.2

Assume that the pole-zero plot of Fig. 13.15 refers to an impedance, $\mathbf{Z}(s)$. Find:
  (a) The critical frequencies
  (b) $\mathbf{Z}(s)$
  (c) The input/output equation in the time domain
  (d) The complex natural frequencies
  (e) The general form of the complementary response
  (f) The particular response for an input $i(t) = 3 \cos(4t + 17°)$ A

■ **SOLUTION**

(a) The critical frequencies are the poles $s_{p1} = -2$, $s_{p2} = -3 + j3$, and $s_{p3} = -3 - j3$ and the zeros $s_{z1} = 0$, $s_{z2} = -1 + j2$, and $s_{z3} = -1 - j2$.
(b) From Eq. (13.41) and the answer to (a),

$$\mathbf{Z}(s) = 10 \frac{(s - s_{z1})^2(s - s_{z2})(s - s_{z3})}{(s - s_{p1})^2(s - s_{p2})(s - s_{p3})} = 10 \frac{s^2(s + 1 - j2)(s + 1 + j2)}{(s + 2)^2(s + 3 - j3)(s + 3 + j3)}$$

$$= 10 \frac{s^4 + 2s^3 + 5s^2}{s^4 + 10s^3 + 46s^2 + 96s + 72}$$

(c) Cross-multiply

$$\frac{\mathbf{V}}{\mathbf{I}} = 10 \frac{s^4 + 2s^3 + 5s^2}{s^4 + 10s^3 + 46s^2 + 96s + 72}$$

to obtain

$$s^4\mathbf{V} + 10s^3\mathbf{V} + 46s^2\mathbf{V} + 96s\mathbf{V} + 72\mathbf{V} = 10s^4\mathbf{I} + 20s^3\mathbf{I} + 50s^2\mathbf{I}$$

The corresponding time-domain equation is

$$\frac{d^4}{dt^4}v + 10\frac{d^3}{dt^3}v + 46\frac{d^2}{dt^2}v + 96\frac{d}{dt}v + 72v = 10\frac{d^4}{dt^4}i + 20\frac{d^3}{dt^3}i + 50\frac{d^2}{dt^2}i$$

(This can be confirmed if we substitute $\mathbf{V}e^{st}$ and $\mathbf{I}e^{st}$ and simplify.)
(d) The complex natural frequencies are the poles $s_{p1} = -2$, $s_{p2} = -3 + j3$, and $s_{p3} = -3 - j3$. Notice that s_{p1} is a second-order pole.
(e) The general form of the complementary response is given by Theorem 4 of Chapter 8.

$$v_c(t) = (A_{1,0} + A_{1,1}t)e^{-2t} + A_{2,0}e^{(-3+j3)t} + A_{3,0}e^{(-3-j3)t}$$

This expression can be put in the form

$$v_c(t) = (A_{1,0} + A_{1,1}t)e^{-2t} + e^{-3t}(A \cos 3t + B \sin 3t)$$

where the constants are determined from initial conditions.
(f) Note that $3 \cos(4t + 17°) = \mathcal{R}e\{\mathbf{I}e^{st}\}$, where $\mathbf{I} = 3\underline{/17°}$ and $s = j4$. The particular response, therefore, is given by $v_p(t) = \mathcal{R}e\{\mathbf{V}e^{st}\}$ where $\mathbf{V} = \mathbf{Z}(s)\mathbf{I}$ and

$s = j4$. The impedance function $\mathbf{Z}(s)$ was determined in (b). By evaluating the second expression for $\mathbf{Z}(s)$ from (b) with $s = j4$, we obtain

$$\mathbf{Z}(j4) = 10\,\frac{(j4)^2(1 + j2)(1 + j6)}{(2 + j4)^2(3 + j)(3 + j7)} = 4.52\underline{/111.8°}\ \Omega$$

The output voltage phasor is, accordingly,

$$\mathbf{V} = \mathbf{ZI} = (4.52\underline{/111.8°})(3\underline{/17°}) = 13.56\underline{/128.8°}\ \text{V}$$

Therefore the particular response to

$$i(t) = 3\cos(4t + 17°)\ \text{A}$$

is

$$v_p(t) = 13.56\cos(4t + 128.8°)\ \text{V}\ \blacksquare$$

Notice that the pole-zero plot of Fig. 13.15 has mirror symmetry about the σ axis. We can show that pole-zero plots describing real circuits *always* have mirror symmetry about the σ axis. To establish this property, we use the fact that the coefficients of the polynomial $\mathscr{A}(s) = a_n s^n + a_{n-1}s^{n-1} + \cdots + a_0$ are real. Since a_n, $a_{n-1}, \ldots,$ are real,

$$\mathscr{A}^*(s) = (a_n s^n)^* + (a_{n-1}s^{n-1})^* + \cdots + a_0^*$$

$$= a_n(s^*)^n + a_{n-1}(s^*)^{n-1} + \cdots + a_0 = \mathbf{A}(s^*) \tag{13.42}$$

where the asterisk denotes complex conjugation. The ith pole of $\mathbf{H}(s)$ satisfies

$$\mathscr{A}(s_{pi}) = 0 \tag{13.43}$$

If we conjugate Eq. (13.43), we obtain

$$\mathscr{A}^*(s_{pi}) = 0 \tag{13.44}$$

which, with the aid of Eq. (13.42), becomes

$$\mathscr{A}(s_{pi}^*) = 0 \tag{13.45}$$

We conclude that if s_{pi} is a pole, then so is s_{pi}^*. Therefore complex poles always occur in conjugate pairs. The same arguments apply to the zeros of $\mathbf{H}(s)$. *The result is that all pole-zero plots describing real LLTI systems have mirror symmetry about the σ axis, as illustrated by Fig. 13.15.*

Pole-zero plots can be used to represent both stable and unstable circuits. If the circuit is stable, then all the poles will reside in the left half of the s plane. It is easy to see why. If a pole were to occur on the $j\omega$ axis or in the right half of the s plane, then the damping coefficient of that pole would not be positive, and the circuit's complementary response would not decay asymptotically to zero. The circuit would not be stable. If a circuit is not stable, then at least one pole resides on the $j\omega$ axis or in the right half of the s plane. The zeros of a stable circuit can lie anywhere in the s plane.

A final noteworthy feature of pole-zero plots follows from the Cramer's rule solution for the node voltages of an arbitrary circuit. The key result was given in Eq. (11.37), which we repeat below.

$$\mathbf{V}_i = \frac{\Delta_{1i}}{\Delta}\,\mathbf{I}_{s1} + \frac{\Delta_{2i}}{\Delta}\,\mathbf{I}_{s2} + \cdots + \frac{\Delta_{(M-1)i}}{\Delta}\,\mathbf{I}_{s(M-1)} + \frac{\Delta_{Mi}}{\Delta}\,\mathbf{V}_{sM} + \cdots + \frac{\Delta_{Ni}}{\Delta}\,\mathbf{V}_{sN} \tag{13.46}$$

In the above equation, the phasors $\mathbf{I}_{s1}, \mathbf{I}_{s2}, \ldots, \mathbf{V}_{sN}$ are independent source phasors, and $\mathbf{V}_i$ is the output-voltage phasor for node i. (The general Cramer's rule solution for $\mathbf{V}_i$ included the possibility that there is more than one independent source). The factors Δ_{1i}/Δ, Δ_{2i}/Δ, ..., Δ_{Ni}/Δ represent transfer impedances and voltage transfer functions between the independent sources and the output $\mathbf{V}_i$. Observe that regardless of the location of the output node (that is, regardless of the value of i), the denominator of the transfer function is always the same function $\Delta = \Delta(s)$. Since the denominators of the transfer functions are the same, the poles of the transfer functions are also the same. Thus, if we were to draw pole-zero plots of all the possible voltage transfer functions and transfer impedances of a circuit, the poles in the various plots would all be the same.*

As should now be evident, pole-zero plots are very useful tools to analyze both the natural response and the sinusoidal steady-state response of LLTI circuits. In the following section we describe how the pole-zero plot provides the basis for the rubber sheet analogy.

REMEMBER

A transfer function $\mathbf{H}(s) = \mathscr{B}(s)/\mathscr{A}(s)$ can be represented by a pole-zero plot. The poles and zeros are the roots of $\mathscr{A}(s)$ and $\mathscr{B}(s)$, respectively. The poles are the complex natural frequencies that appear in the circuit's complementary response.

EXERCISES

19. In the circuit shown, $R = 1$ MΩ and $C = 1$ μF.
    (a) Use any method to find the transfer function

$$\mathbf{H}(s) = \frac{\mathbf{V}_o}{\mathbf{V}_s}$$

    (b) Draw the pole-zero plot.
    (c) Find the complex natural frequencies.
    (d) What is the form of the complementary response?
    (e) Find $v_{op}(t)$ if $v_s(t) = 10 \cos (t + 20°)$ V.

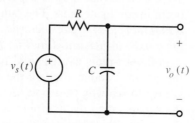

20. Repeat Exercise 17 for the following circuit, where $R = 1$ MΩ and $C = 1$ μF.

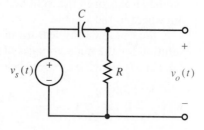

* Exceptions arise where the numerator Δ_{ki} contains a zero that cancels a particular pole in the denominator Δ.

21. The differential equation relating the input $i(t)$ to the output $v(t)$ of a circuit

is
$$\frac{d^2}{dt^2}v + 15\frac{d}{dt}v + 50v = \frac{d^2}{dt^2}i + 7\frac{d}{dt}i$$

Assume that the input is $i(t) = \mathbf{I}e^{st}$ where $\mathbf{I} = I_m\underline{/\phi_\mathbf{I}}$ and $s = \sigma + j\omega$.

(a) Show that the particular solution has the form $v_p(t) = \mathbf{V}e^{st}$.

(b) Using your analysis of (a), write $\mathbf{V}$ as $\mathbf{V} = \mathbf{Z}(s)\mathbf{I}$. Give the expression for $\mathbf{Z}(s)$ as the ratio of two polynomials in s.

(c) Express $\mathbf{Z}(s)$ in terms of the circuit's critical frequencies and sketch the corresponding pole-zero plot.

(d) Evaluate $\mathbf{Z}(s)$ at $s = j5$.

(e) Evaluate $\mathbf{V}$ for $s = j5$ and $I = 10\underline{/0°}$.

(f) Find $v_p(t)$ for $i(t) = 10 \cos 5t$.

(g) What is the form of the complementary response?

22. Repeat Exercise 21 for the equation

$$\frac{d^2}{dt^2}v + 15\frac{d}{dt}v + 50v = \frac{d^2}{dt^2}i + 6\frac{d}{dt}i + 34i$$

The Rubber Sheet Analogy

We saw that the amplitude characteristic $|\mathbf{H}(s)|$ of a resonant RLC circuit resembles a rubber sheet above the complex-frequency plane (Fig. 13.11). Additional examples of the rubber sheet analogy are shown in Figs. 13.16 and 13.17.

Figures 13.16 and 13.17 depict three-dimensional plots of the amplitude characteristics $|\mathbf{H}(s)|$ of the high-pass RL filter and the low-pass RC filter described at the beginning of this chapter. We see again that the amplitude characteristics have

FIGURE 13.16

Magnitude of the transfer function of the high-pass RL circuit of Fig. 13.1

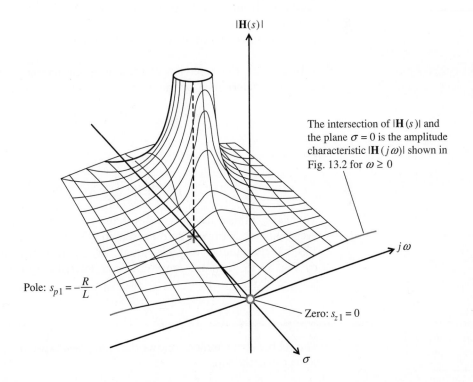

$|\mathbf{H}(s)|$

The intersection of $|\mathbf{H}(s)|$ and the plane $\sigma = 0$ is the amplitude characteristic $|\mathbf{H}(j\omega)|$ shown in Fig. 13.2 for $\omega \geq 0$

$j\omega$

Pole: $s_{p1} = -\dfrac{R}{L}$

Zero: $s_{z1} = 0$

σ

FIGURE 13.17
Magnitude of the transfer
function of the low-pass *RC*
circuit of Fig. 13.4

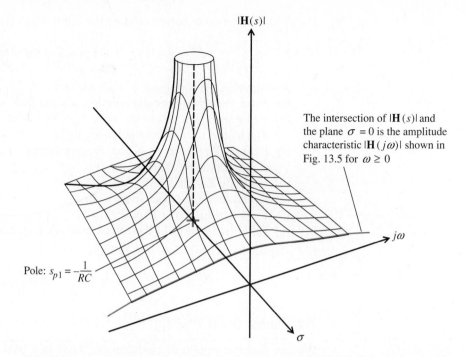

The intersection of $|\mathbf{H}(s)|$ and
the plane $\sigma = 0$ is the amplitude
characteristic $|\mathbf{H}(j\omega)|$ shown in
Fig. 13.5 for $\omega \geq 0$

Pole: $s_{p1} = -\dfrac{1}{RC}$

the appearance of thin rubber sheets stretched above the *s* plane. The sheets appear
to be *supported by poles* at the complex natural frequencies (or poles) and *held
down by nails* at the zeros. The amplitude characteristic $|\mathbf{H}(j\omega)|$ can be visualized
as the intersection of the rubber sheet with the $\sigma = 0$ plane.

A significant feature of the rubber sheet analogy is that it illustrates the funda-
mental importance of the poles and zeros in determining the shape of $|\mathbf{H}(j\omega)|$. We
see clearly from the analogy that the poles are responsible for the peaks in $|\mathbf{H}(j\omega)|$,
whereas the zeros are responsible for the dips. Thus, we have found the answer to
the question we asked at the start of Section 13.2.

In general, the behavior of the amplitude characteristic (or rubber sheet) as
$|s| \to \infty$ is determined by the degrees of $\mathscr{A}(s)$ and $\mathscr{B}(s)$. As $|s| \to \infty$, the expression

$$\mathbf{H}(s) = \frac{\mathscr{B}(s)}{\mathscr{A}(s)} = \frac{b_m s^m + \cdots + b_0}{a_n s^n + \cdots + a_0} \tag{13.47}$$

becomes

$$\mathbf{H}(s) = \frac{b_m}{a_n} s^{m-n} = K s^{m-n} \tag{13.48}$$

Therefore, if there are more poles than zeros ($n > m$), the amplitude characteristic
eventually tends toward zero as $|s|$ increases. This is illustrated by Figs. 13.11 and
13.17. If the number of poles equals the number of zeros ($n = m$), as illustrated
by Fig. 13.16, then the amplitude characteristic approaches the constant K as
$|s| \to \infty$.

REMEMBER

The rubber sheet analogy shows that the poles cause the peaks in $|\mathbf{H}(j\omega)|$, and the
zeros cause the dips.

23. Use the rubber sheet analogy to sketch the approximate amplitude characteristic $|\mathbf{H}(j\omega)|$ corresponding to the following pole-zero plot.

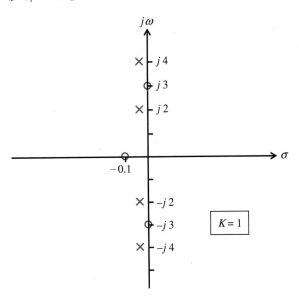

13.5 Band-Pass Filter

Band-pass filters are very important in radio and television systems, where they are used to separate one broadcast channel from another. The most important (and the simplest) examples of band-pass filters are provided by the series and parallel *RLC* circuits. We consider the series *RLC* circuit first.

Series *RLC* Circuit

To use a series *RLC* circuit as a band-pass filter, we drive it with a voltage source and take the output voltage across the resistance, as illustrated in Fig. 13.18.

FIGURE 13.18
Series *RLC* band-pass filter

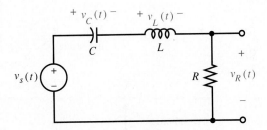

The voltage transfer function is given by

$$\mathbf{H}_R(s) = \frac{\mathbf{V}_R}{\mathbf{V}_s} = \frac{R}{R + sL + 1/sC} = \frac{sR/L}{s^2 + (R/L)s + 1/LC} = \frac{sR/L}{(s - s_{p1})(s - s_{p2})} \quad (13.49)$$

(We use a subscript R on $\mathbf{H}_R(s)$ to emphasize that the output voltage is taken across the resistance.) The poles are given by

$$s_{p1} = -\alpha + j\omega_d$$
$$s_{p2} = -\alpha - j\omega_d$$

(13.50)

if $\alpha < \omega_0$ (underdamped circuit),

$$s_{p1} = -\alpha$$
$$s_{p2} = -\alpha$$

(13.51)

if $\alpha = \omega_0$ (critically damped circuit)

and

$$s_{p1} = -\alpha + \beta$$
$$s_{p2} = -\alpha - \beta$$

(13.52)

if $\alpha > \omega_0$ (overdamped circuit).

The quantity

$$\alpha = \frac{R}{2L}$$

(13.53)

is the damping coefficient,

$$\omega_0 = \frac{1}{\sqrt{LC}}$$

(13.54)

is the undamped natural frequency

$$\omega_d = \sqrt{\omega_0^2 - \alpha^2}$$

(13.55)

is the damped natural frequency,

and

$$\beta = \sqrt{\alpha^2 - \omega_0^2}$$

(13.56)

The pole-zero plot is shown in Fig. 13.19.

FIGURE 13.19
Pole-zero plot of $\mathbf{H}_R(s) =$ $(sR/L)/(s - s_{p1})\ (s - s_{p2})$
(a) Underdamped $(\alpha < \omega_0)$
(b) critically damped $(\alpha = \omega_0)$
(c) overdamped $(\alpha > \omega_0)$

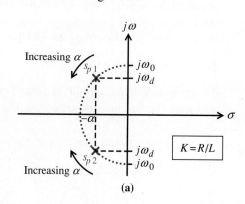

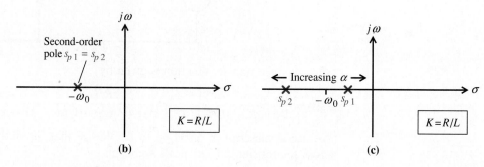

As the damping coefficient α is varied from 0 to ω_0 with ω_0 held constant, the poles move toward each other along the circular path defined by Eq. (13.55) and shown in Fig. 13.19a. The poles lie on the $j\omega$ axis at $\pm j\omega_0$ when $\alpha = 0$ and meet on the σ axis at $\sigma = -\omega_0$ when $\alpha = \omega_0$. When the poles are on the circular path, the damping ratio $\zeta = \alpha/\omega_0$ lies in the range $0 \le \zeta < 1$, which means that the circuit is underdamped. When the poles meet on the σ axis, the damping ratio is unity, and the circuit is critically damped (Fig. 13.19b). As α increases beyond ω_0, the natural frequencies separate as indicated, with s_{p1} approaching zero and s_{p2} approaching minus infinity (Fig. 13.19c). The damping ratio exceeds unity, $\zeta > 1$, and the circuit is overdamped for $\alpha > \omega_0$. We use the *underdamped* series *RLC* circuit for the band-pass filter application.

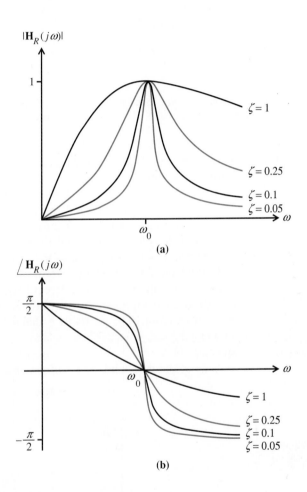

FIGURE 13.20
Frequency response of series *RLC* band-pass filter. Damping ratio $\zeta = \alpha/\omega_0$. (*a*) Amplitude; (*b*) phase

Plots of $\mathbf{H}_R(j\omega)$ are shown in Fig. 13.20 for $0 \le \zeta \le 1$. Notice that the peak in the amplitude characteristic, found by solving

$$\frac{d|\mathbf{H}_R(j\omega)|}{d\omega} = 0 \qquad (13.57)$$

occurs at $\omega = \omega_0$. It is instructive to compare the frequency-domain characteristics with the circuit's natural response. The natural response that is shown in Fig. 13.21 results from a nonzero initial current $i_L(0^-)$ in the inductance. The

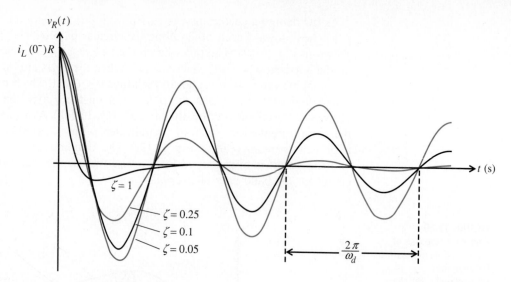

FIGURE 13.21 Natural response of series RLC band-pass filter. The waveform shown occurs when $v_s(t) = 0$ and $v_C(0^-) = 0$. $i_L(0^-)$ is a nonzero initial current in the inductance. The response is a damped sinusoid with frequency ω_d and damping coefficient α for $0 \leq \zeta < 1$

⊞ *See Problem 13.4 in the PSpice manual.*

natural response is a damped sinusoid with damping coefficient α and damped natural frequency $\omega_d = \omega_0 \sqrt{1 - \zeta^2}$. We see that when the damping ratio is small, so that the poles lie close to the $j\omega$ axis (that is, $\alpha \ll \omega_0$), the amplitude characteristic is sharply peaked, and the natural response is highly oscillatory. As the damping ratio increases, the poles move away from the $j\omega$ axis, so the amplitude characteristic is less peaked and the natural response is less oscillatory. The natural response is not oscillatory for $\zeta \geq 1$.

In the band-pass filter application, the input voltage $v_s(t)$ contains two or more sinusoidal components with different frequencies. Only the sinusoidal components in $v_s(t)$ that have frequencies near ω_0 can pass to the output, $v_R(t)$, without significant attenuation. The *passband* is defined as the band of frequencies between ω_L and ω_U, $\omega_L < \omega < \omega_U$, where ω_L and ω_U are the lower and upper half-power frequencies of the filter shown in Fig. 13.22. The *stopband* includes all frequencies outside of the passband. The *half-power bandwidth* is given by

$$\text{BW} = \omega_U - \omega_L \tag{13.58}$$

FIGURE 13.22
Illustration of half-power frequencies ω_L and ω_U

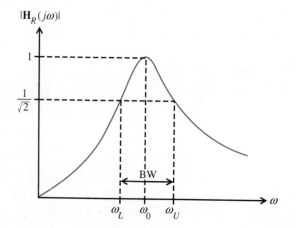

We will now derive formulas for ω_L, ω_U, and BW. The frequencies ω_L and ω_U are the positive solutions to the equation

$$|\mathbf{H}_R(j\omega)| = \frac{1}{\sqrt{2}} \tag{13.59}$$

where $\mathbf{H}_R(j\omega)$ is given by Eq. (13.49) with $s = j\omega$. Thus Eq. (13.59) becomes

$$\left| \frac{R}{R + j\left(\omega L - \dfrac{1}{\omega C}\right)} \right| = \frac{1}{\sqrt{2}} \tag{13.60}$$

After we square both sides, Eq. (13.60) becomes

$$\frac{R^2}{R^2 + \left(\omega L - \dfrac{1}{\omega C}\right)^2} = \frac{1}{2} \tag{13.61a}$$

and we see that

$$\left(\omega L - \frac{1}{\omega C}\right)^2 = R^2 \tag{13.61b}$$

Therefore either

$$\omega L - \frac{1}{\omega C} = +R \tag{13.62}$$

or

$$\omega L - \frac{1}{\omega C} = -R \tag{13.63}$$

The solutions to Eq. (13.62) are

$$\omega_1 = \frac{R}{2L} + \sqrt{\frac{1}{LC} + \left(\frac{R}{2L}\right)^2} \tag{13.64a}$$

and

$$\omega_2 = \frac{R}{2L} - \sqrt{\frac{1}{LC} + \left(\frac{R}{2L}\right)^2} \tag{13.64b}$$

and the solutions to Eq. (13.63) are

$$\omega_3 = -\frac{R}{2L} + \sqrt{\frac{1}{LC} + \left(\frac{R}{2L}\right)^2} \tag{13.65a}$$

and

$$\omega_4 = -\frac{R}{2L} - \sqrt{\frac{1}{LC} + \left(\frac{R}{2L}\right)^2} \tag{13.65b}$$

Roots ω_2 and ω_4 are negative and can be discarded. The half-power frequencies ω_L and ω_U are given by ω_3 and ω_1, respectively:

$$\omega_L = -\frac{R}{2L} + \sqrt{\frac{1}{LC} + \left(\frac{R}{2L}\right)^2}$$

$$= -\alpha + \sqrt{\omega_0^2 + \alpha^2} \qquad (13.66)$$

and

$$\omega_U = +\frac{R}{2L} + \sqrt{\frac{1}{LC} + \left(\frac{R}{2L}\right)^2}$$

$$= +\alpha + \sqrt{\omega_0^2 + \alpha^2} \qquad (13.67)$$

By subtracting ω_L from ω_U, we see that the half-power bandwidth, Eq. (13.58), is given by the simple formula

$$BW = 2\alpha = \frac{R}{L} \qquad (13.68)$$

⊡ *See Example 3.8 in the PSpice manual.*

We have seen that the series RLC circuit can be used as a band-pass filter. Band-pass filters without inductance can be constructed with the use of op amps. A typical op-amp band-pass filter is described in Problem 51.

REMEMBER

The amplitude characteristic $|\mathbf{H}_R(j\omega)|$ of a resonant series RLC circuit has a peak at $\omega_0 = 1/\sqrt{LC}$ and a half-power bandwidth $BW = 2\alpha = R/L$.

EXERCISE

24. (a) Use the equivalent source concept to represent the circuit shown as a series RLC circuit.
    (b) Use your answer to (a) to determine the transfer function $\mathbf{H}_o(s) = \mathbf{V}_o/\mathbf{V}_1$.
    (c) Find an expression for the half-power bandwidth BW in terms of R_1, R_2, R, and L.

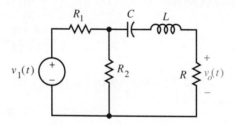

Parallel *RLC* Circuit

To use a parallel RLC circuit as a band-pass filter, we drive it with a current and take the output as the voltage across the resistance, as shown in Fig. 13.23. Thus, the band-pass transfer function is simply the driving-point impedance. If instead we took the output as the current through the resistance, then the analysis of the parallel circuit would be the exact dual of the analysis of the series circuit of

Fig. 13.18. Because of this close connection between the series and parallel *RLC* bandpass filters, we leave the development of the parallel filter to the following exercises.

FIGURE 13.23
Parallel *RLC* band-pass filter

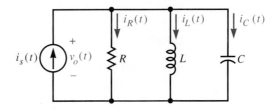

EXERCISES

25. For the circuit of Fig. 13.23:
    (a) Use the current divider relation to find $\mathbf{H}_G(s) = \mathbf{I}_R/\mathbf{I}_s$.
    (b) Use duality and Eq. (13.49) to check your result in (a).
    (c) Use duality to show that $|\mathbf{H}_G(j\omega)|$ has a peak when $\omega = \omega_0$, where $\omega_0 = 1/\sqrt{LC}$.
    (d) Use duality to show that the half-power bandwidth of $|\mathbf{H}_G(j\omega)|$ is given by $\mathrm{BW} = 1/RC$.

26. Assume that $C = 1\ \mu\mathrm{F}$ and $L = 2.5\ \mathrm{mH}$ in the band-pass filter of Fig.13.23. The output is the voltage $\mathbf{V}_o$.
    (a) What value of R is required for a half-power bandwidth of 1 krad/s?
    (b) Sketch the amplitude and phase characteristics for the value of R determined in (a).

13.6 Unity-Power-Factor Resonance

There is a traditionally accepted way to define resonance that does not explicitly refer to the network's peak response. In this definition, resonance occurs when the power factor of the network, $\cos(/\underline{\mathbf{Z}})$, equals unity. This is *unity-power-factor resonance*. Unity-power-factor resonance occurs when the network appears purely resistive to the source. Thus no reactive power is supplied to a network operating at unity-power-factor resonance.

The unity-power-factor resonance frequency is determined by finding $\omega = \omega_{upf}$ for which

$$\mathscr{Im}\{\mathbf{Z}(j\omega)\} = 0 \qquad (13.69a)$$

or, equivalently,

$$\mathscr{Im}\{\mathbf{Y}(j\omega)\} = 0 \qquad (13.69b)$$

where $\mathbf{Z}$ and $\mathbf{Y}$ are, respectively, the network's driving-point impedance and admittance.

EXAMPLE 13.3

Refer to the series *RLC* circuit of Fig. 13.18.
(a) Determine the unity-power-factor resonance frequency.
(b) Determine $i(t)$, $v_L(t)$, and $v_C(t)$ when $\omega = \omega_{upf}$.
(c) Analyze the flow of stored energy in the circuit when $\omega = \omega_{upf}$.

■ SOLUTION

(a) The driving-point impedance $\mathbf{Z}(j\omega)$ is

$$\mathbf{Z}(j\omega) = R + j\left(\omega L - \frac{1}{\omega C}\right)$$

and therefore the unity-power-factor resonance frequency is found by solving

$$\omega L - \frac{1}{\omega C} = 0$$

This yields
$$\omega_{upf} = \sqrt{\frac{1}{LC}}$$

$$= \omega_0$$

which is recognized as the maximum-response resonance frequency associated with $\mathbf{H}_R(j\omega)$ (see Fig. 13.20).

(b) The driving-point impedance at unity-power-factor resonance is

$$\mathbf{Z}(j\omega_0) = R$$

This results in a driving-point current phasor

$$\mathbf{I} = \frac{\mathbf{V}_s}{\mathbf{Z}(j\omega_0)} = \frac{\mathbf{V}_s}{R}$$

a voltage phasor

$$\mathbf{V}_L = j\omega_0 L \mathbf{I} = j\omega_0 L \frac{\mathbf{V}_s}{R} = j\frac{1}{2\zeta}\mathbf{V}_s$$

across the inductance, and a voltage phasor

$$\mathbf{V}_C = \frac{1}{j\omega_0 C}\mathbf{I} = -j\omega_0 L \frac{\mathbf{V}_s}{R} = -j\frac{1}{2\zeta}\mathbf{V}_s$$

across the capacitance. These quantities are summarized in Fig. 13.24. Note that because $\mathbf{V}_L + \mathbf{V}_C = 0$, the voltage phasor $\mathbf{V}_R$ must equal the source phasor $\mathbf{V}_s$.

FIGURE 13.24
Series *RLC* at unity-power-factor resonance

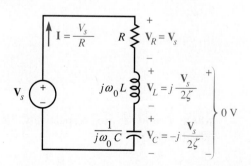

The waveforms associated with $\mathbf{I}$, $\mathbf{V}_L$, and $\mathbf{V}_s$ are

$$i(t) = \mathcal{R}e\left\{\mathbf{I}e^{j\omega_0 t}\right\} = \frac{V_{sm}}{R}\cos{(\omega_0 t + \phi_{\mathbf{V}})}$$

$$v_L(t) = \mathcal{R}e\left\{\frac{j}{2\zeta}\mathbf{V}_s e^{j\omega_0 t}\right\} = \frac{V_{sm}}{2\zeta}\cos{(\omega_0 t + \phi_{\mathbf{V}} + 90°)}$$

and $$v_C(t) = \mathcal{R}e\left\{\frac{-j}{2\zeta}\mathbf{V}_s e^{j\omega_0 t}\right\} = \frac{V_{sm}}{2\zeta}\cos{(\omega_0 t + \phi_{\mathbf{V}} - 90°)}$$

The amplitudes of $v_L(t)$ and $v_C(t)$ will be larger than the amplitude of the voltage source if the damping ratio is less than 0.5. However, because $v_L(t)$ and $v_C(t)$ have equal amplitudes and differ by 180° in phase, the combined voltage drop across the inductance and the capacitance is zero. It is this fact that makes the circuit appear resistive to the source for $\omega = \omega_0$.

(c) No reactive power enters the RLC network at unity-power-factor resonance. Nevertheless, the network does contain stored energy. With the use of Eqs. (2.30) and (2.43), we find that the instantaneous energies stored in the inductance and the capacitance are, respectively,

$$w_L(t) = \frac{1}{2}Li^2(t) = \frac{1}{2}L\left(\frac{V_{sm}}{R}\right)^2\cos^2{(\omega_0 t + \phi_{\mathbf{V}})}$$

$$= \frac{1}{4}L\left(\frac{V_{sm}}{R}\right)^2 + \frac{1}{4}L\left(\frac{V_{sm}}{R}\right)^2\cos{(2\omega_0 t + 2\phi_{\mathbf{V}})}$$

and

$$w_C(t) = \frac{1}{2}Cv_C^2(t)$$

$$= \frac{1}{2}C\left(\frac{V_{sm}}{2\zeta}\right)^2\cos^2{(\omega_0 t + \phi_{\mathbf{V}} - 90°)}$$

$$= \frac{1}{4}C\left(\frac{V_{sm}}{2\zeta}\right)^2 + \frac{1}{4}C\left(\frac{V_{sm}}{2\zeta}\right)^2\cos{(2\omega_0 t + 2\phi_{\mathbf{V}} - 180°)}$$

Now, with the use of $\zeta = \alpha/\omega_0$, where $\alpha = R/2L$ and $\omega_0 = 1/\sqrt{LC}$, we can verify that

$$\frac{1}{4}C\left(\frac{V_{sm}}{2\zeta}\right)^2 = \frac{1}{4}L\left(\frac{V_{sm}}{R}\right)^2$$

which indicates that $w_C(t)$ has the same amplitude as $w_L(t)$. The instantaneous energies are plotted in Fig. 13.25. From this figure and the preceding formulas,

FIGURE 13.25
Instantaneous energies stored in inductance and capacitance of series RLC circuit when $\omega = \omega_{upf}$

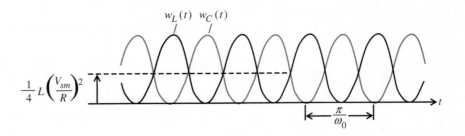

we can conclude that the stored energy is periodically swapped between the capacitance and the inductance; the average energies contained in the inductance and the capacitance are equal;

$$W_L = W_C = \frac{1}{4} L \left(\frac{V_{sm}}{R} \right)^2 \qquad (13.70)$$

and the total instantaneous stored energy is a constant:

$$w_L(t) + w_C(t) = \frac{1}{2} L \left(\frac{V_{sm}}{R} \right)^2 \blacksquare$$

Equation (13.70) is a special case of an important general result. Recall that the instantaneous power delivered to an arbitrary RLC circuit by a sinusoidal source is, by Eq. (12.15),

$$p(t) = \underbrace{P[1 + \cos(2\omega t + 2\phi_1)]}_{\substack{\text{Energy flow into} \\ \text{the network}}} + \underbrace{Q \cos\left(2\omega t + 2\phi_1 + \frac{\pi}{2}\right)}_{\substack{\text{Energy borrowed and} \\ \text{returned by the network}}} \qquad (13.71)$$

where from Eq. (12.62b),

$$Q = 2\omega(W_{\text{tot ave magnetic}} - W_{\text{tot ave electric}}) \qquad (13.72)$$

When unity-power-factor resonance occurs, the reactive power Q supplied to a network equals zero. By setting the right-hand side of Eq. (13.72) to zero, we obtain the following important result.

> When unity-power-factor resonance occurs in any circuit made up of Rs, Ls, and Cs, the total average energy stored in the magnetic fields equals the total average energy stored in the electric fields.

REMEMBER

At unity-power-factor resonance, a network appears purely resistive to the source.

EXERCISES

27. Show that the unity-power-factor resonance frequency of a parallel RLC circuit is $\omega_0 = 1/\sqrt{LC}$. Use duality to check your result.
28. Show that when the parallel RLC circuit shown is operated at its unity-

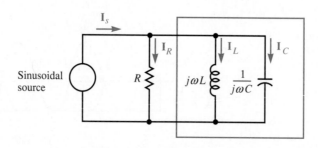

power-factor resonance frequency, the phasor currents are $\mathbf{I}_R = \mathbf{I}_s$, $\mathbf{I}_L = -j\omega_0 RC\mathbf{I}_s$, and $\mathbf{I}_C = +j\omega_0 RC\mathbf{I}_s$. Explain why no current enters the part of the circuit enclosed by red when $\omega = \omega_{upf}$.

13.7 Quality Factor

The *quality factor* Q provides a figure of merit for ac circuits and circuit components.* Q is a dimensionless quantity that has been defined somewhat differently by various authors. We define Q as follows:

DEFINITION

Quality Factor Q

$$Q = 2\pi \frac{\text{peak energy stored in the circuit}}{\text{energy dissipated by the circuit in one period}} \qquad (13.73)$$

The value of Q depends on the frequency ω at which it is evaluated; hence

$$Q = Q(\omega) \qquad (13.74)$$

Series and Parallel *RL* and *RC* Circuits

For illustration, consider the series *RL* circuit of Fig. 13.26a. To determine the Q of this circuit, we must assume that the circuit is operating in the sinusoidal steady state at some frequency ω. This means that a sinusoidal current

$$i(t) = I_m \cos{(\omega t + \phi_\mathbf{I})} \qquad (13.75)$$

is flowing through the resistance and the inductance, as illustrated in Fig. 13.26b.

FIGURE 13.26
Series *RL* circuit: (*a*) alone;
(*b*) with sinusoidal source

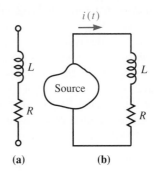

(a)   (b)

We can evaluate the expression for Q, Eq. (13.73) by recognizing that the inductance L stores energy and that the resistance R dissipates energy. The instantaneous value of the energy stored is

---

* Quality factor Q should not be confused with reactive power (also symbolized as Q). They are totally different quantities.

$$w_L(t) = \frac{1}{2} Li^2(t)$$

$$= \frac{1}{2} LI_m^2 \cos^2 (\omega t + \phi_1) \tag{13.76}$$

Therefore

$$\text{Peak stored energy} = w_L(t)\Big|_{\text{peak}}$$

$$= \frac{1}{2} LI_m^2 \tag{13.77}$$

The average power dissipated by the resistance is

$$P = \frac{1}{2} I_m^2 R \tag{13.78}$$

The energy loss in one period is the product of the average power dissipated times the period $T = 2\pi/\omega$. Therefore

$$\text{Energy loss in one period} = \frac{1}{2} I_m^2 R \frac{2\pi}{\omega} \tag{13.79}$$

The substitution of Eqs. (13.77) and (13.79) into Eq. (13.73) yields the result

$$Q = 2\pi \frac{(1/2)LI_m^2}{(1/2)I_m^2 R(2\pi/\omega)} = \frac{\omega L}{R} \tag{13.80}$$

Observe that this result has the form

$$Q = \frac{|X|}{R} \tag{13.81}$$

where $X = \omega L$ is the reactance of the inductance.

The series RL circuit of Fig. 13.26 is sometimes used to model a physical inductor. In the model, R represents the internal or *winding resistance* of the inductor and L represents the inductance. According to this model, an inductor whose inductive reactance is large compared with its winding resistance has a high Q. Also according to this model, the Q of the inductor decreases with decreasing frequency ω. Because winding resistance is unavoidable, it is difficult to construct inductors that have high Qs at low frequencies.

A second illustration is provided by the parallel RC circuit of Fig. 13.27. One way to determine the Q of this circuit is to apply the definition in Eq. (13.73). The Q of the parallel RC circuit can also be determined easily if we recognize that this circuit is the dual of the series RL circuit. Substituting C for L and $1/R$ for R in Eq. (13.80) leads immediately to

$$Q = \omega CR \tag{13.82}$$

or

$$Q = \frac{R}{|X|} \tag{13.83}$$

where $X = -1/\omega C$ is the reactance of the capacitance.

The parallel RC circuit is sometimes used to model a physical capacitor. In the model, R represents the capacitor's *leakage resistance* and C represents the capacitance. Leakage resistance arises from charge carriers in the dielectric material

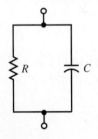

FIGURE 13.27
Parallel RC circuit

separating the plates of the capacitor. According to the model, capacitors whose capacitive reactances are small compared with their leakage resistances have high Qs.

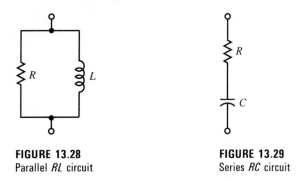

FIGURE 13.28
Parallel *RL* circuit

FIGURE 13.29
Series *RC* circuit

Additional illustrations are provided by the parallel RL and series RC circuits of Figs. 13.28 and 13.29. Derivations similar to the preceding ones reveal that the Q of the parallel RL circuit is

$$Q = \frac{R}{\omega L} = \frac{R}{|X|} \tag{13.84}$$

where $X = \omega L$, and that the Q of its dual series RC circuit is

$$Q = \frac{1}{\omega C R} = \frac{|X|}{R} \tag{13.85}$$

where $X = -1/\omega C$.

The results for the four circuits considered so far are conveniently summarized in Fig. 13.30.

FIGURE 13.30
Series and parallel networks. The boxes contain *only* an inductance or a capacitance. (*a*) Series network $Q_s = |X_s|/R_s$; (*b*) parallel network $Q_p = R_p/|X_p|$

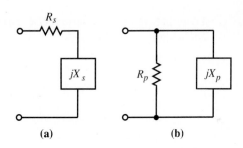

(a)　　　　(b)

In this figure, the subscripts s and p refer to series and parallel circuits, respectively, and X refers to the reactance of either an inductance or a capacitance, as appropriate. The results indicated in the figure for the series circuit follow from Eqs. (13.81) and (13.85), which indicate that the expression for the Q of either the RC or the RL series circuit can be written as

$$Q_s = \frac{|X_s|}{R_s} \tag{13.86}$$

The results indicated in the figure for the parallel circuit follow from Eqs. (13.83) and (13.84), which indicate that the expression for the Q of either the RC or the RL parallel circuit can be written as

$$Q_p = \frac{R_p}{|X_p|} \tag{13.87}$$

REMEMBER

The quality factors of the series and parallel RC or RL networks of Fig. 13.30 are given by Eqs. (13.86) and (13.87), respectively.

Series *RLC* Circuit

A final and especially important illustration of Q is provided by the series RLC circuit. The Q of this circuit is ordinarily evaluated at the circuit's unity-power-factor resonance frequency $\omega_{upf} = \sqrt{1/LC} = \omega_0$, where it is denoted as Q_0. As was shown in Example 13.3, the total instantaneous energy stored in this circuit is constant at unity-power-factor resonance. The peak value of a constant is just that constant. Therefore, from Example 13.3,

$$\text{Peak stored energy} = \frac{1}{2} L \left(\frac{V_{sm}}{R} \right)^2 \tag{13.88}$$

The energy loss in one period is simply the average power dissipated $P = V_{sm}^2/2R$ times the period $T_0 = 2\pi/\omega_0$:

$$\text{Energy loss in one period} = \frac{V_{sm}^2}{2R} \frac{2\pi}{\omega_0} \tag{13.89}$$

Therefore, the Q of a series RLC circuit at the unity-power-factor resonance frequency, $\omega_0 = 1/\sqrt{LC}$, is

$$Q_0 = 2\pi \frac{\frac{1}{2} L \left(\frac{V_{sm}}{R} \right)^2}{\frac{V_{sm}^2}{2R} \frac{2\pi}{\omega_0}}$$

$$= \frac{\omega_0 L}{R} = \frac{\omega_0}{2\alpha} = \frac{1}{2\zeta} \tag{13.90}$$

See Problems 13.3 and 13.5 in the PSpice manual.

Notice that $1/2\zeta$ is the factor appearing in $\mathbf{V}_L$ and $\mathbf{V}_C$ of Fig. 13.24. At unity-power-factor resonance the voltage amplitudes across the inductance and capacitance are each Q_0 times the voltage amplitude of the source.

The quality factor Q_0 is also an indicator of the sharpness of the peak occurring in the amplitude characteristic of Fig. 13.20. Qualitatively speaking, the higher the value of Q_0, the sharper the resonance peak. Remember that the half-power bandwidth associated with $|\mathbf{H}_R(j\omega)|$ is exactly 2α [Eq. (13.68)]. If we combine Eqs. (13.68) and (13.90), we obtain

$$Q_0 = \frac{\omega_0}{\text{BW}} \tag{13.91}$$

or

$$\text{BW} = \frac{\omega_0}{Q_0} \tag{13.92}$$

We see from Eq. (13.92) that the half-power bandwidth is small compared with the "center frequency" ω_0 when Q_0 is large ($Q_0 \gg 1$).

REMEMBER

The quality factor of a series RLC circuit at angular frequency $\omega_0 = 1/\sqrt{LC}$ is $Q_0 = \omega_0/\text{BW}$ where $\text{BW} = 2\alpha = R/L$.

EXERCISES

29. A series RLC circuit has a Q of 10 at its unity-power-factor resonance frequency of 100 kHz, and it dissipates 2 mW average power when driven by a 4-V rms 100-kHz sinusoidal voltage source.
    (a) What is the half-power bandwidth?
    (b) Determine the numerical values of R, L, and C.

30. A parallel RLC circuit has a Q of 100 at its unity-power-factor resonance frequency of 1 kHz, and it dissipates 1 W average power when driven by a 1-A rms 1-kHz sinusoidal current source.
    (a) What is the half-power bandwidth?
    (b) Determine the numerical values of R, L, and C.

13.8 The Universal Resonance Characteristic

In this section we will show that all resonance peaks caused by isolated simple poles located close to the $j\omega$ axis have approximately the same shape. The reason for this important fact can be understood if we consider a typical transfer function, for example,

$$\mathbf{H}(s) = \mathbf{K}\frac{(s - s_{z1})^2(s - s_{z2})(s - s_{z3})}{(s - s_{p1})(s - s_{p2})(s - s_{p3})(s - s_{p4})} \tag{13.93}$$

An exact calculation of the transfer function involves the lengths and the angles of the vectors from the poles and zeros of $\mathbf{H}(s)$ to the point s as illustrated in Fig. 13.31. In this illustration, s is assumed to be close to the isolated simple pole at s_{p1}.

If the point s is moved about in the vicinity of s_{p1}, the long vectors (in black) will remain nearly constant both in length and in angle, but the short vector $s - s_{p1}$ (in red) will vary by a relatively large amount in length and angle. Thus, for s sufficiently close to s_{p1}, each of the long vectors can be approximated with reasonable

FIGURE 13.31
Graphical evaluation of $\mathbf{H}(s)$

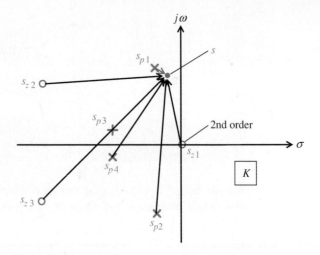

accuracy as constants. A standard choice for the constant vectors is shown in Fig. 13.32, where each long vector has been approximated by a vector starting at the corresponding critical frequency and ending at s_{p1}. By means of this approximation, we see that

$$\mathbf{H}(s) \simeq K \frac{(s_{p1} - s_{z1})^2(s_{p1} - s_{z2})(s_{p1} - s_{z3})}{(s - s_{p1})(s_{p1} - s_{p2})(s_{p1} - s_{p3})(s_{p1} - s_{p4})} \tag{13.94}$$

for $s \simeq s_{p1}$.* Notice that only one factor in Eq. (13.94) depends on s. Therefore we can write

$$\mathbf{H}(s) \simeq \frac{k_{11}}{s - s_{p1}} \qquad \text{for } s \simeq s_{p1} \tag{13.95}$$

FIGURE 13.32
Approximations leading to
Eq. (13.95)

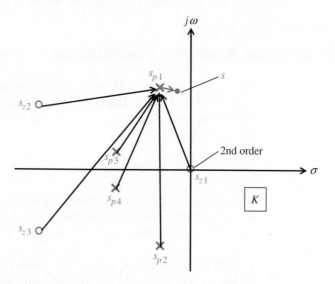

* A complex quantity z is approximately equal to another complex quantity z_1 if $|z - z_1| < \varepsilon$, where ε is a small positive number. Therefore the region of the s plane for which $s \simeq s_{p1}$ is a circle centered at s_{p1} with radius ε.

The constant

$$k_{11} = K \frac{(s_{p1} - s_{z1})^2(s_{p1} - s_{z2})(s_{p1} - s_{z3})}{(s_{p1} - s_{p2})(s_{p1} - s_{p3})(s_{p1} - s_{p4})} \qquad (13.96)$$

is called the *residue* of $\mathbf{H}(s)$ at the simple pole s_{p1} and can be expressed as

$$k_{11} = (s - s_{p1})\mathbf{H}(s)\Big|_{s=s_{p1}} \qquad (13.97)$$

The technique leading to Eqs. (13.95) and (13.97) can be used to determine the approximate form of *any* transfer function in the vicinity of an isolated simple pole. The result will always have the form of Eq. (13.95), where the residue k_{11} is given by Eq. (13.97). It follows that if the isolated pole is close to the $j\omega$ axis, then the frequency response in the vicinity of the resonance peak is approximately

$$\mathbf{H}(j\omega) \simeq \frac{k_{11}}{j\omega - s_{p1}} \qquad (13.98)$$

Another form for the approximation in Eq. (13.98) can be obtained if we set

$$s_{p1} = -\alpha + j\omega_d \qquad (13.99)$$

where α and ω_d are the damping coefficient and the damped natural frequency associated with s_{p1}. The pole s_{p1} is close to the $j\omega$ axis if and only if $\alpha \ll \omega_d$.* Using Eq. (13.99), Eq. (13.98) becomes

$$\mathbf{H}(j\omega) \simeq \frac{k_{11}}{j(\omega - \omega_d) + \alpha} \qquad (13.100)$$

This approximation to $\mathbf{H}(j\omega)$ can be simplified further by momentarily setting $\omega = \omega_d$, which shows that

$$\mathbf{H}(j\omega_d) \simeq \frac{k_{11}}{\alpha} \qquad (13.101)$$

By substituting Eq. (13.101) into Eq. (13.100), we obtain the final result:

$$\mathbf{H}(j\omega) \simeq \frac{1}{1 + j\gamma} \mathbf{H}(j\omega_d) \qquad (13.102)$$

where

$$\gamma = \frac{\omega - \omega_d}{\alpha} \qquad (13.103)$$

The factor $1/(1 + j\gamma)$ is referred to as the *normalized universal resonance characteristic*. The normalized amplitude and phase characteristics are plotted in Fig. 13.33. These plots describe the approximate shapes of the amplitude and phase characteristics of *every* circuit in the vicinity of a resonance peak caused by an

---

* We assume that $\alpha > 0$ (stable circuit).

FIGURE 13.33
The normalized universal resonance characteristic.
(*a*) Amplitude; (*b*) phase

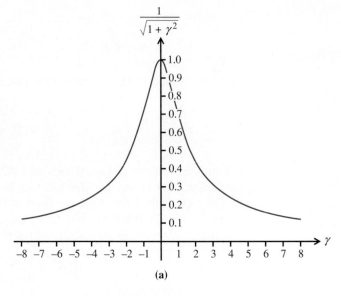

(a)

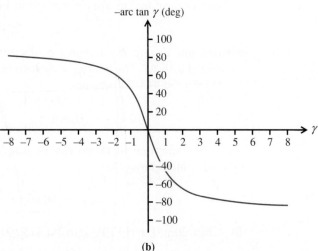

(b)

isolated simple pole near the $j\omega$ axis of the s plane. The peak in the universal amplitude characteristic occurs when $\gamma = 0$, or, equivalently, when $\omega = \omega_d$. The half-power frequencies occur when $\gamma = 1$, or, equivalently, when $\omega = \omega_d \pm \alpha$. Therefore the half-power bandwidth associated with the universal resonance peak is BW $= 2\alpha$. The horizontal axis in Fig. 13.33 can be interpreted as the number of *half bandwidths*, $\alpha = \frac{1}{2}$BW, by which ω exceeds the maximum-response resonance frequency. For example, if ω is three half bandwidths above ω_d, then $\omega = \omega_d + 3\alpha$ and $\gamma = 3$.

The normalized universal resonance characteristic is important because it describes the *approximate* shape of *all* resonance peaks caused by isolated simple poles close to the $j\omega$ axis. The universal resonance characteristic does not apply, however, for frequencies that are far removed from resonance. To show this, consider the following voltage transfer functions associated with the series *RLC* circuit of Fig. 13.18.

$$\mathbf{H}_R(s) = \frac{\mathbf{V}_R}{\mathbf{V}_s} = \frac{sR/L}{(s - s_{p1})(s - s_{p2})} \tag{13.104}$$

$$\mathbf{H}_C(s) = \frac{\mathbf{V}_C}{\mathbf{V}_s} = \frac{1/LC}{(s - s_{p1})(s - s_{p2})} \tag{13.105}$$

$$\mathbf{H}_L(s) = \frac{\mathbf{V}_L}{\mathbf{V}_s} = \frac{s^2}{(s - s_{p1})(s - s_{p2})} \tag{13.106}$$

The subscripts R, L, and C denote the circuit component across which the output voltage is measured. Recall that $\mathbf{H}_R(s)$ and $\mathbf{H}_C(s)$ were described previously in Sections 13.5 and 13.3 [see Eq. (13.22)], respectively. All three of these transfer functions can easily be derived by inspection of Fig. 13.18. The poles are given by Eqs. (13.50) through (13.56). Exact plots of the amplitude characteristics associated with the three transfer functions are shown in Fig. 13.34 for a damping ratio $\zeta = \alpha/\omega_0 = 0.05$. We can see from this figure that the locations, bandwidths, and shapes of the three resonance peaks are nearly equal. However, the amplitude characteristics are substantially different for frequencies far removed from resonance. The exact expressions for the maximum-response resonance frequencies, shown in the figure, were obtained as usual by setting the derivatives of the amplitude characteristics with respect to ω to zero. (We leave the derivations as

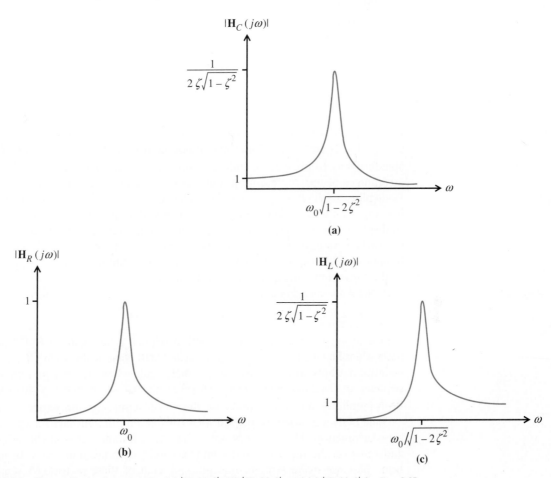

FIGURE 13.34 The amplitude characteristics (a) $|\mathbf{H}_C(j\omega)|$, (b) $|\mathbf{H}_R(j\omega)|$, and (c) $|\mathbf{H}_L(j\omega)|$ for $\zeta = 0.05$

an exercise.) As predicted by the universal resonance characteristic, all three maximum-response resonance frequencies approximately equal the damped natural frequency $\omega_d = \omega_0\sqrt{1 - \zeta^2}$ if the damping ratio is small.

REMEMBER

All resonance peaks caused by isolated simple poles close to the $j\omega$ axis have approximately the same shape.

EXERCISES

31. The amplitude characteristic of an unknown circuit has a peak of 20 at $\omega_r = 1$ krad/s, with associated half-power bandwidth of 4 krad/s. Assuming that the resonance is due to an isolated simple pole, what is the value of the amplitude characteristic at $\omega = 6$ krad/s?

32. Repeat Exercise 31 for $\omega = 8$ krad/s.

33. Derive the expressions for the three maximum-response resonance frequencies shown in Fig. 13.34.

34. Refer again to Fig. 13.34. Explain in simple circuit terms why:
    (a) $H_C(0) = 1$
    (b) $H_L(\infty) = 1$
    (c) $H_R(0) = H_R(\infty) = 0$

13.9 Bode Plots

Pocket calculators and personal computers make it easy to accurately compute and plot a circuit's amplitude and phase characteristics $|H(j\omega)|$ and $/H(j\omega)$. A standard way to present this information is by means of Bode plots. The primary advantages of our representing amplitude and phase information in the form of Bode plots are (1) Bode plots can be sketched rapidly with the use of only straight-line approximations, (2) the Bode amplitude and frequency scales make it possible to display a large range of numerical values compactly, and (3) Bode plots are an industrial standard. Another advantage, which is by no means insignificant, is that human sensory response to both audio and video stimuli seems naturally matched to the logarithmic measures used in Bode plots.

In the *Bode amplitude plot*,

$$|H(j\omega)|_{dB} = 20 \log_{10} |H(j\omega)| \qquad (13.107)$$

is plotted versus frequency. The function $|H(j\omega)|_{dB}$ has a *decibel (dB) amplitude* scale when $H(j\omega)$ is dimensionless, a *decibel (dB) ohms* scale when $H(j\omega)$ is an impedance in ohms, and a *decibel (dB) siemens* scale when $H(j\omega)$ is an admittance in siemens. In the *Bode phase plot*, $/H(j\omega)$ is plotted (in degrees) versus frequency. Both amplitude and phase plots are drawn on semilog graph paper.

Bode plots are based on the fact that only seven different factors appear in any transfer function. These factors are a gain K, poles and zeros at the origin, poles and zeros on the real axis, and complex-conjugate poles and zeros. In making a Bode plot, we make straight-line plots of each of these factors. (A straight-line plot is just a plot of straight-line segments.) We then construct a straight-line plot of the entire transfer function by *adding* the straight-line magnitude plots and *adding* the straight-line phase plots. The result is a straight-line approximation to the actual amplitude and phase characteristics. The final step is to use a table of correction factors to obtain the actual characteristics.

We will explain how to draw Bode plots by starting with a few simple examples. It is important to study the following three examples carefully, because they provide the information you will need to progress easily to the general case.

EXAMPLE 13.4

Construct a Bode plot for the low-pass *RC* filter of Fig. 13.35.

FIGURE 13.35
Low-pass filter

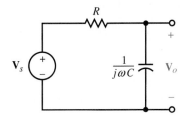

■ SOLUTION

See Problem 13.6 in the PSpice manual.

The voltage transfer function is

$$\mathbf{H}_{LP}(s) = \frac{\dfrac{1}{sC}}{R + \dfrac{1}{sC}} = \frac{1}{s\tau + 1} = \frac{1/\tau}{s + 1/\tau}$$

where $\tau = RC$ is the time constant. We see that $\mathbf{H}_{LP}(s)$ contains a real-axis pole at $s_{p1} = -1/\tau$. By setting $s = j\omega$, we obtain

$$\mathbf{H}_{LP}(j\omega) = \frac{1}{j\omega\tau + 1} = \frac{1}{|j\omega\tau + 1|} \underline{/-\text{arc tan }\omega\tau}$$

The dB amplitude characteristic, Eq. (13.107), is

$$|\mathbf{H}_{LP}(j\omega)|_{dB} = 20 \log |\mathbf{H}_{LP}(j\omega)| = -20 \log |j\omega\tau + 1|$$

In Bode's method, we start by approximating the dB amplitude characteristic by straight lines. The straight-line approximation to $|\mathbf{H}_{LP}(j\omega)|_{dB}$ is obtained from the fact that

$$|j\omega\tau + 1| = \sqrt{(\omega\tau)^2 + 1} \simeq \begin{cases} 1 & \text{for } \omega \ll \dfrac{1}{\tau} \\[2ex] \omega\tau & \text{for } \omega \gg \dfrac{1}{\tau} \end{cases}$$

Therefore

$$|\mathbf{H}_{LP}(j\omega)|_{dB} \simeq \begin{cases} 0 & \text{for } \omega \ll \dfrac{1}{\tau} \\[2ex] -20 \log \omega\tau & \text{for } \omega \gg \dfrac{1}{\tau} \end{cases}$$

The functions appearing on the right-hand side of this equation are the low-frequency and high-frequency asymptotes of $|\mathbf{H}_{LP}(j\omega)|_{dB}$. Both are straight lines when plotted versus log ω and are shown as black lines in Fig. 13.36a.

The frequency at which the two asymptotes meet is called a *corner frequency*. The corner frequency can be found if we solve

$$-20 \log \omega\tau = 0$$

which yields

$$\omega = \frac{1}{\tau}$$

Thus the corner frequency is just the reciprocal of the time constant.

The slope of the asymptote $-20 \log \omega\tau$ can be obtained if we notice that if ω is increased by a factor of 2, then $-20 \log \omega\tau$ decreases by $20 \log 2 = 6.02$ dB:

$$-20 \log 2\omega\tau = -20 \log \omega\tau - 20 \log 2 = -20 \log \omega\tau - 6.02 \text{ dB}$$

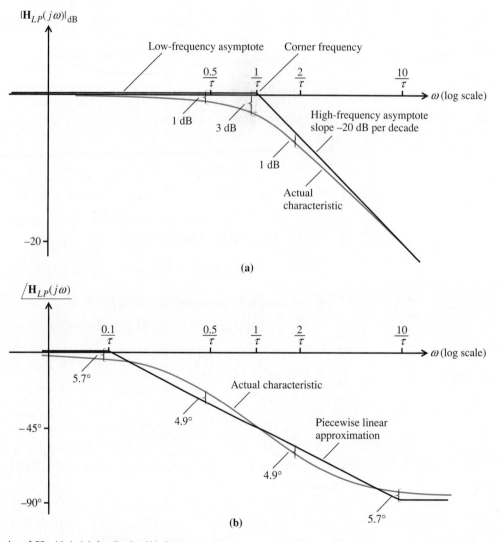

FIGURE 13.36 Bode plot of $\mathbf{H}_{LP}(j\omega)$. (*a*) Amplitude; (*b*) phase

A factor of 2 in frequency is called an *octave*. Therefore the asymptote $-20 \log \omega\tau$ has a slope of approximately -6 dB *per octave*. Similarly, if we increase ω by a factor of 10 (a *decade*), then $-20 \log \omega\tau$ decreases by $20 \log 10 = 20$ dB:

$$-20 \log 10\omega\tau = -20 \log \omega\tau - 20 \log 10 = -20 \log \omega\tau - 20 \text{ dB}$$

Therefore 6.02 dB per octave is equivalent to 20 dB per decade.

To examine how good the straight-line approximation to $|\mathbf{H}_{LP}(j\omega)|_{dB}$ is, we have plotted the exact amplitude characteristic as the red curve in Fig. 13.36a. We see from this figure that the actual characteristic lies below the approximate characteristic. The approximation error is *geometrically symmetrical* about the corner frequency $\omega = 1/\tau$. This means that the error at $\omega = \beta/\tau$ is the same as the error at $\omega = 1/\beta\tau$, where β is any positive number. The approximation error is 3 dB at $\omega = 1/\tau$, 2 dB at $\omega = 0.76/\tau$ and $\omega = 1.31/\tau$, and 1 dB at $\omega = 0.5/\tau$ and $2/\tau$. These numbers are recorded in Table 13.2. We will use this table to correct similar piecewise-linear characteristics in the future.

$\tau = RC$

TABLE 13.2
Magnitudes of corrections for corners due to first-order* real poles or zeros

$\omega\tau$	$\pm$ AMPLITUDE CORRECTION (dB)	$\pm$ PHASE CORRECTION (DEGREES)
1	3	0
0.76 and 1.31	2	2.4
0.5 and 2	1	4.9
0.1 and 10	0.04	5.7

The phase characteristic of the low-pass circuit is given by

$$\underline{/\mathbf{H}_{LP}(j\omega)} = -\arctan \omega\tau$$

Bode's straight-line phase plot is based on the fact that

$$-\arctan \omega\tau \simeq \begin{cases} 0° & \text{for } \omega < \dfrac{0.1}{\tau} \\[2mm] -90° & \text{for } \omega > \dfrac{10}{\tau} \end{cases}$$

It follows from this that for $\omega < 0.1/\tau$ and for $\omega > 10/\tau$, we may approximate $\underline{/\mathbf{H}_{LP}(j\omega)}$ by the constants $0°$ and $-90°$, respectively. These constants are plotted as the horizontal black lines in Fig. 13.36b. These lines are the low- and high-frequency asymptotes of the phase characteristic. To complete the straight-line phase plot, we join the two horizontal asymptotes with a line that passes through $-45°$ at $\omega = 1/\tau$ and intersects the asymptotes at $\omega = 0.1/\tau$ and $\omega = 10/\tau$. The slope of this line is $-45°$ per decade. The exact phase characteristic is plotted as the red curve in Fig. 13.36b.

By examining Fig. 13.36b, we see that the approximation error is geometrically symmetrical about $\omega = 1/\tau$. The magnitude of the error is $5.7°$ at the two *breakpoints*, $\omega = 0.1/\tau$ and $\omega = 10/\tau$, and $4.9°$ at $\omega = 0.5/\tau$ and $\omega = 2/\tau$. These numbers are included in Table 13.2. ∎

* Multiply the corrections by N for an Nth-order real pole or zero.

EXAMPLE 13.5

Construct a Bode plot for the high-pass RL filter of Fig. 13.37

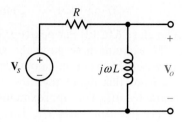

■ SOLUTION

The voltage transfer function is

$$\mathbf{H}_{HP}(s) = \frac{sL}{R + sL} = \frac{s\tau}{s\tau + 1} = \frac{s}{s + 1/\tau}$$

where $\tau = L/R$ is the time constant. Notice that the basic difference between $\mathbf{H}_{HP}(s)$ and $\mathbf{H}_{LP}(s)$ of the previous example is the numerator factor, $s\tau$, in $\mathbf{H}_{HP}(s)$. This factor means, of course, that $\mathbf{H}_{HP}(s)$ contains a zero at the origin. By setting $s = j\omega$, we have

$$\mathbf{H}_{HP}(j\omega) = \frac{j\omega\tau}{j\omega\tau + 1} = \frac{\omega\tau}{|j\omega\tau + 1|} \left| \frac{\pi}{2} - \text{arc tan } \omega\tau \right.$$

The dB amplitude characteristic, Eq. (13.107), is

$$\left| \mathbf{H}_{HP}(j\omega) \right|_{dB} = 20 \log \frac{\omega\tau}{|j\omega\tau + 1|} = 20 \log \omega\tau - 20 \log |j\omega\tau + 1|$$

Since the logarithm of a *product* equals the *sum* of logarithms, the dB amplitude characteristic equals the sum of two functions. We have already plotted the function $-20 \log |j\omega\tau + 1|$ in Fig. 13.36a. The other function, $20 \log \omega\tau$, is a straight line with a slope of 6.02 dB per octave (20 dB per decade) when plotted versus $\log \omega$. The location of this line can be determined from the fact that $\log \omega\tau = 0$ at $\omega = 1/\tau$. The line $20 \log \omega\tau$ is plotted in Fig. 13.38 along with the straight-line approximation to $-20 \log |j\omega\tau + 1|$ obtained in Example 13.4.

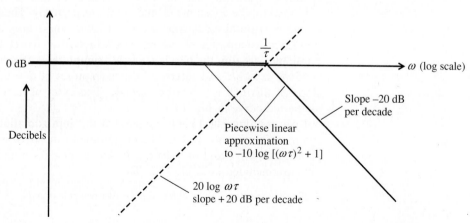

We can obtain a straight-line approximation to $|\mathbf{H}(j\omega)|_{dB}$ by simply adding the two straight-line plots of Fig. 13.38. This sum is plotted as black lines on Fig. 13.39a. The exact amplitude characteristic is plotted as the red curve.

We can see from Fig. 13.39 that the approximation error is 3 dB at the corner frequency $\omega = 1/\tau$ and geometrically decreases symmetrically on either side of $\omega = 1/\tau$ in agreement with the entries in Table 13.2.

The phase characteristic of the high-pass filter is

$$\underline{/\mathbf{H}_{HP}(j\omega)} = \frac{\pi}{2} - \text{arc tan } \omega\tau$$

which is the sum of the constant $\pi/2$ plus the function $-\tan^{-1}\omega\tau$ described in the preceding example. The straight-line approximation to $\underline{/\mathbf{H}_{HP}(j\omega)}$ is shown in Fig. 13.39b along with the exact phase curve. The difference between the two is in agreement with Table 13.2. ■

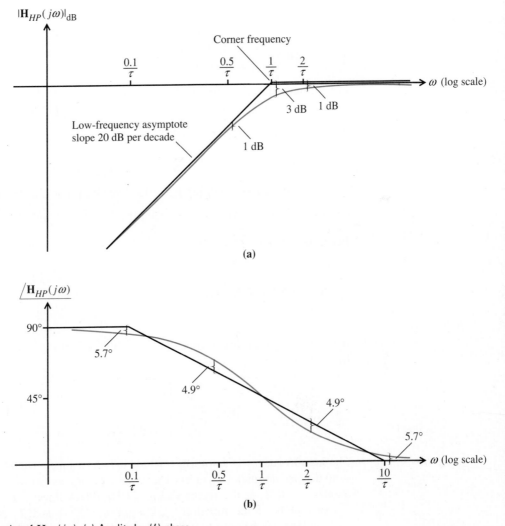

FIGURE 13.39 Bode plot of $\mathbf{H}_{HP}(j\omega)$. (*a*) Amplitude; (*b*) phase

Now that we have considered the Bode plots of simple low-pass and high-pass filters, let us consider that of a resonant *RLC* circuit.

EXAMPLE 13.6

■ *See Problem 13.8 in the PSpice manual.*

Construct a Bode plot of the voltage transfer function for the *RLC* circuit of Fig. 13.40. Assume that the circuit is underdamped.

FIGURE 13.40
Circuit for Example 13.6

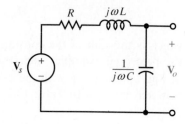

■ SOLUTION

The transfer function is given by

$$\mathbf{H}(s) = \frac{1/sC}{R + sL + (1/sC)} = \frac{1}{s^2LC + sRC + 1} = \frac{1/LC}{(s - s_{p1})(s - s_{p2})}$$

For an underdamped circuit, the poles are complex and are given by $-\alpha \pm j\omega_d$, where $\alpha = R/2L$ and $\omega_d = \sqrt{\omega_0^2 - \alpha^2}$ with $\omega_0 = 1/\sqrt{LC}$. If we set $s = j\omega$ and define $T = \sqrt{LC}$, the above becomes

$$\mathbf{H}(j\omega) = \frac{1}{1 - \omega^2 T^2 + j\omega2\zeta T}$$

where $\zeta = RC/2T$ is the damping ratio. The amplitude characteristic in decibels is

$$|\mathbf{H}(j\omega)|_{dB} = -20 \log |1 - \omega^2 T^2 + j\omega2\zeta T|$$

Because

$$|1 - \omega^2 T^2 + j\omega2\zeta T| = \sqrt{(1 - \omega^2 T^2)^2 + (\omega2\zeta T)^2} \simeq \begin{cases} 1 & \text{for } \omega \ll \dfrac{1}{T} \\[2mm] \omega^2 T^2 & \text{for } \omega \gg \dfrac{1}{T} \end{cases}$$

then $|\mathbf{H}(j\omega)|_{dB}$ is an asymptotically linear function of $\log \omega$ for both low and high frequencies:

$$|\mathbf{H}(j\omega)|_{dB} \simeq \begin{cases} 0 & \text{for } \omega \ll \dfrac{1}{T} \\[2mm] -40 \log \omega T & \text{for } \omega \gg \dfrac{1}{T} \end{cases}$$

The high-frequency asymptote, $-40 \log \omega T$, has slope -12.04 dB per octave (or -40 dB per decade) and meets the low-frequency asymptote at the corner frequency $\omega = 1/T$, as illustrated in Fig. 13.41a (black lines).

Exact plots of the function $-20 \log |1 - \omega^2 T^2 + j\omega2\zeta T|$ are shown in Fig. 13.42a; these can be used to correct the straight-line approximation. We will illus-

FIGURE 13.41
Bode plot of $\mathbf{H}(j\omega)$.
(a) Amplitude; (b) phase.
The black lines depict
the piecewise-linear
approximations. The red
curves are the exact
characteristics for $\zeta = 0.1$

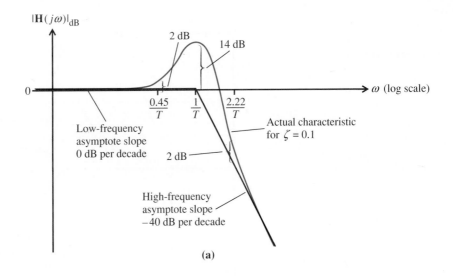

(a)

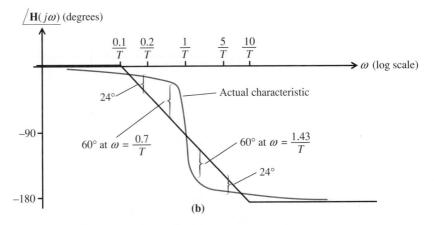

(b)

trate this process by assuming that in the present example, $T = 0.01$ and $\zeta = 0.1$. We then find from Fig. 13.42a that the actual characteristic is approximately 14 dB higher than the piecewise-linear approximation at $\omega = 100$, 9 dB higher at $\omega = 80$ and 125, and 2 dB higher at $\omega = 45$ and 222. The red curve in Fig. 13.41a results when these corrections are added to the piecewise-linear approximation.

The Bode straight-line approximation to the phase characteristic

$$\underline{/\mathbf{H}(j\omega)} = -\arctan\left(\frac{\omega 2\zeta T}{1 - \omega^2 T^2}\right)$$

is given by the black lines in Fig. 13.41b. The outer two black lines in the approximation are based on the fact that

$$-\arctan\left(\frac{\omega 2\zeta T}{1 - \omega^2 T^2}\right) \simeq \begin{cases} 0° & \text{for } \omega < \dfrac{0.1}{T} \\ -180° & \text{for } \omega > \dfrac{10}{T} \end{cases}$$

The constants $0°$ and $-180°$ are, respectively, the low-frequency and high-frequency asymptotes of $-\arctan\left[\omega 2\zeta T/(1 - \omega^2 T^2)\right]$. The middle segment of the

FIGURE 13.42
Bode plot of $\mathbf{H}(j\omega) = 1/(1 - \omega^2 T^2 + j\omega 2\zeta T)$
(a) Amplitude; (b) phase

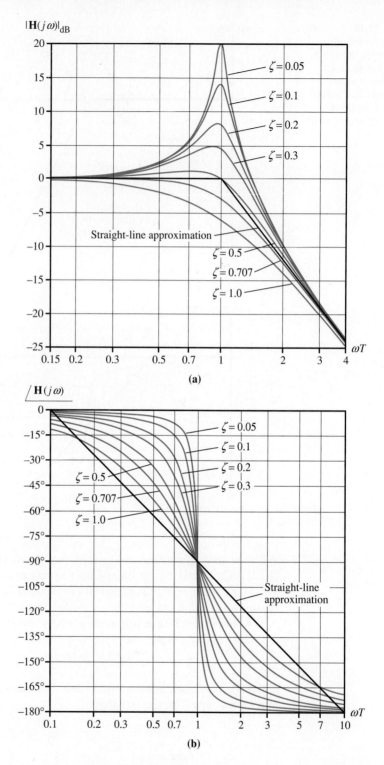

(a)

(b)

straight-line approximation intersects the outer horizontal asymptotes at the breakpoints $\omega = 0.1/T$ and $\omega = 10/T$. Thus, it has a slope of $-90°$ per decade.

Exact plots of $\underline{/\mathbf{H}(j\omega)}$ are shown in Fig. 13.42b. To illustrate the use of Fig. 13.42b, we again assume that $T = 0.01$ and $\zeta = 0.1$. We see from Fig. 13.42b that the magnitude of the phase correction is approximately 24° at $\omega = 20$ and 500, 40° at $\omega = 30$ and 333, and 60° at $\omega = 70$ and 143. The corrected phase characteristic is shown as the red curve in Fig. 13.41b. ∎

In the preceding three examples we showed how to construct Bode plots for transfer functions with a real-axis pole, a real-axis pole and a zero at the origin, and a pair of complex-conjugate poles. Equipped with this experience, we now progress to the general case.* The transfer function under consideration is

$$\mathbf{H}(s) = \frac{\mathscr{A}(s)}{\mathscr{B}(s)} \tag{13.108}$$

The amplitude characteristic is drawn in four steps.

⊞ *See Problem 13.7 in the PSpice manual.*

Step 1

Write $\mathbf{H}(s)$ as a product of factors of the forms
(a) Constant factor K
(b) Poles or zeros at the origin $s^{\pm N}$
(c) Real poles or zeros not at the origin $(s\tau + 1)^{\pm N}$
(d) Complex-conjugate poles or zeros $(s^2 T^2 + s2\zeta T + 1)^{\pm N}$

Step 2

Recognize that the amplitude characteristic in decibels, $|\mathbf{H}(j\omega)|_{dB} = 20 \log |\mathbf{H}(j\omega)|$, is the *sum* of terms corresponding to forms a through d above.
(a) Constant: $20 \log |K|$
(b) Straight line: $\pm 20N \log |j\omega| = \pm 20N \log \omega$
(c) Curves: $\pm 20N \log |j\omega\tau + 1|$
(d) Curves: $\pm 20N \log |-\omega^2 T^2 + 1 + j\omega 2\zeta T|$

In b through d, the + is associated with zeros, whereas the − is associated with poles. The curves in b through d are the same as those described in the three preceding examples except for the factor $\pm N$ and are readily approximated by straight-line characteristics. Draw the straight-line approximations for the individual forms a through d in Step 1.

Step 3

Draw the actual curves with the forms a through d in Step 1 by applying the corrections of Table 13.2 to the straight-line approximations or by using Table 13.3.

Step 4

Obtain the straight-line and the actual characteristic for $|\mathbf{H}(j\omega)|_{dB}$ by adding the straight-line and the actual curves, respectively, from Step 3.

A similar procedure is used for the construction of the phase characteristic. Table 13.3 summarizes the straight-line amplitude and phase approximations needed.

---

* We limit our description to the most common class of transfer functions, called *minimum-phase* transfer functions. The zeros of a minimum-phase transfer function all lie in the left half of the s plane.

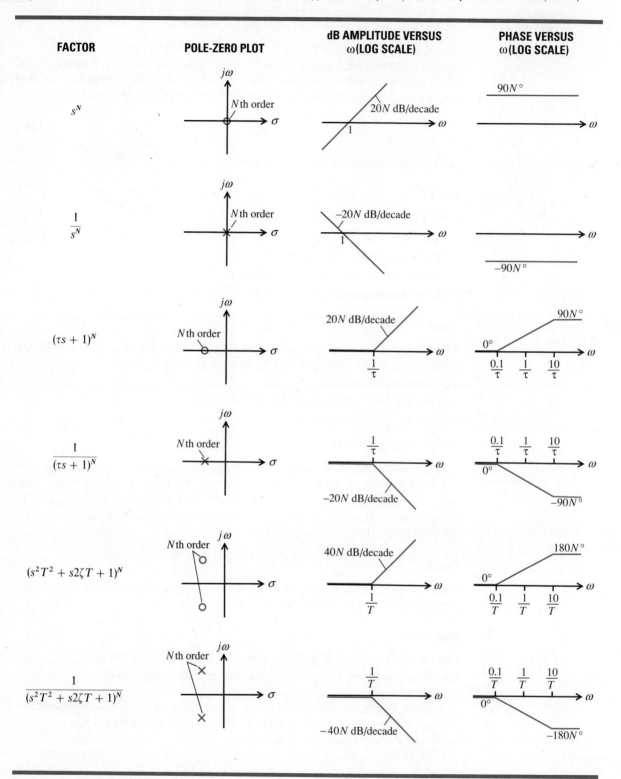

The following example demonstrates the procedure for an op-amp circuit whose transfer function contains real poles and zeros.

EXAMPLE 13.7

Draw the Bode plots for the voltage amplifier of Fig. 13.43. Assume that $R_1 = 400 \,\Omega$, $R_F = 10 \text{ k}\Omega$, $C_1 = 100 \,\mu\text{F}$, $C_F = 2 \,\mu\text{F}$, and $A = \infty$ (ideal op amp).

FIGURE 13.43
Circuit for Example 13.7

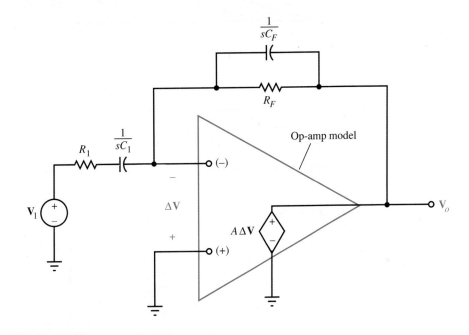

SOLUTION

Step 1

The first step is to obtain the transfer function and to write it in the form of Step 1 of the procedure. The KCL equation at the $(-)$ op-amp terminal is

$$\frac{-\Delta \mathbf{V} - \mathbf{V}_1}{R_1 + \dfrac{1}{sC_1}} + \frac{-\Delta \mathbf{V} - \mathbf{V}_o}{R_F \,\|\, \dfrac{1}{sC_F}} = 0$$

where $\Delta \mathbf{V} = \mathbf{V}_o/A = 0$ because $A \to \infty$ for an ideal op amp. If we set $\Delta \mathbf{V} = 0$, the above rearranges to

$$\frac{\mathbf{V}_o}{\mathbf{V}_1} = \mathbf{H}(s) = -\left(\frac{sC_1}{s\tau_1 + 1}\right)\left(\frac{R_F}{s\tau_F + 1}\right)$$

where $\tau_1 = R_1 C_1$ and $\tau_F = R_F C_F$. For the numbers given, the transfer function is

$$\mathbf{H}(s) = -\frac{s}{(0.02s + 1)(0.04s + 1)}$$

Step 2

We now write the transfer function as

FIGURE 13.44
Bode plot components for Example 13.7. (*a*) Individual terms in $|\mathbf{H}(j\omega)|_{dB}$; (*b*) individual terms in $\underline{/\mathbf{H}(j\omega)}$

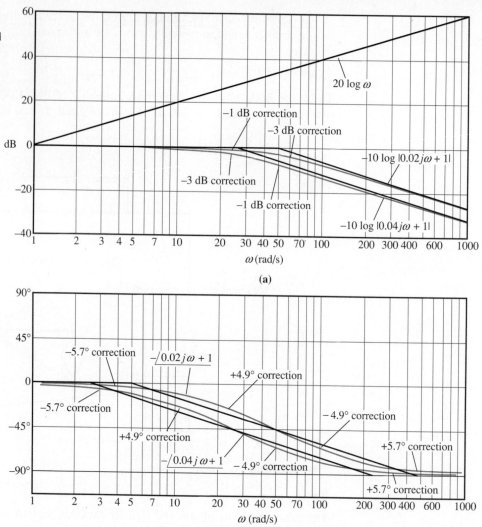

$$|\mathbf{H}(j\omega)|_{dB} = 20 \log \omega - 10 \log |0.02j\omega + 1| - 10 \log |0.04j\omega + 1|$$

and

$$\underline{/\mathbf{H}(j\omega)} = -90° - \underline{/0.02j\omega + 1} - \underline{/0.04j\omega + 1}$$

The straight-line approximations (black) and exact curves (red) for each of the terms in $|\mathbf{H}(j\omega)|_{dB}$ and $\underline{/\mathbf{H}(j\omega)}$ are shown in Fig. 13.44. We obtain the straight-line approximations to $|\mathbf{H}(j\omega)|_{dB}$ and $\underline{/\mathbf{H}(j\omega)}$ by adding the straight-line approximations from Fig. 13.44. The results are given by the black line in Fig. 13.45. The actual amplitude and phase characteristics, shown red in Fig. 13.45, are obtained by adding the red curves from Fig. 13.44. ∎

There is another procedure for drawing Bode amplitude plots that is often faster than the one we just discussed. The alternative procedure uses the simple fact that

FIGURE 13.45
(a) Bode amplitude plot;
(b) Bode phase plot

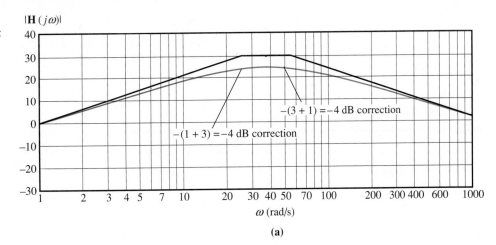

(a)

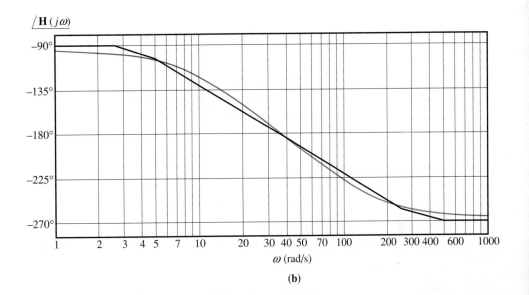

(b)

the *zeros* in a transfer function cause *upward* turns at corner frequencies, whereas the *poles* cause *downward* turns. The amounts of upward and downward turn depend on the orders of the zeros and poles, and whether they are real or complex. Nth-order real zeros cause upward turns of $20N$ dB per decade. Nth-order real poles cause downward turns of $20N$ dB per decade. Nth-order complex zeros and poles cause, respectively, upward and downward turns of $40N$ dB per decade. The details are best learned from an example.

EXAMPLE 13.8

Draw the Bode amplitude plot corresponding to

$$\mathbf{H}(s) = \frac{120s(s + 10^3)^2}{(s + 30)(s + 2 \times 10^4)^2[(3 \times 10^5)^{-2}s^2 + 0.2 \times (3 \times 10^5)^{-1}s + 1]}$$

■ **SOLUTION**

We draw the straight-line approximation first. The corner frequencies are determined by the poles and zeros of $\mathbf{H}(s)$. We make a *mental* note of the facts tabulated in Table 13.4.

TABLE 13.4 Table for Example 13.8

FACTOR	TYPE	CORNER FREQUENCY (RAD/S)	SLOPE CHANGE (dB PER DECADE)
s	1st-order real zero at $s = 0$	—	$+20$
$s + 30$	1st-order real pole at $s = -30$	30	-20
$(s + 10^3)^2$	2nd-order real zero at $s = -10^3$	10^3	$+40$
$(s + 2 \times 10^4)^2$	2nd-order real pole at $s = -2 \times 10^4$	2×10^4	-40
$(3 \times 10^5)^{-2}s^2 + 0.2 \times (3 \times 10^5)^{-1}s + 1$	1st-order complex pole	3×10^5	-40

We begin the plot to the left of all the corner frequencies. Since the smallest corner frequency is 30 rad/s, we begin the plot with $\omega \ll 30$. For $|s| \ll 30$, we can neglect all the s's in $\mathbf{H}(s)$ except the one not in parentheses. That is, we approximate

$$\mathbf{H}(s) \simeq \frac{120s(10^3)^2}{(30)(2 \times 10^4)^2} = 0.01s \qquad \text{for } |s| \ll 30$$

and therefore

$$|\mathbf{H}(j\omega)|_{\text{dB}} = 20 \log |0.01j\omega| = -40 + 20 \log \omega \qquad \text{for } \omega \ll 30$$

We can conclude from this result that the zero at $s = 0$ causes a $+20$ dB per decade asymptote to the left of $\omega = 30$ rad/s as shown by the black line in Fig. 13.46.

FIGURE 13.46
Bode amplitude plot for
Example 13.8

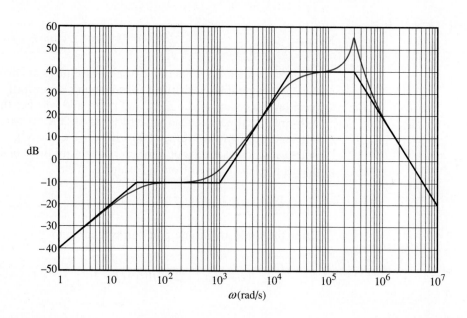

We can find the absolute level of this asymptote by substituting a convenient value of ω. Putting $\omega = 1$ gives

$$\left|\mathbf{H}(j1)\right|_{dB} = -40$$

To continue the straight-line plot, we simply increase ω, and change the slope of the line segments in accordance with the factors in $\mathbf{H}(s)$ as noted in Table 13.4. Thus, at $\omega = 30$ rad/s, there is a slope change of -20 dB per decade due to the first-order real pole at $s = -30$. This slope change causes a leveling off of the amplitude characteristic for $30 < \omega < 10^3$, as seen in Fig. 13.46. At $\omega = 10^3$ rad/s there is an upward turn of 40 dB per decade due to the second-order real zero at $s = -10^3$. Then, at $\omega = 2 \times 10^4$ there is a downward turn of 40 dB per decade due to the second-order real pole at $s = -2 \times 10^4$. A final slope change of -40 dB per decade occurs at $\omega = 3 \times 10^5$ rad/s due to the first-order complex pole. This completes the straight-line plot.

We next draw the actual amplitude characteristic by incorporating the corrections of Table 13.2 and Fig. 13.42a with $\zeta = 0.1$. Since the corner frequencies are all separated by more than a decade, the corrections are easy.

At $\omega = 30$, the actual characteristic is 3 dB below the straight-line approximation. At $\omega = \frac{30}{2} = 15$ and $\omega = 2 \times 30 = 60$, the actual characteristic is 1 dB below the straight-line approximation. We have a $2 \times 3 = 6$-dB correction at the corner frequency $\omega = 10^3$ because the associated zero is second-order. Also, the corrections are $2 \times 1 = 2$ dB at one octave to either side of the corner (at $\omega = 10^3/2 = 500$ and $\omega = 2 \times 10^3$). Similar corrections are made at the corner frequency $\omega = 2 \times 10^4$. We use Fig. 13.42a to help draw the actual characteristic in the vicinity of the $\omega = 3 \times 10^5$ rad/s corner frequency. We see from Fig. 13.42a with $\zeta = 0.1$ that the required correction is approximately 14 dB at the corner frequency and approximately 2 dB at $0.45 \times 3 \times 10^5 = 1.35 \times 10^5$ rad/s and $2.22 \times 3 \times 10^5 = 6.66 \times 10^5$ rad/s. The actual amplitude characteristic is given by the red curve in Fig. 13.46. ∎

The procedure we have just described is faster than the one on page 595 if we draw only the straight-line approximation to the Bode amplitude plot. It is also faster when we draw the actual amplitude characteristic *provided* the corner frequencies are all separated by more than a decade. When the corner frequencies are closer than a decade, however, as was the case in Example 13.7, it is usually faster to use the procedure on page 595 because we must then add together corrections from adjacent corners.

EXERCISES

35. Draw the Bode amplitude and phase plots associated with the circuit shown. Assume that $R_1 = R_2 = 10$ kΩ and $C = 1000$ μF.

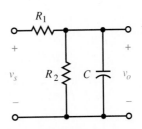

36. Repeat Exercise 34 for the circuit shown.

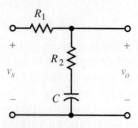

37. Repeat Exercise 35 for the circuit shown.

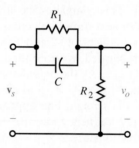

38. Repeat Exercise 35 for the circuit shown.

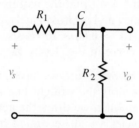

39. (a) Use an ideal op-amp model and determine the voltage transfer function V_o/V_1 of the amplifier shown. Assume that $R = 1\ \text{k}\Omega$ and $C = 1\mu\text{F}$.
    (b) Sketch the Bode amplitude and phase plots.

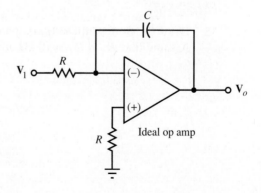

40. Draw the Bode amplitude and phase plots associated with the *RLC* circuit of Fig. 13.6. Assume that $R = 0.1\ \Omega$, $L = 1\ \mu H$, and $C = 1\ \mu F$.

41. Repeat Exercise 40 with $\mathbf{V}_o$ taken across the resistance instead of the capacitance.

42. Repeat Exercise 40 with $\mathbf{V}_o$ taken across the inductance instead of the capacitance.

13.10 Frequency Response with PSpice

The .AC and .PLOT commands; see Sections B.2.1 and B.2.8 of the PSpice manual.

The use of asymptotic approximations and the related correction factors reduces the labor required for manual calculation of a frequency response to an acceptable level. Nevertheless, the modern circuit designer often calculates the frequency response with a program such as PSpice. We do this with the aid of the sweep parameter in the .AC solution control statement. The inclusion of a .PLOT statement* or, as we see later, a .PROBE statement† provides a plot of the frequency response. The following example demonstrates the output obtained with a .PLOT statement.

EXAMPLE 13.9

Determine the frequency response of the *RC* low-pass filter shown in Fig. 13.47. Calculate five points per decade over the range from 0.01 to 10 Hz. Use the .PLOT statement to plot the log of the amplitude of the output on one graph, and the phase on another.

FIGURE 13.47
An *RC* low-pass filter

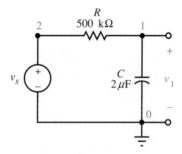

■ SOLUTION

We first write the circuit file shown on page 604 and store it on disk under the file name EX13-9.CIR. Although we can print both the amplitude and the phase of v_1 on the same plot, we have chosen to use two .PLOT statements for a clearer presentation. Now run PSpice and use this circuit file as the input. After we edit out some unneeded information, the output file (EX13-9.OUT by default) consists of the two plots shown in Figs. 13.48 and 13.49. Because we chose v_s to have an amplitude of 1 and a phase of 0, these plots give the frequency response directly. ■

---

* Refer to Chapter 4, page 176, for a description of the .PLOT statement.
† Refer to Chapter 9, page 394, for a description of the .PROBE statement.

```
CIRCUIT FILE FOR THE RC CIRCUIT OF EXAMPLE 13.9

R         2      1     500K

C         1      0     2UF

VS        2      0     AC        1        0

.AC       DEC    5     .01       10

.PLOT     AC     VM(1)

.PLOT     AC     VP(1)

.END
```

FIGURE 13.48
Computer output amplitude characteristic for Example 13.9

```
   LEGEND:

   *: VM(1)

    FREQ        VM(1)


   (*)----------   1.0000E-02   1.0000E-01   1.0000E+00   1.0000E+01   1.0000E+02
   1.000E-02  9.980E-01 - - - - - - . - - - - - - . - - - - -*- . - - - - - - . - - - - -
   1.585E-02  9.951E-01 .            .            *            .            .
   2.512E-02  9.878E-01 .            .            *            .            .
   3.981E-02  9.701E-01 .            .            *            .            .
   6.310E-02  9.296E-01 .            .            *            .            .
   1.000E-01  8.467E-01 .            .            *            .            .
   1.585E-01  7.086E-01 .            .        *.            .            .
   2.512E-01  5.352E-01 .            .     *            .            .
   3.981E-01  3.712E-01 .            .   *            .            .
   6.310E-01  2.446E-01 .            .  *            .            .
   1.000E+00  1.572E-01 .            . *            .            .
   1.585E+00  9.992E-02 .          *            .            .
   2.512E+00  6.323E-02 .        *.            .            .
   3.981E+00  3.995E-02 .      *            .            .
   6.310E+00  2.522E-02 .    *            .            .
   1.000E+01  1.591E-02 .  *            .            .            .
                        - - - - - - . - - - - - - . - - - - - - . - - - - - - . - - - - -
```

FIGURE 13.49
Computer output phase characteristic for Example 13.9

```
    FREQ        VP(1)


   (*)----------  -1.5000E+02  -1.0000E+02  -5.0000E+01  -7.1054E-15   5.0000E+01
   1.000E-02 -3.595E+00 .            .            .            *.  - - - - . - - - - -
   1.585E-02 -5.687E+00 .            .            .            *.           .
   2.512E-02 -8.969E+00 .            .            .           *.            .
   3.981E-02 -1.404E+01 .            .            .          *            .
   6.310E-02 -2.163E+01 .            .            .        *            .
   1.000E-01 -3.214E+01 .            .            .      *            .
   1.585E-01 -4.488E+01 .            .            .   *.            .
   2.512E-01 -5.764E+01 .            .            *            .
   3.981E-01 -6.821E+01 .            .        *            .
   6.310E-01 -7.584E+01 .            .      *            .
   1.000E+00 -8.096E+01 .            .    *            .
   1.585E+00 -8.427E+01 .            .  *            .
   2.512E+00 -8.637E+01 .            . *            .
   3.981E+00 -8.771E+01 .            . *            .
   6.310E+00 -8.856E+01 .            .*            .
   1.000E+01 -8.909E+01 .            .*            .
                        - - - - - - . - - - - - - . - - - - - - . - - - - - - . - - - - -
```

The general form for the .AC solution control statement is

```
.AC       SWEEP       NPOINTS       FSTART       FSTOP
```

As we have seen in the preceding example, this statement lets us calculate the response of the circuit for a range of values. The SWEEP parameter specifies the type of sweep, NPOINTS determines the number of points in the sweep, and FSTART and FSTOP specify the beginning and ending frequency in hertz for the sweep. The three sweep types available are

LIN   The frequency is swept linearly from FSTART to FSTOP. NPOINTS is the *total* number of points in the sweep (equally spaced on a linear scale).

OCT   The frequency is swept logarithmically by octaves from FSTART to FSTOP. NPOINTS is the number of points *per octave* (equally spaced on a logarithmic scale).

DEC   The frequency is swept logarithmically by decades from FSTART to FSTOP. NPOINTS is the number of points *per decade* (equally spaced on a logarithmic scale).

The .PROBE statement can be used in place of, or in addition to, the .PLOT statement to provide a much higher quality graphical output. The following example demonstrates the use of .PROBE.

EXAMPLE 13.10

Add the .PROBE statement to the circuit file of Example 13.9 and obtain a plot of the amplitude of node voltage v_1 in decibels versus frequency.

■ SOLUTION

We will retain the two .PLOT statements used in Example 13.9 so that the frequency response plots shown will be placed in our circuit file, but this is not necessary for the use of the .PROBE statement. The circuit file is as follows:

```
CIRCUIT FILE FOR THE RC CIRCUIT OF EXAMPLE 13.10

R          2     1          500K

C          1     0          2UF

VS         2     0          AC      1      0

.AC        DEC   5          .01     10

.PLOT      AC    VM(1)

.PLOT      AC    VP(1)

.PROBE

.END
```

When the program is run with this circuit file, the .PLOT statements place plots of the amplitude and phase of node voltage v_1 in the output file. The .PROBE statement fills the computer screen* with a rectangular grid. A list of options appears under the grid. Highlight "Add trace," and press the enter key. In response to the request for a variable, type

```
VDB(1)
```

and the amplitude of node voltage v_1 in decibels is plotted as shown in Fig. 13.50.

FIGURE 13.50
Computer output amplitude characteristic for Example 13.10

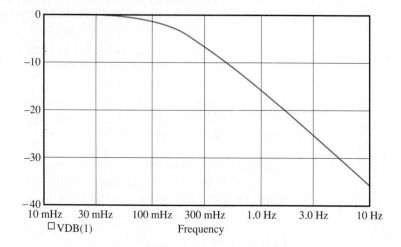

Highlight "Remove trace," and press the enter key. Highlight "Remove all," and press the enter key. The plot will disappear. We can now construct a new plot. For example, we can plot the amplitude of node voltage v_1 on a linear amplitude scale by the following steps. Highlight "Add trace," and press the enter key. In response for a variable, type

```
VM(1)
```

Highlight "Hard copy," and press the enter key to obtain an output that is a printer plot. An output similar to that shown is available with a print screen command.

PSpice is very user-friendly. Do not be afraid to experiment. For example, if you wish the frequency axis to be linear, rather than logarithmic, after you obtain a display using .PROBE, simply select the X axis in the menu that is below the plot. A new menu will appear. Now select Linear, and your new graph will appear with a linear frequency scale. ■

REMEMBER

Frequency-response plots can be obtained with the use of the .AC solution control statement. The frequency sweep can be specified as linear (LIN), logarithmic by octave (OCT), or logarithmic by decade (DEC). The .PLOT statement gives a

---

* A file called EX13-10.DAT is stored on the disk by the .PROBE command. .PROBE uses this data. In some versions of PSpice, you must open this file to obtain the rectangular grid.

printer plot output. A higher-quality graphical output is available with the use of the .PROBE statement.

EXERCISES

43. Change the solution control statement in Example 13.9 to

    ```
    .AC    LIN    15    0.1    1
    ```

    Run PSpice with this input file. Print out the two graphs that are in the output file.

44. Run PSpice with the circuit file from Example 13.9.
    (a) Use .PROBE to generate a plot of the amplitude of node voltage v_1 as a function of the logarithm of frequency.
    (b) Use .PROBE to generate a plot of the amplitude of node voltage v_1 in decibels versus the logarithm of frequency.
    (c) Use .PROBE to generate a plot of the phase of node voltage v_1 as a function of the logarithm of frequency.

13.11 Summary

In this chapter we saw that circuits can be used as filters. We described three types of filters: high-pass, low-pass, and band-pass.

To provide a better understanding of filters, we introduced the concept of complex frequency, $s = \sigma + j\omega$, and we examined the transfer function $\mathbf{H}(s) = \mathscr{B}(s)/\mathscr{A}(s)$. We showed that $\mathbf{H}(s)$ can be determined from the circuit's LLTI equation and also from the circuit itself. We used spiraling phasors to show that a damped sinusoidal input $v_s(t) = V_{sm}e^{\sigma t} \cos(\omega t + \phi_s)$ produces a damped sinusoidal particular response $v_{op}(t) = V_{sm}|\mathbf{H}(s)|e^{\sigma t} \cos(\omega t + \phi_s + \underline{/\mathbf{H}(s)})$.

We defined the poles and the zeros of $\mathbf{H}(s)$ to be the roots to $\mathscr{A}(s) = 0$ and $\mathscr{B}(s) = 0$, respectively, and we showed that $\mathbf{H}(s)$ is completely specified by its pole-zero plot. By using the rubber sheet analogy, we saw that the amplitude characteristic $|\mathbf{H}(j\omega)|$ is given by the intersection of the two-dimensional surface $|\mathbf{H}(s)|$ and the plane $\sigma = 0$. Poles near the $j\omega$ axis cause resonance peaks in the amplitude characteristic, whereas zeros cause dips. Since the poles are the roots to $\mathscr{A}(s) = 0$, the poles also determine the form of the circuit's natural response. Thus we saw that a filter's frequency response $\mathbf{H}(j\omega)$ is closely related to its natural response.

By referring to pole-zero diagrams, we showed that all resonance peaks caused by isolated simple poles near the $j\omega$ axis have approximately the same shape. An isolated simple pole, $s_{p1} = -\alpha + j\omega_d$, is near the $j\omega$ axis when $\alpha \ll \omega_d$. The maximum-response resonance frequency ω_{mr} associated with such a pole is always *approximately* ω_d, and the half-power bandwidth BW of the resonance peak is always *approximately* 2α. We showed that for a series RLC band-pass filter, ω_{mr} is *exactly* equal to $\omega_0 = 1/\sqrt{LC}$, and the half-power bandwidth is *exactly* equal to 2α. We also showed that for the series RLC circuit, $BW = \omega_0/Q_0$, where Q_0 is the quality factor of the circuit evaluated at ω_0.

We defined the unity-power-factor resonance frequency ω_{upf} as the frequency at which the circuit's driving-point impedance is purely real. In general, $\omega_{upf} \neq \omega_{mr}$. When $\omega = \omega_{upf}$, the total average energy stored in the magnetic and the electric fields of a circuit composed of Rs, Ls, and Cs are equal.

We concluded this chapter with descriptions of Bode plots and computer-generated frequency response plots using PSpice. The Bode plot format, in which the amplitude characteristic is plotted in decibels and the frequency axis is logarithmic, has become an industrial standard. (See Table 13.3 for a summary of Bode's straight-line amplitude and phase approximations.) Computer-generated frequency response plots often use the Bode plot format.

KEY FACTS

♦   Complex frequency s is defined by $s = \sigma + j\omega$, where $\sigma = \mathcal{R}e\{s\}$ is the neper frequency and $\omega = \mathcal{I}m\{s\}$ is the angular frequency.

♦   The poles of a transfer function

$$H(s) = \frac{\mathcal{B}(s)}{\mathcal{A}(s)}$$

are values of s for which $\mathcal{A}(s) = 0$. The poles are the roots of the characteristic equation.

♦   The zeros of $H(s)$ are the values of s for which $\mathcal{B}(s) = 0$.

♦   When the complex exponential $\mathbf{V}_s e^{st}$ is applied to the LLTI equation $\mathcal{A}(p)v_o(t) = \mathcal{B}(p)v_s(t)$, the particular response is $\mathbf{V}_o e^{st}$, where $\mathbf{V}_o = \mathbf{H}(s)\mathbf{V}_s$ provided that the complex input frequency s is not a pole.

♦   A complex exponential $\mathbf{V}e^{st}$, where $\mathbf{V} = V_m\underline{/\phi_\mathbf{V}}$, may be interpreted as a spiraling phasor. The projection of the spiraling phasor onto the real axis is the damped sinusoid $V_m e^{\sigma t} \cos(\omega t + \phi_\mathbf{V})$.

♦   When the damped sinusoid $V_{sm} e^{\sigma t} \cos(\omega t + \phi_{v_s})$ is applied to a LLTI circuit, the particular response is $V_{om} e^{\sigma t} \cos(\omega t + \phi_{v_o})$, where the amplitudes and phases are related by $\mathbf{V}_o = \mathbf{H}(s)\mathbf{V}_s$, with $s = \sigma + j\omega$.

♦   The rubber sheet analogy provides a quick way to visualize the function $|\mathbf{H}(s)|$ as a surface above the s plane. A circuit's amplitude characteristic $|\mathbf{H}(j\omega)|$ is the intersection of the function $|\mathbf{H}(s)|$ with the $\sigma = 0$ plane.

♦   Every LLTI circuit's transfer function in the vicinity of an isolated simple pole s_{p1} is given approximately by

$$\mathbf{H}(s) \simeq \frac{k_{11}}{s - s_{p1}}$$

where

$$k_{11} = (s - s_{p1})\mathbf{H}(s)\Big|_{s = s_{p1}}$$

is the residue of $\mathbf{H}(s)$ at the pole.

♦   If the isolated pole $s_{p1} = -\alpha + j\omega_d$ is close to the $j\omega$ axis ($\alpha \ll \omega_d$), the circuit's frequency response characteristic is given approximately by

$$\mathbf{H}(j\omega) \simeq \frac{k_{11}}{j(\omega - \omega_d) + \alpha} \qquad \text{for } \omega \simeq \omega_d$$

We have called this characteristic the *universal resonance characteristic*.

- The half-power bandwidth of *every* resonance peak caused by an isolated pole $s_{p1} = -\alpha + j\omega_d$ near the $j\omega$ axis is given approximately by

$$BW \simeq 2\alpha$$

- A circuit's maximum-response resonance frequency is the frequency at which the amplitude characteristic $|\mathbf{H}(j\omega)|$ is maximum.

- A circuit's unity-power-factor resonance frequency is the frequency at which the driving-point impedance is purely real.

- At unity-power-factor resonance, the total average energies stored in the magnetic and the electric fields of a circuit composed of Rs, Ls, and Cs are equal.

- Quality factor is defined by

$$Q(\omega) = 2\pi \frac{\text{peak energy stored in the circuit}}{\text{energy dissipated by the circuit in one period}}$$

- For a band-pass series *RLC* circuit, the maximum-response and unity-power-factor resonance frequencies both occur at $\omega_0 = 1/\sqrt{LC}$ rad/s. The half-power bandwidth is given by $BW = \omega_0/Q_0$ rad/s, where $Q_0 = \omega_0 L/R$ is the quality factor evaluated at $\omega = \omega_0$.

PROBLEMS

Section 13.1

1. Make a reasonably accurate sketch of the amplitude and phase characteristics of $\mathbf{Z}(j\omega)$ for the following circuit.

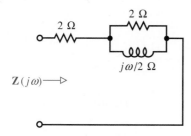

2. Plot the voltage transfer function amplitude and phase characteristics for the circuit shown.

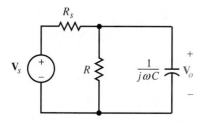

3. Repeat Problem 2 for the following circuit.

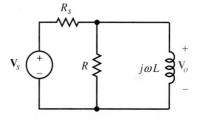

4. Repeat Problem 2 for the circuit shown.

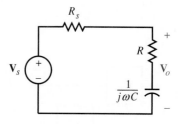

5. Repeat Problem 2 for the circuit shown.

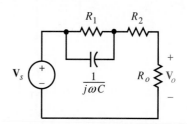

6. The frequency response characteristics of an *ideal low-pass filter* are depicted in Fig. 13.51, where K is the gain, ω_0 is the angular cutoff frequency or bandwidth, and τ is the *signal delay*. (The ideal low-pass filter is an analytically convenient but fictitious system whose transfer function can only be approximated using physical elements.)

   (a) Find the steady-state response to an input $v_1(t) = V_{1m} \cos(\omega_1 t + \phi_{V1})$. Assume that $\omega_1 > \omega_0$.

   (b) Show that if $\omega_1 < \omega_0$, the steady-state response to the input from (a) is $v_o(t) = KV_{1m} \cos[\omega_1(t - \tau) + \phi_{V1}]$.

   (c) Assume that $K = 1$, $\omega_0 = 1000$ rad/s, and $\tau = 1$ ms. Find the steady-state response to the input $v_1(t) = 6 + 2\cos(300t + 20°) + 4\sin 800t + 17\cos(1100t + 45°) - 5\cos 2000t$.

7. Let $\mathbf{Z}_a = \mathbf{Z}_1 = 100\ \Omega$, $\mathbf{Z}_b = 1/j\omega C$, and $\mathbf{Z}_2 = R$ in the ideal op-amp circuit in Fig. 13.52.

   (a) Choose R and C such that the amplitude characteristic approximates that of the

ideal low-pass filter shown in Problem 6 with $K = 20$ and bandwidth 1000 rad/s.

   (b) Plot the amplitude and phase characteristics of your op-amp filter and compare them with those shown in Problem 6 for $K = 20$ and $\tau = 1$ ms.

   (c) Find the steady-state response of your filter to the input $v_1(t) = 6 + 2\cos(300t + 20°) + 4\sin 800t + 17\cos(1100t + 45°) - 5\cos 2000t$ and compare this with the ideal low-pass filter response for $\tau = 1$ ms.

8. The frequency response characteristics of an *ideal high-pass filter* are depicted in Fig. 13.53, where K is the gain, ω_0 is the angular cutoff frequency, and τ is the signal delay. (As with the ideal low-pass filter, the ideal high-pass filter is not physically realizable.)

   (a) Find the steady-state response to an input $v_1(t) = V_{1m} \cos(\omega_1 t + \phi_{V_1})$ for (i) $\omega_1 < \omega_0$ and (ii) $\omega_1 > \omega_0$.

   (b) Assume that $K = 1$, $\omega_0 = 1000$ rad/s, and $\tau = 1$ ms. Find the steady-state response to the input $v_1(t) = 6 + 2\cos(300t + 20°) +$

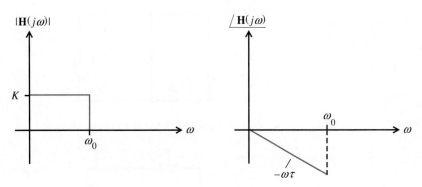

FIGURE 13.51

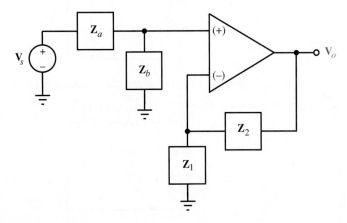

FIGURE 13.52

$$4 \sin 800t + 17 \cos (1100t + 45°) - 5 \cos 2000t.$$

9. For the circuit configuration of Problem 7,
   (a) Use an ideal op amp, two 1-kΩ resistances, a resistance R_x, and a capacitance C_x to design a high-pass filter. Choose R_x and C_x such that $\mathbf{H}(j\infty) = 20$ and $|\mathbf{H}(j1000)| = 20/\sqrt{2}$.

   (b) Plot the amplitude and phase characteristics for the circuit you designed.
10. Hypothetical source and amplifier models are shown in the circuit in Fig. 13.54. A circuit designer sets $G = 10$ and connects the source model directly to the amplifier as shown in an effort to amplify $v_1(t)$ by a factor of 10.
    (a) Determine the voltage transfer function $\mathbf{H}(j\omega) = \mathbf{V}_o/\mathbf{V}_1$.

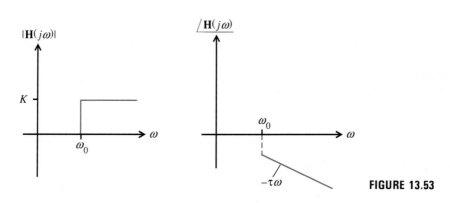

FIGURE 13.53

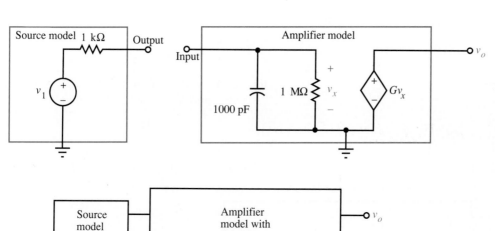

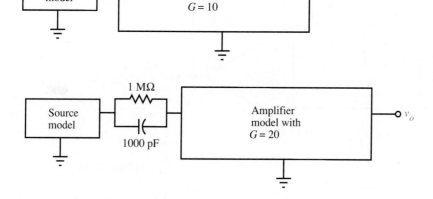

FIGURE 13.54

(b) Plot the amplitude and phase characteristics for $\mathbf{H}(j\omega)$.

(c) What is the half-power bandwidth in hertz?

(d) For what values of $f = \omega/2\pi$ is the approximation $v_o(t) \simeq 10v_1(t)$ reasonable? Discuss.

To improve the performance, the circuit designer inserts a parallel RC circuit in the line connecting the source to the amplifier and sets $G = 20$ as shown.

(e) Determine and plot the amplitude and phase characteristics of the revised circuit.

(f) In what way, if any, has the approximation $v_o(t) \simeq 10v_1(t)$ been improved?

Input impedance compensation of the type shown in the circuit in Fig. 13.54 is frequently used in CRT oscilloscope voltage probes in which the compensating capacitance is adjustable. (Also see Problem 92.)

Section 13.2

11. Draw the dual of the series RLC circuit of Fig. 13.6 and specify the dual transfer function $\mathbf{H}(j\omega) = \mathbf{I}_o/\mathbf{I}_s$.

12. For the circuit shown, determine and plot the amplitude and phase characteristics of $\mathbf{V}_R/\mathbf{V}_s$ and $\mathbf{V}_{LC}/\mathbf{V}_s$.

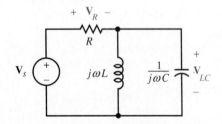

13. In the circuit shown, $L = 1\ \mu\text{H}$, $C = 1\ \mu\text{F}$, and $r = 1\ \text{k}\Omega$.

(a) Choose the value of R such that $\mathbf{V}_o/\mathbf{V}_s = 0.5$ for $\omega = 1/\sqrt{LC}$.

(b) Use the value of R determined in (a), and calculate and compare the currents $\mathbf{I}_s$, $\mathbf{I}_L$, $\mathbf{I}_C$, and $\mathbf{I}_R$ for $\omega = 1/\sqrt{LC}$ if $\mathbf{V}_s = 10\underline{/0^\circ}$ V. What is happening here?

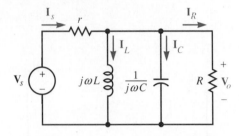

14. Repeat part (b) of Problem 13 with $\omega = 0$.

15. Repeat part (b) of Problem 13 with $\omega = \infty$.

16. For the ideal op-amp circuit shown in Fig. 13.55,

(a) Write the node-voltage equations.

(b) Solve the node-voltage equations for $\mathbf{V}_o$ to show that the voltage transfer function is given by

$$\mathbf{H} = \frac{\mathbf{V}_0}{\mathbf{V}_s} = \frac{-\mathbf{Y}_1\mathbf{Y}_3}{\mathbf{Y}_5(\mathbf{Y}_1 + \mathbf{Y}_2 + \mathbf{Y}_3 + \mathbf{Y}_4) + \mathbf{Y}_3\mathbf{Y}_2}$$

(c) Let $\mathbf{Y}_1 = R_1^{-1}$, $\mathbf{Y}_2 = G_2$, $\mathbf{Y}_3 = R_3^{-1}$, $\mathbf{Y}_4 = j\omega C_4$, and $\mathbf{Y}_5 = j\omega C_5$. Show that

$$\mathbf{H}(j\omega) =$$

$$\frac{-1}{(j\omega)^2 C_4 C_5 R_1 R_3 + j\omega C_5(R_1 + R_3 + R_1 G_2 R_3) + R_1 G_2}$$

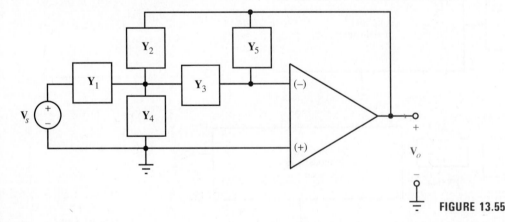

FIGURE 13.55

(d) Let $C_4C_5R_1R_2 = 10^{-12}$ s^2, and $C_5(R_1 + R_3 + R_1G_2R_3) = 10^{-7}$ s and $R_1G_2 = 1$. Plot the amplitude and phase characteristics and compare them with Fig. 13.7.

Section 13.3

17. The complex functions $\mathbf{i}(t)$ and $\mathbf{v}(t)$ satisfy the equation

$$\frac{d^2}{dt^2}\mathbf{v} + 15\frac{d}{dt}\mathbf{v} + 50\mathbf{v} = \frac{d^2}{dt^2}\mathbf{i} + 7\frac{d}{dt}\mathbf{i}$$

(a) Without solving the equation, show that the real functions $i_R(t) = \mathscr{R}e\,\{\mathbf{i}(t)\}$ and $v_R(t) = \mathscr{R}e\,\{\mathbf{v}(t)\}$ must also satisfy the same equation.
(b) Without solving the equation, show that the real functions $i_I(t) = \mathscr{I}m\,\{\mathbf{i}(t)\}$ and $v_I(t) = \mathscr{I}m\,\{\mathbf{v}(t)\}$ must also satisfy the same equation.

18. Find the impedance $\mathbf{Z}(s)$ associated with the LLTI equation of Problem 17.

19. The driving-point impedance of a network is

$$\mathbf{Z}(s) = \frac{s}{s+1}\ \Omega$$

A current $i(t) = 2e^{-4t}\cos(3t + 15°)$ A is applied to the network. Determine the particular response.

20. Repeat Problem 19 if $i(t) = 2e^{-4t}$ A.
21. Repeat Problem 19 if $i(t) = 2$ A.
22. The driving-point impedance of a network is

$$\mathbf{Z}(s) = 1 + \frac{2s}{1 + 2s}\ \Omega$$

(a) Find the differential equation relating input $i(t)$ to output $v(t)$.
(b) Find the complete response to an input $i(t) = 2e^{-4t}\cos(3t + 15°)u(t)$ A. Assume that the circuit is initially at rest.

23. The following circuit depicts a simple model of a one-stage transistor circuit.
(a) Draw the s-domain circuit.
(b) Find $\mathbf{H}(s) = \mathbf{V}_o/\mathbf{V}_s$.
(c) Let $v_s(t) = 7e^{-100t}\cos(2000t + 30°)$ V. What is the value of complex frequency s for this input?
(d) Find the particular response $v_{op}(t)$ in terms of r, R, C, and β for the input of (c) by first determining $\mathbf{V}_o$.

(e) Use $\mathbf{H}(s)$ to determine the LLTI equation relating $v_o(t)$ to $v_s(t)$.

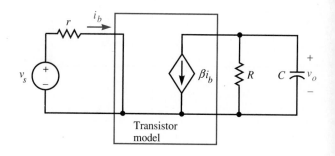

Transistor model

Section 13.4

24. The source in the circuit shown below is a current or voltage source.
(a) Determine the driving-point impedance $\mathbf{Z}(s)$, and sketch the corresponding pole-zero plot.
(b) Determine the form of the complementary (natural) response if the source is an ideal *current* source.
(c) Determine the form of the complementary response if the source is an ideal *voltage* source.
(d) Determine the particular response $v_p(t)$ if the input is $i(t) = 10e^{-t/2}\cos(\sqrt{3}t/2 + 20°)$ A.
(e) Determine the particular response $i_p(t)$ if the input is $v(t) = 25 + e^{-t}$ V.

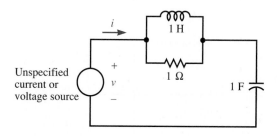

For the pole-zero plots in Problems 25 through 27,
(a) Find the natural response.
(b) Express $\mathbf{H}(s)$ as the ratio of two polynomials.
(c) Find the value of the transfer function at a frequency $\omega = 3$ rad/s (neper frequency $\sigma = 0$).

25.

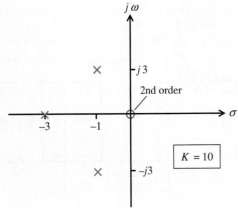

$K = 10$

26.

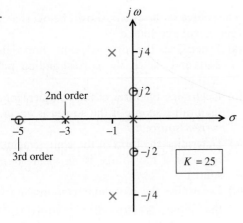

$K = 25$

27.

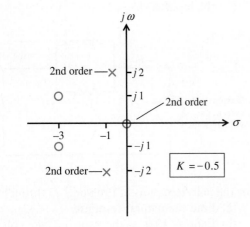

$K = -0.5$

28. Find $\mathbf{H}(s) = \mathbf{V}_2/\mathbf{V}_1$ and draw the corresponding pole-zero plot for the following circuit.

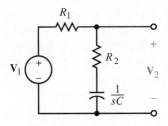

29. Repeat Problem 28 for the following circuit.

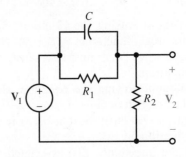

30. Repeat Problem 28 for the following circuit.

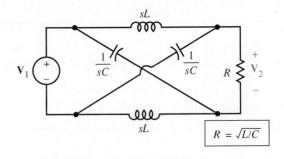

$R = \sqrt{L/C}$

31. (a) Determine the s-domain transfer functions $\mathbf{H}_R(s) = \mathbf{V}_R/\mathbf{V}_s$ and $\mathbf{H}_{LC}(s) = \mathbf{V}_{LC}/\mathbf{V}_s$ for the circuit of Problem 12.
(b) Draw the pole-zero plots, and explain how they differ.

32. For the following circuit,
(a) Draw the s-domain version of the circuit, and write the node-voltage equations for $\mathbf{V}_1$, $\mathbf{V}_2$, and $\mathbf{V}_3$.
(b) Use Cramer's rule to solve for $\mathbf{V}_1$, $\mathbf{V}_2$, and $\mathbf{V}_3$. Evaluate all determinants, and write your answers in the form $\mathbf{V}_i = \mathbf{Z}_{iA}(s)\mathbf{I}_A + \mathbf{Z}_{iB}(s)\mathbf{I}_B$ for $i = 1, 2, 3$, where the numerator and denominator polynomials of $\mathbf{Z}_{iA}(s)$ and $\mathbf{Z}_{iB}(s)$ are written in factored form.

(c) Do the complementary response voltages $v_{1c}(t)$, $v_{2c}(t)$, and $v_{3c}(t)$ contain the same complex natural frequencies? If not, explain why in terms of the circuit.

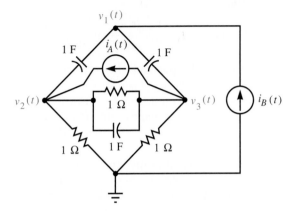

33. Use the rubber sheet analogy to make an approximate sketch of the amplitude response characteristic $|\mathbf{H}(j\omega)|$ associated with each of pole-zero constellations of Problems 25 through 27.

34. Use the rubber sheet analogy to make an approximate sketch of the amplitude characteristic $|\mathbf{H}(j\omega)|$ for the circuit of Problem 28.

35. The input to the circuit shown below is $v_s(t) = \mathbf{V}_s e^{st}$. For what value of s is the particular response, $v_{op}(t)$ zero for all t?

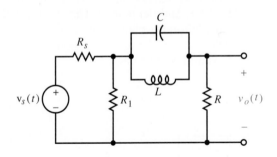

36. For what complex frequencies is a series RC circuit equivalent to (a) a short circuit, and (b) an open circuit?

37. Repeat Problem 36 for a series RL circuit.
38. Repeat Problem 36 for a series LC circuit.
39. Repeat Problem 36 for a series RLC circuit.
40. Repeat Problem 36 for a parallel RC circuit.
41. Repeat Problem 36 for a parallel RL circuit.
42. Repeat Problem 36 for a parallel LC circuit.
43. Repeat Problem 36 for a parallel RLC circuit.

44. For the circuit shown,
(a) Find the driving-point impedance $\mathbf{Z}(s)$.
(b) For what value(s) of s is $\mathbf{Z}(s)$ equal to zero?
(c) For what value(s) of s is $\mathbf{Z}(s)$ equal to infinity?
(d) What is the form of the complementary response if the input is a current?
(e) What is the form of the complementary response if the input is a voltage?

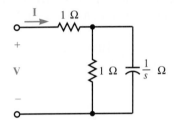

45. For the circuit shown,
(a) Write the driving-point admittance as the ratio of two polynomials in s.
(b) Indicate the location of the poles and zeros of $\mathbf{Y}(s)$ in the s plane in terms of $\alpha = R/2L$ and $\omega_0 = 1/\sqrt{LC}$.
(c) Write an expression for the complementary solution when the circuit is driven by a current source.

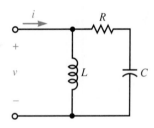

46. For what values of r is the circuit shown unstable?

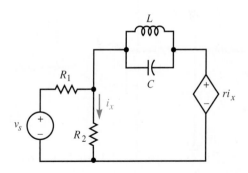

47. An *nth-order Butterworth filter* is a low-pass filter with the transfer function

$$H(s) = \frac{k\omega_0^n}{(s - s_1)(s - s_2) \cdots (s - s_n)}$$

where k is a real constant and where the poles are symmetrically spaced about a semicircle of radius ω_0 in the left half of the s plane, as shown below for $n = 5$. The amplitude characteristic of an nth-order Butterworth filter is called *maximally flat*, meaning that

$$\frac{d^i}{d\omega^i}|H(j\omega)|\Big|_{\omega = 0} = 0$$

for $i = 1, 2, \ldots, n - 1$.

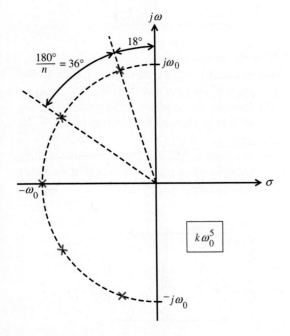

(a) Show that

$$|H(j\omega)| = \frac{|k|\omega_0^n}{\sqrt{\omega^{2n} + \omega_0^{2n}}}$$

Hint:

$$H(s)H(-s) = \frac{k^2\omega_0^{2n}}{s^{2n} + \omega_0^{2n}}$$

because the poles of $H(s)$ are the left-half s-plane roots to $s^{2n} + \omega_0^{2n} = 0$.

(b) Using your result from (a), show that ω_0 is the half-power bandwidth in radians per second. That is, show that $|H(j\omega_0)| = (1/\sqrt{2})|H(0)|$.

(c) Sketch $|H(j\omega)|$ in the limit $n \to \infty$.

(d) Plot the amplitude characteristic for a fifth-order Butterworth filter. ▪ 13.9

48. Butterworth filters can be synthesized by cascading op-amp circuits as shown in Fig. 13.56 for a fourth-order filter.

(a) Assuming ideal op amps, show that the voltage transfer function V_2/V_1 is given by

$$H(s) = \frac{-k_1}{a_1 s^2 + b_1 s + 1} \cdot \frac{-k_2}{a_2 s^2 + b_2 s + 1}$$

For $i = 1, 2$, give the formulas for the constants k_i and for a_i and b_i in terms of the circuit elements.

(b) Choose k_i, a_i, and b_i, $i = 1, 2$, such that $H(0) = K$ and the poles of $H(s)$ are located at the left-half s-plane roots of $s^{2n} + \omega_0^{2n} = 0$ where $n = 4$. Write your answers in terms of ω_0 and K. ▪ 13.9

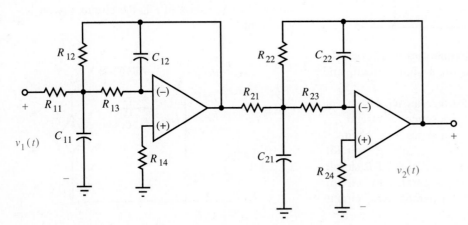

FIGURE 13.56

(c) Draw a circuit with three-op-amps that can be used to synthesize a sixth-order Butterworth filter.

(d) Draw a circuit using three op amps that can be used to synthesize a fifth-order Butterworth filter.

Section 13.5

49. For the circuit shown,
(a) Show that the transfer impedance is

$$H(s) = \frac{V_o}{I_s} = \frac{\dfrac{sRR_s}{L}}{s^2 + s\left(\dfrac{R + R_s}{L}\right) + \dfrac{1}{LC}}$$

(b) Find the form of the complementary response $v_c(t)$.

(c) Find the exact expression for the maximum-response resonance frequency ω_{mr} associated with $|H(j\omega)|$.

(d) Find the exact expression for the half-power bandwidth associated with $|H(j\omega)|$.

(e) Specify your answers to (a) through (d) using $R_s = 6\ \Omega$, $R = 20\ \Omega$, $L = 13$ mH, and $C = 2.2\ \mu$F.

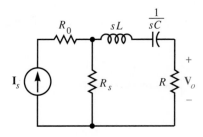

50. The frequency response characteristics of an *ideal band-pass filter* are depicted below, where

K is the gain, ω_L and ω_U are the lower and upper angular cutoff frequencies, respectively, and τ is the *signal delay*. (As with the ideal low-pass and high-pass filters, the ideal band-pass filter is an analytically convenient but physically unrealizable system.)

(a) Find the steady-state response to the input $v_1(t) = V_{1m} \cos(\omega_1 t + \phi_{v1})$ for (i) $\omega_1 < \omega_L$, (ii) $\omega_L < \omega_1 < \omega_U$, and (iii) $\omega_U < \omega_1$.

(b) Assume that $K = 1$, $\omega_L = 500$ rad/s, $\omega_U = 1200$ rad/s, and $\tau = 1$ ms. Find the steady-state response to the input

$$v_1(t) = 6 + 2 \cos(300t + 20°) + 4 \sin 800t$$
$$+ 17 \cos(1100t + 45°) - 5 \cos 2000t$$

51. A band-pass filter employing an ideal op amp is shown below. Show that the voltage transfer function has the form

$$H(s) = \frac{V_o}{V_s} = \frac{-\alpha A \omega_0 s}{s^2 + 2\alpha s + \omega_0^2}$$

Express α, ω_0, and A in terms of the circuit elements.

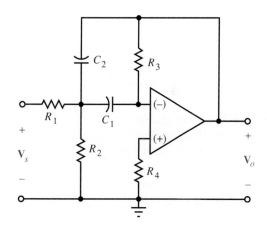

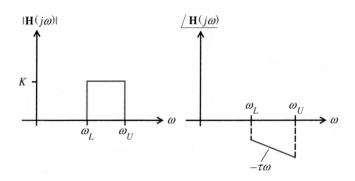

FIGURE 13.57

52. (a) Explain technically, but in your own words, why an *RLC* band-pass filter that has a small damping ratio $\zeta = \alpha/\omega_0$ must have a highly oscillatory natural response. Use mathematics and plots to support your explanation.
   (b) Find a formula that gives the half-power bandwidth of the band-pass filter in terms of the damping ratio and the pass-band "center frequency" ω_0.

Section 13.6

53. Compare the meanings of unity-power-factor resonance and maximum power transfer.
54. For the circuit shown,
   (a) Determine the unity-power-factor resonance frequency.
   (b) Determine the input impedance at unity-power-factor resonance.
   (c) Show that $W_{L,\text{ave}} = W_{C,\text{ave}}$ at the frequency determined in (a).

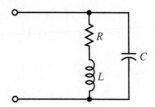

55. For the circuit shown,
   (a) Use the fact that $W_{L,\text{ave}} = W_{C,\text{ave}}$ to determine the unity-power-factor resonance frequency.
   (b) Verify your answer to (a) by showing that $\mathbf{V}/\mathbf{I}$ is purely real when $\omega = \omega_{upf}$.

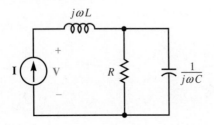

56. Find the unity-power-factor resonance frequency for the circuit shown, where $L = 1$ mH, $C = 1000$ μF, $R = 100$ Ω, and $r = 0.5$ Ω.

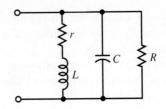

57. For the circuit shown,
   (a) Determine the expression for the unity-power-factor resonance frequency.
   (b) Determine the expression for the driving-point impedance at the unity-power-factor resonance frequency.

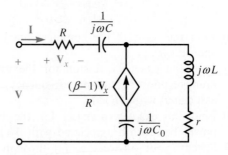

Section 13.7

58. The circuit shown below is a model of a capacitor. Determine its Q.

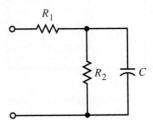

59. In the two circuits shown below, the reactances X_s and X_p are due to either an inductance or a capacitance.

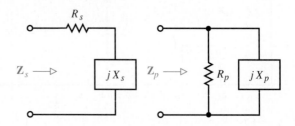

(a) Suppose we want to design the parallel circuit so that it has the same impedance as the series circuit at a fixed angular frequency ω. Show that $\mathbf{Z}_p$ will equal $\mathbf{Z}_s$ if we use the following design equations:

$$R_p = (1 + Q_s^2)R_s \quad \text{and} \quad X_p = \frac{1 + Q_s^2}{Q_s^2}X_s$$

where $Q_s = |X_s|/R_s$ is the quality factor of the series circuit.

(b) Assume instead that we want to design the series circuit so that it has the same impedance as the parallel circuit at a fixed frequency ω. Show that $\mathbf{Z}_s$ will equal $\mathbf{Z}_p$ if we use the following design equations:

$$R_s = \frac{1}{1 + Q_p^2}R_p \quad \text{and} \quad X_s = \frac{Q_p^2}{1 + Q_p^2}X_p$$

where $Q_p = R_p/|X_p|$ is the quality factor of the parallel circuit.

(c) Compare (a) with (b), to show that if $\mathbf{Z}_s$ and $\mathbf{Z}_p$ are equal, then $Q_s = Q_p$.

(d) Use the results of (a) through (c) to find the series RC circuit that has the same impedance at $\omega = 1$ krad/s as a parallel RC circuit in which $R_p = 100$ kΩ and $C_p = 1$ μF. Evaluate $\mathbf{Z}_p$, $\mathbf{Z}_s$, Q_p, and Q_s.

(e) Find the parallel RL circuit that has the same impedance at $\omega = 20$ Mrad/s as a series RL circuit in which $R_s = 100$ Ω, $L = 100$ μH. Evaluate $\mathbf{Z}_p$, $\mathbf{Z}_s$, Q_p, and Q_s.

Section 13.8

60. Use approximations similar to those in Fig. 13.32 to explain why the maximum-response resonance frequencies and the half-power bandwidths associated with $\mathbf{H}_R(j\omega)$, $\mathbf{H}_C(j\omega)$, and $\mathbf{H}_L(j\omega)$ of Fig. 13.34 are nearly equal for $\zeta \ll 1$.

61. The amplitude response of an unknown circuit has a peak of 100 at 1 Mrad/s and equals 10 at 200 krad/s. Assuming that the peak is due to an isolated simple pole, sketch the (approximate) amplitude and phase characteristics.

62. Measurements of the amplitude response of a certain LLTI system reveal that $|\mathbf{H}(j100)| = 316$, $|\mathbf{H}(j110)| = 447$, and $|\mathbf{H}(j120)| = 707$. Assuming that the response for these frequencies is caused primarily by a simple isolated pole, estimate:
    (a) The maximum response frequency
    (b) The peak value of $|\mathbf{H}|$

Section 13.9

63. The Bode straight-line amplitude plot asymptotes associated with a certain circuit are shown in Fig. 13.58.
    (a) Determine the poles of the circuit and their order.
    (b) Determine the zeros of the circuit and their order.
    (c) What is the transfer function of the circuit?
    (d) Are the solutions to (a) through (c) necessarily unique? That is, is a circuit's transfer

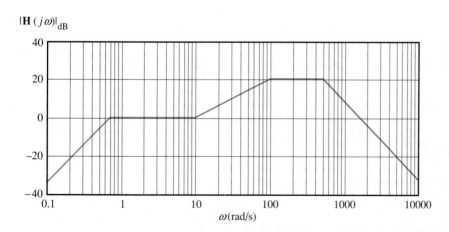

FIGURE 13.58

function completely specified by the associated straight-line amplitude plot? Explain.

64. The circuit shown in Fig. 13.59 represents three identical RC low-pass filters separated by buffer amplifiers.
    (a) Show that the voltage transfer function of the circuit is

$$\frac{V_o}{V_s} = H(s) = H_1^3(s)$$

    where

$$H_1(s) = \frac{1}{1 + sRC}$$

    (b) Draw the Bode amplitude and phase plots associated with $H(s)$.

65. (a) Determine the transfer function $H(s) = V_o/V_s$ for the circuit in Fig. 13.60.
    (b) Draw the Bode amplitude and phase plots for $R = 1\ k\Omega$ and $C = 1\ \mu F$.

66. In Example 13.7, we assumed that the op-amp open-loop gain was infinite, $A = \infty$. A practical op-amp open-loop gain will be finite. Redraw the Bode amplitude and phase plots for the amplifier of Example 13.7 when $A = 10,000$.

67. Part (a) of Fig. 13.61 shows a model of a single-stage amplifier circuit, where $g \gg 1$, $R_o \ll R_L$, and C represents a small unavoidable (parasitic) capacitance existing between output and input.
    (a) Derive the exact formula for $H(j\omega) = V_o/V_1$.
    (b) Show that $H(j\omega)$ is approximately constant, $H(j\omega) \simeq -g_0$, for $0 \le \omega \le 1/RC$, where $R = R_L \| R_o$.
    (c) Show that the input admittance of the amplifier is given by $I_1/V_1 = Y_{in} = G_1 + [1 - H(j\omega)]j\omega C$ for all ω, where $G_1 = 1/R_1$.
    (d) Using the results from (b) and (c), conclude that $Y_{in} \simeq G_1 + j\omega C_M$ over the useful bandwidth of the amplifier, where $C_M = (1 + g)C$. Since $Y_{in} \simeq G_1 + j\omega C_M$, C_M is, in effect, in parallel with R_1. Therefore C_M reduces the input impedance of the amplifier. C_M is much larger than C because $g \gg 1$. This is the *Miller effect*. The Miller effect decreases the transmission bandwidth when the amplifier is driven by a voltage source with internal resistance R_s.
    (e) The approximate amplifier model shown in (b) of the illustration is valid for $0 \le \omega \le 1/RC$, where $R = R_o \| R_L$. Use this approximate model to find V_o/V_s. Assume that $R_s = 1\ k\Omega$, $R_1 = 49\ k\Omega$, $C = 100\ pF$, $g =$

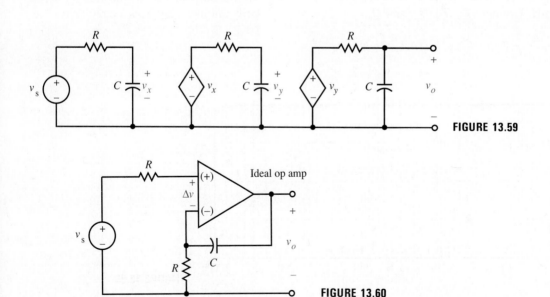

FIGURE 13.59

FIGURE 13.60

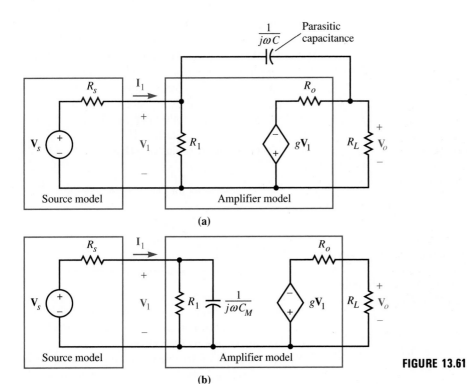

FIGURE 13.61

(a)

(b)

10^4, $R_o = 10\ \Omega$, and $R_L = 1\ \text{k}\Omega$. Plot the Bode plots for $|\mathbf{V}_o/\mathbf{V}_s|$ and $/\mathbf{V}_o/\mathbf{V}_s$ for $0 \le \omega \le 1/RC$. Comment on the result.

Section 13.10

For Problems 68 through 77 use a computer to generate plots. Use the .PROBE command if your program is PSpice. *Experiment* with frequency range and number of points to obtain best results. Use Bode amplitude and frequency scales if indicated by your instructor.

68. Plot the frequency response of the *RL* high-pass filter of Fig. 13.3.

69. Plot the frequency response of the *RLC* circuit of Fig. 13.6.

70. Plot the frequency response, $\mathbf{Z}(j\omega)$, of the circuit of Problem 1.

71. Plot the frequency response, $\mathbf{Z}(j\omega)$, of the circuit of Problem 24.

72. Use PSpice to generate a Bode plot of $\mathbf{H}(s) = \mathbf{V}_{cd}/\mathbf{V}_{ab}$ of the "lag network" shown. Then state the reason for the name.

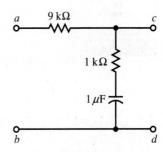

73. Repeat Problem 72 for the "lead network" shown.

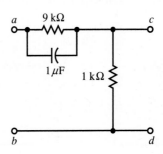

74. Repeat Problem 72 for the "lead-lag network" shown.

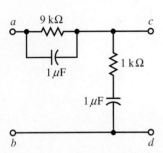

75. For fixed angular frequency ω, a circuit's transfer function, $\mathbf{H}(j\omega)$, is a complex number that can be plotted as a point in the complex plane. By varying ω, we can plot several such points in the complex plane and identify each point by the associated value of ω. Use PSpice to develop a plot of $\mathbf{H}(j\omega) = |\mathbf{H}(j\omega)|\ \underline{/\mathbf{H}(j\omega)}$ in the complex plane for the lag network of Problem 72. Label the values of ω that correspond to several points on the curve.

76. Repeat Problem 75 for the lead network of Problem 73.

77. Repeat Problem 75 for the lead-lag network of Problem 74.

Comprehensive Problems

78. (a) Determine the transfer function $\mathbf{H}_1(s) = \mathbf{V}_2/\mathbf{V}_1$ for circuit (a) in Fig. 13.62, if $R_1 = 1\ \mathrm{k\Omega}$ and $C_1 = 1\ \mu\mathrm{F}$.
    (b) Determine the transfer function $\mathbf{H}_2(s) = \mathbf{V}_2/\mathbf{V}_1$ and $\mathbf{H}_3(s) = \mathbf{V}_3/\mathbf{V}_2$ for circuit (b) in the illustration, for $R_1 = R_2 = 1\ \mathrm{k\Omega}$ and $C_1 = C_2 = 1\ \mu\mathrm{F}$.
    (c) Does $\mathbf{H}_2(s) = \mathbf{H}_1(s)$? Explain.
    (d) Does $\mathbf{H}_3(s) = \mathbf{H}_1^2(s)$? Explain.

79. Use PSpice and .PROBE to plot the amplitude response in decibels and the phase response for $\mathbf{H}_1(s)$, $\mathbf{H}_2(s)$, and $\mathbf{H}_3(s)$ found in Problem 78.

80. Refer to Problem 78. Alter the value of resistance R_2 and capacitance C_2 so that $\mathbf{H}_2(s) \simeq \mathbf{H}_1(s)$ and $\mathbf{H}_3(s) \simeq \mathbf{H}_1^2(s)$. Use PSpice to plot the amplitude response in decibels and phase response to check your design.

81. (a) Determine the transfer function $\mathbf{H}(s) = \mathbf{V}_x/\mathbf{V}$ for the following circuit.
    (b) Give the formula for $|\mathbf{H}(j\omega)|$.
    (c) Give the formula for $\underline{/\mathbf{H}(j\omega)}$.

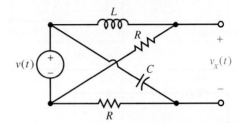

82. In general, what must be the relationship between the components, R, L, and C in Problem 81 for $|\mathbf{H}(j\omega)|$ to be constant?

83. The circuit of Problem 81 can be used as a *crossover network* for an audio system in which the resistance R connected to the inductance is a model for the resistance of a low-range speaker (woofer), and the resistance R connected to the capacitance models the resistance of the midrange speaker. Assuming that $R = 4\ \Omega$, determine inductance and capacitance values so that the sum of the average powers delivered to the two speakers is constant, and the average power delivered to each speaker is the same at $\omega = 1000\ \mathrm{rad/s}$.

Problems 84 through 89 refer to the following circuit.

84. Determine the transfer function $\mathbf{H}(s) = \mathbf{V}_2/\mathbf{V}_1$.
85. If $R_1 = 10\ \Omega$, $R_2 = 20\ \Omega$, and $R_3 = 5\ \Omega$, what is the value of
    (a) $\mathbf{H}(0)$?
    (b) $\mathbf{H}(j\infty)$?

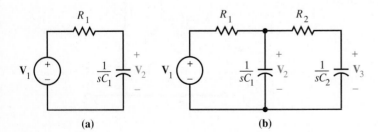

(a)   (b)   **FIGURE 13.62**

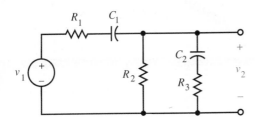

86. The values of resistance are as given in Problem 85, and $C_1 = 10\mu F$ and $C_2 = 0.1\ \mu F$. Use PSpice and .PROBE to make a Bode plot for a range of frequencies that clearly shows how the transfer function makes the transition from the value $\mathbf{H}(0)$ to $\mathbf{H}(j\infty)$.

87. Make a low-frequency equivalent circuit for the network with the assumption that C_2 is an open circuit. Use PSpice to make a Bode plot for this equivalent circuit. Use the same range of frequencies used in Problem 86.

88. Make a high-frequency equivalent circuit for the network with the assumption that C_1 is a short circuit. Use PSpice to make a Bode plot for this equivalent circuit. Use the same range of frequencies used in Problem 86.

89. Combine the Bode plots obtained in Problems 87 and 88 by using the low-frequency portion of Problem 87 and the high-frequency portion of Problem 88. How does the combined plot compare with the exact result obtained in Problem 86?

90. For the circuit shown,

(a) Use PSpice to plot the magnitude and phase response $\mathbf{H}(s) = \mathbf{V}_{ab}/\mathbf{V}_s$ if $R = 1000\ \Omega$ and $C = 1\ \mu F$.
(b) What is the frequency at which there is zero phase shift between the input and output voltages?
(c) What is the magnitude of the transfer function at this frequency?

91. Repeat Problem 90 without the aid of PSpice.

92. The circuit shown illustrates a *scope probe attenuator* and oscilloscope input circuit model. $v_2(t)$ is the voltage that will be amplified and appear on the oscilloscope screen, and $v_1(t)$ is the voltage input to the probe. Capacitance C_2 models the input capacitance of the oscilloscope and coaxial cable, and R_2 is the input resistance of the oscilloscope.

(a) Determine the transfer function $\mathbf{H}(s) = \mathbf{V}_2/\mathbf{V}_1$.
(b) If R_1 is αR_2, what is the value of $\mathbf{H}(0)$? (Typically $\alpha = 9$.)
(c) What value must C_1 have so that $\mathbf{H}(s)$ is a constant?
(d) What equivalent capacitance and parallel resistance is seen by the source for the condition in (c)?
(e) Do you see why an attenuator probe is often used instead of a simple coaxial cable? Explain.

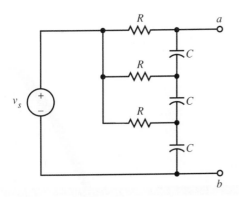

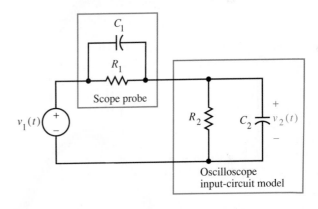

93. Find the quality of factor $Q(\omega)$ of a series RLC circuit for an arbitrary angular frequency ω. Plot $Q(\omega)$ versus ω.

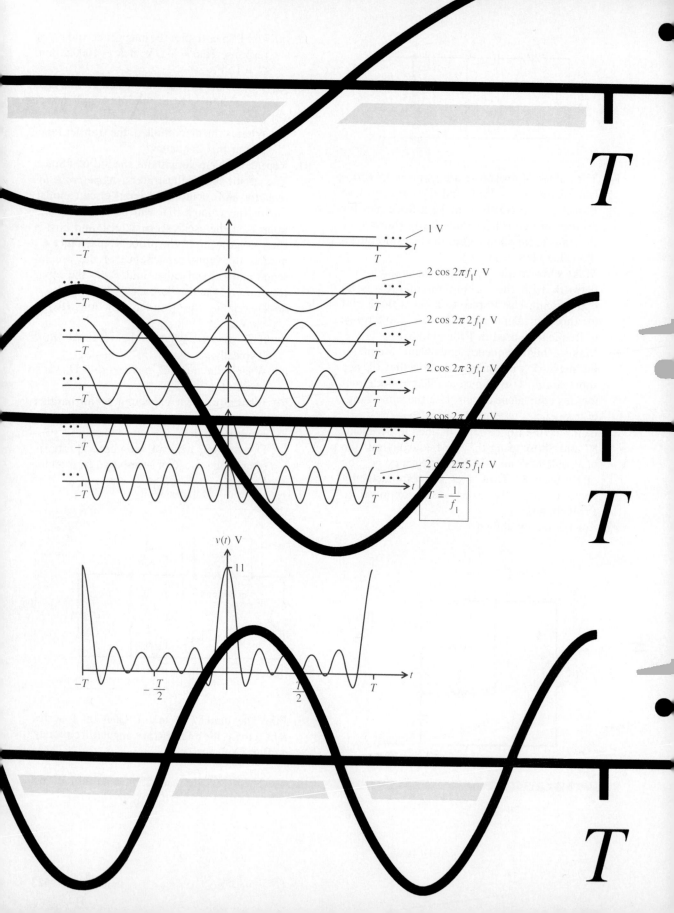

SERIES AND TRANSFORM METHODS

In Part Four, we will focus our attention on LLTI circuit response to *nonsinusoidal* inputs. Our analysis will be based on the insight of a nineteenth-century mathematician, physicist, and statesman, Jean Baptiste Joseph Fourier. Fourier's insight was that any physical signal can be represented as either a sum or an integral of sinusoidal signals. If we represent a nonsinusoidal input as a sum or an integral of sinusoids, we can use ac circuit analysis and superposition to find the response.

CHAPTER
14

Fourier Series

In this chapter we will describe the Fourier series and show how to use it in circuit analysis. In the Fourier series, a periodic function with period T is represented by a sum of sinusoidal functions whose frequencies are multiples of $1/T$. We use the Fourier series to help us find the response of a circuit to a periodic input waveform. The response of an LLTI circuit to a periodic input waveform is found by adding the circuit's responses to the individual sinusoidal terms in the Fourier-series representation of the input.

The essential advantage of a Fourier series is that it represents a periodic input waveform by a sum of signals, *each of which is simpler to work with than the input periodic waveform*. The method is analogous to the technique used by physicists in which a force vector is decomposed into a vector sum of perpendicular force-vector components to simplify a problem in mechanics. In a Fourier series, the input waveform is decomposed into a sum of mutually orthogonal sinusoidal-waveform components. We shall see that the mutual orthogonality of the component waveforms simplifies the problem of finding the series representation of the input waveform.

14.1 Outline of the Fourier-Series Method

⬇ The .FOUR command; see Section B.2.4 of the PSpice manual.

We will now outline the basic strategy used in determining LLTI response to a periodic input. For illustration, we will assume that the triangular wave of Fig. 14.1 is applied to an LLTI circuit as shown in Fig. 14.2. The problem is to determine the complete response $v_o(t)$. The solution is easy with use of the Fourier series. The first step is to represent the input triangular wave by its Fourier series, which is a sum of sinusoidal terms. This is the *Fourier analysis* step. The Fourier series for the triangular wave shown in Fig. 14.1 is given in Fig. 14.2. Later sections will provide formulas to obtain this and other Fourier series. The second step is to determine the particular response of the LLTI circuit to the *individual* sinusoidal terms shown in Fig. 14.2. This is an easy step because we can use ac circuit analysis. The third and final step is to apply superposition. Because the circuit is *linear*, the particular response to the triangular wave—a sum of sinusoids—is the sum of the particular responses to each sinusoid acting alone. Step three is the *Fourier synthesis* step, in which we construct the output waveform by adding together its sinusoidal terms.

FIGURE 14.1
Triangular wave. According to Fourier, this, and every other periodic signal, can be represented by a sum of sinusoidal components

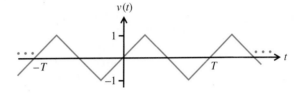

FIGURE 14.2
(a) Triangular wave input; (b) LLTI circuit. Once it is recognized that $v_s(t)$ is equivalent to a sum of sinusoidal components, the output $v_o(t)$ can be determined with the use of ac circuit analysis and superposition

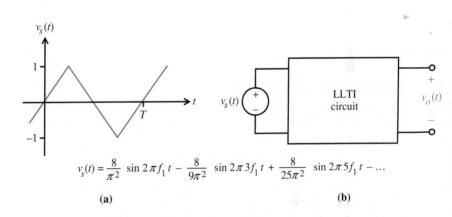

$$v_s(t) = \frac{8}{\pi^2}\sin 2\pi f_1 t - \frac{8}{9\pi^2}\sin 2\pi 3f_1 t + \frac{8}{25\pi^2}\sin 2\pi 5f_1 t - \dots$$

(a)   **(b)**

We can see from the preceding discussion that the circuit response $v_o(t)$ to a periodic input will be represented analytically by an infinite series (sum) of sinusoids—just as the input was represented by an infinite series (sum) of sinusoids. This is a general property of the method of Fourier series: The solution is usually described by an infinite series and *not* by a closed form. As you gain experience in circuit analysis, you may well regard this as a strength—not a weakness—of the method. Very often a closed-form solution is not pertinent, and a series solution provides exactly the kind of information you want. In any case, it is not difficult to program a computer to add the first hundred or so sinusoidal components of a Fourier series to obtain computer-generated plots approximating $v_o(t)$. Examples of computer-generated plots are given in Sections 14.8 and 14.10.

1. The triangular voltage waveform of Fig. 14.1 can be represented as the infinite sum of sinusoids:

$$v(t) = \frac{8}{\pi^2} \sin 2\pi f_1 t - \frac{8}{9\pi^2} \sin 2\pi 3 f_1 t + \frac{8}{25\pi^2} \sin 2\pi 5 f_1 t - \cdots$$

$$= \sum_{n=1,3,5,\ldots}^{\infty} (-1)^{(n-1)/2} \frac{8}{(\pi n)^2} \sin 2\pi n f_1 t$$

   (a) Calculate and plot the sum of the first three nonzero terms of the above formula,

$$v_5(t) = \frac{8}{\pi^2} \sin 2\pi f_1 t - \frac{8}{9\pi^2} \sin 2\pi 3 f_1 t + \frac{8}{25\pi^2} \sin 2\pi 5 f_1 t$$

   where $f_1 = 1$ Hz. To make the plot, calculate and plot $v_5(t)$ for $t = 0, 0.05$, $0.10, 0.15, \ldots, 0.95, 1.0$. Then connect your plotted points with a smooth curve. Compare your result with the triangular wave of Fig. 14.1. Notice that $v_5(t)$ is approximately equal to the triangular wave.
   (b) Suppose that $v_5(t)$ is applied to a series RL circuit where $R = 1\,\Omega$, $L = 2$ H, and $f_1 = 1$ Hz. Determine the particular response voltage across the inductance by adding the particular responses due to the individual terms on the right-hand side of the above equation. Express the solution as a sum of three sinusoids.
   (c) Plot your answer to (b).

2. Use a calculator or PSpice to help you sum and plot the first seven nonzero terms of the summation in Exercise 1. Compare your result with Fig. 14.1.

14.2 Interference Patterns

To understand Fourier series, it is helpful to first understand the phenomenon of interference. Interference occurs whenever sinusoidal waveforms with different frequencies are added. The result of interference is always a nonsinusoidal waveform. As an illustration, consider the sum

$$v(t) = 1 + 2 \cos 2\pi f_1 t + 2 \cos 2\pi 2 f_1 t + \cdots + 2 \cos 2\pi 5 f_1 t \qquad (14.1)$$

We will call the individual sinusoidal terms in this sum *components*. Notice that the components have frequencies $f_0 = 0/T$ (dc), $f_1 = 1/T$, $f_2 = 2/T, \ldots, f_5 = 5/T$ that are multiples of $1/T$. The components are plotted individually in Fig. 14.3a. The sum of the components, $v(t)$, is plotted in Fig. 14.3b.

   We see from Fig. 14.3b that $v(t)$ is periodic with period T. This periodicity may also be seen from the trigonometry:

$$v(t + T) = 1 + 2 \cos 2\pi f_1 (t + T) + 2 \cos 2\pi 2 f_1 (t + T) + \cdots$$
$$+ 2 \cos 2\pi 5 f_1 (t + T)$$
$$= 1 + 2 \cos 2\pi f_1 t + 2 \cos 2\pi 2 f_1 t + \cdots + 2 \cos 2\pi 5 f_1 t$$
$$= v(t) \qquad (14.2)$$

FIGURE 14.3
A typical interference
pattern and its sinusoidal
components. An interference
pattern is produced when
two or more sinusoids with
different frequencies
are added. (*a*) Sinusoidal
components; (*b*) interference
pattern

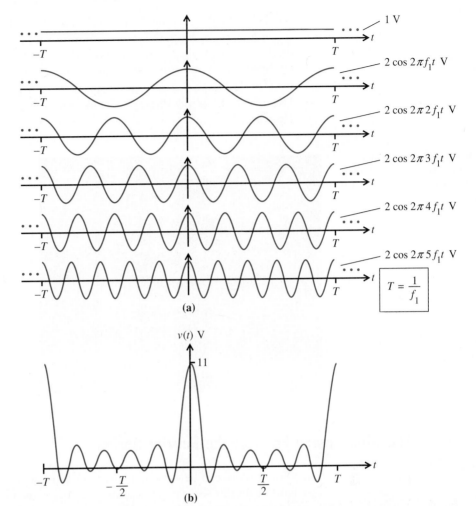

(a)

(b)

We also see from Fig. 14.3b that $v(t)$ is decidedly *not* sinusoidal. Large values of $v(t)$ are produced for t in the vicinity of kT, $k = 0, \pm 1, \pm 2, \ldots$, where many or all of the numerical values of the components have the same sign, thus reinforcing one another in the sum. This is called *constructive interference*. Smaller values of $v(t)$ are produced for t not in the vicinity of kT, $k = 0, \pm 1, \pm 2, \ldots$, where many or all of the numerical values of the components do not have the same sign, thus more or less canceling one another in the sum. This is called *destructive interference*.

Therefore the phenomenon of interference causes the waveform $v(t)$ to be non-sinusoidal. It was Fourier who first asserted that *every* periodic waveform can be obtained as a result of constructive and destructive interference of sinusoidal components. He claimed, for example, that we can regard the periodic waveform in the photograph on page 630 as an interference pattern.

EXERCISES

3. (a) The voltage $v(t)$ plotted in Fig. 14.3b is applied to a series *RC* circuit. Determine the particular response voltage across the capacitor by adding

the particular responses due to the individual components in $v(t)$ shown in Fig. 14.3a. Express the solution as a sum of six terms.

(b) Explain why your answer to (a) is the complete response.

(c) Specialize your solution to the case $f_1 = 1$ kHz, $R = 1$ kΩ, and $C = 2\pi\ \mu$F.

4. Use a calculator or PSpice to help you sum and plot your solution to Exercise 3.

Oscilloscope trace of a periodic waveform. Courtesy of John Fluke Mfg. Co., Inc.

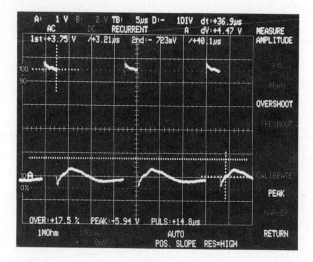

14.3 The Sine-Cosine Form of the Fourier Series

The Fourier series can be written in any of three basic forms, all of which are used by engineers. In this section, we describe the sine-cosine form of the Fourier series. Let us now consider the problem of representing an arbitrary real periodic waveform $v(t)$ by a sum of sinusoidal terms or *components* as follows:

DEFINITION

Sine-Cosine Form of Fourier Series

$$v(t) = a_0 + a_1 \cos 2\pi f_1 t + \cdots + a_n \cos 2\pi n f_1 t + \cdots + b_1 \sin 2\pi f_1 t + \cdots$$
$$+ b_n \sin 2\pi n f_1 t + \cdots$$

$$= a_0 + \sum_{n=1}^{\infty} (a_n \cos 2\pi n f_1 t + b_n \sin 2\pi n f_1 t) \tag{14.3}$$

---

where $f_1 = 1/T$ and T is the period of $v(t)$. We now ask, "If we *can* express $v(t)$ as in Eq. (14.3), how *must* the coefficients a_0 and a_n, b_n, $n = 1, 2, 3, \ldots$, be related to the function $v(t)$?"

Part of the answer to this question follows from the observation that a_0 must be the average value of $v(t)$. This is because the average values of all the other terms on the right-hand side are zero. Therefore,

$$a_0 = \frac{1}{T} \int_{t_0}^{t_0+T} v(t)\, dt \qquad (14.4)$$

where t_0 is arbitrary. [If you are not convinced, substitute the right-hand side of Eq. (14.3) into Eq. (14.4) and integrate. The result will be a_0.]

A formula for a_n ($n \geq 1$) can be obtained if we multiply both sides of Eq. (14.3) by $\cos 2\pi n f_1 t$ and integrate over an interval of duration T:

$$\int_{t_0}^{t_0+T} v(t) \cos 2\pi n f_1 t\, dt = \int_{t_0}^{t_0+T} a_0 \cos 2\pi n f_1 t\, dt$$

$$+ \sum_{k=1}^{\infty} a_k \int_{t_0}^{t_0+T} \cos 2\pi k f_1 t \cos 2\pi n f_1 t\, dt$$

$$+ \sum_{k=1}^{\infty} b_k \int_{t_0}^{t_0+T} \sin 2\pi k f_1 t \cos 2\pi n f_1 t\, dt \qquad (14.5)$$

Observe that the first integral on the right-hand side of Eq. (14.5) is zero because the area under an integer number of cycles of a cosine function is zero. Also observe that each of the remaining integrals on the right-hand side *except* the one involving the $\cos^2 2\pi n f_1 t$ (this is the term for $k = n$ in the first summation) is also zero. It is easy to see why. Because

$$\cos 2\pi k f_1 t \cos 2\pi n f_1 t = \frac{1}{2} \cos 2\pi (k+n) f_1 t + \frac{1}{2} \cos 2\pi (k-n) f_1 t$$

the integral

$$\int_{t_0}^{t_0+T} \cos 2\pi k f_1 t \cos 2\pi n f_1 t\, dt$$

is equal to one-half the area under $k + n$ cycles of a cosine function plus one-half the area under $k - n$ cycles of a cosine function. Both of these are zero if $k \neq n$.

Similar remarks apply to the integrals involving the products

$$\sin 2\pi k f_1 t \cos 2\pi n f_1 t$$

Because

$$\sin 2\pi k f_1 t \cos 2\pi n f_1 t = \frac{1}{2} \sin 2\pi (k+n) f_1 t + \frac{1}{2} \sin 2\pi (k-n) f_1 t$$

the integral

$$\int_{t_0}^{t_0+T} \sin 2\pi k f_1 t \cos 2\pi n f_1 t\, dt$$

equals zero for all integer values of k. Since all but one integral on the right-hand side of Eq. (14.5) equal zero, Eq. (14.5) becomes simply

$$\int_{t_0}^{t_0+T} v(t) \cos 2\pi n f_1 t\, dt = a_n \int_{t_0}^{t_0+T} \cos^2 2\pi n f_1 t\, dt$$

$$= a_n \frac{T}{2} \qquad (14.6)$$

and this yields

$$a_n = \frac{2}{T} \int_{t_0}^{t_0+T} v(t) \cos 2\pi n f_1 t\, dt \qquad \text{for } n = 1, 2, \ldots \qquad (14.7)$$

The coefficient b_n can be obtained similarly if we multiply both sides of Eq. (14.3) by $\sin 2\pi n f_1 t$ and integrate over an interval T. All the integrals on the right-hand side except the one involving $\sin^2 2\pi n f_1 t$ will be zero. This leads to

$$b_n = \frac{2}{T} \int_{t_0}^{t_0 + T} v(t) \sin 2\pi n f_1 t \, dt \qquad \text{for } n = 1, 2, \ldots \qquad (14.8)$$

The a_n and b_n uniquely determined by Eqs. (14.4), (14.7), and (14.8) are the *real Fourier coefficients* of $v(t)$. Their units are the same as those of $v(t)$. With a_n and b_n so determined, the right-hand side of Eq. (14.3) is called the *sine-cosine* form of the *Fourier series* of $v(t)$.

Notice that the formulas for the Fourier coefficients given by Eq. (14.4), (14.7), and (14.8) are *extremely simple. Each* coefficient is computed directly from $v(t)$ *independently* of the other coefficients. This independence is a consequence of the fact that the terms on the right-hand side of Eq. (14.3) are *orthogonal* over an interval T. This means that the terms satisfy Eq. (7.65), which we repeat here:

$$\int_{t_0}^{t_0 + T} x_1(t) x_2(t) \, dt = 0 \qquad (14.9)$$

where $x_1(t)$ and $x_2(t)$ denote any two distinct terms (a_0, $a_k \cos 2\pi k f_1 t$, $b_k \sin 2\pi k f_1 t$) on the right-hand side of Eq. (14.3). As a consequence of Eq. (14.9), all but one integral on the right-hand side of Eq. (14.5) equaled zero—a significant simplification.

We use Eqs. (14.4), (14.7), and (14.8) to compute the values of the Fourier coefficients of $v(t)$. Once we have computed the values, we substitute them into Eq. (14.3) to obtain the Fourier-series representation of $v(t)$.

Parseval's Theorem for Real Periodic Waveforms

Recall from Chapter 7 that the mean square value of a waveform is numerically equal to the normalized power in the waveform. The normalized power is the average power the waveform dissipates when applied to a 1-Ω resistance. Parseval's theorem shows us that the normalized power in a periodic waveform can be expressed directly in terms of the real Fourier coefficients. We can derive Parseval's theorem by writing the normalized power in $v(t)$ as

$$\frac{1}{T} \int_{t_0}^{t_0 + T} v^2(t) \, dt = \frac{1}{T} \int_{t_0}^{t_0 + T} \left(a_0 + \sum_{n=1}^{\infty} a_n \cos 2\pi n f_1 t + \sum_{n=1}^{\infty} b_n \sin 2\pi n f_1 t \right)^2 dt$$

$$(14.10)$$

When we square the quantity in parentheses on the right-hand side of Eq. (14.10) as indicated, we obtain many cross products. All the cross products integrate to zero because the components in a Fourier series are orthogonal. The integrals of the non-cross-product terms are easily obtained. The result is

Parseval's Theorem for Real Periodic Waveforms

$$\frac{1}{T} \int_{t_0}^{t_0 + T} v^2(t) \, dt = a_0^2 + \sum_{n=1}^{\infty} \frac{1}{2} a_n^2 + \sum_{n=1}^{\infty} \frac{1}{2} b_n^2 \qquad (14.11)$$

It is easily shown that the quantity a_0^2 in Eq. (14.11) is the normalized power in the dc component a_0, $\frac{1}{2}a_n^2$ is the normalized power in the component $a_n \cos 2\pi n f_1 t$, and $\frac{1}{2}b_n^2$ is the normalized power in $b_n \sin 2\pi n f_1 t$. Therefore Parseval's theorem tells us that *the total normalized power in v(t) equals the sum of the normalized powers in each of the Fourier-series components of v(t)*. This means that if $v(t)$ were applied across a resistance, then the total average power supplied would equal the sum of the average powers dissipated by the individual Fourier-series components.

REMEMBER

A real periodic waveform can be represented by the sine-cosine form of the Fourier series given by Eq. (14.3). The real Fourier coefficients are given by Eqs. (14.4), (14.7), and (14.8).

EXERCISES

5. Find the period and the real Fourier-series coefficients of the periodic waveform $v(t) = 3 + 5 \cos 2\pi 10^3 t + 7 \sin 6\pi 10^3 t$.

6. A *periodic impulse train* is defined as

$$v(t) = \sum_{n=-\infty}^{+\infty} \delta(t - nT)$$

   (a) Show that $v(t)$ consists of unit impulses that occur periodically every T seconds.
   (b) Sketch $v(t)$.
   (c) Use Eqs. (14.4), (14.7), and (14.8) to determine the real Fourier coefficients of $v(t)$.
   (d) Write the real Fourier series of $v(t)$, and compare your answer with Eq. (14.1).

14.4 Symmetry Properties

We now investigate the relationship between a waveform's symmetry and its Fourier series. The three symmetry properties described below are important because they tell us when certain Fourier coefficients equal zero *without the need to evaluate the integrals* in Eqs. (14.4), (14.7), and (14.8).

Symmetry Property 1

If $v(t)$ is an even function, that is, if $v(-t) = v(t)$, then the Fourier series for $v(t)$ contains only even component functions.

The even component functions are the dc term a_0 and the cosine terms $a_n \cos 2\pi n f_1 t$; $n = 1, 2, \ldots$. Therefore, if $v(t)$ is an even function, then $b_n = 0$ for $n = 1, 2, 3, \ldots$. An example of an even function is the trapezoidal pulse train of Fig. 14.4.

FIGURE 14.4
A function with even
symmetry

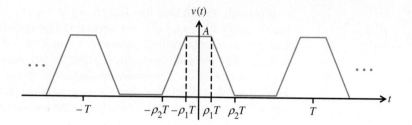

PROOF

Write Eq. (14.3) as

$$v(t) = v_a(t) + v_b(t) \tag{14.12}$$

where

$$v_a(t) = a_0 + \sum_{n=1}^{\infty} a_n \cos 2\pi n f_1 t \tag{14.13}$$

is an even function, and

$$v_b(t) = \sum_{n=1}^{\infty} b_n \sin 2\pi n f_1 t \tag{14.14}$$

is an odd function. We know that $v(t)$ is an even function. Therefore the right-hand side of Eq. (14.12) must be even:

$$v_a(-t) + v_b(-t) = v_a(t) + v_b(t) \tag{14.15a}$$

Since $v_a(t)$ is even and $v_b(t)$ is odd, Eq. (14.15a) can be written as

$$v_a(t) - v_b(t) = v_a(t) + v_b(t) \tag{14.15b}$$

from which follows the result

$$v_b(t) = 0 \tag{14.16}$$

It then follows from Eqs. (14.14) and (14.16) that

$$b_n = \frac{2}{T} \int_{t_0}^{t_0 + T} v_b(t) \sin 2\pi n f_1 t \, dt = 0 \qquad n = 1, 2, \ldots \tag{14.17}$$

Symmetry Property 2

If $v(t)$ is an odd function, $v(-t) = -v(t)$, then its Fourier series contains only odd component functions.

The odd component functions are the sine terms $b_n \sin 2\pi n f_1 t$, $n = 1, 2, \ldots$. Therefore, if $v(t)$ is an odd function, then $a_n = 0$ for $n = 0, 1, 2, \ldots$. An example of an odd function is the sawtooth wave of Fig. 14.5.

Symmetry Property 2 can be proved in a manner similar to the proof of Symmetry Property 1. The details are in Problem 7.

FIGURE 14.5
A function with odd symmetry

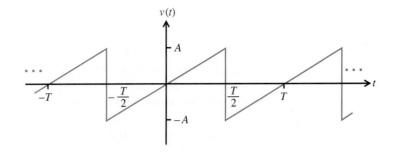

FIGURE 14.6
A function with half-wave symmetry

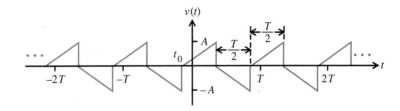

A periodic function satisfying $v(t) = -v(t - T/2)$ is said to be *half-wave symmetric*. An example of a half-wave-symmetric waveform is the double sawtooth wave shown in Fig. 14.6.

Symmetry Property 3

If $v(t)$ is half-wave symmetric, then its Fourier series contains only half-wave-symmetric component functions.

The half-wave-symmetric component functions are $a_n \cos 2\pi n f_1 t$ and $b_n \sin 2\pi n f_1 t$ for $n = 1, 3, 5, \ldots$. Only frequencies that are *odd* multiples of $f_1 = 1/T$ are present in a half-wave-symmetric function. Thus $a_0 = a_n = b_n = 0$ for $n = 2, 4, 6, \ldots$.

See Problem 8 for a proof of Symmetry Property 3. The following example illustrates how the symmetry properties can be used to simplify the problem of finding a Fourier series.

EXAMPLE 14.1

Find the Fourier series of the square wave illustrated in Fig. 14.7.

FIGURE 14.7
A square wave

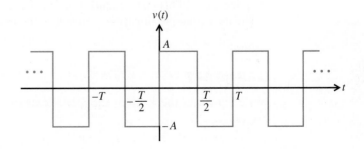

■ *See Problems 14.1
and 14.2 in the PSpice
manual.*

■ **SOLUTION**

We can see by inspection of Fig. 14.7 that $v(t)$ is an odd function. By Symmetry Property 2, $a_n = 0$ for $n = 0, 1, \ldots$. We can also see from Fig. 14.7 that $v(t)$ is half-wave symmetric. By Symmetry Property 3, $a_n = b_n = 0$ for n even. Therefore, it suffices to evaluate Eq. (14.8) for $n = 1, 3, 5, \ldots$. It is convenient here to choose $t_0 = -T/2$. For the square wave of Fig. 14.7, Eq. (14.8) becomes

$$b_n = \frac{2}{T} \int_{-T/2}^{T/2} v(t) \sin 2\pi n f_1 t \, dt$$

$$= -\frac{2}{T} \int_{-T/2}^{0} A \sin 2\pi n f_1 t \, dt + \frac{2}{T} \int_{0}^{T/2} A \sin 2\pi n f_1 t \, dt$$

The change of variable t to $-t$ in the first integral shows that the first integral is a disguised version of the second integral:

$$-\frac{2}{T} \int_{-T/2}^{0} A \sin 2\pi n f_1 t \, dt = \frac{2}{T} \int_{0}^{T/2} A \sin 2\pi n f_1 t \, dt$$

Therefore

$$b_n = \frac{4}{T} \int_{0}^{T/2} A \sin 2\pi n f_1 t \, dt$$

$$= \frac{4A}{2\pi n f_1 T} (1 - \cos \pi n f_1 T)$$

$$= \frac{4A}{n\pi} \qquad n = 1, 3, 5, \ldots$$

The Fourier series of the square wave is therefore

$$v(t) = \frac{4A}{\pi} \sin 2\pi f_1 t + \frac{4A}{3\pi} \sin 2\pi 3 f_1 t + \frac{4A}{5\pi} \sin 2\pi 5 f_1 t + \cdots$$

$$= \sum_{n=1,3,5,\ldots}^{\infty} \frac{4A}{n\pi} \sin 2\pi n f_1 t \quad ■$$

No computer can sum the infinite number of terms appearing in the Fourier-series representation of a waveform. To obtain a computer-generated plot, therefore, a computer is programmed to calculate the *partial* Fourier series of $v(t)$. The partial Fourier series $\hat{v}_N(t)$ is simply the sum of all the terms of the Fourier series up to and including those with frequencies $n f_1$ where $n = N$. Computer plots of the partial Fourier series of the square wave of Example 14.1 are shown in Fig. 14.28. Notice that $\hat{v}_N(t)$ in Fig. 14.28 appears to converge to the square wave of Fig. 14.6 as N increases. This convergence is discussed in more detail in Section 14.10.

The next example illustrates Parseval's theorem.

EXAMPLE 14.2

Write Parseval's theorem for the special case that $v(t)$ is the square wave shown in Fig. 14.7.

The mean square value of the square wave is, by Parseval's theorem and the results of Example 14.1,

$$\frac{1}{T} \int_{-T/2}^{T/2} v^2(t)\, dt = \frac{1}{2}\left(\frac{4A}{\pi}\right)^2 + \frac{1}{2}\left(\frac{4A}{3\pi}\right)^2 + \frac{1}{2}\left(\frac{4A}{5\pi}\right)^2 + \cdots$$

$$= \sum_{n=1,3,5,\ldots}^{\infty} \frac{1}{2}\left(\frac{4A}{n\pi}\right)^2 = A^2$$

Since $v^2(t) = A^2$ for all t (see Fig. 14.7), the evaluation of the integral tells us that the total normalized power in $v(t)$ is A^2. This is the term on the right-hand side of the preceding equation. The series expression shows the normalized powers contained in each of the Fourier components in $v(t)$. Parseval's theorem tells us that the sum of these normalized powers equals A^2. ■

Frequently, electronic circuits may add a dc voltage to a waveform that has either odd or half-wave symmetry. Strictly speaking, this will destroy the symmetry involved. However, the addition of a dc voltage to any periodic waveform changes only the dc coefficient, a_0 of the Fourier series. If you encounter a waveform that *would be* either odd symmetric or half-wave symmetric after the *subtraction* of a dc voltage, all the Fourier coefficients of that waveform except a_0 will satisfy the conditions given in symmetry properties 2 and 3. Exercise 7 illustrates this simple point.

REMEMBER

An even symmetric periodic waveform contains only dc and cosine components. An odd symmetric periodic waveform contains only sine components. A half-wave symmetric waveform contains only frequencies that are odd multiples of $1/T$.

EXERCISES

⊟ *See Problem 14.3 of the PSpice manual.*

7. A dc voltage of A V is added to the odd, half-wave symmetric waveform $v(t)$ of Example 14.1, shown in Fig. 14.7, to produce a new waveform $v'(t) = v(t) + A$.
   (a) Sketch $v'(t)$.
   (b) Does $v'(t)$ have odd symmetry?
   (c) Does $v'(t)$ have half-wave symmetry?
   (d) Add A to the Fourier series for $v(t)$ derived in Example 14.1 to obtain the Fourier series for $v'(t)$.
   (e) Discuss the difference between the Fourier-series coefficients of $v'(t)$ and $v(t)$.

8. State which of the Fourier series of the waveforms shown in the following illustration contain (a) no sine components, (b) no cosine components, and (c) only frequencies that are odd multiples of $f_1 = 1/T$.

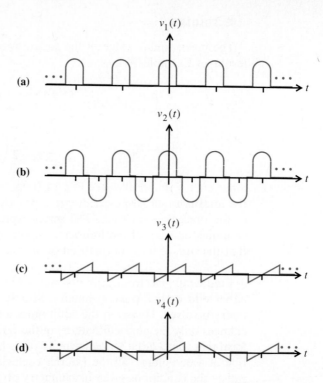

(a)

(b)

(c)

(d)

9. Derive the real Fourier series of the triangular wave of Fig. 14.1.

10. Derive the real Fourier series of the sawtooth wave $v(t)$ of Fig. 14.5.

14.5 The Amplitude-Phase Form of the Fourier Series

The form of the Fourier series described in the previous sections is not the only form used. Another useful form of the Fourier series of a real periodic waveform can be obtained from the identity

$$A_n \cos (2\pi n f_1 t + \phi_n) = A_n \cos \phi_n \cos 2\pi n f_1 t - A_n \sin \phi_n \sin 2\pi n f_1 t \quad (14.18)$$

If we set

$$a_n = A_n \cos \phi_n \qquad n = 1, 2, \ldots \quad (14.19a)$$

$$b_n = -A_n \sin \phi_n \qquad n = 1, 2, \ldots \quad (14.19b)$$

In Eqs. (14.3) and use Eq. (14.18), Eq. (14.3) becomes

DEFINITION

Amplitude-Phase Form of Fourier Series

$$v(t) = A_0 + A_1 \cos (2\pi f_1 t + \phi_1) + \cdots + A_n \cos (2\pi n f_1 t + \phi_n) + \cdots$$

$$= A_0 + \sum_{n=1}^{\infty} A_n \cos (2\pi n f_1 t + \phi_n) \quad (14.20)$$

---

in which

$$A_0 = a_0 \qquad \text{(14.21a)}$$

$$A_n = \sqrt{a_n^2 + b_n^2} \qquad n = 1, 2, \dots \qquad \text{(14.21b)}$$

and

$$\phi_n = \tan^{-1}\left(\frac{-b_n}{a_n}\right) \qquad n = 1, 2, \dots \qquad \text{(14.21c)}$$

Engineers call Eq. (14.20) the *amplitude-phase* form of the Fourier series of $v(t)$. The A_n and ϕ_n are the *Fourier amplitudes* and *phases*. The sinusoidal components in Eq. (14.20) are called *tones* or *harmonics*: $A_1 \cos(2\pi f_1 t + \phi_1)$ is the *fundamental* or *first harmonic*, $A_2 \cos(2\pi 2 f_1 t + \phi_2)$ is the *second harmonic*, and so forth.* Doubling of the frequency raises the harmonic by an *octave*. Thus the second harmonic is one octave above the first, and the fourth harmonic is two octaves above the first.

The terms *octave* and *harmonics* used by engineers are identical to those used by musicians. This common terminology is helpful in analyzing or designing an audio amplifier.

One-Sided Amplitude, Phase, and Power Spectra

A plot of the Fourier amplitudes versus frequency, as illustrated in Fig. 14.8a, is called the *one-sided amplitude spectrum* of $v(t)$. The corresponding plot of the Fourier phases in Fig. 14.8b is called the *one-sided phase spectrum* of $v(t)$. The plots indicate that at each frequency $f = nf_1$, $n = 0, 1, 2, \dots$, there are sinusoidal components in $v(t)$. The amplitude spectrum tells us that the amplitude of the component with frequency nf_1 is A_n (designated by the height of the corresponding line). The phase spectrum tells us that the phase of the component with frequency nf_1 is ϕ_n (again designated by the height of the corresponding line). The amplitude and phase spectra are called *line* spectra because they consist of lines.

FIGURE 14.8
A typical one-sided spectrum:
(a) amplitude; (b) phase

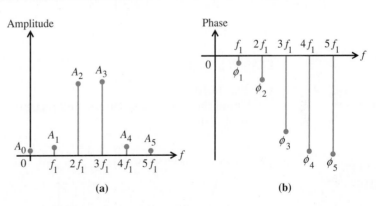

(a) (b)

* The dc term A_0 is occasionally called the zeroth harmonic.

Restatement of Parseval's Theorem for Real Periodic Waveforms

The terms on the right-hand side of Eq. (14.20), like those in Eq. (14.3), are orthogonal over an interval T. Consequently, the normalized power in $v(t)$,

$$\frac{1}{T} \int_{t_0}^{t_0+T} v^2(t)\, dt = \frac{1}{T} \int_{t_0}^{t_0+T} \left[A_0 + \sum_{n=1}^{\infty} A_n \cos\left(2\pi n f_1 t + \phi_n\right)\right]^2 dt \qquad (14.22)$$

equals the sum of the normalized powers in the individual components.

Parseval's Theorem for Real Periodic Waveforms

$$\frac{1}{T} \int_{t_0}^{t_0+T} v^2(t)\, dt = A_0^2 + \sum_{n=1}^{\infty} \frac{1}{2} A_n^2 \qquad (14.23)$$

Equation (14.23) is a restatement of Parseval's theorem, Eq. (14.11). A plot of the normalized powers in the components in Eq. (14.20) versus frequency as illustrated in Fig. 14.9 is called the *one-sided power spectrum* of $v(t)$. The height of the line at $f = nf_1$ in the power spectrum is numerically equal to the average power that would be dissipated if the nth-harmonic voltage were applied across the terminals of a 1-Ω resistance. According to Parseval's theorem, the sum of the heights of all the lines in the power spectrum is numerically equal to the average power that would be dissipated if $v(t)$ were applied across the terminals of a 1-Ω resistance.

FIGURE 14.9
A typical one-sided power spectrum

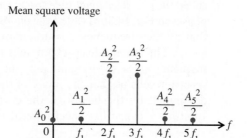

EXAMPLE 14.3

See Problem 14.4 in the PSpice manual.

For Fig. 14.10:
 (a) Find the Fourier amplitudes and phases for the periodic voltage pulse train (where $A > 0$).

FIGURE 14.10
Periodic pulse train

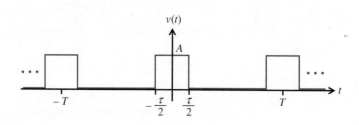

(b) Sketch the one-sided amplitude, phase, and power spectra for $A = 10$ V, $\tau = 0.2$ ms, and $T = 1$ ms.

(c) Assume that the pulse train is applied across the terminals of a 1-Ω resistance. Determine the percentage of power dissipation due to harmonics 0 through 5. Assume A, τ, and T as in (b).

■ SOLUTION

(a) First determine the Fourier coefficients. Since $v(t)$ is even, $b_n = 0$ for $n = 1, 2, \ldots$. Choose $t_0 = -T/2$ for convenience. Then, from Eqs. (14.4) and (14.7),

$$a_0 = \frac{1}{T} \int_{-T/2}^{T/2} v(t)\, dt = \frac{1}{T} \int_{-\tau/2}^{\tau/2} A\, dt = A \frac{\tau}{T}$$

$$a_n = \frac{2}{T} \int_{-T/2}^{T/2} v(t) \cos 2\pi n f_1 t\, dt$$

$$= \frac{2}{T} \int_{-\tau/2}^{\tau/2} A \cos 2\pi n f_1 t\, dt$$

$$= 2A \frac{\tau}{T} \frac{\sin \pi n f_1 \tau}{\pi n f_1 \tau} \qquad \text{for } n = 1, 2, \ldots$$

where $f_1 = 1/T$. Then, by Eq. (14.21a),

$$A_0 = A \frac{\tau}{T}$$

and by Eq. (14.21b),

$$A_n = 2A \frac{\tau}{T} \left| \frac{\sin \pi n f_1 \tau}{\pi n f_1 \tau} \right| \qquad \text{for } n = 1, 2, \ldots$$

Because there are no sine components, the Fourier phases can equal either $0°$ or $\pm 180°$ depending on the sign of the corresponding a_n: If $a_n > 0$, $\phi_n = 0°$, and if $a_n < 0$, $\phi_n = \pm 180°$. We arbitrarily choose the minus sign in front of the $180°$. Then we write

$$\phi_n = \begin{cases} 0 & \dfrac{\sin \pi n f_1 \tau}{\pi n f_1 \tau} > 0 \\[2ex] -180° & \dfrac{\sin \pi n f_1 \tau}{\pi n f_1 \tau} < 0 \end{cases}$$

(b) The amplitude, phase, and power spectra are illustrated in Fig. 14.11 for $A = 10$ V, $\tau = 0.2$ ms, and $T = 1$ ms.

(c) The total average power dissipated in the 1-Ω resistance is

$$P = \frac{1}{T} \int_{-T/2}^{+T/2} v^2(t)\, dt = \frac{\tau}{T} A^2 = 20 \text{ W}$$

The average power dissipation due to the nth harmonic is

$$P_n = \begin{cases} \left(A \dfrac{\tau}{T} \right)^2 & \text{for } n = 0 \\[3ex] \dfrac{1}{2} \left(\dfrac{2A\tau}{T} \dfrac{\sin \pi n f_1 \tau}{\pi n f_1 \tau} \right)^2 & \text{for } n \neq 0 \end{cases}$$

which becomes, for the given values of A, T, and τ,

$$
\begin{array}{ll}
P_0: & 4.00 \text{ W} \\
P_1: & 7.00 \\
P_2: & 4.58 \\
P_3: & 2.04 \\
P_4: & 0.44 \\
P_5: & 0.0 \\
\hline
\text{Total} & 18.06 \text{ W}
\end{array}
$$

Therefore harmonics 0 through 5 account for $100 \times (18.06/20) \simeq 90\%$ of the total average power dissipation. ■

FIGURE 14.11
(*a*) One-sided amplitude, (*b*) phase, and (*c*) power spectra of a periodic pulse train

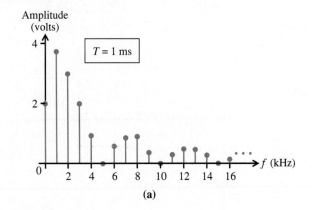

(a)

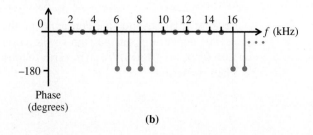

(b)

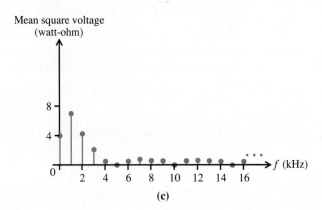

(c)

A real periodic waveform can be represented by the amplitude-phase form of the Fourier series given by Eqs. (14.20) and (14.21). This form of the Fourier series is used to define the one-sided amplitude, phase, and power spectra of the waveform.

11. (a) Sketch the one-sided amplitude, phase, and power spectra of the triangular waveform $v(t)$ of Fig. 14.1. Assume that $T = 1$ ms. [*Hint:* The real Fourier series of $v(t)$ is given in Exercise 1.]
    (b) Assume that $v(t)$ is applied across the terminals of a 1-Ω resistance. What is the average power dissipated?
    (c) Determine the percentage of power dissipation due to harmonics 0 through 4.

14.6 The Complex Form of the Fourier Series

The *complex form* of the Fourier series is a representation of a periodic waveform by a sum of complex exponentials. This form is sometimes more convenient to use than either of the real forms. In addition, it provides us with an analytical stepping stone to the Fourier and Laplace transforms of Chapters 15 and 16.

The complex form of the Fourier series of a real periodic waveform can be obtained directly from the real form of the Fourier series. However, when we discuss the complex form of the Fourier series, it is conventional to generalize to include *complex* periodic waveforms. Therefore, we will make this generalization now. Consider an arbitrary real *or* complex periodic function $v(t)$, and write it in the complex Fourier series:

DEFINITION

Complex Form of Fourier Series

$$v(t) = \sum_{n=-\infty}^{+\infty} V_n e^{j2\pi n f_1 t} \qquad (14.24)$$

---

where $f_1 = 1/T$ and T is the period. The quantities V_n, $n = 0, \pm 1, \pm 2, \ldots$, are phasors.* They are called *complex Fourier coefficients*. Observe that the terms in the summation are rotating phasors. Those corresponding to $n > 0$ rotate counterclockwise, and those corresponding to $n < 0$ rotate clockwise. The sum of the rotating phasors is $v(t)$. A formula for the kth phasor V_k can be obtained if we multiply both sides of Eq. (14.24) by $e^{-j2\pi k f_1 t}$ and integrate from t_0 to $t_0 + T$, where t_0 is arbitrary. This yields

---

* For notational simplicity we will drop the boldface convention for phasors when we write the Fourier series or its relatives, the Fourier and Laplace transforms.

$$\int_{t_0}^{t_0+T} v(t)e^{-j2\pi k f_1 t}\, dt = \int_{t_0}^{t_0+T} \sum_{n=-\infty}^{+\infty} V_n e^{j2\pi n f_1 t} e^{-j2\pi k f_1 t}\, dt$$

$$= \sum_{n=-\infty}^{+\infty} V_n \int_{t_0}^{t_0+T} e^{j2\pi(n-k)f_1 t}\, dt \qquad (14.25)$$

By using Euler's identity, we can write

$$\int_{t_0}^{t_0+T} e^{j2\pi(n-k)f_1 t}\, dt = \int_{t_0}^{t_0+T} \cos 2\pi(n-k)f_1 t\, dt + j \int_{t_0}^{t_0+T} \sin 2\pi(n-k)f_1 t\, dt \qquad (14.26)$$

Note that if $n \neq k$, then each of the two integrals on the right-hand side of Eq. (14.26) equals zero, because the area under an integer number of cycles of any sinusoid is zero. If $n = k$, the second integral on the right-hand side is still zero, because $\sin 0 = 0$. However, since $\cos 0 = 1$, the value of the first integral on the right-hand side is T for $n = k$:

$$\int_{t_0}^{t_0+T} e^{j2\pi(n-k)f_1 t}\, dt = \begin{cases} T & n = k \\ 0 & n \neq k \end{cases} \qquad (14.27)$$

The substitution of Eq. (14.27) into Eq. (14.25) yields

$$\int_{t_0}^{t_0+T} v(t)e^{-j2\pi k f_1 t}\, dt = V_k T \qquad (14.28)$$

where V_k is the kth complex Fourier coefficient. By substituting n for k in Eq. (14.28) and solving for V_n, we obtain

$$V_n = \frac{1}{T} \int_{t_0}^{t_0+T} v(t)e^{-j2\pi n f_1 t}\, dt \qquad n = 0, \pm 1, \pm 2, \ldots \qquad (14.29)$$

Equation (14.29) is the formula for the nth complex Fourier series coefficient, V_n, of an arbitrary real or complex periodic waveform $v(t)$. Special results occur when $v(t)$ is real. These are explained in the following subsection.

Relationship Between Real and Complex Forms of the Fourier Series of a Real Signal

We will now examine the relationship between the real and the complex forms of the Fourier series. With application of Euler's identity, Eq. (14.29) can be written as

$$V_n = \frac{1}{T} \int_{t_0}^{t_0+T} v(t) \cos 2\pi n f_1 t\, dt - j \frac{1}{T} \int_{t_0}^{t_0+T} v(t) \sin 2\pi n f_1 t\, dt \qquad (14.30)$$

for $n = 0, \pm 1, \pm 2, \ldots$. Comparing Eq. (14.30) with Eq. (14.7), (14.8), and (14.4), we see that for *real* $v(t)$,

$$V_n = \frac{1}{2} a_n - j \frac{1}{2} b_n \qquad \text{for } n = 1, 2, \ldots \qquad (14.31a)$$

and

$$V_0 = a_0 \qquad (14.31b)$$

It follows that the *real* Fourier coefficients a_0, a_n, and b_n, $n = 1, 2, \ldots$, of a *real* periodic waveform $v(t)$ can be obtained from the complex Fourier-series coefficients V_n of that waveform with the use of the following formulas:

$$a_n = 2 \mathscr{R}e \left\{ V_n \right\} \qquad n = 1, 2, \ldots$$

$$b_n = -2 \mathscr{I}m \left\{ V_n \right\} \qquad n = 1, 2, \ldots \qquad \text{(14.32)}$$

$$a_0 = V_0$$

The preceding equations are very useful. A convenient feature of Eq. (14.29) is that it replaces three integrals, Eqs. (14.4), (14.7), and (14.8), with one. Thus, to compute the real Fourier-series coefficients of a real periodic waveform $v(t)$, we can now evaluate *one* integral, Eq. (14.29), and then apply Eq. (14.32).

We can also obtain the Fourier amplitudes A_n, $n = 0, 1, 2, \ldots$, and phases $\phi_n = 1, 2, 3, \ldots$, of a real periodic waveform $v(t)$ from the complex Fourier-series coefficients V_n of that waveform. To derive the appropriate formulas, we substitute Eqs. (14.19a), (14.19b), and (14.21a) into Eqs. (14.31a) and (14.31b), to obtain

$$V_n = \frac{1}{2} A_n \cos \phi_n + j \frac{1}{2} A_n \sin \phi_n$$

$$= \frac{1}{2} A_n e^{j\phi_n} \qquad n = 1, 2, 3, \ldots \qquad \text{(14.33)}$$

and

$$V_0 = A_0 \qquad \text{(14.34)}$$

This yields the convenient formulas

$$A_0 = V_0$$

$$A_n = 2|V_n| \qquad n = 1, 2, 3, \ldots \qquad \text{(14.35)}$$

$$\phi_n = \underline{/V_n} \qquad n = 1, 2, 3, \ldots$$

A final observation regarding the complex Fourier coefficients of a real periodic waveform follows from Eq. (14.29) with n replaced by $-n$:

$$V_{-n} = \frac{1}{T} \int_{t_0}^{t_0 + T} v(t) e^{+j2\pi n f_1 t} \, dt \qquad \text{(14.36)}$$

Note that the substitution of $-n$ for n has changed the sign of the j in the exponent in the integral in Eq. (14.29). If $v(t)$ is real, then the change in the sign of the j in the exponent is equivalent to conjugation of the entire right-hand side. That is, for real $v(t)$,

$$V_{-n} = \left[\frac{1}{T} \int_{t_0}^{t_0 + T} v(t) e^{-j2\pi n f_1 t} \, dt \right]^* \qquad \text{(14.37)}$$

which, by comparison with Eq. (14.29), shows that

$$V_{-n} = V_n^* \tag{14.38}$$

Therefore, for $v(t)$ real, we can obtain the negatively indexed complex Fourier coefficients by conjugating the positively indexed coefficients as in Eq. (14.38).

Parseval's Theorem for Real or Complex Periodic Waveforms

The most general form of Parseval's theorem applies to both real and complex waveforms. We obtain the general form by conjugating both sides of Eq. (14.24), multiplying the result by $v(t)$, and integrating from t_0 to $t_0 + T$. This yields

$$\int_{t_0}^{t_0+T} v(t)v^*(t)\,dt = \int_{t_0}^{t_0+T} v(t) \sum_{n=-\infty}^{+\infty} V_n^* e^{-j2\pi n f_1 t}\,dt \tag{14.39}$$

By interchanging the order of summation and integration we obtain

$$\int_{t_0}^{t_0+T} v(t)v^*(t)\,dt = \sum_{n=-\infty}^{+\infty} V_n^* \int_{t_0}^{t_0+T} v(t)e^{-j2\pi n f_1 t}\,dt \tag{14.40}$$

which, with the aid of Eq. (14.29), becomes

$$\frac{1}{T}\int_{t_0}^{t_0+T} v(t)v^*(t)\,dt = \sum_{n=-\infty}^{+\infty} V_n V_n^* \tag{14.41}$$

or

Parseval's Theorem for Real or Complex Periodic Waveforms

$$\frac{1}{T}\int_{t_0}^{t_0+T} |v(t)|^2\,dt = \sum_{n=-\infty}^{+\infty} |V_n|^2 \tag{14.42}$$

This is Parseval's theorem for real or complex periodic waveforms. Of course, if $v(t)$ is real, then $|v(t)|^2 = v^2(t)$.

Two-Sided Amplitude, Phase, and Power Spectra

Plots of the magnitudes and phases of the complex Fourier coefficients versus frequency as illustrated in Fig. 14.12a, b are called the *two-sided amplitude spectrum* and the *two-sided phase spectrum* of $v(t)$, respectively. The *spectrum* of $v(t)$ is the pair of amplitude and phase spectra. Because $v(t)$ can be synthesized from its spectrum by means of Eq. (14.24), the spectrum provides a complete frequency-domain representation of $v(t)$.

You will notice that in a two-sided spectrum, the spectral lines exist for both positive and negative frequencies. Negative frequencies do not physically exist. The concept of negative frequency arises because we represent sinusoids as rotating phasors. Consider the simplest example illustrated in Fig. 14.13. Since $\cos 2\pi ft = \mathscr{R}e\,\{e^{j2\pi ft}\} = \frac{1}{2}e^{j2\pi ft} + \frac{1}{2}e^{-j2\pi ft}$ the process of taking the real part of $e^{j2\pi ft}$ is equivalent to adding the counterclockwise rotating phasor $\frac{1}{2}e^{j2\pi ft}$ to the clockwise rotating phasor $\frac{1}{2}e^{-j2\pi ft}$. We see therefore, that negative frequencies simply refer

FIGURE 14.12
A typical two-sided spectrum:
(a) amplitude; (b) phase

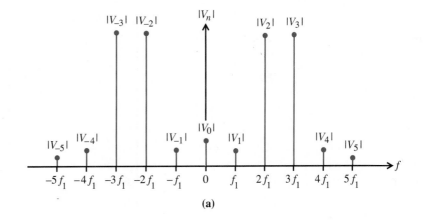

(a)

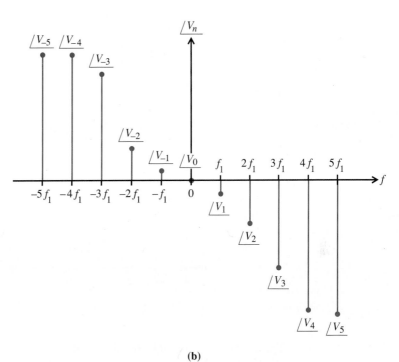

(b)

to phasors that rotate clockwise. The lines plotted on the positive frequency axis in the two-sided spectrum of Fig. 14.12 refer to phasors in Eq. (14.24) that rotate counterclockwise ($n > 0$). The lines plotted on the *negative* frequency axis correspond to phasors that rotate clockwise ($n < 0$). Equation (14.38) applies for real $v(t)$. Therefore, for a real periodic waveform the amplitude spectrum has even symmetry, $|V_n| = |V_{-n}|$, and the phase spectrum has odd symmetry $\underline{/V_n} = -\underline{/V_{-n}}$.

A plot of $|V_n|^2$ versus frequency, as illustrated in Fig. 14.14, is called the *two-sided power spectrum* of $v(t)$. It follows from Eqs. (14.35) and (14.38) that for real $v(t)$,

$$|V_n|^2 + |V_{-n}|^2 = \frac{1}{2} A_n^2 \qquad n = 1, 2, 3, \ldots \qquad (14.43)$$

FIGURE 14.13
Rotating phasor
representations of $\cos 2\pi ft$

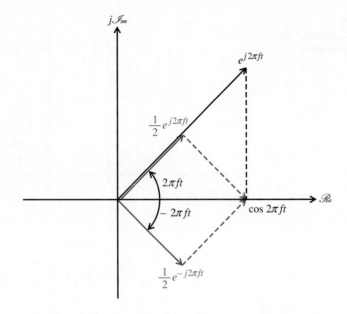

FIGURE 14.14
A typical two-sided power
spectrum

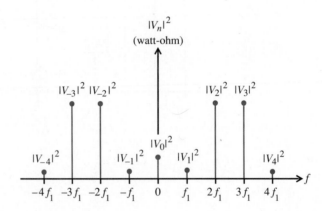

The quantity on the right-hand side of Eq. (14.43) is the normalized power in the nth Fourier component $A_n \cos (2\pi n f_1 t + \phi_n)$. Therefore, we find the normalized power carried by the nth harmonic of $v(t)$, $n \geq 1$, by adding $|V_n|^2$ and $|V_{-n}|^2$. The normalized power carried by the dc component is $|V_0|^2$. The total normalized power in $v(t)$ is, according to Parseval's theorem, Eq. (14.42), equal to the sum of all the line heights of the power spectrum.

EXAMPLE 14.4

From Fig. 14.15,
- (a) Determine the complex Fourier coefficients of the waveform.
- (b) Use your result from (a) to find the real Fourier coefficients a_n and b_n and the Fourier amplitudes and phases A_n and ϕ_n.
- (c) Write the first several terms of each form of the Fourier series with the assumption that $T = 1$ ms, $\tau = 1/2\pi$ ms, and $A = 100$ V.

FIGURE 14.15
Periodically repeated
exponential waveform

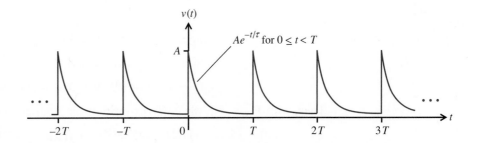

$v(t)$ ... $Ae^{-t/\tau}$ for $0 \le t < T$ ... A ... $-2T$ ... $-T$ ... 0 ... T ... $2T$ ... $3T$ ... t

■ **SOLUTION**

(a) A convenient interval of integration for Eq. (14.29) is $0 \le t < T$:

$$V_n = \frac{1}{T}\int_0^T v(t)e^{-j2\pi nf_1 t}\,dt$$

$$= \frac{1}{T}\int_0^T Ae^{-t/\tau}e^{-j2\pi nf_1 t}\,dt$$

$$= \frac{A}{T}\frac{e^{[-(t/\tau)-j2\pi nf_1 t]}}{-(1/\tau)-j2\pi nf_1}\Bigg|_0^T$$

Because $f_1 T = 1$, this becomes

$$V_n = A\frac{\tau}{T}\frac{1-e^{-T/\tau}}{1+j2\pi nf_1\tau}$$

$$= A\frac{\tau}{T}\frac{1-e^{-T/\tau}}{\sqrt{1+(2\pi nf_1\tau)^2}}\underline{/-\tan^{-1}2\pi nf_1\tau}$$

(b) From Eqs. (14.32) and (14.35),

$$a_0 = V_0 = 2A\frac{\tau}{T}(1-e^{-T/\tau})$$

$$a_n = 2\mathscr{R}e\{V_n\} = 2A\frac{\tau}{T}\frac{1-e^{-T/\tau}}{1+(2\pi nf_1\tau)^2}$$

$$b_n = -2\mathscr{I}m\{V_n\} = 2A\frac{\tau}{T}\frac{(1-e^{-T/\tau})2\pi nf_1\tau}{1+(2\pi nf_1\tau)^2}\qquad n = 1,2,\dots$$

$$A_0 = V_0 = A\frac{\tau}{T}(1-e^{-T/\tau})$$

$$A_n = 2|V_n| = 2A\frac{\tau}{T}\frac{1-e^{-T/\tau}}{\sqrt{1+(2\pi nf_1\tau)^2}}\qquad n = 1,2,\dots$$

$$\phi_n = \underline{/V_n} = -\tan^{-1}2\pi nf_1\tau \qquad n = 1,2,\dots$$

(c) Numerical evaluation of the first six complex Fourier coefficients in (b) with the use of $A = 100$ V, $T = 1$ ms, and $\tau = 1/2\pi$ ms $\simeq 0.159$ ms yields

$$V_0 \simeq 15.89 \text{ V}$$

$$V_1 \simeq 11.23\underline{/-45^\circ}\text{ V} = 7.94 - j7.94 \text{ V}$$

$$V_2 \simeq 7.10\underline{/-63.4°}\ \text{V} = 3.12 - j6.35\ \text{V}$$
$$V_3 \simeq 5.02\underline{/-71.6°}\ \text{V} = 1.58 - j4.76\ \text{V}$$
$$V_4 \simeq 3.85\underline{/-76.0°}\ \text{V} = 0.93 - j3.74\ \text{V}$$
$$V_5 \simeq 3.12\underline{/-78.7°}\ \text{V} = 0.61 - j3.06\ \text{V}$$

Then, with $V_{-n} = V_n^*$, the complex form of the Fourier series is

$$
\begin{aligned}
v(t) \simeq\ & 15.89 + (11.23\underline{/-45°})e^{j2\pi f_1 t} + (11.23\underline{/+45°})e^{-j2\pi f_1 t} \\
& + (7.10\underline{/-63.4°})e^{j2\pi 2 f_1 t} + (7.10\underline{/+63.4°})e^{-j2\pi 2 f_1 t} \\
& + (5.02\underline{/-71.6°})e^{j2\pi 3 f_1 t} + (5.02\underline{/+71.6°})e^{-j2\pi 3 f_1 t} \\
& + (3.85\underline{/-76.0°})e^{j2\pi 4 f_1 t} + (3.85\underline{/+76.0°})e^{-j2\pi 4 f_1 t} \\
& + (3.12\underline{/-78.7°})e^{j2\pi 5 f_1 t} + (3.12\underline{/+78.7°})e^{-j2\pi 5 f_1 t} + \cdots\ \text{V}
\end{aligned}
$$

where $f_1 = 1000$ Hz. This leads to the real forms

$$
\begin{aligned}
v(t) \simeq\ & 15.89 + 22.46 \cos(2\pi 1000t - 45°) \\
& + 14.20 \cos(2\pi 2000t - 63.4°) \\
& + 10.04 \cos(2\pi 3000t - 71.6°) \\
& + 7.70 \cos(2\pi 4000t - 76.0°) \\
& + 6.24 \cos(2\pi 5000t - 78.7°) + \cdots\ \text{V}
\end{aligned}
$$

and

$$
\begin{aligned}
v(t) \simeq\ & 15.89 + 15.88 \cos 2\pi 1000t + 6.24 \cos 2\pi 2000t \\
& + 3.16 \cos 2\pi 3000t + 1.86 \cos 2\pi 4000t \\
& + 1.22 \cos 2\pi 5000t + \cdots \\
& + 15.88 \sin 2\pi 1000t + 12.7 \sin 2\pi 2000t \\
& + 9.52 \sin 2\pi 3000t + 7.48 \sin 2\pi 4000t \\
& + 6.12 \sin 2\pi 5000t + \cdots\ \text{V}\ \blacksquare
\end{aligned}
$$

REMEMBER

The complex form of the Fourier series applies to both complex and real periodic waveforms. When $v(t)$ is real, we can find its real Fourier series by first finding V_n by the use of Eq. (14.29) and then applying Eq. (14.32) or (14.35). The complex form of the Fourier series is used to define the two-sided amplitude, phase, and power spectra of the waveform. If $v(t)$ is real, then $V_{-n} = V_n^*$.

EXERCISES

12. (a) Use Eq. (14.29) to determine the complex Fourier-series coefficients of the periodic pulse train of Fig. 14.10.

    (b) Check your answer by referring to the results for a_n and b_n in Example 14.3 and applying Eq. (14.32).

13. (a) Show that if $v(t)$ is a real, even function of t, then V_n is a real, even function of n.

    (b) Show that if $v(t)$ is a real, odd function of t, then V_n is an imaginary, odd function of n.

14. Use Eq. (14.29) to determine the complex Fourier-series coefficients of a periodic impulse train

$$v(t) = \sum_{i=-\infty}^{+\infty} \delta(t - iT)$$

15. Use the formula in Exercise 1 to determine the complex Fourier-series coefficients of the triangular waveform $v(t)$ of Fig. 14.1.

14.7 Fourier-Series Tables

Electrical engineers frequently use tables to save them from needlessly rederiving results that are already known. One such table is Table 14.1, which lists the complex Fourier-series coefficients for several periodic signals found in engineering. The corresponding results for the real series can be obtained easily from the table with the use of Eqs. (14.32) and (14.35).

Table 14.2 can also save you time. It lists four basic properties of the complex Fourier series. These properties can be helpful to derive the Fourier coefficients of waveforms not shown in Table 14.1.

Linearity (P1)

Property P1 states that the nth complex Fourier coefficient of the weighted sum of two periodic waveforms $ax(t) + by(t)$, where $x(t)$ and $y(t)$ have period T, is simply $aX_n + bY_n$, where X_n and Y_n are the nth complex Fourier coefficients of $x(t)$ and $y(t)$, respectively. This property can be established by simply observing that

$$ax(t) + by(t) = \sum_{n=-\infty}^{+\infty} aX_n e^{j2\pi n f_1 t} + \sum_{n=-\infty}^{+\infty} bY_n e^{j2\pi n f_1 t}$$

$$= \sum_{n=-\infty}^{+\infty} (aX_n + bY_n)e^{j2\pi n f_1 t} \qquad (14.44)$$

Therefore $aX_n + bY_n$ is the nth complex Fourier coefficient of $ax(t) + by(t)$.

Reversal (P2)

Property P2 states that the effect of reversing a periodic waveform in time, $v(t) \rightarrow v(-t)$, is to exchange V_n with V_{-n}. Property P2 is illustrated in Fig. 14.16. To establish Property P2, we compute the nth complex Fourier coefficient G_n of $g(t) = v(-t)$:

$$G_n = \frac{1}{T} \int_{t_0}^{t_0+T} v(-t)e^{-j2\pi n f_1 t}\, dt \qquad (14.45)$$

By a change of variables, λ for $-t$, this becomes

$$G_n = \frac{1}{T} \int_{-t_0-T}^{-t_0} v(\lambda)e^{j2\pi n f_1 \lambda}\, d\lambda = \frac{1}{T} \int_{t_0}^{t_0+T} v(\lambda)e^{j2\pi n f_1 \lambda}\, d\lambda \qquad (14.46)$$

where the last step follows from the fact that $v(\lambda)$ has period T. But the nth complex Fourier coefficient of $v(t)$ is

$$V_n = \frac{1}{T} \int_{t_0}^{t_0+T} v(\lambda)e^{-j2\pi n f_1 \lambda}\, d\lambda \qquad (14.47)$$

Therefore, by comparing Eqs. (14.46) and (14.47), we obtain

$$G_n = V_{-n} \tag{14.48}$$

Time Shift (P3)

According to Property P3, the nth complex Fourier coefficient of the time-shifted periodic waveform $v(t - t_d)$ equals V_n times $e^{-j2\pi n f_1 t_d}$. This property is illustrated in Fig. 14.17. To establish Property P3, we observe that since

$$v(t) = \sum_{n=-\infty}^{+\infty} V_n e^{j2\pi n f_1 t} \tag{14.49}$$

then

$$v(t - t_d) = \sum_{n=-\infty}^{+\infty} V_n e^{j2\pi n f_1(t - t_d)}$$

$$= \sum_{n=-\infty}^{+\infty} (V_n e^{-j2\pi n f_1 t_d}) e^{j2\pi n f_1 t} \tag{14.50}$$

Therefore $V_n e^{-j2\pi n f_1 t_d}$ is the nth complex Fourier coefficient of $v(t - t_d)$.

TABLE 14.1   Some familiar periodic waveforms and their complex Fourier-series coefficients

	NAME	FIGURE	COMPLEX FOURIER COEFFICIENT V_n
FS1	Square wave		$-j\dfrac{2A}{n\pi}\cdot$ n odd 0 n even
FS2	Periodic rectangular pulse (rectangular wave)		$\dfrac{A}{\pi n}\sin\dfrac{n\pi\tau}{T}$ $n \neq 0$ $A\dfrac{\tau}{T}$ $n = 0$
FS3	Periodically repeated exponential		$A\dfrac{\tau}{T}\dfrac{1 - e^{-T/\tau}}{1 + j2\pi n\tau/T}$
FS4	Sawtooth		$j\dfrac{A}{n\pi}(-1)^n$ $n \neq 0$ 0 $n = 0$

TABLE 14.1 (*continued*)

$\dfrac{2k\pi x}{L}$

NAME	FIGURE	COMPLEX FOURIER COEFFICIENT V_n
FS5 Double sawtooth	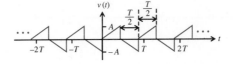	$-\dfrac{2A}{(\pi n)^2} - j\dfrac{A}{\pi n}$ n odd 0 n even
FS6 Triangular wave	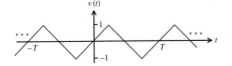	$j\dfrac{4}{\pi^2 n^2}(-1)^{(n+1)/2}$ n odd 0 n even
FS7 Trapezoidal wave	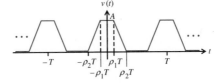	$\dfrac{A\tau}{T}\dfrac{\sin(\pi n t_0/T)}{\pi n}\dfrac{\sin(\pi n\tau/T)}{\pi n}$ where $t_0 = (\rho_1 + \rho_2)T$ and $\tau = (\rho_2 - \rho_1)T$
FS8 Half-rectified cosine	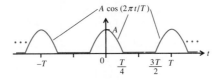	$\dfrac{(A/\pi)\cos n\pi/2}{1-n^2}$ $n \neq \pm 1$ $\dfrac{A}{4}$ $n = \pm 1$
FS9 Full-rectified cosine (where frequency n/T_o is associated with V_n)	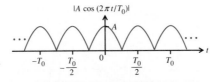	$\dfrac{(2A/\pi)\cos n\pi/2}{1-n^2}$ $n \neq \pm 1$ 0 $n = \pm 1$
FS10 Periodic impulse train		$\dfrac{1}{T}$

TABLE 14.2
Four properties of the complex Fourier series

PROPERTY	WAVEFORM	COMPLEX FOURIER COEFFICIENT
P1 Linearity	$ax(t) + by(t)$	$aX_n + bY_n$
P2 Reversal	$v(-t)$	V_{-n}
P3 Time shift	$v(t - t_d)$	$V_n e^{-j2\pi n f_1 t_d}$
P4 Differentiation	$\dfrac{d}{dt}v(t)$	$j2\pi n f_1 V_n$

FIGURE 14.16
Illustration of time reversal
property: (*a*) periodic
waveform $v(t)$; (*b*) time-
reversed waveform $v(-t)$

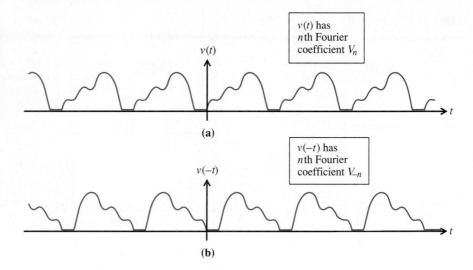

(a)

(b)

FIGURE 14.17
Illustration of time-shift
property: (*a*) periodic
waveform $v(t)$; (*b*) time-
shifted waveform $v(t - t_d)$

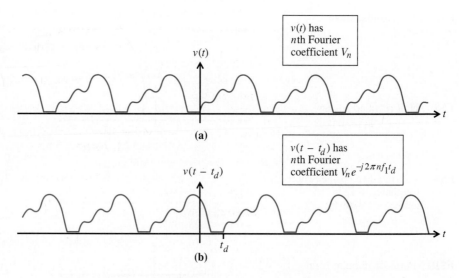

(a)

(b)

FIGURE 14.18
Illustration of differentiation
property: (*a*) periodic
waveform $v(t)$;
(*b*) differentiated waveform
dv/dt

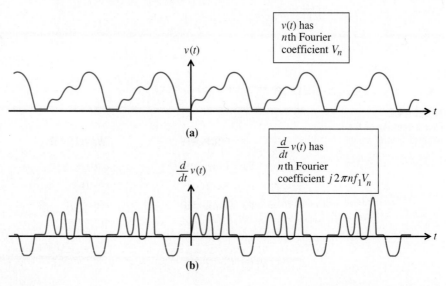

(a)

(b)

Differentiation (P4)

Property P4 states that the nth complex Fourier coefficient of the derivative of $v(t)$ is the product $j2\pi nf_1\, V_n$, where V_n is the nth complex coefficient of $v(t)$. Property P4 is illustrated in Fig. 14.18. Property P4 can be established simply by differentiating both sides of Eq. (14.24) with respect to t.

Another important property of the complex Fourier series is developed in Problem 35.

EXERCISES

16. Let $v(t)$ be the sum of the square wave and the half-rectified cosine wave, FS1 and FS8 in Table 14.1. Assume that their amplitudes A and their periods T are identical.
    (a) Sketch $v(t)$.
    (b) Determine the complex Fourier-series coefficients of $v(t)$. (*Hint*: Use the linearity Property P1 of Table 14.2.)
    (c) Sketch the two-sided amplitude and phase spectra.

17. Let $g(t)$ be the time-reversed version of a periodically repeated exponential wave: $g(t) = v(-t)$, where $v(t)$ is the waveform of Fig. 14.15.
    (a) Sketch $g(t)$.
    (b) Use Tables 14.1 and 14.2 to determine the complex Fourier-series coefficients of $g(t)$.
    (c) Sketch the two-sided amplitude and phase spectra of $v(t)$ and $g(t)$.

18. Let $v_d(t) = v(t - t_d)$ denote the time-shifted version of the half-wave-rectified cosine wave, FS8 of Table 14.1. Assume that the time delay is $t_d = T/2$.
    (a) Sketch $v_d(t)$.
    (b) Use Property P3 of Table 14.2 to determine the complex Fourier-series coefficients of $v_d(t)$.
    (c) Use your result from (b) and Property P1 to verify entry FS9 of Table 14.1.
    (d) Sketch the two-sided amplitude and phase spectra of entry FS9.

19. Differentiate the triangular wave of Fig. 14.1, and apply Property P4 of Table 14.2 to FS6 of Table 14.1 to determine the complex Fourier series of a square wave. Explain why your answer differs from FS1 of Table 14.1.

14.8 Circuit Response to Periodic Input

See Problems 14.5 and 14.6 in the PSpice manual.

Now that we know how to represent a periodic waveform by its Fourier series, we are prepared to determine the particular response of a stable LLTI circuit to a periodic input. Our method for obtaining the particular response uses the principle of superposition and ac circuit analysis. The Fourier series enables us to represent the input as a sum of sinusoidal components. Because the circuit is linear, the particular response to this sum is given by the sum of the particular responses to the individual components. Because the circuit is also time-invariant, the individual responses can be determined with the use of methods in Chapter 11. Because the circuit is also stable, the complete response to the periodic input will be equal to the particular response. Details are given in the following subsections.

Application of Real Fourier Series

We will show how to apply the real Fourier series to circuit analysis by considering a specific example. Assume that the particular response $v_o(t)$ of the band-pass RLC circuit of Fig. 14.19 is to be determined for the periodically repeated exponential waveform of Fig. 14.15 in which $A = 100$, $T = 1$ ms, and $\tau = 1/2\pi$ ms.

FIGURE 14.19
Band-pass RLC circuit

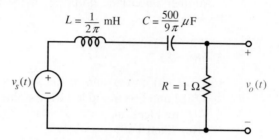

The first step in the solution is to express $v_s(t)$ in its Fourier-series form. Either the real or the complex form of the series can be used. Suppose that the real form of Eq. (14.20) is used. Then, according to Example 14.4,

$$v_s(t) = A_0 + A_1 \cos(2\pi f_1 t + \phi_1) + A_2 \cos(2\pi 2 f_1 t + \phi_2) + \cdots$$
$$= 15.89 + 22.46 \cos(2\pi 1000 t - 45°) + 14.20 \cos(2\pi 2000 t - 63.4°)$$
$$+ 10.04 \cos(2\pi 3000 t - 71.6°) + 7.70 \cos(2\pi 4000 t - 76.6°)$$
$$+ 6.24 \cos(2\pi 5000 t - 78.7°) + \cdots \tag{14.51}$$

The second step is to determine the circuit's particular response to the *individual* components in Eq. (14.51). This step is easy because, by ac circuit analysis, the particular response to the individual harmonic $A_n \cos(2\pi n f_1 t + \phi_n)$ is simply $A_n |H(j2\pi n f_1)| \cos(2\pi n f_1 t + \phi_n + \underline{/H(j2\pi n f_1)}$, where $H(j\omega)$ is the transfer function of the circuit:*

$$H(j\omega) = \cfrac{R}{R + j\left(\omega L - \cfrac{1}{\omega C}\right)}$$

$$= \frac{1}{1 + j(0.001f - 9/0.001f)} \tag{14.52}$$

The third step is to apply superposition. When the input is the sum in Eq. (14.51), the particular response is the sum

$$v_o(t) = A_0 H(0) + A_1 |H(j2\pi f_1 t)| \cos(2\pi f_1 t + \phi_1 + \underline{/H(j\omega_1)}$$
$$+ A_2 |H(j2\pi 2 f_1)| \cos(2\pi 2 f_1 t + \phi_2 + \underline{/H(j2\omega_1)}$$
$$+ \cdots + A_n |H(j2\pi n f_1)| \cos(2\pi n f_1 t + \phi_1 + \underline{/H(j2\pi n f_1)}) + \cdots$$

$$= (15.89)(0) + (22.46)(0.124) \cos(2\pi 1000 t - 45° + 82.87°)$$
$$+ (14.20)(0.371) \cos(2\pi 2000 t - 63.4° + 68.20°)$$
$$+ (10.04)(1) \cos(2\pi 3000 t - 71.6° + 0°)$$
$$+ (7.70)(0.496) \cos(2\pi 4000 t - 76.0° - 60.26°)$$
$$+ (6.24)(0.298) \cos(2\pi 5000 t - 78.7° - 72.65°) + \cdots$$

---

* Again, for notational simplicity we drop the boldface notation, $\mathbf{H}(j\omega)$, that was used in Chapters 10 through 13.

$$= 2.79 \cos (2\pi 1000t + 37.87°)$$
$$+ 5.27 \cos (2\pi 2000t + 4.8°)$$
$$+ 10.04 \cos (2\pi 3000t - 71.6°)$$
$$+ 3.82 \cos (2\pi 4000t - 136.26°)$$
$$+ 1.86 \cos (2\pi 5000t - 151.35°) + \cdots \qquad (14.53)$$

This step is illustrated in Fig. 14.20.

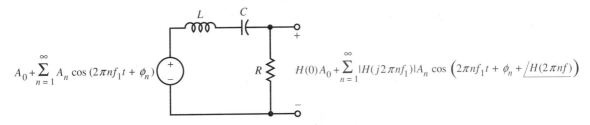

$$A_0 + \sum_{n=1}^{\infty} A_n \cos (2\pi nf_1 t + \phi_n) \qquad\qquad H(0)A_0 + \sum_{n=1}^{\infty} |H(j2\pi nf_1)|A_n \cos \left(2\pi nf_1 t + \phi_n + \underline{/H(2\pi nf)}\right)$$

FIGURE 14.20 A sum of sinusoidal inputs yields a sum of sinusoidal outputs. Here $H(j\omega) = 1/\{R + j[\omega L - (1/\omega C)]\}$

A good way to view the relationship between the input and output is illustrated in Fig. 14.21. This figure contains plots of the one-sided spectrum of the input $v_s(t)$ and the transfer function of the circuit, all plotted versus f. Because the independent variable in these plots is f, it has been convenient to introduce the more economical notation

$$\mathrm{H}(f) = H(j2\pi f) \qquad (14.54)$$

as indicated in Fig. 14.21. As is evident in Eq. (14.53), the amplitude of each sinusoidal component in $v_o(t)$ is given by the product $A_n|H(j2\pi nf_1)|$. Therefore, *the output waveform $v_o(t)$ has an amplitude spectrum that is the product of the amplitude spectrum of the input waveform and the amplitude characteristic of the filter.* It can

FIGURE 14.21
The product of (*a*) the input amplitude spectrum and (*b*) the filter's amplitude characteristic yields (*c*) the output amplitude spectrum. The sum of (*d*) the input phase spectrum and (*e*) the filter's phase characteristic at nf_1 yields (*f*) the output phase spectrum

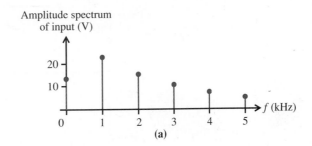

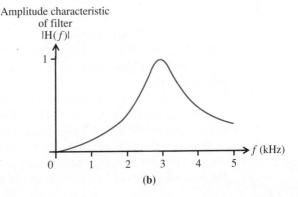

FIGURE 14.21
Continued

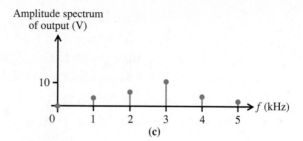

Amplitude spectrum of output (V)

(c)

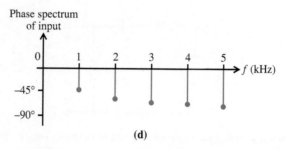

Phase spectrum of input

(d)

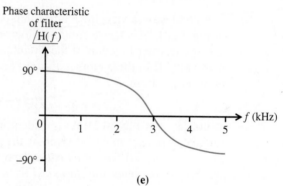

Phase characteristic of filter

$\underline{/H(f)}$

(e)

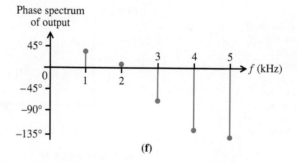

Phase spectrum of output

(f)

also be seen from Eq. (14.53), that the phase of each sinusoidal component in $v_o(t)$ is given by the sum $\phi_n + \underline{/H(j2\pi nf_1)}$. Therefore, *the phase spectrum of $v_o(t)$ equals the sum of the phase spectrum of $v_s(t)$ and the phase characteristic of the filter at frequencies nf_1*.

Figure 14.22 contains plots of *one period* of the partial Fourier series $v_{oN}(t)$ of the output waveform of Eq. (14.53). This figure was generated by a computer, which summed the first N terms in Eq. (14.53) for $N = 1, 2, 3, 4, 5, 10$, and 15. (The dc term is zero and is not included in this count.) It can be seen from the figure that the convergence is rapid; there is very little change in the appearance of the plots for $N > 10$. A plot of two periods of the partial Fourier series of $v_o(t)$ with $N = 25$

FIGURE 14.22
One period of the partial Fourier series $\hat{v}_{oN}(t)$ with $N =$ (a) 1, (b) 2, (c) 3, (d) 4, (e) 5, (f) 10, and (g) 15

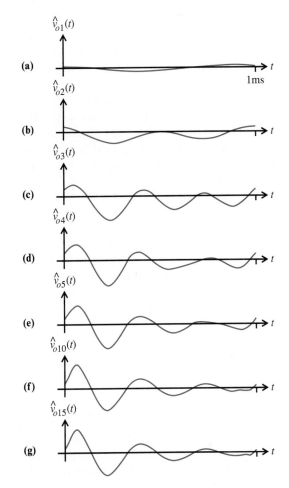

FIGURE 14.23
Two periods of the partial Fourier series $\hat{v}_{oN}(t)$ with $N = 25$

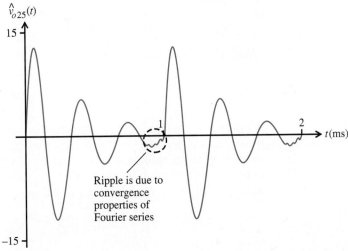

is given in Fig. 14.23. Except for the small ripple (enclosed in the dotted region), this plot can be considered to be the actual output $v_o(t)$. The ripple is related to the Gibbs phenomenon, which concerns the way in which a partial Fourier series converges. The Gibbs phenomenon is described in Section 14.10.

Recognize that the output shown in Fig. 14.23 is what we would expect. This can be explained from two points of view. The first point of view is a frequency-domain one. Notice from the *RLC* circuit's amplitude characteristic in Fig. 14.21 that the circuit is a band-pass filter. The circuit passes the third ($f = 3$ kHz) harmonic of the input waveform while attenuating the other harmonics. We see that the output waveform of Fig. 14.23 exhibits three major cycles or swings each millisecond. These three cycles each millisecond are caused by the relatively large amplitude of the 3-kHz harmonic in the output. The output also contains harmonics at 1, 2, 4, 5, ... kHz, but these are smaller than the 3-kHz harmonic. The presence of the smaller-amplitude harmonics at 1, 2, 4, 5, ... kHz causes $v_o(t)$ to have the nonsinusoidal shape shown in the figure. If the values of the *RLC* circuit had been chosen so that the filter was more frequency-selective (that is, had a narrower pass band centered at 3 kHz), then $v_o(t)$ would look much more like a pure 3-kHz sinusoid.

The second point of view for understanding the appearance of $v_o(t)$ (Fig. 14.23) is to consider the time domain. For the values given, the input waveform of Fig. 14.15, $v_s(t)$, resembles a sequence of narrow pulses that occur every $T = 1$ ms. We can show that the response of the band-pass *RLC* circuit to a single narrow pulse is approximately a damped sinusoid with frequency 3 kHz. The response to the sequence of pulses in $v_s(t)$ is, by superposition, approximately a sequence of damped 3-kHz sinusoids occurring every 1 ms, as seen in Fig. 14.23. This point of view can be formulated rigorously with the use of the concepts of *convolution* and *impulse response*, which are described in the following chapter.

Application of Complex Fourier Series

We stated earlier that we can also find the response of an LLTI circuit to a periodic input by using the complex form of the Fourier series. This method is illustrated in Fig. 14.24, in which the real periodic input waveform $v_s(t)$ is represented as the sum

$$v_s(t) = \sum_{n=-\infty}^{+\infty} V_{sn} e^{j2\pi n f_1 t} \tag{14.55}$$

where V_{sn} is the complex Fourier coefficient

$$V_{sn} = \frac{1}{T} \int_{t_0}^{t_0+T} v_s(t) e^{-j2\pi n f_1 t} \, dt \tag{14.56}$$

As shown in Chapter 10, the individual complex exponential $V_{sn} e^{j2\pi n f_1 t}$ produces a complex exponential particular response $V_{on} e^{j2\pi n f_1 t}$, where $V_{on} = H(j2\pi n f_1)V_{sn}$, or in terms of the economical notation of Eq. (14.54), $V_{on} = H(nf_1)V_{sn}$. From the principle of superposition, it follows that the response to the sum of complex exponentials $v_s(t)$ of Eq. (14.55) is simply the sum

$$v_o(t) = \sum_{n=-\infty}^{+\infty} V_{on} e^{jn\omega_1 t} \tag{14.57}$$

where

$$V_{on} = H(jn\omega_1)V_{sn} \tag{14.58}$$

or, equivalently, where

$$V_{on} = H(nf_1)V_{sn} \tag{14.59}$$

FIGURE 14.24
A sum of complex
exponential inputs yields a
sum of complex exponential
outputs

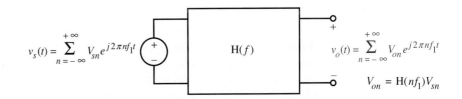

$$v_s(t) = \sum_{n=-\infty}^{+\infty} V_{sn} e^{j2\pi nf_1 t} \qquad \mathrm{H}(f) \qquad v_o(t) = \sum_{n=-\infty}^{+\infty} V_{on} e^{j2\pi nf_1 t}$$

$$V_{on} = \mathrm{H}(nf_1)V_{sn}$$

The computation of the two-sided output spectrum is illustrated in Fig. 14.25. The height of each line in the output amplitude spectrum is the product of the height of the corresponding line of the input amplitude spectrum and the value of the amplitude characteristic $|\mathrm{H}(f)|$ at the associated frequency. The output phase spectrum is the sum of the input phase spectrum and the phase characteristic of the filter at frequencies nf_1.

It follows directly from Eq. (14.58) that the output power spectrum is equal to the product of the input power spectrum and the squared magnitude of the transfer function

$$|V_{on}|^2 = |H(jn\omega_1)|^2 |V_{sn}|^2 \tag{14.60}$$

or

$$|V_{on}|^2 = |\mathrm{H}(nf_1)|^2 |V_{sn}|^2 \tag{14.61}$$

FIGURE 14.25
Computation of two-sided
output amplitude and phase
spectra: (a) input spectrum;
(b) filter transfer function;
(c) output spectrum

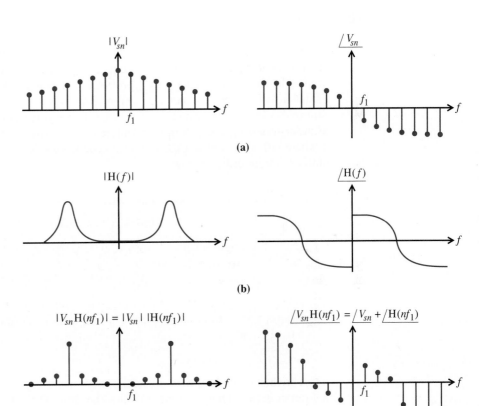

When the input to an LLTI circuit is a periodic waveform, the output is also a periodic waveform. The amplitude spectrum of the output is given by the product of the amplitude spectrum of the input and the amplitude characteristic of the circuit. The output phase spectrum is given by the sum of the phase spectrum of the input and the phase characteristic of the circuit.

EXERCISES

20. The pulse train of Fig. 14.10 is applied to a low-pass RC circuit. The parameter values are $A = 10$ V, $T = 1$ ms, $\tau = 0.1$ ms, $R = 50$ Ω, and $C = 1$ μF.
    (a) Plot the two-sided amplitude and phase characteristics of the circuit and the two-sided amplitude and phase spectra of the pulse train input.
    (b) Plot the two-sided amplitude and phase spectra of the output.

21. Repeat Exercise 20 with the low-pass RC circuit replaced by a high-pass RC circuit with the same values of R and C. (For a high-pass RC circuit, the output voltage is the voltage across the R.)

22. Repeat Exercise 20 with the low-pass RC circuit replaced by the band-pass RLC circuit of Fig. 14.19.

14.9 Existence of Fourier Series

Up to this point it has been assumed that the trigonometric series on the right-hand side of Eq. (14.3) actually can equal the arbitrary real periodic function $v(t)$ on the left-hand side. This was a problem that confounded the greatest mathematicians of Fourier's time: *Can any* periodic function be equated to a sum of sinusoids? It is now known that if $v(t)$ satisfies three conditions, then it has a Fourier-series representation that will be *equal* to $v(t)$ except at isolated values of t where $v(t)$ is discontinuous. The three conditions, known as the Dirichlet conditions, are described below.

Dirichlet Conditions

1. $v(t)$ is absolutely integrable over a period; that is,

$$\int_{-T/2}^{+T/2} |v(t)|\, dt < \infty$$

2. $v(t)$ has a finite number of maxima and minima within a period.
3. $v(t)$ has a finite number of discontinuities within a period. Each discontinuity must be finite.

The Dirichlet conditions are sufficient but not necessary conditions for $v(t)$ to have a Fourier series. That is, if $v(t)$ satisfies the three Dirichlet conditions, then it is guaranteed to have a Fourier-series representation that *equals* $v(t)$ except at the isolated discontinuities of $v(t)$. If $v(t)$ does not satisfy one or more of the Dirichlet conditions, then it *may or may not* have a Fourier-series representation that equals $v(t)$ except at the isolated discontinuities of $v(t)$.

Virtually all periodic waveforms that can exist in a physical circuit satisfy the Dirichlet conditions and, therefore, can be represented by a Fourier series. Conditions 2 and 3 are violated only by functions that have a distinctly pathological character. Examples of waveforms that violate conditions 2 and 3 are illustrated in Fig. 14.26. Notice that it is impossible to completely draw one period of these exemplifying functions either by hand or by any other physical means.

FIGURE 14.26
Functions not satisfying the Dirichlet conditions. (*a*) $v(t)$ has an infinite number of discontinuities in one period; (*b*) $v(t)$ has an infinite number of maxima and minima in one period. Each function has period 1

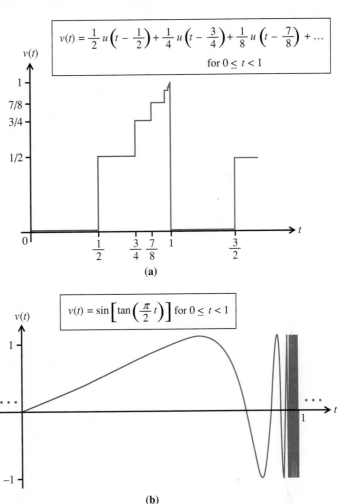

$$v(t) = \frac{1}{2} u\left(t - \frac{1}{2}\right) + \frac{1}{4} u\left(t - \frac{3}{4}\right) + \frac{1}{8} u\left(t - \frac{7}{8}\right) + \cdots$$

$$\text{for } 0 \le t < 1$$

(a)

$$v(t) = \sin\left[\tan\left(\frac{\pi}{2} t\right)\right] \text{ for } 0 \le t < 1$$

(b)

EXERCISES

In Exercises 23 through 29, the functions are portions of signals that repeat every T seconds outside the given interval. Sketch two periods of each signal, and state which Dirichlet conditions, if any, are violated.

23. $e^{-3t}, 0 \le t < T$

24. $e^{+3t}, 0 \le t < T$

25. $1/t, 0 \le t < T$

26. $1/t^2, 0 < t \le T$

27. $\cos(1/t), 0 \le t < T$

28. $t^{-0.5}, 0 < t \le T$

29. $u(t) + u(t - \frac{1}{2}T) + u[t - \frac{3}{4}T] + u[t - \frac{7}{8}T] + \cdots, 0 \le t < T$

14.10 Convergence of Fourier Series

An understanding of the way the Fourier series converges to $v(t)$ can be obtained from the problem illustrated in Fig. 14.27. In this figure, $v(t)$ is a real periodic waveform with period T, and $\hat{v}_N(t)$ is an approximation to $v(t)$ given by the *finite* trigonometric series

$$\hat{v}_N(t) = a_0 + \sum_{n=1}^{N} (a_n \cos 2\pi n f_1 t + b_n \sin 2\pi n f_1 t) \qquad (14.62)$$

where $f_1 = 1/T$. The problem is to adjust the a_n's and b_n's so that the approximation error

$$e_N(t) = v(t) - \hat{v}_N(t) \qquad (14.63)$$

is as small as possible. How do we choose the a_n and b_n to make the *waveform* $e_N(t)$ as small as possible?

The solution to this problem depends on what is meant by a small waveform. *One* way to define the size of a waveform is by the waveform's mean

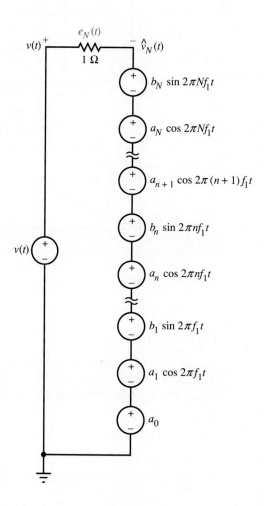

FIGURE 14.27
How should the dc voltage a_0 and the amplitudes a_n and b_n $(n = 1, 2, \ldots, N)$ be chosen to make the waveform $e_N(t)$ as "small" as possible?

square value. In the problem at hand, both $v(t)$ and $\hat{v}_N(t)$ have period T. Therefore $e_N(t)$ also has period T, and its mean square value is given by

$$\varepsilon(N) = \frac{1}{T} \int_{t_0}^{t_0+T} e_N^2(t)\, dt \qquad (14.64)$$

Recall that the mean square value of $e_N(t)$ is numerically equal to the average power that $e_N(t)$ supplies to the 1-Ω resistance of Fig. 14.27. In many engineering applications, it is reasonable to regard $\hat{v}_N(t)$ as the best possible approximation to $v(t)$ when the average power dissipated in the resistance is a minimum, that is, when $\varepsilon(N)$ is a minimum. This is the celebrated *minimum mean square error fidelity criterion*. To find the values of a_n that minimize $\varepsilon(N)$, we differentiate

$$\varepsilon(N) = \frac{1}{T} \int_{t_0}^{t_0+T} [v(t) - \hat{v}_N(t)]^2\, dt \qquad (14.65)$$

with respect to a_n, $n = 1, 2, \ldots, N$, to obtain

$$\frac{d\varepsilon(N)}{da_n} = \frac{-2}{T} \int_{t_0}^{t_0+T} [v(t) - \hat{v}_N(t)] \frac{d\hat{v}_N(t)}{da_n}\, dt$$

$$= \frac{-2}{T} \int_{t_0}^{t_0+T} [v(t) - \hat{v}_N(t)] \cos 2\pi n f_1\, dt \qquad (14.66)$$

Then by setting the right-hand side to zero, we find

$$\frac{2}{T} \int_{t_0}^{t_0+T} \hat{v}_N(t) \cos 2\pi n f_1\, dt = \frac{2}{T} \int_{t_0}^{t_0+T} v(t) \cos 2\pi n f_1 t\, dt \qquad (14.67a)$$

or

$$\frac{2}{T} \int_{t_0}^{t_0+T} \left(a_0 + \sum_{m=1}^{N} a_m \cos 2\pi m f_1 t + \sum_{m=1}^{N} b_m \sin 2\pi m f_1 t \right) \cos 2\pi n f_1 t\, dt$$

$$= \frac{2}{T} \int_{t_0}^{t_0+T} v(t) \cos 2\pi n f_1 t\, dt \qquad (14.67b)$$

All the terms in the left-hand integral integrate to zero except the one involving $a_m \cos 2\pi m f_1 t \cos 2\pi n f_1 t$ where $m = n$. By integration this term, Eq. (14.67b) becomes

$$a_n = \frac{2}{T} \int_{t_0}^{t_0+T} v(t) \cos 2\pi n f_1 t\, dt \qquad (14.68)$$

which is recognized as the Fourier coefficient given by Eq. (14.7). To see whether the Fourier coefficient results in a minimum or a maximum value for $\varepsilon(N)$, we use Eq. (14.66) to compute

$$\frac{d^2\varepsilon(N)}{da_n^2} = \frac{2}{T} \int_{t_0}^{t_0+T} \frac{d\hat{v}_N(t)}{da_n} \cos 2\pi n f_1 t\, dt$$

$$= \frac{2}{T} \int_{t_0}^{t_0+T} \cos^2 2\pi n f_1 t\, dt$$

$$= 1 \qquad n = 1, 2, \ldots, N \qquad (14.69)$$

Because the second derivative is positive, the mean square error $\varepsilon(N)$ obtained by the use of Eq. (14.68) is a minimum.

By differentiating $\varepsilon(N)$ with respect to a_0 and b_n, we similarly find that $\varepsilon(N)$ is minimized by

$$a_0 = \frac{1}{T} \int_{t_0}^{t_0 + T} v(t)\, dt \tag{14.70}$$

which is the Fourier coefficient of Eq. (14.4), and

$$b_n = \frac{2}{T} \int_{t_0}^{t_0 + T} v(t) \sin 2\pi n f_1 t\, dt \tag{14.71}$$

which is the Fourier coefficient of Eq. (14.8). Therefore

The Fourier coefficients make $\hat{v}_N(t)$ the best possible approximation to $v(t)$ in the sense that they cause the mean square approximation error $\varepsilon(N)$ to be a minimum.

In other words, we minimize the average power dissipation in the resistance in Fig. 14.27 by making the dc voltage and the sinusoidal voltage amplitudes equal to the Fourier coefficients [Eqs. (14.68), (14.70), and (14.71)].

The interference pattern $\hat{v}_N(t)$ is called a *partial Fourier series* when the a_n and b_n are Fourier coefficients. An alternative form of the partial Fourier series can be obtained with the use of Eq. (14.21):

$$\hat{v}_N(t) = A_0 + A_1 \cos(2\pi f_1 t + \phi_1) + \cdots + A_N \cos(2\pi N f_1 t + \phi_N) \tag{14.72}$$

We obtain the mean square approximation error arising from the partial Fourier series by substituting Eq. (14.62) into Eq. (14.65) and simplifying using Eqs. (14.68), (14.70), and (14.71). After some tedious but routine algebra, we obtain the following result:

$$\varepsilon_{\min}(N) = \frac{1}{T} \int_{t_0}^{t_0 + T} v^2(t)\, dt - \left(a_0^2 + \sum_{n=1}^{N} \frac{1}{2} a_n^2 + \sum_{n=1}^{N} \frac{1}{2} b_n^2 \right) \tag{14.73a}$$

or equivalently,

$$\varepsilon_{\min}(N) = \frac{1}{T} \int_{t_0}^{t_0 + T} v^2(t)\, dt - \left(A_0^2 + \sum_{n=1}^{N} \frac{1}{2} A_n^2 \right) \tag{14.73b}$$

and

$$\varepsilon_{\min}(N) = \frac{1}{T} \int_{t_0}^{t_0 + T} v^2(t)\, dt - \frac{1}{T} \int_{t_0}^{t_0 + T} \hat{v}_N^2(t)\, dt \tag{14.73c}$$

Therefore

The minimum mean square error $\varepsilon_{\min}(N)$ equals the difference between the mean square values of $v(t)$ and $\hat{v}_N(t)$, where $\hat{v}_N(t)$ is the partial Fourier series of $v(t)$.

It can be shown that if $v(t)$ has finite power, then

$$\lim_{N \to \infty} \varepsilon_{\min}(N) = 0 \tag{14.74}$$

This result is known as *Plancherel's theorem*. The implication of Eq. (14.74) is that the average power delivered to the 1-Ω resistance of Fig. 14.27 vanishes in the limit $N \rightarrow \infty$. *This means that there is no physical way to distinguish between any periodic finite power signal $v(t)$ and its (infinite) Fourier series.*

The nature of the convergence of $\hat{v}_N(t)$ to $v(t)$ is illustrated in Fig. 14.28, which contains plots of the partial Fourier series

$$\hat{v}_N(t) = \sum_{n=1,3,5,\ldots}^{N} \frac{4A}{\pi n} \sin 2\pi n f_1 t \qquad (14.75)$$

of the square wave of Fig. 14.7. Figure 14.28 also indicates the mean square approximation errors $\varepsilon_{\min}(N)$ associated with each partial Fourier series. It can be seen from this figure that as N increases, $\hat{v}_N(t)$ approaches $v(t)$ in appearance and $\varepsilon_{\min}(N)$ gets smaller. An interesting feature of the partial Fourier series occurs in the vicinity of the discontinuities of $v(t)$, where $\hat{v}_N(t)$ overshoots and oscillates. The amount of overshoot is approximately 9 percent of the amplitude of discontinuity of $v(t)$ for large but finite N *and does not decrease as N increases.* As N increases, the oscillations become more rapid and are grouped closer to the discontinuity. This oscillatory behavior of the partial Fourier series at the discontinuities of $v(t)$ can be seen in computer-generated plots of $\hat{v}_N(t)$, where N is necessarily finite. This feature of the partial Fourier series is called the *Gibbs phenomenon*. Because the square wave in this example satisfies the Dirichlet conditions, in the limit $N \rightarrow \infty$, $\hat{v}_N(t)$ converges to $v(t)$ for every value of t not corresponding to a discontinuity of $v(t)$. At a discontinuity, $\hat{v}_N(t)$ converges to a point halfway between the values on either side of the discontinuity.

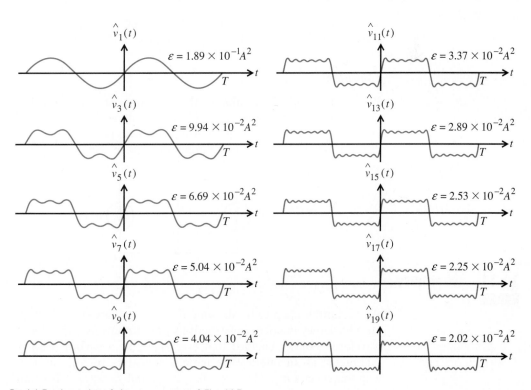

FIGURE 14.28 Partial Fourier series of the square wave of Fig. 14.7

FIGURE 14.28
Continued

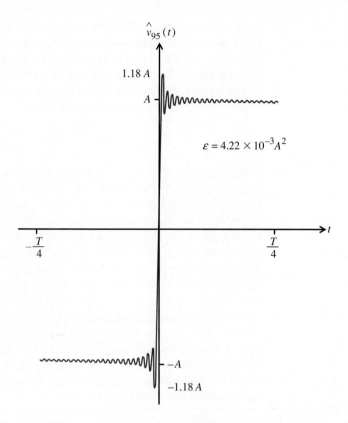

$$\varepsilon = 4.22 \times 10^{-3} A^2$$

REMEMBER

The Fourier coefficients make $\hat{v}_N(t)$ the best possible approximation to $v(t)$ in the sense that they cause the mean square approximation error $\varepsilon(N)$ to be a minimum.

EXERCISE

30. (a) Determine the value of the constant α in the circuit below so that the average power dissipated in the resistance is minimized. The current $i_s(t)$ is an arbitrary periodic waveform with period T.
    (b) What is the resulting minimum average power dissipation?

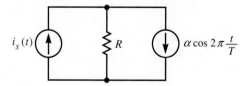

14.11 Summary

We began this chapter by showing that when sinusoids with different frequencies are added, they interfere constructively or destructively to produce a nonsinusoidal waveform. Fourier claimed that every periodic waveform with period T can be interpreted as an interference pattern consisting of real sinusoidal components with frequencies nf_1, where $n = 0, 1, 2, \ldots$ and $f_1 = 1/T$. This sum of sinusoids is the Fourier-series representation of the periodic waveform. The real form of the

Fourier series consists of sines and cosines. The coefficients of the sines and cosines are called Fourier coefficients. We showed that even symmetric waveforms contain only dc and cosine Fourier components, odd symmetric waveforms contain only sine components, and half-wave symmetric waveforms contain only components whose frequencies are odd harmonics, nf_1, $n = 1, 3, 5, \ldots$. By means of standard trigonometric identities, we gave an alternative form of the Fourier series with terms of the form $A_n \cos (2\pi nf_1 t + \phi_n)$, where A_n was called a Fourier amplitude and ϕ_n was called a Fourier phase. Finally, we described the complex form of the Fourier series. The complex form applies to both complex and real periodic waveforms. The real series Fourier coefficients, a_n and b_n, and the Fourier amplitudes and phases, A_n and ϕ_n, are related to the complex series Fourier coefficients V_n by very simple formulas, Eqs. (14.32) and (14.35). All three forms of the Fourier series are used in practice. The Fourier series provides a convenient method to compute the particular response of an LLTI circuit to a periodic input. An important topic of this chapter is that of the existence of the Fourier series. By referring to certain conditions established by Dirichlet, we were able to conclude that every periodic waveform that has engineering significance can be represented by a Fourier series. Finally, we discussed the minimum mean square error convergence property of the Fourier series and the Gibbs phenomenon.

KEY FACTS

◆   Any real periodic waveform found in nature or engineering can be represented by a sum of sinusoidal components whose frequencies are multiples of $f_1 = 1/T$, where T is the period.

◆   There are two basic forms of the real Fourier series: A *sine-cosine* form

$$v(t) = a_0 + \sum_{n=1}^{\infty} a_n \cos 2\pi nf_1 t + \sum_{n=1}^{\infty} b_n \sin 2\pi nf_1 t$$

where the *Fourier coefficients* a_n and b_n are given by

$$a_0 = \frac{1}{T} \int_{t_0}^{t_0 + T} v(t) \, dt$$

$$a_n = \frac{2}{T} \int_{t_0}^{t_0 + T} v(t) \cos 2\pi nf_1 t \, dt \qquad \text{for } n = 1, 2, \ldots$$

$$b_n = \frac{2}{T} \int_{t_0}^{t_0 + T} v(t) \sin 2\pi nf_1 t \, dt \qquad \text{for } n = 1, 2, \ldots$$

and an *amplitude-phase* form

$$v(t) = A_0 + \sum_{n=1}^{\infty} A_n \cos (2\pi nf_1 t + \phi_n)$$

where the *Fourier amplitudes and phases*, A_n and ϕ_n, are given by

$$A_0 = a_0$$

$$A_n = \sqrt{a_n^2 + b_n^2} \qquad n = 1, 2, \ldots$$

$$\phi_n = \tan^{-1}\left(\frac{-b_n}{a_n}\right) \qquad n = 1, 2, \ldots$$

- If $v(t)$ is even, then $b_n = 0$ for $n = 1, 2, \ldots$.

- If $v(t)$ is odd, then $a_n = 0$ for $n = 0, 1, 2, \ldots$.

- If $v(t)$ is half-wave symmetric, then $a_0 = a_n = b_n = 0$ for $n = 2, 4, 6, \ldots$.

- The *one-sided amplitude spectrum* is a plot of the Fourier amplitudes A_n versus frequency, $f \geq 0$.

- The *one-sided phase spectrum* is a plot of the Fourier phases ϕ_n versus frequency, $f \geq 0$.

- The *one-sided power spectrum* of $v(t)$ is a plot of $A_0^2, \frac{1}{2}A_1^2, \frac{1}{2}A_2^2, \ldots$ versus frequency, $f \geq 0$.

- According to *Parseval's theorem*, for real periodic signals,

$$\frac{1}{T} \int_{t_0}^{t_0 + T} v^2(t)\, dt = a_0^2 + \sum_{n=1}^{\infty} \frac{1}{2} a_n^2 + \sum_{n=1}^{\infty} \frac{1}{2} b_n^2$$

$$= A_0^2 + \sum_{n=1}^{\infty} \frac{1}{2} A_n^2$$

- A real or complex periodic waveform $v(t)$ can be represented by the *complex form* of the Fourier series,

$$v(t) = \sum_{n=-\infty}^{+\infty} V_n e^{j2\pi n f_1 t}$$

where the complex Fourier coefficient V_n is given by

$$V_n = \frac{1}{T} \int_{t_0}^{t_0 + T} v(t) e^{-j2\pi n f_1 t}\, dt \qquad n = 0, \pm 1, \pm 2, \ldots$$

- If $v(t)$ is real, then $V_{-n} = V_n^*$.

- The *two-sided amplitude spectrum* is a plot of $|V_n|$ versus frequency, $-\infty < f < \infty$.

- The *two-sided phase spectrum* is a plot of $\underline{/V_n}$ versus frequency, $-\infty < f < \infty$.

- According to *Parseval's theorem*, for real or complex periodic signals,

$$\frac{1}{T} \int_{t_0}^{t_0 + T} |v(t)|^2\, dt = \sum_{n=-\infty}^{+\infty} |V_n|^2$$

- By representing a periodic waveform by its Fourier series, the response of any LLTI circuit to that periodic waveform may be obtained using ac circuit analysis and superposition.

- The amplitude spectrum of an LLTI circuit's output is given by the *product* of the amplitude spectrum of the input and the amplitude characteristic of the circuit.

- The phase spectrum of the output is given by the *sum* of the phase spectrum of the input and the phase characteristic of the circuit.

- The *Dirichlet conditions* are sufficient conditions for the Fourier series of $v(t)$ to *equal* $v(t)$ except at isolated discontinuities.

◆ The Fourier coefficients make the partial Fourier series $\hat{v}_N(t)$ the best possible approximation to $v(t)$ in the sense of minimum mean square error.

◆ A 9 percent overshoot of a partial Fourier series occurs in the vicinity of a discontinuity of $v(t)$.

■ PROBLEMS ■

Section 14.2

1. The purpose of this problem is to derive a compact formula for the waveform $v(t)$ of Eq. (14.1), shown in Fig. 14.3b.
   (a) Define the finite geometric sum

   $$S_n = 1 + z + z^2 + \cdots + z^n$$

   Multiply S_n by z, and subtract the resulting product from S_n to show that

   $$S_n = \frac{1 - z^{n+1}}{1 - z}$$

   (b) Set $z = e^{j\omega_1 t}$ to establish the identity

   $$\frac{1 - e^{j(n+1)\omega_1 t}}{1 - e^{j\omega_1 t}}$$

   $$= 1 + e^{j\omega_1 t} + e^{j2\omega_1 t} + \cdots + e^{jn\omega_1 t}$$

   (c) Add the equation from (b) to its conjugate, then rearrange the result to show that

   $$v(t) = 1 + 2 \cos \omega_1 t + 2 \cos 2\omega_1 t + \cdots$$
   $$+ 2 \cos n\omega_1 t$$
   $$= \frac{\sin\left[(2n+1)\omega_1 t/2\right]}{\sin\left(\omega_1 t/2\right)}$$

   (d) The result from (c) applies for any integer $n \geq 0$. Let $n = 5$ to obtain the compact formula for the waveform of Eq. (14.1).

Section 14.3

2. A periodic waveform is given by $v(t) = 10 + 7 \cos 6\pi 10^3 t + 5 \cos (8\pi 10^3 t + 45°)$. Find the period T and the real Fourier-series coefficients.
3. Repeat Problem 2 for the waveform

   $$v(t) =$$
   $$[10 + 7 \cos 6\pi 10^3 t + 5 \cos (8\pi 10^3 t + 45°)]^2$$

4. Consider the time-shifted periodic impulse train

$$v(t) = \sum_{n=-\infty}^{+\infty} A\delta(t - t_d - nT)$$

where T is the period and t_d is an arbitrary time shift.
   (a) Sketch $v(t)$ for $A = 5$, $T = 1$ ms, and $t_d = 0.25$ ms.
   (b) Use Eqs. (14.4), (14.7), and (14.8) to determine the real Fourier coefficients of $v(t)$. (Assume that A, T, and t_d are arbitrary.)
   (c) Write the real Fourier series of $v(t)$.
   (d) Use the trigonometric identity $\cos (u - v) = \cos u \cos v + \sin u \sin v$ to rewrite the real Fourier series of (c) in terms of only cosine functions. How does the parameter t_d appear in your result?

5. The periodic waveform

$$v(t) = x_0\sqrt{2} \cos 2\pi t + y_0\sqrt{2} \sin 2\pi t + z_0\sqrt{2} \cos 4\pi t$$

is obviously determined by the values of x_0, y_0, and z_0. The quantity $\mathbf{v} = (x_0, y_0, z_0)$ can be regarded as a vector in a three-dimensional cartesian coordinate system. Find a relationship between the mean square value of $v(t)$ and the length of the vector $\mathbf{v}$.

6. Let $\{\phi_i(t), i = 1, 2, \ldots\}$ be any set of real waveforms having the property that

$$\int_{t_0}^{t_0 + T} \phi_i(t)\phi_k(t)\, dt = \begin{cases} c_i & i = k \\ 0 & i \neq k \end{cases}$$

where c_i is a constant that depends on i. Assume that

$$v(t) = \sum_{i=1}^{\infty} \alpha_i \phi_i(t) \qquad t_0 \leq t \leq t_0 + T \quad (14.76)$$

   (a) Show that if Eq. (14.76) is true, then

$$\alpha_i = \frac{1}{c_i} \int_{t_0}^{t_0 + T} v(t)\phi_i(t)\, dt \qquad (14.77)$$

(b) Show that if Eq. (14.76) is true, then

$$\int_{t_0}^{t_0+T} v^2(t)\,dt = \sum_{i=1}^{\infty} c_i \alpha_i^2 \qquad (14.78)$$

(c) Show that Eqs. (14.3), (14.4), (14.7), (14.8), and (14.11) of the text are special cases of Eqs. (14.76), (14.77), and (14.78).

Section 14.4

7. To prove Symmetry Property 2, write the Fourier series of an arbitrary real periodic waveform $v(t)$ as $v(t) = v_a(t) + v_b(t)$, where

$$v_a(t) = a_0 + \sum_{n=1}^{\infty} a_n \cos 2\pi n f_1 t$$

and

$$v_b(t) = \sum_{n=1}^{\infty} b_n \sin 2\pi n f_1 t$$

(a) Show that $v_a(t)$ is an even function and $v_b(t)$ is an odd function.
(b) Use your result from (a) to show that $v(t)$ is an odd function if and only if $v_a(t) = 0$ for all t.
(c) Use your result from (b) to show that $v(t)$ is an odd function if and only if $a_n = 0$ for $n = 0, 1, 2, \ldots$.

8. To prove Symmetry Property 3, write the Fourier series of an arbitrary real periodic waveform as $v(t) = v_1(t) + v_2(t)$, where

$$v_1(t) =$$

$$a_0 + \sum_{n=2,4,6,\ldots}^{\infty} (a_n \cos 2\pi n f_1 t + b_n \sin 2\pi n f_1 t)$$

and

$$v_2(t) = \sum_{n=1,3,5,\ldots}^{\infty} (a_n \cos 2\pi n f_1 t + b_n \sin 2\pi n f_1 t)$$

(a) Show that $v_1(t)$ has the property that $v_1(t) = v_1(t - T/2)$, and $v_2(t)$ has the property that $v_2(t) = -v_2(t - T/2)$.
(b) Use your result from (a) to show that $v(t)$ is half-wave symmetric if and only if $v_1(t) = 0$ for all t.
(c) Use your result from (b) to show that $v(t)$ is half-wave symmetric if and only if $a_0 = a_n = b_n = 0$ for $n = 2, 4, 6, \ldots$.

9. Suppose that A V dc is added to the odd symmetric waveform $v(t)$ of Fig. 14.5 to produce a new waveform $v'(t) = A + v(t)$.
(a) Sketch $v'(t)$.
(b) Is $v'(t)$ odd symmetric?
(c) Show why $v'(t)$ contains no *cosine* functions.

10. Suppose that A V dc is added to the half-wave symmetric waveform $v(t)$ of Fig. 14.6 to produce a new waveform $v'(t) = A + v(t)$.
(a) Sketch $v'(t)$.
(b) Is $v'(t)$ half-wave symmetric?
(c) Show why the Fourier coefficients of $v'(t)$ satisfy $a_0 \neq 0$, $a_n = b_n = 0$ for $n = 2, 4, 6, \ldots$.

11. A *full-wave rectified cosine* wave is defined by

$$v(t) = |A \cos 2\pi t/T_0|$$

(a) Sketch $v(t)$ versus t. What is the period?
(b) Determine the real Fourier series.

Section 14.5

12. Plot the one-sided amplitude, phase, and power spectra of a periodic impulse train

$$v(t) = \sum_{n=-\infty}^{+\infty} A\delta(t - nT)$$

13. Repeat Problem 12 for the delayed periodic impulse train

$$v(t) = \sum_{n=-\infty}^{+\infty} A\delta(t - t_d - nT)$$

where t_d is an arbitrary constant.

14. The periodic pulse train of Fig. 14.10 is applied across the terminals of a 3-kΩ resistance. Find the percentage of the total average power dissipation caused by the dc component. Write your answer in terms of τ and T.

15. The *duty factor* of the periodic pulse train $v(t)$ of Fig. 14.10 is defined by DF $= (\tau/T) \times 100\%$.
(a) Sketch $v(t)$ for $A = 5$ and a duty factor of 50%.
(b) Plot the one-sided amplitude, phase, and power spectra of $v(t)$ of (a).
(c) Comment on any interesting features of your plots.

Sketch the periodic pulse train of Example 14.3 and its amplitude, phase, and power spectra for the following two special cases.

16. $\tau = T$
17. $\tau = \varepsilon$ and $A = \varepsilon^{-1}$, where ε is arbitrarily small ($\varepsilon \to 0$)
18. Plot the one-sided amplitude, phase, and power spectra of the periodic waveform

$$v(t) = 5 \cos 100\pi t + 17 - 3 \sin 300\pi t$$
$$+ 4 \sin 100\pi t$$

19. (a) Use Eq. (14.20) to show that the effect of replacing a periodic waveform $v(t)$ by $v'(t) = v(t - t_d)$ is to replace the Fourier phases ϕ_n, $n = 1, 2, \ldots$, by $\phi'_n = \phi_n - 2\pi n f_1 t_d$.
   (b) Illustrate your result from (a) by assuming that the periodic pulse train of Fig. 14.10 is delayed by $t_d = T/4$, where $A = 10$ V, $\tau = 0.2$ ms, and $T = 1$ ms. Plot the new waveform and its associated one-sided amplitude, phase, and power spectra.

Section 14.6

20. (a) Use the fact that $\cos\theta = \frac{1}{2}e^{j\theta} + \frac{1}{2}e^{-j\theta}$ to find the complex Fourier series of the periodic voltage $v(t)$ given by Eq. (14.1).
   (b) Plot the two-sided amplitude, phase, and power spectra.
21. A periodic waveform $v(t)$ is obtained by repeating a truncated exponential pulse $p(t) = Ae^{at}[u(t) - u(t - T)]$ every T seconds, that is

$$v(t) = \sum_{n=-\infty}^{+\infty} p(t - nT)$$

   (a) Sketch $p(t)$ for $A = 5$, $a = 1$, and $T = 2$ s.
   (b) Sketch $v(t)$ for the parameters used in (a).
   (c) Find the complex Fourier series of $v(t)$. Express your answer in terms of A, a, T, and $f_1 = 1/T$.
22. (a) Derive the complex Fourier coefficients of the waveform shown.

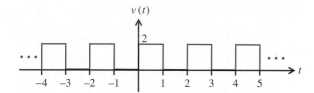

23. What percentage of the total power of the waveform $v(t)$ of Fig. 14.15 is contained in the nth harmonic? Express your answer in terms of n, τ, and T.
24. (a) Use Eq. (14.29) to derive the complex Fourier coefficients of a time-shifted periodic impulse train

$$v(t) = \sum_{n=-\infty}^{+\infty} \delta(t - nT - t_d)$$

   where t_d is arbitrary.
   (b) Plot the two-sided amplitude, phase, and power spectra.
   (c) Use your answer from (a) to determine the Fourier amplitudes and phases A_n and ϕ_n.
   (d) Plot the one-sided amplitude, phase, and power spectra.
   (e) Use your answer to (a) to obtain the real Fourier coefficients a_n and b_n.
   (f) Write the three forms of the Fourier series of $v(t)$.
25. Repeat Problem 24 with $v(t)$ equal to the saw-tooth wave of Fig. 14.5.

Section 14.7

26. A waveform

$$v_a(t) = \sum_{n=-\infty}^{+\infty} V_{an} e^{j2\pi n f_1 t}$$

   with period $T = 1/f_1$ is time-delayed $T/2$ s to produce another waveform $v_b(t) = v_a(t - T/2)$. The waveform $v_b(t)$ is then subtracted from $v_a(t)$ to produce a third waveform $v_c(t) = v_a(t) - v_b(t)$. An illustration of this procedure is given in Fig. 14.29 for the special case that $v_a(t)$ is a periodically repeated exponential signal.
   (a) Use the time-shift property P3 to show that, in general, the nth complex Fourier-series coefficient of $v_c(t)$ is given by

$$V_{cn} = \begin{cases} 2V_{an} & n \text{ odd} \\ 0 & n \text{ even} \end{cases}$$

   (b) Does $v_c(t)$ (in general) have half-wave symmetry?
   (c) Show that the nth complex Fourier-series coefficient of the alternating exponential

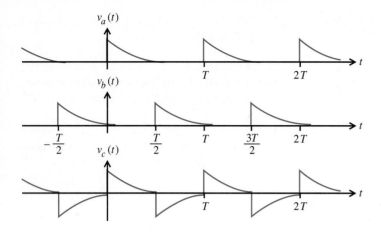

FIGURE 14.29

waveform $v_c(t)$ shown in the preceding figure is

$$V_{cn} = \begin{cases} \dfrac{2A\tau}{T} \dfrac{1 - e^{-T/\tau}}{1 + j2\pi n\tau/T} & n \text{ odd} \\ 0 & n \text{ even} \end{cases}$$

27. Use any method to verify or derive FS1 of Table 14.1.
28. Use any method to verify or derive FS2 of Table 14.1.
29. Use any method to verify or derive FS5 of Table 14.1.
30. Use any method to verify or derive FS6 of Table 14.1.
31. Use any method to verify or derive FS7 of Table 14.1.
32. Use any method to verify or derive FS8 of Table 14.1.
33. Use any method to verify or derive FS9 of Table 14.1.
34. Let $x(t)$ and $y(t)$ be two periodic functions with period T. Show that

$$\frac{1}{T} \int_{t_0}^{t_0 + T} x(t)y^*(t)\,dt = \sum_{n=-\infty}^{+\infty} X_n Y_n^*$$

where X_n and Y_n are the complex Fourier-series coefficients $x(t)$ and $y(t)$, respectively.

35. Let $x(t)$ and $y(t)$ be two periodic functions with period T. Define a new function

$$z(t) = \frac{1}{T} \int_{t_0}^{t_0 + T} x(\lambda)y(t - \lambda)\,d\lambda$$

where t_0 is arbitrary. The function $z(t)$ is referred to as the *convolution* of the periodic functions $x(t)$ and $y(t)$.
   (a) Show that $z(t)$ is periodic with period T.
   (b) Show that the nth complex Fourier-series coefficient of $z(t)$, Z_n, is given by $Z_n = X_n Y_n$, where X_n and Y_n are the nth complex Fourier-series coefficients of $x(t)$ and $y(t)$, respectively.

Section 14.8

36. A low-pass RC circuit has a periodic pulse train input, where $A = 1$ V, $\tau = 0.3$ ms, and $T = 1$ ms. The resistance is $R = 1$ kΩ. Choose C so that the fundamental component of the output waveform is 100 times smaller than the dc component of the output waveform.
37. A series band-pass RLC filter uses a resistance $R = 2.2$ kΩ.
   (a) Choose the values of L and C to pass the fifth harmonic of a 10-kHz input triangular wave without attenuation and so that the seventh output harmonic is 25 times smaller in amplitude than the fifth.
   (b) Make a rough sketch of the output to illustrate the predominant shape of the output waveform.
38. The full-wave-rectified cosine $v(t)$ of FS9 (Table 14.1) is applied to a low-pass RC filter.
   (a) Use FS9 to write the real Fourier series of $v(t)$. Put your result in the form of Eq. (14.20). Assume that $A = 120\sqrt{2}$ and $T_0 = \frac{1}{60}$ s, and give the numerical values of the

dc term and the amplitudes and phases of the first five harmonics.

(b) Plot the one-sided amplitude and phase spectra of $v(t)$ for $A = 120\sqrt{2}$ and $T_0 = \frac{1}{60}$ s.

(c) Find the voltage transfer function of the filter and plot the filter's amplitude and phase characteristics. Assume for your plots that $RC = 1/(120\pi)$ s.

(d) Plot the one-sided amplitude and phase spectra of the output $v_0(t)$ for the numerical values of (a) through (c).

(e) Write the series expression for the filter output, giving the numerical values of the dc term and the amplitudes and phases of the first five harmonics.

39. A half-rectified cosine voltage with a peak of 1 V and period 1 ms is input to a low-pass RC filter.

(a) Give the formula for the nth complex Fourier coefficient of the output.

(b) Evaluate your answer to (a) for $n = 0, \pm1, \pm2,$ and ±3 and $RC = 10$ ms.

(c) Give a rough sketch of the output. State basically what the circuit has done to the input.

40. A hypothetical circuit called an *ideal low-pass filter* has the voltage transfer function illustrated in the figure. The input voltage is the half-rectified cosine of Table 14.1, where $A = 1$ V and $T = 1$ ms. Determine and sketch the output voltage for $W = 500$ Hz.

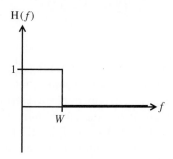

41. Repeat Problem 40 for $W = 1500$ Hz.

Section 14.10

42. (a) Determine the values of α and ϕ in the following circuit to minimize the average power dissipated in the resistance. The cur-

rent $i_s(t)$ is an arbitrary periodic waveform with period T.

(b) What is the resulting minimum average power dissipation?

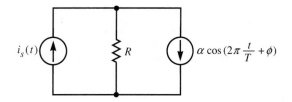

43. In what sense is
(a) Plancherel's theorem *more general* than the Dirichlet conditions?

(b) Plancherel's theorem weaker than the Dirichlet conditions?

44. Let $\{\phi_i(t), i = 1, 2, \ldots, I\}$ be any set of real functions satisfying

$$\int_{t_0}^{t_0+T} \phi_i(t)\phi_k(t)\, dt = \begin{cases} c_i & i = k \\ 0 & i \neq k \end{cases}$$

where c_i is a constant that depends on i. Let $v(t)$ be any waveform, and let $\hat{v}_I(t)$ be an approximation to $v(t)$:

$$\hat{v}_I(t) = \sum_{i=1}^{I} \alpha_i \phi_i(t)$$

(a) Show that the coefficients α_i minimizing the integral squared error

$$\varepsilon = \int_{t_0}^{t_0+T} [v(t) - \hat{v}_I(t)]^2\, dt$$

are

$$\alpha_i = \frac{1}{c_i} \int_{t_0}^{t_0+T} v(t)\phi_i(t)\, dt \qquad i = 1, 2, \ldots, I$$

(b) Show that

$$\varepsilon_{\min} = \int_{t_0}^{t_0+T} v^2(t)\, dt - \sum_{i=1}^{I} c_i \alpha_i^2$$

(c) Show that

$$\varepsilon_{\min} = \int_{t_0}^{t_0+T} v^2(t)\, dt - \int_{t_0}^{t_0+T} \hat{v}_I^2(t)\, dt$$

Comprehensive Problems

45. Frequencies ω_1 and ω_2 are present in the waveform

$$v_s(t) = a \cos(\omega_1 t + \phi_1) + b \cos(\omega_2 t + \phi_2)$$

(a) Assume that $v_s(t)$ is the input to an LLTI circuit. Show that the same frequencies ω_1 and ω_2 are present in the output $v_o(t)$.

(b) Assume that $v_s(t)$ is the input to a *square-law device* whose output is $v_o(t) = cv_s^2(t)$, where c is a real nonnegative constant. What frequencies are present in $v_o(t)$?

(c) Assume that $v_s(t)$ appears at the input to a modulator (multiplier), shown schematically below. The modulator output is $v_o(t) = v_s(t) \cos \omega_c t$. What frequencies are present in $v_o(t)$?

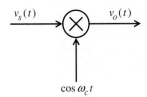

(d) Assume that $v_s(t)$ is input to the switching circuit shown below. The switch opens and closes periodically every T seconds, as shown in the following graph. The output of the switching circuit is

$$v_o(t) = \begin{cases} v_s(t) & mT - \dfrac{T}{4} \le t \le mT + \dfrac{T}{4} \\ 0 & mT + \dfrac{T}{4} < t < mT + \dfrac{3T}{4} \end{cases}$$

where m is an integer. What frequencies are present in $v_o(t)$? [*Hint:* $v_o(t) = v_s(t)v(t)$, where $v(t)$ is the periodic rectangular pulse of Fig. 14.10 with $\tau = T/2$.]

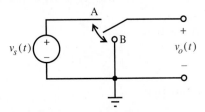

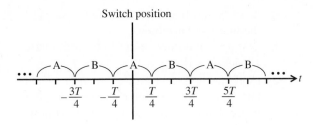

Switch position

46. As stated in this chapter, it is usually impossible to recognize a closed form for the Fourier-series solution of the output of a circuit. The following is an exception to this rule.

The square wave of Fig. 14.7 is input to a series RL circuit. Let A' (rather than A) denote the amplitude of the square wave. The output is the inductance voltage $v_L(t)$.

(a) Use Eq. (14.59) and FS1 to show that $v_L(t)$ has the complex Fourier series

$$v_L(t) = \sum_{n=-\infty}^{+\infty} V_{Ln} e^{j2\pi n f_1 t}$$

where

$$V_{Ln} = \begin{cases} \dfrac{4A'f_1 L}{R + j2\pi n f_1 L} & n \text{ odd} \\ 0 & n \text{ even} \end{cases}$$

(b) Show that by appropriate choice of constants A and τ, V_{Ln} of (a) can be put into the form

$$V_{Ln} = \begin{cases} \dfrac{2A\tau}{T} \dfrac{1 - e^{-T/\tau}}{1 + j2\pi n \tau/T} & n \text{ odd} \\ 0 & n \text{ even} \end{cases}$$

(c) Compare the result from (b) with the result in Problem 26. Conclude that $v_L(t)$ is the alternating exponential waveform of Problem 26. Write the *closed form* expressions for $v_L(t)$ for $0 \le t < T/2$ and $T/2 \le t < T$.

(d) Sketch the circuit and its input and output signals.

CHAPTER
15

The Fourier Transform

In Chapter 14 we saw that a waveform with period T can be represented by a discrete sum of complex exponential components whose frequencies are integer multiples of $1/T$. This sum was called the Fourier-series representation of the periodic waveform. This chapter describes the limit of the Fourier series as $T \to \infty$. This limit is the *inverse Fourier transform*. The inverse Fourier transform is a representation of an *aperiodic* (nonperiodic) signal $v(t)$ as an *integral* of complex exponential components whose frequencies range *continuously* from minus to plus infinity.

The motivation to represent an aperiodic signal $v(t)$ as an integral of complex exponential components is the same as that to represent a periodic signal by a Fourier series: When $v(t)$ is input to a stable LLTI circuit, the output can be found by simple addition (integration) of the responses to the individual complex exponential components in $v(t)$.

Just as there is a formula to compute the Fourier coefficients of a periodic waveform, there is a formula to compute the frequency-domain representation of an aperiodic signal $v(t)$. This formula is called the *direct Fourier transform* of the signal. The direct Fourier transform transforms the time-domain signal $v(t)$ into its frequency-domain representation $V(f)$. The *inverse Fourier transform* transforms $V(f)$ back into $v(t)$.

In studying the Fourier transform, we enter the general theory of "signals and systems." Most of the results presented here apply not only to lumped linear time-invariant (LLTI) systems but also more generally to linear time-invariant (LTI) systems. Indeed, the Fourier transform is the most useful of all analytical aids in the theory of LTI systems. Therefore, many of the examples and problems of this chapter deal with signals and systems in a more general way than other chapters did.

15.1 Definitions

We begin with the definition of the Fourier transform:

DEFINITION

Fourier Transform

The **direct Fourier transform** of a signal $v(t)$ is defined by

$$V(f) = \int_{-\infty}^{+\infty} v(t)e^{-j2\pi ft}\, dt \qquad (15.1)$$

The direct Fourier transform $V(f)$ is a complete frequency-domain description of the time-domain signal $v(t)$. It is a *complete* description because we can always obtain the original signal, $v(t)$, from $V(f)$. [In mathematical jargon, the transformation between $v(t)$ and $V(f)$ is said to be *one-to-one**: For each $v(t)$ there is a unique $V(f)$.]

The formula used to recover $v(t)$ from $V(f)$ is called the *inverse Fourier transform*:

Inverse Fourier Transform

$$v(t) = \int_{-\infty}^{+\infty} V(f)e^{j2\pi ft}\, df \qquad (15.2)$$

The *inverse* Fourier transform, Eq. (15.2), should be compared with Eq. (14.24), which is the corresponding representation for periodic signals and which is reproduced below.

$$v(t) = \sum_{n=-\infty}^{+\infty} V_n e^{j2\pi n f_1 t}$$

By comparing Eq. (15.2) with Eq. (14.24), we can acquire an intuitive understanding of Eqs. (15.1) and (15.2). The quantity $V(f)e^{j2\pi ft}\, df = [V(f)\, df]e^{j2\pi ft}$ appearing in Eq. (15.2) is a complex exponential with frequency f and infinitesimal complex amplitude $V(f)\, df$. It corresponds to the complex exponential $V_n e^{j2\pi n f_1 t}$ of Eq. (14.24), which has frequency nf_1 and finite complex amplitude V_n. Thus $V(f)$ of

---

* There are mathematical qualifications to this statement that are best ignored here.

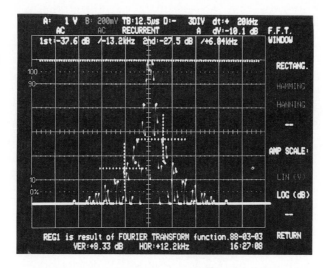

The Fourier transform can be approximately obtained using a digital computer algorithm known as the fast Fourier transform (FFT). The result of a typical FFT computation is recorded on a storage oscilloscope. Courtesy of John Fluke Mfg. Co., Inc.

Eq. (15.1) tells us the essential information regarding the infinitesimal complex amplitudes in Eq. (15.2). Although the sum in Eq. (14.24) is taken over only discrete frequencies nf_1, where n is an integer, the sum in Eq. (15.2) is taken over all f, $-\infty < f < +\infty$. If $v(t)$ has the unit volts, then $V(f)\,df$ also has the unit volts, and, in turn, the unit of $V(f)$ is volt-seconds or volts per hertz (V·s or V/Hz). $V(f)$ is often referred to as the *voltage density spectrum** of $v(t)$.

The density spectrum $V(f)$ is generally a complex function of frequency f. The magnitude of $V(f)$, $|V(f)|$, is called the *amplitude density spectrum*, and the angle of $V(f)$, $\underline{/V(f)}$, is called the *phase spectrum*. The transformation from $v(t)$ to $V(f)$ is called *Fourier analysis* or *spectrum analysis*, whereas that from $V(f)$ to $v(t)$ is called *Fourier synthesis*.

Convenient notations for the transformations of Eqs. (15.1) and (15.2) are

$$V(f) = \mathscr{F}\{v(t)\} \tag{15.3a}$$

$$v(t) = \mathscr{F}^{-1}\{V(f)\} \tag{15.3b}$$

To emphasize the fact that the transformations of Eqs. (15.1) and (15.2) are one-to-one, we write

$$v(t) \leftrightarrow V(f) \tag{15.3c}$$

The functions $v(t)$ and $V(f)$ taken together are called a *Fourier transform pair*.

EXAMPLE 15.1

Determine and plot the density spectrum of the one-sided decaying exponential

$$v(t) = e^{-t/\tau}u(t)$$

(where τ is positive), illustrated in Fig. 15.1.

■ SOLUTION

For notational simplicity, we set $a = -1/\tau$. Note that a is negative because τ is positive. Substitution of $e^{at}u(t)$ into Eq. (15.1) yields

---

* $V(f)$ is referred to as simply the *spectrum* (rather than density spectrum) of $v(t)$ in much of the electrical engineering literature. In this book, the use of the term *spectrum* is restricted to the Fourier-series coefficients described in Chapter 14.

FIGURE 15.1
One-sided decaying
exponential waveform

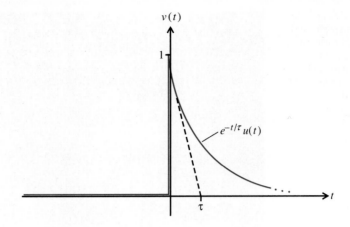

$$V(f) = \int_{-\infty}^{+\infty} e^{at} u(t) e^{-j2\pi ft} \, dt$$

$$= \int_{0}^{+\infty} e^{(a-j2\pi f)t} \, dt$$

$$= \frac{1}{a - j2\pi f} e^{(a-j2\pi f)t} \Big|_{0}^{\infty}$$

$$= \frac{1}{a - j2\pi f} e^{at} e^{-j2\pi ft} \Big|_{\infty} - \frac{1}{a - j2\pi f} e^{at} e^{-j2\pi ft} \Big|_{0}$$

Because a is negative, the factor e^{at} will equal zero for $t = \infty$. Therefore, the first term on the right-hand side of the above equation equals zero. This leaves

$$V(f) = -\frac{1}{a - j2\pi f} e^{at} e^{-j2\pi ft} \Big|_{0} = \frac{1}{-a + j2\pi f}$$

or

$$V(f) = \frac{1}{(1/\tau) + j2\pi f}$$

Plots of $|V(f)|$ and $\underline{/V(f)}$ are given in Fig. 15.2. ∎

FIGURE 15.2
Density spectrum of the
one-sided decaying
exponential: (a) amplitude
density; (b) phase

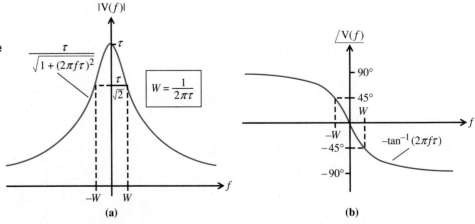

Analogous to Eq. (14.38) for Fourier coefficients, the Fourier transform of a real signal has the property that $V(-f) = V^*(f)$. It follows that the amplitude density spectrum of a real signal is an even function, $|V(-f)| = |V(f)|$, and the phase spectrum is an odd function, $\underline{/V(-f)} = -\underline{/V(f)}$. This property is proved in Problem 3.

EXAMPLE 15.2

Determine and plot the density spectrum of the rectangular pulse

$$v(t) = \Pi\left(\frac{t}{\tau}\right)$$

illustrated in Fig. 15.3.

FIGURE 15.3
Rectangular pulse

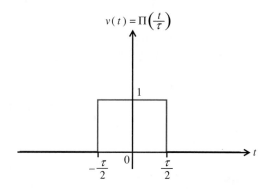

$$v(t) = \Pi\left(\frac{t}{\tau}\right)$$

■ **SOLUTION**

Substitution of $\Pi(t/\tau)$ into Eq. (15.1) yields

$$V(f) = \int_{-\infty}^{+\infty} \Pi\left(\frac{t}{\tau}\right) e^{-j2\pi ft}\, dt = \int_{-\tau/2}^{+\tau/2} e^{-j2\pi ft}\, dt$$

$$= \frac{1}{-j2\pi f} e^{-j2\pi ft}\Big|_{-\tau/2}^{+\tau/2}$$

$$= \frac{1}{-j2\pi f}\left(e^{-j\pi f\tau} - e^{+j\pi f\tau}\right)$$

$$= \tau\,\frac{\sin \pi f\tau}{\pi f\tau}$$

The function $\sin \pi x/\pi x$ is very important in engineering and is called the sinc function:

$$\mathrm{sinc}\, x = \frac{\sin \pi x}{\pi x} \tag{15.4}$$

The sinc function is plotted in Fig. 15.4. Notice that $\mathrm{sinc}\, x = 1$ for $x = 0^*$ and $\mathrm{sinc}\, x = 0$ for $x = \pm 1, \pm 2, \ldots$.

---

* Direct substitution of $x = 0$ into the right-hand side of Eq. 15.4 results in the indeterminant form $0/0$. The quantity sinc 0 is *defined* as $\mathrm{sinc}\, 0 = \lim_{x \to 0} (\sin \pi x)/\pi x = 1$.

FIGURE 15.4
The function sinc (x)

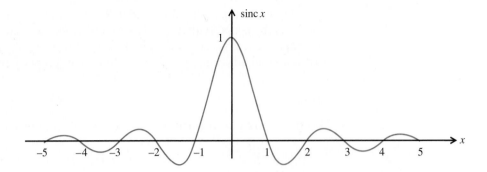

In terms of Eq. (15.4),

$$\Pi\left(\frac{t}{\tau}\right) \leftrightarrow \tau \operatorname{sinc} f\tau$$

The density spectrum is plotted in Fig. 15.5. Because $V(f)$ is purely real, there is no need to plot the amplitude density and phase spectra separately. ∎

FIGURE 15.5
Density spectrum of the
rectangular pulse

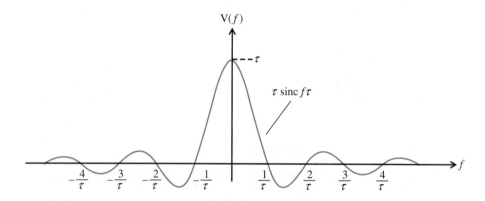

REMEMBER

For a real signal $v(t)$, the amplitude density spectrum is an even function of f, and the phase spectrum is an odd function of f.

EXERCISES

Use Eq. (15.1) to determine the density spectra of the waveforms in Exercises 1 through 3. Sketch each waveform and its corresponding density spectrum.

1. $v(t) = \delta(t)$
2. $v(t) = e^{bt}u(-t),\ b > 0$
3. $v(t) = e^{at}u(t) + e^{bt}u(-t),\ a < 0,\ b > 0$

Use Eq. (15.2) to determine the waveforms with the density spectra in Exercises 4 through 6. Sketch each waveform and its corresponding density spectrum.

4. $V(f) = \delta(f)$
5. $V(f) = \frac{1}{2}A\delta(f - f_0) + \frac{1}{2}A\delta(f + f_0)$

6. $V(f) = \Pi(f/2W)$

7.   Show that (a) $\mathcal{F}\{v_1(t) + v_2(t)\} = \mathcal{F}\{v_1(t)\} + \mathcal{F}\{v_2(t)\}$ and (b) $\mathcal{F}\{av(t)\} = a\mathcal{F}\{v(t)\}$. What can you conclude about the Fourier operator $\mathcal{F}\{\cdot\}$?

8.   Find $v(t)$ if $V(f) = (B/2j)\delta(f - f_0) - (B/2j)\delta(f + f_0)$.

15.2   Relation of the Direct and Inverse Fourier Transforms to Fourier Series

The inverse Fourier transform is the limiting form of the complex Fourier series as $T \to \infty$. This can be shown in four simple steps. The first step is to define the *periodic version* $v_T(t)$ of the aperiodic signal $v(t)$. Figure 15.6 illustrates the definition of $v_T(t)$ when $v(t)$ is a rectangular pulse. The periodic version $v_T(t)$ is simply the waveform $v(t)$ repeated every T seconds and we can write it as a Fourier series:

$$v_T(t) = \sum_{n=-\infty}^{+\infty} V_n e^{j2\pi n f_1 t} \tag{15.5a}$$

where $f_1 = 1/T$ and

$$V_n = \frac{1}{T} \int_{-T/2}^{T/2} v_T(t) e^{-j2\pi n f_1 t}\, dt \tag{15.5b}$$

FIGURE 15.6
(a) An aperiodic signal $v(t)$ and (b) its periodic version $v_T(t)$

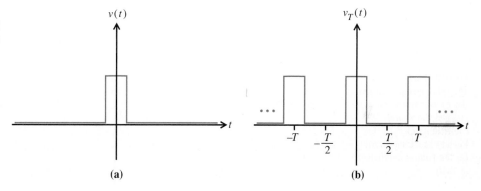

(a)

(b)

Notice from Fig. 15.6 that

$$v(t) = \begin{cases} v_T(t) & \text{for } |t| \le \dfrac{T}{2} \\ 0 & \text{otherwise} \end{cases} \tag{15.6}$$

The third step is to use Eq. (15.6) to rewrite the nth complex Fourier coefficient V_n in Eq. (15.5b) as

$$V_n = \frac{1}{T} \int_{-\infty}^{+\infty} v(t) e^{-j2\pi n f_1 t}\, dt \tag{15.7a}$$

or equivalently,

$$V_n = \frac{1}{T} V(n f_1) \tag{15.7b}$$

where $V(f)$ is the Fourier transform of $v(t)$ defined in Eq. (15.1). The relationship between the Fourier transform of $v(t)$ and the Fourier coefficients of $v_T(t)$ expressed by Eq. (15.7b) is illustrated in Fig. 15.7. Observe that except for the scale factor $1/T$, the complex Fourier coefficients V_n of the periodic signal $v_T(t)$ can be regarded as "samples" (represented by the vertical lines) of the density spectrum $V(f)$ of the nonperiodic signal $v(t)$ at the discrete frequencies nf_1, $n = 0, \pm 1, \pm 2, \ldots$. The substitution of Eq. (15.7b) into Eq. (15.5b) yields

$$v_T(t) = \frac{1}{T} \sum_{n=-\infty}^{+\infty} V(nf_1)e^{j2\pi n f_1 t} \tag{15.8}$$

The final step in the derivation is the key conceptual step. As we can see by referring to Fig. 15.6, the periodic signal $v_T(t)$ equals the nonperiodic signal $v(t)$ if $T = \infty$. That is,

$$v(t) = \lim_{T \to \infty} v_T(t)$$

$$= \lim_{T \to \infty} \frac{1}{T} \sum_{n=-\infty}^{+\infty} V(nf_1)e^{j2\pi n f_1 t} \tag{15.9}$$

In taking the limit in Eq. (15.9), it is useful to recognize that f_1 is not only the frequency of the fundamental component of the Fourier series, it is also the frequency interval

$$\Delta f = f_1 = \frac{1}{T} \tag{15.10}$$

separating the lines in the spectrum of $v_T(t)$ as shown in Fig. 15.7. Observe that

FIGURE 15.7
The relationship between the Fourier transform of $v(t)$ and the Fourier coefficients of $v_T(t)$. (a) Density spectrum of $v(t)$; (b) the Fourier coefficients of $v_T(t)$

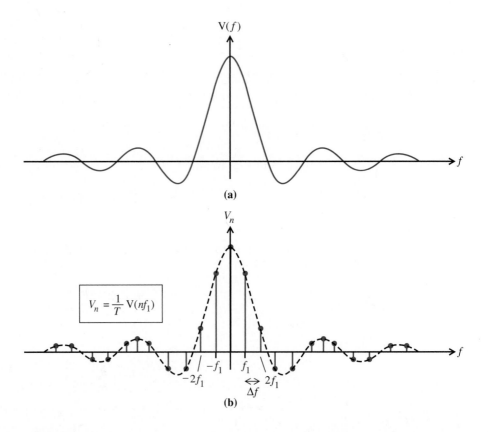

$V(f)$

(a)

V_n

$$V_n = \frac{1}{T} V(nf_1)$$

$-2f_1$ $-f_1$ f_1 $2f_1$

Δf

(b)

as T increases to infinity, this frequency interval shrinks to zero. The substitution of Eq. (15.10) into Eq. (15.9) yields

$$v(t) = \lim_{\Delta f \to 0} \sum_{n=-\infty}^{+\infty} V(n\,\Delta f)e^{j2\pi n\,\Delta f t}\,\Delta f$$

$$= \lim_{\substack{\Delta f \to 0 \\ f = n\,\Delta f \\ n=-\infty}} \sum_{n=-\infty}^{+\infty} V(f)e^{j2\pi ft}\,\Delta f \tag{15.11}$$

In the limit as $\Delta f \to 0$, Eq. (15.11) becomes, by the definition of an integral,

$$v(t) = \int_{-\infty}^{+\infty} V(f)e^{j2\pi ft}\,df \tag{15.12}$$

which is the inverse Fourier transform of $V(f)$.

The preceding analysis has far-reaching implications. Because the inverse Fourier transform is a limiting form of the Fourier series, all the properties of the Fourier series can be extended to the inverse Fourier transform. We will see this extension in the following section.

15.3 Properties and Tables

In this section, we develop tables of important Fourier transform properties and pairs. We will use these tables later in this chapter to solve circuit and system problems of practical interest.

Important Fourier Transform Properties

Ten useful properties of the Fourier transform are listed in Table 15.1.

Linearity (P1) Property P1 states that if a scaled waveform $a_2v_2(t)$ is added to a scaled waveform $a_1v_1(t)$, then the effect is to add the scaled density spectrum

TABLE 15.1
Fundamental properties of the Fourier transform

NUMBER	PROPERTY	SIGNAL	DENSITY SPECTRUM
P1	Linearity	$a_1v_1(t) + a_2v_2(t)$	$a_1V_1(f) + a_2V_2(f)$
P2	Differentiation	$\dfrac{d^nv}{dt^n}$	$(j2\pi f)^nV(f)$
P3	Integration	$\displaystyle\int_{-\infty}^{t} v(\lambda)\,d\lambda$	$\dfrac{1}{2}V(0)\delta(f) + \dfrac{1}{j2\pi f}V(f)$
P4	Reversal	$v(-t)$	$V(-f)$
P5	Scaling	$v\left(\dfrac{t}{a}\right)$	$aV(af),\ a > 0$
P6	Duality	$V(t)$	$v(-f)$
P7	Time shift	$v(t - t_d)$	$V(f)e^{-j2\pi ft_d}$
P8	Frequency shift	$v(t)e^{j2\pi f_c t}$	$V(f - f_c)$
P9	Multiplication	$v_1(t)v_2(t)$	$\displaystyle\int_{-\infty}^{+\infty} V_1(\xi)V_2(f - \xi)\,d\xi$
P10	Convolution	$v_1(t){*}v_2(t) =$ $\displaystyle\int_{-\infty}^{+\infty} v_1(\lambda)v_2(t - \lambda)\,d\lambda$	$V_1(f)V_2(f)$

$a_2V_2(f)$ to the scaled density spectrum $a_1V_1(f)$. In other words, the density spectrum of the sum of scaled time-domain signals equals the sum of the scaled density spectra. It is easy to prove this property if we simply substitute $a_1v_1(t) + a_2v_2(t)$ into Eq. (15.1).

Differentiation (P2)   Property P2 states that differentiation in the time domain corresponds to multiplication by $j2\pi f$ in the frequency domain. This property is proved by differentiation of both sides of Eq. (15.2) with respect to t n times, with the derivative operator moved inside the integral each time. This leads to

$$\frac{d^n v}{dt^n} = \int_{-\infty}^{+\infty} (j2\pi f)^n V(f)e^{j2\pi ft}\, df \tag{15.13a}$$

The right-hand side of Eq. (15.13a) is the inverse Fourier transform of $(j2\pi f)^n V(f)$. Consequently,

$$\frac{d^n v}{dt^n} \leftrightarrow (j2\pi f)^n V(f) \tag{15.13b}$$

Integration (P3)   Property P3 states that if we integrate $v(\lambda)$ over the interval $-\infty < \lambda < t$, the result has a density spectrum $\frac{1}{2}V(0)\delta(f) + (1/j2\pi f)V(f)$. The impulse at $f = 0$ occurs whenever the density spectrum of $v(t)$ is finite at $f = 0$. A derivation of Property P3 is outlined in Problem 13.

Reversal (P4)   Property P4 states that the effect of reversing $v(t)$ about the time axis is to reverse $V(f)$ about the frequency axis. Property P4 can be derived by simply changing variables of integration in Eq. (15.2).

Scaling (P5)   Property P5 states that if we expand $v(t)$ along the time axis, then we compress $V(f)$ along the frequency axis and multiply it by the expansion factor, a. This result is obtained easily by a change of variables of integration in Eq. (15.2).

Duality (P6)   Property P6 states that if we replace f by t in a density function $V(f)$, where $V(f) = \mathscr{F}\{v(t)\}$, then $\mathscr{F}\{V(t)\} = v(-f)$. This shows us, for example that since $\text{sinc}(f) = \mathscr{F}\{\Pi(t)\}$, then $\mathscr{F}\{\text{sinc}(t)\} = \Pi(t)$.

Property P6 can be proved if we change variables $t \to -f, f \to +t$ in

$$v(t) = \int_{-\infty}^{+\infty} V(f)e^{j2\pi ft}\, df = \mathscr{F}^{-1}\{V(f)\} \tag{15.14}$$

to obtain

$$v(-f) = \int_{-\infty}^{+\infty} V(t)e^{-j2\pi ft}\, dt = \mathscr{F}\{V(t)\} \tag{15.15}$$

Time Shift (P7)   Property P7 states that delaying a waveform by t_d seconds has the effect of multiplying its density spectrum by $e^{-j2\pi ft_d}$. This can be proved if we replace t by $t - t_d$ in Eq. (15.2). This leads to

$$v(t - t_d) = \int_{-\infty}^{+\infty} V(f)e^{-j2\pi ft_d}e^{j2\pi ft}\, df$$

$$= \mathscr{F}^{-1}\{V(f)e^{-j2\pi ft_d}\} \tag{15.16}$$

Frequency Shift (P8)   Property P8 states that we can translate any density spectrum by an amount f_c along the frequency axis by multiplying the time-domain signal by $e^{j2\pi f_c t}$. To prove Property P8, we replace f by $f - f_c$ in Eq. (15.1) and recognize that the resulting integral is the direct Fourier transform of $v(t)e^{j2\pi f_c t}$.

A practical application of Property P8 is found in communication systems in which a message signal $v(t)$ modulates (multiplies) a "carrier" $\cos 2\pi f_c t$. Because $\cos \theta = (e^{j\theta} + e^{-j\theta})/2$, the transmitted signal

$$v_x(t) = v(t) \cos 2\pi f_c t \qquad (15.17)$$

can be written as

$$v_x(t) = \frac{1}{2} v(t)e^{j2\pi f_c t} + \frac{1}{2} v(t)e^{-j2\pi f_c t} \qquad (15.18)$$

It follows from Properties P1 and P8 that the density spectrum of the transmitted signal is given by

$$V_x(f) = \frac{1}{2} V(f - f_c) + \frac{1}{2} V(f + f_c) \qquad (15.19)$$

where $V(f)$ is the density spectrum of $v(t)$. Typical density spectra of the message and transmitted signals are illustrated in Fig. 15.8. As can be seen, $V_x(f)$ contains "bands" of frequency components just above and below f_c. For this reason, communications engineers refer to $v_x(t)$ as a *double sideband* signal.

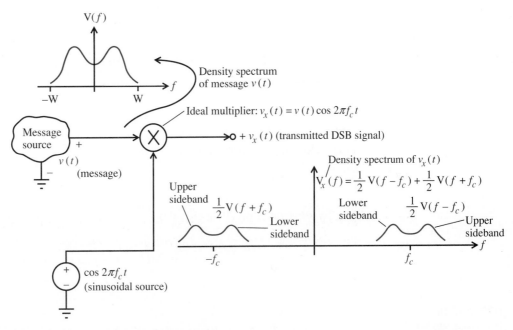

FIGURE 15.8 Double-sideband (DSB) transmitter. The message signal $v(t)$ is multiplied by $\cos 2\pi f_c t$ to produce the transmitted signal $v_x(t)$. Notice that $V_x(f)$ consists of upper and lower "sidebands" located about the center frequency f_c

The derivations of Properties P9 and P10 are outlined in Problem 15. The following examples illustrate how the properties in Table 15.1 can be used.

EXAMPLE 15.3

Find and sketch the density spectrum of the signal

$$v(t) = Ae^{-|t|/\tau}$$

shown in Fig. 15.9.

FIGURE 15.9
The function $Ae^{-|t|/\tau}$

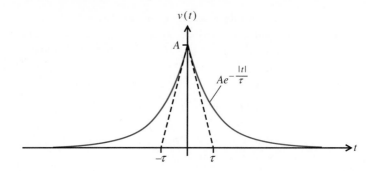

■ SOLUTION

Consider $v(t)$ to be the sum of the two waveforms:

$$v(t) = v_1(t) + v_2(t)$$

where

$$v_1(t) = Ae^{-t/\tau}u(t)$$

and

$$v_2(t) = v_1(-t)$$

We know from Example 15.1 and linearity that

$$\mathscr{F}\{v_1(t)\} = \frac{A\tau}{1 + j2\pi f\tau}$$

and therefore (by Property P4),

$$\mathscr{F}\{v_2(t)\} = \frac{A\tau}{1 - j2\pi f\tau}$$

Then, by superposition (Property P1),

$$\mathscr{F}\{v_1(t) + v_2(t)\} = \frac{A\tau}{1 + j2\pi f\tau} + \frac{A\tau}{1 - j2\pi f\tau} = \frac{2A\tau}{1 + (2\pi f\tau)^2}$$

FIGURE 15.10
The density spectrum of
$Ae^{-|t|/\tau}$:
(a) amplitude density;
(b) phase

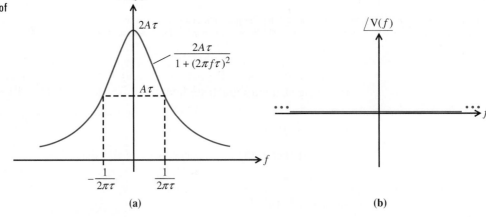

(a)　　　　　　　　　　　　　　　　　　(b)

which gives the result

$$Ae^{-|t|/\tau} \leftrightarrow \frac{2A\tau}{1 + (2\pi f\tau)^2}$$

Plots of $|V(f)|$ and $\underline{/V(f)}$ are given in Fig. 15.10. ■

EXAMPLE 15.4

Find and sketch the density spectrum of the signal

$$v(t) = Ae^{-|t - t_d|/\tau}$$

shown in Fig. 15.11.

FIGURE 15.11
Time-shifted signal

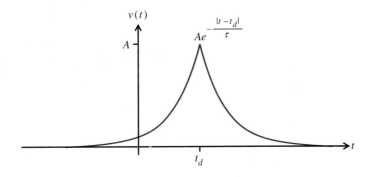

■ **SOLUTION**

Use Property P7 (time shift) and the result of the preceding example to obtain

$$V(f) = \frac{2A\tau}{1 + (2\pi f\tau)^2} e^{-j2\pi f t_d}$$

The density spectrum is plotted in Fig. 15.12. By comparing Figs. 15.10 and 15.12, we can see that the time-shifted signal $v(t - t_d)$ has the same amplitude density spectrum as $v(t)$, but the phase spectrum of $v(t - t_d)$ differs from that of $v(t)$ by a linear phase factor $\underline{/-2\pi f t_d}$. ■

FIGURE 15.12
The density spectrum of a time-shifted signal: (*a*) amplitude density; (*b*) phase

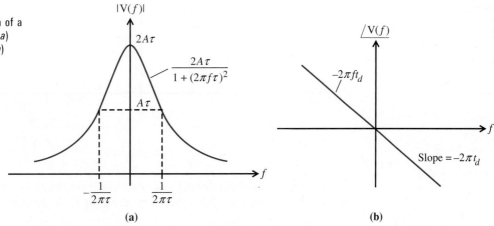

EXAMPLE 15.5

(a) Find and sketch the Fourier transform of the voltage

$$v(t) = A\Pi\left(\frac{t}{\tau}\right) \cos 2\pi f_c t$$

illustrated in Fig. 15.13. This signal is called a *gated tone* or a *toneburst*.
(b) Find and sketch the limiting forms of $v(t)$ and $V(f)$ for $\tau \to \infty$.

FIGURE 15.13
A toneburst

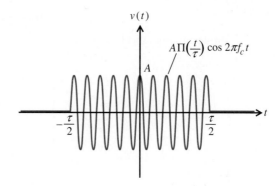

■ **SOLUTION**

(a) From Example 15.2,

$$A\Pi\left(\frac{t}{\tau}\right) \leftrightarrow A\tau \text{ sinc } f\tau$$

Then, according to Eqs. (15.17) through (15.19),

$$A\Pi\left(\frac{t}{\tau}\right) \cos 2\pi f_c t \leftrightarrow \frac{A\tau}{2} \text{ sinc } (f - f_c)\tau + \frac{A\tau}{2} \text{ sinc } (f + f_c)\tau$$

The density spectrum is shown in Fig. 15.14.

FIGURE 15.14
The density spectrum of a toneburst

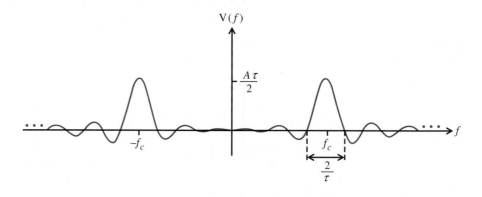

Notice that most of the density spectrum of the toneburst is concentrated within bandwidths of roughly $2/\tau$ Hz centered about $\pm f_c$. A measure of the degree of concentration is the ratio of center frequency to the bandwidth: $f_c \div (2/\tau) = f_c\tau/2$. The product $f_c\tau$ is the number of oscillations occurring within the toneburst (see Fig. 15.13). Therefore a toneburst containing many oscillations will have a voltage density spectrum that is highly concentrated.

(b) In the limit $\tau \to \infty$, $v(t)$ becomes the sinusoid $A \cos 2\pi f_c t$. As τ increases, $\tau \operatorname{sinc}(f - f_c)\tau$ and $\tau \operatorname{sinc}(f + f_c)\tau$ become narrower and higher. In the limit $\tau \to \infty$, they become unit impulses occurring at $f = f_c$ and $f = -f_c$, respectively. Thus

$$A \cos 2\pi f_c t \leftrightarrow \frac{A}{2}\delta(f - f_c) + \frac{A}{2}\delta(f + f_c)$$

We can easily confirm this result by substituting $V(f) = (A/2)\delta(f - f_c) + (A/2)\delta(f + f_c)$ into Eq. (15.1). This yields, as in Exercise 5,

$$v(t) = \int_{-\infty}^{+\infty} \left[\frac{A}{2}\delta(f - f_c) + \frac{A}{2}\delta(f + f_c) \right] e^{j2\pi ft}\, df$$

$$= \frac{A}{2} \int_{-\infty}^{+\infty} \delta(f - f_c) e^{j2\pi ft}\, df + \frac{A}{2} \int_{-\infty}^{+\infty} \delta(f + f_c) e^{j2\pi ft}\, df$$

$$= \frac{A}{2} e^{j2\pi f_c t} + \frac{A}{2} e^{-j2\pi f_c t} = A \cos 2\pi f_c t$$

The cosine waveform and its density spectrum are illustrated in Fig. 15.15. Notice what has happened here. We have obtained the Fourier transform of the *periodic* signal $A \cos 2\pi f_c t$. It is interesting to compare the voltage density spectrum of the sinusoid of Fig. 15.15 with that of the toneburst of Figs. 15.13 and 15.14. Whereas the voltage density spectrum of the toneburst is continuous for all f, $-\infty < f < \infty$, that of the sinusoid consists solely of impulses located at $f = \pm f_c$. In Example 15.6, we will obtain the Fourier transform of an arbitrary periodic signal. ■

FIGURE 15.15
(a) The cosine waveform and (b) its density spectrum

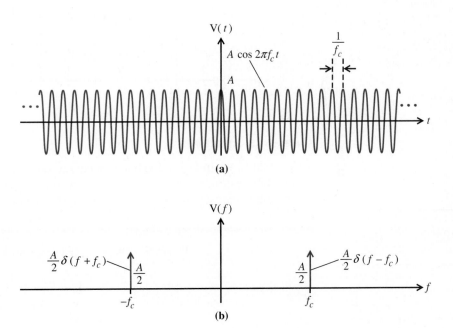

EXAMPLE 15.6

Find the Fourier transform of the periodic waveform

$$v(t) = \sum_{n=-\infty}^{+\infty} V_n e^{j2\pi n f_1 t}$$

where the V_n are the complex Fourier coefficients of $v(t)$.

■ SOLUTION

The Fourier transform of $v(t)$,

$$V(f) = \mathscr{F}\left\{ \sum_{n=-\infty}^{+\infty} V_n e^{j2\pi n f_1 t} \right\}$$

is, by Property P1,

$$V(f) = \sum_{n=-\infty}^{+\infty} V_n \mathscr{F}\{e^{j2\pi n f_1 t}\}$$

which is

$$V(f) = \sum_{n=-\infty}^{+\infty} V_n \delta(f - nf_1)$$

where the last step follows from Property P8 and the fact that $1 \leftrightarrow \delta(f)$, as shown in Exercise 15.4. The density spectrum of a periodic waveform is illustrated in Fig. 15.16. Compare this result with Fig. 14.12, which illustrates the (line) spectrum of the same periodic waveform. ■

FIGURE 15.16
Density spectrum of a typical periodic waveform; the Fourier coefficients V_n are the complex areas of the impulses

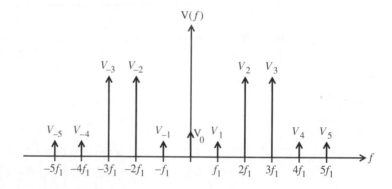

Parseval's Theorem

Recall that in Chapter 7 we defined an energy signal as a voltage or current that would deliver a finite amount of energy if it were applied to a resistance.

The following version of Parseval's theorem applies to energy signals.

Parseval's Theorem for Energy Signals

$$\int_{-\infty}^{+\infty} |v(t)|^2 \, dt = \int_{-\infty}^{+\infty} |V(f)|^2 \, df \qquad (15.20)$$

The proof of Eq. (15.20) is straightforward. By conjugating both sides of Eq. (15.2), we can write the left-hand side of Eq. (15.20) as

$$\int_{-\infty}^{+\infty} |v(t)|^2 \, dt = \int_{-\infty}^{+\infty} v(t)v^*(t) \, dt$$

$$= \int_{-\infty}^{+\infty} v(t) \left\{ \int_{-\infty}^{+\infty} V^*(f)e^{-j2\pi ft} \, df \right\} dt \qquad (15.21)$$

Changing the orders of integration then leads to

$$\int_{-\infty}^{+\infty} |v(t)|^2 \, dt = \int_{-\infty}^{+\infty} V^*(f) \left\{ \int_{-\infty}^{+\infty} v(t)e^{-j2\pi ft} \, dt \right\} df \qquad (15.22)$$

The term in braces is $V(f)$. This leads directly to Eq. (15.20).

The integral on the left-hand side of Eq. (15.20) is numerically equal to the total energy dissipated in a 1-Ω resistance with terminal voltage $v(t)$. The right-hand side of Eq. (15.20) expresses this energy as a (Riemann) sum of infinitesimal quantities $|V(f)|^2 \, df$. These infinitesimal quantities have the unit volt^2-seconds and represent the energy contributions from the individual complex sinusoidal components in $v(t)$. The function $|V(f)|^2$ has the unit of volt^2-seconds per hertz and is called the *energy density spectrum* of $v(t)$. According to Eq. (15.20), the total energy dissipated in a 1-Ω resistance with terminal voltage $v(t)$ is numerically equal to the area under the energy density spectrum. The contribution to the total energy dissipated due to the infinitesimal complex sinusoids with frequencies in the range $f_1 \le f \le f_2$ equals the area under the energy density spectrum between f_1 and f_2.

Important Fourier Transform Pairs

Table 15.2 lists 15 important Fourier transform pairs. The derivations of most of the pairs listed in Table 15.2 appear within this chapter. We will use Table 15.2 in several examples and problems in subsequent sections.

EXERCISES

9. Derive FT11 of Table 15.2 two ways: (a) by direct evaluation of Eq. (15.2), and (b) by applying the duality Property P6 of Table 15.1 to FT9.

10. Derive FT12 of Table 15.2.

11. Use Tables 15.1 and 15.2 to determine the Fourier transform of $v(t) = e^{-\alpha t} \cos(2\pi f_c t) u(t)$.

12. Commercial AM radio stations broadcast waveforms of the form $v_x(t) = A[1 + av(t)] \cos 2\pi f_c t$, where A and a are nonnegative constants.
    (a) Assume that $v(t) \leftrightarrow V(f)$, where $V(f)$ is illustrated in Fig. 15.8, and determine and sketch $V_x(f)$.
    (b) How does the voltage density spectrum of the transmitted AM signal differ from that of a DSB signal?

13. Use Tables 15.1 and 15.2 to determine the Fourier transform of $v(t) = 1 - \delta(t - 3)$.

TABLE 15.2 Fifteen important Fourier transform pairs

		SIGNAL	DENSITY SPECTRUM
FT1	Unit impulse	$\delta(t)$	1
FT2	DC signal	1	$\delta(f)$
3	Complex exponential	$e^{j2\pi f_c t}$	$\delta(f - f_c)$
4	Sinusoid	$\cos(2\pi f_c t + \phi)$	$\frac{1}{2}e^{j\phi}\delta(f - f_c) + \frac{1}{2}e^{-j\phi}\delta(f + f_c)$
FT5	Modulated sinusoid	$m(t)\cos(2\pi f_c t + \phi)$	$\frac{1}{2}e^{j\phi}M(f - f_c) + \frac{1}{2}e^{-j\phi}M(f + f_c)$
FT6	Unit step	$u(t)$	$\frac{1}{2}\delta(f) + \frac{1}{j2\pi f}$
FT7	One-sided decaying exponential	$e^{-t/\tau}u(t),\ \tau > 0$	$\dfrac{1}{(1/\tau) + j2\pi f}$
FT8	Symmetrical two-sided exponential	$e^{-|t|/\tau}$	$\dfrac{2\tau}{1 + (2\pi f\tau)^2}$
FT9	Rectangular pulse	$\Pi\left(\dfrac{t}{\tau}\right)$	$\tau\,\text{sinc}\,f\tau$
FT10	Triangular pulse with height 1 and base 2τ	$\left(1 - \dfrac{|t|}{\tau}\right)\Pi\left(\dfrac{t}{2\tau}\right)$	$\tau\,\text{sinc}^2 f\tau$
FT11	Sinc function	$f_s\,\text{sinc}\,f_s t$	$\Pi\left(\dfrac{f}{f_s}\right)$
FT12	Half-cycle cosine pulse	$\Pi\left(\dfrac{t}{\tau}\right)\cos\dfrac{\pi t}{\tau}$	$\dfrac{2\tau\cos\pi f\tau}{\pi[1 - (2f\tau)^2]}$
FT13	Raised cosine pulse	$\Pi\left(\dfrac{t}{\tau}\right)\left(\dfrac{1}{2} + \dfrac{1}{2}\cos\dfrac{2\pi t}{\tau}\right)$	$\dfrac{\sin\pi f\tau}{2\pi f[1 - (f\tau)^2]}$
FT14	Unit area gaussian pulse	$\dfrac{1}{\sqrt{2\pi}\,\tau}e^{-t^2/2\tau^2},\ \tau > 0$	$e^{-(2\pi f\tau)^2/2}$
FT15	General periodic signal	$\displaystyle\sum_{n=-\infty}^{+\infty} V_n e^{j2\pi n f_1 t}$	$\displaystyle\sum_{n=-\infty}^{+\infty} V_n\delta(f - nf_1),\ V_n = \dfrac{1}{T}\int_{t_0}^{t_0+T} v(t)e^{-j2\pi n f_1 t}\,dt$ $Tf_1 = 1$

15.4 Circuit and System Analysis Using the Fourier Transform

If an arbitrary signal $v_s(t)$ is input to a stable LTI system, then the density spectrum of the output signal $v_o(t)$ is

$$\mathbf{V}_o(f) = \mathbf{H}(f)\mathbf{V}_s(f) \qquad (15.23)$$

where $\mathbf{V}_s(f)$ is the density spectrum of $v_s(t)$ and $\mathbf{H}(f)$ is the transfer function of the system. This relationship, which is proved in Appendix 15.1, is the principal reason for the use of Fourier transforms in linear system analysis. The profound implication of Eq. (15.23) is that the transfer function of an LTI system provides a *complete* description of the input-output properties of that system. This implication was encountered previously in Example 11.5, where it was shown that a circuit's differential equation could be obtained from its transfer function. In Eq. (15.23), the transfer function is seen to be the link between the input density spectrum $V_s(f)$ and the output density spectrum $V_o(f)$.

It also follows immediately from Eq. (15.23) that the energy density function $|V_o(f)|^2$ of the system's output is given by

$$|V_o(f)|^2 = |H(f)|^2 |V_s(f)|^2 \tag{15.24}$$

where $|V_s(f)|^2$ is the energy density function of the input.

The following example illustrates an important application of Eq. (15.23) in systems theory. In this example we see why engineers cannot build ideal filters.

EXAMPLE 15.7

An *ideal low-pass filter* is an LTI system with a transfer function $H(f) = \Pi(f/2W)$ that is exactly unity for $|f| < W$ and exactly zero for $|f| > W$. The constant W is the *filter bandwidth* or *cutoff frequency* in hertz. Show that it is impossible to physically construct an ideal low-pass filter.

■ SOLUTION

Consider the response to an input $v_s(t) = \delta(t)$. Since $V_s(f) = \mathscr{F}\{\delta(t)\} = 1$, then

$$V_o(f) = H(f)V_s(f) = H(f)$$

Thus the response of the ideal low-pass filter to an impulse input is

$$v_o(t) = \mathscr{F}^{-1}\left\{\Pi\left(\frac{f}{2W}\right)\right\}$$

By FT11 of Table 15.2 we obtain the result

$$v_o(t) = 2W \operatorname{sinc}(2Wt)$$

We see that $v_o(t)$ is nonzero for $t < 0$. This implies that an ideal low-pass filter *responds to* an impulse $\delta(t)$ before the impulse occurs at $t = 0$—clearly a physical impossibility. Since physical systems must be *causal*, that is, they must obey the rules of cause and effect, we see that we cannot build an ideal low-pass filter. ■

The next example concerns *distortionless* transmission, a topic of great importance to designers of high-fidelity communications systems.

EXAMPLE 15.8

A *distortionless system* is a system that only multiplies its input, $v_s(t)$, by a factor $k > 0$ and (possibly) delays the input. Thus the output of a distortionless system is given by

$$v_o(t) = kv_s(t - t_d)$$

where t_d is the time delay. Find the transfer function of a distortionless system.

SOLUTION

The Fourier transform of the above equation is, by Properties P1 and P7 of Table 15.1,

$$V_o(f) = kV_s(f)e^{-j2\pi ft_d}$$

By comparing this with Eq. (15.23), we see that the transfer function must be

$$H(f) = ke^{-j2\pi ft_d} = k\underline{/-2\pi ft_d}$$

for all f for which $V_s(f) \neq 0$. It follows that a distortionless system has a constant or *flat* amplitude characteristic and a *linear* phase characteristic for all frequencies for which $V_s(f)$ is nonzero. The height of the amplitude characteristic determines the factor k. The negative of the slope of the phase characteristic determines the delay. ■

In the next example, we show how high-fidelity music can be transmitted through a low-fidelity communications channel without loss of fidelity.

EXAMPLE 15.9

High-fidelity music recordings are to be transmitted over a hypothetical low-fidelity communication channel, as illustrated in Fig. 15.17. The density spectrum of the recorded music, also illustrated in Fig. 15.17, lies in the frequency range $30 \leq |f| \leq 15,000$ Hz and is approximately zero outside this range. The frequency response of the communication circuit is unity for $0 \leq |f| \leq 1000$ Hz and falls rapidly to zero outside this range. Devise a system to transmit the music over the communication circuit without loss in fidelity.

SOLUTION

The recording can be transmitted over the channel without loss in fidelity if we slow it down by a factor of 15. This is a consequence of Property P5 of Table

FIGURE 15.17
High-fidelity music is to be transmitted through a low-fidelity communication circuit. The music $v_s(t)$ has a density spectrum $V_s(f)$ that is nonzero for 30 Hz $\leq |f| \leq$ 15 kHz, but the communication circuit bandwidth is only 1 kHz. How can we transmit the music without losing fidelity?

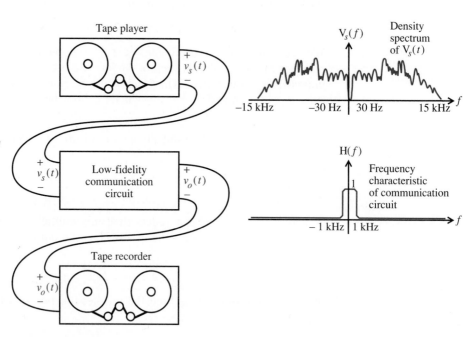

15.1. A slowdown of $v(t)$ by 15, corresponding to $a = 15$ in Property P5, results in a 15-fold compression along the frequency axis of the density spectrum. Thus the highest recorded frequency of 15 kHz will be scaled down to 1 kHz and will be passed by the communications circuit. The slowed music must be recorded at the receiver and then played 15 times faster to reproduce the original recording in full fidelity. ■

The preceding three examples demonstrated the usefulness and elegance of the Fourier transform for solving problems involving general signals and systems. We conclude this section with two examples showing how the Fourier transform may also be used to solve detailed circuit problems—an area for which it is less suited. As we will see in Chapter 16, we can solve the same circuit problems more easily using the Laplace transform.

EXAMPLE 15.10

The input to the high-pass RL filter of Fig. 15.18 is the unit step function $v_s(t) = u(t)$.
   (a) Determine the voltage density spectra of the input and the output.
   (b) Find $v_o(t)$ from $V_o(f)$.

FIGURE 15.18
High-pass RL filter with input $v_s(t) = u(t)$

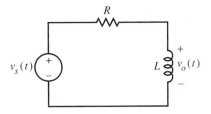

■ **SOLUTION**

The Fourier transform of the unit step input is given in Table 15.2:

$$V_s(f) = \frac{1}{2}\delta(f) + \frac{1}{j2\pi f}$$

Multiplication of this by the transfer function of the filter,

$$H(f) = \frac{j2\pi f L}{R + j2\pi f L}$$

gives the output density spectrum

$$V_o(f) = \left\{ \frac{1}{2}\delta(f) + \frac{1}{j2\pi f} \right\} \frac{j2\pi f L}{R + j2\pi f L}$$

$$= \delta(f)\frac{j\pi f L}{R + j2\pi f L} + \frac{L}{R + j2\pi f L}$$

The factor multiplying $\delta(f)$ equals zero at $f = 0$, and $\delta(f) = 0$ for $f \neq 0$. Therefore we suspect that the first term of $V_o(f)$ may equal zero for all f. We can show that this is the case by taking the inverse Fourier transform of the first term

$$\mathscr{F}^{-1}\left\{\delta(f)\frac{j\pi fL}{R + j2\pi fL}\right\} = \int_{-\infty}^{+\infty} \delta(f)\underbrace{\frac{j\pi fL}{R + j\pi fL}\ e^{j2\pi f}\ df}_{g(f)} = 0$$

[The last step follows from the *definition* of the unit impulse, $\int_{-\infty}^{+\infty} \delta(f)g(f)\,df = g(0)$, in Eq. (7.7).] The direct Fourier transform of 0 is 0. Therefore our suspicion that the first term of $V_o(f)$ equals zero is confirmed.* Consequently

$$V_o(f) = \frac{L}{R + j2\pi fL}$$

(*b*) To determine $v_o(t)$, we can rewrite $V_o(f)$ as

$$V_o(f) = \frac{1}{\dfrac{R}{L} + j2\pi f}$$

and find from Table 15.2, that

$$v_o(t) = e^{-(R/L)t}u(t)\ \blacksquare$$

In the preceding example, the density spectrum of the input waveform was found from Table 15.2. The same table was used to find the output waveform $v_o(t)$. Occasionally a needed transform pair cannot be found in a table. When this happens, the required transform pair can often be obtained with the help of the Fourier transform properties in Table 15.1.

EXAMPLE 15.11

Find the response $v_o(t)$ of the high-pass RL filter of Fig. 15.18 to a unit impulse $v_s(t) = \delta(t)$.

■ SOLUTION

The density spectrum of a unit impulse is $V_s(f) = 1$ (Table 15.2). The voltage density spectrum of the output waveform is therefore

$$V_o(f) = H(f) \cdot 1$$

$$= \frac{j2\pi fL}{R + j2\pi fL}$$

This is not listed in Table 15.2. It can be written as

$$V_o(f) = j2\pi f\left(\frac{1}{R/L + j2\pi f}\right)$$

where the inverse Fourier transform of the quantity in parentheses is in Table 15.2:

$$\mathscr{F}^{-1}\left\{\frac{1}{R/L + j2\pi f}\right\} = e^{-(R/L)t}u(t)$$

---

* This roundabout argument was made necessary by the fact that the unit impulse is not a function in the ordinary sense. We can obtain the same result by treating $\delta(f)$ as the limit of the pulse $(1/\Delta f)\Pi(f/\Delta f)$ as $\Delta f \to 0$. The *Laplace* transform, to be studied in the next chapter, has the significant advantage of bypassing such complicated arguments.

According to Property P2, multiplication by $j2\pi f$ in the frequency domain corresponds to differentiation with respect to t in the time domain. Consequently,

$$v_o(t) = \frac{d}{dt}\left[e^{-(R/L)t}u(t)\right]$$

$$= \left(\frac{d}{dt}\,e^{-(R/L)t}\right)u(t) + e^{-(R/L)t}\frac{d}{dt}\,u(t)$$

$$= -\frac{R}{L}\,e^{-(R/L)t}u(t) + e^{-(R/L)t}\delta(t)$$

and, since $e^{-(R/L)t} = 1$ at $t = 0$,

$$v_o(t) = \delta(t) - \frac{R}{L}\,e^{-(R/L)t}u(t) \quad\blacksquare$$

REMEMBER

The density spectrum of the output of an LTI system is given by the product of the density spectrum of the input and the system's transfer function.

EXERCISES

14. (a) Show that the "message" $v(t)$ of Fig. 15.8 can be recovered (at least approximately) from $v_x(t) = v(t)\cos 2\pi f_c t$ if we multiply $v_x(t)$ by $2\cos 2\pi f_c t$ and put the result into a low-pass filter. (*Hint*: Use the identity $2\cos^2 2\pi f_c t = 1 + \cos 2\pi 2 f_c t$. Then find and sketch the density spectrum of the filter input.)
    (b) Under what conditions is the filter output *exactly* equal to $v(t)$?

15. A low-pass series RC circuit has input $v_s(t) = \delta(t)$.
    (a) Determine the voltage across C for $t > 0$ with the use of Fourier transforms.
    (b) Use your answer from (a) to determine the voltage across R for $t > 0$.

15.5   Impulse Response and Convolution

The impulse response of a system is defined as the output of the system resulting from a unit impulse input. This definition is illustrated in Fig. 15.19 for an LTI system. The input is a unit impulse. By definition, the output is the impulse response, denoted as $h(t)$.

FIGURE 15.19
The response to an impulse $\delta(t)$ is the impulse response $h(t)$

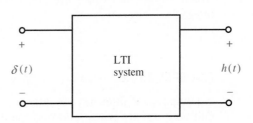

The impulse response of an LTI system has the following two fundamental (and remarkable) properties.

FUNDAMENTAL PROPERTY 1

Relation Between Input and Output of Any LTI System

The response $v_o(t)$ of an LTI system to an arbitrary input $v_s(t)$ is given by

$$v_o(t) = \int_{-\infty}^{+\infty} v_s(\lambda)h(t - \lambda)\, d\lambda = v_s(t) * h(t) \qquad (15.25a)$$

where $h(t)$ is the system's impulse response.

The special integral appearing in Eq. (15.25a) is called a *convolution* integral and is frequently denoted by the *convolution operator* $*$ as shown. By replacing λ with $t - \sigma$, we can rewrite Eq. (15.25a) as

$$v_o(t) = \int_{-\infty}^{+\infty} v_s(t - \sigma)h(\sigma)\, d\sigma = h(t) * v_s(t) \qquad (15.25b)$$

According to Fundamental Property 1, the input-output relation of an LTI system is completely specified by the system's response to a unit impulse. Knowledge of $h(t)$ is all that is required to find the system's response to an *arbitrary* input $v_s(t)$.

FUNDAMENTAL PROPERTY 2

Relation Between Impulse Response and Transfer Function

The Fourier transform of an LTI system's impulse response $h(t)$ is the system's transfer function $H(f)$: $h(t) \leftrightarrow H(f)$.

According to Fundamental Property 2, the impulse response of an LTI system can be determined from the system's transfer function and vice versa.

These two properties are among the most important results in the theory of linear systems. They are used extensively in both the analysis and the design of communication and control systems, and in signal processing. Because they are so important, we give two derivations of each in Appendix 15.2. Example 15.12, which follows, illustrates how to solve a circuit problem using the convolution integral.

EXAMPLE 15.12

A low-pass *RC* filter has the input $v_s(t)$ shown in Fig. 15.20. Assume that the circuit was initially at rest and find the output $v_o(t)$ with the use of convolution.

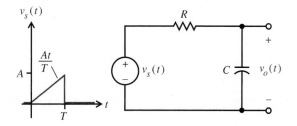

FIGURE 15.20
Low-pass *RC* filter with sawtooth pulse input

SOLUTION

The first step is to find the impulse response of the circuit. This can be determined from the transfer function:

$$H(f) = \frac{\dfrac{1}{j2\pi fC}}{R + \dfrac{1}{j2\pi fC}} = \frac{1}{\tau} \cdot \frac{1}{(1/\tau) + j2\pi f}$$

where $\tau = RC$. It follows from FT7 of Table 15.2 and Property P1 of Table 15.1 that

$$h(t) = \frac{1}{\tau} e^{-t/\tau} u(t)$$

The convolution

$$v_o(t) = \int_{-\infty}^{+\infty} v_s(\lambda) h(t - \lambda) \, d\lambda$$

is interpreted graphically in Fig. 15.21, in which $v_s(\lambda)$, $h(t - \lambda)$, and $v_s(\lambda)h(t - \lambda)$ have been plotted versus λ. As is evident from Fig. 15.21, the integrand $v_s(\lambda)h(t - \lambda)$ assumes three distinct forms, depending on the value of t. As indicated in Fig. 15.21a, if $t < 0$, then

$$v_s(\lambda)h(t - \lambda) = 0 \qquad \text{for all } \lambda$$

It follows that

$$v_o(t) = \int_{-\infty}^{+\infty} 0 \cdot d\lambda = 0 \qquad \text{if } t < 0$$

As indicated in Fig. 15.21b, if t is in the range $0 \le t < T$, then

$$v_s(\lambda)h(t - \lambda) = \begin{cases} \dfrac{A\lambda}{T} \cdot \dfrac{1}{\tau} e^{-(t-\lambda)/\tau} & \text{for } 0 \le \lambda \le t \\ 0 & \text{otherwise} \end{cases}$$

Therefore

$$v_o(t) = \int_0^t \frac{A\lambda}{T} \cdot \frac{1}{\tau} e^{-(t-\lambda)/\tau} \, d\lambda \qquad \text{if } 0 \le t < T$$

which is (integrating by parts)

$$v_o(t) = A\left[\frac{t}{T} - \left(\frac{\tau}{T}\right)(1 - e^{-t/\tau})\right] \qquad \text{if } 0 \le t < T$$

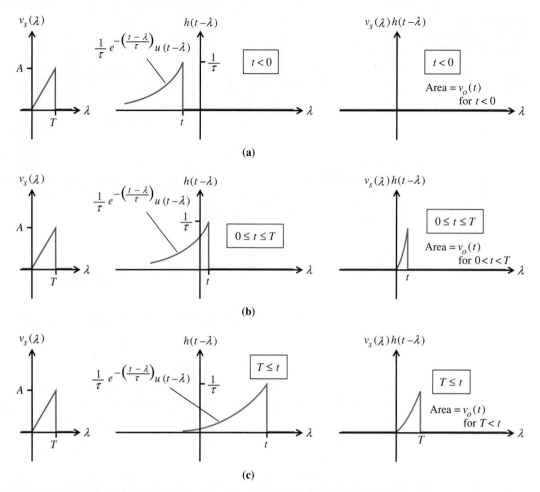

FIGURE 15.21 Graphical interpretation of convolution. As t increases, $h(t - \lambda)$ "slides" along the λ axis. This causes $v_s(\lambda)h(t - \lambda)$ (and its area) to vary. (a) $t < 0$; (b) $0 \le t < T$; (c) $T \le t$

As indicated in Fig. 15.21c, if t is greater than T, then

$$v_s(\lambda)h(t - \lambda) = \begin{cases} \dfrac{A\lambda}{T} \cdot \dfrac{1}{\tau}e^{-(t-\lambda)/\tau} & \text{for } 0 \le \lambda \le T \\ 0 & \text{otherwise} \end{cases}$$

Therefore

$$v_o(t) = \int_0^T \frac{A\lambda}{T} \cdot \frac{1}{\tau}e^{-(t-\lambda)/\tau}\, d\lambda \qquad \text{if } T < t$$

which is (again integrating by parts)

$$v_o(t) = A\left[1 - \frac{\tau}{T}(1 - e^{-T/\tau})\right]e^{-(t-T)/\tau} \qquad \text{if } T < t$$

The output waveform is plotted in Fig. 15.22. ∎

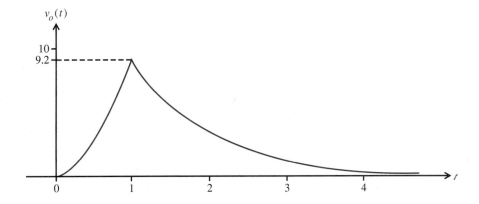

FIGURE 15.22
Plot of $v_o(t)$ from Example 15.12 versus t. The plot assumes that $\tau = T = 1$ and $A = 25$

REMEMBER

The output of an LTI system is given by the convolution of the input and the system's impulse response. The transfer function of an LTI system is given by the Fourier transform of the system's impulse response.

EXERCISES

The convolution operator $*$ was defined in Eq. (15.25a). Show that the convolution operator has the following properties, where $x_1(t)$, $x_2(t)$, and $x_3(t)$ are arbitrary waveforms.

16. Commutation: $x_1(t) * x_2(t) = x_2(t) * x_1(t)$
17. Distribution: $x_1(t) * [x_2(t) + x_3(t)] = x_1(t) * x_2(t) + x_1(t) * x_3(t)$
18. Association: $x_1(t) * [x_2(t) * x_3(t)] = [x_1(t) * x_2(t)] * x_3(t)$

For Exercises 19 through 22, let $x(t)$ be an arbitrary waveform and show that the statement is true.

19. $x(t) * \delta(t) = x(t)$
20. $x(t) * \delta(t - t_0) = x(t - t_0)$
21. $x(t) * u(t) = \int_{-\infty}^{t} x(\lambda) \, d\lambda$
22. $x(t) * \delta'(t) = x'(t)$ where prime denotes differentiation with respect to t

Determine the impulse response and the transfer function of each of the following systems.

23. An ideal differentiator: $v_o(t) = (d/dt)v_s(t)$
24. An ideal integrator: $v_o(t) = \int_{-\infty}^{t} v_s(\lambda) \, d\lambda$
25. An ideal delay line: $v_o(t) = v_s(t - t_0)$

15.6 Existence of the Fourier Transform

The topics of the existence and convergence of the Fourier transform are both practically important and theoretically subtle. In this section, we present two well-known and alternative theorems.

Dirichlet Conditions

Dirichlet has given sufficient conditions for the existence of the Fourier transform.

Dirichlet Conditions

If a signal $v(t)$ satisfies *each* of the following conditions, then $v(t)$ has a Fourier transform $V(f) = \mathscr{F}\{v(t)\}$. Furthermore, the inverse Fourier transform $\mathscr{F}^{-1}\{V(f)\}$ *equals* $v(t)$ except at isolated discontinuities of $v(t)$.

1. $v(t)$ is absolutely integrable, that is,

$$\int_{-\infty}^{+\infty} |v(t)|\, dt < \infty \qquad (15.26)$$

2. $v(t)$ has a finite number of maxima and minima within every finite interval of time.
3. $v(t)$ has a finite number of discontinuities within every finite interval of time. Each discontinuity must be finite.

The Dirichlet conditions are *sufficient but not necessary* conditions for $v(t)$ to have a frequency-domain representation $V(f)$.

Most signals in circuit theory satisfy the Dirichlet conditions. Examples are the decaying one-sided exponential $v(t) = e^{at}u(t)$ $(a < 0)$ and the rectangular pulse $v(t) = \Pi(t)$. The density spectra of these signals were derived in Examples 15.1 and 15.2.

Dirichlet conditions 2 and 3 are violated only by mathematical functions that have a markedly pathological character, analogous to those discussed in Section 14.9. There is little reason to consider functions of this type in circuit theory.

Some signals that are important in circuit theory do not satisfy Dirichlet condition 1. Examples of them are the rising exponential $v(t) = e^{at}u(t)$, $a > 0$, the unit step function $u(t)$, and the dc signal $v(t) = 1$. Each of these signals satisfies conditions 2 and 3 but does not satisfy condition 1. As we show in Examples 15.13 through 15.15, the rising exponential does not have a density spectrum, but the unit step function and dc waveforms do. Examples 15.14 and 15.15 show that special care is needed to derive the density spectra of the unit step function and dc waveforms.

EXAMPLE 15.13

Try to find the density spectrum of the exponential waveform $v(t) = e^{at}u(t)$ with $a > 0$.

■ SOLUTION

The substitution of $e^{at}u(t)$ into Eq. (15.1) yields

$$V(f) = \int_{-\infty}^{+\infty} e^{at}u(t)e^{-j2\pi ft}\, dt$$

$$= \frac{1}{a - j2\pi f}\, e^{at}e^{-j2\pi ft}\Big|_{\infty} - \frac{1}{a - j2\pi f}\, e^{at}e^{-j2\pi ft}\Big|_{0}$$

If $a > 0$, then $e^{a\infty} = \infty$ and

$$V(f) = \infty - \frac{1}{a - j2\pi f}$$

This indicates that the integral defining $V(f)$ diverges for $a > 0$. We say that $V(f)$ *does not exist* for $a > 0$. ∎

EXAMPLE 15.14

Try to obtain the density spectrum $U(f)$ of the unit step function (a) by direct substitution of $u(t)$ into Eq. (15.1), and (b) by taking the limit $a \to 0$ of the density spectrum of $e^{at}u(t)$ $(a < 0)$.

■ SOLUTION

(a) Direct substitution of $u(t)$ into Eq. (15.1) yields

$$V(f) = \int_{-\infty}^{+\infty} u(t)e^{-j2\pi ft}\,dt$$

$$= \int_{0}^{\infty} e^{-j2\pi ft}\,dt$$

$$= \frac{1}{-j2\pi ft}\,e^{-j2\pi ft}\Big|_{0}^{\infty}$$

$$= \frac{1}{-j2\pi ft}\,e^{-j2\pi f\infty} - \frac{1}{-j2\pi ft}$$

But $e^{-j2\pi f\infty}$ is not a mathematically well-defined quantity.* Consequently we have failed to find $U(f)$ by this method.

(b) As was shown in Example 15.1, if $a < 0$, then

$$\mathscr{F}\{e^{at}u(t)\} = \frac{1}{-a + j2\pi f} = V(f)$$

In the limit $a \to 0$, $e^{at}u(t)$ becomes the unit step $u(t)$. This suggests that we can find the density spectrum of $u(t)$ by taking the limit of $V(f)$ above as $a \to 0$. We can most easily accomplish this by expressing $V(f)$ in rectangular form:

$$V(f) = \frac{-a}{a^2 + (2\pi f)^2} + j\,\frac{-2\pi f}{a^2 + (2\pi f)^2}$$

The real and imaginary parts of $V(f)$ are sketched in Fig. 15.23.[†]
The limit of the imaginary part follows easily:

$$\lim_{a \to 0} \mathscr{I}m\,\{V(f)\} = \lim_{a \to 0} \frac{-2\pi f}{a^2 + (2\pi f)^2}$$

$$= -\frac{1}{2\pi f}$$

---

* The function $e^{-j2\pi ft} = \cos 2\pi ft - j\sin 2\pi ft$ has oscillatory real and imaginary parts that do not approach a well-defined limit as $t \to \infty$.

[†] You may be tempted to take the limit $a \to 0$ by simply setting $a = 0$ in $V(f) = 1/(-a + j2\pi f)$, thereby obtaining the incorrect, purely imaginary, result $1/j2\pi f$. This method fails because, as shown in Fig. 15.23, the real part of $V(f)$ actually does not vanish as $a \to 0$.

FIGURE 15.23
The density spectrum of
the one-sided decaying
exponential $e^{at}u(t)$, where
$a < 0$: (a) real part;
(b) imaginary part

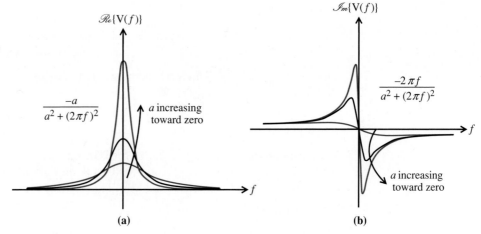

According to Fig. 15.23a, $\mathcal{R}e\,\{V(f)\}$ is a "pulse" that becomes narrower and higher as $a \to 0$. The area under this "pulse" is given by a standard integral:

$$\int_{-\infty}^{+\infty} \frac{-a}{a^2 + (2\pi f)^2}\, df = \frac{1}{2} \qquad \text{where } a < 0$$

Thus, as $a \to 0$, $\mathcal{R}e\,\{V(f)\}$ approaches an impulse with area 1/2, and

$$U(f) = \lim_{\tau \to \infty} V(f) = \frac{1}{2}\delta(f) + \frac{1}{j2\pi f}$$

Figure 15.24 illustrates the unit step function $u(t)$ and its density spectrum $U(f)$. ∎

FIGURE 15.24
(a) The unit step function;
(b) its density spectrum

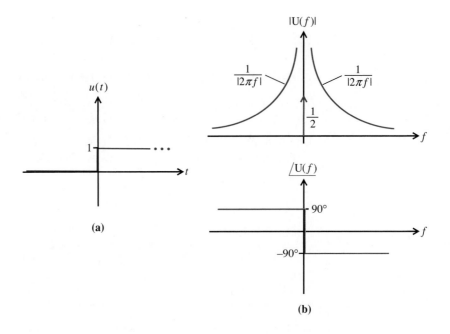

EXAMPLE 15.15

Try to obtain the density spectrum of the dc waveform $v(t) = 1$ (a) by direct substitution of $v(t) = 1$ into Eq. (15.1), and (b) by taking the limit $\tau \to \infty$ of the density spectrum of $\Pi(t/\tau)$.

■ SOLUTION

(a) The direct substitution of $v(t) = 1$ into Eq. (15.1) yields

$$V(f) = \int_{-\infty}^{+\infty} e^{-j2\pi ft}\, dt$$

$$= \frac{1}{-j2\pi f} e^{-j2\pi ft}\Big|_{-\infty}^{+\infty}$$

$$= \frac{1}{-j2\pi f} e^{-j2\pi f\infty} - \frac{1}{-j2\pi f} e^{j2\pi f\infty}$$

Neither of the above two terms is a mathematically well-defined quantity. Therefore we have failed to find $V(f)$ by this method.

(b) As shown in Example 15.2,

$$\mathscr{F}\left\{\Pi\left(\frac{t}{\tau}\right)\right\} = \tau \operatorname{sinc} f\tau = V(f)$$

A sketch of $V(f)$ is given in Fig. 15.5. Notice that $V(f)$ is a "pulse" that becomes narrower and higher as $\tau \to \infty$.

We can determine the area under $V(f)$ by setting $t = 0$ in Eq. (15.2):

$$\int_{-\infty}^{+\infty} V(f)\, df = v(0)$$

which is, for the case at hand,

$$\int_{-\infty}^{+\infty} \tau \operatorname{sinc} f\tau \, df = v(0) = 1$$

As τ increases, $V(f)$ becomes narrower and higher, but its area remains constant. In the limit $\tau \to \infty$, $\Pi(t/\tau)$ becomes a 1-V dc waveform and $V(f)$ becomes a unit impulse:

$$\lim_{\tau \to \infty} \tau \operatorname{sinc} f\tau = \delta(f)$$

as illustrated in Fig. 15.25. ■

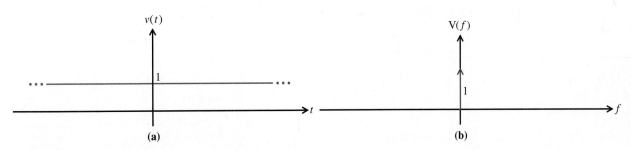

(a)   (b)

FIGURE 15.25   (a) The unit dc signal; (b) its density spectrum

Examples 15.13 through 15.15 illustrated the fact that a signal that satisfies Dirichlet conditions 2 and 3, but does not satisfy condition 1, *may or may not* have a Fourier transform. As illustrated by Examples 15.14 and 15.15, the use of the unit impulse $\delta(f)$ sometimes allows us to obtain $V(f)$ for those cases in which condition 1 is not satisfied.

Plancherel's Theorem

Plancherel gave a sufficient condition for the existence of the Fourier transform that provides an alternative to the Dirichlet conditions. Plancherel's sufficient condition is given in the following theorem.

Plancherel's Theorem

If $v(t)$ is an energy signal, that is,

$$\int_{-\infty}^{+\infty} |v(t)|^2 \, dt < \infty \tag{15.27}$$

then $V(f) = \mathscr{F}\{v(t)\}$ exists for all values of f except for a set of total length zero. Furthermore, $\mathscr{F}^{-1}\{V(f)\}$ is equal to $v(t)$ in the integral square sense.

Equality in the integral square sense means simply that there is no *energy* in the difference

$$e_B(t) = v(t) - \int_{-B}^{+B} V(f)e^{j2\pi ft} \, df$$

in the limit $B \to \infty$. Of course, if there is no energy in the difference $e_B(t)$ in the limit $B \to \infty$, then *there is no physical way to distinguish between $v(t)$ and $\mathscr{F}^{-1}\{V(f)\}$.* Thus, according to Plancherel, we can use Fourier transforms to analyze all physical signals.

An illuminating example of Plancherel's theorem is provided by the signal $v(t) = \text{sinc } t$. This signal does not satisfy Dirichlet condition 1, Eq. (15.26), but does satisfy Eq. (15.27). Thus sinc t is Fourier transformable according to Plancherel's theorem. Indeed, we know from FT11 of Table 15.2 that $\mathscr{F}\{\text{sinc } t\} = \Pi(f)$. This example illustrates the fact that the Fourier transform of a signal $v(t)$ is guaranteed to exist if *either* Plancherel's condition Eq. (15.27), is satisfied *or* the Dirichlet conditions are satisfied.

EXERCISES

Sketch each signal in Exercises 26 through 35, and state whether the signal satisfies all three Dirichlet conditions.

26. e^{-3t}

27. $e^{-3t}u(t)$

28. $t^{-1}u(t)$

29. $\text{sinc } (t)$

30. $\cos \omega_o t$

31. $t^{-0.5}[u(t) - u(t-1)]$

32. $\cos (\omega_o t)\Pi(t/\tau)$

33. $(1 - e^{-t/\tau})u(t)$

34. $\cos (\omega_o t)u(t)$

35. $\sin 1/t$

36. Which of the signals in Exercises 26 through 35 satisfy Plancherel's theorem?

In this chapter we described the Fourier transform and showed how to use it to solve problems involving LTI systems. The Fourier transform

$$V(f) = \int_{-\infty}^{+\infty} v(t)e^{-j2\pi ft}\,dt \tag{15.28}$$

provides the means to represent any physical signal $v(t)$ as an integral of complex exponential components,

$$v(t) = \int_{-\infty}^{+\infty} V(f)e^{j2\pi ft}\,df \tag{15.29}$$

We showed that the inverse Fourier transform, Eq. (15.29), is the limiting form of the complex Fourier series as $T \to \infty$. The energy density function $|V(f)|^2$ describes the distribution of energy with respect to frequency. Parseval's theorem states that if $v(t)$ appears across a 1-Ω resistance, the total energy dissipated is

$$\int_{-\infty}^{+\infty} |v(t)|^2\,dt = \int_{-\infty}^{+\infty} |V(f)|^2\,df \tag{15.30}$$

Although the Fourier relations of Eqs. (15.28) and (15.29) were originally derived for aperiodic signals, we showed that they also apply to signals with period T. For a periodic signal $v(t)$, $V(f)$ consists of impulses located at the harmonics nf_1, $n = 0, \pm1, \pm2, \ldots$, of the fundamental frequency $f_1 = 1/T$. The area of the impulse at $f = nf_1$ is the nth complex Fourier-series coefficient of $v(t)$.

The application of the Fourier transform to LTI system analysis arises from the fundamental relationship

$$V_o(f) = H(f)V_s(f) \tag{15.31}$$

where $V_s(f)$ is the Fourier transform of the input and $H(f)$ is the system's transfer function. The system's response $v_o(t)$ is the inverse Fourier transform of $V_o(f)$. Remarkably, $H(f)$ is the Fourier transform of the system's impulse response $h(t)$. The time-domain form of Eq. (15.31) is

$$v_o(t) = \int_{-\infty}^{\infty} h(t - \sigma)v_s(\sigma)\,d\sigma \tag{15.32}$$

Equations (15.28), (15.29), (15.31), and (15.32) provide means to compute a circuit's output without the use of differential equations.

In addition to being one of the most useful of all analytical tools in LTI systems theory, the Fourier transform leads us directly to the Laplace transform, which is described in Chapter 16.

KEY FACTS

◆ The Fourier transform (density function) of a signal $v(t)$ is defined as

$$\mathscr{F}\{v(t)\} = V(f) = \int_{-\infty}^{+\infty} v(t)e^{-j2\pi ft}\,dt$$

◆ The inverse Fourier transform of $V(f)$ is given by

$$\mathscr{F}^{-1}\{V(f)\} = v(t) = \int_{-\infty}^{+\infty} V(f)e^{j2\pi ft}\,df$$

- The output of an LTI system is given by the convolution of the input with the impulse response of the system:

$$v_o(t) = h(t) * v_s(t)$$

- The Fourier transform of the output of an LTI system is given by

$$V_o(f) = H(f)V_s(f)$$

where $V_s(f)$ is the Fourier transform of the input and $H(f)$ is the transfer function of the system.

- The transfer function of an LTI system equals the Fourier transform of the system's impulse response:

$$H(f) = \mathscr{F}\{h(t)\}$$

- The *energy density function* of a signal $v(t)$ is defined as $|V(f)|^2$, where $V(f) = \mathscr{F}\{v(t)\}$.

- Important Fourier transform properties and pairs are given in Tables 15.1 and 15.2, respectively.

- Parseval's theorem for energy signals states that

$$\int_{-\infty}^{+\infty} |v(t)|^2 \, dt = \int_{-\infty}^{+\infty} |V(f)|^2 \, df$$

Appendix 15.1

PROOFS OF EQ. (15.23)

In this appendix we provide two proofs of Eq. (15.23). The first proof is rigorous but applies only to LLTI systems. The second proof is not rigorous, but it does provide a great deal of insight into Eq. (15.23). It applies to LTI systems in general. (The second proof can be made rigorous by the inclusion of concepts that are outside the scope of this text.) Equation (15.23) states that the density spectrum of the output of a stable LTI circuit is given by

$$V_o(f) = H(f)V_s(f) \tag{15.23}$$

where $V_s(f)$ is the density spectrum of the input and $H(f)$ is the circuit's transfer function.

PROOF 1

This proof applies only to LLTI systems. It is based on Properties P1 and P2 in Table 15.1. The output and input of a stable LLTI system are related by

$$a_n \frac{d^n v_o}{dt^n} + a_{n-1} \frac{d^{n-1} v_o}{dt^{n-1}} + \cdots + a_0 v_o = b_m \frac{d^m v_s}{dt^m} + b_{m-1} \frac{d^{m-1} v_s}{dt^{m-1}} + \cdots + b_0 v_s \tag{15.33}$$

or

$$\mathscr{A}(p)v_o = \mathscr{B}(p)v_s \tag{15.34}$$

where the operators $\mathscr{A}(p)$ and $\mathscr{B}(p)$ were defined by Eqs. (8.5) and (8.6). Both sides of Eq. (15.33) are functions of t. Because these functions of t are equal, their Fourier transforms must also be equal. Therefore

$$\mathscr{F}\left\{a_n \frac{d^n v_o}{dt^n} + a_{n-1} \frac{d^{n-1} v_o}{dt^{n-1}} + \cdots + a_0 v_o\right\}$$

$$= \mathscr{F}\left\{b_m \frac{d^m v_s}{dt^m} + b_{m-1} \frac{d^{m-1} v_s}{dt^{m-1}} + \cdots + b_0 v_s\right\} \qquad (15.35)$$

or

$$\mathscr{F}\{\mathscr{A}(p)v_o\} = \mathscr{F}\{\mathscr{B}(p)v_s\} \qquad (15.36)$$

By the linearity property (Property P1), Eq. (15.35) becomes

$$a_n \mathscr{F}\left\{\frac{d^n v_o}{dt^n}\right\} + a_{n-1} \mathscr{F}\left\{\frac{d^{n-1} v_o}{dt^{n-1}}\right\} + \cdots + a_0 \mathscr{F}\{v_o\}$$

$$= b_m \mathscr{F}\left\{\frac{d^m v_s}{dt^m}\right\} + b_{m-1} \mathscr{F}\left\{\frac{d^{m-1} v_s}{dt^{m-1}}\right\} + \cdots + b_0 \mathscr{F}\{v_s\} \qquad (15.37)$$

and by the differentiation property (Property P2), Eq. (15.37) becomes

$$a_n(j\omega)^n V_o(f) + a_{n-1}(j\omega)^{n-1} V_o(f) + \cdots + a_0 V_o(f)$$

$$= b_m(j\omega)^m V_s(f) + b_{m-1}(j\omega)^{m-1} V_s(f) + \cdots + b_0 V_s(f) \qquad (15.38)$$

where $\omega = 2\pi f$, $V_o(f) = \mathscr{F}\{v_o(t)\}$, and $V_s(f) = \mathscr{F}\{v_s(t)\}$. Equation (15.38) can be rewritten as

$$[a_n(j\omega)^n + a_{n-1}(j\omega)^{n-1} + \cdots + a_0] V_o(f)$$

$$= [b_m(j\omega)^m + b_{m-1}(j\omega)^{m-1} + \cdots + b_0] V_s(f) \qquad (15.39)$$

or

$$\mathscr{A}(j\omega) V_o(f) = \mathscr{B}(j\omega) V_s(f) \qquad (15.40)$$

or

$$V_o(f) = \left[\frac{\mathscr{B}(j\omega)}{\mathscr{A}(j\omega)}\right] V_s(f) \qquad (15.41)$$

The quantity in the brackets is, by definition, the transfer function of the circuit and can be written as either $H(j\omega)$ or $H(f)$. The latter notation appears in Eq. (15.23).

PROOF 2

This proof applies to all LTI systems. It is based on superposition. An arbitrary waveform $v(t)$ is, according to Eq. (15.2), the Riemann sum of a continuum of infinitesimal complex exponentials:

$$v_s(t) = \int_{-\infty}^{+\infty} [V_s(f) df] e^{j2\pi f t} \qquad (15.42)$$

The complex exponential with frequency f is a rotating phasor $[V_s(f) df] e^{j2\pi f t}$. If this rotating phasor were to be applied individually to an LTI system as illustrated in Fig. 15.26, then the particular response would be the rotating phasor

$H(f)[V_s(f)\,df]e^{j2\pi ft}$, where $H(f)$ is the value of the system's transfer function at frequency f. This result is well known for LLTI systems from Chapter 10. We can establish it for all LTI systems by a straightforward evaluation of the convolution integral Eq. (15.25b) for the rotating phasor input $\mathbf{V}_s e^{j2\pi ft}$, where $\mathbf{V}_s = V_s(f)\,df$. We leave this evaluation as an exercise [also see Eqs. (15.55) through (15.58)].

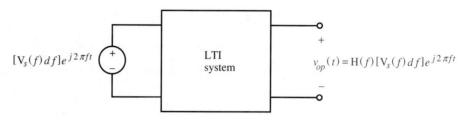

FIGURE 15.26   If the input to a stable LTI system is $[V_s(f)\,df]e^{j2\pi ft}$, then the particular response is $H(f)[V_s(f)\,df]e^{j2\pi ft}$

Figure 15.27 depicts the same system with the input of Eq. (15.42). Because the system is linear, it follows immediately that the particular response is the sum

$$v_{op}(t) = \int_{-\infty}^{+\infty} H(f)[V_s(f)\,df]e^{j2\pi ft}$$

$$= \int_{-\infty}^{+\infty} H(f)V_s(f)e^{j2\pi ft}\,df \tag{15.43}$$

The complementary response $v_c(t)$ is zero because the circuit is stable and each sinusoidal input component in (15.42) began at $t = -\infty$.

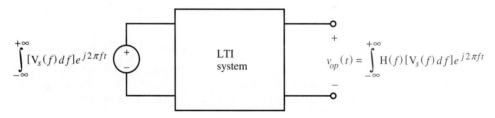

FIGURE 15.27   The particular response to a sum of complex exponentials is the sum of the particular responses to the individual complex exponentials

Thus the complete response is

$$v_o(t) = \int_{-\infty}^{+\infty} [H(f)V_s(f)]e^{j2\pi ft}\,df$$

$$= \int_{-\infty}^{+\infty} V_o(f)e^{j2\pi ft}\,df \tag{15.44}$$

which implies Eq. (15.23).

Appendix 15.2

DERIVATIONS OF PROPERTIES 1 AND 2

In this appendix we provide two separate derivations of Properties 1 and 2 of Section 15.6. The first derivation is rigorous. The second is not rigorous, but it conveys new insights. First we restate these properties.

Property 1

The response $v_o(t)$ of an LTI system to an arbitrary input $v_s(t)$ is given by

$$v_o(t) = \int_{-\infty}^{+\infty} v_s(\lambda) h(t - \lambda) \, d\lambda = v_s(t) * h(t) \tag{15.25a}$$

where $h(t)$ is the system's impulse response.

Property 2

The Fourier transform of a system's impulse response $h(t)$ is the system's transfer function $H(f)$: $h(t) \leftrightarrow H(f)$.

In each of the following derivations, Properties 1 and 2 are obtained together as part of a single development.

Derivation 1 of Properties 1 and 2

We begin with the result found in Appendix 15.1. As was shown in Appendix 15.1, the density spectrum $V_o(f)$ of the output of an LTI system is related to the density spectrum $V_s(f)$ of the input by

$$V_o(f) = H(f) V_s(f) \tag{15.45}$$

where $H(f)$ is the system's transfer function. The inverse Fourier transform of $V_o(f)$ is

$$v_o(t) = \int_{-\infty}^{+\infty} H(f) V_s(f) e^{j2\pi ft} \, df \tag{15.46}$$

By substituting

$$V_s(f) = \int_{-\infty}^{+\infty} v_s(\lambda) e^{-j2\pi f\lambda} \, d\lambda \tag{15.47}$$

into Eq. (15.46) and changing the order of integration, we find that

$$v_o(t) = \int_{-\infty}^{+\infty} H(f) \left[\int_{-\infty}^{+\infty} v_s(\lambda) e^{-j2\pi f\lambda} \, d\lambda \right] e^{j2\pi ft} \, df$$

$$= \int_{-\infty}^{+\infty} v_s(\lambda) \left[\int_{-\infty}^{+\infty} H(f) e^{j2\pi f(t-\lambda)} \, df \right] d\lambda \tag{15.48}$$

The quantity in the brackets can be simplified by defining

$$h(t) = \int_{-\infty}^{+\infty} H(f)e^{j2\pi ft}\, df \tag{15.49}$$

It follows that

$$h(t - \lambda) = \int_{-\infty}^{+\infty} H(f)e^{j2\pi f(t-\lambda)}\, df \tag{15.50}$$

which is the term in the brackets in Eq. (15.48). Therefore, Eq. (15.48) can be written as

$$v_o(t) = \int_{-\infty}^{+\infty} v_s(\lambda)h(t - \lambda)\, d\lambda \tag{15.51}$$

So far, our main results are Eqs. (15.49) and (15.51). To establish Properties 1 and 2, it remains to show that $h(t)$ is the system's impulse response. To show this, we simply substitute $v_s(t) = \delta(t)$ into Eq. (15.51). This yields

$$v_o(t) = \int_{-\infty}^{+\infty} \delta(\lambda)h(t - \lambda)\, d\lambda$$

$$= h(t) \tag{15.52}$$

which shows that $h(t)$ is indeed the impulse response of the system.

Derivation 2 of Properties 1 and 2

In this derivation, we begin by denoting the system's impulse response as $h(t)$, as illustrated in Fig. 15.28a. Next, we use the system's properties of linearity and time invariance. Because the system is time-invariant, its response to a delayed impulse $\delta(t - \lambda)$ is the delayed impulse response $h(t - \lambda)$, as illustrated in Fig. 15.28b. Because the system is also linear, its response to the scaled and delayed impulse $[v_s(\lambda)\, d\lambda]\delta(t - \lambda)$ must be the scaled and delayed impulse response $[v_s(\lambda)\, d\lambda]h(t - \lambda)$, as illustrated in Fig. 15.28c. Now an arbitrary input $v_s(t)$ can be expressed as the Riemann sum of scaled and delayed impulses

$$v_s(t) = \int_{-\infty}^{+\infty} v_s(\lambda)\delta(t - \lambda)\, d\lambda \tag{15.53}$$

The response to such a sum of scaled and delayed impulse responses is, by superposition, the sum of scaled and delayed impulse responses

$$v_o(t) = \int_{-\infty}^{+\infty} v_s(\lambda)h(t - \lambda)\, d\lambda \tag{15.54}$$

as illustrated in Fig. 15.28d. This establishes Property 1. It remains to be shown that the Fourier transform of $h(t)$ is the system's transfer function $H(f)$. If $v_s(t) = V_s e^{j2\pi ft}$, then, according to Eq. (15.54),

$$v_o(t) = \int_{-\infty}^{+\infty} v_s(t - \sigma)h(\sigma)\, d\sigma$$

$$= \int_{-\infty}^{+\infty} V_s e^{j2\pi f(t-\sigma)}h(\sigma)\, d\sigma$$

$$= \left[\int_{-\infty}^{+\infty} h(\sigma)e^{-j2\pi f\sigma}\, d\sigma \right] V_s e^{j2\pi ft} \tag{15.55}$$

FIGURE 15.28
Pertinent to second
derivation of Properties 1
and 2. (*a*) The response to
$\delta(t)$ is $h(t)$; (*b*) the
response to $\delta(t - \lambda)$ is
$h(t - \lambda)$ (time invariance);
(*c*) the response to
$[v_s(\lambda)\, d\lambda]\delta(t - \lambda)$ is
$[v_s(\lambda)\, d\lambda]h(t - \lambda)$ (linearity
and time invariance); (*d*)
superposition leads to
convolution integral

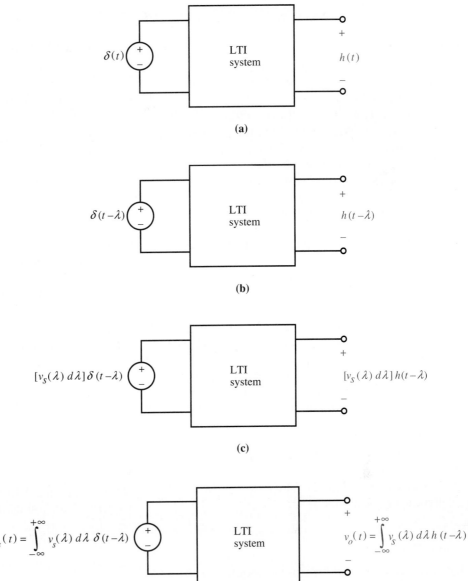

Thus $v_o(t)$ has the form $\qquad v_o(t) = \mathbf{V}_o e^{j2\pi ft}$ (15.56)

where $\qquad\qquad\qquad \mathbf{V}_o = \left[\int_{-\infty}^{+\infty} h(\sigma)e^{-j2\pi f\sigma}\, d\sigma \right]\mathbf{V}_s$ (15.57)

The ratio of output to input phasors, $\mathbf{V}_o/\mathbf{V}_s$, is by definition the transfer function $H(j\omega)$ or $H(f)$. Therefore

$$H(f) = \int_{-\infty}^{+\infty} h(t)e^{-j2\pi ft}\, dt = \mathscr{F}\{h(t)\} \qquad (15.58)$$

which establishes Property 2.

Section 15.1

1. (a) Find $v(t)$ if $V(f) = \Pi(f/2W)e^{-j2\pi f t_d}$.
   (b) Sketch $v(t)$, $|V(f)|$, and $\underline{/V(f)}$ for $W = 1$ kHz and $t_d = 5$ ms.

2. (a) Find $V(f)$ if $v(t) = e^{-t/\tau}[u(t) - u(t - T)]$.
   (b) Sketch $v(t)$, $|V(f)|$, and $\underline{/V(f)}$ for $\tau = T = 1$ ms.

3. Show that if $v(t)$ is a real function, then $V(-f) = V^*(f)$. [*Hint*: This important result follows directly from Eq. (15.1).]

4. (a) Show that $V(f)$ is real and even if $v(t)$ is real and even.
   (b) Show that $V(f)$ is imaginary and odd if $v(t)$ is real and odd.

Section 15.2

5. A rectangular pulse, $v(t) = \Pi(t/\tau)$, is repeated every Ts, where $T = 5\tau$, to yield a periodic signal $v_T(t)$.
   (a) Write down the Fourier transform of $v(t)$ and the Fourier series of $v_T(t)$. [*Hint*: Use Eq. (15.7b) and the Fourier transform pair of Example 15.2.]
   (b) Sketch $v(t)$, $V(f)$, $v_T(t)$, and the spectrum of $v_T(t)$.
   (c) Discuss your results.

6. Repeat Problem 5 for $T = 2\tau$.

7. Repeat Problem 5 for $T = \tau$.

Section 15.3

Use any method to find the Fourier transforms of the signals in Problems 8 through 11.

8. $5 \cos 100t \sin 150t$

9. $7e^{-2|t-1|}u(t)$

10. $\Pi(t) * \cos 2\pi t + \delta(t) \sin t$

11. According to Property P10 of Table 15.1, the Fourier transform of the convolution of two signals equals the product of the Fourier transforms of the signals, that is, $x_1(t) * x_2(t) \leftrightarrow X_1(f)X_2(f)$.
    (a) Apply this property of the special case $x_1(t) = x_2(t) = \Pi(t)$ to derive the Fourier transform of a triangular pulse.
    (b) Apply time-scaling Property P5 and linearity Property P1 to your result from (a)

to obtain the Fourier transform of a triangular pulse with arbitrary amplitude and width.

12. (a) Show that the Fourier transform of a periodic impulse train

$$v(t) = \sum_{n=-\infty}^{+\infty} \delta(t - nT)$$

is a periodic impulse train

$$V(f) = \frac{1}{T} \sum_{n=-\infty}^{+\infty} \delta(f - n/T).$$

   (b) What property does the gaussian pulse function share with the periodic impulse train?
   (c) Can you think of any other functions that have the property of (b)?

13. Derive Property P3 of Table 15.1. [*Hint*: First find the function $g(t)$ that has the property that

$$g(t) * v(t) = \int_{-\infty}^{+t} v(\lambda)\,d\lambda$$

Then use Property P10.]

14. Consider the signal $v(t) = \sqrt{2}\ 10^3$ sinc $(2 \times 10^6 t)\ \mu$V. Use Parseval's theorem to find the energy dissipated if $v(t)$ is applied across the terminals of a 1-Ω resistance.

15. Derive Properties P10 and P9 of Table 15.1. To derive P10, you need to show that

$$\mathscr{F}\left\{ \int_{-\infty}^{+\infty} v_1(\lambda)v_2(t-\lambda)\,d\lambda \right\} = V_1(f)V_2(f)$$

Start with the definition

$$\mathscr{F}\left\{ \int_{-\infty}^{+\infty} v_1(\lambda)v_2(t-\lambda)\,d\lambda \right\}$$

$$= \int_{-\infty}^{+\infty}\left\{ \int_{-\infty}^{+\infty} v_1(\lambda)v_2(t-\lambda)\,d\lambda \right\}e^{-j2\pi ft}\,dt$$

Then interchange orders of integration to obtain

$$\mathscr{F}\left\{ \int_{-\infty}^{+\infty} v_1(\lambda)v_2(t-\lambda)\,d\lambda \right\}$$

$$= \int_{-\infty}^{+\infty} v_1(\lambda)\left\{ \int_{-\infty}^{+\infty} v_2(t-\lambda)e^{-j2\pi ft}\,dt \right\}d\lambda$$

You should now be able to simplify the right-hand side to finish the derivation. Property P9 can be derived in a similar way.

16. A pulse $v_1(t)$ is repeated to yield a signal $v(t) = \sum_{n=-N}^{+N} v_1(t - nT)$ consisting of $2N+1$ pulses.
   (a) Show that $V(f) = V_1(f) \sum_{n=-N}^{+N} e^{-j2\pi nT}$.
   (b) Use the identity

   $$1 + 2 \cos \theta + 2 \cos 2\theta + \cdots + 2 \cos N\theta$$
   $$= \frac{\sin [(2N + 1)\theta/2]}{\sin (\theta/2)}$$

   to put your answer from (a) in the form

   $$V(f) = V_1(f) \frac{\sin [(2N + 1)\pi fT]}{\sin \pi fT}$$

   (c) Sketch $[\sin (2N + 1)\pi fT]/(\sin \pi fT)$ for $N = 5$.
   (d) Demonstrate the significance of the result from (b) by sketching $v_1(t)$, $V_1(f)$, $v(t)$, and $V(f)$. Assume for your sketch that $v_1(t) = e^{-|t|/\tau}$ (see FT8 of Table 15.2) with $\tau \ll T$, and $N = 5$. What effect does repeating the pulse have on the density spectrum?
   (e) Determine and sketch the limiting forms of $v(t)$ and $V(f)$ for $N \to \infty$. Hint:

   $$\lim_{N \to \infty} \frac{\sin [(2N + 1)\pi ft]}{\sin \pi fT}$$
   $$= \frac{1}{T} \sum_{n=-\infty}^{+\infty} \delta\left(f - \frac{n}{T}\right)$$

   What is the relationship between the density spectrum $V(f)$ and the Fourier-series coefficients of $v(t)$ when $N = \infty$?

17. A novel and frequently useful method to derive the Fourier transform $V(f)$ of a waveform $v(t)$ is to first determine an equation that $v(t)$ satisfies, then transform that equation into one for $V(f)$. This is illustrated by the following example.
   If $v(t) = \Pi(t/\tau)$, then $dv/dt = \delta(t + \tau/2) - \delta(t - \tau/2)$. We take the Fourier transform of each side of the latter equation with the aid of Property P2 of Table 15.1 to obtain

   $$j2\pi f V(f) = e^{+j\pi f\tau} - e^{-j\pi f\tau}$$
   $$= 2j \sin \pi f\tau$$

   We then solve for $V(f)$ and use Eq. (15.4) to obtain $V(f) = \tau \operatorname{sinc} f\tau$. Use this method on (a) through (c) below.
   (a) If $v(t) = e^{at}u(t)$, then $dv/dt = av(t) + \delta(t)$. Transform this equation and solve for $V(f)$. (Assume that $a < 0$.)

   (b) Assume that $v(t)$ is a triangular pulse $v(t) = (1 - |t|/\tau)\Pi(t/2\tau)$. Find an equation that $v(t)$ satisfies. [Hint: Consider d^2v/dt^2.] Transform this equation and solve for $V(f)$.
   (c) Determine $\mathscr{F}\{\cos (\pi t/\tau)\Pi(t/\tau)\}$ with the use of the method.

18. Show that

   $$\int_{-\infty}^{+\infty} x(t)y^*(t)\, dt = \int_{-\infty}^{+\infty} X(f)Y^*(f)\, df$$

   where $X(f) = \mathscr{F}\{x(t)\}$ and $Y(f) = \mathscr{F}\{y(t)\}$. The above equality is the *generalized Parseval's theorem*.

19. A real current $i(t)$ is applied to a circuit. The driving-point impedance of the circuit is $Z(j\omega) = R(\omega) + jX(\omega)$. Show that the total energy dissipated is given by the formula

   $$E = \int_{-\infty}^{+\infty} |I(f)|^2 R(\omega)\, df$$

   where $I(f) = \mathscr{F}\{i(t)\}$.

Section 15.4

20. The signal $v_s(t) = Ae^{-t/\tau}u(t)$ is input to a filter with transfer function $H(f) = \Pi(f/2W)$ to yield the output $v_o(t)$.
   (a) Sketch $V_s(f)$, $H(f)$, and $V_o(f)$.
   (b) Assume that $v_o(t)$ is applied across the terminals of a 1-Ω resistance. Use Parseval's theorem to determine the energy E_o dissipated.
   (c) Note that E_o depends on the product $W\tau$, $E_o = E_o(W\tau)$. Plot the function $E_o(W\tau)/E_o(\infty)$ versus $W\tau$. What is the physical significance of this plot?

21. A transmitted commercial AM signal has the form $v_x(t) = A_c[1 + av(t)] \cos 2\pi f_c t$, where A_c and a are nonnegative constants and $v(t)$ is the message. Assume that $V(f) = (1/2W)\Pi(f/2W)$, $a \ll 1$, and $f \gg W$.
   (a) Sketch $V_x(f)$.
   (b) Let $y(t) = v_x^2(t)$. Sketch $Y(f)$.
   (c) Show that $v(t)$ can be recovered (at least approximately) by putting $y(t)$ into a low-pass filter. Draw the system, including any additional operations that are necessary to obtain $v(t)$ from $y(t)$.

(d) Under what conditions will the output of the system you described in (c) be equal to $v(t)$?

22. In upper-single-sideband transmission, the DSB signal $v_x(t)$ of Fig. 15.8 is input to a high-pass filter that passes only the upper sideband in $V_x(f)$ before transmission. The transfer function of the ideal filter is $H(f) = 1 - \Pi(f/2f_c)$.
    (a) Sketch $H(f)$.
    (b) Let $v_x'(t)$ denote the transmitted signal. Sketch $V_x'(f) = V_x(f)H(f)$.
    (c) Show that $v(t)$ can be recovered (at least approximately) from $v_x'(t)$ by multiplying $v_x'(t)$ by $4 \cos 2\pi f_c t$ and low-pass filtering the product. (*Hint*: Use Property P9 from Table 15.1 and FT4 from Table 15.2. Sketch the spectral density of the low-pass filter input signal.)
    (d) Under what conditions is the low-pass filter output exactly equal to $v(t)$?

23. Repeat Problem 22 with the high-pass filter replaced with an ideal low-pass filter whose transfer function is $H(f) = \Pi(f/2f_c)$.

Section 15.5

24. The Fourier transform of a *gaussian pulse*

$$v(t) = \frac{1}{\sqrt{2\pi\tau^2}}\, e^{-t^2/2\tau^2}$$

is

$$V(f) = e^{-(2\pi f\tau)^2/2}$$

(a) Sketch $v(t)$. Indicate the value of $v(t)$ at $t = 0$ and $t = \tau$ on your sketch.
(b) Sketch $V(f)$. Indicate the value of $V(f)$ at $f = 0$ and $f = 1/\tau$.
(c) The gaussian pulse $v(t)$ of (a) is input to an LTI system whose impulse response is

$$h(t) = \frac{1}{\sqrt{2\pi\sigma^2}}\, e^{-t^2/2\sigma^2}$$

Determine the output $v_o(t)$. Sketch $v_o(t)$ and $V_o(f)$.
(d) State two special properties of a gaussian pulse.

25. The signal $x(t) = A \cos(\omega_s t)u(t)$ is applied to a low-pass filter whose transfer function is $H(f) = 1/(1 + j2\pi f\tau)$ (where $\tau > 0$).
    (a) Find the filter output $y(t)$ by convolving

$x(t)$ with $h(t)$ where $h(t)$ is the impulse response of the filter.
    (b) Find the density spectrum $Y(f)$ of the filter output.
    (c) Check your answer from (a) by solving the filter's differential equation.
    (d) Sketch $y(t)$, indicating the influence of the filter half-power frequency $1/\tau$ rad/s on your sketch. Approximately how long does it take for the output to reach steady-state?

26. Repeat Problem 25 for a high-pass filter with transfer function $H(f) = j2\pi f/(1 + j2\pi f\tau)$.

27. The function sinc x was defined in Eq. (15.4).
    (a) Show that $(\tau^{-1} \operatorname{sinc} t/\tau) * \tau^{-1} \operatorname{sinc} t/\tau = \tau^{-1} \operatorname{sinc} t/\tau$.
    (b) Show that $x(t) * x(t) = x(t)$ where $x(t) = 2\tau^{-1} \operatorname{sinc}(t/\tau)\cos \omega_o t$ with $\omega_o > \pi/\tau$.

28. An *ideal low-pass filter* is defined by the transfer function $H(f) = \Pi(f/2W)$, where W is the bandwidth.
    (a) Sketch $H(f)$.
    (b) Determine the output of the filter for an input $v_s(t) = A \cos 2\pi f_s t$. Assume that $f_s < W$.
    (c) Repeat (b) for $f_s > W$.
    (d) Determine and sketch $h(t)$.
    (e) Determine the output of the filter to an input $v_s(t) = \operatorname{sinc} 2f_a t$. Assume that $f_a < W$.
    (f) Repeat (e) for $f_a > W$.
    (g) Determine the output of the filter for an arbitrary input $v_s(t)$ with the property that $V_s(f) = 0$ for $|f| > W$.

29. An *ideal band-pass filter* is defined by the transfer function

$$H(f) = \Pi\left(\frac{f - f_c}{2W}\right) + \Pi\left(\frac{f + f_c}{2W}\right)$$

where f_c is the *center frequency* and $2W$ is the bandwidth.
    (a) Sketch $H(f)$.
    (b) Determine the output of the filter for an input $v_s(t) = A \cos 2\pi f_s t$. Assume that $f_s < f_c - W$.
    (c) Repeat (b) for $f_c - W < f_s < f_c + W$.
    (d) Repeat (b) for $f_c + W < f_s$.
    (e) Determine and sketch $h(t)$.
    (f) Determine the output of the filter for an arbitrary input $v_s(t)$ with the property that $V_s(f) = 0$ for $|f - f_c| > W$.

30. According to the analysis of Chapter 10, the response of a stable LLTI circuit with impedance $\mathbf{Z}(j\omega)$ to the input $i(t) = I_m \cos(\omega t + \phi_I)$ is $v(t) = V_m \cos(\omega t + \phi_V)$, where $V_m = |\mathbf{Z}(j\omega)|I_m$ and $\phi_V = \underline{/\mathbf{Z}(j\omega)} + \phi_I$. Rederive this result with the use of Fourier transforms.

31. According to the principle of superposition as stated in Chapter 11, the voltage response of a stable LLTI circuit to an input current $i(t) = I_{1m} \cos(\omega_1 t + \phi_1) + I_{2m} \cos(\omega_2 t + \phi_2)$ is
$v(t) = I_{1m}|\mathbf{Z}(j\omega_1)| \cos[\omega_1 t + \phi_1 + \underline{/\mathbf{Z}(j\omega_1)}]$
$+ I_{2m}|\mathbf{Z}(j\omega_2)| \cos[\omega_2 t + \phi_2 + \underline{/\mathbf{Z}(j\omega_2)}]$,
where $\mathbf{Z}(j\omega)$ is the impedance of the circuit. Rederive this result with the use of Fourier transforms.

32. Show that an LTI system is causal if and only if its impulse response equals zero for $t < 0$.

33. The driving-point impedance of an arbitrary real LLTI circuit is $Z(f) = R(f) + jX(f)$.
    (a) Show that the resistance $R(f)$ is an even function of f.
    (b) Show that the reactance $X(f)$ is an odd function of f.
    (c) Do the properties of (a) and (b) also apply for transfer functions?

34. In Example 15.7 we showed that it is impossible to physically construct an ideal low-pass filter. A student proposes than an approximation to an ideal low-pass filter can be constructed by allowing delay. Specifically, the student proposes to construct an LTI system having impulse response

$$h(t) = 2W \text{ sinc } [2W(t - t_d)]u(t)$$

where t_d is a positive constant and $u(t)$ is a unit step function.
    (a) Is the proposed system causal?
    (b) Sketch $h(t)$. Assume in your sketch that $t_d \gg \frac{1}{2}W$.
    (c) Show that if we replace $u(t)$ by unity then the proposed system is equivalent to an ideal low-pass filter followed by a delay t_d.
    (d) What is the effect of increasing t_d in the proposed system?
    (e) What do you think of the student's idea? Is it basically sound? Do you foresee any problems?

35. (a) The voltage density spectrum $V(f)$ of a typical *band-limited* signal $v(t)$ is shown in the figure below (top). (The term *band-limited* means that $V(f) = 0$ for $|f| > f_h$ for some finite f_h as shown in the first figure below.) The signal $v(t)$ is multiplied by a periodic impulse train $p(t) = \sum_{n=-\infty}^{+\infty} \delta(t - nT)$ to yield the "sample-carrying" signal

$$v_s(t) = v(t) \sum_{n=-\infty}^{+\infty} \delta(t - nT)$$

Show that $v_s(t) = \sum_{n=-\infty}^{+\infty} v(nT)\delta(t - nT)$, and sketch $v_s(t)$. Observe that only the "samples" $v(nT)$, $-\infty \le n \le \infty$, appear in $v_s(t)$.

    (b) Find and sketch the voltage density spectrum of the periodic impulse train $p(t) = \sum_{n=-\infty}^{+\infty} \delta(t - nT)$. (*Hint*: Use FT15.)

    (c) Show that the voltage density spectrum of $v_s(t)$ is $V_s(f) = (1/T) \sum_{m=-\infty}^{+\infty} V(f - m/T)$. [*Hint*: Use Property P9 and your result from (b).]

    (d) The sample carrying signal $v_s(t)$ is input to a low-pass filter whose frequency response characteristic is shown in part (b) of the figure below. Assume that the *sampling frequency* $f_s = 1/T$ satisfies $f_s \ge 2f_h$, where f_h is the highest frequency of $V(f)$ as indicated in part (a) of the figure. The output of the filter is $v_o(t)$. Sketch the voltage density spectra of $v_s(t)$ and $v_o(t)$ and conclude that $V_o(f) = V(f)$.

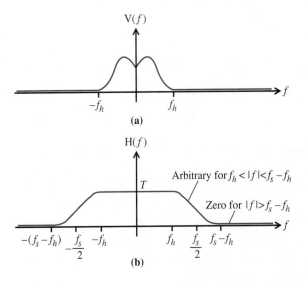

(e) Conclude that $v_o(t) = v(t)$ provided that $f_s > 2f_h$. Note, therefore, that *any band-limited signal can be determined from its samples* $v(nT)$, $-\infty \le n \le \infty$, *provided that the sampling rate* $f_s = 1/T$ *is at least twice the highest frequency* f_h *in* $V(f)$. This startling and important result is known as *Shannon's sampling theorem*. The minimum permitted sampling rate, $2f_h$, is called the *Nyquist rate*.

(f) Show that the general formula to reconstruct the band-limited signal $v(t)$ from its samples is

$$v(t) = \sum_{n=-\infty}^{+\infty} v_s(nT)h(t - nT)$$

where $h(t) = \mathscr{F}^{-1}\{H(f)\}$ is the impulse response of the filter in part (b) of the figure and $f_s > 2f_h$.

(g) Assume as a special case of the result from (f) that $H(f) = T\Pi(f/f_s)$ to obtain the reconstruction formula

$$v(t) = \sum_{n=-\infty}^{+\infty} v_s(nT)\operatorname{sinc}\left[f_s(t - nT)\right]$$

where $f_s > 2f_h$.

(h) Use the result from (g) to determine and sketch $v(t)$ and $V(f)$ if $v_s(t)$ has only three nonzero samples:

$$v_s(-2T) = 3, \qquad v_s(0) = 6, \qquad v_s(T) = 3$$

The Laplace Transform

One of the easiest ways to solve LLTI differential equations and LLTI circuit problems involving initial conditions is by means of the Laplace transform. The principal property of the Laplace transform is that it transforms differential operators into algebraic operators. By exploiting this property, we can obtain the *complete* response of an LLTI circuit to almost any input signal by using only algebra and table look-up. In addition to its value in the solution of differential equations and circuit problems, the Laplace transform provides valuable theoretical insight into the interaction of signals and LLTI systems. A third strength of the Laplace transform is that it can be applied to virtually every signal encountered in circuit theory.

16.1 Definitions

The Laplace transform* of a signal $v(t)$ is defined as follows:

DEFINITION

Laplace Transform

$$V(s) = \int_{0^-}^{\infty} v(t)e^{-st}\, dt \qquad \mathcal{R}e\,\{s\} > \sigma_0 \qquad\qquad (16.1)$$

---

where the notation 0^- indicates that the point $t = 0$ is included in the integration. The integral in Eq. (16.1) converges only for values of s that satisfy $\mathcal{R}e\,\{s\} > \sigma_0$, where σ_0 is a constant determined by $v(t)$. The region $\mathcal{R}e\,\{s\} > \sigma_0$ is the *region of convergence*, and σ_0 is the *abscissa of convergence*. We say that $V(s)$ *exists* in the region of convergence and is *undefined* outside of the region of convergence. When we study specific examples of the Laplace transform, we will see that the condition $\mathcal{R}e\,\{s\} > \sigma_0$ arises naturally as part of the development.

Associated with the *direct* Laplace transform, Eq. (16.1), is the *inverse* Laplace transform, which is given by:

Inverse Laplace Transform

$$v(t) = \int_{c-j\infty}^{c+j\infty} V(s)e^{st}\, \frac{ds}{2\pi j} \qquad c > \sigma_0 \qquad\qquad (16.2)$$

Notice that the inverse Laplace transform represents $v(t)$ as a Riemann sum (that is, an integral) of infinitesimal exponential signals, each with the form $V(s)e^{st}(ds/2\pi j)$, where $s = \sigma + j\omega$. The integration is usually performed along the straight-line path $(c + j\omega,\ -\infty < \omega < \infty)$, called the *Bromwich path*, which is illustrated in Fig. 16.1.

FIGURE 16.1
Illustration of the abscissa of convergence, the Bromwich path, and the region of convergence

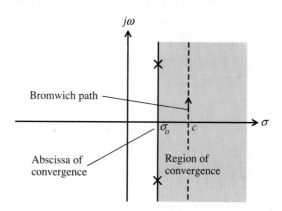

---

* The Laplace transform comes in two varieties, the *unilateral* or *one-sided* Laplace transform and the *bilateral* or *two-sided* Laplace transform. Only the unilateral Laplace transform is considered in this book, where it is referred to as simply the Laplace transform.

In future work we will not have to evaluate the integral in Eq. (16.2) *analytically*. Instead, we will simply use a look-up table. We develop the look-up table in Section 16.3.

It is important to note that the direct Laplace transform $V(s)$, defined by Eq. (16.1), depends only on $v(t)$ for $t \geq 0$. As is shown in Appendix 16.1, the right-hand side of Eq. (16.2) converges to $v(t)$ for $t \geq 0$ but to 0 for $t < 0$. This result means that the transformation in Eq. (16.1) is one-to-one* for waveforms $v(t)$ that equal zero for $t < 0$. Waveforms of this type are called *causal* waveforms. Causal waveforms occur frequently in circuit analysis problems. Moreover, many noncausal waveforms can be made causal by the simple use of a time-shift operation, $t' = t - t_0$. Unless stated to the contrary, we will assume throughout this chapter that $v(t)$ is a causal waveform, and write Eqs. (16.1) and (16.2), respectively, as

$$V(s) = \mathscr{L}\{v(t)\} \tag{16.3a}$$

and

$$v(t) = \mathscr{L}^{-1}\{V(s)\} \tag{16.3b}$$

or

$$v(t) \leftrightarrow V(s) \tag{16.3c}$$

The functions $v(t)$ and $V(s)$ taken together are called a *Laplace transform pair*.

To develop a feeling for Eq. (16.1), we now derive the Laplace transforms of three simple but important waveforms. Study Example 16.1, and notice how the condition $\mathscr{R}e\{s\} > \sigma_0$ arises.

EXAMPLE 16.1

Determine the Laplace transform of $v(t) = e^{at}u(t)$, where a is an arbitrary real number.

■ SOLUTION

Substitution of $v(t) = e^{at}u(t)$ into Eq. (16.1) yields

$$V(s) = \int_{0^-}^{\infty} e^{at}e^{-st}\, dt$$

$$= \int_{0^-}^{\infty} e^{(a-s)t}\, dt = \frac{1}{a-s}\, e^{(a-s)t}\Big|_{0^-}^{\infty}$$

which is

$$V(s) = \frac{1}{a-s}\, e^{(a-s)t}\Big|_{t=\infty} - \frac{1}{a-s}\, e^{(a-s)t}\Big|_{t=0^-}$$

With $\sigma + j\omega$ substituted for s in the exponent, this becomes

$$V(s) = \frac{1}{a-s}\, e^{(a-\sigma)t}e^{-j\omega t}\Big|_{t=\infty} - \frac{1}{a-s}\, e^{(a-\sigma)t}e^{-j\omega t}\Big|_{t=0^-}$$

---

* As with Fourier series and Fourier transforms, exceptions of a purely mathematical nature exist. We will ignore the mathematical exceptions, because they do not correspond to physically measurable quantities.

The second term is simply

$$-\frac{1}{a-s}e^{(a-\sigma)0^-}e^{-j\omega0^-} = -\frac{1}{a-s}$$

Evaluation of the first term requires care. The term $\left.[1/(a-s)]e^{(a-\sigma)t}e^{-j\omega t}\right|_{t=\infty}$ should be interpreted as the limit $\lim_{t\to\infty}[1/(a-s)]e^{(a-\sigma)t}e^{-j\omega t}$. If $\sigma \leq a$, then this limit is not a well-defined mathematical quantity. If $\sigma > a$, then the limit is well defined because the factor $e^{(a-\sigma)\infty}$ equals zero for $a-\sigma < 0$. Therefore

$$V(s) = \frac{1}{s-a}$$

provided that $\mathscr{R}e\ \{s\} > a$, and we have the result

$$e^{at}u(t) \leftrightarrow \frac{1}{s-a} \qquad \mathscr{R}e\ \{s\} > a$$

The abscissa of convergence is $\sigma_0 = a$.

   In Section 15.6 we found that the causal exponential waveform $v(t) = e^{at}u(t)$ does not have a Fourier transform if a is positive. We have just shown that $v(t) = e^{at}u(t)$ has a Laplace transform, $V(s) = 1/(s-a)$, regardless of the value of a. The value of a affects only the region of convergence of $V(s)$. This region is illustrated in Fig. 16.2a for $a < 0$ and in Fig. 16.2b for $a > 0$.

FIGURE 16.2
The region of convergence (shaded) of $V(s) =$ $\mathscr{L}\{e^{at}u(t)\} = 1/(s-a)$. Note that $V(s)$ has a pole $s = a$. (a) $a < 0$; (b) $a > 0$

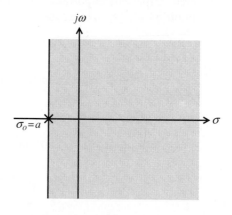

(a)

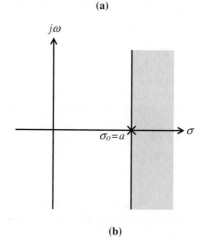

(b)

EXAMPLE 16.2

Determine the Laplace transform of a unit step function $v(t) = u(t)$.

■ **SOLUTION**

This function is a special case of Example 16.1 with $a = 0$. Consequently

$$u(t) \leftrightarrow \frac{1}{s} \qquad \sigma > 0$$

The region of convergence is illustrated in Fig. 16.3.

FIGURE 16.3
The region of convergence
(shaded) of $V(s) =$
$\mathscr{L}\{u(t)\} = 1/s$

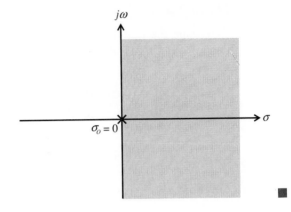

EXAMPLE 16.3

Determine the Laplace transform of the unit impulse $v(t) = \delta(t)$.

■ **SOLUTION**

Substitution of $\delta(t)$ into Eq. (16.1) yields

$$V(s) = \int_{0^-}^{\infty} \delta(t)e^{-st}\,dt = 1 \qquad \text{for all } s$$

Thus
$$\delta(t) \leftrightarrow 1$$

The abscissa of convergence is $\sigma_0 = -\infty$, because $V(s)$ exists for all s. The region of convergence is illustrated in Fig. 16.4.

FIGURE 16.4
The region of convergence
(shaded) of $V(s)$ for
$v(t) = \delta(t)$

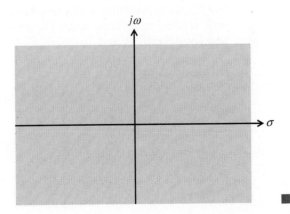

In the preceding examples, we have found the Laplace transforms of only three of an infinite number of possible waveforms. These three waveforms, however, are certainly among the most important in circuit analysis, and they will be encountered repeatedly in this chapter.

REMEMBER

The Laplace transform of $e^{at}u(t)$, where a is any real number, is $1/(s - a)$, where $\mathscr{R}e\,\{s\} > a$.

EXERCISES

Determine the Laplace transform of the signals in Exercises 1 through 5. Show the poles and zeros of $V(s)$ in the s plane, and shade the region of convergence of $V(s)$.

1. $v(t) = u(t - t_0)$, where $t_0 > 0$
2. $v(t) = u(t) - u(t - t_0)$, where $t_0 > 0$
3. $v(t) = e^{-3t}u(t) + e^{-5t}u(t)$
4. $v(t) = e^{-3t}u(t) + e^{5t}u(t)$
5. $v(t) = tu(t)$
6. Find $v(t)$ if

$$V(s) = 1 + \frac{1}{s} + \frac{1}{s + 2} \qquad \mathscr{R}e\,\{s\} > 0$$

(*Hint*: Use the results of Example 16.1 through 16.3.)

16.2 Properties of the Laplace Transform

As was stated in the introduction to this chapter, the Laplace transform provides a means to solve LLTI circuit problems through the use of algebra and table look-up only. In this section we describe the Laplace transform properties that make this method of solution possible.

Table 16.1 lists several important properties of the Laplace transform. Most of these properties can be established by proofs very similar to those used to establish the corresponding properties of the Fourier transform (Table 15.1). Properties P2 and P3 are particularly useful to solve circuit problems and differential equations with nonzero initial conditions. The proofs of Properties P2, P3, P8, and P9 involve new considerations and are therefore shown here.

Differentiation (P2)

Property 2 states that the Laplace transform of the nth derivative of $v(t)$ is given by

$$\mathscr{L}\left\{\frac{d^n v}{dt^n}\right\} = -v^{(n)}(0^-) - sv^{(n-1)}(0^-) - \cdots - s^{n-1}v(0^-) + s^n V(s)$$

where

$$v^{(k)}(0^-) = \left.\frac{d^k v}{dt^k}\right|_{t=0^-} \qquad \text{for } k = 1, 2, \ldots \tag{16.4}$$

TABLE 16.1   Properties of the Laplace transform

PROPERTY		SIGNAL	LAPLACE TRANSFORM
P1	Linearity	$a_1 v_1(t) + a_2 v_2(t)$	$a_1 V_1(s) + a_2 V_2(s)$
P2	Differentiation	$\dfrac{d^n v}{dt^n}$	$-v^{(n)}(0^-) - sv^{(n-1)}(0^-) - \cdots$ $\quad - s^{n-1} v(0^-) + s^n V(s)$
	Special case, $n=1$	$\dfrac{dv}{dt}$	$-v(0^-) + sV(s)$
	Special case, $n=2$	$\dfrac{d^2 v}{dt^2}$	$-v^{(1)}(0^-) - sv(0^-) + s^2 V(s)$
P3	Integration	$\displaystyle\int_{-\infty}^{t} v(\lambda)\,d\lambda,\ t>0$	$\dfrac{1}{s} V(s) + \dfrac{1}{s}\displaystyle\int_{-\infty}^{0^-} v(\lambda)\,d\lambda$
P4	Scaling	$v\!\left(\dfrac{t}{a}\right)$	$aV(as),\ a>0$
P5	Time delay	$v(t-t_d)u(t-t_d),\ t_d \ge 0$	$e^{-st_d} V(s)$
P6	Complex frequency shift	$v(t)e^{s_0 t}$	$V(s-s_0)$
P7	Convolution	$\displaystyle\int_{0}^{t} v_1(\lambda)v_2(t-\lambda)\,d\lambda$	$V_1(s)V_2(s)$
P8	Initial value theorem	$v(0^+) = \displaystyle\lim_{\mathcal{Re}\,\{s\}\to\infty} sV(s)$	Provided the limit exists
P9	Final value theorem	$\displaystyle\lim_{t\to\infty} v(t) = \lim_{s\to 0} sV(s)$	Provided the limit exists

(handwritten annotations in margin: $V_R = IR = I$; $\dfrac{d\,v_c}{dt}$; $\dfrac{1}{s}\left(V_c(0^-)\right) + sV(s)$; $V_c(s)$)

This property is important in the solution of LLTI equations. We can prove Property P2 for $n=1$ by integrating

$$\mathcal{L}\left\{\frac{dv}{dt}\right\} = \int_{0^-}^{\infty} \frac{dv}{dt} e^{-st}\,dt \tag{16.5}$$

by parts. If we set $y = e^{-st}$ and $dx = (dv/dt)\,dt$, the identity

$$\int_{0^-}^{\infty} y\,dx = xy\Big|_{0^-}^{\infty} - \int_{0^-}^{\infty} x\,dy \tag{16.6}$$

yields

$$\mathcal{L}\left\{\frac{dv}{dt}\right\} = v(t)e^{-st}\Big|_{0^-}^{\infty} + s\int_{0^-}^{\infty} v(t)e^{-st}\,dt$$

$$= v(t)e^{-st}\Big|_{t=\infty} - v(t)e^{-st}\Big|_{t=0^-} + s\mathcal{L}\{v(t)\} \tag{16.7}$$

It follows from Eq. (16.77) of Appendix 16.1 that

$$v(t)e^{-st}\Big|_{t=\infty} = v(t)e^{-\sigma t}e^{-j\omega t}\Big|_{t=\infty} = 0 \tag{16.8}$$

Therefore Eq. (16.7) becomes

$$\mathcal{L}\left\{\frac{dv}{dt}\right\} = -v(0^-) + s\mathcal{L}\{v(t)\} \tag{16.9}$$

Equation (16.9) establishes Property P2 for $n = 1$. To establish Property P2 for $n = 2$, we simply replace $v(t)$ with dv/dt in Eq. (16.9) to obtain

$$\mathcal{L}\left\{\frac{d^2v}{dt^2}\right\} = -\left.\frac{dv}{dt}\right|_{t=0^-} + s\mathcal{L}\left\{\frac{dv}{dt}\right\} \tag{16.10}$$

Substitution of Eq. (16.9) into Eq. (16.10) yields

$$\mathcal{L}\left\{\frac{d^2v}{dt^2}\right\} = -\left.\frac{dv}{dt}\right|_{t=0^-} - sv(0^-) + s^2\mathcal{L}\{v(t)\} \tag{16.11}$$

which establishes Property P2 for $n = 2$. If we repeat the process we obtain Property P2 for any n:

$$\mathcal{L}\left\{\frac{d^nv}{dt^n}\right\} = -v^{(n)}(0^-) - sv^{(n-1)}(0^-) + \cdots - s^{n-1}v(0^-) + s^nV(s) \tag{16.12}$$

Integration (P3)

According to Property P3, the Laplace transform of the integral of $v(t)$ is given by

$$\mathcal{L}\left\{\int_{-\infty}^t v(\lambda)\, d\lambda\right\} = \frac{1}{s}V(s) + \frac{1}{s}\int_{-\infty}^{0^-} v(\lambda)\, d\lambda$$

where $t > 0$. This property can be obtained from Eq. (16.9) by defining

$$v(t) = \int_{-\infty}^t g(\lambda)\, d\lambda \tag{16.13}$$

where $g(t)$ is not necessarily zero for $t < 0$. It follows that

$$\frac{dv}{dt} = g(t) \tag{16.14}$$

and

$$v(0^-) = \int_{-\infty}^{0^-} g(\lambda)\, d\lambda \tag{16.15}$$

so that when Eqs. (16.14) and (16.15) are substituted into Eq. (16.9), we obtain

$$\mathcal{L}\{g(t)\} = -\int_{-\infty}^{0^-} g(\lambda)\, d\lambda + s\mathcal{L}\left\{\int_{-\infty}^t g(\lambda)\, d\lambda\right\} \tag{16.16}$$

which can be solved to give

$$\mathcal{L}\left\{\int_{-\infty}^t g(\lambda)\, d\lambda\right\} = \frac{1}{s}\mathcal{L}\{g(t)\} + \frac{1}{s}\int_{-\infty}^{0^-} g(\lambda)\, d\lambda \tag{16.17}$$

A purely notational change then yields

$$\mathcal{L}\left\{\int_{-\infty}^t v(\lambda)\, d\lambda\right\} = \frac{1}{s}V(s) + \frac{1}{s}\int_{-\infty}^{0^-} v(\lambda)\, d\lambda \tag{16.18}$$

which is Property P3. Of course, the integral on the right-hand side of Eq. (16.18) will equal zero whenever $v(t)$ is causal.

The initial and final value theorems, Properties P8 and P9, are useful because they enable us to compute initial and final values of $v(t)$ without computing the inverse transform of $V(s)$.

The Initial Value Theorem (P8)

The *initial value theorem* states that the value of $v(0^+)$ is given by

$$v(0^+) = \lim_{\mathscr{R}e\,\{s\} \to \infty} sV(s)$$

provided that the limit exists. To prove the initial value theorem, we begin with the differentiation Property P2 specialized to $n = 1$:

$$\int_{0^-}^{\infty} \frac{dv}{dt} e^{-st}\, dt = sV(s) - v(0^-) \qquad (16.19)$$

The left-hand side can be broken up into an integral from 0^- to 0^+ and an integral from 0^+ to ∞. This yields

$$\int_{0^-}^{\infty} \frac{dv}{dt} e^{-st}\, dt = v(0^+) - v(0^-) + \int_{0^+}^{\infty} \frac{dv}{dt} e^{-st}\, dt \cdot \qquad (16.20)$$

provided that $v(t)$ does not contain an impulse or derivatives of an impulse at $t = 0$. Substitution of Eq. (16.20) into Eq. (16.19) yields

$$v(0^+) + \int_{0^+}^{\infty} \frac{dv}{dt} e^{-st}\, dt = sV(s) \qquad (16.21)$$

By using inequality (16.77) in Appendix 16.1, we can show that the integral in Eq. (16.21) vanishes in the limit $\mathscr{R}e\,\{s\} \to \infty$. Therefore, if we take the limit $\mathscr{R}e\,\{s\} \to \infty$, Eq. (16.21) becomes

$$v(0^+) = \lim_{\mathscr{R}e\,\{s\} \to \infty} sV(s) \qquad (16.22)$$

provided the limit exists. This is Property P8 of Table 16.1.

The Final Value Theorem (P9)

The *final value theorem* gives the following formula for the final value of $v(t)$:

$$\lim_{t \to \infty} v(t) = \lim_{s \to 0} sV(s)$$

provided $v(t)$ *has* a final value.*

To prove the final value theorem, we again begin with the differentiation Property P2 where $n = 1$, namely, Eq. (16.19). In the limit $s \to 0$, e^{-st} becomes unity. Thus, with $s \to 0$, Eq. (16.19) becomes

$$\int_{0^-}^{\infty} \frac{dv}{dt}\, dt = \lim_{s \to 0} sV(s) - v(0^-) \qquad (16.23)$$

But

$$\int_{0^-}^{\infty} \frac{dv}{dt}\, dt = \lim_{t \to \infty} v(t) - v(0^-) \qquad (16.24)$$

By substituting Eq. (16.24) into Eq. (16.23), we get

$$\lim_{t \to \infty} v(t) = \lim_{s \to 0} sV(s) \qquad (16.25)$$

provided the limit exists. This is Property P9 of Table 16.1.

---

* For example, the final value theorem cannot be used to find the final value of $v(t) = \cos t$ because $\cos t$ oscillates between $+1$ and -1 forever. However, the initial value theorem *can* be used to find the value of $\cos 0^+$ because $\lim_{\substack{t \to 0 \\ t > 0}} \cos t$ is well defined.

We will use the properties in Table 16.1 to solve circuit problems in later sections of this chapter. Examples 16.4 and 16.5 illustrate how to use the properties to derive new Laplace transform pairs.

EXAMPLE 16.4

Use Properties P1 and P6 and the results of Example 16.2 to find the Laplace transform of $v(t) = \cos(\omega_0 t)u(t)$.

■ SOLUTION

By using Euler's formula, we can write $v(t)$ as

$$v(t) = \frac{1}{2}e^{j\omega_0 t}u(t) + \frac{1}{2}e^{-j\omega_0 t}u(t)$$

It follows by the linearity Property (P1) that

$$V(s) = \frac{1}{2}\mathscr{L}\{e^{j\omega_0 t}u(t)\} + \frac{1}{2}\mathscr{L}\{e^{-j\omega_0 t}u(t)\}$$

We know from Example 16.2 that

$$u(t) \leftrightarrow \frac{1}{s} \qquad \mathscr{R}e\{s\} > 0$$

Therefore, from Property P6,

$$\mathscr{L}\{e^{j\omega_0 t}u(t)\} = \frac{1}{s - j\omega_0} \qquad \mathscr{R}e\{s\} > 0$$

and

$$\mathscr{L}\{e^{-j\omega_0 t}u(t)\} = \frac{1}{s + j\omega_0} \qquad \mathscr{R}e\{s\} > 0$$

By combining the above results, we obtain

$$V(s) = \frac{1}{2}\frac{1}{s - j\omega_0} + \frac{1}{2}\frac{1}{s + j\omega_0} \qquad \mathscr{R}e\{s\} > 0$$

which simplifies to

$$V(s) = \frac{s}{s^2 + \omega_0^2} \qquad \mathscr{R}e\{s\} > 0$$

Therefore

$$\cos(\omega_0 t)u(t) \leftrightarrow \frac{s}{s^2 + \omega_0^2} \qquad \mathscr{R}e\{s\} > 0 \quad ■$$

In the preceding example, the region of convergence of each Laplace transform we encountered was $\mathscr{R}e\{s\} > 0$. In Example 16.5, we work with Laplace transforms that have different regions of convergence.

EXAMPLE 16.5

Assume that

$$v(t) = 2\delta(t) + 3e^{-2t}u(t) + \cos(2t)u(t) + 4e^{3t}u(t)$$

(a) Find $V(s)$.

(b) Plot the poles of $V(s)$ in the complex plane and show the region of convergence.

■ **SOLUTION**

We can find $V(s)$ using the linearity property (Property P1) and the results of Examples 16.1, 16.3, and 16.4.

By Property P1, we know that

$$V(s) = 2\mathscr{L}\{\delta(t)\} + 3\mathscr{L}\{e^{-2t}u(t)\} + \mathscr{L}\{\cos(2t)u(t)\} + 4\mathscr{L}\{e^{3t}u(t)\}$$

It follows from Examples 16.1, 16.3, and 16.4 that

$$\mathscr{L}\{\delta(t)\} = 1 \qquad \text{for all } s$$

$$\mathscr{L}\{e^{-2t}u(t)\} = \frac{1}{s+2} \qquad \text{for } \mathscr{R}e\,\{s\} > -2$$

$$\mathscr{L}\{\cos(2t)u(t)\} = \frac{s}{s^2+4} \qquad \text{for } \mathscr{R}e\,\{s\} > 0$$

and

$$\mathscr{L}\{e^{3t}u(t)\} = \frac{1}{s-3} \qquad \text{for } \mathscr{R}e\,\{s\} > 3$$

Therefore

$$V(s) = 2 + \frac{3}{s+2} + \frac{s}{s^2+4} + \frac{4}{s-3}$$

$$= \frac{2s^4 + 6s^3 - 6s^2 + 14s - 52}{(s+2)(s^2+4)(s-3)}$$

(b) The poles are the values of s that cause the denominator of $V(s)$ to equal zero. Therefore the poles of $V(s)$ are given by $s = -2$, $s = j2$, $s = -j2$, and $s = 3$. These are plotted in Fig. 16.5. We can determine the region of convergence of $V(s)$

FIGURE 16.5
The region of convergence (shaded) for $V(s) = (2s^4 + 6s^3 - 6s^2 + 14s - 52)/[(s+2)(s^2+4)(s-3)]$. Notice that the region of convergence lies to the immediate right of the poles. The abscissa of convergence is $\sigma_0 = 3$

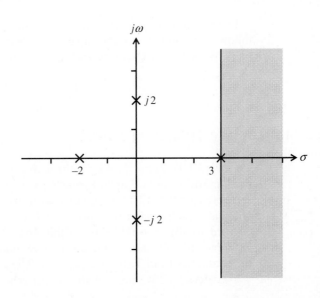

by writing $V(s)$ in its original form and noting the region of convergence of each term:

$$V(s) = \underbrace{2}_{\text{All } s} + \underbrace{\frac{3}{s+2}}_{\mathcal{R}e\,\{s\} > -2} + \underbrace{\frac{s}{s^2+4}}_{\mathcal{R}e\,\{s\} > 0} + \underbrace{\frac{4}{s-3}}_{\mathcal{R}e\,\{s\} > 3}$$

$V(s)$ exists if and only if *all four* terms exist. We see that all four terms exist if and only if $\mathcal{R}e\,\{s\} > 3$. The region of convergence of $V(s)$ is therefore given by $\mathcal{R}e\,\{s\} > 3$, which is the shaded region in Fig. 16.5. Note that the region of convergence of $V(s)$ lies just to the right of the poles of $V(s)$. ∎

In Examples 16.1 through 16.5, the region of convergence was given by $\mathcal{R}e\,\{s\} > \sigma_0$, which was the region just to the right of the poles of $V(s)$. By using methods similar to those in Example 16.5b, we can show that the region of convergence of *every* Laplace transform $V(s)$ lies to the immediate right of the poles of $V(s)$. This result applies to all subsequent examples in this chapter. The formal proof is based on *partial fractions*, which are described in Section 16.4.

Example 16.6 illustrates the initial and final value theorems.

EXAMPLE 16.6

Let $V_o(s) = (5s - 10,000)/(s^2 + 1000s)$ for $\mathcal{R}e\,\{s\} > 0$. Use the initial and final value theorems to find (a) $v(0^+)$ and (b) $\lim\limits_{t \to \infty} v(t)$.

■ **SOLUTION**

(a) According to the initial value theorem,

$$v(0^+) = \lim_{\mathcal{R}e\,\{s\} \to \infty} sV(s) = \lim_{\mathcal{R}e\,\{s\} \to \infty} \left\{ \frac{5s^2 - 10,000s}{s^2 + 1000s} \right\} = 5$$

(b) According to the final value theorem,

$$\lim_{t \to \infty} v(t) = \lim_{s \to 0} sV(s) = \lim_{s \to 0} \left\{ \frac{5s^2 - 10,000s}{s^2 + 1000s} \right\} = -10 \quad ■$$

In this section we presented a table of Laplace transform properties. In the following section we use this table to develop a look-up table of Laplace transform pairs. We will defer the exercises for Table 16.1 until we have developed Table 16.2.

16.3 Important Laplace Transform Pairs

To solve LLTI circuit problems through the use of algebra and table look-up, we need a table of basic Laplace transform pairs in addition to a table of properties. Table 16.2 contains 12 Laplace transform pairs that are used pervasively in circuit analysis. In subsequent sections we will see how to use Table 16.2. This section is devoted totally to establishing Table 16.2. Transform pair LT1 was derived in Example 16.2. We can derive all the remaining entries in Table 16.2 by applying the properties from Table 16.1 to LT1. The essential steps are as follows.

To derive LT2 we begin with the fact that

$$\int_{-\infty}^{t} \delta(\lambda)\, d\lambda = u(t) \tag{16.26}$$

and

$$\int_{-\infty}^{0^-} \delta(\lambda)\, d\lambda = 0 \tag{16.27}$$

Therefore, if we set $v(t) = \delta(t)$ and $V(s) = 1$ and apply Property P3, we obtain

$$u(t) \leftrightarrow \frac{1}{s} \tag{16.28}$$

which is LT2.

We derive LT3 by repeating the above steps. We start with

$$\int_{-\infty}^{t} u(\lambda)\, d\lambda = tu(t) \tag{16.29}$$

and

$$\int_{-\infty}^{0^-} u(\lambda)\, d\lambda = 0 \tag{16.30}$$

TABLE 16.2
Laplace transform pairs*

	SIGNAL $v(t)$	LAPLACE TRANSFORM $V(s)$
LT1	$\delta(t)$	1
LT2	$u(t)$	$\dfrac{1}{s}$
LT3	$tu(t)$	$\dfrac{1}{s^2}$
LT4	$\dfrac{1}{n!}\, t^n u(t)$	$\dfrac{1}{s^{n+1}}$
LT5	$e^{s_0 t} u(t)$	$\dfrac{1}{s - s_0}$
LT6	$t e^{s_0 t} u(t)$	$\dfrac{1}{(s - s_0)^2}$
LT7	$\dfrac{1}{n!}\, t^n e^{s_0 t} u(t)$	$\dfrac{1}{(s - s_0)^{n+1}}$
LT8	$|\mathbf{K}_0| \cos(\omega_0 t + \underline{/\mathbf{K}_0}) u(t)$	$\dfrac{(1/2)\mathbf{K}_0}{s - j\omega_0} + \dfrac{(1/2)\mathbf{K}_0^*}{s + j\omega_0}$
LT9	$|\mathbf{K}_0| e^{\sigma_0 t} \cos(\omega_0 t + \underline{/\mathbf{K}_0}) u(t)$	$\dfrac{(1/2)\mathbf{K}_0}{s - s_0} + \dfrac{(1/2)\mathbf{K}_0^*}{s - s_0^*}$
LT10	$\dfrac{|\mathbf{K}_0| t^{n-1}}{(n-1)!}\, e^{\sigma_0 t} \cos(\omega_0 t + \underline{/\mathbf{K}_0}) u(t)$	$\dfrac{(1/2)\mathbf{K}_0}{(s - s_0)^n} + \dfrac{(1/2)\mathbf{K}_0^*}{(s - s_0^*)^n}$
LT11	$e^{-\alpha t} \cos(\omega_d t) u(t)$	$\dfrac{s + \alpha}{(s + \alpha)^2 + \omega_d^2}$
LT12	$e^{-\alpha t} \sin(\omega_d t) u(t)$	$\dfrac{\omega_d}{(s + \alpha)^2 + \omega_d^2}$

* The region of convergence of each transform listed is the region to the immediate right of the poles of $V(s)$. For entries LT5 through LT10, $s_0 = \sigma_0 + j\omega_0$ denotes an arbitrary point in the complex plane. For entries LT8 through LT10, $\mathbf{K}_0$ is a complex constant.

Property P3 yields [with $v(t) = u(t)$ and $V(s) = 1/s$]

$$tu(t) \leftrightarrow \frac{1}{s^2} \tag{16.31}$$

Iterating n times yields LT4:

$$\frac{1}{n!} t^n u(n) \leftrightarrow \frac{1}{s^{n+1}} \tag{16.32}$$

LT5 follows by applying the complex frequency shift Property P6 to LT2. This shows immediately that

$$e^{s_0 t} u(t) \leftrightarrow \frac{1}{s - s_0} \tag{16.33}$$

Observe that this result is a generalization of the special case in which s_0 is a real constant $s_0 = a$, derived in Example 16.1.

LT7 is obtained by application of the complex frequency shift property to LT4. This yields

$$\frac{1}{n!} t^n e^{s_0 t} u(t) \leftrightarrow \frac{1}{(s - s_0)^{n+1}} \tag{16.34}$$

LT6 is, of course, a special case of LT7 where $n = 1$. LT10 is obtained from LT7 and Property P1 (linearity). If we replace s_0 with s_0^* in Eq. (16.34) and apply Property P1, we find that

$$\frac{\mathbf{K}_0}{(n-1)!} t^{n-1} e^{s_0 t} u(t) + \frac{\mathbf{K}_0^*}{(n-1)!} t^{n-1} e^{s_0^* t} u(t) \leftrightarrow \frac{\mathbf{K}_0}{(s - s_0)^n} + \frac{\mathbf{K}_0^*}{(s - s_0^*)^n} \tag{16.35}$$

where $\mathbf{K}_0$ is a constant. With $s_0 = \sigma_0 + j\omega_0$, the left-hand side of Eq. (16.35) becomes

$$\frac{\mathbf{K}_0}{(n-1)!} t^{n-1} e^{s_0 t} u(t) + \frac{\mathbf{K}_0^*}{(n-1)!} t^{n-1} e^{s_0^* t} u(t) = \frac{2|\mathbf{K}_0|}{(n-1)!} t^{n-1} e^{\sigma_0 t} \cos (\omega_0 t + \underline{/\mathbf{K}_0})$$

$$\tag{16.36}$$

Therefore

$$\frac{2|\mathbf{K}_0|}{(n-1)!} t^{n-1} e^{\sigma_0 t} \cos (\omega_0 t + \underline{/\mathbf{K}_0}) \leftrightarrow \frac{\mathbf{K}_0}{(s - s_0)^n} + \frac{\mathbf{K}_0^*}{(s - s_0^*)^n} \tag{16.37}$$

which is LT10. LT8 and LT9 are special cases of this result.

Transform pairs LT11 and LT12 are special cases of LT9. We leave the details to Exercises 18 and 19.

We have listed only the most basic Laplace transform pairs in Table 16.2. However, using Tables 16.1 and 16.2, we can derive the Laplace transform of virtually every signal encountered in everyday circuit analysis. We shall encounter several pertinent examples in Sections 16.5 through 16.7.

EXERCISES

Use Tables 16.1 and 16.2 to obtain the Laplace transforms of the following signals.

7. $tu(t)$

8. $(t - t_0)u(t - t_0)$ where $t_0 > 0$

9. $tu(t - t_0)$ where $t_0 > 0$

10. $5te^{7t}u(t) + 2\cos(\omega_0 t + 45°)u(t)$

11. $\Pi\left(\dfrac{t - 0.5\tau}{\tau}\right)$

Use Tables 16.1 and 16.2 to determine the inverse Laplace transforms of the following functions.

12. $\dfrac{1}{s}e^{-s}$

13. $\dfrac{1}{s - j30 + 10}$

14. $\dfrac{s - 2}{s^2 - 2s + 2}$

15. Use Tables 16.1 and 16.2 to determine the inverse Laplace transform of $V(s) = (s + 1)/[(s + 2)(s + 3)]$. [*Hint:* First write $V(s)$ in partial fraction form, $V(s) = a/(s + 2) + b/(s + 3)$, where a and b are constants to be determined by you.]

16. Use LT11 of Table 16.2 to show that

$$\cos(\omega_0 t)u(t) \leftrightarrow \frac{s}{s^2 + \omega_0^2}$$

17. Use LT12 of Table 16.2 to show that

$$\sin(\omega_0 t)u(t) \leftrightarrow \frac{\omega_0}{s^2 + \omega_0^2}$$

18. Use LT9 to show that

$$e^{-\alpha t}\cos(\omega_d t)u(t) \leftrightarrow \frac{s + \alpha}{(s + \alpha)^2 + \omega_d^2}$$

19. Repeat Exercise 18 for

$$e^{-\alpha t}\sin(\omega_d t)u(t) \leftrightarrow \frac{\omega_d}{(s + \alpha)^2 + \omega_d^2}$$

Find the initial and final values of the signals with the following Laplace transforms.

20. $\dfrac{1}{s}e^{-3s}$

21. $\dfrac{s}{(s + 1)(s + 2)}$

22. $\dfrac{s + 3}{s^2 + 6s + 18}e^{-s}$

16.4 Partial Fraction Expansions

Partial fractions provide the key to the table look-up technique to determine inverse Laplace transforms. In most problems, inverse Laplace transforms are required for functions of the form

$$G(s) = \frac{N(s)}{D(s)} \tag{16.38}$$

where $N(s)$ and $D(s)$ are polynomials in s. The function $G(s)$ is called a *rational algebraic* function of s. A specific example of a rational algebraic function is

$$G(s) = \frac{s+1}{(s+2)^2(s-1)}$$ (16.39a)

for which the numerator polynomial is

$$N(s) = s + 1$$ (16.39b)

and the denominator polynomial is

$$D(s) = (s+2)^2(s-1) = s^3 + 3s^2 - 4$$ (16.39c)

Arbitrary rational algebraic functions like Eq. (16.39a) do not appear in Table 16.2. However, rational algebraic functions can be written as a sum of partial fractions, each of which does appear in Table 16.2. Therefore, by relying on the linearity of the Laplace transform, we can determine the inverse transform of a rational algebraic function using table look-up. The first step to determine the partial fraction of $G(s)$ in Eq. (16.38) is to write down its *form*.

Partial Fraction Form

The form of the partial fraction expansion of $G(s)$ depends on the degrees of the numerator and denominator polynomials in Eq. (16.38). In what follows, we denote the degrees of $N(s)$ and $D(s)$ by $deg\{N(s)\}$ and $deg\{D(s)\}$, respectively.

If the degree of the numerator polynomial is less than the degree of the denominator polynomial, $deg\{N(s)\} < deg\{D(s)\}$, then the partial fraction expansion of $G(s) = N(s)/D(s)$ will have the form

$$G(s) = L_1(s) + L_2(s) + \cdots + L_l(s)$$ (16.40a)

where

$$L_i(s) = \frac{k_{i1}}{s - s_i} + \frac{k_{i2}}{(s - s_i)^2} + \cdots + \frac{k_{im_i}}{(s - s_i)^{m_i}}$$ (16.40b)

for $i = 1, 2, \ldots, l$. In Eq. (16.40), s_i is the ith distinct root of the equation $D(s) = 0$ and is called the ith distinct *pole* of $G(s)$. l is the number of distinct poles. The quantity m_i is the *order* of s_i. The constant k_{i1} has special significance and is called the *residue* of $G(s)$ at the pole s_i.

EXAMPLE 16.7

Determine the *form* of the partial fraction expansion of

$$G(s) = \frac{s}{(s+2)(s+4)(s+6)} = \frac{s}{s^3 + 12s^2 + 44s + 48}$$

■ SOLUTION

The degree of the numerator, $deg\{s\} = 1$, is less than the degree of the denominator, $deg\{s^3 + 12s^2 + 44s + 48\} = 3$. Therefore, Eq. (16.40) applies, where $s_1 = -2$, $s_2 = -4$, and $s_3 = -6$ are all first-order poles. The partial fraction expansion of $G(s)$ has the form

$$G(s) = \frac{s}{(s+2)(s+4)(s+6)} = \frac{k_{11}}{s+2} + \frac{k_{21}}{s+4} + \frac{k_{31}}{s+6}$$

where k_{11}, k_{21}, and k_{31} are the residues of $G(s)$ at -2, -4, and -6, respectively, and are still to be determined. ∎

EXAMPLE 16.8

Determine the *form* of the partial fraction expansion of

$$G(s) = \frac{s+1}{(s+2)^2(s-1)} = \frac{s+1}{s^3 + 3s^2 - 4}$$

■ SOLUTION

The degree of the numerator, $deg\{s+1\} = 1$, is less than the degree of the denominator, $deg\{s^3 + 3s^2 - 4\} = 3$. Therefore, Eq. (16.40) applies, where $s_1 = -2$ with $m_1 = 2$ and $s_2 = 1$ with $m_2 = 1$. Therefore, the partial fraction expansion of $G(s)$ has the form

$$G(s) = \frac{s+1}{(s+2)^2(s-1)} = \underbrace{\frac{k_{11}}{s+2} + \frac{k_{12}}{(s+2)^2}}_{L_1(s)} + \underbrace{\frac{k_{21}}{s-1}}_{L_2(s)}$$

where k_{11}, k_{12}, and k_{21} are constants to be determined. The residues of $G(s)$ at -2 and $+1$ are k_{11} and k_{21}, respectively. ∎

If the degree of $N(s)$ equals or exceeds the degree of $D(s)$, $deg\{N(s)\} \geq deg\{D(s)\}$, then the partial fraction expansion of $G(s) = N(s)/D(s)$ will have the form

$$G(s) = p(s) + L_1(s) + L_2(s) + \cdots + L_l(s) \tag{16.41}$$

where $p(s)$ is a polynomial in s with degree

$$deg\{p(s)\} = deg\{N(s)\} - deg\{D(s)\} \tag{16.42}$$

and the form of the $L_i(s)$ is as shown in Eq. (16.40b).

EXAMPLE 16.9

Determine the *form* of the partial fraction expansion of

$$G(s) = \frac{s^4 + 2s^3 + s^2 + 1}{s(s+3)} = \frac{s^4 + 2s^3 + s^2 + 1}{s^2 + 3s}$$

■ SOLUTION

The degree of the numerator exceeds the degree of the denominator by 2. Therefore, Eq. (16.41) applies, in which $p(s)$ has degree 2, and $G(s)$ has the form

$$G(s) = \frac{s^4 + 2s^3 + s^2 + 1}{s(s+3)} = \underbrace{\alpha_2 s^2 + \alpha_1 s + \alpha_0}_{p(s)} + \underbrace{\frac{k_{11}}{s}}_{L_1(s)} + \underbrace{\frac{k_{21}}{s+3}}_{L_2(s)}$$

where α_0, α_1, α_2, k_{11}, and k_{21} are constants to be determined. The constants k_{11} and k_{21} are residues of $G(s)$ at 0 and -3, respectively. ∎

Evaluation of Partial Fraction Constants

After the form of a partial fraction expansion has been determined, the next step is to evaluate the constants. A general formulation of the solution is straightforward, but notationally cumbersome. Therefore it is better to proceed with examples.

EXAMPLE 16.10

Determine the partial fraction expansion of

$$G(s) = \frac{s}{(s + 2)(s + 4)(s + 6)} = \frac{k_{11}}{s + 2} + \frac{k_{21}}{s + 4} + \frac{k_{31}}{s + 6}$$

■ **SOLUTION**

An easy way to determine k_{11} is to first multiply the above equation by $s + 2$ to obtain

$$(s + 2)G(s) = \frac{s}{(s + 4)(s + 6)} = k_{11} + \frac{(s + 2)k_{21}}{(s + 4)} + \frac{(s + 2)k_{31}}{s + 6}$$

Now if we set $s = -2$, the above becomes

$$(s + 2)G(s)\Big|_{s=-2} = \frac{-2}{(-2 + 4)(-2 + 6)} = -\frac{1}{4} = k_{11}$$

The above steps can be written compactly as

$$k_{11} = (s + 2)G(s)\Big|_{s=-2}$$

Similarly, we have

$$k_{21} = (s + 4)G(s)\Big|_{s=-4} = \frac{-4}{(-4 + 2)(-4 + 6)} = 1$$

and

$$k_{31} = (s + 6)G(s)\Big|_{s=-6} = \frac{-6}{(-6 + 2)(-6 + 4)} = -\frac{3}{4} \quad ■$$

The method of Example 16.10 worked because the poles in $G(s)$ had order 1. We will show later that if $G(s)$ is *any* rational algebraic function with a simple (that is, first-order) pole at $s = s_i$, then the residue of $G(s)$ at s_i is given by

$$k_{i1} = (s - s_i)G(s)\Big|_{s=s_i} \tag{16.43}$$

The following example will illustrate how to proceed when the poles of $G(s)$ are not simple.

EXAMPLE 16.11

Determine the partial fraction expansion of

$$G(s) = \frac{s + 1}{(s + 2)^2(s - 1)}$$

■ **SOLUTION**

According to the result of Example 16.8, the form of the partial fraction expansion is

$$\frac{s+1}{(s+2)^2(s-1)} = \frac{k_{11}}{s+2} + \frac{k_{12}}{(s+2)^2} + \frac{k_{21}}{s-1} \qquad (16.44)$$

We will describe three methods for finding the values of the constants. Methods 1 and 2 have the advantage of being straightforward and applicable in general. Method 3 has the advantage of being faster, but it yields values only for k_{12} and k_{21}.

Method 1

Multiply both sides of Eq. (16.44) by $(s+2)^2(s-1)$ to obtain

$$s + 1 = k_{11}(s+2)(s-1) + k_{12}(s-1) + k_{21}(s+2)^2$$

and write both sides as second-degree polynomials in s:

$$0s^2 + 1s + 1 = (k_{11} + k_{21})s^2 + (k_{11} + k_{12} + 4k_{21})s - 2k_{11} - k_{12} + 4k_{21}$$

If equality is to hold for all values of s, then the coefficients of like powers of s on either side of this equation must be equal. Thus

$$k_{11} + k_{21} = 0$$

$$k_{11} + k_{12} + 4k_{21} = 1$$

$$-2k_{11} - k_{12} + 4k_{21} = 1$$

This system can be solved to yield $k_{11} = -\frac{2}{9}$, $k_{12} = \frac{1}{3}$, and $k_{21} = \frac{2}{9}$. The conclusion is

$$\frac{s+1}{(s+2)^2(s-1)} = -\frac{2/9}{s+2} + \frac{1/3}{(s+2)^2} + \frac{2/9}{(s-1)}$$

Method 2

Substitute three convenient values for s to obtain three equations in three unknowns. For example, with $s = 0$, Eq. (16.44) becomes

$$-\frac{1}{4} = \frac{1}{2}k_{11} + \frac{1}{4}k_{12} - k_{21}$$

and with $s = -1$, Eq. (16.44) becomes

$$0 = k_{11} + k_{12} - \frac{1}{2}k_{21}$$

Finally, with $s = 2$, Eq. (16.44) becomes

$$\frac{3}{16} = \frac{1}{4}k_{11} + \frac{1}{16}k_{12} + k_{21}$$

The solution to the above system of equations is $k_{11} = -\frac{2}{9}$, $k_{12} = \frac{1}{3}$, and $k_{21} = \frac{2}{9}$, which agrees with the previous result, as it must.

Method 3

Observe that if we multiply both sides of Eq. (16.44) by $s - 1$,

$$(s-1)G(s) = \frac{s+1}{(s+2)^2} = (s-1)\frac{k_{11}}{s+2} + (s-1)\frac{k_{12}}{(s+2)^2} + k_{21}$$

and if we set $s = 1$, we obtain

$$(s - 1)G(s)\Big|_{s=1} = \frac{2}{9} = k_{21}$$

Similarly, if we multiply both sides of Eq. (16.44) by $(s + 2)^2$,

$$(s + 2)^2 G(s) = \frac{s + 1}{s - 1} = (s + 2)k_{11} + k_{12} + (s + 2)^2 \frac{k_{21}}{s - 1}$$

and if we set $s = -2$, we have

$$(s + 2)^2 G(s)\Big|_{s=-2} = \frac{1}{3} = k_{12}$$

Therefore Eq. (16.44) becomes

$$\frac{s + 1}{(s + 2)^2(s - 1)} = \frac{k_{11}}{s + 2} + \frac{1/3}{(s + 2)^2} + \frac{2/9}{s - 1}$$

This method does not work for k_{11}. However, k_{11} can still be found by method 1 or 2. Perhaps the easiest approach is to set $s = -1$ in the above equation, to get

$$0 = k_{11} + \frac{1}{3} - \frac{1}{9}$$

which reveals that

$$k_{11} = -\frac{2}{9} \quad \blacksquare$$

Example 16.12 will illustrate how to proceed when the degree of $N(s)$ exceeds that of $D(s)$.

EXAMPLE 16.12

Determine the partial fraction expansion of

$$G(s) = \frac{s^4 + 2s^3 + s^2 + 1}{s(s + 3)}$$

■ SOLUTION

According to Example 16.9, the form of the partial fraction expansion is

$$G(s) = \underbrace{\alpha_2 s^2 + \alpha_1 s + \alpha_0}_{p(s)} + \frac{k_{11}}{s} + \frac{k_{21}}{s + 3}$$

Again, there are several ways to obtain the values of the constants. Methods 1 and 2 of Example 16.11 will work, but they should be avoided because they are too lengthy. (In using method 2, we would have to substitute five values for s and solve the resulting system of five equations for α_2, α_1, α_0, k_{11}, and k_{21}.) The recommended approach is to find α_2, α_1, and α_0 first by long division, stopping the division process just before a term with a negative exponent is obtained:

$$
\begin{array}{r}
s^2 - s + 4 \\
s^2 + 3s \overline{)\, s^4 + 2s^3 + s^2 + 0s + 1} \\
\underline{s^4 + 3s^3} \\
-s^3 + s^2 + 0s + 1 \\
\underline{-s^3 - 3s^2} \\
4s^2 + 0s + 1 \\
\underline{4s^2 + 12s} \\
-12s + 1 \quad \text{Remainder} \\
\text{---STOP---}
\end{array}
$$

It follows that

$$
G(s) = \frac{s^4 + 2s^3 + s^2 + 1}{s(s+3)} = \underbrace{s^2 - s + 4}_{p(s)} + \frac{-12s + 1}{s(s+3)}
$$

which shows that $\alpha_2 = 1$, $\alpha_1 = -1$, and $\alpha_0 = 4$. The partial fraction expansion of the fractional term [call it $G_1(s)$] can now be taken

$$
G_1(s) = \frac{-12s + 1}{s(s+3)} = \frac{k_{11}}{s} + \frac{k_{21}}{s+3}
$$

Equation (16.43) works beautifully for determining k_{11} and k_{21}. To find k_{11}, multiply both sides of $G_1(s)$ by s and set $s = 0$. Thus

$$
k_{11} = sG_1(s)\Big|_{s=0} = \frac{-12s + 1}{s+3}\bigg|_{s=0} = \frac{1}{3}
$$

Similarly,

$$
k_{21} = (s+3)G_1(s)\Big|_{s=-3} = \frac{-12s + 1}{s}\bigg|_{s=-3} = -\frac{37}{3}
$$

Therefore,

$$
G_1(s) = \frac{-12s + 1}{s(s+3)} = \frac{1/3}{s} - \frac{37/3}{s+3}
$$

and, combining the above with $p(s)$, we obtain the result

$$
\frac{s^4 + 2s^3 + s^2 + 1}{s(s+3)} = s^2 - s + 4 + \frac{1/3}{s} - \frac{37/3}{s+3}
$$

A partial (but reassuring) check can be made by setting $s = 1$. This yields

$$
\frac{1 + 2 + 1 + 1}{4} = 1 - 1 + 4 + \frac{1}{3} - \frac{37}{12}
$$

or

$$
\frac{5}{4} = \frac{15}{12}
$$

which is, of course, a true statement. ∎

Examples 16.10 through 16.12 have illustrated methods that can be applied to any rational algebraic function $G(s) = N(s)/D(s)$. The methods apply for both real and complex poles.* Whenever $deg\{N(s)\} \geq deg\{D(s)\}$, we can use long division as in Example 16.12 to find the constants appearing in the polynomial $p(s)$ of Eq. (16.41). The remaining fractional term $G_1(s) = [\text{Remainder}] \div D(s)$ will always be a rational algebraic function whose numerator has a smaller degree than the denominator. Thus $G(s)$ is easily put in the form $G(s) = p(s) + G_1(s)$, where $deg\{p(s)\} = deg\{N(s)\} - deg\{D(s)\}$. The function $G_1(s)$ can then be put in partial fraction form, Eq. (16.40) and the constants k_{ij} may be evaluated by any of the three methods of Example 16.11. Whenever $deg\{N(s)\} < deg\{D(s)\}$, long division is no longer necessary, since $p(s) = 0$.

Heaviside's Theorem

We have stated that whenever a pole s_i is simple (has order $m_i = 1$), then the residue k_{i1} at that simple pole can be found from Eq. (16.43), which we repeat below:

$$k_{i1} = (s - s_i)G(s)\Big|_{s=s_i} \qquad (16.43)$$

This formula applies regardless of the relative degrees of $N(s)$ and $D(s)$ and can be established by writing

$$G(s) = \frac{N(s)}{c(s-s_1)^{m_1}(s-s_2)^{m_2}\cdots(s-s_i)^1\cdots(s-s_l)^{m_l}}$$

Simple pole s_i

$$= p(s) + L_1(s) + \cdots + \frac{k_{i1}}{s-s_i} + \cdots + L_l(s) \qquad (16.45)$$

Present if $\qquad\qquad G_1(s)$
$deg\{N(s)\} \geq deg\{D(s)\}$

By multiplying both sides by $(s - s_i)$, we obtain

$$(s-s_i)G(s) = \frac{(s-s_i)N(s)}{c(s-s_1)^{m_1}\cdots(s-s_i)\cdots(s-s_l)^{m_l}}$$

$$= (s-s_i)p(s) + (s-s_i)L_1(s) + \cdots + k_{i1} + \cdots + (s-s_i)L_l(s) \quad (16.46)$$

Equation (16.43) is a direct consequence of the fact that all terms on the right-hand side of Eq. (16.45) except k_{i1} equal zero for $s = s_i$. Equation (16.43) is sometimes referred to as *Heaviside's theorem*.

You may recall that the expression, Eq. (16.43), for the residue at a simple pole of a filter arose naturally in connection with the universal resonance characteristic described in Section 13.8 [see Eq. 13.97]. There the residue at the simple pole was seen to have a simple graphical interpretation illustrated in the pole-zero plot of Fig. 13.32. Except for a proportionality constant, the residue k_{i1} is equal to the product of the vectors drawn from the zeros of $G(s)$ to the simple pole s_i divided by the product of the vectors drawn from the poles of $G(s)$ to s_i.

For completeness, we state the following general formula for obtaining any of the coefficients k_{ij} appearing in Eq. (16.40b).

* A helpful shortcut for complex poles is described in Example 16.15.

$$k_{ij} = \frac{1}{(m_i - j)!} \frac{d^{m_i - j}}{ds^{m_i - j}} \left[(s - s_i)^{m_i} G(s) \right] \Big|_{s = s_i} \qquad (16.47)$$

for $j = 1, 2, \ldots, m_i$, where m_i is the order of s_i. Equation (16.47) can be verified if we multiply both sides of Eq. (16.40a) by $(s - s_i)^{m_i}$ and differentiate as indicated. Although Eq. (16.47) gives a complete theoretical solution to the problem of determining the partial fraction expansion constants k_{ij}, it is often less convenient to apply than methods 1 through 3 of Example 16.11.

EXERCISES

Determine the partial fraction expansion of the rational algebraic functions in Exercises 23 through 28.

23. $\dfrac{1}{s^2 + 1}$

24. $\dfrac{s}{(s + 1)^2}$

25. $\dfrac{s^2 + 2s + 2}{s^2 + 7s + 12}$

26. $\dfrac{s^3 + 3s^2 + 2s + 1}{s + 2}$

27. $\dfrac{(s^3 + 3s^2 + 2s + 1)(s^2 + 2s + 2)}{(s + 2)(s^2 + 7s + 12)}$

28. $\dfrac{s(s + 3 + j)(s + 3 - j)}{(s + 1 + 3j)(s + 1 - 3j)(s + 5)}$

29. Suppose that $G_a(s) = k_a/(s - s_a)$ and $G_b(s) = k_b/(s - s_b)$. Determine the partial fraction expansion of $G_a(s)G_b(s)$ if $s_a \neq s_b$.

30. Repeat Exercise 29 for $s_a = s_b$.

Determine the partial fraction expansions of the following rational algebraic functions.

31. $\dfrac{s}{s^2 + \omega_0^2}$

32. $\dfrac{\omega_0}{s^2 + \omega_0^2}$

33. $\dfrac{s + \alpha}{(s + \alpha)^2 + \omega_d^2}$

34. $\dfrac{\omega_d}{(s + \alpha)^2 + \omega_d^2}$

16.5 Solution of LLTI Equations

One of the easiest ways to solve linear differential equations with constant coefficients and arbitrary initial conditions is to use the Laplace transform. Examples 16.13 through 16.16 show how it is done.

EXAMPLE 16.13

The input to the RC circuit of Fig. 16.6 is $v_s(t) = 5u(t)$ V. The circuit is at rest for $t < 0$. Find the output $v_o(t)$ for $t \geq 0$.

FIGURE 16.6
RC circuit

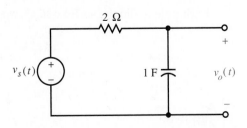

■ SOLUTION

The differential equation relating output to input is

$$2\frac{dv_o}{dt} + v_o(t) = v_s(t)$$

where $v_o(0^-) = 0$ because the circuit is at rest for $t < 0$. Obviously, the quantities on both sides of this equation are functions of t. Because these functions of t are equal, so must be their Laplace transforms. Thus

$$\mathcal{L}\left\{2\frac{dv_o}{dt} + v_o(t)\right\} = \mathcal{L}\{v_s(t)\}$$

which, by the linearity property (Property P1 of Table 16.1), becomes

$$2\mathcal{L}\left\{\frac{dv_o}{dt}\right\} + \mathcal{L}\{v_o(t)\} = \mathcal{L}\{v_s(t)\}$$

By using the differentiation Property P2, we can put this in the form

$$2[sV_o(s) - v_o(0^-)] + V_o(s) = V_s(s)$$

Solving for $V_o(s)$ gives

$$V_o(s) = \frac{1}{2s+1}V_s(s) + \frac{2v_o(0^-)}{2s+1}$$

$$= \frac{0.5}{s+0.5}V_s(s)$$

where we have used the fact that $v_o(0^-) = 0$. It follows from LT2 of Table 16.2 and Property P1 that

$$V_s(s) = \mathcal{L}\{5u(t)\} = 5\mathcal{L}\{u(t)\} = \frac{5}{s}$$

so

$$V_o(s) = \frac{0.5}{s+0.5} \cdot \frac{5}{s} = \frac{2.5}{s(s+0.5)}$$

The complete response for $t \geq 0$ is the inverse Laplace transform of $V_o(s)$. The easiest way to find the inverse transform of $V_o(s)$ is through the use of Tables 16.1 and 16.2. To do this, we need to first express $V_o(s)$ in partial fraction form. Since the degree of the numerator polynomial is less than that of the denominator polynomial, this form is given by Eq. (16.40):

$$V_o(s) = \frac{2.5}{s(s+0.5)} = \frac{k_{11}}{s} + \frac{k_{21}}{s+0.5}$$

Each pole is simple, so Eq. (16.43) applies to each residue k_{11} and k_{21}:

$$k_{11} = \left.\frac{2.5}{s+0.5}\right|_{s=0} = 5$$

$$k_{21} = \left.\frac{2.5}{s}\right|_{s=-0.5} = -5$$

Therefore

$$V_o(s) = \frac{5}{s} - \frac{5}{s + 0.5}$$

It follows from the above and linearity that

$$v_o(t) = \mathscr{L}^{-1}\{V_o(s)\} = 5\mathscr{L}^{-1}\left\{\frac{1}{s}\right\} - 5\mathscr{L}^{-1}\left\{\frac{1}{s + 0.5}\right\}$$

which, by LT2 and LT5, yields the final result:

$$v_o(t) = 5u(t) - 5e^{-0.5t}u(t)$$
$$= 5(1 - e^{-0.5t})u(t) \text{ V} \quad \blacksquare$$

The next example is slightly more challenging. The techniques are the same as in Example 16.13.

EXAMPLE 16.14

The input to the RC circuit of Fig. 16.6 is $v_s(t) = 5tu(t)$ V, and the capacitance voltage is -3 V at $t = 0^-$. Find the output $v_o(t)$ for $t \geq 0$.

■ SOLUTION

The differential equation relating output to input is, for $t > 0$,

$$2\frac{dv_o}{dt} + v_o(t) = v_s(t)$$

By taking the Laplace transform of both sides, we obtain (in a manner identical to that of Example 16.13)

$$2[sV_o(s) - v_o(0^-)] + V_o(s) = V_s(s)$$

which yields

$$V_o(s) = \frac{1}{2s + 1} V_s(s) + \frac{2v_o(0^-)}{2s + 1}$$

$$= \frac{0.5}{s + 0.5} V_s(s) - \frac{3}{s + 0.5}$$

It follows from LT3 and Property P1 that $V_s(s) = 5/s^2$. Therefore

$$V_o(s) = \frac{2.5}{(s + 0.5)s^2} - \frac{3}{s + 0.5}$$

The complete response for $t \geq 0$ is given by the inverse Laplace transform of $V_o(s)$. Again, we will use partial fraction expansion and table look-up. The partial fraction expansion of the first term in the above equation is

$$G(s) = \frac{2.5}{(s + 0.5)s^2}$$

$$= \frac{k_{11}}{s} + \frac{k_{12}}{s^2} + \frac{k_{13}}{s + 0.5}$$

where

$$k_{13} = (s + 0.5)G(s)\Big|_{s=-0.5} = \frac{2.5}{s^2}\Big|_{s=-0.5} = 10$$

$$k_{12} = s^2 G(s)\Big|_{s=0} = \frac{2.5}{s + 0.5}\Big|_{s=0} = 5$$

By substituting these values into the expression for $G(s)$ and setting $s = 1$, we find that $k_{11} = -10$. Thus

$$G(s) = \frac{2.5}{(s + 0.5)s^2} = -\frac{10}{s} + \frac{5}{s^2} + \frac{10}{s + 0.5}$$

which, when combined with the second term in $V_o(s)$, yields

$$V_o(s) = -\frac{10}{s} + \frac{5}{s^2} + \frac{7}{s + 0.5}$$

We now have $V_o(s)$ in a form where we can use table look-up to obtain $v_o(t)$. It follows directly from LT1, LT2, and Property P1 that

$$v_o(t) = \mathcal{L}^{-1}\left\{-\frac{10}{s} + \frac{5}{s^2} + \frac{7}{s + 0.5}\right\}$$

$$= -10\mathcal{L}^{-1}\left\{\frac{1}{s}\right\} + 5\mathcal{L}^{-1}\left\{\frac{1}{s^2}\right\} + 7\mathcal{L}^{-1}\left\{\frac{1}{s + 0.5}\right\}$$

$$= -10u(t) + 5tu(t) + 7e^{-0.5t}u(t) \text{ V} \blacksquare$$

In the next example, the input to the RC circuit is a causal sinusoid. The method of solution is the same as that of Examples 16.13 and 16.14.

EXAMPLE 16.15

The input to the RC circuit in Fig. 16.6 is $v_s(t) = A \cos(\omega_0 t + \phi)u(t)$. The initial voltage on the capacitance is $v_o(0^-)$. Find $v_o(t)$ for $t \geq 0$ in terms of R, C, $v_o(0^-)$, A, ω_0, and ϕ.[†]

■ **SOLUTION**

The differential equation relating output to input is

$$RC\frac{dv_o}{dt} + v_o(t) = v_s(t)$$

Proceeding as in Examples 16.13 and 16.14, we take the Laplace transform of both sides to obtain

$$RC[sV_o(s) - v_o(0^-)] + V_o(s) = V_s(s)$$

Therefore

$$V_o(s) = \frac{\dfrac{1}{RC}}{s + \dfrac{1}{RC}}V_s(s) + \frac{v_o(0^-)}{s + \dfrac{1}{RC}}$$

[†] The use of symbols rather than numerical values can make a simple analysis appear complicated. The result, however, is often well worth the extra effort. The major advantage of expressing a solution in terms of symbols rather than numerical values arises in design problems where we must choose the values of one or more elements to produce a desired result.

which, by LT8, becomes

$$V_o(s) = \frac{\dfrac{A}{2RC}\underline{/\phi}}{\left(s + \dfrac{1}{RC}\right)(s - j\omega_0)} + \frac{\dfrac{A}{2RC}\underline{/-\phi}}{\left(s + \dfrac{1}{RC}\right)(s + j\omega_0)} + \frac{v_o(0^-)}{s + \dfrac{1}{RC}}$$

As in Examples 16.13 and 16.14, the next step is to write $V_o(s)$ in partial fraction expansion form. The partial fraction expansion of the first term is

$$\frac{\dfrac{A}{2RC}\underline{/\phi}}{\left(s + \dfrac{1}{RC}\right)(s - j\omega_0)} = \frac{k_{11}}{s + \dfrac{1}{RC}} + \frac{k_{21}}{s - j\omega_0}$$

where

$$k_{11} = \frac{\dfrac{A}{2RC}\underline{/\phi}}{s - j\omega_0}\Bigg|_{s = -1/RC} = -\frac{\dfrac{A}{2RC}\underline{/\phi}}{\dfrac{1}{RC} + j\omega_0}$$

$$= \frac{-A/2}{\sqrt{1 + (\omega_0 RC)^2}}\underline{/\phi - \tan^{-1}\omega_0 RC}$$

and

$$k_{21} = \frac{\dfrac{A}{2RC}\underline{/\phi}}{s + \dfrac{1}{RC}}\Bigg|_{s = j\omega_0} = \frac{\dfrac{A}{2RC}\underline{/\phi}}{\dfrac{1}{RC} + j\omega_0}$$

$$= \frac{A/2}{\sqrt{1 + (\omega_0 RC)^2}}\underline{/\phi - \tan^{-1}\omega_0 RC}$$

We can obtain the partial fraction expansion of the second term in $V_o(s)$ in the same way. However, there is a shortcut that is often helpful when we deal with complex poles. Since the constants in the second term in $V_o(s)$ are the complex conjugates of those in the first term, we can immediately write the result

$$\frac{\dfrac{A}{2RC}\underline{/-\phi}}{\left(s + \dfrac{1}{RC}\right)(s + j\omega_0)} = \frac{k_{11}^*}{s + \dfrac{1}{RC}} + \frac{k_{21}^*}{s + j\omega_0}$$

where k_{11} and k_{21} are given above. Substitution of the partial fraction expansions into the original expression for $V_o(s)$ yields, after a little algebra,

$$V_o(s) = \frac{k_{11} + k_{11}^* + v_o(0^-)}{s + \dfrac{1}{RC}} + \frac{k_{21}}{s - j\omega_0} + \frac{k_{21}^*}{s + j\omega_0}$$

It follows by LT5, LT8, and Property P1 that

$$v_o(t) = [v_o(0^-) + k_{11} + k_{11}^*]e^{-t/RC}u(t) + 2|k_{21}| \cos (\omega_0 t + \underline{/k_{21}})u(t)$$

where k_{11} and k_{21} are given above. By substituting the expressions for k_{11} and k_{21}, we obtain the final result.

$$v_o(t) = \left[v_o(0^-) - \frac{A}{\sqrt{1 + (\omega_0 RC)^2}} \cos (\phi - \tan^{-1} \omega_0 RC) \right] e^{-t/RC}u(t)$$

$$+ \frac{A}{\sqrt{1 + (\omega_0 RC)^2}} \cos (\omega_0 t + \phi - \tan^{-1} \omega_0 RC)u(t) \quad \blacksquare$$

In this section we have shown how the Laplace transform may be used to solve the differential equations that arise in LLTI circuit problems. In summary, the method consists of the following four steps:

Step 1

Take the Laplace transform of both sides of the time-domain equation to obtain the corresponding complex-frequency-domain equation.

Step 2

Solve the complex-frequency-domain equation for $V_o(s)$.

Step 3

Use partial fraction expansions to write $V_o(s)$ as a sum of terms of the forms included in Table 16.2.

Step 4

Use Tables 16.1 and 16.2 to inverse-transform $V_o(s)$ into the time domain.

Section 16.7 contains a deeper theoretical look at each step. In Section 16.6, we will describe a way to use the Laplace transform that *bypasses* the differential equation of the circuit.

EXERCISES

35. Start from the appropriate differential equation, and use Laplace transforms to determine $v_R(t)$ for $v_s(t) = u(t)$ V in the circuit shown. The voltage $v_C(0^-)$ is zero.

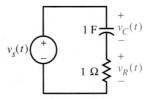

36. Repeat Exercise 35 for $v_s(t) = 10tu(t)$ V and $v_C(0^-) = 6$ V.
37. Start from the appropriate differential equation, and use Laplace transforms to determine $v_L(t)$ for $v_s(t) = u(t)$ in the following circuit. The voltage $v_R(0^-)$ is arbitrary.

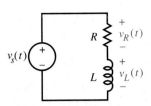

38. Repeat Exercise 37 for $v_s(t) = A \cos(\omega_0 t + \phi)u(t)$.

39. Start from the appropriate differential equation, and determine $v_0(t)$ when $i_s(t) = \delta(t)$ in the circuit shown below. Assume that the circuit is at rest for $t < 0$ and that the response is underdamped.

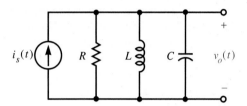

16.6 Laplace Transform Circuit Analysis

In this section we will see that the Laplace transform can be used to transform a circuit from the time domain to the frequency domain. This transformation includes the terminal equations of the circuit components R, L, and C, the terminal equations of the sources, and the underlying circuit laws KCL and KVL. The motivation is similar to that in Chapters 11 and 13: the transformation from time to frequency replaces the integro-differential representation of the circuit by an algebraic representation, which greatly simplifies the task of circuit analysis. As we will see, an advantage of the Laplace transform is that it not only transforms the circuit but also transforms the circuit's initial conditions. Once this is done, we can obtain the circuit's complete response by algebra and table look-up.

The s-domain representations of the circuit elements obtained from the Laplace transform are somewhat different from those obtained in Chapter 13. The reason is that the Laplace transform accounts for initial conditions. To illustrate, the terminal equation of inductance with initial condition $i(0^-) = I_o$ (Fig. 16.7a) is

$$v(t) = L\frac{di}{dt} \qquad \text{given } i(0^-) = I_o \qquad (16.48)$$

FIGURE 16.7
Time- and frequency-domain representations of inductance with initial conditions:
(a) time domain; (b), (c) frequency domain

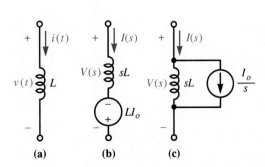

By taking the Laplace transform, we have, according to Properties P1 and P2,

$$V(s) = sLI(s) - Li(0^-)$$
$$= sLI(s) - LI_o \qquad (16.49)$$

which is an algebraic equation relating frequency-domain terminal voltage $V(s)$ and current $I(s)$ to the inductance's impedance sL and initial current I_o. Equation (16.49) can be rearranged to read

$$I(s) = \frac{1}{sL} V(s) + \frac{I_o}{s} \qquad (16.50)$$

Equations (16.49) and (16.50) are represented by Fig. 16.7b, c, which we can verify using KVL and KCL.

Frequency-domain representations of coupled inductances and capacitances can be obtained similarly. The results are shown in Figs. 16.8 and 16.9 along with the corresponding time-domain circuits. As in Chapter 13, the s-domain representation for a resistance R is simply that resistance R.

FIGURE 16.8
Time- and frequency-domain representations of coupled inductances with initial conditions: (a) time domain; (b), (c) frequency domain

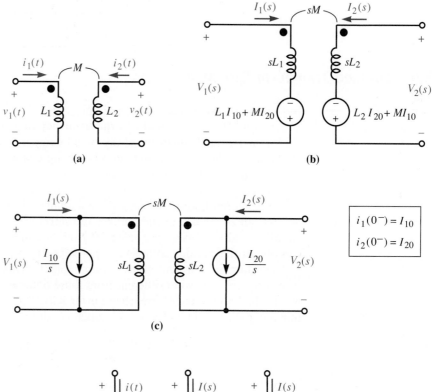

FIGURE 16.9
Time- and frequency-domain representations of capacitance with initial conditions: (a) time domain; (b), (c) frequency domain

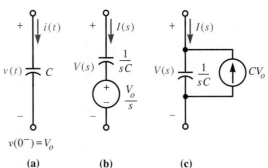

Usually the equivalent circuits in Figs. 16.7b, 16.8b, and 16.9b are easier to use in circuit analysis involving mesh-current equations, whereas those in Figs. 16.7c, 16.8c, and 16.9c are easier to use when writing node-voltage equations.

It is a straightforward exercise to transform the basic circuit theory laws KCL and KVL into the frequency domain. For example, by taking the Laplace transform of Kirchhoff's current equation

$$\sum_{\substack{\text{Closed} \\ \text{surface}}} i_n(t) = 0 \tag{16.51}$$

we obtain

$$\mathscr{L}\left\{\sum_{\substack{\text{Closed} \\ \text{surface}}} i_n(t)\right\} = 0 \tag{16.52}$$

This becomes, by Property P1,

$$\sum_{\substack{\text{Closed} \\ \text{surface}}} \mathscr{L}\{i_n(t)\} = 0 \tag{16.53}$$

which is, by definition,

$$\sum_{\substack{\text{Closed} \\ \text{surface}}} I_n(s) = 0 \tag{16.54}$$

Thus, the sum of the Laplace transforms of the currents leaving any closed surface in the network equals zero. A similar result applies to KVL, namely,

$$\sum_{\substack{\text{Closed} \\ \text{path}}} V_n(s) = 0 \tag{16.55}$$

Now that we have Laplace-transformed the circuit element terminal equations and the axioms KCL and KVL into the s domain, we are prepared to find the complete response of LLTI circuits using only algebra and table look-up. The method consists of the following three steps.

Step 1

Draw the Laplace-transformed circuit.

Step 2

Use any convenient frequency-domain circuit analysis method to relate the Laplace transform output [for example, $V_o(s)$] to the Laplace transform input(s).

Step 3

Find the time-domain output [for example, $v_o(t)$] by taking the inverse Laplace transform of $V_o(s)$.

Examples 16.16 through 16.19 illustrate the method.

EXAMPLE 16.16

Solve the problem in Example 16.14 with the use of Laplace transform circuit analysis.

■ SOLUTION

The first step is to determine the frequency-domain (that is, Laplace-transformed) circuit, using Figs. 16.7, 16.8, and 16.9 for help. Since the *RC* circuit of Fig. 16.6 is a series circuit, it is convenient to select the series frequency-domain representation shown in Fig. 16.9b for the capacitance, setting $V_o = -3$ V. The frequency-domain representation of the source voltage $v_s(t) = 5tu(t)$ follows from LT3 and Property P1, as $V_s(s) = 5/s^2$ V. The resulting frequency-domain circuit is shown in Fig. 16.10. The next step is to determine the equation relating $V_o(s)$ to $V_s(s)$. We will leave it to you as an easy exercise in circuit analysis to derive the equation

$$V_o(s) = \left(\frac{1/s}{2 + 1/s}\right)V_s(s) + \left(\frac{-3/s}{2 + 1/s}\right)2$$

The substitution $V_s(s) = 5/s^2$ leads to

$$V_o(s) = \frac{2.5}{(s + 0.5)s^2} - \frac{3}{s + 0.5}$$

FIGURE 16.10
Laplace-transformed circuit

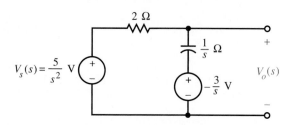

$V_s(s) = \dfrac{5}{s^2}$ V $2\ \Omega$ $\dfrac{1}{s}\ \Omega$ $-\dfrac{3}{s}$ V $V_o(s)$

Notice that this result is identical to that in Example 16.14, which was derived from the circuit's differential equation. We can now determine the complete response $v_o(t)$ by taking the inverse Laplace transform of $V_o(s)$. These steps in the analysis were presented in Example 16.14. ■

EXAMPLE 16.17

The input to the ideal op-amp circuit of Fig. 16.11 is $v_1(t) = 10u(t)$ V, and the capacitance has an initial voltage $v_C(0^-) = 5$ V. Find and sketch $v_o(t)$ for $t \geq 0$ if $R_1 = R_2 = 1$ kΩ and $C = 1\ \mu$F.

FIGURE 16.11
Op-amp circuit with step input and initial condition $v_C(0^-) = 5$ V

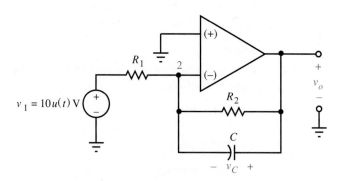

$v_1 = 10u(t)$ V R_1 R_2 C v_o $- v_C +$

■ **SOLUTION**

The Laplace-transformed circuit is shown in Fig. 16.12.

FIGURE 16.12
Laplace-transformed circuit

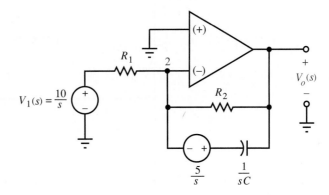

Application of KCL at node 2 yields

$$\frac{V_1(s)}{R_1} + sC\left[V_o(s) - \frac{5}{s}\right] + \frac{V_o(s)}{R_2} = 0$$

By setting $V_1(s) = 10/s$ and rearranging, we obtain

$$V_o(s) = \frac{5}{s + \dfrac{1}{R_2C}} - \frac{10}{sR_1C\left(s + \dfrac{1}{R_2C}\right)}$$

which becomes, for the given numerical values,

$$V_o(s) = \frac{5}{s + 1000} - \frac{10{,}000}{s(s + 1000)}$$

By expanding the second term into partial fractions and rearranging, we obtain

$$V_o(s) = \frac{15}{s + 1000} - \frac{10}{s}$$

Therefore, from Tables 16.1 and 16.2,

$$v_o(t) = 15e^{-1000t}u(t) - 10u(t) \text{ V}$$

This result assumes that the units of t are seconds. If we convert the units of t to milliseconds, we obtain

$$v_o(t) = 15e^{-t}u(t) - 10u(t) \text{ V}$$

A sketch of $v_o(t)$ is given in Fig. 16.13. ■

FIGURE 16.13
Output of op-amp circuit for
$t \geq 0$

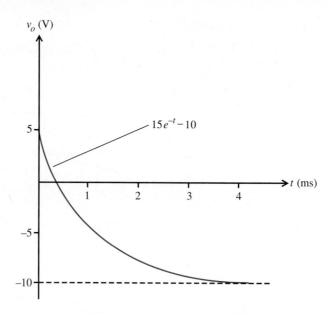

Examples 16.18 and 16.19 are slightly more challenging.

EXAMPLE 16.18

Steady-state conditions exist in the circuit of Fig. 16.14 at $t = 0^-$, and switch S1 closes at $t = 0$. Determine $i_1(t)$ and $i_2(t)$ for $t \geq 0$.

FIGURE 16.14
Circuit with switch S1

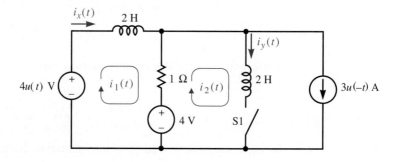

■ **SOLUTION**

By inspection, $i_y(0^-) = 0$ and $i_x(0^-) = -1$ A. The frequency domain circuit is shown in Fig. 16.15.

FIGURE 16.15
Frequency-domain circuit

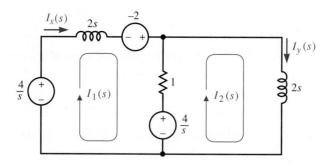

Notice that the $3u(-t)$-A current source has been transformed to an open circuit and that the 4-V dc source has the same transform, $4/s$, as the source $4u(t)$ V. This is because the (unilateral) Laplace transform of a function is determined only by the value of that function for $t \geq 0$.

The mesh-current equations are

$$\begin{bmatrix} 2s+1 & -1 \\ -1 & 2s+1 \end{bmatrix} \begin{bmatrix} I_1(s) \\ I_2(s) \end{bmatrix} = \begin{bmatrix} -2 \\ \dfrac{4}{s} \end{bmatrix}$$

Cramer's rule yields the solution

$$I_1(s) = \frac{\begin{vmatrix} -2 & -1 \\ \dfrac{4}{s} & 2s+1 \end{vmatrix}}{\begin{vmatrix} 2s+1 & -1 \\ -1 & 2s+1 \end{vmatrix}} = \frac{-4s-2+(4/s)}{4s^2+4s+1-1}$$

$$I_2(s) = \frac{\begin{vmatrix} 2s+1 & -2 \\ -1 & \dfrac{4}{s} \end{vmatrix}}{\begin{vmatrix} 2s+1 & -1 \\ -1 & 2s+1 \end{vmatrix}} = \frac{6+(4/s)}{4s^2+4s+1-1}$$

Simplifying and taking partial fractions, we have

$$I_1(s) = \frac{-2s^2-s+2}{2s^2(s+1)} = \frac{1/2}{s+1} - \frac{3/2}{s} + \frac{1}{s^2}$$

$$I_2(s) = \frac{3s+2}{2s^2(s+1)} = \frac{-1/2}{s+1} + \frac{1/2}{s} + \frac{1}{s^2}$$

which yield for $t \geq 0$,

$$i_1(t) = \frac{1}{2}e^{-t} - \frac{3}{2} + t \text{ A}$$

$$i_2(t) = -\frac{1}{2}e^{-t} + \frac{1}{2} + t \text{ A} \quad \blacksquare$$

EXAMPLE 16.19

The circuit of Fig. 16.16 is at rest at $t = 0^-$. Find $v_C(t)$ for $t \geq 0$.

FIGURE 16.16
Circuit for Example 16.19

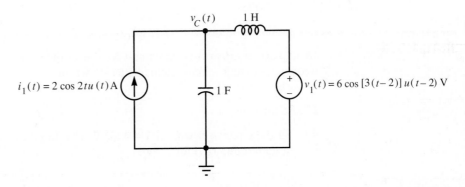

$i_1(t) = 2\cos 2t\, u(t)\,\text{A}$

$v_C(t)$ 1 H

1 F

$v_1(t) = 6\cos[3(t-2)]\,u(t-2)\,\text{V}$

■ **SOLUTION**

The transformed circuit is shown in Fig. 16.17.

FIGURE 16.17
Laplace-transformed circuit

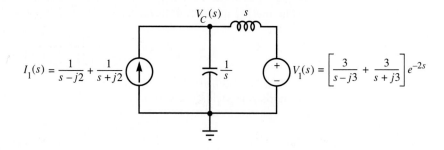

$$I_1(s) = \frac{1}{s - j2} + \frac{1}{s + j2} \qquad \qquad V_1(s) = \left[\frac{3}{s - j3} + \frac{3}{s + j3} \right] e^{-2s}$$

By KCL,

$$\frac{V_C(s) - V_1(s)}{s} + sV_C(s) = I_1(s)$$

which rearranges to

$$V_C(s) = \frac{V_1(s)}{s^2 + 1} + \frac{s}{s^2 + 1} I_1(s)$$

$$= \frac{s}{(s + j)(s - j)(s - j2)} + \frac{s}{(s + j)(s - j)(s + j2)}$$

$$+ \left[\frac{3}{(s + j)(s - j)(s - j3)} + \frac{3}{(s + j)(s - j)(s + j3)} \right] e^{-2s}$$

Using partial fraction expansions and combining terms, we obtain

$$V_C(s) = \frac{j1/3}{s - j} - \frac{j1/3}{s + j} - \frac{j2/3}{s - j2} + \frac{j2/3}{s + j2} + \left[\frac{3/8}{s - j} + \frac{3/8}{s + j} - \frac{3/8}{s - j3} - \frac{3/8}{s + j3} \right] e^{-2s}$$

which, by Property P5 of Table 16.1 and LT8 of Table 16.2, has the inverse transform

$$v_C(t) = \left[\frac{2}{3} \cos (t + 90°) + \frac{4}{3} \cos (2t - 90°) \right] u(t)$$

$$+ \left\{ \frac{3}{4} \cos (t - 2) + \frac{3}{4} \cos [3(t - 2) + 180°] \right\} u(t - 2) \text{ V} \quad ■$$

REMEMBER

In Laplace transform circuit analysis, the initial conditions are included in the frequency-domain circuit elements as shown in Figs. 16.7 through 16.9.

EXERCISES

40. In the circuit shown, $v_1(t) = 4u(t)$ V. Use Laplace transform circuit analysis to determine $i_R(t)$ for $t \geq 0$.

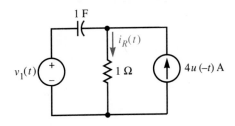

41. Repeat Exercise 40 for $v_1(t) = 4$ V.

42. Repeat Exercise 40 for $v_1(t) = 4u(t-1)$ V.

43. Repeat Exercise 40 for $v_1(t) = 8\cos(t+30°)u(t)$.

44. Switch S1 in the circuit shown opens at $t = 0$. Use Laplace transform circuit analysis to determine $v_1(t)$ and $v_2(t)$ for $t \geq 0$.

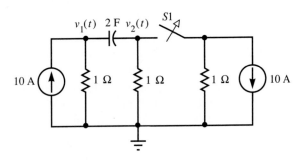

16.7 A Deeper Look at the Solution of LLTI Equations by Laplace Transforms

In this section we use the Laplace transform to develop some new theoretical insights into circuit response. As stated in Section 16.5, there are four basic steps involved in the solution of the differential equation

$$\mathscr{A}(p)v_o(t) = \mathscr{B}(p)v_s(t) \tag{16.56}$$

by Laplace transforms. In the following it is assumed that $v_o(0^-)$, $v_o^{(1)}(0^-), \ldots,$ $v_o^{(n-1)}(0^-)$ are given initial conditions, $v_s(t)$ is the given input, and the complete solution $v_o(t)$ is sought for $t \geq 0$.

Step 1

Take the Laplace transform of both sides of the time-domain equation to obtain the corresponding complex-frequency-domain equation.

The Laplace transform of the left-hand side of Eq. (16.56) is

$$\mathscr{L}\{\mathscr{A}(p)v_o(t)\} = \mathscr{L}\left\{a_n \frac{d^n v_o}{dt^n} + a_{n-1}\frac{d^{n-1}v_o}{dt^{n-1}} + \cdots + a_0 v_o(t)\right\}$$

$$= a_n \mathscr{L}\left\{\frac{d^n v_o}{dt^n}\right\} + a_{n-1}\mathscr{L}\left\{\frac{d^{n-1}v_o}{dt^{n-1}}\right\} + \cdots + a_0 \mathscr{L}\{v_o(t)\} \tag{16.57}$$

According to Property P2, the derivatives are given by

$$\mathcal{L}\left\{\frac{d^n v_o}{dt^n}\right\} = s^n V_o(s) - \sum_{k=0}^{n-1} s^{n-1-k} v_o^{(k)}(0^-)$$

$$\mathcal{L}\left\{\frac{d^{n-1} v_o}{dt^{n-1}}\right\} = s^{n-1} V_o(s) - \sum_{k=0}^{n-2} s^{n-2-k} v_o^{(k)}(0^-)$$

$$\vdots \qquad \qquad \vdots$$

$$\mathcal{L}\left\{\frac{d v_o}{dt}\right\} = s V_o(s) - v_o(0^-)$$

$$\mathcal{L}\{v_o(t)\} = V_o(s) \tag{16.58}$$

Combining Eqs. (16.57) and (16.58), we obtain

$$\mathcal{L}\{\mathcal{A}(p) v_o(t)\} = \mathcal{A}(s) V_o(s) + E(s) \tag{16.59}$$

where

$$E(s) = -a_n \sum_{k=0}^{n-1} s^{n-1-k} v_o^{(k)}(0^-) - a_{n-1} \sum_{k=0}^{n-2} s^{n-2-k} v_o^{(k)}(0^-) - \cdots - a_1 v_o(0^-) \tag{16.60}$$

is an $(n-1)$st-degree polynomial that depends on the initial conditions $v_o(0^-)$, $v_o^{(1)}(0^-)$, $v_o^{(n-1)}(0^-)$ and the coefficients $a_1, a_2, \ldots, a_n$. The Laplace transform of the right-hand side of Eq. (16.56) can be obtained similarly. The result is

$$\mathcal{L}\{\mathcal{B}(p) v_s(t)\} = \mathcal{B}(s) V_s(s) + F(s) \tag{16.61}$$

where

$$F(s) = -b_m \sum_{k=0}^{m-1} s^{m-1-k} v_s^{(k)}(0^-) - b_{m-1} \sum_{k=0}^{m-2} s^{m-2-k} v_s^{(k)}(0^-) - \cdots - b_1 v_s(0^-) \tag{16.62}$$

is an $(m-1)$st-degree polynomial that depends on "initial" input values $v_s(0^-)$, $v_s^{(1)}(0^-), \ldots, v_s^{(m-1)}(0^-)$ and the coefficients $b_1, b_2, \ldots, b_m$. It follows from Eqs. (16.59) and (16.61) that the frequency-domain version (Laplace transform) of LLTI Eq. (16.56) is

$$\mathcal{A}(s) V_o(s) + E(s) = \mathcal{B}(s) V_s(s) + F(s) \tag{16.63}$$

Step 2

Solve the complex-frequency-domain equation for $V_o(s)$.

The solution follows easily from Eq. (16.63):

$$V_o(s) = \frac{\mathcal{B}(s)}{\mathcal{A}(s)} V_s(s) + \frac{F(s) - E(s)}{\mathcal{A}(s)} \tag{16.64}$$

Before proceeding to Step 3, recall that the ratio $\mathcal{B}(s)/\mathcal{A}(s)$ is, by definition, the circuit's transfer function $H(s)$:

$$H(s) = \frac{\mathscr{B}(s)}{\mathscr{A}(s)} \qquad (16.65)$$

Also observe that the second term on the right-hand side of Eq. (16.64) is a function of the initial conditions $v_o(0^-), v_o^{(1)}(0^-), \ldots, v_o^{(n-1)}(0^-)$ and $v_s(0^-), v_s^{(1)}(0^-), \ldots,$ $v_s^{(m-1)}(0^-)$. This term is the *stored-energy response*, which represents that portion of the output $v_o(t), t \geq 0$, caused by the energy that is stored in the circuit at $t = 0^-$. The first term on the right-hand side of Eq. (16.64) is the *forced response*, which represents that portion of the output caused by the input $v_s(t)$ for $t \geq 0$. Thus

$$V_o(s) = \underbrace{H(s)V_s(s)}_{\substack{\text{Forced} \\ \text{response}}} + \underbrace{\frac{F(s) - E(s)}{\mathscr{A}(s)}}_{\substack{\text{Stored-energy} \\ \text{response}}} \qquad (16.66)$$

The forced response is the inverse Laplace transform of $V_s(s)H(s)$. According to Property P7, multiplication in the frequency domain corresponds to convolution in the time domain. Hence, in general, $v_o(t)$ consists of two components:

$$v_o(t) = \underbrace{\int_{0^-}^{t} h(t - \lambda)v_s(\lambda)\,d\lambda}_{\substack{\text{Forced} \\ \text{response}}} + \underbrace{\mathscr{L}^{-1}\left\{\frac{F(s) - E(s)}{\mathscr{A}(s)}\right\}}_{\substack{\text{Stored-energy} \\ \text{response}}} \qquad (16.67)$$

The function $h(t) = \mathscr{L}^{-1}\{H(s)\}$ is the impulse response of the circuit. We can easily establish this result by evaluating Eqs. (16.66) and (16.67) for the case that $v_s(t)$ is a unit impulse. If $v_s(t) = \delta(t)$, then $V_s(s) = 1$ and the first terms on the right-hand sides of Eqs. (16.66) and (16.67), respectively, become $H(s)$ and $h(t)$.

Step 3

Use partial fraction expansions to write $V_o(s)$ as a sum of terms of the forms included in Table 16.2.

A general mathematical formulation of this step is notationally messy. However, it is worth noting that the poles of the function $H(s)V_s(s)$ include both the poles of $H(s)$ and the poles of $V_s(s)$. Assuming that the poles of $H(s)$ are distinct from those of $V_s(s)$, the partial fraction expansion of $H(s)V_s(s)$ consists of two groups of terms; one group contains the poles $V_s(s)$, and the other group contains the poles of $H(s)$. The significance of each group of terms is shown in Eq. (16.68):

$$
\begin{aligned}
V_o(s) &= \overset{\substack{\text{Forced} \\ \text{response}}}{\underbrace{V_s(s)H(s)}} + \overset{\substack{\text{Stored-energy} \\ \text{response}}}{\underbrace{\frac{F(s) - E(s)}{\mathscr{A}(s)}}} \\[2mm]
&= \underbrace{G_1(s)}_{\substack{\text{Particular} \\ \text{response}}} + \underbrace{G_2(s) + G_3(s)}_{\substack{\text{Complementary} \\ \text{response}}}
\end{aligned} \qquad (16.68)
$$

where

$G_1(s) =$ that portion of the partial fraction expansion of $V_s(s)H(s)$ containing the poles of $V_s(s)$

$G_2(s) =$ that portion of the partial fraction expansion of $V_s(s)H(s)$ containing the poles of $H(s)$

$G_3(s) =$ the partial fraction expansion of $[F(s) - E(s)]/\mathscr{A}(s)$

Because $H(s) = \mathscr{B}(s)/\mathscr{A}(s)$, the poles in $G_2(s)$ are the same as those in $G_3(s)$.

Step 4

Use Tables 16.1 and 16.2 to inverse-transform Eq. (16.68) into the time domain.

The result of this procedure is the complete response $v_o(t)$. As will be illustrated by Example 16.20, the inverse Laplace transform of $G_1(s)$ is the particular response of the differential equation, and the inverse Laplace transform of $G_2(s) + G_3(s)$ is the complementary response. Thus the particular response involves the poles of $V_s(s)$ and the complementary response involves the poles of $H(s)$ [the roots of $\mathscr{A}(s) = 0$]. Observe that the task of finding the constants of integration as described in Chapter 9 has been replaced by that of determining the partial fraction expansion in Eq. (16.68).

The connection between the Laplace transform formulation of Eq. (16.66) and the sinusoidal steady-state analysis of Chapter 10 can be established if we assume that $v_s(t) = \mathbf{V}_1 e^{j\omega_0 t}u(t)$, where $\mathbf{V}_1 = V_{1m}\underline{/\phi_1}$ is a complex constant. This implies, by LT5, that $V_s(s) = \mathbf{V}_1/(s - j\omega_0)$. For this case, Eq. (16.66) becomes

$$V_o(s) = \frac{\mathbf{V}_1}{(s - j\omega_0)} H(s) + \frac{F(s) - E(s)}{\mathscr{A}(s)} \qquad (16.69)$$

Assuming that $j\omega_0$ is not a pole of $H(s)$, the partial fraction expansion of $V_o(s)$ is given by Eq. (16.68), where

$$G_1(s) = \frac{k_{11}}{s - j\omega_0} \qquad (16.70)$$

is that portion of the partial fraction expansion containing the pole at $s = j\omega_0$. The residue k_{11} can be determined using Heaviside's theorem [Eq. (16.43)]:

$$k_{11} = (s - j\omega_0) \frac{\mathbf{V}_1}{(s - j\omega_0)} H(s)\Big|_{s = j\omega_0}$$

$$= \mathbf{V}_1 H(j\omega_0) \qquad (16.71)$$

Substitution of Eq. (16.71) into Eq. (16.70) yields

$$V_o(s) = \underbrace{\frac{\mathbf{V}_1 H(j\omega_0)}{s - j\omega_0}}_{\text{Particular response}} + \underbrace{G_2(s) + G_3(s)}_{\text{Complementary response}} \qquad (16.72)$$

It follows directly from LT5 that

$$v_o(t) = \underbrace{\mathbf{V}_1 H(j\omega_0)e^{j\omega_0 t}u(t)}_{\text{Particular response}} + \text{complementary response} \qquad (16.73)$$

Therefore, as in Chapter 10, the particular response to the rotating phasor input $\mathbf{V}_1 e^{j\omega_0 t}$ is the rotating phasor $\mathbf{V}_0 e^{j\omega_0 t}$, where $\mathbf{V}_0 = H(j\omega_0)\mathbf{V}_1$.

EXAMPLE 16.20

The input-output equation of the network shown in Fig. 16.18 is

$$\tau_2 \frac{dv_o}{dt} + v_o = \tau_1 \frac{dv_s}{dt} + v_s$$

where $\tau_1 = R_2 C$ and $\tau_2 = (R_1 + R_2)C$. Find $v_o(t)$ for $t \geq 0$ if the input is $v_s(t) = Au(t)$, where A is a real constant, and the capacitance has initial voltage $v_C(0^-)$.

FIGURE 16.18
Circuit for Example 16.20

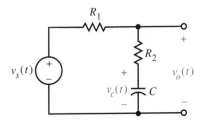

■ SOLUTION

The Laplace transform of the input-output equation is (Step 1):

$$\tau_2 s V_o(s) - \tau_2 v_o(0^-) + V_o(s) = \tau_1 s V_s(s) - \tau_1 v_s(0^-) + V_s(s)$$

which can be solved for $V_o(s)$ (Step 2):

$$V_o(s) = \underbrace{\frac{s\tau_1 + 1}{s\tau_2 + 1} V_s(s)}_{\substack{\text{Forced} \\ \text{response}}} + \underbrace{\frac{\tau_2 v_o(0^-) - \tau_1 v_s(0^-)}{s\tau_2 + 1}}_{\substack{\text{Stored-energy} \\ \text{response}}}$$

Rearranging with $\tau_1 = R_2 C$ and $\tau_2 = (R_1 + R_2)C$ we obtain

$$V_o(s) = \underbrace{\left(\frac{R_2}{R_1 + R_2}\right)\left(\frac{s + \tau_1^{-1}}{s + \tau_2^{-1}}\right) V_s(s)}_{\text{Forced response}} + \underbrace{\frac{v_o(0^-) - \dfrac{R_2}{R_1 + R_2} v_s(0^-)}{s + \tau_2^{-1}}}_{\text{Stored-energy response}}$$

This equation has the form of Eq. 16.66. Observe that this equation specifies $V_o(s)$ in terms of initial conditions on the output $v_o(0^-)$ and the input $v_s(0^-)$. Because the only "givens" are $v_s(t)$ and the initial voltage on the capacitance $v_C(0^-)$, it is necessary to determine $v_o(0^-)$. This can be done by writing a single KCL equation at the node connecting R_1 and R_2:

$$\frac{v_o(t) - v_s(t)}{R_1} + \frac{v_o(t) - v_C(t)}{R_2} = 0$$

which yields

$$v_o(t) = \frac{R_1}{R_1 + R_2} v_C(t) + \frac{R_2}{R_1 + R_2} v_s(t)$$

and therefore

$$v_o(0^-) = \frac{R_1}{R_1 + R_2} v_C(0^-) + \frac{R_2}{R_1 + R_2} v_s(0^-)$$

Substitution of the above expression for $v_o(0^-)$ into that for $V_o(s)$ yields

$$V_o(s) = \left(\frac{R_2}{R_1 + R_2}\right)\left(\frac{s + \tau_1^{-1}}{s + \tau_2^{-1}}\right)V_s(s) + \left(\frac{R_1}{R_1 + R_2}\right)\frac{v_C(0^-)}{s + \tau_2^{-1}}$$

The input is a step function, $v_s(t) = Au(t)$. Therefore, from LT2 and Property P1, $V_s(s) = A/s$ and

$$V_o(s) = \left(\frac{R_2}{R_1 + R_2}\right)\left(\frac{s + \tau_1^{-1}}{s + \tau_2^{-1}}\right)\frac{A}{s} + \left(\frac{R_1}{R_1 + R_2}\right)\frac{v_C(0^-)}{s + \tau_2^{-1}}$$

It is again helpful to use partial fraction expansions (Step 3):

$$\left(\frac{s + \tau_1^{-1}}{s + \tau_2^{-1}}\right)\frac{1}{s} = \frac{k_{11}}{s} + \frac{k_{21}}{s + \tau_2^{-1}}$$

Because each pole is simple, one can use Heaviside's theorem to find k_{11} and k_{21}:

$$k_{11} = \left.\frac{s + \tau_1^{-1}}{s + \tau_2^{-1}}\right|_{s=0} = \frac{\tau_2}{\tau_1} = \frac{R_1 + R_2}{R_2}$$

$$k_{21} = \left.\frac{s + \tau_1^{-1}}{s}\right|_{s=-\tau_2^{-1}} = \frac{-\tau_2^{-1} + \tau_1^{-1}}{-\tau_2^{-1}} = -\frac{R_1}{R_2}$$

Therefore

$$\left(\frac{s + \tau_1^{-1}}{s + \tau_2^{-1}}\right)\frac{1}{s} = \left(\frac{R_1 + R_2}{R_2}\right)\frac{1}{s} - \frac{R_1/R_2}{s + \tau_2^{-1}}$$

and $V_o(s)$ can be written as

$$V_o(s) = \overbrace{\frac{R_2}{R_1 + R_2} \frac{s + \tau_1^{-1}}{s + \tau_2^{-1}} V(s)}^{\text{Forced response}} + \overbrace{\frac{R_1}{R_1 + R_2} \frac{v_C(0^-)}{s + \tau_2^{-1}}}^{\text{Stored-energy response}}$$

$$= \underbrace{\frac{A}{s}}_{\text{Particular response}} - \left(\frac{R_1}{R_1 + R_2}\right)\frac{A}{s + \tau_2^{-1}} + \underbrace{\left(\frac{R_1}{R_1 + R_2}\right)\frac{v_C(0^-)}{s + \tau_2^{-1}}}_{\text{Complementary response}}$$

Observe that the stored-energy response depends only on the circuit parameters and the initial voltage across the capacitance. The time-domain output can now be obtained directly from LT2, LT5, and Property P1. The result is

$$v_o(t) = Au(t) - \left(\frac{R_1}{R_1 + R_2}\right)Ae^{-t/\tau_2}u(t) + \left(\frac{R_1}{R_1 + R_2}\right)v_C(0^-)e^{-t/\tau_2}u(t)$$

$$= Au(t) + [v_C(0^-) - A]\left(\frac{R_1}{R_1 + R_2}\right)e^{-t/\tau_2}u(t) \quad \blacksquare$$

An LLTI circuit's complete response consists of the sum of the forced response and the stored-energy response. The complete response also consists of the sum of the particular response and the complementary response. The relationships among the responses are summarized by Eq. (16.68).

EXERCISES

45. Identify the forced response, stored-energy response, particular response, and complementary response terms in the solution to Example 16.14.

46. Identify the forced response, stored-energy response, particular response, and complementary response terms in the solution to Example 16.15.

47. Determine the response $i(t)$ in the circuit shown for $t > 0$ with $v_s(t) = Au(t)$. Identify the forced response, stored-energy response, particular response, and complementary response terms in your solution. The current $i_L(0^-)$ is arbitrary.

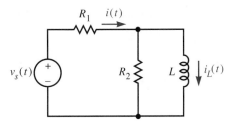

16.8   Summary

This chapter described one of the most useful of all analytical aids to determine the response of an LLTI circuit with arbitrary input and initial conditions. We defined the Laplace transform of a signal $v(t)$ as

$$V(s) = \int_{0^-}^{\infty} v(t)e^{-st}\,dt \qquad \mathscr{R}e\,\{s\} > \sigma_0 \tag{16.74}$$

where $s = \sigma + j\omega$ and σ_0 is the abscissa of convergence. The Laplace transform $V(s)$ is a frequency-domain representation of $v(t)$ that exists for virtually every signal that arises in circuit theory and engineering practice. The transformation of Eq. (16.74) is one-to-one when $v(t)$ is causal (that is, when $v(t) = 0$ for $t < 0$). The inverse transform is

$$v(t) = \int_{c-j\infty}^{c+j\infty} V(s)e^{st}\,\frac{ds}{2\pi j} \qquad c > \sigma_0 \tag{16.75}$$

We saw that the Laplace transform can be applied to LLTI circuit analysis with the use of one of two methods. In the first method, the Laplace transform is applied to the differential equation (or the system of differential equations) relating the input and output. This process yields an algebraic equation (or system of algebraic equations) from which the frequency-domain representation (the Laplace transform) of the output can be obtained. In the second method, the Laplace transform is applied to the circuit itself to obtain a frequency-domain circuit from which the Laplace transform of the response can be obtained. In both methods, all initial conditions are included in the analysis from the start, and purely algebraic manipulations are all that are required to obtain the frequency-domain form of the response. The time-domain response is obtained by taking the inverse Laplace transform. We showed that this inversion can be done using partial fraction expansions and table look-up. In Section 16.7, we saw that the Laplace transform provides insight into the relation between a circuit's forced and stored-energy responses and its particular and complementary responses.

KEY FACTS

◆ The Laplace transform of a signal $v(t)$ is defined as

$$V(s) = \int_{0^-}^{\infty} v(t)e^{-st}\, dt \qquad \mathscr{R}e\,\{s\} > \sigma_0$$

where σ_0 is the abscissa of convergence.

◆ The region of convergence, $\mathscr{R}e\,\{s\} > \sigma_0$, of every rational algebraic $V(s)$ lies to the immediate right of the poles of $V(s)$.

◆ The Laplace transform exists for virtually every signal encountered in circuit theory.

◆ The Laplace transform is one-to-one if $v(t) = 0$ for $t < 0$.

◆ The inverse Laplace transform of $V(s)$ is given by

$$v(t) = \int_{c-j\infty}^{c+j\infty} V(s)e^{st}\, \frac{ds}{2\pi j} \qquad c > \sigma_0$$

◆ Laplace transform properties and pairs are given in Tables 16.1 and 16.2, respectively.

◆ A rational algebraic function of s may be expanded into *partial fractions*. This step simplifies the task of finding the inverse Laplace transform.

◆ There are two useful techniques to apply the Laplace transform to LLTI circuit analysis. The first is to apply the Laplace transform to the LLTI equation relating output to input. The second is to use the Laplace transform to transform the circuit itself into the frequency domain.

◆ After the algebraic equation relating the Laplace transform of the output to the Laplace transform of the input has been obtained, the time-domain output can be obtained by table look-up.

Appendix 16.1

DERIVATION OF THE LAPLACE TRANSFORM

In this appendix we show that the Laplace transform can be derived from the Fourier transform.

We begin the derivation by considering any signal $v(t)$ (causal or noncausal) that satisfies Dirichlet conditions 2 and 3 (page 704) but may or may not satisfy condition 1. Thus, we assume that $\int_{-\infty}^{+\infty} |v(t)|\, dt$ may be infinite. Consequently,

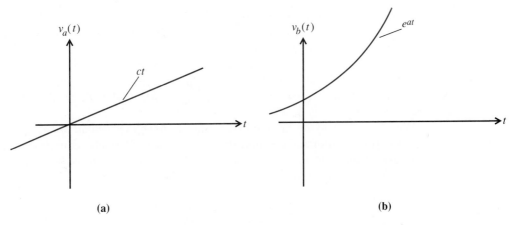

(a)   (b)

FIGURE 16.19   Two signals that do not satisfy Dirichlet condition 1; (a) $v_a(t) = ct$; (b) $v_b(t) = e^{at}$

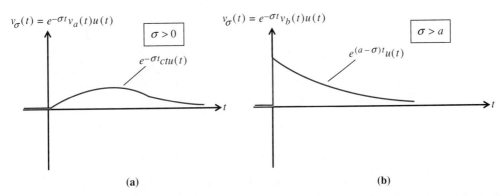

(a)   (b)

FIGURE 16.20   By multiplying the functions $v_a(t)$ and $v_b(t)$ in Fig. 16.89 by $e^{-\sigma t}u(t)$, we can obtain functions $v_\sigma(t)$ that satisfy Dirichlet condition 1

the Fourier transform of $v(t)$ may not exist. Signals of this type are illustrated in Fig. 16.19.

To derive the Laplace transform of $v(t)$, we define an "exponentially weighted causal version" of $v(t)$ as follows:

$$v_\sigma(t) = v(t)e^{-\sigma t}u(t) = \begin{cases} v(t)e^{-\sigma t} & t \geq 0 \\ 0 & t < 0 \end{cases} \tag{16.76}$$

The function $v_\sigma(t)$ is illustrated in Fig. 16.20, which should be compared with Fig. 16.19.

As illustrated by the waveforms in Figs. 16.19 and 16.20, $\int_{-\infty}^{\infty} |v_\sigma(t)|\, dt$ will be finite provided that $|v(t)|$ does not increase faster than exponentially with t and provided that σ is sufficiently large. Specifically, the integral $\int_{-\infty}^{\infty} |v_\sigma(t)|\, dt$ will be finite for all $\sigma > \sigma_0$ if

$$|v(t)| \leq Me^{\sigma_0 t} \qquad \text{for all } t > 0 \tag{16.77}$$

where M and σ_0 are finite. This statement can be proven by observing that if Eq. (16.77) is true, then

$$\int_{-\infty}^{\infty} |v_\sigma(t)| \, dt = \int_{0-}^{\infty} |v(t)| e^{-\sigma t} \, dt$$

$$\leq M \int_{0-}^{\infty} e^{\sigma_0 t} e^{-\sigma t} \, dt$$

$$= \frac{M}{\sigma - \sigma_0} < \infty \qquad \sigma > \sigma_0 \qquad (16.78)$$

Waveforms satisfying Eq. (16.77) are said to have *exponential order*.

If $v(t)$ has exponential order, then $v_\sigma(t)$ satisfies Dirichlet condition 1, and the Fourier transform of $v_\sigma(t)$ exists. With the aid of Eq. (16.76), we can write the Fourier transform of $v_\sigma(t)$ as

$$\mathbf{V}_\sigma(f) = \int_{-\infty}^{+\infty} v_\sigma(t) e^{-j2\pi f t} \, dt = \int_{0-}^{\infty} v(t) e^{-(\sigma + j2\pi f)t} \, dt \qquad \sigma > \sigma_0 \qquad (16.79)$$

where the condition $\sigma > \sigma_0$ is needed to guarantee that the integral exists. Now by setting $\sigma + j2\pi f = s$, we can write Eq. (16.79) as

$$\mathbf{V}_\sigma(f) = V(s) = \int_{0-}^{\infty} v(t) e^{-st} \, dt \qquad \sigma > \sigma_0 \qquad (16.80)$$

$V(s)$ is, by definition, the unilateral Laplace transform of $v(t)$. We see that $V(s)$ exists for $\sigma > \sigma_0$, where σ_0 is the smallest number satisfying Eq.(16.78).

Let us now derive the inverse Laplace transform. The inverse *Fourier* transform of $\mathbf{V}_\sigma(f)$ is, by Eq. (16.79), $v_\sigma(t)$. Consequently, we can write

$$v_\sigma(t) = v(t) e^{-\sigma t} u(t) = \int_{-\infty}^{+\infty} \mathbf{V}_\sigma(f) e^{+j2\pi f t} \, df \qquad (16.81)$$

for $\sigma > \sigma_0$. The rightmost equality can be rearranged to yield

$$v(t) u(t) = \int_{-\infty}^{+\infty} \mathbf{V}_\sigma(f) e^{(\sigma + j2\pi f)t} \, df \qquad \sigma > \sigma_0 \qquad (16.82)$$

We can substitute $\mathbf{V}_\sigma(f) = V(s)$ and $s = \sigma + j2\pi f$ into Eq. (16.82). If we hold σ equal to a constant c, then $ds = j2\pi \, df$ and Eq. (16.82) becomes

$$v(t) u(t) = \int_{c-j\infty}^{c+j\infty} V(s) e^{st} \frac{ds}{2\pi j} \qquad c > \sigma_0 \qquad (16.83)$$

which is the formula for the inverse Laplace transform.

Our derivation of the Laplace transform has been based on the Fourier transform. It follows that, excluding the pathological waveforms that do not satisfy Dirichlet conditions 2 and 3, every waveform satisfying Eq. (16.77) has a Laplace transform.

EXERCISES

For each of the following signals, determine whether it is (a) absolutely integrable, (b) of exponential order, and (c) Laplace transformable.

48. $e^{2t} u(t)$ 49. $t^3 u(t)$

50. $t^{-1} u(t)$ 51. $\cos(t) u(t)$

52. $e^{t^2} u(t)$

53. Determine the abscissa of convergence for each of the Laplace-transformable signals in Exercises 48 through 52.

PROBLEMS

Section 16.1

1. Determine the Laplace transform of the following signals. Show the poles and zeros of $V(s)$ in the s plane, and shade the region of convergence of $V(s)$.

$v(t) = e^{a(t-t_0)}u(t - t_0)$ for $a < 0$ where $t_0 > 0$

$v(t) = e^{a(t-t_0)}u(t - t_0)$ for $a > 0$ where $t_0 > 0$

2. Determine the Laplace transform of $v(t) = Ve^{s_0 t}u(t)$ where $s_0 = \sigma_0 + j\omega_0$ is an arbitrary complex number. Show the poles and zeros of $V(s)$ in the s plane and shade the region of convergence.

3. Determine the Laplace transform of $v(t) = V_m e^{\sigma_0 t} \cos(\omega_0 t + \phi_0)u(t)$, where σ_0, ω_0, and ϕ_0 are arbitrary real numbers. Show the poles and zeros of $V(s)$ in the s plane, and shade the region of convergence. [*Hint*: Express $v(t)$ as the sum of two causal exponential signals.]

For Problems 4 through 6, explain why the unilateral Laplace transforms of the pairs of functions of t are identical.

4. $u(t)$ and 1
5. $u(-t)$ and 0
6. $\cos \omega t$ and $\cos \omega t + t^2 u(-t)$
7. (a) Let $v(t)$ be a causal signal with Fourier transform $V(f)$ and Laplace transform $V(s)$. Show that $V(j2\pi f) = V(f)$. This important result provides a link between the Fourier and Laplace transforms.
   (b) Show that the result of (a) holds for $v(t) = e^{at}u(t)$ for $a < 0$.
   (c) Why doesn't the result of (a) hold for $v(t) = e^{at}u(t)$ for $a \geq 0$?
8. Use the results of Problem 7 to show that the Fourier transform of a causal signal exists if and only if the region of convergence of the Laplace transform of the signal includes the $j\omega$ axis of the s plane.

Sections 16.2 and 16.3

For Problems 9 through 12, sketch each signal and use Tables 16.1 and 16.2 to find their Laplace transforms.

9. $\Pi\left(\dfrac{t - t_0}{\tau}\right)$, where $t_0 > 0.5\tau$

10. $\cos(\omega_0 t)\Pi\left(\dfrac{t - t_0}{\tau}\right)$, where $t_0 > 0.5\tau$

11. $\cos[\omega_0(t - t_0)]\Pi\left(\dfrac{t - t_0}{\tau}\right)$, where $t_0 > 0.5\tau$

12. $tu(t) + (2 - 2t)u(t - 1) + (t - 2)u(t - 2)$

13. Prove Property P1 of Table 16.1.
14. Prove Property P4 of Table 16.1.
15. Prove Property P5 of Table 16.1.
16. Prove Property P6 of Table 16.1.
17. Prove Property P7 of Table 16.1.
18. Consider a causal signal $v_o(t)$. Assume that $v_o(t) \leftrightarrow V_o(s)$ for $\sigma > \sigma_0$ and let $v(t) = \sum_{n=0}^{\infty} v_0(t - nT)$, where $T > 0$. Determine $V(s) = \mathscr{L}\{v(t)\}$ and its abscissa of convergence. [*Hint*: Use Property P5 of Table 16.1 and the fact that $\sum_{n=0}^{\infty} z^{-n} = z/(z - 1)$ for $|z| > 1$.]
19. Verify that the initial value theorem gives the correct result for $v(t) = Ae^{-\alpha t}u(t)$.
20. Verify that the final value theorem gives the correct result for $v(t) = A(1 - e^{-\alpha t})u(t)$.
21. Show that the final value theorem gives an incorrect result for $v(t) = A\cos(\omega_0 t)u(t)$. Discuss.

Section 16.4

22. Find $g(t) = \mathscr{L}^{-1}\{G(s)\}$, where

$$G(s) = \frac{2s^2 + 5s + 4}{(s + 1)(s^2 + 2s + 2)}$$

23. Find $g(t) = \mathscr{L}^{-1}\{G(s)\}$, where

$$G(s) = \frac{s^3 + 3s^2 + 2s + 2}{s^2 + 2s + 1}$$

24. The Laplace transform of a particular function $g(t)$ is $G(s) = N(s)/D(s)$, where $deg\{N(s)\} < deg\{D(s)\}$ and where all the roots $s_1, s_2, \ldots, s_n$ of the equation $D(s) = 0$ are distinct.
   (a) Use Eq. (16.40) to show that the partial fraction expansion of $G(s)$ is given by

$$G(s) = \frac{k_{11}}{s - s_1} + \frac{k_{21}}{s - s_2} + \cdots + \frac{k_{n1}}{s - s_n}$$

$$= \sum_{i=1}^{n} \frac{k_{i1}}{s - s_i}$$

where k_{i1} is the residue of $G(s)$ at s_i.

(b) Identify the entries in Tables 16.1 and 16.2 that enable you to conclude that the inverse Laplace transform of $G(s)$ is given by

$$g(t) = k_{11}e^{s_1 t}u(t) + k_{21}e^{s_2 t}u(t)$$
$$+ \cdots + k_{n1}e^{s_n t}u(t)$$
$$= \sum_{i=1}^{n} k_{i1}e^{s_i t}u(t)$$

25. (a) Let $G(s) = N(s)/D(s)$ be a rational algebraic function with a simple pole at s_1. Assume that $deg\{N(s)\} < deg\{D(s)\}$. Show that the residue of $G(s)$ at s_1 is given by

$$k_{11} = \frac{N(s_1)}{D'(s_1)}$$

where $D'(s_1)$ denotes the derivative of $D(s)$ evaluated at s_1. [*Hint:* Write $D(s)$ as $D(s) = (s - s_1) Q(s)$ and note that by Eq. (16.43), $k_{11} = N(s_1)/Q(s_1)$. Then show that $D'(s_1) = Q(s_1)$.]

(b) Generalize the results from (a) to show that if all the poles $s_1, s_2, \ldots, s_n$ of the function $G(s) = N(s)/D(s)$ are simple, then the partial fraction expansion of $G(s)$ is

$$G(s) = \sum_{i=1}^{n} \frac{N(s_i)/D'(s_i)}{s - s_i}$$

This result is known as *Heaviside's expansion theorem*.

26. Use the Heaviside expansion theorem of Problem 25 to find the partial fraction expansion of

$$G(s) = \frac{2s^2 + 5s + 4}{(s + 3)(s^2 + 3s + 2)}$$

Then find $g(t) = \mathscr{L}^{-1}\{G(s)\}$.

27. Prove that the region of convergence of the Laplace transform lies to the immediate right of the poles of $V(s)$. {*Hint:* Write $V(s)$ in partial fraction form. If you are stuck, consider a simple example like $V(s) = 1/[(s + 1)(s - 3)]$, then generalize.}

28. Find $v(t)$ if

$$V(s) = \frac{s + 1}{(s + 2)(s + 3)} e^{-s4}$$

Use the Laplace transform to find the complete response of the following differential equations for $t > 0$.

29. $\dfrac{dv}{dt} + 4v = 100,\ v(0^-) = 20$

30. $\dfrac{dv}{dt} + 4v = 100t,\ v(0^-) = 20$

31. $\dfrac{dv}{dt} + 4v = 100 \cos t,\ v(0^-) = 20$

32. $\dfrac{dv}{dt} + 4v = 48e^{-t},\ v(0^-) = 20$

33. $\dfrac{d^2v}{dt^2} + v = \cos t,\ v(0^-) = 2,\ v'(0^-) = 0$

34. $\dfrac{d^2v}{dt^2} + v = e^{-t} \cos t,\ v(0^-) = 2,\ v'(0^-) = 0$

35. $\dfrac{d^2v}{dt^2} + 3\dfrac{dv}{dt} + 2v = 4 \cos t + 8 \cos 2t$, where

$$v(0^-) = 0 \text{ and } \left.\frac{dv}{dt}\right|_{t=0^-} = 0$$

Section 16.5

36. In the circuit shown, $R = 1\ \Omega,\ C = 1$ F, $v_1(t) = 10e^{\sigma t} \cos \omega t\ u(t)$ V and $v_C(0^-) = -10$ V.
    (a) Find $v_R(t)$ and $v_C(t)$ for $t \geq 0$ if $\sigma = \omega = 0$.
    (b) Repeat (a) with $\sigma = 0$, and $\omega = 1$ rad/s.
    (c) Repeat (a) with $\sigma = -1$ Np/s, and $\omega = 0$.
    (d) Repeat (a) with $\sigma = -1$ Np/s and $\omega = 1$ rad/s.

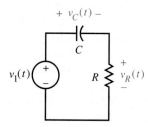

37. (a) Use Laplace transforms to determine $v_R(t)$ and $v_C(t)$ for the circuit of Problem 36 if $v_1(t) = Atu(t)$ and $v_C(0^-) = 0$.
    (b) Sketch $v_R(t)$ and $v_C(t)$.

38. In the circuit shown, $R = 1\ \Omega,\ L = 1$ H, $v_1(t) = 10e^{\sigma t} \cos \omega t\ u(t)$ V and $v_R(0^-) = -10$ V.
    (a) Find $v_R(t)$ and $v_L(t)$ for $t \geq 0$ if $\sigma = \omega = 0$.

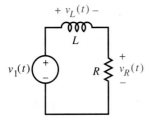

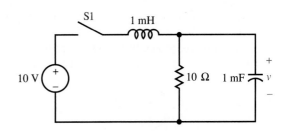

(b) Repeat (a) with $\sigma = 0$, and $\omega = 1$ rad/s.
(c) Repeat (a) with $\sigma = -1$ Np/s, and $\omega = 0$.
(d) Repeat (a) with $\sigma = -1$ Np/s and $\omega = 1$ rad/s.

39. (a) Use Laplace transforms to determine $v_R(t)$ and $v_L(t)$ for the circuit of Problem 38 if $v_1(t) = Atu(t)$ and $v_R(0^-) = 0$.
    (b) Sketch $v_R(t)$ and $v_L(t)$.

40. In the circuit shown, $L = 1$ H, $C = 1$ F, $v_1(t) = 10e^{\sigma t}\cos\omega t\, u(t)$ V, and there is no energy stored in the circuit for $t < 0$.
    (a) Find $v_L(t)$ and $v_C(t)$ for $t \geq 0$ if $\sigma = \omega = 0$.
    (b) Repeat (a) with $\sigma = 0$, and $\omega = 1$ rad/s.
    (c) Repeat (a) with $\sigma = -1$ Np/s, and $\omega = 0$.
    (d) Repeat (a) with $\sigma = -1$ Np/s and $\omega = 1$ rad/s.

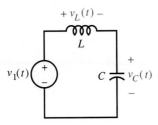

41. (a) Use Laplace transforms to determine $v_L(t)$ and $v_C(t)$ for the circuit of Problem 40 if $v_1(t) = Atu(t)$ and $i_L(0^-) = v_C(0^-) = 0$.
    (b) Sketch $v_L(t)$ and $v_C(t)$.

42. Rework Example 9.11 with the use of Laplace transforms. Start with the equation $(p + 3)v = 12 + 24e^{-t}$, $t > 0$ where $v(0^-) = v(0^+) = 6$ V.

Section 16.6

43. In the network shown, switch S1 has been closed for a very long time. At time $t = 0$, the switch is opened. Use Laplace transform circuit analysis to find v for $t > 0$.

44. Repeat Problem 43 if switch S1 has been open for a very long time and closes at time $t = 0$.

45. (a) Switch S1 closes at $t = 0$ in the circuit shown. Determine $i_1(t)$ and $i_2(t)$ for $t \geq 0$ with the use of Laplace transform circuit analysis.
    (b) Use your answer from (a) to determine $v_1(t)$ for $t \geq 0$.
    (c) Show how your answers from (a) and (b) would be modified if switch S1 closes at $t = t_0$, where t_0 is arbitrary.

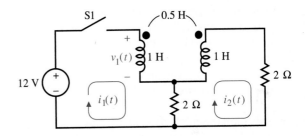

46. Assume that switch S1 has been closed for a sufficiently long time for steady-state conditions to exist in the circuit of Problem 45. Use Laplace transform circuit analysis to determine the currents $i_1(t)$ and $i_2(t)$ and voltage $v_1(t)$ that result if the switch opens at $t = 0$.

47. In the circuit shown, switch S1 has been in position B for a long time. At $t = 0$ it switches to position A, where it remains until $t = 1$ ms, when it switches to position B. Determine and plot v for $t \geq 0$. Assume that the capacitances are uncharged for $t < 0$.

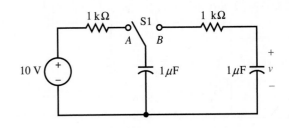

48. Repeat Problem 47 with the capacitances replaced with 1-mH inductances.
49. In the ideal op-amp circuit shown, we are given $v_C(0^-) = 5$ V and $v_1 = 10u(t)$V. Find and sketch v_o for $t > 0$.

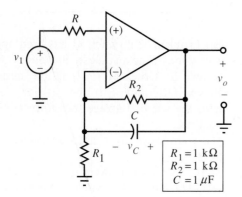

$R_1 = 1$ kΩ
$R_2 = 1$ kΩ
$C = 1\,\mu$F

50. Repeat Problem 49 with $v_1 = tu(t)$ V.
51. Assume that the input to the ideal op-amp circuit of Problem 49 is $v_1 = A\cos(1000t)u(t)$. If $v_C(0^-) = 5$ V, what value of A causes the output v_o to be purely sinusoidal for $t > 0$?
52. Use Laplace transform circuit analysis to find the mesh currents for $t \geq 0$ in the circuit shown in Fig. 16.21.
53. Solve for the mesh currents i_1 and i_2 for $t \geq 0$ in the circuit shown in Fig. 16.22. The circuit is at rest for $t < 0$.

54. In the circuit shown, the inductance has an initial current $i(0^-) = 3$ mA.
    (a) Draw the Laplace-transformed circuit.
    (b) Show that $I(s) = H(s)0.003/s$, where $0.003/s$ represents the initial current and $H(s)$ is a transfer function.
    (c) Sketch the pole-zero plots of $H(s)$ for $\beta = 0, -1, -2$.
    (d) Determine and sketch $i(t) = \mathscr{L}^{-1}\{I(s)\}$ for $\beta = 0, -1, -2$. Discuss your results.

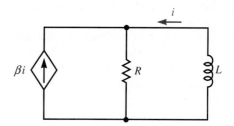

55. In the circuit shown, the capacitance has an initial voltage $v(0^-) = 2$ V.
    (a) Draw the Laplace-transformed circuit.

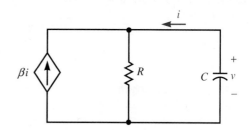

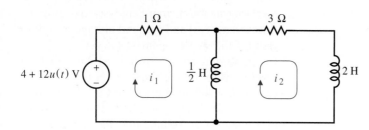

FIGURE 16.21

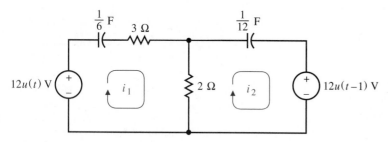

FIGURE 16.22

(b) Show that $I(s) = Y(s)2/s$, where $2/s$ represents the initial voltage and $Y(s)$ is an admittance.

(c) Sketch the pole-zero constellation of $Y(s)$ for $\beta = 0, -1, -2$.

(d) Determine and sketch $i(t) = \mathcal{L}^{-1}\{I(s)\}$ for $\beta = 0, -1, -2$. Discuss your results.

56. In the circuit shown, $v(0^-) = 2$ V. There is no current in the inductance at $t = 0^-$.

(a) Draw the Laplace-transformed circuit.

(b) Determine $I(s)$. Show that it can be written in the form $I(s) = Y(s)2/s$.

(c) Sketch the pole-zero plots of $Y(s)$ for $\beta = 0, -1, -2$.

(d) Determine and sketch $i(t) = \mathcal{L}^{-1}\{I(s)\}$ for $\beta = 0, -1, -2$. Discuss your results.

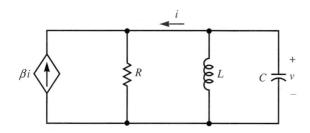

In Problems 57 through 61, use Laplace transform circuit analysis to solve the indicated problems from Chapter 9. In each case, draw and work directly from the Laplace-transformed circuit.

57. Problem 78
58. Problem 79
59. Problem 80
60. Problem 81
61. Problem 82
62. In the ideal op-amp circuit shown, $v_C(0^-) = 2$ V and $v_1 = 5u(t)$ V. Find and sketch v_o for $t \geq 0$.

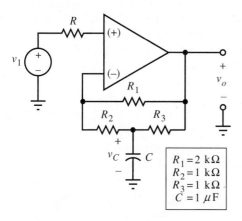

63. Repeat Problem 62 for $v_1(t) = tu(t)$.

64. Assume that the input to the ideal op-amp circuit of Problem 62 is $v_1 = 5\cos(1000t)u(t)$ V. Find $v_C(0^-)$ such that the output v_o is purely sinusoidal for $t > 0$.

Section 16.7

65. The input to a parallel RL circuit is $i_s(t) = 3e^{-t}u(t)$ A. Determine the resulting inductance current $i_L(t)$ for $t > 0$ if $R = 2\,\Omega$, $L = 1$ H, and $i_L(0^-) = -5$ A. Identify the forced response, stored-energy response, particular response, and complementary response terms in your solution.

66. Repeat Problem 65 for a parallel RC circuit where $R = 2\,\Omega$, $C = 1$ F, and $v_C(0^-) = 5$ V.

67. The expression

$$V_o(s) = \frac{1}{(s+1)(s+2)}$$

can be written as $V_o(s) = H(s)V_1(s)$ in four distinct ways, depending on which factors are assigned to $H(s)$ and $V(s)$. List each possibility and solve for $h(t)$, $v_1(t)$, and $v_o(t)$ in each case.

68. (a) An LLTI network containing no independent sources is at rest at $t = 0^-$. An impulse is applied at $t = 0$ and the response is $e^{-2t}\cos(3t)u(t)$. Use Laplace transforms to determine the complete response of the same system to the input $\cos(2t)u(t)$.

(b) What is the differential equation relating the input and output?

Comprehensive Problems

69. In Example 15.10, we needed an impulse, $\delta(f)$, to solve a differential equation by the Fourier transform method. Repeat this example with the use of the Laplace transform, and explain why the impulse is no longer needed.

70. Op amps are frequently configured to act as *buffers*. The purpose of a buffer is to prevent one circuit from drawing current from another. In part (a) of the figure, the ideal op amp is the buffer.

(a) Show that the voltage transfer function of the circuit of Fig. 16.23a is $H(s) = H_1(s)H_2(s)$, where $H_1(s) = 1/(sR_1C_1 + 1)$ and $H_2(s) = 1/(sR_2C_2 + 1)$.

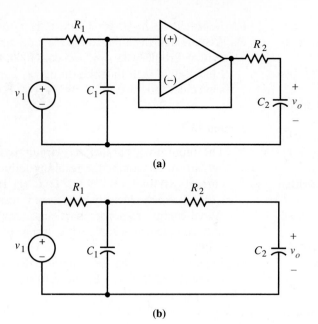

(a)

(b)

FIGURE 16.23

(b) Assume that $R_1 = R_2 = 1 \text{ k}\Omega$, and $C_1 = C_2 = 1 \ \mu\text{F}$, and $v_{c_1}(0^-) = v_{c_2}(0^-) = 0$. Evaluate the response of the circuit to $v_1(t) = \delta(t)$.

(c) Find the voltage transfer function of the circuit of Fig. 16.23b. How does your result compare to that of (a)?

(d) Evaluate the response of the circuit in Fig. 16.23b to $v_1(t) = \delta(t)$, with $R_1 = R_2 = 1 \text{ k}\Omega$, $C_1 = C_2 = 1 \ \mu\text{F}$, and $v_{c_1}(0^-) = v_{c_2}(0^-) = 0$.

71. The circuit shown is an important part of multivibrator circuits discussed in electronics courses. The switch S1 has been open for all

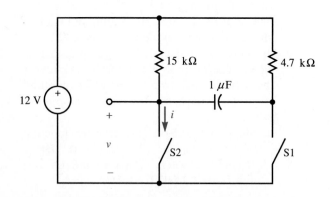

$t < 0$. Switch S2 is a controlled switch that remains closed if, and only if, the current through it remains positive: $i > 0$.

(a) Show why switch S2 is closed for $t < 0$.

(b) At $t = 0$, switch S1 closes. Determine and plot v for $t \geq 0$.

(c) Find the value of t when $v(t)$ just reaches 0 V.

72. The transfer function

$$H_R(s) = \frac{sR/L}{(s - s_1)(s - s_2)}$$

of a band-pass RLC filter was examined in detail in Chapter 13. Use partial fraction expansions and Tables 16.1 and 16.2 to show that the impulse response $h(t)$ is as given below. [Refer to Eqs. (13.49) through (13.56) for the definitions.)

(a) Overdamped circuit:

$$h(t) = \frac{R}{L \cosh \theta} \cosh (\beta t - \theta) e^{-\alpha t} u(t)$$

where $\theta = \text{arc tanh } \alpha/\beta$.

(b) Critically damped circuit:

$$h(t) = \frac{R}{L} (1 - \alpha t) e^{-\alpha t} u(t)$$

(c) Underdamped circuit:

$$h(t) = \frac{R}{L \cos \phi} \cos (\omega_d t + \phi)e^{-\alpha t}u(t)$$

where $\phi = \arctan \alpha/\omega_d$.

73. Repeat Problem 49 with the use of PSpice and PROBE.

74. Repeat Problem 62 with the use of PSpice and PROBE.

75. Repeat Problem 63 with the use of PSpice and PROBE.

76. Repeat Example 16.18 with the use of PSpice. Use PROBE to plot $i_1(t)$ and $i_2(t)$ for $0 \leq t \leq$ 4 s.

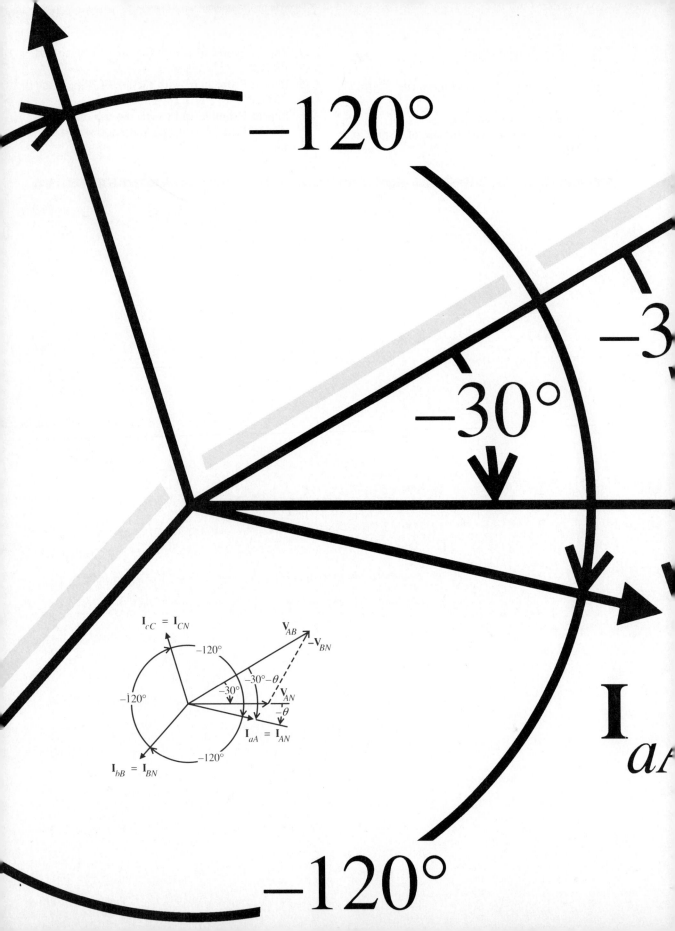

SELECTED TOPICS

Each chapter in Part Five describes a topic in circuit analysis that lies somewhat outside of the principal development of Chapters 1 through 16. These topics may be covered in any order. Chapter 17 requires Chapter 11 as a prerequisite. Chapters 18 and 19 require Chapter 12 as a prerequisite.

Equivalent Circuits for Three-Terminal Networks and Two-Port Networks

In Chapter 11 we showed that an arbitrary two-terminal network of sources and impedances can be represented by a Thévenin or Norton equivalent. In this chapter we will generalize the equivalent-circuit concept further to include arbitrary *three-terminal* networks composed of sources and impedances, and an important class of four-terminal networks called *two-port* networks.

The equivalent circuits developed here are used in electronics as linear models for physical devices. They are also used to simplify the analysis of circuits used in communication and power-transmission systems.

17.1 Equivalent Circuits for Three-Terminal Networks

Figure 17.1 depicts an arbitrary three-terminal LLTI ac network or device containing impedances, dependent sources, and independent sources. Phasor currents I_1 and I_2 enter at terminals 1 and 2. Phasor voltage drops V_1 and V_2 appear from terminals 1 and 2, respectively, to terminal 0.

FIGURE 17.1
A general three-terminal network

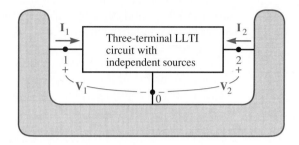

We can develop equivalent circuits for the three-terminal network of Fig. 17.1 by regarding various pairs of the variables I_1, I_2, V_1, and V_2 as independent variables (inputs) and the remaining pairs as dependent variables (outputs). The possible combinations are listed in Table 17.1 along with the equivalent-circuit parameters with which they are associated.* The derivations of the first three equivalent circuits listed are in Sections 17.2, 17.3, and 17.6. The hybrid-**g** equivalent circuit, listed last, is developed in Problem 20.

TABLE 17.1
Viewpoints leading to three-terminal equivalent circuits

VARIABLES TREATED AS INPUTS	VARIABLES TREATED AS OUTPUTS	RESULTING EQUIVALENT CIRCUIT PARAMETERS
1. I_1, I_2	V_1, V_2	Impedance
2. V_1, V_2	I_1, I_2	Admittance
3. I_1, V_2	I_2, V_1	Hybrid-**h**
4. I_2, V_1	I_1, V_2	Hybrid-**g**

17.2 Three-Terminal Impedance-Parameter Equivalent Circuits

To derive the impedance-parameter equivalent network of the three-terminal LLTI circuit of Fig. 17.1, we take the viewpoint that voltages V_1 and V_2 are dependent variables or "outputs." The independent variables are taken to be the currents I_1 and I_2 (these are the "inputs") and the independent sources within the rectangle. Since the network is linear, we know from the superposition principle that outputs V_1 and V_2 can be obtained by linearly superimposing the voltage

---

* No source exists that can simultaneously apply independent values of both driving-point voltage and driving-point current to a network. Therefore the pairs (I_1, V_1) and (I_2, V_2) are excluded from the first column of Table 17.1.

contributions caused by the independent sources acting one at a time. This means that we can write $\mathbf{V}_1$ and $\mathbf{V}_2$ as

$$\begin{bmatrix} \mathbf{V}_1 \\ \mathbf{V}_2 \end{bmatrix} = \begin{bmatrix} \mathbf{z}_{11} & \mathbf{z}_{12} \\ \mathbf{z}_{21} & \mathbf{z}_{22} \end{bmatrix} \begin{bmatrix} \mathbf{I}_1 \\ \mathbf{I}_2 \end{bmatrix} + \begin{bmatrix} \mathbf{V}_{1oc} \\ \mathbf{V}_{2oc} \end{bmatrix} \tag{17.1}$$

where the $\mathbf{z}_{ij}$ are proportionality constants and $\mathbf{V}_{1oc}$ and $\mathbf{V}_{2oc}$ represent the contributions from the independent sources inside the box. (The meaning of the subscripts will soon be clear.) We can derive formulas for the proportionality constants and $\mathbf{V}_{1oc}$ and $\mathbf{V}_{2oc}$ by recalling the following simple fact: The response caused by each independent source acting alone is found by setting the other independent sources to zero.

The components of $\mathbf{V}_1$ and $\mathbf{V}_2$ caused by the independent sources *within* the network can be found if we set $\mathbf{I}_1 = \mathbf{I}_2 = 0$ (open terminals 1 and 2), as illustrated in Fig. 17.2. The resulting values of $\mathbf{V}_1$ and $\mathbf{V}_2$ are called *open-circuit voltages*:

$$\mathbf{V}_{1oc} = \mathbf{V}_1 \Big|_{\substack{\mathbf{I}_1 = 0 \\ \mathbf{I}_2 = 0}} \tag{17.2}$$

$$\mathbf{V}_{2oc} = \mathbf{V}_2 \Big|_{\substack{\mathbf{I}_1 = 0 \\ \mathbf{I}_2 = 0}} \tag{17.3}$$

FIGURE 17.2
$\mathbf{V}_{1oc}$ and $\mathbf{V}_{2oc}$ are determined by setting $\mathbf{I}_1 = \mathbf{I}_2 = 0$

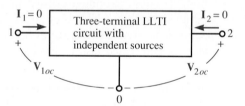

The components of $\mathbf{V}_1$ and $\mathbf{V}_2$ due to the current $\mathbf{I}_1$ are determined by setting $\mathbf{I}_2 = 0$ (open terminal 2) and setting all independent sources inside the network to zero, as illustrated in Fig. 17.3. It follows from an inspection of this figure that the contribution to $\mathbf{V}_1$ due to $\mathbf{I}_1$ is

$$\mathbf{V}_1 \Big|_{\substack{\mathbf{I}_2 = 0 \\ \text{IIS} = 0}} = \mathbf{z}_{11} \mathbf{I}_1 \tag{17.4}$$

where IIS = 0 denotes that the independent internal sources are set to zero and where $\mathbf{z}_{11}$ is the network's driving-point impedance looking into terminals 1 and

FIGURE 17.3
The open-circuit parameters $\mathbf{z}_{11}$ and $\mathbf{z}_{21}$ are determined by setting $\mathbf{I}_2$ and the internal independent sources to zero

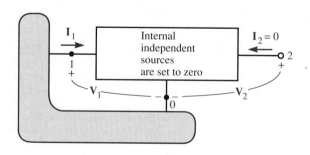

0 when terminal 2 is open-circuited. We can now solve Eq. (17.4) for $\mathbf{z}_{11}$. We can calculate the remaining proportionality constants in a similar way to obtain:

$$\mathbf{z}_{11} = \left.\frac{\mathbf{V}_1}{\mathbf{I}_1}\right|_{\substack{\mathbf{I}_2=0 \\ \text{IIS}=0}} \qquad (17.5)$$

$$\mathbf{z}_{21} = \left.\frac{\mathbf{V}_2}{\mathbf{I}_1}\right|_{\substack{\mathbf{I}_2=0 \\ \text{IIS}=0}} \qquad (17.6)$$

$$\mathbf{z}_{12} = \left.\frac{\mathbf{V}_1}{\mathbf{I}_2}\right|_{\substack{\mathbf{I}_1=0 \\ \text{IIS}=0}} \qquad (17.7)$$

$$\mathbf{z}_{22} = \left.\frac{\mathbf{V}_2}{\mathbf{I}_2}\right|_{\substack{\mathbf{I}_1=0 \\ \text{IIS}=0}} \qquad (17.8)$$

We will show how to apply Eqs. (17.5) through (17.8) in Example 17.1.

The parameters $\mathbf{z}_{11}$, $\mathbf{z}_{21}$, $\mathbf{z}_{12}$, and $\mathbf{z}_{22}$ are called the *open-circuit impedance or z parameters* of the three-terminal network. The parameters are impedances since each parameter equals the ratio of phasor voltage to phasor current. They are *open-circuit* parameters because each parameter is found by setting either $\mathbf{I}_1$ or $\mathbf{I}_2$ equal to zero (open circuit). With the obvious definitions, Eq. (17.1) can be written as a matrix equation:

$$\mathbf{V} = \mathbf{Z}\mathbf{I} + \mathbf{V}_{oc} \qquad (17.9)$$

As far as external measurements are concerned, the three-terminal network of Fig. 17.1 is completely characterized by its open-circuit voltages $\mathbf{V}_{1oc}$ and $\mathbf{V}_{2oc}$ and its impedance parameters $\mathbf{z}_{11}$, $\mathbf{z}_{12}$, $\mathbf{z}_{21}$, and $\mathbf{z}_{22}$.

Consider now the circuit of Fig. 17.4. We can see by inspection that the terminal equations of this circuit are also given by Eq. (17.1). Consequently, the network of Fig. 17.1 can be replaced by the network of Fig. 17.4, called the *three-terminal z-parameter equivalent*.

FIGURE 17.4
Three-terminal impedance-parameter equivalent network for a network containing independent sources

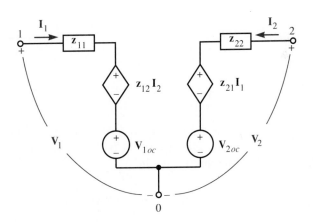

An important special case arises when the network of Fig. 17.1 contains no independent sources. For this special case, $\mathbf{V}_{oc} = \mathbf{0}$, and the network of Fig. 17.4 assumes the form shown in Fig. 17.5.

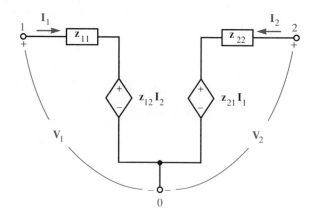

For reciprocal three-terminal networks, $\mathbf{z}_{12} = \mathbf{z}_{21}$. All three-terminal networks composed of only *independent* sources, resistances, capacitances, inductances, and coupled inductances are reciprocal.

As illustrated by the following example, there is more than one way to determine an impedance-parameter equivalent circuit.

EXAMPLE 17.1

Find $\mathbf{V}_{1oc}$, $\mathbf{V}_{2oc}$, and the z parameters of the network in Fig. 17.6 by (a) direct application of Eqs. (17.2), (17.3), and (17.5) through (17.8) and (b) mesh-current analysis.

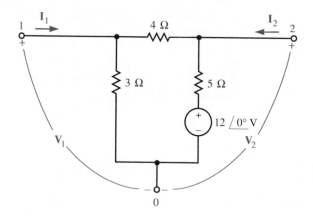

■ **SOLUTION**

(a) We will first find $\mathbf{V}_{1oc}$ and $\mathbf{V}_{2oc}$, which are the values of $\mathbf{V}_1$ and $\mathbf{V}_2$, respectively, when $\mathbf{I}_1 = \mathbf{I}_2 = 0$. We set $\mathbf{I}_1 = \mathbf{I}_2 = 0$ in Fig. 17.6 and use a voltage divider equation to obtain

$$\mathbf{V}_{1oc} = \frac{3}{3 + 4 + 5}\, 12\underline{/0^\circ} = 3\underline{/0^\circ}\ \mathrm{V}$$

and
$$\mathbf{V}_{2oc} = \frac{3 + 4}{3 + 4 + 5} \; 12\underline{/0^\circ} = 7\underline{/0^\circ} \; \text{V}$$

The impedance parameters follow from Eqs. (17.5) through (17.8). The values of $\mathbf{z}_{11}$ and $\mathbf{z}_{21}$ can be found from Eqs. (17.5) and (17.6) and inspection of Fig. 17.7:

$$\mathbf{z}_{11} = \frac{\mathbf{V}_1}{\mathbf{I}_1}\bigg|_{\substack{\mathbf{I}_2 = 0 \\ \text{IIS} = 0}} = 3 \; \| \; (4 + 5) = \frac{9}{4} = 2\frac{1}{4} \; \Omega$$

$$\mathbf{z}_{21} = \frac{\mathbf{V}_2}{\mathbf{I}_1}\bigg|_{\substack{\mathbf{I}_2 = 0 \\ \text{IIS} = 0}} = \frac{\mathbf{V}_2}{\mathbf{V}_1}\frac{\mathbf{V}_1}{\mathbf{I}_1}\bigg|_{\substack{\mathbf{I}_2 = 0 \\ \text{IIS} = 0}} = \left(\frac{5}{4 + 5}\right)\left(\frac{9}{4}\right) = 1\frac{1}{4} \; \Omega$$

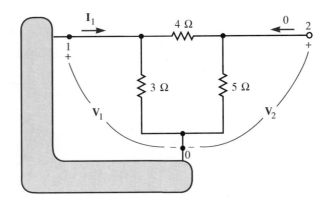

FIGURE 17.7
Circuit for finding $\mathbf{z}_{11}$ and $\mathbf{z}_{21}$

Similarly, from Fig. 17.8 and Eqs. (17.7) and (17.8),

$$\mathbf{z}_{22} = \frac{\mathbf{V}_2}{\mathbf{I}_2}\bigg|_{\substack{\mathbf{I}_1 = 0 \\ \text{IIS} = 0}} = (3 + 4) \; \| \; 5 = \frac{35}{12} = 2\frac{11}{12} \; \Omega$$

and

$$\mathbf{z}_{12} = \frac{\mathbf{V}_1}{\mathbf{I}_2}\bigg|_{\substack{\mathbf{I}_1 = 0 \\ \text{IIS} = 0}} = \frac{\mathbf{V}_1}{\mathbf{V}_2}\frac{\mathbf{V}_2}{\mathbf{I}_2}\bigg|_{\substack{\mathbf{I}_1 = 0 \\ \text{IIS} = 0}} = \left(\frac{3}{3 + 4}\right)\left(\frac{35}{12}\right) = 1\frac{1}{4} \; \Omega$$

Notice that $\mathbf{z}_{12} = \mathbf{z}_{21}$ as expected for a network not containing dependent sources.

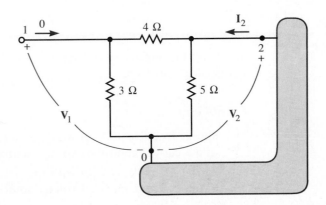

FIGURE 17.8
Circuit for finding $\mathbf{z}_{22}$ and $\mathbf{z}_{12}$

(b) We define mesh currents $\mathbf{I}_1$, $\mathbf{I}_2$, and $\mathbf{I}_3$ as shown in Fig. 17.9. The mesh-current equations are

$$3\mathbf{I}_1 + 0\mathbf{I}_2 - 3\mathbf{I}_3 = \mathbf{V}_1$$

$$0\mathbf{I}_1 + 5\mathbf{I}_2 + 5\mathbf{I}_3 = \mathbf{V}_2 - 12\underline{/0°}$$

$$-3\mathbf{I}_1 + 5\mathbf{I}_2 + 12\mathbf{I}_3 = -12\underline{/0°}$$

FIGURE 17.9
Mesh-current definitions

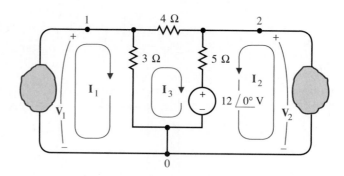

By elimination of $\mathbf{I}_3$ these become

$$\mathbf{V}_1 = \frac{9}{4}\mathbf{I}_1 + \frac{5}{4}\mathbf{I}_2 + 3\underline{/0°}$$

$$\mathbf{V}_2 = \frac{5}{4}\mathbf{I}_1 + \frac{35}{12}\mathbf{I}_2 + 7\underline{/0°}$$

which is, in matrix form

$$\begin{bmatrix} \mathbf{V}_1 \\ \mathbf{V}_2 \end{bmatrix} = \begin{bmatrix} \dfrac{9}{4} & \dfrac{5}{4} \\ \dfrac{5}{4} & \dfrac{35}{12} \end{bmatrix} \begin{bmatrix} \mathbf{I}_1 \\ \mathbf{I}_2 \end{bmatrix} + \begin{bmatrix} 3\underline{/0°} \\ 7\underline{/0°} \end{bmatrix}$$

Since this has the form of Eq. (17.9), we can read the impedance parameters and open-circuit voltages directly from it. We find as in (a) that $\mathbf{z}_{11} = \frac{9}{4}\,\Omega$, $\mathbf{z}_{12} = \mathbf{z}_{21} = \frac{5}{4}\,\Omega$, $\mathbf{z}_{22} = \frac{35}{12}\,\Omega$, $\mathbf{V}_{1oc} = 3\underline{/0°}$ V, and $\mathbf{V}_{2oc} = 7\underline{/0°}$ V. ∎

REMEMBER

We derive the impedance-parameter equivalent circuit by regarding $\mathbf{I}_1$ and $\mathbf{I}_2$ as independent variables.

EXERCISES

1. Repeat Example 17.1 (a) with the $+12\underline{/0°}$-V independent source replaced with a current-controlled voltage source with terminal equation $\mathbf{V} = 2\mathbf{I}_1$.

2. Repeat Example 17.1 (b) with the $+12\underline{/0°}$-V independent source replaced with a current-controlled voltage source with terminal equation $\mathbf{V} = 2\mathbf{I}_1$.

3. Determine the three-terminal z-parameter equivalent circuit of the network shown by direct application of Eqs. (17.2), (17.3), and (17.5) through (17.8).

4. Repeat Exercise 3 using mesh-current analysis.

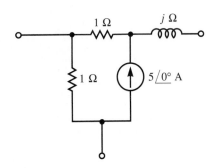

17.3 Three-Terminal Admittance-Parameter Equivalent Circuits

Another way to obtain terminal equations for the original three-terminal network of Fig. 17.1 is to regard the currents I_1 and I_2 as the outputs caused by two independent voltage sources V_1 and V_2 and the independent sources inside the network. Here we use the principle of superposition to write I_1 and I_2 as a linear combination of current components caused by the external voltage sources and the independent sources inside the network. That is, we write I_1 and I_2 as

$$\begin{bmatrix} I_1 \\ I_2 \end{bmatrix} = \begin{bmatrix} y_{11} & y_{12} \\ y_{21} & y_{22} \end{bmatrix} \begin{bmatrix} V_1 \\ V_2 \end{bmatrix} - \begin{bmatrix} I_{1sc} \\ I_{2sc} \end{bmatrix} \qquad (17.10)$$

where the y_{ij} are proportionality constants and where I_{1sc} and I_{2sc} represent *outward* current contributions due to the independent sources inside the box.

The components arising from the independent sources within the network are obtained by setting voltages $V_1 = V_2 = 0$ (connect terminals 1 and 2 to terminal 0) as illustrated in Fig. 17.10. The resulting currents flowing from terminals 1 and 2 to terminal 0 are called *short-circuit currents* I_{1sc} and I_{2sc}, respectively. Thus, comparing the sign convention of Fig. 17.1,

$$I_{1sc} = -I_1 \Big|_{\substack{V_1 = 0 \\ V_2 = 0}} \qquad (17.11)$$

$$I_{2sc} = -I_2 \Big|_{\substack{V_1 = 0 \\ V_2 = 0}} \qquad (17.12)$$

FIGURE 17.10
I_{1sc} and I_{2sc} are determined by setting $V_1 = V_2 = 0$

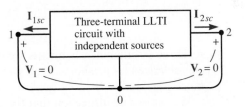

The components in terminal currents I_1 and I_2 caused by the voltage source V_1 can be determined if we set $V_2 = 0$ (connect terminal 2 to terminal 0) and set the

FIGURE 17.11
The short-circuit parameters.
y_{11} and y_{21} are determined
by setting V_2 and the
internal independent sources
to zero

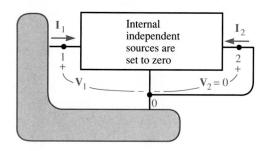

independent sources inside the network to zero (IIS = 0) as shown in Fig. 17.11. Then, by inspection of Fig. 17.11, the contribution to I_1 from V_1 is

$$\left. I_1 \right|_{\substack{V_2=0 \\ IIS=0}} = y_{11}V_1 \qquad (17.13)$$

where y_{11} is the network's driving-point admittance looking into terminals 1 and 0 when terminal 2 is shorted to terminal 0. Equation (17.13) can easily be solved for y_{11}. We can calculate the remaining admittance parameters in a similar way to obtain

$$y_{11} = \left. \frac{I_1}{V_1} \right|_{\substack{V_2=0 \\ IIS=0}} \qquad (17.14)$$

$$y_{21} = \left. \frac{I_2}{V_1} \right|_{\substack{V_2=0 \\ IIS=0}} \qquad (17.15)$$

$$y_{12} = \left. \frac{I_1}{V_2} \right|_{\substack{V_1=0 \\ IIS=0}} \qquad (17.16)$$

$$y_{22} = \left. \frac{I_2}{V_2} \right|_{\substack{V_1=0 \\ IIS=0}} \qquad (17.17)$$

The parameters y_{11}, y_{12}, y_{21}, and y_{22} are called the *short-circuit admittance parameters* of the three-terminal network of Fig. 17.1. For a reciprocal network, $y_{12} = y_{21}$. With the obvious definitions, Eq. (17.10) has the form

$$I = YV - I_{sc} \qquad (17.18)$$

As far as external measurements are concerned, the original network is completely characterized by its short-circuit currents I_{1sc} and I_{2sc} and its short-circuit admittance parameters.

We can see by inspection that the terminal equations of the circuit of Fig. 17.12 are also given by Eqs. (17.11) and (17.12). Consequently, the network of Fig. 17.1 can be replaced by the network of Fig. 17.12, called the *three-terminal y-parameter equivalent network*.

FIGURE 17.12
Three-terminal admittance-parameter equivalent for a network containing independent sources

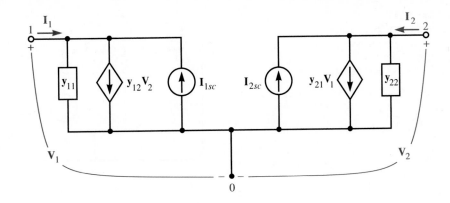

As with the three-terminal impedance parameter equivalent, an important special case arises if the network of Fig. 17.1 contains no independent sources. For this special case, $\mathbf{I}_{sc} = \mathbf{0}$ and the network of Fig. 17.12 assumes the form shown in Fig. 17.13.

FIGURE 17.13
Three-terminal admittance-parameter equivalent network for a network not containing independent sources

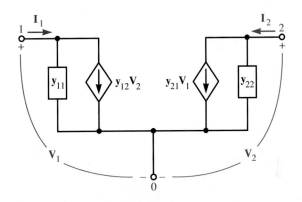

EXAMPLE 17.2

Find $\mathbf{I}_{1sc}$, $\mathbf{I}_{2sc}$, and the admittance parameters of the network of Example 17.1 by (a) direct application of Eqs. (17.11), (17.12), and (17.14) through (17.17) and (b) node-voltage analysis.

■ SOLUTION

(a) We will first find the short-circuit currents. It follows from Eqs. (17.11) and (17.12) and Fig. 17.14 that

$$\mathbf{I}_{1sc} = -\mathbf{I}_1 \Big|_{\substack{\mathbf{V}_1=0 \\ \mathbf{V}_2=0}} = 0 \text{ A}$$

and

$$\mathbf{I}_{2sc} = -\mathbf{I}_2 \Big|_{\substack{\mathbf{V}_1=0 \\ \mathbf{V}_2=0}} = \frac{12}{5} \underline{/0°} \text{ A}$$

FIGURE 17.14
Circuit for finding I_{1sc} and I_{2sc}

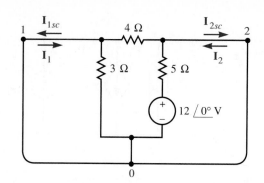

The values of $\mathbf{y}_{11}$ and $\mathbf{y}_{21}$ can be obtained by an inspection of Fig. 17.15 and Eqs. (17.14) and (17.15):

$$\mathbf{y}_{11} = \left.\frac{\mathbf{I}_1}{\mathbf{V}_1}\right|_{\substack{\mathbf{V}_2=0 \\ \text{IIS}=0}}$$

$$= \frac{1}{3} + \frac{1}{4} = \frac{7}{12}\ \text{S}$$

and

$$\mathbf{y}_{21} = \left.\frac{\mathbf{I}_2}{\mathbf{V}_1}\right|_{\substack{\mathbf{V}_2=0 \\ \text{IIS}=0}}$$

$$= -\frac{1}{4}\ \text{S}$$

FIGURE 17.15
Circuit for finding $\mathbf{y}_{11}$ and $\mathbf{y}_{21}$

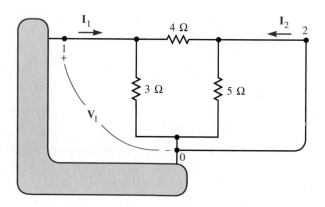

Similarly, we find from Fig. 17.16 and Eqs. (17.16) and (17.17):

$$\mathbf{y}_{12} = \left.\frac{\mathbf{I}_1}{\mathbf{V}_2}\right|_{\substack{\mathbf{V}_1=0 \\ \text{IIS}=0}} = -\frac{1}{4}\ \text{S}$$

and

$$\mathbf{y}_{22} = \left.\frac{\mathbf{I}_2}{\mathbf{V}_2}\right|_{\substack{\mathbf{V}_1=0 \\ \text{IIS}=0}} = \frac{1}{5} + \frac{1}{4} = \frac{9}{20}\ \text{S}$$

FIGURE 17.16
Circuit for finding y_{12} and y_{22}

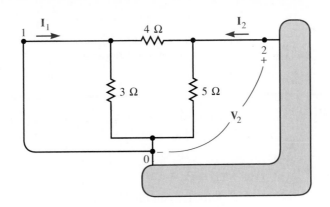

(b) The node-voltage definitions are shown in Fig. 17.17.

FIGURE 17.17
Node-voltage definitions

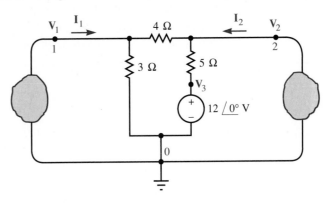

The node-voltage equations are

$$\left(\frac{1}{3}+\frac{1}{4}\right)\mathbf{V}_1 - \frac{1}{4}\mathbf{V}_2 + 0\mathbf{V}_3 = \mathbf{I}_1$$

$$-\frac{1}{4}\mathbf{V}_1 + \left(\frac{1}{4}+\frac{1}{5}\right)\mathbf{V}_2 - \frac{1}{5}\mathbf{V}_3 = \mathbf{I}_2$$

and
$$\mathbf{V}_3 = 12\,\underline{/0^\circ}$$

The elimination of $\mathbf{V}_3$ leads to

$$\frac{7}{12}\mathbf{V}_1 - \frac{1}{4}\mathbf{V}_2 = \mathbf{I}_1$$

and
$$-\frac{1}{4}\mathbf{V}_1 + \frac{9}{10}\mathbf{V}_2 - \frac{12}{5}\,\underline{/0^\circ} = \mathbf{I}_2$$

Therefore

$$\begin{bmatrix}\mathbf{I}_1 \\ \mathbf{I}_2\end{bmatrix} = \begin{bmatrix} \dfrac{7}{12} & -\dfrac{1}{4} \\[2mm] -\dfrac{1}{4} & \dfrac{9}{20} \end{bmatrix}\begin{bmatrix}\mathbf{V}_1 \\ \mathbf{V}_2\end{bmatrix} - \begin{bmatrix} 0 \\[2mm] \dfrac{12}{5}\,\underline{/0^\circ} \end{bmatrix}$$

which has the form of Eq. (17.18). Consequently, $y_{11} = \frac{7}{12}$ S, $y_{12} = y_{21} = -\frac{1}{4}$ S, $y_{22} = \frac{9}{20}$ S, $\mathbf{I}_{1sc} = 0$, and $\mathbf{I}_{2sc} = \frac{12}{5}\,\underline{/0^\circ}$ A as in (a). ∎

Up to now we have obtained the impedance- and admittance-parameter equivalents of a three-terminal circuit. In the next section, we will show how these two equivalent circuits are related.

REMEMBER

We derive the admittance-parameter equivalent circuit by regarding V_1 and V_2 as independent variables.

EXERCISES

5. Repeat Example 17.2(a) with the $+12\underline{/0°}$-V independent source replaced with a current-controlled voltage source with terminal equation $V = 2I_1$.

6. Repeat Example 17.2(b) with the $+12\underline{/0°}$-V independent source replaced with a current-controlled voltage source with terminal equation $V = 2I_1$.

7. Determine the three-terminal **y**-parameter equivalent circuit of the network shown by direct application of Eqs. (17.11), (17.12), and (17.14) through (17.17).

8. Repeat Exercise 7 with node-voltage analysis.

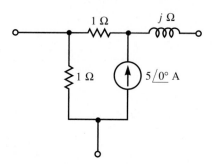

17.4 Relationship Between Impedance-Parameter and Admittance-Parameter Equivalent Circuits

Equations (17.9) and (17.18) and their corresponding circuits of Figs. 17.4 and 17.12 provide alternative equivalent circuits for a given three-terminal network. This equivalency implies that there is a relationship between **Z**, V_{oc}, **Y**, and I_{sc}. To obtain this relationship, solve Eq. (17.9), repeated below, for I_1 and I_2:

$$\begin{bmatrix} V_1 \\ V_2 \end{bmatrix} = \begin{bmatrix} z_{11} & z_{12} \\ z_{21} & z_{22} \end{bmatrix} \begin{bmatrix} I_1 \\ I_2 \end{bmatrix} + \begin{bmatrix} V_{1oc} \\ V_{2oc} \end{bmatrix} \qquad (17.19)$$

Cramer's rule leads quickly to the solution

$$\begin{bmatrix} I_1 \\ I_2 \end{bmatrix} = \begin{bmatrix} \dfrac{z_{22}}{|Z|} & -\dfrac{z_{12}}{|Z|} \\ -\dfrac{z_{21}}{|Z|} & \dfrac{z_{11}}{|Z|} \end{bmatrix} \begin{bmatrix} V_1 \\ V_2 \end{bmatrix} - \begin{bmatrix} \dfrac{z_{22}}{|Z|} & -\dfrac{z_{12}}{|Z|} \\ -\dfrac{z_{21}}{|Z|} & \dfrac{z_{11}}{|Z|} \end{bmatrix} \begin{bmatrix} V_{1oc} \\ V_{2oc} \end{bmatrix} \qquad (17.20)$$

where $|\mathbf{Z}|$ denotes the determinant* of $\mathbf{Z}$. We can write Eq. (17.20) as

$$\mathbf{I} = \mathbf{Z}^{-1}\mathbf{V} - \mathbf{Z}^{-1}\mathbf{V}_{oc} \qquad (17.21)$$

where it has been natural to define the "$\mathbf{Z}$ inverse" matrix

$$\mathbf{Z}^{-1} = \begin{bmatrix} \dfrac{\mathbf{z}_{22}}{|\mathbf{Z}|} & -\dfrac{\mathbf{z}_{12}}{|\mathbf{Z}|} \\[2ex] -\dfrac{\mathbf{z}_{21}}{|\mathbf{Z}|} & \dfrac{\mathbf{z}_{11}}{|\mathbf{Z}|} \end{bmatrix} \qquad (17.22)$$

Observe that $\mathbf{Z}^{-1}$ exists if and only if $|\mathbf{Z}| \neq 0$. We can see that Eq. (17.20) has the same form as Eq. (17.18), namely,

$$\mathbf{I} = \mathbf{YV} - \mathbf{I}_{sc} \qquad (17.23)$$

Therefore, by comparing Eqs. (17.21) and (17.23), we see that

$$\mathbf{Y} = \begin{bmatrix} \mathbf{y}_{11} & \mathbf{y}_{12} \\ \mathbf{y}_{21} & \mathbf{y}_{22} \end{bmatrix} = \begin{bmatrix} \dfrac{\mathbf{z}_{22}}{|\mathbf{Z}|} & -\dfrac{\mathbf{z}_{12}}{|\mathbf{Z}|} \\[2ex] -\dfrac{\mathbf{z}_{21}}{|\mathbf{Z}|} & \dfrac{\mathbf{z}_{11}}{|\mathbf{Z}|} \end{bmatrix} = \mathbf{Z}^{-1} \qquad (17.24)$$

and

$$\mathbf{I}_{sc} = \begin{bmatrix} \mathbf{I}_{1sc} \\ \mathbf{I}_{2sc} \end{bmatrix} = \begin{bmatrix} \dfrac{\mathbf{z}_{22}}{|\mathbf{Z}|} & -\dfrac{\mathbf{z}_{12}}{|\mathbf{Z}|} \\[2ex] -\dfrac{\mathbf{z}_{21}}{|\mathbf{Z}|} & \dfrac{\mathbf{z}_{11}}{|\mathbf{Z}|} \end{bmatrix} \begin{bmatrix} \mathbf{V}_{1oc} \\ \mathbf{V}_{2oc} \end{bmatrix} = \mathbf{Z}^{-1}\mathbf{V}_{oc} \qquad (17.25)$$

Equations (17.24) and (17.25) are formulas for computing the short-circuit admittance and current matrices $\mathbf{Y}$ and $\mathbf{I}_{sc}$ from the open-circuit impedance and voltage matrices $\mathbf{Z}$ and $\mathbf{V}_{oc}$

The reverse calculation can be obtained if we solve Eq. (17.10) for $\mathbf{V}_1$ and $\mathbf{V}_2$. Cramer's rule leads quickly to the result

$$\mathbf{V} = \mathbf{Y}^{-1}\mathbf{I} + \mathbf{Y}^{-1}\mathbf{I}_{sc} \qquad (17.26)$$

This has the form $\mathbf{V} = \mathbf{ZI} + \mathbf{V}_{oc}$ where

$$\mathbf{Z} = \mathbf{Y}^{-1} = \begin{bmatrix} \dfrac{\mathbf{y}_{22}}{|\mathbf{Y}|} & -\dfrac{\mathbf{y}_{12}}{|\mathbf{Y}|} \\[2ex] -\dfrac{\mathbf{y}_{21}}{|\mathbf{Y}|} & \dfrac{\mathbf{y}_{11}}{|\mathbf{Y}|} \end{bmatrix} \qquad (17.27)$$

and

$$\mathbf{V}_{oc} = \mathbf{Y}^{-1}\mathbf{I}_{sc} = \mathbf{ZI}_{sc} \qquad (17.28)$$

* Determinants are described in Appendix A.

The matrix $\mathbf{Y}^{-1}$, ($\mathbf{Y}$ inverse), exists if and only if $|\mathbf{Y}| \neq 0$, where $|\mathbf{Y}|$ is the determinant of $\mathbf{Y}$. Equations (17.24), (17.25), (17.27), and (17.28) provide the means for converting a z-parameter equivalent circuit to a y-parameter equivalent circuit and vice versa. For example, if we wish to convert a y-parameter equivalent to a z-parameter equivalent, we simply use Eq. (17.27) to compute the z parameters and Eq. (17.28) to compute $\mathbf{V}_{oc}$.

The conversion from one equivalent to another is not possible, however, if $|\mathbf{Z}|$ or $|\mathbf{Y}|$ equals zero. We shall examine this issue in the next section.

EXERCISE

9. We found that the z-parameters of the network in Example 17.1 are $z_{11} = \frac{9}{4}$, $z_{12} = z_{21} = \frac{5}{4}$, and $z_{22} = \frac{35}{12}$ ohms. The open-circuit voltages were found to be $\mathbf{V}_{1oc} = 3\underline{/0^\circ}$ V and $\mathbf{V}_{2oc} = 7\underline{/0^\circ}$ V. Use Eqs. (17.24) and (17.25) to find the y parameters and the short-circuit currents of the same network. Do your results agree with those determined in Example 17.2?

17.5 Existence of Three-Terminal z- and y-Parameter Equivalent Circuits

It is possible for *theoretical* circuits to have impedance-parameter equivalent circuits but not admittance-parameter equivalent circuits and vice versa. The reason for this can be understood in terms of the simple example shown in Fig. 17.18.

FIGURE 17.18
This circuit has a z-parameter equivalent circuit but no y-parameter equivalent circuit

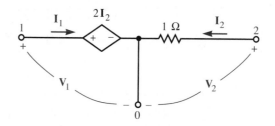

It follows by an inspection of the circuit shown that

$$\mathbf{V}_1 = 0\mathbf{I}_1 + 2\mathbf{I}_2 \tag{17.29}$$

and

$$\mathbf{V}_2 = 0\mathbf{I}_1 + 1\mathbf{I}_2 \tag{17.30}$$

It follows from the above equations that

$$\mathbf{Z} = \begin{bmatrix} 0 & 2 \\ 0 & 1 \end{bmatrix} \tag{17.31}$$

and

$$\mathbf{V}_{oc} = 0 \tag{17.32}$$

It follows from Eq. (17.31) and Appendix A that $|\mathbf{Z}| = 0$. Therefore $\mathbf{Y} = \mathbf{Z}^{-1}$ does not exist. As a consequence, the circuit cannot be characterized by a y-parameter equivalent circuit. The mathematical reason for this is that $\mathbf{V}_1$ and $\mathbf{V}_2$ cannot be chosen independently. The impossibility of a **y**-parameter equivalent for Fig. 17.18 can also be explained by examination of the circuit. Remember that circuit theory is based on KVL, KCL, and component definitions. The dependent source voltage in Fig. 17.18 is, by *definition*, $2I_2$ V. If terminal 1 is connected to terminal 0, as depicted in Fig. 17.19, KVL is violated, which is impossible in the context of circuit theory. (The sum of voltage drops around the red path in Fig. 17.19 is not zero for arbitrary values of V_2.) In addition, terminal 2 *cannot*, within the context of circuit theory, be shorted to terminal 0 if an arbitrary voltage V_1 is applied at terminal 1 as depicted in Fig. 17.20.

FIGURE 17.19
Shorting terminal 1 to terminal 0 results in an impossible circuit (KVL is violated)

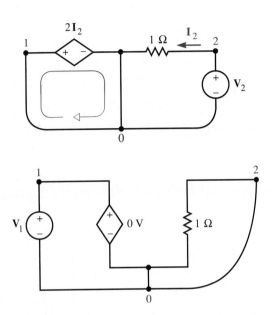

FIGURE 17.20
Shorting terminal 2 to terminal 0 results in an impossible circuit

It is similarly possible for a circuit to have a **y**-parameter equivalent but not have a **z**-parameter equivalent. An example is shown in Fig. 17.21. Note that I_1 and I_2 cannot be chosen independently.

FIGURE 17.21
This circuit has a **y**-parameter equivalent circuit but no **z**-parameter equivalent circuit

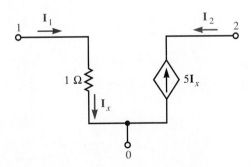

Finally, it is possible for a circuit to have neither **y**- nor **z**-parameter equivalent circuits. An example is shown in Fig. 17.22, where neither voltage V_1 nor current I_2 can be independent variables.

FIGURE 17.22
This circuit has neither a
z- nor a **y**-parameter
equivalent circuit

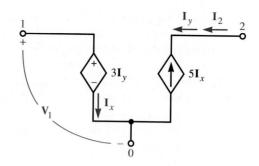

EXERCISES

In Exercises 10 through 13, give an example of the circuits specified.

10. A circuit that has both **z**- and **y**-parameter equivalent circuits.

11. A circuit that has a **z**-parameter equivalent but not a **y**-parameter equivalent.

12. A circuit that has a **y**-parameter equivalent but not a **z**-parameter equivalent.

13. A circuit that has neither a **z**- nor a **y**-parameter equivalent.

17.6   Three-Terminal Hybrid-h-Parameter Equivalent Circuits

A third equivalent circuit for the network in Fig. 17.1 can be obtained if we assume that I_1 and V_2 are independent variables and V_1 and I_2 are dependent variables. Superposition indicates that V_1 and I_2 can then be expressed as linear combinations of I_1, V_2 and the independent sources in the network. Thus

$$\begin{bmatrix} V_1 \\ I_2 \end{bmatrix} = \begin{bmatrix} h_{11} & h_{12} \\ h_{21} & h_{22} \end{bmatrix} \begin{bmatrix} I_1 \\ V_2 \end{bmatrix} + \begin{bmatrix} V_{1occs} \\ -I_{2occs} \end{bmatrix} \qquad (17.33)$$

The quantities h_{11}, h_{12}, h_{21}, and h_{22} are the *hybrid-h-parameters* of the network of Fig. 17.1. The significance of the **h**-parameters can be determined directly from Eq. (17.33) if we observe that

$$h_{11} = \left. \frac{V_1}{I_1} \right|_{\substack{V_2 = 0 \\ \text{IIS} = 0}} \qquad (17.34)$$

$$h_{21} = \left. \frac{I_2}{I_1} \right|_{\substack{V_2 = 0 \\ \text{IIS} = 0}} \qquad (17.35)$$

$$h_{12} = \left. \frac{V_1}{V_2} \right|_{\substack{I_1 = 0 \\ \text{IIS} = 0}} \qquad (17.36)$$

$$h_{22} = \left. \frac{I_2}{V_2} \right|_{\substack{I_1 = 0 \\ \text{IIS} = 0}} \qquad (17.37)$$

As before, IIS = 0 denotes that the independent internal sources are set to zero. Observe that the units of h_{11} and h_{22} are ohms and siemens, respectively, whereas

h_{12} and h_{21} are dimensionless. The fact that the four parameters do not all have the same units is the reason these parameters are referred to as *hybrid parameters*. It also follows from Eq. (17.33) that V_{1ocsc} and I_{2ocsc} are given by

$$V_{1ocsc} = V_1 \Big|_{\substack{I_1 = 0 \\ V_2 = 0}} \qquad (17.38)$$

$$I_{2ocsc} = -I_2 \Big|_{\substack{I_1 = 0 \\ V_2 = 0}} \qquad (17.39)$$

The equivalent circuit associated with Eq. (17.33) is shown in Fig. 17.23. The special case in which there are no independent sources inside the network is shown in Fig. 17.24. The networks shown in Figs. 17.23 and 17.24 are called *three-terminal hybrid-**h**-parameter equivalent networks*.

FIGURE 17.23
Three-terminal hybrid-h-parameter equivalent network for a network containing independent sources

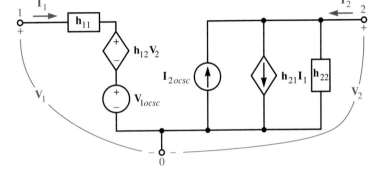

FIGURE 17.24
Three-terminal hybrid-h-parameter equivalent network for a network not containing independent sources

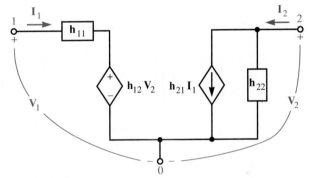

The elements in the **h**-parameter equivalent circuit can be related to those of the **z**- and **y**-parameter equivalent circuits by straightforward manipulations of the corresponding equations. A summary of the results is contained in Tables 17.2 through 17.5 of Section 17.8.

As with **z** and **y** parameters, there are several ways to determine the **h**-parameter equivalent circuit of a network. The following example illustrates one approach.

EXAMPLE 17.3

Find V_{1ocsc}, I_{2ocsc}, and the hybrid-**h** parameters for the circuit of Example 17.1.

We find $\mathbf{V}_{1ocsc}$ and $\mathbf{I}_{2ocsc}$ by setting $\mathbf{I}_1 = 0$ and $\mathbf{V}_2 = 0$ as shown in Fig. 17.25. It follows from Eqs. (17.38) and (17.39) and an inspection of Fig 17.25 that

$$\mathbf{V}_{1ocsc} = \mathbf{V}_1 \Big|_{\substack{\mathbf{I}_1 = 0 \\ \mathbf{V}_2 = 0}} = 0 \text{ V}$$

and

$$\mathbf{I}_{2ocsc} = -\mathbf{I}_2 \Big|_{\substack{\mathbf{I}_1 = 0 \\ \mathbf{V}_2 = 0}} = \frac{12}{5} \underline{/0^\circ} \text{ A}$$

FIGURE 17.25
Circuit for finding $\mathbf{V}_{1ocsc}$ and $\mathbf{I}_{2ocsc}$

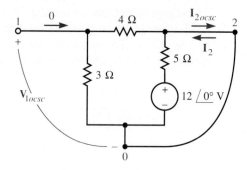

The parameters $\mathbf{h}_{11}$ and $\mathbf{h}_{21}$ can be obtained from Eqs. (17.34) and (17.35) and an inspection of Fig. 17.26:

FIGURE 17.26
Circuit for finding $\mathbf{h}_{11}$ and $\mathbf{h}_{21}$

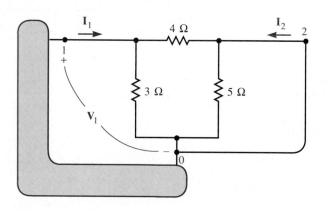

$$\mathbf{h}_{11} = \frac{\mathbf{V}_1}{\mathbf{I}_1} \Big|_{\substack{\mathbf{V}_2 = 0 \\ \text{IIS} = 0}} = 3 \parallel 4 = \frac{12}{7} \ \Omega$$

$$\mathbf{h}_{21} = \frac{\mathbf{I}_2}{\mathbf{I}_1} \Big|_{\substack{\mathbf{V}_2 = 0 \\ \text{IIS} = 0}} = -\frac{3}{3+4} = -\frac{3}{7}$$

Similarly, from Eqs. (17.36) and (17.37) and Fig. 17.27 we obtain

$$\mathbf{h}_{12} = \frac{\mathbf{V}_1}{\mathbf{V}_2} \Big|_{\substack{\mathbf{I}_1 = 0 \\ \text{IIS} = 0}} = \frac{3}{3+4} = \frac{3}{7}$$

$$\mathbf{h}_{22} = \frac{\mathbf{I}_2}{\mathbf{V}_2} \Big|_{\substack{\mathbf{I}_1 = 0 \\ \text{IIS} = 0}} = \frac{1}{5} + \frac{1}{7} = \frac{12}{35} \text{ S} \ ■$$

FIGURE 17.27
Circuit for finding $\mathbf{h}_{12}$ and
$\mathbf{h}_{22}$

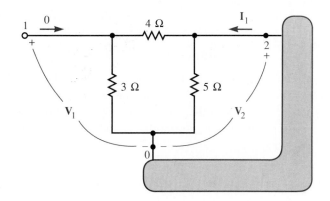

Hybrid-**h** equivalent circuits are frequently used to model bipolar junction transistors. In the next example we will see how a simple but remarkably useful hybrid-**h** transistor model is used to analyze a transistor amplifier.

EXAMPLE 17.4

Figure 17.28 depicts a single-stage transistor amplifier where B is a dc voltage used to supply power to the transistor. The capacitors are employed simply to prevent the dc voltages caused by B from reaching the source $v_s(t)$ and the load resistance R_L. Their values are large enough so that impedances, $1/j\omega C_1$ and $1/j\omega C_2$, are negligible. Use the simplified hybrid-**h** transistor model of Fig. 17.29,

FIGURE 17.28
Transistor amplifier

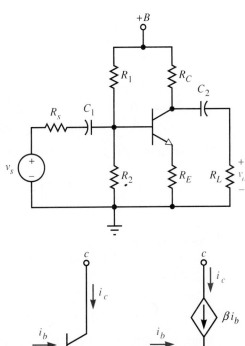

FIGURE 17.29
(a) Transistor and
(b) simplified hybrid-**h** model.
A transistor is composed of a
base b, an emitter e, and a
collector c

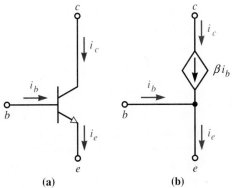

(a) (b)

in which $\mathbf{h}_{21} = \beta$ is the only nonzero parameter, to determine the voltage transfer function $\mathbf{V}_o/\mathbf{V}_s$.

■ SOLUTION

The ac circuit model is shown in Fig. 17.30. Since we are interested only in the portion of the output due to the source v_s, we have set voltage B to zero and have replaced the capacitances by short circuits. This places R_1 and R_2 in parallel and R_C and R_L in parallel as indicated.

FIGURE 17.30
Circuit model

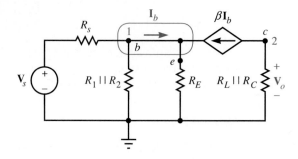

The application of KCL at the surface enclosing node 1 yields

$$\frac{\mathbf{V}_1 - \mathbf{V}_s}{R_s} + \frac{\mathbf{V}_1}{R_1 \| R_2} + \frac{\mathbf{V}_1}{R_E} = \beta\mathbf{I}_b$$

where

$$\mathbf{V}_1 = (\beta + 1)\mathbf{I}_b R_E$$

The KCL equation at node 2 is

$$\frac{\mathbf{V}_o}{R_L \| R_C} + \beta\mathbf{I}_b = 0$$

By combining the previous three equations, we obtain

$$\frac{\mathbf{V}_o}{\mathbf{V}_s} = -\left\{ \left(\frac{\beta}{\beta + 1}\right) \left[\frac{1}{1 + \dfrac{R_s}{R_1 \| R_2} + \dfrac{R_s}{(\beta + 1)R_E}} \right] \right\} \frac{R_L \| R_C}{R_E}$$

This expression simplifies considerably for typical circuit values for which $\beta \gg 1$, $R_s \ll R_1 \| R_2$, and $R_s \ll (\beta + 1)R_E$. For these values, the factor inside the braces is approximately 1. Thus

$$\frac{\mathbf{V}_o}{\mathbf{V}_s} \simeq -\frac{R_L \| R_C}{R_E}$$

which is a very convenient design formula. ■

We have now derived three of the four possible three-terminal equivalent circuits listed in Table 17.1. At this point you should be well prepared to derive the remaining (hybrid-**g**) equivalent. (See Tables 17.2 through 17.5 and Problem 20.) We turn to two-port networks next.

We derive the hybrid-**h**-parameter equivalent circuit by regarding $\mathbf{I}_1$ and $\mathbf{V}_2$ as independent variables.

EXERCISES

14. Show that the bipolar transistor model in Fig. 17.29 is an example of the three-terminal hybrid-**h**-parameter equivalent network of Fig. 17.23.

15. Determine the three-terminal hybrid-**h**-parameter equivalent circuit of the network shown by direct application of Eqs. (17.34) through (17.39).

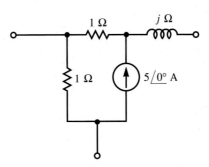

17.7 Two-Port Networks

The definition of a two-port network can be understood from Fig. 17.31, which depicts a four-terminal network with terminals a and a' and b and b' grouped into pairs. Notice that the four-terminal network shown has the special property that the *net current entering each terminal pair is zero*. In other words, the currents exiting at terminals a' and b', $\mathbf{I}_1$ and $\mathbf{I}_2$, equal the corresponding currents entering at terminals a and b. This is called the *two-port* property. It is the key property that distinguishes a two-port network from an arbitrary four-terminal network.

FIGURE 17.31
Two-port network

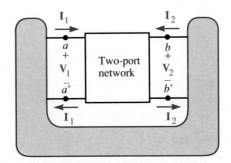

The special constraint on the currents entering a two-port network is *sometimes* the result of the circuit inside the rectangle of Fig. 17.31. The simplest example of this is provided by the coupled inductances of Fig. 17.32. Frequently, however, the constraints on the currents are *not* a result of the circuit itself, but of the way it is connected to other circuits. This is illustrated in Fig. 17.33.

FIGURE 17.32
Coupled inductances as
two-port network

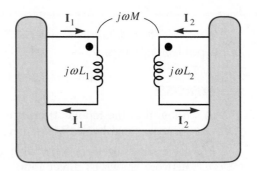

FIGURE 17.33
The arbitrary circuit enclosed
in the rectangle is a two port

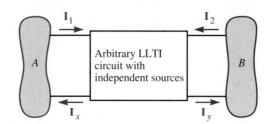

By KCL, the net current entering each circuit, A and B, is zero. Therefore $I_x = I_1$, $I_y = I_2$, and the four-terminal network in the middle is a two-port. A more general configuration is shown in Fig. 17.34. Observe that each of the networks I, II, III, and IV is properly called a two-port network because the two-port property must be satisfied for each network. (Show why this is so.)

FIGURE 17.34
Networks I, II, III, and IV
are two ports

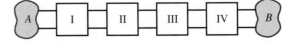

Notice that in the defining circuit of Fig. 17.31, *no statement is made* about voltage $V_{a'b'}$ or V_{ab}. The only variables of interest in defining the two-port network are the currents I_1 and I_2 and the voltages $V_{aa'} = V_1$ and $V_{bb'} = V_2$.

Three-terminal networks are frequently connected as four-terminal networks, as illustrated in Fig. 17.35. As a result, two-port networks and three-terminal networks are often confused. The distinction is as follows:

When a three-terminal network is connected as shown in Fig. 17.35, then, of course, the equations and equivalent circuits developed in Sections 17.2, 17.3, and 17.6 for the three-terminal network (inner rectangle) still apply. Therefore the relationships between the variables I_1, I_2, V_1, and V_2 of the four-terminal network (outer rectangle) are the same as those for the three-terminal network. It does *not* follow, however, that $I_x = I_1$ and $I_y = I_2$. The only constraint is that imposed by KCL, which states that $I_1 + I_2 = I_x + I_y$. The values of I_x and I_y are determined both by the sum $I_1 + I_2$ and by the external circuit connected at terminals a, a', b, and b'. The external circuit determines whether or not the four-terminal network is a two-port.

FIGURE 17.35
A three-terminal network
connected as a four-terminal
network

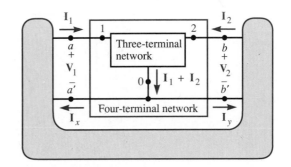

REMEMBER   The net current entering each terminal pair of a two-port network equals zero.

EXERCISES

16. Prove that network III in Fig. 17.34 is a two-port network.

17. Network A shown below is an arbitrary four-terminal network. Is there a value of β that forces network A to satisfy the two-port property? If so, state the value(s) of β.

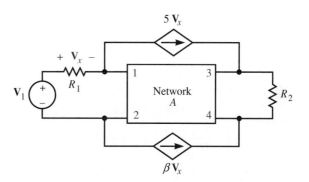

17.8 Equivalent Circuits for Two-Port Networks

Two-port networks are characterized by the equations relating $\mathbf{I}_1$, $\mathbf{I}_2$, $\mathbf{V}_1$, and $\mathbf{V}_2$. The derivations of the terminal equations of a two-port network are identical to those of three-terminal networks. The results are given in Tables 17.2 through 17.5. We must be cautious when defining equivalent circuits for two-port networks because the equations given in Tables 17.2 through 17.4 *make no statement* about the values of $\mathbf{V}_{ab}$ and $\mathbf{V}_{a'b'}$ in the two-port network of Fig. 17.31. The circuits of Fig. 17.36a–h are frequently used as equivalent circuits for two-port networks even though they may not specify the correct values for $\mathbf{V}_{ab}$ and $\mathbf{V}_{a'b'}$. Example 17.5 illustrates this point.

TABLE 17.2
Summary of terminal
equations for three-
terminal and two-port
networks

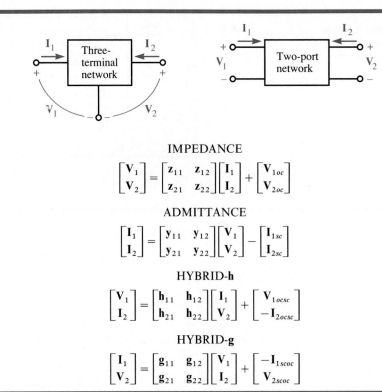

$$\text{IMPEDANCE}$$

$$\begin{bmatrix} V_1 \\ V_2 \end{bmatrix} = \begin{bmatrix} z_{11} & z_{12} \\ z_{21} & z_{22} \end{bmatrix} \begin{bmatrix} I_1 \\ I_2 \end{bmatrix} + \begin{bmatrix} V_{1oc} \\ V_{2oc} \end{bmatrix}$$

$$\text{ADMITTANCE}$$

$$\begin{bmatrix} I_1 \\ I_2 \end{bmatrix} = \begin{bmatrix} y_{11} & y_{12} \\ y_{21} & y_{22} \end{bmatrix} \begin{bmatrix} V_1 \\ V_2 \end{bmatrix} - \begin{bmatrix} I_{1sc} \\ I_{2sc} \end{bmatrix}$$

$$\text{HYBRID-}h$$

$$\begin{bmatrix} V_1 \\ I_2 \end{bmatrix} = \begin{bmatrix} h_{11} & h_{12} \\ h_{21} & h_{22} \end{bmatrix} \begin{bmatrix} I_1 \\ V_2 \end{bmatrix} + \begin{bmatrix} V_{1ocsc} \\ -I_{2ocsc} \end{bmatrix}$$

$$\text{HYBRID-}g$$

$$\begin{bmatrix} I_1 \\ V_2 \end{bmatrix} = \begin{bmatrix} g_{11} & g_{12} \\ g_{21} & g_{22} \end{bmatrix} \begin{bmatrix} V_1 \\ I_2 \end{bmatrix} + \begin{bmatrix} -I_{1scoc} \\ V_{2scoc} \end{bmatrix}$$

TABLE 17.3
Summary of formulas for
three-terminal and two-port
networks

IMPEDANCE

$$z_{11} = \left. \frac{V_1}{I_1} \right|_{\substack{I_2 = 0 \\ IIS = 0}} \qquad z_{12} = \left. \frac{V_1}{I_2} \right|_{\substack{I_1 = 0 \\ IIS = 0}}$$

$$z_{21} = \left. \frac{V_2}{I_1} \right|_{\substack{I_2 = 0 \\ IIS = 0}} \qquad z_{22} = \left. \frac{V_2}{I_2} \right|_{\substack{I_1 = 0 \\ IIS = 0}}$$

$$V_{1oc} = \left. V_1 \right|_{\substack{I_1 = 0 \\ I_2 = 0}} \qquad V_{2oc} = \left. V_2 \right|_{\substack{I_1 = 0 \\ I_2 = 0}}$$

ADMITTANCE

$$y_{11} = \left. \frac{I_1}{V_1} \right|_{\substack{V_2 = 0 \\ IIS = 0}} \qquad y_{12} = \left. \frac{I_1}{V_2} \right|_{\substack{V_1 = 0 \\ IIS = 0}}$$

$$y_{21} = \left. \frac{I_2}{V_1} \right|_{\substack{V_2 = 0 \\ IIS = 0}} \qquad y_{22} = \left. \frac{I_2}{V_2} \right|_{\substack{V_1 = 0 \\ IIS = 0}}$$

$$I_{1sc} = \left. -I_1 \right|_{\substack{V_1 = 0 \\ V_2 = 0}} \qquad I_{2sc} = \left. -I_2 \right|_{\substack{V_1 = 0 \\ V_2 = 0}}$$

HYBRID-h

$$h_{11} = \left. \frac{V_1}{I_1} \right|_{\substack{V_2 = 0 \\ IIS = 0}} \qquad h_{12} = \left. \frac{V_1}{V_2} \right|_{\substack{I_1 = 0 \\ IIS = 0}}$$

$$h_{21} = \left. \frac{I_2}{I_1} \right|_{\substack{V_2 = 0 \\ IIS = 0}} \qquad h_{22} = \left. \frac{I_2}{V_2} \right|_{\substack{I_1 = 0 \\ IIS = 0}}$$

$$V_{1ocsc} = \left. V_1 \right|_{\substack{I_1 = 0 \\ V_2 = 0}} \qquad I_{2ocsc} = \left. -I_2 \right|_{\substack{I_1 = 0 \\ V_2 = 0}}$$

HYBRID-g

$$g_{11} = \left. \frac{I_1}{V_1} \right|_{\substack{I_2 = 0 \\ IIS = 0}} \qquad g_{12} = \left. \frac{I_1}{I_2} \right|_{\substack{V_1 = 0 \\ IIS = 0}}$$

$$g_{21} = \left. \frac{V_2}{V_1} \right|_{\substack{I_2 = 0 \\ IIS = 0}} \qquad g_{22} = \left. \frac{V_2}{I_2} \right|_{\substack{V_1 = 0 \\ IIS = 0}}$$

$$I_{1scoc} = \left. -I_1 \right|_{\substack{V_1 = 0 \\ I_2 = 0}} \qquad V_{2scoc} = \left. V_2 \right|_{\substack{V_1 = 0 \\ I_2 = 0}}$$

TABLE 17.4
Summary of relationships among parameters for three-terminal and two-port networks

	Z	Y	H	G
Z	$\begin{bmatrix} z_{11} & z_{12} \\ z_{21} & z_{22} \end{bmatrix}$	$\begin{bmatrix} \dfrac{y_{22}}{|Y|} & -\dfrac{y_{12}}{|Y|} \\ -\dfrac{y_{21}}{|Y|} & \dfrac{y_{11}}{|Y|} \end{bmatrix}$	$\begin{bmatrix} \dfrac{|H|}{h_{22}} & \dfrac{h_{12}}{h_{22}} \\ -\dfrac{h_{21}}{h_{22}} & \dfrac{1}{h_{22}} \end{bmatrix}$	$\begin{bmatrix} \dfrac{1}{g_{11}} & -\dfrac{g_{12}}{g_{11}} \\ \dfrac{g_{21}}{g_{11}} & \dfrac{|G|}{g_{11}} \end{bmatrix}$
Y	$\begin{bmatrix} \dfrac{z_{22}}{|Z|} & -\dfrac{z_{12}}{|Z|} \\ -\dfrac{z_{21}}{|Z|} & \dfrac{z_{11}}{|Z|} \end{bmatrix}$	$\begin{bmatrix} y_{11} & y_{12} \\ y_{21} & y_{22} \end{bmatrix}$	$\begin{bmatrix} \dfrac{1}{h_{11}} & -\dfrac{h_{12}}{h_{11}} \\ \dfrac{h_{21}}{h_{11}} & \dfrac{|H|}{h_{11}} \end{bmatrix}$	$\begin{bmatrix} \dfrac{|G|}{g_{22}} & \dfrac{g_{12}}{g_{22}} \\ -\dfrac{g_{21}}{g_{22}} & \dfrac{1}{g_{22}} \end{bmatrix}$
H	$\begin{bmatrix} \dfrac{|Z|}{z_{22}} & \dfrac{z_{12}}{z_{22}} \\ -\dfrac{z_{21}}{z_{22}} & \dfrac{1}{z_{22}} \end{bmatrix}$	$\begin{bmatrix} \dfrac{1}{y_{11}} & -\dfrac{y_{12}}{y_{11}} \\ \dfrac{y_{21}}{y_{11}} & \dfrac{|Y|}{y_{11}} \end{bmatrix}$	$\begin{bmatrix} h_{11} & h_{12} \\ h_{21} & h_{22} \end{bmatrix}$	$\begin{bmatrix} \dfrac{g_{22}}{|G|} & -\dfrac{g_{12}}{|G|} \\ -\dfrac{g_{21}}{|G|} & \dfrac{g_{11}}{|G|} \end{bmatrix}$
G	$\begin{bmatrix} \dfrac{1}{z_{11}} & -\dfrac{z_{12}}{z_{11}} \\ \dfrac{z_{21}}{z_{11}} & \dfrac{|Z|}{z_{11}} \end{bmatrix}$	$\begin{bmatrix} \dfrac{|Y|}{y_{22}} & \dfrac{y_{12}}{y_{22}} \\ -\dfrac{y_{21}}{y_{22}} & \dfrac{1}{y_{22}} \end{bmatrix}$	$\begin{bmatrix} \dfrac{h_{22}}{|H|} & -\dfrac{h_{12}}{|H|} \\ -\dfrac{h_{21}}{|H|} & \dfrac{h_{11}}{|H|} \end{bmatrix}$	$\begin{bmatrix} g_{11} & g_{12} \\ g_{21} & g_{22} \end{bmatrix}$

TABLE 17.5
Summary of relationships among open-circuit voltages and short-circuit currents for three-terminal and two-port networks

$$\begin{bmatrix} V_{1oc} \\ V_{2oc} \end{bmatrix} = \begin{bmatrix} z_{11} & z_{12} \\ z_{21} & z_{22} \end{bmatrix}\begin{bmatrix} I_{1sc} \\ I_{2sc} \end{bmatrix} = \begin{bmatrix} 1 & -\dfrac{h_{12}}{h_{22}} \\ 0 & -\dfrac{1}{h_{22}} \end{bmatrix}\begin{bmatrix} V_{1ocsc} \\ -I_{2ocsc} \end{bmatrix} = \begin{bmatrix} -\dfrac{1}{g_{11}} & 0 \\ -\dfrac{g_{21}}{g_{11}} & 1 \end{bmatrix}\begin{bmatrix} -I_{1scoc} \\ V_{2scoc} \end{bmatrix}$$

$$\begin{bmatrix} I_{1sc} \\ I_{2sc} \end{bmatrix} = \begin{bmatrix} y_{11} & y_{12} \\ y_{21} & y_{22} \end{bmatrix}\begin{bmatrix} V_{1oc} \\ V_{2oc} \end{bmatrix} = \begin{bmatrix} \dfrac{1}{h_{11}} & 0 \\ \dfrac{h_{21}}{h_{11}} & -1 \end{bmatrix}\begin{bmatrix} V_{1ocsc} \\ -I_{2ocsc} \end{bmatrix} = \begin{bmatrix} -1 & \dfrac{g_{12}}{g_{22}} \\ 0 & \dfrac{1}{g_{22}} \end{bmatrix}\begin{bmatrix} -I_{1scoc} \\ V_{2scoc} \end{bmatrix}$$

$$\begin{bmatrix} V_{1ocsc} \\ -I_{2ocsc} \end{bmatrix} = \begin{bmatrix} 1 & -\dfrac{z_{12}}{z_{22}} \\ 0 & -\dfrac{1}{z_{22}} \end{bmatrix}\begin{bmatrix} V_{1oc} \\ V_{2oc} \end{bmatrix} = \begin{bmatrix} y_{11} & 0 \\ \dfrac{y_{21}}{y_{11}} & -1 \end{bmatrix}\begin{bmatrix} I_{1sc} \\ I_{2sc} \end{bmatrix} = -\begin{bmatrix} g_{11} & g_{12} \\ g_{21} & g_{22} \end{bmatrix}\begin{bmatrix} -I_{1scoc} \\ V_{2scoc} \end{bmatrix}$$

$$\begin{bmatrix} -I_{1scoc} \\ V_{2scoc} \end{bmatrix} = \begin{bmatrix} -\dfrac{1}{z_{11}} & 0 \\ -\dfrac{z_{21}}{z_{11}} & 1 \end{bmatrix}\begin{bmatrix} V_{1oc} \\ V_{2oc} \end{bmatrix} = \begin{bmatrix} -1 & \dfrac{y_{12}}{y_{22}} \\ 0 & \dfrac{1}{y_{22}} \end{bmatrix}\begin{bmatrix} I_{1sc} \\ I_{2sc} \end{bmatrix} = \begin{bmatrix} h_{11} & h_{12} \\ h_{21} & h_{22} \end{bmatrix}\begin{bmatrix} V_{1ocsc} \\ -I_{2ocsc} \end{bmatrix}$$

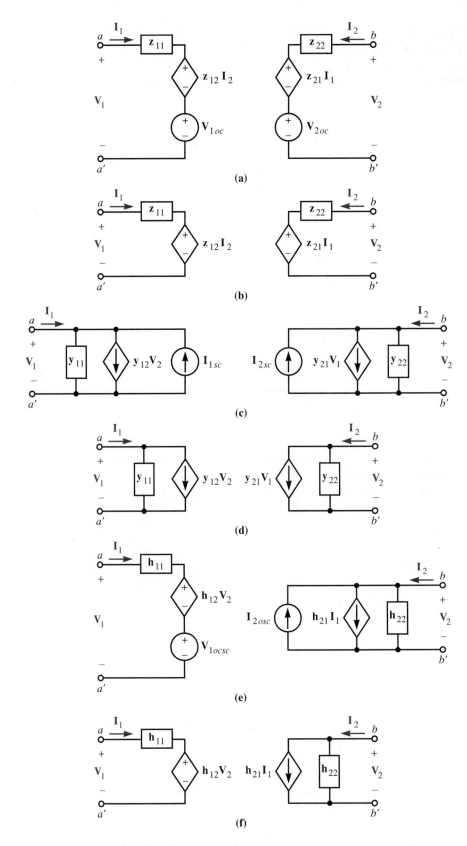

FIGURE 17.36
Equivalent circuits for two ports: (a) **z**-parameter for two port containing independent sources; (b) **z**-parameter for two port not containing independent sources; (c) **y**-parameter for two port containing independent sources; (d) **y**-parameter for two port not containing independent sources; (e) hybrid-**h** for two port not containing independent sources; (f) hybrid-**h** for two port not containing independent sources; (g) hybrid-**g** for two port containing independent sources; (h) hybrid-**g** for two port not containing independent sources

FIGURE 17.36
Continued

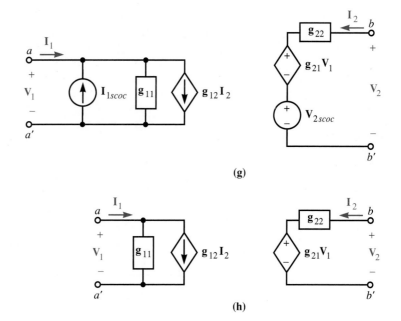

(g)

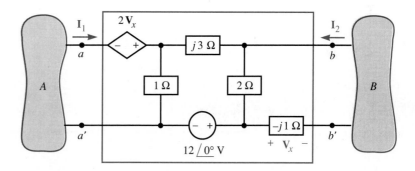

(h)

EXAMPLE 17.5

Find a **z**-parameter equivalent circuit for the two-port network enclosed in the rectangle in Fig. 17.37.

FIGURE 17.37
Two-port network

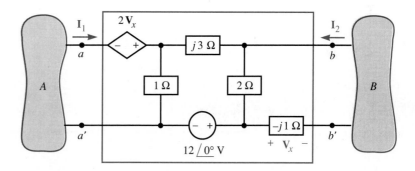

■ SOLUTION

Observe first that the network in the rectangle does satisfy the two-port property because the net current entering network A or B must be zero. Therefore Table 17.2 applies. *Assuming* that we are not interested in the voltages $\mathbf{V}_{ab}$ and $\mathbf{V}_{a'b'}$, we can use the equivalent circuit of Fig. 17.36a. Let us now evaluate the parameters of the equivalent circuit.

$\mathbf{V}_{oc}$ is determined by setting $\mathbf{I}_1 = \mathbf{I}_2 = 0$. The condition $\mathbf{I}_2 = 0$ leads immediately to $\mathbf{V}_x = 0$, and inspection yields

$$\mathbf{V}_{1oc} = \frac{1}{3 + j3}\, 12\underline{/0^\circ} = 2.83\underline{/-45^\circ}\ \mathbf{V}$$

$$\mathbf{V}_{2oc} = -\frac{2}{3 + j3}\, 12\underline{/0^\circ} = 5.66\underline{/135^\circ}\ \mathbf{V}$$

Parameter z_{11} is the ratio V_1/I_1 when $I_2 = 0$ and the 12-V internal source is set to zero. With $I_2 = 0$, and $V_x = 0$, an inspection of Fig. 17.38 reveals that

$$z_{11} = 1 \,\|\, (2 + j3) = 0.85\underline{/11.3°}\ \Omega$$

Similarly,

$$z_{21}\Big|_{\substack{I_2 = 0 \\ \text{IIS} = 0}} = 2\,\frac{1}{3 + j3} = 0.47\underline{/-45°}\ \Omega$$

FIGURE 17.38
I_2 and the internal independent source are set to zero

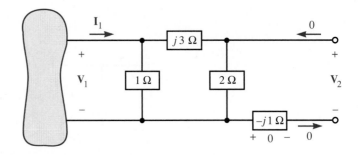

The configuration for computing z_{22} and z_{12} is shown in Fig. 17.39. Again, we find by inspection that

$$z_{12} = \frac{V_1}{I_2}\Big|_{\substack{I_1 = 0 \\ \text{IIS} = 0}} = 1\,\frac{2}{3 + j3} + j2 = 1.7\underline{/78.7°}\ \Omega$$

and

$$z_{22} = \frac{V_2}{I_2}\Big|_{\substack{I_1 = 0 \\ \text{IIS} = 0}} = 2\,\|\,(1 + j3) - j1 = 1.37\underline{/-14.0°}\ \Omega$$

FIGURE 17.39
I_1 and the internal independent source are set to zero

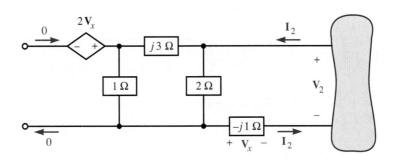

Therefore, as in the top row of Table 17.2, we have

$$\begin{bmatrix} V_1 \\ V_2 \end{bmatrix} = \begin{bmatrix} 0.85\underline{/11.3°} & 1.7\underline{/78.7°} \\ 0.47\underline{/-45°} & 1.37\underline{/-14.0°} \end{bmatrix}\begin{bmatrix} I_1 \\ I_2 \end{bmatrix} + \begin{bmatrix} 2.83\underline{/-45°} \\ 5.66\underline{/135°} \end{bmatrix}\ \blacksquare$$

REMEMBER

The circuits of Fig. 17.36 for the two-port network of Fig. 17.31 make no statement about V_{ab} and $V_{a'b'}$.

18. Give a possible **z**, **y**, and hybrid-**h**-parameter equivalent circuit for the two-port network shown. State why the term "equivalent circuit" should be used with caution.

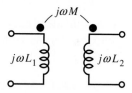

19. Repeat Exercise 18 for the circuit shown.

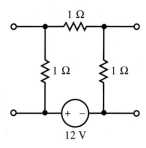

20. Repeat Exercise 18 for the circuit shown.

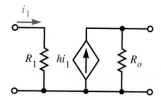

17.9   Summary

This chapter was an introduction to multiterminal network analysis. We considered the problem of finding equivalent circuits for three-terminal networks and two-port networks. We found that we could obtain four equivalent circuits for three-terminal networks by rearranging the equations relating terminal voltages and currents. We showed that for certain theoretical circuits, one or more of the equivalents might not exist. We then defined two-port networks and compared two-port networks with three-terminal networks. We showed that, with certain limitations, two-port networks can be represented by equivalent circuits that are similar to the three-terminal equivalents.

KEY FACTS

◆   Four equivalent circuits are available to model three-terminal and two-port networks. The four equivalent circuits are summarized in Tables 17.1 through 17.5 and in Figs. 17.4, 17.12, 17.13, 17.23, and 17.24.

• For certain theoretical circuits, one or more of the equivalent circuits may not exist.

• Two-port networks are four terminal networks with the following special property: The net current entering the input terminal pair equals zero and the net current entering the output pair equals zero.

• The equivalent circuits for two-port networks are limited because they do not describe the voltages $\mathbf{V}_{ab}$ and $\mathbf{V}_{a'b'}$ (Fig. 17.31).

■ PROBLEMS ■

Section 17.2

1. Find the impedance-parameter equivalent of the ladder network shown.

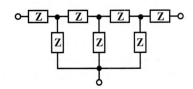

2. Find the impedance-parameter equivalent of the circuit shown.

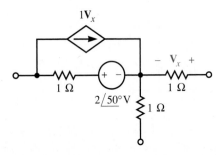

3. Find the impedance-parameter equivalent of the bridged-T circuit shown.

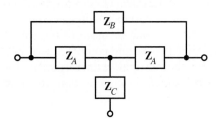

4. A three-terminal network is driven by a voltage source with source impedance $\mathbf{Z}_s$ as shown. Use the impedance-parameter equivalent network of Fig. 17.4, and determine the Thévenin equiv-

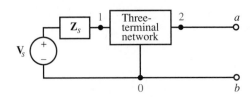

alent looking into terminals a and b. (Find $\mathbf{V}_{Th}$ and $\mathbf{Z}_{Th}$.)

5. Assume that the three-terminal network shown in Problem 4 does not contain independent sources. Assume also that a load impedance $\mathbf{Z}_L$ is connected across terminals a and b. Determine the transfer admittance $\mathbf{I}_{ab}/\mathbf{V}_s$ in terms of $\mathbf{Z}_L$, $\mathbf{Z}_s$, and the impedance parameters.

6. Find the impedance-parameter equivalent circuit of the circuit shown.

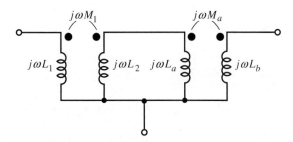

7. An engineer knows the open-circuit voltages $\mathbf{V}_{1oc}$ and $\mathbf{V}_{2oc}$ and the short-circuit currents $\mathbf{I}_{1sc}$ and $\mathbf{I}_{2sc}$ of a certain three-terminal network. By using only this information, can the engineer determine *any* of the impedance parameters? Explain.

Section 17.3

8. Find the admittance-parameter equivalent circuit of the network of Problem 1.

9. Find the admittance-parameter equivalent of the circuit of Problem 2.

10. Find the admittance-parameter equivalent of the parallel-T network shown.

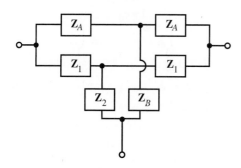

11. Use the admittance-parameter equivalent circuit of Fig. 17.13, and determine the Norton equivalent of the circuit shown looking into terminals a and b.

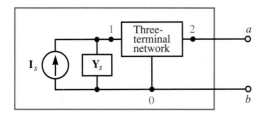

Section 17.4

12. A certain three-terminal network has admittance parameters $y_{11} = 2\underline{/0^\circ}$ S, $y_{22} = 1\underline{/0^\circ}$ S, $y_{12} = 1\underline{/45^\circ}$ S and $y_{21} = 1\underline{/-45^\circ}$ S. Find the impedance parameters.
13. The open-circuit voltages of the network of Problem 12 are given by

$$\mathbf{V}_{oc} = \begin{bmatrix} 3\underline{/60^\circ} \\ 1+j \end{bmatrix} \text{V}$$

Find the short-circuit currents.
14. Discuss the similarities and the differences between Eqs. (17.27) and (17.28) and the corresponding results for Thévenin and Norton equivalents of two-terminal networks.

Section 17.5

15. We have shown that certain circuit models are not possible. On the other hand, at our own peril, we can interconnect physical components in any way we wish. Give some examples to show why it can be unwise to try to physically construct a circuit corresponding to an impossible circuit model.

Section 17.6

16. Find the hybrid-**h**-parameter equivalent of the ladder network of Problem 1.
17. Find the hybrid-**h**-parameter equivalent of the network of Problem 2.
18. Find the hybrid-**h**-parameter equivalent of the coupled inductances shown below.

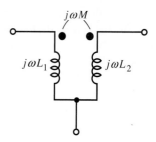

19. Use the simplified hybrid-**h** transistor model shown in Fig. 17.29 to determine the relationship between i_1 and i_2 in the circuit shown.

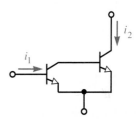

20. Hybrid-**g** equivalent circuits are frequently used to model field-effect transistors (FETs). The terminal equations for the hybrid-**g** representation of an ac network containing independent sources have the form

$$\mathbf{I}_1 = \mathbf{g}_{11}\mathbf{V}_1 + \mathbf{g}_{12}\mathbf{I}_2 - \mathbf{I}_{1scoc}$$
$$\mathbf{V}_2 = \mathbf{g}_{21}\mathbf{V}_1 + \mathbf{g}_{22}\mathbf{I}_2 + \mathbf{V}_{2scoc}$$

(a) Proceed in a manner similar to that of Sections 17.2, 17.3, and 17.6 to develop formulas for the constants appearing in the above equations.
(b) Draw the **g** equivalent circuit.
(c) Show that

$$\begin{bmatrix} \mathbf{g}_{11} & \mathbf{g}_{12} \\ \mathbf{g}_{21} & \mathbf{g}_{22} \end{bmatrix} = \begin{bmatrix} \mathbf{h}_{11} & \mathbf{h}_{12} \\ \mathbf{h}_{21} & \mathbf{h}_{22} \end{bmatrix}^{-1}$$

where the $\mathbf{h}_{ij}$'s are the hybrid-**h** parameters of the circuit.

(d) Show that

$$\begin{pmatrix} -\mathbf{I}_{1scoc} \\ \mathbf{V}_{2scoc} \end{pmatrix} = -\begin{bmatrix} g_{11} & g_{12} \\ g_{21} & g_{22} \end{bmatrix} \begin{pmatrix} \mathbf{V}_{1ocsc} \\ -\mathbf{I}_{2ocsc} \end{pmatrix}$$

where $\mathbf{V}_{1ocsc}$ and $\mathbf{I}_{2ocsc}$ appear in the hybrid-**h** model of Fig. 17.21.

Section 17.7

21. In your own words, explain the difference between an ordinary four-terminal network and a two-port network.

22. In your own words, explain whether or not a three-terminal network can be treated as a two-port network.

Section 17.8

23. Give the **z**-, **y**-, and **h**-parameter equivalents for the circuit shown.

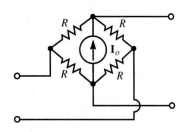

24. Repeat Problem 23 for the circuit shown.

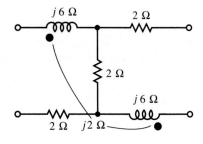

25. Repeat Problem 23 for the circuit shown.

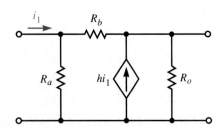

26. Repeat Problem 23 for the circuit shown.

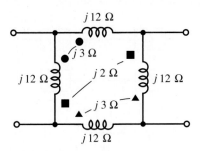

27. The term "equivalent circuit" should be used with special care for two-port networks. Explain why.

28. The figure shown is a *series* connection of two four-terminal networks A and B.

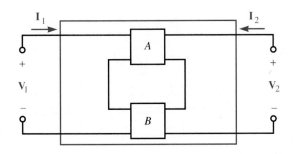

(a) Show that if networks A and B each have the two-port property, then the composite network enclosed in red automatically has the two-port property.

(b) Show that if the two-port requirement is satisfied by the composite network (enclosed in red), then the two-port requirement is *not necessarily* satisfied by either network A or network B.

29. Show that if networks A and B of Problem 28 each satisfy the two-port requirement, then the **z**-parameter matrix of the composite network is given by the sum of the **z**-parameter matrices of networks A and B.

30. The network shown is a *parallel* connection of two four-terminal networks A and B.

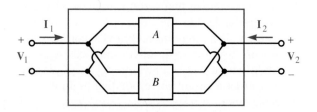

(a) Show that if networks A and B each have the two-port property, then the composite network (enclosed in red) automatically has the two-port property.

(b) Show that if the two-port requirement is satisfied by the composite network enclosed in red, then the two-port requirement is *not necessarily* satisfied by either network A or network B.

31. Show that if networks A and B of Problem 30 each satisfy the two-port requirement, then the **y**-parameter matrix of the composite network is given by the sum of the **y**-parameter matrices of networks A and B.

Comprehensive Problems

32. The three-terminal network in the circuit shown contains no independent sources. Find an expression for the voltage transfer function $\mathbf{V}_o/\mathbf{V}_s$ in terms of the **z** parameters of the three-terminal network.

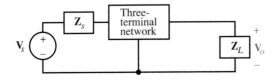

33. Repeat Problem 32 with the use of **y** parameters.
34. Repeat Problem 32 with the use of hybrid-**h** parameters.
35. Use PSpice to find the **z**-parameter equivalent of the circuit of Fig. 17.6.
36. Use PSpice to find the **y**-parameter equivalent of the circuit of Fig. 17.6.
37. Try using PSpice to find the **y**-parameter equivalent of the circuit of Fig. 17.18.
38. Use PSpice to obtain Bode amplitude and phase plots of the voltage transfer function of the transistor amplifier of Fig. 17.28. Assume that $C_1 = C_2 = 1000\ \mu F$, $R_s = 100\ \Omega$, $R_L = 18\ k\Omega$, $R_1 = 48\ k\Omega$, $R_2 = 22\ k\Omega$, $R_E = 2.2\ k\Omega$, and $R_C = 4.1\ k\Omega$. Use the transistor model of Fig. 17.29 with $\beta = 100$. Compare your plots with the design formula given in the solution to Example 17.4.
39. Repeat Problem 38 without the use of PSpice.
40. Assume that the input to the transistor amplifier of Fig. 17.28 is $v_s(t) = 1u(t)$ V. Use the transistor

model of Fig. 17.29, the component values of Problem 38, and PSpice to obtain a plot of the output $v_o(t)$.
41. Repeat Problem 40 with the use of the methods in Chapter 9.
42. Repeat Problem 40 with the use of Laplace transforms.

The networks shown below model important parts or stages in three hypothetical transistor amplifiers. (The numerical values are for illustration only.) Solve Problems 43 through 46 with the use of PSpice.

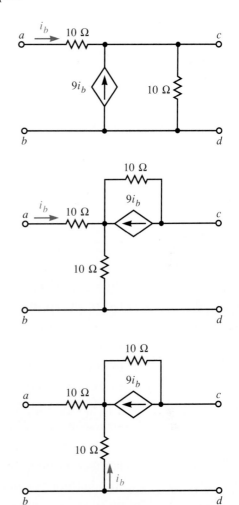

43. Calculate the **z** parameters of each stage.
44. Calculate the **y** parameters of each stage.
45. Calculate the hybrid-**h** parameters of each stage.
46. Calculate the hybrid-**g** parameters of each stage.

CHAPTER 18

Linear Models for Transformers

In Chapter 2 we discussed how magnetic flux caused by current in one coil can link (pass through) the turns of another coil and thus couple energy from one coil to another. Magnetic coupling led to the mutual inductance model.

In this chapter we see how magnetically coupled coils can transform the value of an impedance. When viewed in this manner, we refer to a pair of magnetically coupled coils as a *transformer*. We also show how, in many cases, we can replace a pair of coupled coils by three inductances connected in an equivalent tee (T) or pi (Π).

We introduce leakage inductance and magnetizing inductance, and see how we can use these to develop a transformer model that includes dependent sources. We then replace the dependent sources with a new four-terminal component, which is called an ideal transformer, and see how to analyze circuits that contain these new components. We conclude with a linear model for transformers that includes power absorbed by the ferromagnetic cores.

18.1   Linear Transformers

Electronic circuits, such as radio and television receivers, often use magnetically coupled coils, as shown in Fig. 18.1, to alter the effective input or output imped- ance of a portion of a circuit. These coupled coils *transform* an impedance, and we refer to them as *linear transformers*.* The transformers are linear because they are wound on magnetically linear (magnetic flux is proportional to the current) forms, such as plastic. We will model a linear transformer by coupled inductances with series resistances, as shown in Fig. 18.2.

FIGURE 18.1
Two magnetically coupled coils wound on a plastic tube to form a linear transformer. This is often called an air-core transformer

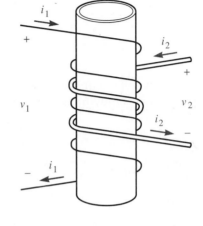

FIGURE 18.2
Time-domain model for a linear transformer

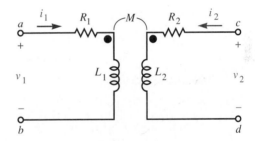

We can easily develop a frequency-domain model for a linear transformer by remembering that differentiation in the time domain corresponds to multiplication by $j\omega$ in the frequency domain. If a linear circuit drives the *primary* coil and a linear load is connected to the *secondary* coil,[†] we can model our circuit in the frequency domain as shown in Fig. 18.3.

Summing the voltages for closed paths on both sides of the transformer gives us

$$\begin{bmatrix} R_1 + j\omega L_1 & j\omega M \\ j\omega M & R_L + j\omega L_2 + \mathbf{Z}_L \end{bmatrix} \cdot \begin{bmatrix} \mathbf{I}_1 \\ \mathbf{I}_2 \end{bmatrix} = \begin{bmatrix} \mathbf{V}_1 \\ 0 \end{bmatrix} \qquad (18.1)$$

---

* Linear transformers are often called air-core transformers.

† Either coil can be called the primary coil and the other the secondary, but we usually refer to the source side as the primary and the load side as the secondary. There can be a source on each side in some applications.

FIGURE 18.3
A circuit with a linear
transformer

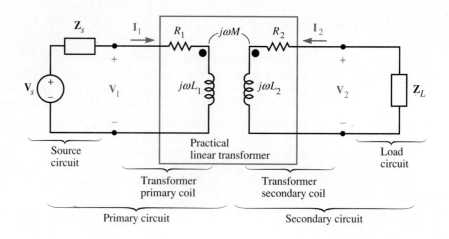

To simplify our notation, we define the following impedances. The *self-impedances* of the transformer primary and secondary sides are

$$\mathbf{Z}_{11} = R_1 + j\omega L_1 \tag{18.2}$$

and

$$\mathbf{Z}_{22} = R_2 + j\omega L_2 \tag{18.3}$$

respectively, and the *mutual impedances* are

$$\mathbf{Z}_{12} = \mathbf{Z}_{21} = j\omega M \tag{18.4}$$

This lets us write Eq. (18.1) as

$$\begin{bmatrix} \mathbf{Z}_{11} & \mathbf{Z}_{12} \\ \mathbf{Z}_{21} & \mathbf{Z}_{22} + \mathbf{Z}_L \end{bmatrix} \cdot \begin{bmatrix} \mathbf{I}_1 \\ \mathbf{I}_2 \end{bmatrix} = \begin{bmatrix} \mathbf{V}_1 \\ 0 \end{bmatrix} \tag{18.5}$$

which is of the form

$$\mathbf{ZI} = \mathbf{V} \tag{18.6}$$

The determinant of the impedance matrix **Z** is

$$\Delta = |\mathbf{Z}| = \mathbf{Z}_{11}(\mathbf{Z}_{22} + \mathbf{Z}_L) - \mathbf{Z}_{12}\mathbf{Z}_{21} \tag{18.7}$$

We can use Cramer's rule to solve for currents $\mathbf{I}_1$ and $\mathbf{I}_2$:

$$\mathbf{I}_1 = \frac{\mathbf{Z}_{22} + \mathbf{Z}_L}{\Delta} \mathbf{V}_1 \tag{18.8}$$

and

$$\mathbf{I}_2 = -\frac{\mathbf{Z}_{21}}{\Delta} \mathbf{V}_1 \tag{18.9}$$

The *input impedance*, $\mathbf{Z}_1$, seen by the source circuit is given by

$$\mathbf{Z}_1 = \frac{\mathbf{V}_1}{\mathbf{I}_1} = \frac{\Delta}{\mathbf{Z}_{22} + \mathbf{Z}_L} = \frac{\mathbf{Z}_{11}(\mathbf{Z}_{22} + \mathbf{Z}_L) - \mathbf{Z}_{12}\mathbf{Z}_{21}}{\mathbf{Z}_{22} + \mathbf{Z}_L}$$

$$= \mathbf{Z}_{11} - \mathbf{Z}_{12}\mathbf{Z}_{21} \frac{1}{\mathbf{Z}_{22} + \mathbf{Z}_L} \tag{18.10}$$

We see from Eq. (18.10) that the input impedance Z_1 is equal to the impedance of the primary circuit Z_{11} in series with an equivalent impedance

$$Z_{e1} = -Z_{12}Z_{21}\frac{1}{Z_2} \tag{18.11}$$

The equivalent impedance Z_{e1} is the total impedance of the secondary circuit Z_2:

$$Z_2 = Z_{22} + Z_L \tag{18.12}$$

reflected to the primary circuit. The mutual impedances Z_{12} and Z_{21} effectively *transform* the impedance of the secondary circuit Z_2 to a new value Z_{e1} and place it in series with the impedance of the primary coil Z_{11} to give the input impedance Z_1 seen by the source network.

This impedance transformation property of coupled inductances is frequently used in electronic amplifiers to perform *impedance matching*. Impedance matching maximizes the power transferred from one part of the amplifier to another and permits maximization of the overall power gain of the amplifier.

We now use Ohm's law for phasors to calculate the terminal voltage V_2 of the secondary side of the transformer (the load side):

$$V_2 = -Z_L I_2$$
$$= \frac{Z_{21}Z_L}{\Delta} V_1 \tag{18.13}$$

Equation (18.13) gives the forward voltage transfer function, which is the ratio of the secondary voltage V_2 to the primary voltage V_1:

$$\frac{V_2}{V_1} = \frac{Z_{21}Z_L}{\Delta} \tag{18.14}$$

From Eqs. (18.8) and (18.9), the forward current transfer function, which is the ratio of secondary current I_2 to primary current I_1, is

$$\frac{I_2}{I_1} = -\frac{Z_{21}}{Z_{22} + Z_L} = -\frac{Z_{21}}{Z_2} \tag{18.15}$$

Equations (18.14) and (18.15) provide an alternative interpretation of the transformer action of coupled inductances. The transformer lets us provide a voltage (or current) at the secondary terminals that is larger or smaller than the voltage (or current) at the primary terminals.

Linear transformers are frequently used in electronics to perform one or more of the following three functions.

1. The transformer action transforms the load impedance into a new value to match the impedance of the source for maximum power transfer.
2. The transformer provides dc voltage *isolation* between the source and load, because there is no dc path between the two coils.
3. When used in conjunction with capacitors, the inductances of the coils can provide a voltage transfer function that passes a sinusoid of one frequency and rejects a sinusoid of another frequency (the circuit is a *filter*).

An example of a circuit where a linear transformer performs all three functions is shown in Fig. 18.4.

FIGURE 18.4
A radio-frequency amplifier
with transformer-coupled
output

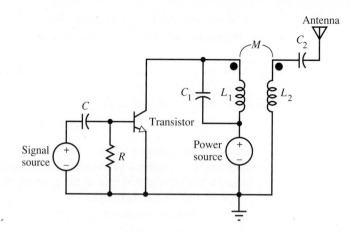

In the circuit of Fig. 18.4, the transformer, consisting of coupled inductances L_1 and L_2 in conjunction with capacitances C_1 and C_2, matches the impedance of the antenna to the output impedance of the transistor. The combined effect of the inductances and capacitances provides an exact impedance match at only one frequency, thereby providing the filtering action. The coils are connected only at the reference (ground). This prevents the dc voltage of the power source from appearing on the antenna side of the transformer. Analysis of this application of transformers is outlined in Problems 51, 52, and 53.

REMEMBER

Coupled coils form a linear transformer that can alter the effective impedance of a device. Impedance matching is often used to maximize power transfer in electronic circuits.

EXERCISES

1. Draw the frequency-domain circuit that corresponds to the time-domain circuit shown below.

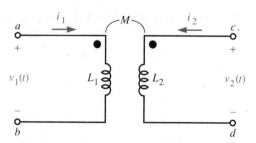

Write the terminal equations in the frequency domain for the preceding network in the form indicated in Exercises 2 and 3.

2. $\mathbf{V} = \mathbf{ZI}$ 3. $\mathbf{I} = \mathbf{YV}$

4. Determine the equivalent impedance measured between terminals a and b of the coupled inductances in the circuit in Exercise 1 if terminals c and d are connected together.

5. Write an equation, similar to Eq. (18.1), that relates the source voltage $\mathbf{V}_s$ to currents $\mathbf{I}_1$ and $\mathbf{I}_2$ in Fig. 18.3.

6. Use the relation $\mathbf{V}_2 = -\mathbf{Z}_L\mathbf{I}_2$ for Fig. 18.3 and write an equation similar to Eq. (18.1) that relates voltages $\mathbf{V}_1$ and $\mathbf{V}_2$ to currents $\mathbf{I}_1$ and $\mathbf{I}_2$. (The equation should not include $\mathbf{Z}_L$.)

7. (a) Find the input impedance $\mathbf{Z}_{in}(j\omega) = \mathbf{V}_1/\mathbf{I}_1$ for the following network.
   (b) Determine the forward voltage transfer function $\mathbf{H}_v(j\omega) = \mathbf{V}_2/\mathbf{V}_1$ for the following circuit.

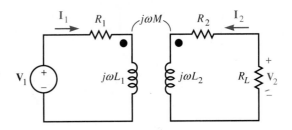

8. Find the forward-current transfer function $\mathbf{H}_I(j\omega) = \mathbf{I}_2/\mathbf{I}_1$ for the preceding circuit.

9. Determine the complex-power gain $\mathbf{G}(j\omega) = -\mathbf{V}_2\mathbf{I}_2^*/\mathbf{V}_1\mathbf{I}_1^*$ for the preceding circuit.

18.2   Equivalent T and Π Networks*

With the exception of dc isolation, we can often duplicate a linear transformer by an equivalent inductive T or Π network. Assume that terminals b and d of two coupled inductances are directly connected together as shown in Fig. 18.5.

FIGURE 18.5
Coupled inductances with a common connection

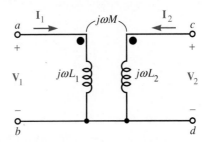

We can write the frequency-domain terminal equations that describe the coupled inductances of Fig. 18.5 as

$$\begin{bmatrix} \mathbf{V}_1 \\ \mathbf{V}_2 \end{bmatrix} = \begin{bmatrix} j\omega L_1 & j\omega M \\ j\omega M & j\omega L_2 \end{bmatrix} \cdot \begin{bmatrix} \mathbf{I}_1 \\ \mathbf{I}_2 \end{bmatrix} \qquad (18.16)$$

---

* This section can be omitted with no loss of continuity.

or in the form

$$\begin{bmatrix} \mathbf{I}_1 \\ \mathbf{I}_2 \end{bmatrix} = \begin{bmatrix} \dfrac{L_2}{j\omega(L_1 L_2 - M^2)} & \dfrac{-M}{j\omega(L_1 L_2 - M^2)} \\[3mm] \dfrac{-M}{j\omega(L_1 L_2 - M^2)} & \dfrac{L_1}{j\omega(L_1 L_2 - M^2)} \end{bmatrix} \cdot \begin{bmatrix} \mathbf{V}_1 \\ \mathbf{V}_2 \end{bmatrix} \qquad (18.17)$$

We can describe the T network (Y network) of Fig. 18.6 by the terminal equations (see Problem 9):

$$\begin{bmatrix} \mathbf{V}_1 \\ \mathbf{V}_2 \end{bmatrix} = \begin{bmatrix} j\omega(L_a + L_c) & j\omega L_c \\ j\omega L_c & j\omega(L_b + L_c) \end{bmatrix} \cdot \begin{bmatrix} \mathbf{I}_1 \\ \mathbf{I}_2 \end{bmatrix} \qquad (18.18)$$

FIGURE 18.6
A T network (Y network)

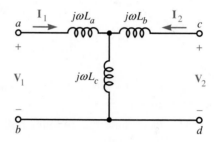

We can design a T network that is equivalent to the coupled inductances of Fig. 18.5 by equating each element in the impedance matrix of Eq. (18.18) to the corresponding element in the impedance matrix of Eq. (18.16). Solving for L_a, L_b, and L_c in terms of L_1, L_2, and M, we obtain

---

Equivalent T Network

$$L_a = L_1 - M \qquad (18.19)$$

$$L_b = L_2 - M \qquad (18.20)$$

$$L_c = M \qquad (18.21)$$

The Π network (Δ network) of Fig. 18.7 is described by the terminal equations (see Problem 10):

$$\begin{bmatrix} \mathbf{I}_1 \\ \mathbf{I}_2 \end{bmatrix} = \begin{bmatrix} \dfrac{1}{j\omega L_A} + \dfrac{1}{j\omega L_C} & -\dfrac{1}{j\omega L_C} \\[3mm] -\dfrac{1}{j\omega L_C} & \dfrac{1}{j\omega L_B} + \dfrac{1}{j\omega L_C} \end{bmatrix} \cdot \begin{bmatrix} \mathbf{V}_1 \\ \mathbf{V}_2 \end{bmatrix} \qquad (18.22)$$

FIGURE 18.7
A Π network (Δ network)

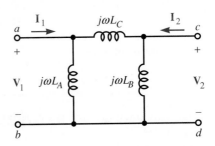

We can design a Π network that is equivalent to the coupled inductances of Fig. 18.5. Equating each element in the admittance matrix of Eq. (18.22) to the corresponding element in the admittance matrix of Eq. (18.17) and solving for L_A, L_B, and L_C in terms of L_1, L_2, and M, we obtain

Equivalent Π Network

$$L_A = \frac{L_1 L_2 - M^2}{L_2 - M} \tag{18.23}$$

$$L_B = \frac{L_1 L_2 - M^2}{L_1 - M} \tag{18.24}$$

$$L_C = \frac{L_1 L_2 - M^2}{M} \tag{18.25}$$

Although the mathematically equivalent T and Π networks always exist, they can be constructed only if the calculated inductances are nonnegative. Also, practical coils always have resistance, so the equivalent T and Π networks have a resistive component in the mutual impedance as well as the self-impedances. Magnetically coupled inductances have purely reactive mutual impedances.

If the circuit is operated with a sinusoidal signal of fixed angular frequency ω_0 and if either reactance in the T network,

$$\omega_0 L_a = \omega_0 (L_1 - M) \tag{18.26}$$

or

$$\omega_0 L_b = \omega_0 (L_2 - M) \tag{18.27}$$

is negative, we can still construct an equivalent T network by using a capacitance in place of the inductance to form the negative reactance. For instance, if the calculated reactance $\omega_0 (L_1 - M)$ is negative, we replace the inductance L_a in the T network by a capacitance of value C_a that satisfies the equation

$$-\frac{1}{\omega_0 C_a} = \omega_0 (L_1 - M) \tag{18.28}$$

Equation (18.28) gives a capacitance value of

$$C_a = \frac{1}{\omega_0^2 (M - L_1)} \tag{18.29}$$

Similar results hold true for the Π network. The choice of whether to use coupled inductances, a T network, or a Π network for impedance matching is usually based on economics.

REMEMBER

We can often replace coupled coils, which have a common terminal, with an equivalent T network or Π network. If the calculated value of an inductance for the equivalent circuit is negative, we cannot construct the equivalent. However, if the circuit operates at a single fixed frequency, we can construct an equivalent that is valid for this frequency by using a capacitance in place of an inductance.

EXERCISES

Replace the coupled inductance network in Fig. 18.5 with an equivalent T network and give numerical values for L_a, L_b, and L_c if $L_1 = 2$ H, $L_2 = 40$ H, and M has the value specified in Exercises 10 through 13.

10. 1 H   11. 2 H

12. 3 H   13. 6 H

Replace the coupled inductance network in Fig. 18.5 with an equivalent Π network and give numerical values for L_A, L_B, and L_C if $L_1 = 2$ H, $L_2 = 40$ H, and M has the value specified in Exercises 14 through 17.

14. 1 H   15. 2 H

16. 3 H   17. 6 H

18.3 Alternative Transformer Models

In principle, we can analyze any linear transformer by the method we used in Section 18.1. We now develop an alternative, but equivalent model. This analysis leads to a new four-terminal network component called an ideal transformer, which is a usable approximation for some physical transformers.

We begin by rewriting the terminal equations for the network of Fig. 18.2 as given in Eq. (18.1):

$$\begin{bmatrix} \mathbf{V}_1 \\ \mathbf{V}_2 \end{bmatrix} = \begin{bmatrix} R_1 + j\omega L_1 & j\omega M \\ j\omega M & R_2 + j\omega L_2 \end{bmatrix} \cdot \begin{bmatrix} \mathbf{I}_1 \\ \mathbf{I}_2 \end{bmatrix} \tag{18.30}$$

where we used

$$\mathbf{V}_2 = -\mathbf{Z}_L \mathbf{I}_2 \tag{18.31}$$

Leakage Inductance and Magnetizing Inductance

To develop our new model, we will use the coupled inductance model shown in Fig. 2.30 and repeated in Fig. 18.8 (with a minor change in notation for voltages

FIGURE 18.8
A simplified model for
coupled coils

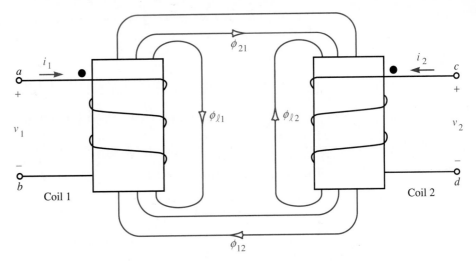

$\phi_{\ell 1}$ The flux due to current i_1, which links only coil 1 (leakage flux)

$\phi_{\ell 2}$ The flux due to current i_2, which links only coil 2 (leakage flux)

ϕ_{21} The flux due to current i_1, which links coil 2

ϕ_{12} The flux due to current i_2, which links coil 1

$\phi_{11} = \phi_{\ell 1} + \phi_{21}$

$\phi_{22} = \phi_{\ell 2} + \phi_{12}$

The total flux due to current i_1, which links coil 1

The total flux due to current i_2, which links coil 2

$\phi_1 = \phi_{\ell 1} + \phi_{21} + \phi_{12}$
$\quad = \phi_{11} + \phi_{12}$

$\phi_2 = \phi_{\ell 2} + \phi_{12} + \phi_{21}$
$\quad = \phi_{22} + \phi_{21}$

The total flux linking coil 1

The total flux linking coil 2

and currents), to express inductances L_1 and L_2 in terms of component inductances. The total flux ϕ_{11} linking coil 1 due to current i_1 is composed of two components,

$$\phi_{11} = \phi_{\ell 1} + \phi_{21} \tag{18.32}$$

as shown in Fig. 18.8.

The *leakage flux* $\phi_{\ell 1}$ is the component of flux ϕ_{11} linking coil 1 that does not link coil 2. The *magnetizing flux* ϕ_{21} is the flux due to current i_1 that passes through both coils. Equation (2.61), which we repeat below,

$$k_1 = \frac{\phi_{21}}{\phi_{11}} \tag{18.33}$$

defines the coupling coefficient k_1 as the fraction of the flux caused by current i_1 that passes through both coils.

Substitution from Eq. (18.33) into Eq. (18.32) gives

$$\phi_{\ell 1} = (1 - k_1)\phi_{11} \tag{18.34}$$

Equation (2.46) gives the *self-inductance* L_1 of coil 1:

$$L_1 = \frac{N_1}{i_1} \phi_{11} \qquad (18.35)$$

Substitution from Eqs. (18.32), (18.33), and (18.34) lets us write Eq. (18.35) as

$$L_1 = \frac{N}{i_1}(1 - k_1)\phi_{11} + \frac{N}{i_1} k_1 \phi_{11}$$

$$= L_{\ell 1} + L_{M1} \qquad (18.36)$$

The *leakage inductance* of coil 1,

$$L_{\ell 1} = \frac{N}{i_1}(1 - k_1)\phi_{11}$$

$$= (1 - k_1)L_1 \qquad (18.37)$$

is the component of the self-inductance L_1 caused by the magnetic flux that passes through coil 1 but not coil 2. The *magnetizing inductance*

$$L_{M1} = \frac{N}{i_1} k_1 \phi_{11}$$

$$= k_1 L_1 \qquad (18.38)$$

of coil 1 is due to the magnetic flux, caused by current i_1, that passes through both coils. We see that Eq. (18.36) is equivalent to

$$L_1 = (1 - k_1)L_1 + k_1 L_1 \qquad (18.39)$$

In a similar fashion we can express the *self-inductance* L_2 of coil 2 as a sum of the leakage inductance of coil 2 plus the magnetizing inductance of coil 2,

$$L_2 = L_{\ell 2} + L_{M2} \qquad (18.40)$$

which we can rewrite as

$$L_2 = (1 - k_2)L_2 + k_2 L_2 \qquad (18.41)$$

The magnetic coefficient of coupling as given by Eq. (2.60) is

$$k = \sqrt{k_1 k_2} \qquad (18.42)$$

We can substitute this into Eq. (2.59), which is

$$M = k\sqrt{L_1 L_2} \qquad (18.43)$$

to obtain an expression for M in terms of magnetizing inductances:

$$M = \sqrt{k_1 k_2}\sqrt{L_1 L_2}$$
$$= (\sqrt{k_1 L_1})(\sqrt{L_2 k_2})$$
$$= (\sqrt{L_{M1} L_{M2}}) \qquad (18.44a)$$

$$= L_{M1}\left(\sqrt{\frac{L_{M2}}{L_{M1}}}\right) \qquad (18.44b)$$

$$= L_{M2}\left(\sqrt{\frac{L_{M1}}{L_{M2}}}\right) \qquad (18.44c)$$

To simplify our notation, we define an *effective turns ratio*

$$\frac{n_1}{n_2} = \sqrt{\frac{L_{M1}}{L_{M2}}} \qquad (18.45)$$

For most transformers, this ratio of n_1 to n_2 is approximately equal to the actual turns ratio

$$\frac{n_1}{n_2} \simeq \frac{N_1}{N_2} \qquad (18.46)$$

of the coils, because, for a coil of fixed configuration, the inductance is proportional to the square of the number of turns.

Transformer Models

We can now use Eqs. (18.36), (18.39), (18.44b), and (18.45) to write the inductances in Eq. (18.30) in terms of magnetizing and leakage inductances:

$$\begin{bmatrix} V_1 \\ V_2 \end{bmatrix} = \begin{bmatrix} R_1 + j\omega(L_{\ell 1} + L_{M1}) & j\omega L_{M1}\dfrac{n_2}{n_1} \\[2ex] j\omega L_{M1}\dfrac{n_2}{n_1} & R_2 + j\omega(L_{\ell 2} + L_{M2}) \end{bmatrix} \cdot \begin{bmatrix} I_1 \\ I_2 \end{bmatrix} \qquad (18.47)$$

We can arrange the two equations contained in Eq. (18.47) in a manner that lets us develop a new equivalent circuit:

$$V_1 = (R_1 + j\omega L_{\ell 1})I_1 + j\omega L_{M1}\left(I_1 + \frac{n_2}{n_1}I_2\right) \qquad (18.48)$$

$$V_2 = (R_2 + j\omega L_{\ell 2})I_2 + j\omega \frac{n_2}{n_1} L_{M1}\left(I_1 + \frac{n_2}{n_1}I_2\right) \qquad (18.49)$$

Careful inspection of Eqs. (18.48) and (18.49) lets us construct a new model for a linear transformer, as shown in Fig. 18.9.

If we write a KVL equation for the primary side of the circuit shown in Fig. 18.9, we obtain Eq. (18.48). A KVL equation for the secondary side of the circuit of Fig. 18.9 gives Eq. (18.49). Therefore, the circuit of Fig. 18.9 has the same terminal equations as the linear transformer of Fig. 18.3. In the following sections we

FIGURE 18.9
A transformer model that uses the primary magnetizing inductance

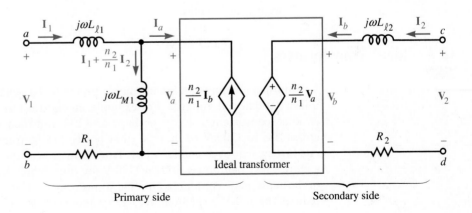

will use the model of Fig. 18.9 to develop an approximate transformer model. This will simplify the analysis of many circuits.

We can use an analysis similar to that which led to Fig. 18.9 and construct an equivalent circuit for linear transformers that uses the magnetizing inductance of the secondary circuit, as shown in Fig. 18.10.

The dependent sources enclosed by the red lines of Figs. 18.9 and 18.10 form an ideal transformer model, which we discuss in the following section.

FIGURE 18.10
A transformer model that uses the secondary magnetizing inductance

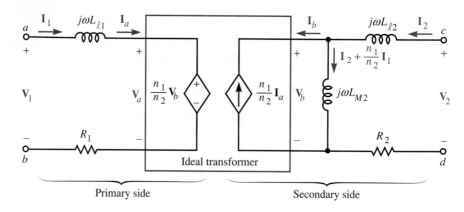

REMEMBER

Leakage inductance is caused by flux that links only one coil of coupled inductances. Increasing the coupling coefficients k_1 and k_2 decreases the leakage inductance. We can use either the primary or secondary magnetizing inductance in the transformer model.

EXERCISES

18. Draw a model like that shown for the linear transformer in Fig. 18.3 if $L_1 = 20$ H, $L_2 = 40$ H, and $M = 10$ H. The resistance values are $R_1 = 1\ \Omega$ and $R_2 = 2\ \Omega$, and the radian frequency is ω.

19. Draw a transformer model like that shown in Fig. 18.9. Use the values specified in Exercise 18. (The model is valid if k replaces k_1 and k_2.)

20. Draw a transformer model like that shown in Fig. 18.10. Use the values specified in Exercise 18.

18.4   Ideal Transformers

Transformers in power systems enable power to be generated, transmitted, and utilized at different voltages. For example, an electrical utility will generate power at some convenient voltage, perhaps 13.8 kV, and then use transformers to increase this to 120 kV or more so that power can be transmitted at high voltage and low current over smaller wires. Transformers near the customer reduce this to a more convenient lower voltage (120 V for your lights). Equipment in a residence or business may contain additional transformers. For example, a transformer reduces the 120-V residential voltage to about 6 V for a model train.

Small transformer.
Photograph by
James Scherer

All transformers used for 60-Hz power and many transformers used in electronics have ferromagnetic cores. The type of construction is often similar to that depicted in Fig. 18.11. This gives a very large inductance for a coil, compared with the value if the core is not ferromagnetic. This typically gives a large ratio of inductive reactance (ωL_1 or ωL_2) to resistance (R_1 or R_2) and a coupling coefficient close to one.*

FIGURE 18.11
A transformer with a ferromagnetic core (a physical transformer may have hundreds of turns of wire in each coil)

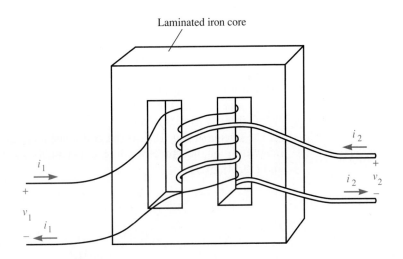

Laminated iron core

The result is that, in many practical applications, the magnitude of the current through the magnetizing reactance wL_{m2} (see Fig. 8.10) is much less than the magnitude of the load current, and the coil resistances (R_1 and R_2) and the leakage reactances (wL_{l1} and wL_{l2}) have little effect on the load voltage.

This means that we can often approximate the linear transformer models of Figs. 18.9 and 18.10 with the ideal transformer models, which we repeat in Fig.

* Transformers with ferromagnetic cores can be constructed with magnetic coefficients of coupling greater than 0.995. Transformers with nonmagnetic cores rarely have a coupling coefficient greater than 0.5.

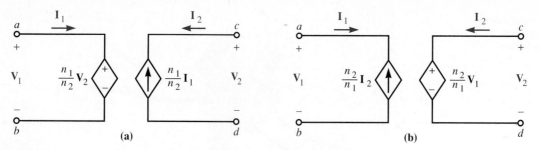

FIGURE 18.12 Two equivalent ideal transformer models that use dependent sources

18.12. Inspection of the ideal transformer models shown in Fig. 18.12 gives the terminal equations:

$$\begin{bmatrix} \mathbf{V}_1 \\ \mathbf{I}_2 \end{bmatrix} = \begin{bmatrix} 0 & \dfrac{n_1}{n_2} \\ -\dfrac{n_1}{n_2} & 0 \end{bmatrix} \cdot \begin{bmatrix} \mathbf{I}_1 \\ \mathbf{V}_2 \end{bmatrix} \tag{18.50}$$

or the equivalent equation

$$\begin{bmatrix} \mathbf{V}_2 \\ \mathbf{I}_1 \end{bmatrix} = \begin{bmatrix} 0 & \dfrac{n_2}{n_1} \\ -\dfrac{n_2}{n_1} & 0 \end{bmatrix} \cdot \begin{bmatrix} \mathbf{I}_2 \\ \mathbf{V}_1 \end{bmatrix} \tag{18.51}$$

Equations (18.48) and (18.49) break down in the limiting case given by Eqs. (18.50) and (18.51). We can, however, justify the ideal transformer model of Eq. (18.50) directly from the coupled inductance model and the equations we developed for the circuit of Fig. 18.3. If the coil resistances are negligible compared with the magnitude of the corresponding coil reactances, Eqs. (18.2), (18.3), (18.4), (18.7) and (18.14) give us

$$\mathbf{V}_1 = \frac{\Delta}{\mathbf{Z}_{21}\mathbf{Z}_L} \mathbf{V}_2 = \frac{\mathbf{Z}_{11}(\mathbf{Z}_{22} + \mathbf{Z}_L) - \mathbf{Z}_{12}\mathbf{Z}_{21}}{\mathbf{Z}_{21}\mathbf{Z}_L} \mathbf{V}_2$$

$$= \frac{j\omega L_1(j\omega L_2 + \mathbf{Z}_L) - (j\omega M)^2}{j\omega M \mathbf{Z}_L} \mathbf{V}_2 \tag{18.52}$$

Using Eqs. (18.41) and (18.43) lets us write the above equation as

$$\mathbf{V}_1 = \frac{(k^2 - 1)\omega^2 L_1 L_2 + j\omega L_1 \mathbf{Z}_L}{j\omega M \mathbf{Z}_L} \mathbf{V}_2 \tag{18.53}$$

For unity coupling coefficient this becomes

$$\mathbf{V}_1 = \frac{j\omega L_1 \mathbf{Z}_L}{j\omega M \mathbf{Z}_L} \mathbf{V}_2 = \frac{L_1}{M} \mathbf{V}_2 = \frac{L_1}{\sqrt{L_1 L_2}} \mathbf{V}_2$$

$$= \sqrt{\frac{L_1}{L_2}} \mathbf{V}_2 = \frac{n_1}{n_2} \mathbf{V}_2 \tag{18.54}$$

which is as given in Eq. (18.50). For $\omega L_2 \gg |\mathbf{Z}_L|$ and negligible coil resistances, Eqs. (18.3), (18.4), (18.15), and (18.43) give

$$\mathbf{I}_2 = -\frac{\mathbf{Z}_{21}}{\mathbf{Z}_2}\mathbf{I}_1 = -\frac{j\omega M}{j\omega L_2}\mathbf{I}_1 = \frac{k\sqrt{L_1 L_2}}{L_2}\mathbf{I}_1 = -k\sqrt{\frac{L_1}{L_2}}\,\mathbf{I}_1 \qquad (18.55)$$

With $k = 1$, this gives

$$\mathbf{I}_2 = -\frac{n_1}{n_2}\mathbf{I}_1 \qquad (18.56)$$

as in Eq. (18.50).

The Ideal Transformer Component

Although the ideal transformer models shown in Fig. 18.12 are useful for circuit analysis, we now define a new four-terminal network component. This new network component has the symbol shown in Fig. 18.13.

COMPONENT SYMBOL

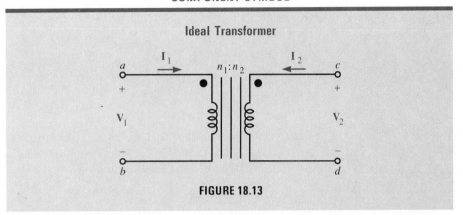

Ideal Transformer

FIGURE 18.13

We define the ideal transformer in the following way (the voltage references are as assigned in Fig. 18.13):

DEFINITION

Ideal Transformer

The complex power absorbed by an **ideal transformer** is zero. That is,

$$\mathbf{S}_1 + \mathbf{S}_2 = 0 \qquad (18.57)$$

where $\mathbf{S}_1$ is the complex power absorbed by side 1 and $\mathbf{S}_2$ is the complex power absorbed by side 2. In addition, the effective turns ratio is constant and equal to the voltage ratio:

$$\frac{n_1}{n_2} = \frac{\mathbf{V}_1}{\mathbf{V}_2} \qquad (18.58)$$

Referring to Fig. 18.13, we can write Eq. (18.57) as (we use rms values):

$$\mathbf{V}_1\mathbf{I}_1^* + \mathbf{V}_2\mathbf{I}_2^* = 0 \qquad (18.59)$$

With the aid of Eq. (18.58), Eq. (18.59) becomes

$$\mathbf{V}_1\mathbf{I}_1^* + \frac{n_2}{n_1}\,\mathbf{V}_1\mathbf{I}_2^* = 0 \qquad (18.60)$$

We can obtain a relationship between currents $\mathbf{I}_1$ and $\mathbf{I}_2$ by taking the complex conjugate of Eq. (18.60):

$$\mathbf{I}_2 = -\left(\frac{n_1}{n_2}\right)^*\mathbf{I}_1 \qquad (18.61)$$

Equations (18.58) and (18.61) are the terminal equations for an ideal transformer with reference directions assigned as in Fig. 18.13:

TERMINAL EQUATIONS

Ideal Transformer

$$\begin{bmatrix} \mathbf{V}_1 \\ \mathbf{I}_2 \end{bmatrix} = \begin{bmatrix} 0 & \dfrac{n_1}{n_2} \\ -\left(\dfrac{n_1}{n_2}\right)^* & 0 \end{bmatrix} \cdot \begin{bmatrix} \mathbf{I}_1 \\ \mathbf{V}_2 \end{bmatrix} \qquad (18.62)$$

Because of considerations that arise when transformers have more than two coils, we take Eqs. (18.57) and (18.58) as the definition of an ideal transformer. We consider Eq. (18.61) a derived relationship.

Reflected Impedance

Connection of a load impedance $\mathbf{Z}_L$ to the secondary terminals (c and d) of an ideal transformer as shown in Fig. 18.14a gives

$$\mathbf{V}_2 = -\mathbf{I}_2\mathbf{Z}_L = -\left[-\left(\frac{n_1}{n_2}\right)^*\mathbf{I}_1\right]\mathbf{Z}_L \qquad (18.63)$$

FIGURE 18.14
(a) An ideal transformer with load and (b) its equivalent circuit

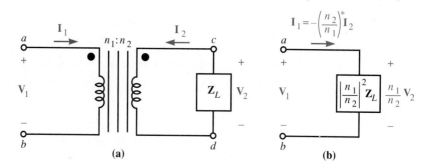

Equations (18.58) and (18.63) yield

$$\mathbf{V}_1 = \frac{n_1}{n_2}\mathbf{V}_2 = \frac{n_1}{n_2}\left(\frac{n_1}{n_2}\right)^{*}\mathbf{I}_1\mathbf{Z}_L$$

$$= \left|\frac{n_1}{n_2}\right|^2 \mathbf{Z}_L \mathbf{I}_1 \qquad (18.64)$$

Equation (18.64) is easily solved for the input impedance (Thévenin impedance) looking into terminals a and b:

$$\mathbf{Z}_1 = \frac{\mathbf{V}_1}{\mathbf{I}_1} = \left|\frac{n_1}{n_2}\right|^2 \mathbf{Z}_L \qquad (18.65)$$

We say that impedance $\mathbf{Z}_1$ is the impedance of $\mathbf{Z}_L$ *reflected* to the primary side. Equations (18.63), (18.64), and (18.65) give the equivalent circuit shown in Fig. 18.14b. In the next section we will see how reflected impedances simplify the analysis of some circuits.

The *effective turns ratio n_1/n_2 must be real for a two-coil transformer*, but in three-phase power system applications, multicoil transformers or combinations of transformers, called phase-shifting transformers, are conveniently modeled using complex turns ratios. With little loss in generality, we consider only real voltage ratios in the remainder of this chapter.

Networks with Ideal Transformers

Analysis of networks containing ideal transformers is no more difficult than the analysis of other linear networks. Example 18.1 illustrates how we can use reflected impedances to analyze some circuits with transformers.

EXAMPLE 18.1

Use the reflected impedances to find the impedance seen by the source in Fig. 18.15.

FIGURE 18.15
An ideal transformer example

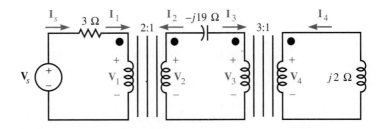

■ **SOLUTION**

If we reflect the inductive load of $j2\ \Omega$ to the primary of the right-hand transformer we obtain the equivalent circuit in Fig. 18.16.

FIGURE 18.16
The first equivalent circuit

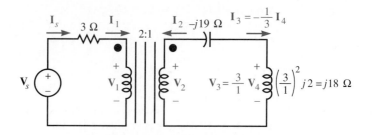

The effective load on the secondary side of the left-hand transformer is

$$\mathbf{Z}'_L = -j19 + j18$$
$$= -j1 \ \Omega$$

Now reflect $\mathbf{Z}'_L$ to the primary side of the left-hand transformer to obtain the circuit of Fig. 18.17.

FIGURE 18.17
The second equivalent circuit

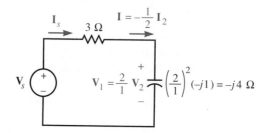

Source $\mathbf{V}_s$ sees impedance

$$\mathbf{Z} = 3 - j4 \ \Omega$$
$$= 5\underline{/-53.13°} \ \Omega \quad \blacksquare$$

Ideal Transformers with More than Two Coils

We defined the ideal transformer by Eqs. (18.57) and (18.58). Equation (18.61) is a derived expression. This distinction is important for a transformer with more than two coils.

DEFINITION

An **ideal transformer with K coils** absorbs zero complex power:

$$\sum_{k=1}^{K} \mathbf{S}_k = 0 \qquad (18.66)$$

and the coil voltages are defined by the turns ratios:

$$\frac{\mathbf{V}_k}{\mathbf{V}_l} = \frac{n_k}{n_l} \qquad k, l = 1, 2, \ldots, K \qquad (18.67)$$

With $\mathbf{I}_k$ defined to be the current entering the dot marked end of coil k, Eqs. (18.66) and (18.67) give

$$\sum_{k=1}^{K} n_k \mathbf{I}_k = 0 \qquad (18.68)$$

where n_k is assumed real.

REMEMBER

We can model an ideal transformer by a dependent current source and a dependent voltage source. The ideal transformer component gives an equivalent model for an ideal transformer. The voltage ratio $\mathbf{V}_1/\mathbf{V}_2$ for an ideal transformer is equal to the effective turns ratio n_1/n_2. The current ratio $-\mathbf{I}_1/\mathbf{I}_2$ is inversely proportional to the turns ratio n_1/n_2. An impedance $\mathbf{Z}_L$ is reflected from the secondary side to the primary side of a transformer by multiplying $\mathbf{Z}_L$ by the turns ratio squared $(n_1/n_2)^2$.

EXERCISES

21. Redraw the transformer model of Fig. 18.9 with the dependent sources replaced by the ideal transformer component of Fig. 18.13.

22. Redraw the transformer model of Fig. 18.10 with the dependent sources replaced by the ideal transformer model of Fig. 18.13.

23. For the following circuit, calculate: (a) $\mathbf{I}_s$, (b) $\mathbf{V}_1$, (c) $\mathbf{I}_1$, (d) $\mathbf{V}_2$, (e) $\mathbf{I}_2$, and (f) the impedance seen by the 360-V source. Use the method of reflected impedance.

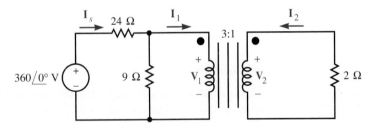

24. Repeat Exercise 23 if the dot on the right-hand side of the transformer is on the lower end of the secondary.

25. Refer to the preceding circuit. Use Thévenin's theorem, reflected impedance, and the terminal characteristics for the ideal transformer to replace all components to the left of the 2-Ω resistance with a voltage source and a series resistance.

18.5 Mesh-Current Analysis with Ideal Transformers

Some circuits contain transformers interconnected in a manner that makes analysis by reflected impedances impractical. An example of such a circuit is shown in Fig. 18.18. We can easily adapt the methods of analysis by node voltages or mesh currents to such problems.

Analysis of networks containing ideal transformers by the method of mesh currents is straightforward. To do this, we assign a voltage variable to each coil of the transformer, and write the mesh-current equations as if the transformer coils were voltage sources. The transformer voltages are unknowns, so we obtain additional equations from the turns-ratio relations between primary and secondary voltages and primary and secondary currents. This technique is illustrated in the following example.

EXAMPLE 18.2

Solve for mesh currents $\mathbf{I}_1$, $\mathbf{I}_2$, and $\mathbf{I}_3$ in the circuit of Fig. 18.18.

FIGURE 18.18
A mesh-current example with an ideal transformer

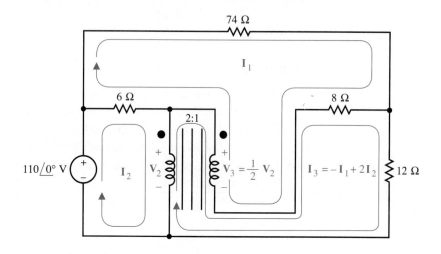

■ SOLUTION

First assign voltages $\mathbf{V}_2$ and $\mathbf{V}_3$ to the transformer as shown on the figure. Summation of the voltages around mesh 1 gives

$$74\mathbf{I}_1 + 8(\mathbf{I}_1 - \mathbf{I}_3) - \mathbf{V}_3 + 6(\mathbf{I}_1 - \mathbf{I}_2) = 0 \tag{18.69}$$

Summation of voltages around mesh 2 gives

$$-110 + 6(\mathbf{I}_2 - \mathbf{I}_1) + \mathbf{V}_2 = 0 \tag{18.70}$$

and KVL applied to mesh 3 gives

$$-\mathbf{V}_2 + \mathbf{V}_3 + 8(\mathbf{I}_3 - \mathbf{I}_1) + 12\mathbf{I}_3 = 0 \tag{18.71}$$

The turns-ratio relation between primary and secondary voltage and primary and secondary current provides the two additional equations required by the introduction of $\mathbf{V}_2$ and $\mathbf{V}_3$. The voltage relationship is

$$\mathbf{V}_3 = \frac{1}{2}\mathbf{V}_2 \tag{18.72}$$

which gives

$$\mathbf{V}_2 - 2\mathbf{V}_3 = 0 \tag{18.72a}$$

and the current relationship is

$$\mathbf{I}_3 - \mathbf{I}_1 = -\frac{2}{1}(\mathbf{I}_2 - \mathbf{I}_3) \tag{18.73}$$

which we write as

$$-\mathbf{I}_1 + 2\mathbf{I}_2 - \mathbf{I}_3 = 0 \tag{18.73a}$$

We can solve these five simultaneous equations for $\mathbf{I}_1$, $\mathbf{I}_2$, $\mathbf{I}_3$, $\mathbf{V}_2$, and $\mathbf{V}_3$, but considerable effort can be saved by substituting

$$\mathbf{V}_3 = \frac{1}{2}\mathbf{V}_2$$

into the first and third equation. The fifth equation is thus eliminated and we can write the remaining four equations as

$$\begin{bmatrix} 88 & -6 & -8 & -\frac{1}{2} \\ -6 & 6 & 0 & 1 \\ -8 & 0 & 20 & -\frac{1}{2} \\ -1 & 2 & -1 & 0 \end{bmatrix} \cdot \begin{bmatrix} \mathbf{I}_1 \\ \mathbf{I}_2 \\ \mathbf{I}_3 \\ \mathbf{V}_2 \end{bmatrix} = \begin{bmatrix} 0 \\ 110 \\ 0 \\ 0 \end{bmatrix}$$

These can be solved directly, but the last equation can be solved for

$$\mathbf{I}_3 = -\mathbf{I}_1 + 2\mathbf{I}_2$$

Thus $\mathbf{I}_3$ can be eliminated from the first three equations to give

$$\begin{bmatrix} 96 & -22 & -\frac{1}{2} \\ -6 & 6 & 1 \\ -28 & 40 & -\frac{1}{2} \end{bmatrix} \cdot \begin{bmatrix} \mathbf{I}_1 \\ \mathbf{I}_2 \\ \mathbf{V}_2 \end{bmatrix} = \begin{bmatrix} 0 \\ 110 \\ 0 \end{bmatrix}$$

We can use a *shortcut* and write these last three equations (the last matrix equation) directly by using the transformer turns ratio to write

$$\mathbf{V}_3 = \frac{1}{2}\mathbf{V}_2$$

and

$$\mathbf{I}_3 = -\mathbf{I}_1 + 2\mathbf{I}_2$$

on the network diagram as noted. We then use these in place of $\mathbf{V}_3$ and $\mathbf{I}_3$ when the KVL equations are written. In either case,

$$\begin{bmatrix} \mathbf{I}_1 \\ \mathbf{I}_2 \\ \mathbf{I}_3 \end{bmatrix} = \begin{bmatrix} 1 \\ 2 \\ 3 \end{bmatrix} \mathrm{A} \quad \text{and} \quad \begin{bmatrix} \mathbf{V}_2 \\ \mathbf{V}_3 \end{bmatrix} = \begin{bmatrix} 104 \\ 52 \end{bmatrix} \mathrm{V}$$

An alternative procedure to analyze transformer circuits with mesh-current equations is indicated in Exercises 26 and 27. ∎

We can adapt the method of analysis by mesh currents to any planar network that contains ideal transformers. We use the turns ratio of the transformer to write the voltage and current of one coil in terms of the voltage and current of the other.

EXERCISES

26. Use the ideal transformer model of Fig. 18.12a to write mesh-current equations for the following network. Solve for the mesh currents.

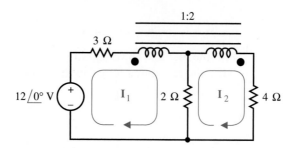

27. Use the ideal transformer model of Fig. 18.12b and repeat Exercise 26.
28. Use the ideal transformer component of Fig. 18.13, and repeat Exercise 26. (Use the shortcut discussed in the last paragraph of Example 18.2.)
29. Move the right-hand dot mark to the other end of the coil in the preceding circuit, and repeat Exercise 28.

18.6 Node-Voltage Analysis with Ideal Transformers

We can easily adapt the method of analysis by node voltages to include ideal transformers. To do this we assign a current variable to each coil of the transformer and write the node-voltage equations as if the transformer coils were current sources. The transformer currents are unknowns, so we obtain additional equations from the turns-ratio relations between primary and secondary voltages and primary and secondary currents. The following example demonstrates the procedure.

EXAMPLE 18.3

Solve for the node voltages in the network of Fig. 18.19.

FIGURE 18.19
A node-voltage example with an ideal transformer

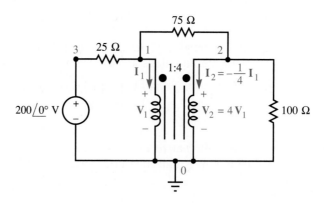

■ **SOLUTION**

We assign currents $\mathbf{I}_1$ and $\mathbf{I}_2$ to the two transformer coils as shown. The KCL equation for a surface enclosing node 1 is

$$\frac{1}{25}(\mathbf{V}_1 - 200) + \frac{1}{75}(\mathbf{V}_1 - \mathbf{V}_2) + \mathbf{I}_1 = 0 \tag{18.74}$$

and that for a surface enclosing node 2 is

$$\frac{1}{75}(\mathbf{V}_2 - \mathbf{V}_1) + \frac{1}{100}\mathbf{V}_2 + \mathbf{I}_2 = 0 \tag{18.75}$$

We obtain the required two additional equations from the transformer turns ratio:

$$\mathbf{I}_2 = -\frac{1}{4}\mathbf{I}_1 \tag{18.76}$$

$$\mathbf{V}_2 = 4\mathbf{V}_1 \tag{18.77}$$

We can solve these four simultaneous equations, or use the latter two to eliminate $\mathbf{I}_2$ and $\mathbf{V}_2$ from the first two equations. Substitution of these expressions for $\mathbf{I}_2$ and $\mathbf{V}_2$ in the first two equations gives

$$\begin{bmatrix} 0 & 1 \\ \dfrac{2}{25} & -\dfrac{1}{4} \end{bmatrix} \begin{bmatrix} \mathbf{V}_1 \\ \mathbf{I}_1 \end{bmatrix} = \begin{bmatrix} 8 \\ 0 \end{bmatrix}$$

We can use a *shortcut* and write this last matrix equation directly. Write

$$\mathbf{I}_2 = -\frac{1}{4}\mathbf{I}_1$$

and

$$\mathbf{V}_2 = 4\mathbf{V}_1$$

on the network diagram. Then use these values in place of $\mathbf{I}_2$ and $\mathbf{V}_2$ to write the KCL equations. Regardless of the method used,

$$\begin{bmatrix} \mathbf{V}_1 \\ \mathbf{V}_2 \end{bmatrix} = \begin{bmatrix} 25 \\ 100 \end{bmatrix} \text{V} \quad \text{and} \quad \begin{bmatrix} \mathbf{I}_1 \\ \mathbf{I}_2 \end{bmatrix} = \begin{bmatrix} 8 \\ -2 \end{bmatrix} \text{A} \quad ■$$

An alternative procedure to write the node-voltage equations is indicated in Exercises 30 and 31.

REMEMBER

We can use the method of analysis by node voltage for any network with ideal transformers. We use the turns ratio of the transformer to write the voltage and current of one coil in terms of the voltage and current of the other.

EXERCISES

30. Use the ideal transformer model of Fig. 18.12a to write a set of node-voltage equations for the following network and solve for the node voltages.

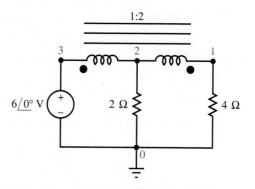

31. Use the ideal transformer model of Fig. 18.12b, and repeat Exercise 30.
32. Use the ideal transformer component of Fig. 18.13, and repeat Exercise 30.
33. Move the right-hand dot mark to the other end of the coil, and repeat Exercise 32.

18.7   Core Losses

Coupled inductances wound on ferromagnetic cores have heating losses in addition to the resistive or Ri^2 losses in the wire. These losses, caused by the changing magnetic flux in the core, are due to two causes. The changing magnetic flux induces voltages in the core material that cause *eddy currents* (the current circulates

FIGURE 18.20
Two equivalent circuits for ferromagnetic-core transformers: (*a*) equivalent circuit; (*b*) approximate equivalent circuit

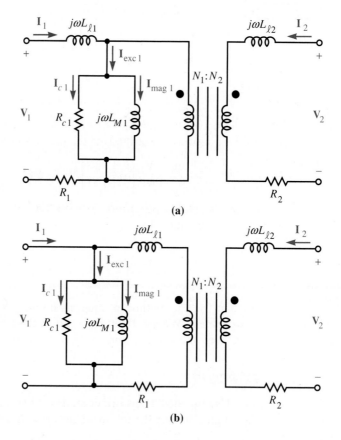

in eddies) in the core. This produces resistive losses called eddy-current losses. Changing flux in the ferromagnetic core also produces *hysteresis* losses in the core. We model these core losses as a resistance R_{c1} in parallel with the magnetizing inductance, as shown in Fig. 18.20a. We call current $\mathbf{I}_{exc1}$ the *excitation current*. Current $\mathbf{I}_{c1}$ accounts for the core losses, and the current $\mathbf{I}_{mag1}$ is the *magnetization current*. Typically the voltage drop across the leakage inductance is much less than the corresponding terminal voltage under normal operating conditions. In this case, the less exact equivalent circuit of Fig. 18.20b is used to simplify calculations.

We often further simplify the model by assuming that the excitation current is zero and the resistances are negligible, in which case the transformer model consists only of the leakage inductances $L_{\ell 1}$ and $L_{\ell 2}$ connected to the ideal transformer. (We frequently use this model for the large power transformers used in power systems.)

Large electrical transformers as used in power systems have their voltage ratio specified as their actual turns ratio, N_1/N_2, which is nearly identical to their effective turns ratio. The true turns ratio is often unknown for small transformers, but we only need the effective turns ratio n_1/n_2. If we refer to Fig. 18.20b, we see that we can make reasonably accurate measurements of the effective turns ratio of a transformer by using

$$\frac{n_1}{n_2} = \left|\frac{\mathbf{V}_1}{\mathbf{V}_2}\right|\Bigg|_{\mathbf{I}_2 = 0} \tag{18.78}$$

if the coupling coefficient is close to 1. Procedures for measuring the parameters for transformer models are presented in most texts on electrical machinery.

Distribution transformer
of a utility pole.
Copyright © Story Litchfield

EXAMPLE 18.4

A voltage of 120 V rms is applied to the primary of a transformer. The secondary is open-circuited. The transformer primary current is measured to be 5 A rms, and the input power is 360 W. Use the approximate model of Fig. 18.20b and calculate the value of R_{c1} and the magnetizing reactance ωL_{mag}.

The core losses are

$$P = \frac{|V_1|^2}{R_{c1}} = 360 \text{ W}$$

Therefore

$$R_c = \frac{120^2}{360} = 40 \ \Omega$$

The power factor angle is

$$\theta = \text{arc cos} \frac{P}{|\mathbf{S}|} = \text{arc cos} \frac{360}{5(120)} = 53.13°$$

The reactive power is

$$Q = |\mathbf{S}| \text{ arc sin } \theta = 5(120) \sin 53.13° = 480 \text{ VAR}$$

$$X_{\text{mag}} = \omega L_{\text{mag}} = \frac{|V_1|^2}{Q} = \frac{120^2}{480} = 30 \ \Omega \quad ■$$

Ferromagnetic material is not linear magnetically, so transformers using ferro-magnetic cores have nonlinear inductance and our linear models are approximate. Nevertheless, a linear model is useful for most analyses. The principal effect of the nonlinear core material is to distort the magnetizing current so that this current is not sinusoidal for a sinusoidal input voltage.

REMEMBER

Ferromagentic cores introduce additional losses, which we call core losses. We account for these core losses by introducing a resistance in parallel with the mag-netizing inductance of our model.

EXERCISES

34. We measure an input current of 5 A rms and a core loss of 100 W for a lin-ear transformer when the rated voltage of 240 V rms is applied to the pri-mary winding (the secondary winding is open-circuited). Draw a model as shown in Fig. 18.20b, and give the values for R_c and ωL_{M1}.

35. The rated voltage for the secondary coil of the transformer described in Exer-cise 34 is 120 V. This gives a turns ratio n_1/n_2 of 2. Construct a model similar to that of Fig. 18.20b, but the magnetizing inductance and core loss resis-tance must appear on the secondary side of the model.

18.8 Summary

This chapter began with a development of the equations that show how coupled inductances transform an impedance in the secondary circuit into an equivalent impedance in the primary circuit. We then developed the relationships for equiv-alent T and Π networks that perform the same transformation. We introduced leakage inductance and magnetizing inductance. This led to the ideal transformer as a model for coupled inductances when the coil inductances approach infinity, and the coupling coefficients approach 1. We demonstrated how mesh-current analysis and node-voltage analysis are modified for use with network models containing ideal transformers. We concluded the chapter with a linear model for transformers with core losses.

KEY FACTS

◆ Coupled inductances form a linear transformer.

◆ The primary and secondary voltages and currents of a linear transformer are related by an equation of the form

$$\begin{bmatrix} \mathbf{V}_1 \\ \mathbf{V}_2 \end{bmatrix} = \begin{bmatrix} \mathbf{Z}_{11} & \mathbf{Z}_{12} \\ \mathbf{Z}_{21} & \mathbf{Z}_{22} \end{bmatrix} \cdot \begin{bmatrix} \mathbf{I}_1 \\ \mathbf{I}_2 \end{bmatrix}$$

◆ A load impedance $\mathbf{Z}_L$ connected between the terminals of the secondary side of a transformer gives an input impedance to the primary side of value

$$\mathbf{Z}_1 = \mathbf{Z}_{11} - \mathbf{Z}_{12}\mathbf{Z}_{21} \frac{1}{\mathbf{Z}_{22} + \mathbf{Z}_L}$$

where we call

$$\mathbf{Z}_{e1} = -\mathbf{Z}_{12}\mathbf{Z}_{21} \frac{1}{\mathbf{Z}_{22} + \mathbf{Z}_L}$$

the impedance of the secondary circuit reflected to the primary circuit.

◆ We can often use Π or T connections of inductances to perform the same impedance transformation as a linear transformer.

◆ We can use an alternative model for coupled inductances. This model includes leakage inductance, magnetizing inductance, and dependent sources.

◆ We can use an ideal transformer component in place of the dependent sources of our alternative transformer model.

◆ An ideal transformer reflects an impedance as the turns ratio squared. That is, an impedance of value $\mathbf{Z}_L$ connected to the secondary terminals appears to be an impedance of

$$\mathbf{Z}_1 = \left(\frac{n_1}{n_2}\right)^2 \mathbf{Z}_L$$

connected across the primary terminals.

◆ The voltage ratio for an ideal transformer is

$$\frac{\mathbf{V}_1}{\mathbf{V}_2} = \frac{n_1}{n_2}$$

◆ The total complex power absorbed by an ideal transformer is zero:

$$\mathbf{S} = \mathbf{S}_1 + \mathbf{S}_2 = 0$$

The current ratio of an ideal transformer is

$$\frac{\mathbf{I}_2}{\mathbf{I}_1} = -\frac{n_1}{n_2}$$

◆ A useful model for many large power transformers consists of an ideal transformer and the leakage inductances. (Inclusion of the resistances improves the model.)

◆ The use of reflected impedances simplifies the analysis of some networks with transformers.

◆ Mesh-current analysis can be used for any planar network that includes transformers.

◆ Node-voltage analysis can be applied to any network with transformers.

■ PROBLEMS ■

Section 18.1

Problems 1 through 8 refer to the following circuit.

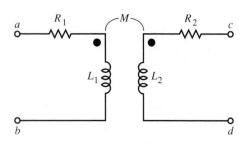

1. A 60-Hz source is connected to terminals a and b. Terminals c and d are open-circuited. We measure $V_{ab} = 85$ V rms, $V_{cd} = 40$ V rms, $I_{ab} = 2$ A rms, and the power absorbed by the transformer $P_{ab} = 25$ W. Determine R_1, L_1, and M.

2. A 60-Hz source is connected to terminals c and d. Terminals a and b are open-circuited. We measure $V_{cd} = 35$ V rms, $V_{ab} = 40$ V rms, $I_{cd} = 2$ A rms, and the power input to the transformer $P_{cd} = 15$ W. Determine R_2, L_2, and M.

3. Use the values of inductances calculated in Problems 1 and 2 to calculate the magnetic coefficient of coupling, $k = M/\sqrt{L_1 L_2}$.

4. The primary impedance of a linear transformer is $\mathbf{Z}_{11} = R_1 + j\omega L_1 = 6 + j40\ \Omega$ and the secondary impedance is $\mathbf{Z}_{22} = 4 + j16\ \Omega$. The mutual impedance is $\mathbf{Z}_{12} = \mathbf{Z}_{21} = j\omega M = j20\ \Omega$. A load of $\mathbf{Z}_L = 8 + j0\ \Omega$ is connected

between terminals c and d. What is the impedance $\mathbf{Z}_1 = \mathbf{V}_{ab}/\mathbf{I}_{ab}$ looking into terminals a and b?

5. Calculate the magnetic coefficient of coupling $k = M/\sqrt{L_1 L_2}$ for the linear transformer described in Problem 4.

6. For the circuit described in Problem 4, calculate $\mathbf{I}_{ab}$, $\mathbf{I}_{cd}$, $\mathbf{V}_{cd}$, and the complex power absorbed at terminals a and b ($\mathbf{S}_{ab} = \mathbf{V}_{ab}\mathbf{I}_{ab}^*$), if $V_{ab} = 100\underline{/0°}$ V rms.

7. For the circuit that precedes Problem 1, $R_1 = 1\ \Omega$, $R_2 = 2\ \Omega$, $L_1 = 10$ H, and $L_2 = 40$ H. A resistance $R_L = 2\ \Omega$ is connected from terminal c to d.
   (a) Calculate the coefficient of coupling k for (i) $M = 1$ H, (ii) $M = 2$ H, (iii) $M = 3$ H, and (iv) $M = 6$ H.
   (b) Plot $|\mathbf{Z}_{in}(j\omega)|$ for $0.002\pi \le \omega \le 0.8\pi$ for the values of k found in part (a).
   (c) Plot $|H(j\omega)| = |\mathbf{V}_{cd}/\mathbf{V}_{ab}|$ for $0.002\pi \le \omega \le 0.8\pi$ for the values of k found in part b, when the input is to terminals a and b. The use of PSpice is suggested.

8. Repeat Problem 7 when a 0.1-F capacitance is placed in series with R_1 to the right of terminal a, and a 0.025-F capacitance is in series with R_2 and to the left of terminal c.

Section 18.2

9. Connect a voltage source of value $\mathbf{V}_1$ to terminals a and b and a voltage source of value

V_2 to terminals c and d of the T network of Fig. 18.6. Write two mesh-current equations with the use of currents I_1 and I_2. Write these KVL equations in matrix form: $V = ZI$. This gives the terminal equations [Eq. (18.18)] for this T network.

10. Connect a current source of value I_1 between terminals a and b and a current source of value I_2 between terminals c and d of the Π network of Fig. 18.7. Write two node-voltage equations with the lower node as the reference node. Write these two KCL equations in matrix form: $I = YV$. This gives the terminal equations [Eq. (18.22)] for this Π network.

Calculate the inductance values for a T network, as shown in Fig. 18.6, which is equivalent to the coupled inductance network of Fig. 18.5. Use the inductance values specified in Problems 11 through 14.

11. $L_1 = 10$ H
    $L_2 = 40$ H
    $M = 2$ H
12. $L_1 = 2$ H
    $L_2 = 8$ H
    $M = 2$ H
13. $L_1 = 4$ H
    $L_2 = 4$ H
    $M = 4$ H
14. $L_1 = 18$ H
    $L_2 = 2$ H
    $M = 3$ H

Calculate the inductance values for a Π network, as shown in Fig 18.7, which is equivalent to the coupled inductance network of Fig. 18.5. Use the inductance values specified in Problems 15 through 18.

15. $L_1 = 10$ H
    $L_2 = 40$ H
    $M = 2$ H
16. $L_1 = 2$ H
    $L_2 = 8$ H
    $M = 2$ H
17. $L_1 = 4$ H
    $L_2 = 4$ H
    $M = 4$ H
18. $L_1 = 18$ H
    $L_2 = 2$ H
    $M = 3$ H

19. Two coupled inductances, as shown in Fig. 18.5, have inductance values $L_1 = 18$ mH, $L_2 = 2$ mH, and $M = 3$ mH. Design and sketch a T network that is equivalent to these coupled inductances at an angular frequency of 10,000 rad/s.

20. Two coupled inductances, as shown in Fig. 18.5, have inductance values $L_1 = 18$ mH, $L_2 = 2$ mH, and $M = 3$ mH. Design and sketch a Π network that is equivalent to these coupled inductances at an angular frequency of 10,000 rad/s.

21. Replace the coupled inductances in the following circuit by an equivalent T network and use parallel and series combinations to calculate the impedance looking into terminals a and b.

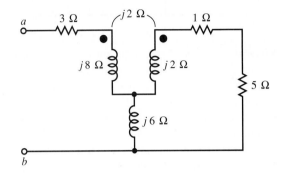

22. Equate the coefficients in the impedance matrix of Eq. (18.16) to the corresponding coefficients in the impedance matrix of Eq. (18.18). Use these equations to derive Eqs. (18.19), (18.20), and (18.21).

23. Equate the coefficients in the admittance matrix of Eq. (18.17) to the corresponding coefficients in the admittance matrix of Eq. (18.22). Use these equations to derive Eqs. (18.23), (18.24), and (18.25).

24. Refer to Figs. 18.5 and 18.6 and determine the upper bound on mutual inductance M, in terms of L_1 and L_2, so that L_a and L_b of the equivalent T network are nonnegative.

25. Refer to Figs. 18.5 and 18.7 and determine the upper bound on mutual inductance M in terms of L_1 and L_2, so that L_A and L_B of the equivalent Π network are nonnegative.

26. Refer to Figs. 18.5 and 18.6 and determine the upper bound, in terms of L_1 and L_2, for the coefficient of coupling k of the coupled inductances so that L_a and L_b of the equivalent T network are nonnegative.

27. Refer to Figs. 18.5 and 18.7 and determine the upper bound, in terms of L_1 and L_2, for the coefficient of coupling k of the coupled inductances so that L_A and L_B of the equivalent Π network are nonnegative.

Section 18.3

Problems 28 and 29 refer to the following circuit. Assume that the primary coupling coefficient k_1 is equal to the secondary coupling coefficient k_2.

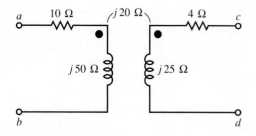

28. Draw a circuit diagram as shown in Fig. 18.9, and give the component values that make this transformer model equivalent to the preceding coupled inductance model.
29. Draw a circuit diagram as shown in Fig. 18.10, and give the component values that make this transformer model equivalent to the preceding coupled inductance model.
30. Draw a circuit diagram as shown in Fig. 18.9, and show the component values that make this model equivalent to the coupled inductances described by Problems 1 and 2.
31. Draw a circuit diagram as shown in Fig. 18.10, and show the component values that make this model equivalent to the coupled inductances described by Problems 1 and 2.

Section 18.4

32. Replace the dependent source model for an ideal transformer, as shown in Fig. 18.9, by the ideal transformer component shown in Fig. 18.13. Calculate the value of all components so that this model is equivalent to the coupled inductances shown below.

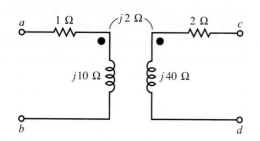

FIGURE 18.21

33. Replace the dependent source model for an ideal transformer, as shown in Fig. 18.10, by the ideal transformer component shown in Fig. 18.13. Calculate the value of all components so that this model is equivalent to the coupled inductances shown above.
34. Repeat Problem 32 for the coupled inductances described by Problems 1 and 2.
35. For the following ideal transformer circuit, determine: (a) the source current $\mathbf{I}$, (b) the complex power $\mathbf{S}$ supplied by the source, (c) the output voltage $\mathbf{V}_o$, and (d) the voltage on the primary winding $\mathbf{V}_1$, if $\mathbf{V}_s = 20\underline{/0°}$ V rms.

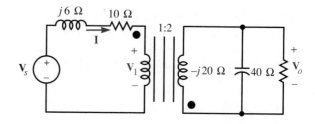

36. For the following circuit, resistance R_L is 250 Ω. Calculate currents $\mathbf{I}_1$ and $\mathbf{I}_2$ and voltages $\mathbf{V}_1$ and $\mathbf{V}_2$, if $\mathbf{V}_s = 120\underline{/0°}$ V rms.

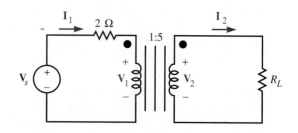

37. For the preceding circuit, calculate the value of load resistance R_L that will absorb the maximum power. What power is absorbed by this load resistance?

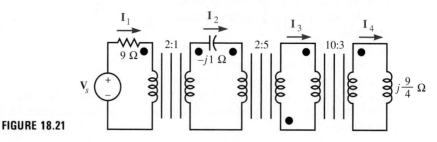

38. Find the phasor currents I_1, I_2, I_3, and I_4, if $V_s = 30\underline{/30°}$ V in the circuit in Fig. 18.21.

39. The connection of two transformer coils in series, as shown in the following circuit, is often called an *autotransformer*. Assume that the transformers are ideal. Calculate: (a) V_{cd}/V_{ad}, (b) I_2/I_1, and (c) $Z_1 = V_{ad}/I_1$.

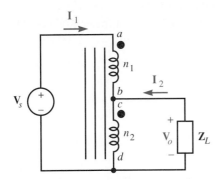

40. Repeat Problem 39 with the dot mark by terminal c moved to the end by terminal d.

41. For the following circuit, determine the turns ratio n_1/n_2 and the reactance so that the load Z_L absorbs maximum power (n_1/n_2 is real, not complex). Calculate the power absorbed by this value of Z_L, if $V_s = 100\underline{/0°}$ V rms.

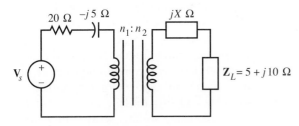

42. Use the dependent source model for an ideal transformer, as shown in Fig. 18.12a, and solve for the mesh currents in the following circuit if $V_s = 6\underline{/0°}$ V.

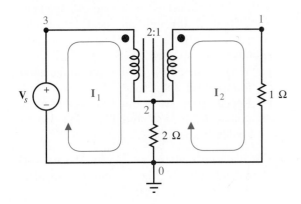

43. Use the dependent source model for an ideal transformer, as shown in Fig. 18.12b, and solve for the mesh currents in the preceding circuit, if $V_s = 6\underline{/0°}$ V.

44. Use the dependent source model for an ideal transformer, as shown in Fig. 18.12a, and solve for the node voltages in the preceding circuit, if $V_s = 6\underline{/0°}$ V.

45. Use the dependent source model for an ideal transformer, as shown in Fig. 18.12b, and solve for the node voltages in the preceding circuit, if $V_s = 6\underline{/0°}$ V.

Section 18.5

46. Use the ideal transformer component and the method described in Example 18.3 to solve for the mesh currents in the following circuit. Use these mesh currents and calculate voltages V_1 and V_2 if $V_s = 120\underline{/0°}$ V.

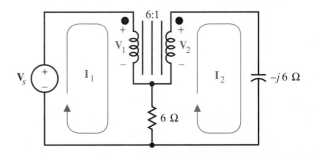

47. Use the ideal transformer component and solve for the mesh currents in the following circuit if $V_s = 20\underline{/0°}$ V.

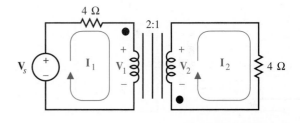

Section 18.6

48. Use the ideal transformer component and the method of node voltages described in Example 18.4 to solve for voltages V_1 and V_2 in the preceding circuit if $V_s = 20\underline{/0°}$ V.

49. Use the ideal transformer component and solve for the node voltages in the following circuit if $\mathbf{V}_s = 24\underline{/0°}$ V.

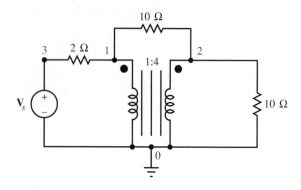

Section 18.7

50. A transformer has a core loss of 20 W when the primary coil operates at its rated voltage of 480 V rms. The secondary is rated at 120 V rms. The coupling coefficient k is very close to 1. What would be the value of R_{c1} in the model of Fig. 18.20b? What would be the value of the resistance R_{c2} that would be used in the model if the excitation circuit, R_{c1} and $j\omega L_{M1}$, were reflected to the other side of the transformer model?

Comprehensive Problems

51. The following network is used as an impedance transformation device at $f_0 = 1000$ Hz.

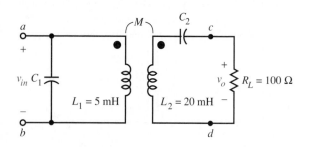

(a) Select capacitance C_2 so that the self-impedance of the secondary is real.
(b) Select M so that the conductance looking into terminals a and b with C_1 disconnected is $\frac{1}{600}$ S.

(c) Now select C_1 so that the driving-point impedance looking into terminals a and b is 600 Ω.
(d) If a practical source, modeled as an ideal voltage source in series with a 600-Ω resistance, is connected to terminals a and b, what is the Thévenin impedance seen by the 100-Ω load?
(e) Determine the transfer function $\mathbf{H}(j\omega_0) = \mathbf{V}_o/\mathbf{V}_{in}$, where $\omega_0 = 2\pi1000$ rps.

52. Use the values of M, C_1, and C_2 calculated in Problem 51, to plot $|\mathbf{H}(j\omega)|$ for $\frac{1}{2}\omega_0 \leq \omega \leq 2\omega_0$.

53. Consider the *doubly-tuned transformer* shown in the following circuit.

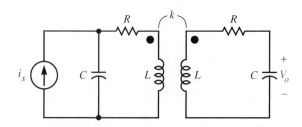

(a) Calculate the transfer function $\mathbf{V}_o/\mathbf{I}_s$.
(b) If $k = 0.04$, $R = 0.02$ Ω, $L = 1$ H, and $C = 1$ F, plot $|\mathbf{H}(j\omega)| = |\mathbf{V}_o/\mathbf{I}_s|$.
(c) Magnitude- and frequency-scale the network so that the resonant frequency of the primary circuit and secondary circuit is 455 kHz and L is 1 mH.
(d) After frequency scaling, what is the bandwidth for which the magnitude of the response is greater than or equal to 0.707 of the value at 455 kHz?

54. A voltage source of internal impedance $\mathbf{Z}_s = 300\underline{/0°}$ Ω is connected to a load of impedance $\mathbf{Z}_L = 50\underline{/0°}$ Ω through an ideal transformer with turns ratio n_1/n_2, where side 1 corresponds to the source.
(a) What value of n_1/n_2 will result in the maximum power delivered to $\mathbf{Z}_L$?
(b) Repeat part (a) if the transformer resistances $R_1 = 36$ Ω and $R_2 = 1$ Ω are included in the model.

55. Calculate currents $\mathbf{I}_1$ and $\mathbf{I}_2$ for the network shown in Fig. 18.22. The transformers are ideal.

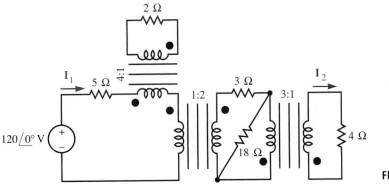

FIGURE 8.22

56. The two coils of the ideal transformer in the following circuit are connected in series to form an autotransformer.

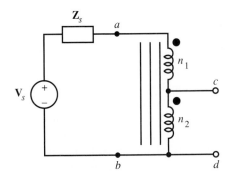

(a) Find the Thévenin impedance with respect to terminals c and d.
(b) Find the Thévenin impedance seen by the source when an impedance of $\mathbf{Z}_L$ is connected between terminals c and d.

57. Repeat Problem 56 with the dot mark on the lower coil at the bottom.

58. The transformer shown in Problem 56 is not ideal. The coil with n_1 turns has inductance L_1, and the coil with n_2 turns has inductance L_2. The coefficient of coupling is 1.
(a) What is the Thévenin impedance looking into terminals c and d?
(b) What is the Thévenin impedance seen by the source when an impedance of $\mathbf{Z}_L$ is connected between terminals c and d?

59. Repeat Problem 58 with the dot mark on the lower coil at the bottom.

60. Consider the ideal autotransformer connected as shown in the following circuit.

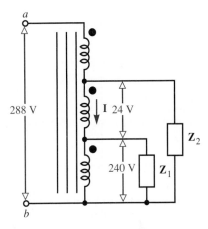

(a) What is the Thévenin impedance looking into terminals a and b?
(b) Find a relation between $\mathbf{Z}_1$ and $\mathbf{Z}_2$ so that current $\mathbf{I}$ is zero.

61. Find the Thévenin impedance looking into terminals a and b for the following network. If a resistance of value R_L is connected between terminals a and b, what resistance is seen between terminals c and d by the voltage sources?

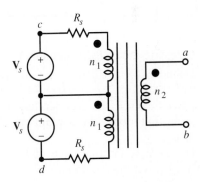

62. For the ideal transformer network below, solve for voltages $\mathbf{V}_1$ and $\mathbf{V}_2$ if $\mathbf{V}_s = 72\underline{/0°}$ V.

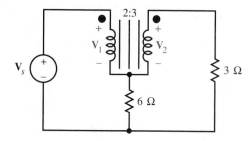

63. Calculate the transfer function $\mathbf{I}_2/\mathbf{I}_1$ and the impedance seen by the current source for the network shown in the following circuit.

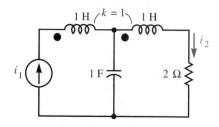

64. Construct a model that includes an ideal transformer for the preceding circuit.

65. The load on the linear transformer shown in the following circuit is a capacitance of value C_2. The voltage $\mathbf{V}_1$ is supplied by a voltage source.

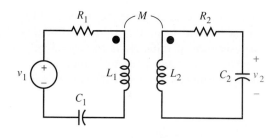

(a) Calculate the transfer function $\mathbf{V}_2/\mathbf{V}_1$.
(b) Define $\omega_0 = 1/\sqrt{L_2C_2} = 1/\sqrt{L_1C_1}$, $Q_1 = \omega_0 L_1/R_1$, $Q_2 = \omega_0 L_2/R_2$ and $k = M/\sqrt{L_1L_2}$. Write the transfer function in a form that does not contain M, L_1, L_2, or C_1.
(c) Define k_c to be the value of k for which $|\mathbf{V}_2/\mathbf{V}_1|$ is a maximum at $\omega = \omega_0$. Develop an expression for k_c in terms of Q_1 and Q_2.

66. A transformer has an effective turns ratio of 20, a primary resistance of $R_1 = 200\ \Omega$, and a secondary resistance of $R_2 = 0.5\ \Omega$. The core losses are negligible. The inductance of the primary with the secondary open-circuited is $L_1 = 5$ H. The inductance of the primary coil with the secondary coil short-circuited is $L_{1sc} = 0.1$ H. Construct a model for the transformer that incorporates an ideal transformer. You can assume that $k = k_1 = k_2$ and $\omega L_1 \gg R_1$.

Use the method introduced in Chapter 17 to calculate the *open-circuit impedance matrix* in the indicated equations.

67. Eq. (18.16)   68. Eq. (18.18)

Use the method introduced in Chapter 17 to calculate the *short-circuit* admittance matrix in the indicated equations.

69. Eq. (18.17)   70. Eq. (18.22)
71. Use PSpice to solve Problem 4.
72. Use the ideal transformer model of Fig. 18.12a, and PSpice, to solve Problem 49.
73. Use PSpice to solve for current $\mathbf{I}_2$ when the angular frequency is 1 rad/s and current $\mathbf{I}_1 = 10\underline{/0°}$ A in the circuit of Problem 63.
74. Use PSpice to solve Example 18.2.
75. Use PSpice to solve Example 18.3.
76. Your fishing cabin is at the end of the rural distribution system. The line voltage is 109 V rms with or without the television set being turned on. The result is that your 12-inch television has an 11-inch picture. You buy a 120/12 V transformer (assume it is ideal) rated at 12 VA. Show how you would connect this as an auto-transformer to supply 120 V to the television set (show the dot marks). The television requires 100 VA. Will this overload the 12 VA transformer?
77. A small transformer has a turns ratio of 5/1 and is designed for a supply voltage of 120 V rms and a maximum output current of 2 A rms. Transformers of this size are typically far from ideal. The parameters of the model shown in Fig. 18.20a are $R_1 = 10\ \Omega$, $R_2 = 0.5\ \Omega$, $R_{c1} = 1.2\ \text{k}\Omega$, $X_{m1} = 5\ \text{k}\Omega$, $X_{\ell1} = 40\ \Omega$, and $X_{\ell2} = 2\ \Omega$. Use PSpice to calculate the load voltage when a 10-Ω resistive load is connected. What is the transformer input current?

78. Repeat Problem 77, but use the equivalent circuit of Fig. 18.20b. Compare your answers to those obtained in Problem 77.

A practical power transformer is rated at 2400/240 V rms ($n_1/n_2 = 10/1$) and 15 kVA at 60 Hz. Refer to the model of Fig. 18.20a. The model parameters are $R_1 = 3\ \Omega$, $R_2 = 0.025\ \Omega$, $R_{C1} = 64\ \text{k}\Omega$, $X_{m1} = 16\ \text{k}\Omega$, $X_{\ell 1} = 9\ \Omega$, and $X_{\ell 2}\ 0.08\ \Omega$. Use this transformer model and PSpice to solve Problems 79 through 83. The load voltage is maintained at 240 V rms.

79. What must be the primary voltage if the rated kVA is delivered to (a) a resistive load, (b) a load with a lagging power factor of 0.8, and (c) a load with a leading power factor of 0.8?

80. Use the approximate equivalent circuit of Fig. 18.20b, and repeat Problem 79. Compare answers for the two problems. Is the approximation reasonable?

81. The transformer has a 15-kVA resistive load connected. The load voltage is to be maintained at 240 V rms in the steady state. A primary voltage of $V_1 = V_a\sqrt{2}\cos(2\pi 60t + \theta)$ V, where V_a is as found in part a of Problem 79, is applied at time zero. Determine the primary peak transient current that will flow for (a) $\theta = 0°$, (b) $\theta = 45°$, and (c) $\theta = 90°$. (Nonlinearities in the magnetic material can cause peak transient currents larger than these values. PSpice can be used to solve such problems, but they are not considered in this book.)

82. Repeat Problem 81 where the load is 15 kVA at a lagging power factor of 0.8 and the rms primary voltage is as found in part b of Problem 79.

83. Repeat Problem 81 where the load is 15 kVA at a leading power factor of 0.8 and the rms primary voltage is as found in part c of Problem 79.

CHAPTER
19

Single- and Three-Phase
Power Circuits

The analyses of electrical generation, transmission, and distribution systems are specialized topics in electrical engineering that can fill several semester courses. Expertise in these subjects is typically required only of engineers employed by electrical utilities and consulting engineering firms and of plant engineers. Nevertheless, all engineers require some knowledge of how electric power is supplied.

We begin by describing three-wire single-phase systems, which supply power to nearly all residences in the United States. We then present a brief introduction to balanced three-phase circuits, which are used in industry and many commercial air conditioning systems. We see that an important property of a balanced three-phase system is that it delivers *constant* power. That is, the power delivered does not fluctuate with time as in a single-phase system.

19.1 Three-Wire Single-Phase Systems

In the United States nearly all residences and many small commercial facilities are supplied power by a 120/240-V 60-Hz three-wire connection to the power system. The power source and load can be modeled as in Fig. 19.1.

FIGURE 19.1
The three-wire single-phase system that supplies residential power

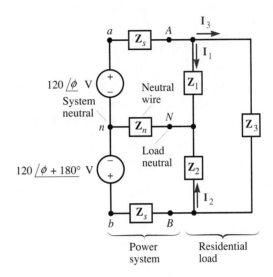

In Fig. 19.1 the line-to-neutral voltages at the source are $\mathbf{V}_{an} = 120\underline{/\phi}$ V and $\mathbf{V}_{bn} = 120\underline{/\phi + 180°}$ V, where the system neutral (n) is typically connected to the earth (earth ground). Thus $\mathbf{V}_{ab} = \mathbf{V}_{an} + \mathbf{V}_{nb} = 240\underline{/\phi}$ V and the system supplies both 120- and 240-V service. The system impedances $\mathbf{Z}_s$ and $\mathbf{Z}_n$ are normally small enough that the load voltages $\mathbf{V}_{AN}$, $\mathbf{V}_{BN}$, and $\mathbf{V}_{AB}$ differ from the system voltages by only a few percent.

Low-power loads such as lighting and small appliances are normally designed for 120-V line voltages and are represented by impedances $\mathbf{Z}_1$ and $\mathbf{Z}_2$. Large loads such as cooking stoves and central air conditioners, which are usually connected to 240 V, are represented by impedance $\mathbf{Z}_3$. Equal 120-V loads $\mathbf{Z}_1$ and $\mathbf{Z}_2$ will result in *line currents* being related by $\mathbf{I}_{aA} = -\mathbf{I}_{bB}$ and *neutral current* $\mathbf{I}_{nN}$ equal to zero. Loads $\mathbf{Z}_1$ and $\mathbf{Z}_2$ are, of course, rarely balanced, but the neutral current is usually less than the line current, which results in reduced power losses. The connection shown in Fig. 19.1 is used for safety as well as efficiency. The system supplies 240-V service without any line having more than 120 V with respect to the system neutral, which is connected to earth ground.

For the unusual case where $\mathbf{Z}_1 = \mathbf{Z}_2$, the analysis can be simplified if we realize that because the neutral line, represented by impedance $\mathbf{Z}_n$, carries no current, we can replace it by an open circuit. Loads $\mathbf{Z}_1$ and $\mathbf{Z}_2$ are effectively in series, and this series combination is in parallel with $\mathbf{Z}_3$. Typically $\mathbf{Z}_1 \neq \mathbf{Z}_2$ and a technique such as mesh-current or node-voltage analysis is appropriate. When numerical values for the impedances and source voltages are given, a circuit analysis program such as PSpice can be used.

Although power is supplied to the load by three wires, this *is not* considered to be a three-phase system, because $\mathbf{V}_{an}$, $\mathbf{V}_{bn}$, and $\mathbf{V}_{ab}$ differ in phase by either $0°$ or $180°$. The corresponding voltages in a balanced three-phase system would differ by $120°$.

EXAMPLE 19.1

Refer to Fig. 19.1. The load represented by $\mathbf{Z}_1$ absorbs a complex power of $\mathbf{S}_1 = 1200\underline{/60°}$ VA, $\mathbf{Z}_2$ absorbs $\mathbf{S}_2 = 600\underline{/0°}$ VA, and $\mathbf{Z}_3$ absorbs $\mathbf{S}_3 = 2400\underline{/-45°}$ VA. Neglect line impedances $\mathbf{Z}_s$ and neutral impedance $\mathbf{Z}_n$ when calculating the currents. Find $\mathbf{I}_1$, $\mathbf{I}_2$, $\mathbf{I}_3$, $\mathbf{I}_{aA}$, $\mathbf{I}_{bB}$, $\mathbf{I}_{nN}$, and the line losses.

■ **SOLUTION**

From the relation[†]

$$\mathbf{S} = \mathbf{VI}^*$$

$$\mathbf{I}_1 = \left(\frac{\mathbf{S}_1}{\mathbf{V}_{AN}}\right)^* = \left(\frac{1200\underline{/60°}}{120\underline{/\phi}}\right)^* = 10\underline{/(\phi - 60°)} \text{ A}$$

$$\mathbf{I}_2 = \left(\frac{\mathbf{S}_2}{\mathbf{V}_{BN}}\right)^* = \left(\frac{600\underline{/0°}}{120\underline{/\phi + 180°}}\right)^* = 5\underline{/(\phi + 180°)} \text{ A}$$

and

$$\mathbf{I}_3 = \left(\frac{\mathbf{S}_3}{\mathbf{V}_{AB}}\right)^* = \left(\frac{2400\underline{/-45°}}{240\underline{/\phi}}\right)^* = 10\underline{/(\phi + 45°)} \text{ A}$$

From KCL,

$$\mathbf{I}_{aA} = \mathbf{I}_1 + \mathbf{I}_3 = 10\underline{/\phi - 60°} + 10\underline{/\phi + 45°} = 12.18\underline{/\phi - 7.5°} \text{ A}$$

$$\mathbf{I}_{bB} = \mathbf{I}_2 - \mathbf{I}_3 = 5\underline{/\phi + 180°} - 10\underline{/\phi + 45°} = 13.99\underline{/\phi - 149.64°} \text{ A}$$

$$\mathbf{I}_{nN} - (\mathbf{I}_1 + \mathbf{I}_2) = -(10\underline{/\phi - 60°} + 5\underline{/\phi + 180°})$$
$$= 8.66\underline{/\phi + 90°} \text{ A}$$

The power lost in the lines for the system shown in Fig. 19.1 is

$$P = |\mathbf{I}_{nN}|^2 \mathscr{R}e\{\mathbf{Z}_n\} + |\mathbf{I}_{aA}|^2 \mathscr{R}e\{\mathbf{Z}_s\} + |\mathbf{I}_{bB}|^2 \mathscr{R}e\{\mathbf{Z}_s\}$$
$$= 75\mathscr{R}e\{\mathbf{Z}_n\} + 148.24\mathscr{R}e\{\mathbf{Z}_s\} + 195.71\mathscr{R}e\{\mathbf{Z}_s\}$$

If the resistances of the lines and neutral wire are the same,

$$R = \mathscr{R}e\{\mathbf{Z}_n\} = \mathscr{R}e\{\mathbf{Z}_s\}$$

then the power loss in the three wires is

$$P = (419.0)R \text{ W} ■$$

[†] The use of effective values for voltage and current is assumed, so a factor of $\frac{1}{2}$ is not included in the equation for complex power. The phasor $\mathbf{I}^*$ is the complex conjugate of $\mathbf{I}$.

In power systems, voltage and current are universally specified in terms of rms (effective) values. A three-wire single-phase system supplies power to most residences. Two lines are 120 V with respect to the neutral wire, and the neutral wire is connected to earth ground. The voltage between the two 120-V lines is 240 V, and voltages are supplied at a frequency of 60 Hz in the United States.

EXERCISES

1. Refer to Fig. 19.1. Assume that the angle ϕ is zero. The impedances are $Z_s = j1\ \Omega$, $Z_n = j2\ \Omega$, $Z_1 = Z_2 = -j10\ \Omega$, and $Z_3 = j10\ \Omega$. Find currents I_1, I_2, and I_3, and voltages V_{AN}, V_{BN}, and V_{Nn}. Show these on a phasor diagram that also includes V_{an} and V_{bn}. Check your answer with PSpice or a similar program, if available.

2. Repeat Exercise 1 when the load impedances are $Z_1 = -j10\ \Omega$, $Z_2 = j10\ \Omega$, and $Z_3 = \infty$.

3. Repeat Exercise 1 when the load impedances are $Z_1 = -j10\ \Omega$, $Z_2 = j10\ \Omega$, and $Z_3 = 10\ \Omega$.

19.2 Power Generation

Although the direct conversion of heat or solar radiation to electrical energy is used in special applications, most electrical energy is supplied from electromechanical generators. More specifically, most electrical energy is generated by *synchronous machines*, called alternators, in which an electromagnet is rotated inside an internally slotted laminated-steel cylinder. The rotating magnetic field induces voltages in wires (conductors) placed in these slots. These conductors supply electrical energy to the connected load.

Actual alternator design and construction is rather involved, since the number and connection of the wires placed in each slot is chosen to make the resultant terminal voltage as nearly sinusoidal as practicable. Nevertheless, the end view of a single-phase two-wire alternator can be represented pictorially as shown in Fig. 19.2a. Terminals a and a' represent the ends of a loop of wire going into and coming out of the page. As the electromagnet rotates, a voltage is generated. This is represented by voltage $v_{aa'}$ or phasor voltage $V_{aa'}$.

A second set of windings displaced by 90°, as shown in Fig. 19.2b, permits generation of a voltage $v_{bb'}$ displaced 90° electrically from $v_{aa'}$ as shown. This would be a *two-phase* alternator. Two-phase generation and utilization of power has some advantages over single-phase power. One advantage is that, unlike single-phase induction motors, two-phase induction motors are self-starting and have an instantaneous power output that is constant.

Electrical utilities no longer generate or supply two-phase power, but some two-phase motors are still used on old equipment. These old motors can still be operated, because two-phase electric power can be supplied from a three-phase system with the proper transformer connections. Two-phase motors with less than $\frac{1}{4}$ hp are occasionally used in control applications, where they are powered by electronic amplifiers.

FIGURE 19.2
Simplified representations
of alternators and the
generated voltages: (a)
single-phase; (b) two-phase;
(c) three-phase

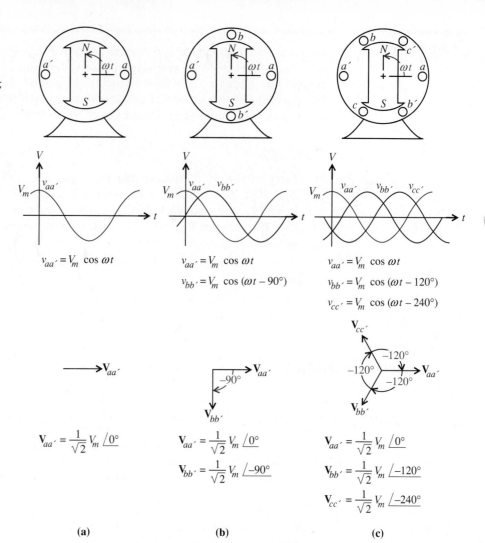

Installation of three sets of windings displaced by 120 mechanical degrees, as depicted in Fig. 19.2c, gives three voltages $v_{aa'}$, $v_{bb'}$, and $v_{cc'}$, each displaced by 120 electrical degrees from each other. If all voltages are sinusoidal and of equal magnitude V_m, this is a *balanced* three-phase set of voltages.

Generation, transmission, and utilization of three-phase power has some advantages over either single- or two-phase power. Three-phase windings in alternators fill all slots uniformly and thus make very efficient use of the space available for conductors. As a result, virtually all large alternators are three-phase. Three-phase alternators are also used in conjunction with rectifiers to supply dc electric power in automobiles.

For reasons that are detailed in power systems texts, three-phase power transmission makes more efficient use of transmission-line materials than does single- or two-phase transmission. For this reason, most long-distance power transmission is by three-phase lines, although some dc lines are in use. In Section 19.7 we will show that a balanced load on a balanced three-phase system absorbs

Three-phase alternator.
Courtesy of Sunstrand
Corporation

constant power. This constant power results in less noise and vibration than the oscillating power in a single-phase system. Three-phase induction motors are also self-starting and economically advantageous for many applications.

The phase voltages for a balanced three-phase generator have equal magnitude and a phase displacement of 120° with respect to each other. Phasor voltages and currents in power systems are expressed in effective (rms) values.

EXERCISE

4. The voltage for one phase of a balanced three-phase generator is known to be $v_{aa'} = 3919 \cos (2\pi 60t + 45°)$ V. What is the phasor value of $\mathbf{v}_{aa'}$? Give possible values for the phasors $\mathbf{v}_{bb'}$ and $\mathbf{v}_{cc'}$.

19.3 Y Connections

Although three-phase power generation requires three windings that have a total of six terminals, these windings are interconnected, and only three or at most four wires are used to supply power. We will use three ideal voltage sources to represent the voltages generated in the alternator windings.

Positive Phase Sequence

A Y source connection is shown in Fig. 19.3. Terminals a', b', and c' are connected to a common point n called the *neutral*.

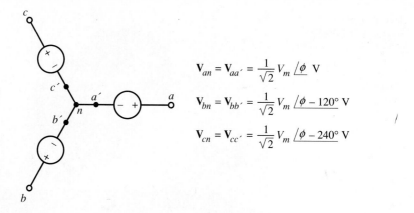

FIGURE 19.3
A balanced Y-connected source with positive (*abc*) phase sequence

$$\mathbf{V}_{an} = \mathbf{V}_{aa'} = \frac{1}{\sqrt{2}} V_m \underline{/\phi} \text{ V}$$

$$\mathbf{V}_{bn} = \mathbf{V}_{bb'} = \frac{1}{\sqrt{2}} V_m \underline{/\phi - 120°} \text{ V}$$

$$\mathbf{V}_{cn} = \mathbf{V}_{cc'} = \frac{1}{\sqrt{2}} V_m \underline{/\phi - 240°} \text{ V}$$

As mentioned earlier, voltages in electric power systems are specified in rms or effective values unless otherwise noted. The rms value of the *line-to-neutral voltage* or *phase voltage* for the Y connection is

$$V_{LN} = \frac{1}{\sqrt{2}} V_m \tag{19.1}$$

where V_m is the maximum value of the sinusoidal line-to-neutral voltage. If all phase voltages have the same magnitude and are displaced one from the other by 120°, we say that the system is *balanced*. If $\mathbf{V}_{bb'}$ lags $\mathbf{V}_{aa'}$ by 120° and $\mathbf{V}_{cc'}$ lags $\mathbf{V}_{aa'}$ by 240°, as shown in Fig. 19.4, we say that the voltages have *positive* or *abc phase sequence*. Voltages $\mathbf{V}_{an}$, $\mathbf{V}_{bn}$, and $\mathbf{V}_{cn}$ are the phasor line-to-neutral voltages. For a Y connection, these are the same as the phase voltages $\mathbf{V}_{aa'}$, $\mathbf{V}_{bb'}$, and $\mathbf{V}_{cc'}$. *Line-to-line* or simply *line voltages* are easily calculated from the line-to-neutral voltages:

$$\begin{aligned}
\mathbf{V}_{ab} &= \mathbf{V}_{an} - \mathbf{V}_{bn} \\
&= \mathbf{V}_{an} - \mathbf{V}_{an}(1\underline{/-120°}) \\
&= \mathbf{V}_{an}[1 - (\cos 120° - j \sin 120°)] \\
&= \mathbf{V}_{an}\left(\frac{3}{2} + j\frac{1}{2}\sqrt{3}\right) \\
&= \sqrt{3}\underline{/30°}\,\mathbf{V}_{an}
\end{aligned} \tag{19.2}$$

In a similar manner we can calculate the other two line voltages to show that:

Line Voltages for a Positive (*abc*) Sequence Y Connection

$$\mathbf{V}_{ab} = \sqrt{3}\underline{/30°}\,\mathbf{V}_{an} \tag{19.3}$$

$$\mathbf{V}_{bc} = \sqrt{3}\underline{/30°}\,\mathbf{V}_{bn} \tag{19.4}$$

$$\mathbf{V}_{ca} = \sqrt{3}\underline{/30°}\,\mathbf{V}_{cn} \tag{19.5}$$

Thus the magnitudes of the line voltages are the square roots of three times the magnitudes of the line-to-neutral voltages. For the positive (*abc*) phase sequence, the line-to-line voltages lead the corresponding line-to-neutral voltages by 30°. This relationship is more easily visualized with reference to the phasor diagram of

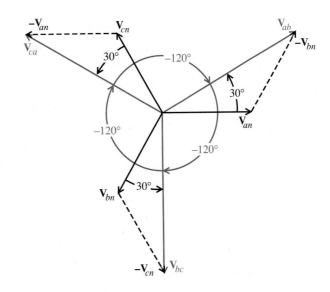

Fig. 19.4, where ϕ, the angle on $\mathbf{V}_{an}$, has been chosen as zero for convenience. If ϕ were unequal to zero, the entire phasor diagram would simply be rotated counterclockwise by angle ϕ.

Voltages for Y-Connected Loads

Consider a Y-connected load and a Y-connected source as shown in Fig. 19.5. Assume that the source is balanced with a positive (*abc*) phase sequence and $\mathbf{V}_{an} = V_{LN}\underline{/\phi}$. The circuit has four nodes and three voltage sources. With n chosen as the reference node,

---

Phase Voltages Equal Line-to-Neutral Voltages for Y-Connected Loads

$$\mathbf{V}_{AN} = \mathbf{V}_{an} = V_{LN}\underline{/\phi} = \frac{1}{\sqrt{3}}\underline{/-30°}\,\mathbf{V}_{ab} \tag{19.6}$$

$$\mathbf{V}_{BN} = \mathbf{V}_{bn} = V_{LN}\underline{/(\phi-120°)} = \frac{1}{\sqrt{3}}\underline{/-30°}\,\mathbf{V}_{bc} \tag{19.7}$$

$$\mathbf{V}_{CN} = \mathbf{V}_{cn} = V_{LN}\underline{/(\phi-240°)} = \frac{1}{\sqrt{3}}\underline{/-30°}\,\mathbf{V}_{ca} \tag{19.8}$$

where a positive (*abc*) phase sequence is assumed.

FIGURE 19.5
A Y-connected source with a Y-connected load

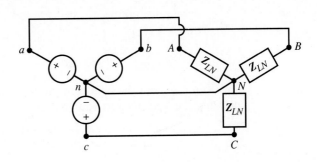

Currents for Y-Connected Loads

We can easily find the *line-to-neutral currents* for a Y-connected load. Examination of Fig. 19.5 reveals that the line currents are the same as the corresponding line-to-neutral currents. Define the magnitude of the line current by

$$I_L = |\mathbf{I}_{aA}|$$
$$= |\mathbf{I}_{AN}|$$
$$= I_{LN} \tag{19.9}$$

where I_{LN} is the magnitude of the line-to-neutral current. The line-to-neutral currents are *phase currents* for a Y-connected load. Defining θ to be the angle on the line-to-neutral impedance (the phase-impedance for a Y-connected load),

$$\mathbf{Z}_{LN} = |\mathbf{Z}_{LN}|\underline{/\theta} \tag{19.10}$$

gives us:

Line Currents Equal Phase Currents for Y-Connected Loads

$$\mathbf{I}_{aA} = \mathbf{I}_{AN} = \frac{1}{\mathbf{Z}_{LN}}\,\mathbf{V}_{AN} = I_L\underline{/(\phi - \theta)} \tag{19.11}$$

$$\mathbf{I}_{bB} = \mathbf{I}_{BN} = \frac{1}{\mathbf{Z}_{LN}}\,\mathbf{V}_{BN} = \mathbf{I}_{AN}(1\underline{/-120°}) = I_L\underline{/(\phi - \theta - 120°)} \tag{19.12}$$

$$\mathbf{I}_{cC} = \mathbf{I}_{CN} = \frac{1}{\mathbf{Z}_{LN}}\,\mathbf{V}_{CN} = \mathbf{I}_{AN}(1\underline{/-240°}) = I_L\underline{/(\phi - \theta - 240°)} \tag{19.13}$$

where a positive (*abc*) phase sequence is assumed.

These currents are shown in Fig. 19.6 for $\phi = 0$ and $\theta > 0$. Voltages $\mathbf{V}_{an}$ and $\mathbf{V}_{AB}$ are included for reference.

FIGURE 19.6
Phasor diagram for the line currents and phase currents in a balanced Y-connected three-phase load in an *abc* phase-sequence system

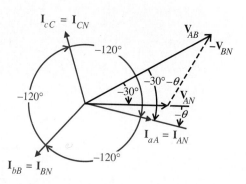

We can calculate the neutral current $\mathbf{I}_{nN}$ by applying KCL to a closed surface enclosing the Y-connected source:

$$\mathbf{I}_{aA} + \mathbf{I}_{bB} + \mathbf{I}_{cC} + \mathbf{I}_{nN} = 0 \tag{19.14}$$

$$\mathbf{I}_{nN} = -\frac{1}{\mathbf{Z}_{LN}}(\mathbf{V}_{AN} + \mathbf{V}_{BN} + \mathbf{V}_{CN})$$

$$= -\frac{1}{\mathbf{Z}_{LN}}\mathbf{V}_{AN}(1 + 1\underline{/-120°} + 1\underline{/-240°})$$

$$= 0 \tag{19.15}$$

The obvious conclusion is that a balanced load does not require a neutral wire connecting n to N. Omission of the neutral wire gives a *three-wire three-phase system*.

Complex Power for Y-Connected Loads

The complex power absorbed by a balanced Y-connected load is the sum of the power absorbed by each of the three phases. For a balanced load, the power absorbed by each phase is the same. Therefore, the total complex power absorbed by the load is

$$\mathbf{S} = \mathbf{S}_A + \mathbf{S}_B + \mathbf{S}_C = 3\mathbf{S}_A \tag{19.16}$$

which gives

Complex Power Absorbed by Y-Connected Loads

$$\mathbf{S} = 3\mathbf{V}_{AN}\mathbf{I}_{AN}^* = 3V_{LN}I_{LN}\underline{/\theta}\ \text{VA} \tag{19.17}$$

Line voltage has a magnitude of

$$V_L = \sqrt{3}\,V_{LN}\ \text{V} \tag{19.18}$$

and line current has a magnitude of

$$I_L = I_{LN}\ \text{A} \tag{19.19}$$

Substitution of Eqs. (19.18) and (19.19) into Eq. (19.17) gives the following relationship for the total three-phase complex power absorbed by the load:

Complex Power Absorbed by Y-Connected Loads

$$\mathbf{S} = \sqrt{3}\,V_L I_L\underline{/\theta}\ \text{VA} \tag{19.20}$$

---

where V_L and I_L are in rms values. We can conveniently express the complex power absorbed by a Y-connected load as

$$\mathbf{S} = 3\mathbf{V}_{AN}\mathbf{I}_{AN}^* = 3\mathbf{V}_{AN}\left(\frac{\mathbf{V}_{AN}}{\mathbf{Z}_{LN}}\right)^*$$

$$= 3\frac{1}{\mathbf{Z}_{LN}^*}V_{LN}^2 = 3\frac{1}{\mathbf{Z}_{LN}^*}\left(\frac{V_L}{\sqrt{3}}\right)^2 \tag{19.21}$$

which gives

EXAMPLE 19.2

Three impedances of $5\underline{/36.87°}\ \Omega$ are Y-connected to a 440-V three-phase line. Find the complex power absorbed by the three-phase load and the magnitude of the rms line current.

■ **SOLUTION**

$$\mathbf{S} = \frac{1}{\mathbf{Z}_{LN}^*} V_L^2 = \frac{440^2}{5\underline{/-36.87°}} = 38{,}720\underline{/36.87°}\ \text{VA}$$

$$= 30{,}976 + j23{,}232\ \text{VA}$$

$$I_L = \frac{|\mathbf{S}|}{\sqrt{3}\,V_L} = 50.81\ \text{A}\ ■$$

Negative Phase Sequence

Interchanging the wire designations b and c and b' and c' would result in phase voltages

$$\mathbf{V}_{aa'} = V_{LN}\underline{/\phi} \qquad (19.23)$$

$$\mathbf{V}_{cc'} = V_{LN}\underline{/\phi - 120°} = V_{LN}\underline{/\phi + 240°} \qquad (19.24)$$

$$\mathbf{V}_{bb'} = V_{LN}\underline{/\phi - 240°} = V_{LN}\underline{/\phi + 120°} \qquad (19.25)$$

as shown in Fig. 19.7. We refer to this arrangement as a *negative* or *acb phase sequence.** Proceeding as before, a Y connection can be shown to give

For a negative phase sequence, the line-to-line voltages are $\sqrt{3}$ times the line-to-neutral voltages, but lag the corresponding line-to-neutral voltages by 30°. This relationship is shown in the phasor diagram of Fig. 19.7, where ϕ is again chosen to be zero for convenience.

The distinction between the positive and negative phase sequences is significant in applications. One important reason is that interchanging any two wires of a three-phase system effectively changes the phase sequence and reverses the direction of rotation for three-phase motors.

---

* The negative phase sequence is also called a *cba* sequence.

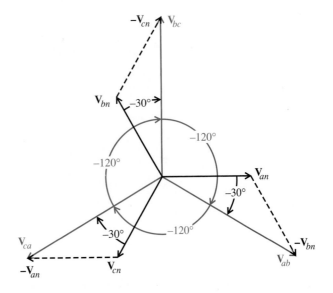

REMEMBER

For a balanced positive (*abc*) phase sequence three-phase system, the line voltage (line-to-line voltage) is $\sqrt{3}$ times the line-to-neutral voltage in magnitude and leads the line-to-neutral voltage by 30°. The phase current for a Y connection is the same as the corresponding line current.

EXERCISES

A balanced positive (*abc*) phase sequence system with a balanced Y-connected load has a line-to-neutral impedance (phase impedance) of $\mathbf{Z}_{LN} = 20\underline{/15°}\ \Omega$. The line voltage is $\mathbf{V}_{AB} = 208\underline{/0°}$ V.

5. Calculate the line voltages $\mathbf{V}_{BC}$ and $\mathbf{V}_{CA}$.
6. Calculate the line-to-neutral voltages $\mathbf{V}_{AN}$, $\mathbf{V}_{BN}$, and $\mathbf{V}_{CN}$.
7. Calculate the line-to-neutral currents $\mathbf{I}_{AN}$, $\mathbf{I}_{BN}$, and $\mathbf{I}_{CN}$. These are also the line currents.
8. Find the complex power absorbed by the load.
9. You wish to design a three-phase system so that 120-V lighting can be operated from the line-to-neutral connections of Y-connected transformers. What will be the three-phase line voltage (line-to-line voltage)? This system is widely used in industry.

19.4 Δ Connections

Balanced three-phase systems do not always use the Y connection shown in Fig. 19.5 (with or without the neutral wire). We will begin by analyzing a Δ-connected load with power supplied by a Y-connected source. This system is pictured in Fig. 19.8.

FIGURE 19.8
A Y-connected source with a
Δ-connected load

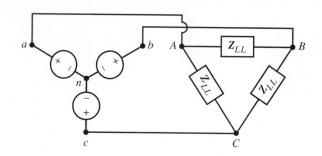

Currents for Δ-Connected Loads

Line-to-line currents or *phase currents* through impedances $\mathbf{Z}_{LL}$ for a Δ-connected load are easily found in terms of line voltages. Define the magnitude of the phase current by

$$I_{LL} = |\mathbf{I}_{AB}| \tag{19.29}$$

Line-to-Line Currents Are Equal to Phase Currents for Δ-Connected Loads

$$\mathbf{I}_{AB} = \frac{1}{\mathbf{Z}_{LL}} \mathbf{V}_{AB} = I_{LL}\underline{/\phi + 30° - \theta} \tag{19.30}$$

$$\mathbf{I}_{BC} = \frac{1}{\mathbf{Z}_{LL}} \mathbf{V}_{BC} = \mathbf{I}_{AB}(1\underline{/-120°}) = I_{LL}\underline{/\phi + 30° - \theta - 120°} \tag{19.31}$$

$$\mathbf{I}_{CA} = \frac{1}{\mathbf{Z}_{LL}} \mathbf{V}_{CA} = \mathbf{I}_{AB}(1\underline{/-240°}) = I_{LL}\underline{/\phi + 30° - \theta - 240°} \tag{19.32}$$

where a positive *(abc)* phase sequence is assumed.

As before, ϕ is the phase angle of the line-to-neutral voltage $\mathbf{V}_{AN}$.

Line currents are easily found in terms of the phase currents of the Δ:

$$\mathbf{I}_{aA} = \mathbf{I}_{AB} - \mathbf{I}_{CA} = \mathbf{I}_{AB}(1 - 1\underline{/-240°})$$

$$= \mathbf{I}_{AB}\left(\frac{3}{2} - j\frac{1}{2}\sqrt{3}\right) = \sqrt{3}\underline{/-30°}\,\mathbf{I}_{AB} \tag{19.33}$$

We can calculate the remaining two line currents in a similar manner to give

---

Line Currents for a Δ-Connected Load

$$\mathbf{I}_{aA} = \sqrt{3}\underline{/-30°}\,\mathbf{I}_{AB} \tag{19.34}$$

$$\mathbf{I}_{bB} = \sqrt{3}\underline{/-30°}\,\mathbf{I}_{BC} \tag{19.35}$$

$$\mathbf{I}_{cC} = \sqrt{3}\underline{/-30°}\,\mathbf{I}_{CA} \tag{19.36}$$

Thus for a positive *(abc)* phase sequence, line currents lag the corresponding phase currents (line-to-line currents) for a Δ by 30° and are larger than the phase currents by a factor of $\sqrt{3}$. These relationships are shown in Fig. 19.9.

FIGURE 19.9
Phasor diagram for the line
currents and phase currents
in a balanced Δ-connected
three-phase load in an *abc*
phase-sequence system

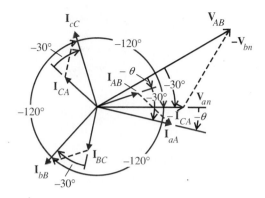

The complex power absorbed by each branch of a balanced Δ-connected load is the same. Therefore, the total power absorbed is three times the power absorbed by one branch of the delta. Therefore

$$\mathbf{S} = 3\mathbf{S}_a = 3\mathbf{V}_{AB}\mathbf{I}^*_{AB}$$

$$= 3\mathbf{V}_{AB}\left(\frac{1}{\sqrt{3}}\,\underline{/30°}\,\mathbf{I}_{aA}\right)^*$$

$$= (3V_L\,\underline{/\phi+30°})\left(\frac{1}{\sqrt{3}}\,I_L\,\underline{/\phi+30°-\theta}\right)^* \tag{19.37}$$

This gives

The Complex Power Absorbed by Δ-Connected Loads

$$\mathbf{S} = \sqrt{3}\,V_L I_L\,\underline{/\theta}\ \text{VA} \tag{19.38}$$

This is *identical* to Eq. (19.20) for a Y-connected load. In terms of the line-to-line impedance $\mathbf{Z}_{LL}$ (phase impedance of a Δ-connected load),

$$\mathbf{S} = 3\mathbf{V}_{AB}\mathbf{I}^*_{AB}$$

$$= 3\mathbf{V}_{AB}\left(\frac{\mathbf{V}_{AB}}{\mathbf{Z}_{LL}}\right)^* \tag{19.39}$$

which gives

The Complex Power Absorbed by a Δ-Connected Load

$$\mathbf{S} = 3\,\frac{1}{\mathbf{Z}^*_{LL}}\,V^2_L\ \text{VA} \tag{19.40}$$

We can obtain a relationship between the impedances of equivalent balanced Y and Δ loads by equating the power absorbed by a Δ-connected load and a Y-connected load:

$$\frac{1}{\mathbf{Z}^*_{LN}}\,V^2_L = 3\,\frac{1}{\mathbf{Z}^*_{LL}}\,V^2_L \tag{19.41}$$

If we take the complex conjugate of both sides of Eq. (19.41) we obtain

for equal complex power absorption from a power system.

EXAMPLE 19.3

Three impedances of $5\underline{/36.87°}\ \Omega$ are connected in Δ to a 440-V three-phase line. Find the complex power absorbed by the three-phase load and the magnitude of the rms line current.

■ **SOLUTION**

$$\mathbf{S} = 3\,\frac{1}{\mathbf{Z}_{LL}^*}\,V_L^2 = 3\,\frac{440^2}{5\underline{/-36.87°}} = 116{,}160\underline{/36.87°}\text{ VA}$$

$$\text{\textit{(handwritten)}} \; j69{,}969 \text{ VA}$$

$$I_L = \frac{|\mathbf{S}|}{\sqrt{3}\,V_L} = 152.42 \text{ A} \quad ■$$

Δ-Connected Sources

Instead of a Y connection, the coils of a balanced three-phase generator can be connected in a Δ, as shown in Fig. 19.10. A quick check will reveal that

$$\mathbf{V}_{ab} + \mathbf{V}_{bc} + \mathbf{V}_{ca} = 0 \qquad\qquad (19.43)$$

so KVL is not violated. (In practice, any slight imbalance is accommodated by the winding impedances, which have been neglected, but some current will circulate in the Δ mesh.) Line voltages are equal to the corresponding phase voltages for a Δ connection. Line voltages rather than line-to-neutral voltages would now be the same magnitude as the generated voltages and would be in phase with them. Given that the phase angle ϕ is zero, the phasor diagram for a Δ-connected source and load would be as in Fig. 19.11. The current $\mathbf{I}_{AB}$ would be the phase current (line-to-line current) for phase a of a Δ load, and current $\mathbf{I}_{aA}$ would be the line current. If the load is Y-connected, the line current is the phase current, and the phase voltage of phase a is $\mathbf{V}_{an}$.

FIGURE 19.10
A Δ-connected source with positive (*abc*) phase sequence

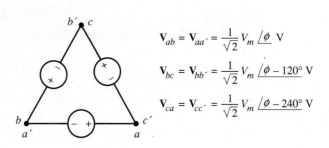

$$\mathbf{V}_{ab} = \mathbf{V}_{aa'} = \frac{1}{\sqrt{2}} V_m \underline{/\phi}\ \text{V}$$

$$\mathbf{V}_{bc} = \mathbf{V}_{bb'} = \frac{1}{\sqrt{2}} V_m \underline{/\phi - 120°}\ \text{V}$$

$$\mathbf{V}_{ca} = \mathbf{V}_{cc'} = \frac{1}{\sqrt{2}} V_m \underline{/\phi - 240°}\ \text{V}$$

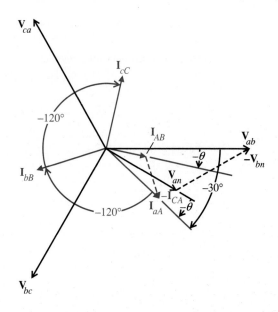

FIGURE 19.11
Phasor diagram for
Δ-connected load with
$V_{ab} = (1/\sqrt{2})V_m\underline{/0°}$ and a
positive phase sequence

REMEMBER

For a Δ-connected load or source in a balanced system, the line voltage (line-to-line voltage) is the same as the corresponding phase voltage. The magnitude of the line current is $\sqrt{3}$ times the magnitude of the corresponding phase current (line-to-line current). The line current lags the corresponding phase current by 30° for a positive (*abc*) phase sequence.

EXERCISES

A balanced positive (*abc*) phase sequence system has a balanced Δ-connected load. The phase impedances of the Δ-connected load each have a value of $Z_{LL} = 20\underline{/15°}\ \Omega$. The line voltage is $V_{AB} = 240\underline{/0°}$ V.

10. Calculate the phase currents (line-to-line currents) I_{AB}, I_{BC}, and I_{CA} for the Δ-connected load.
11. Calculate the line currents I_{aA}, I_{bB}, and I_{cC}.
12. Determine the complex power absorbed by the load.

19.5 Three-Phase Systems

A power system may have both Δ- and Y-connected sources and usually includes both Δ- and Y-connected loads. Care is required when discussing phase voltages and currents, as the relation between line voltages and phase voltages and line currents and phase currents is different for a Δ and a Y.

A summary of the important relationships in balanced positive (*abc*) phase sequence, three-phase circuits follows. These are for convenience and not for memorization. They are easily derived from a phasor diagram and the symmetry properties of a balanced system.

1. **Phase Voltage and Line Voltage**
   (a) A line-to-neutral voltage lags the corresponding line voltage (line-to-line voltage) by 30° and

$$V_{LN} = \frac{1}{\sqrt{3}} V_L \qquad (19.44)$$

   (b) The phase voltage for a Y connection is the line-to-neutral voltage.
   (c) The phase voltage for a Δ connection is the line-to-line voltage (line voltage).

2. **Phase Current and Line Current**
   (a) A line-to-neutral current is equal to the corresponding line current.
   (b) The line-to-line current leads the corresponding line current by 30° and

$$I_{LL} = \frac{1}{\sqrt{3}} I_L \qquad (19.45)$$

   (c) The phase current for a Y connection is the line current.
   (d) The phase current for a Δ connection is the line-to-line current.

3. **Calculating Line Currents**
   (a) For both Δ and Y loads, a line current is shifted in phase from the corresponding line voltage by $-30° - \theta$, where θ is the power factor angle (angle on the load impedance).
   (b) For a Y-connected load, the line current $\mathbf{I}_{aA}$ is

$$\mathbf{I}_{aA} = \frac{1}{\mathbf{Z}_{LN}} \cdot \frac{\mathbf{V}_{AB}}{\sqrt{3}} (1\underline{/-30°}) \qquad (19.46)$$

   and the magnitude of each line current is

$$I_L = \frac{1}{|\mathbf{Z}_{LN}|} \cdot \frac{V_L}{\sqrt{3}} \qquad (19.47)$$

   (c) For a Δ-connected load, the line current $\mathbf{I}_{aA}$ is

$$\mathbf{I}_{aA} = \frac{1}{\mathbf{Z}_{LL}} \cdot \sqrt{3}\,\mathbf{V}_{AB} \cdot (1\underline{/-30°}) \qquad (19.48)$$

   and the magnitude of each line current is

$$I_L = \frac{1}{|\mathbf{Z}_{LL}|} \sqrt{3}\,V_L \qquad (19.49)$$

4. **Calculating Complex Power**
   (a) The complex power absorbed by the load is

$$\mathbf{S} = \sqrt{3}\,V_L I_L \underline{/\theta} \text{ VA} \qquad (19.50)$$

   where V_L and I_L are the line voltage and line current in rms or effective values.
   (b) For a Y-connected load with impedance $\mathbf{Z}_{LN}$ Ω per phase,

$$\mathbf{S} = \frac{1}{\mathbf{Z}_{LN}^*} V_L^2 \text{ VA} \qquad (19.51)$$

The above relationships also apply to balanced negative phase sequence systems if "lags" is replaced by "leads" in relation 1a, "leads" by "lags" in relation 2b, and "$-30°$" by "$30°$" in relations 3a, 3b, and 3c.

EXAMPLE 19.4

A balanced three-phase load draws 80 kW at a lagging power factor of 0.8 and a line voltage of 440 V. Find the complex power and the line current.

■ **SOLUTION**

Equation (19.50) gives

$$I_L = \frac{|S|}{\sqrt{3} V_L}$$

where $V_L = 440$ V. We can write the complex power as

$$S = P + jQ = |S| \cos \theta + j|S| \sin \theta$$

with the power factor angle given by

$$\theta = \text{arc} \cos 0.8 = 36.87°$$

Therefore, the magnitude of the complex power can be calculated from the real power by

$$|S| = \frac{P}{\cos \theta} = \frac{80}{\cos 36.87°} = 100 \text{ kVA}$$

and

$$S = 100 \underline{/36.87°} \text{ kVA}$$

The line current is then

$$I_L = \frac{|S|}{\sqrt{3} V_L} = \frac{100 \times 10^3}{440 \sqrt{3}} = 131.21 \text{ A}$$

We should note that if the complex power is known, we do not need to know whether the load is Δ-connected or Y-connected in order to calculate the line current. ■

EXAMPLE 19.5

Two balanced three-phase loads, one drawing 80 kW at a power factor of 0.8 lagging and the second drawing 75 kVA at a power factor of 0.6 leading are connected to a 440-V three-phase line. What line current is required to supply the combined load?

■ **SOLUTION**

From the previous example the complex power for one load is

$$\mathbf{S}_1 = 100\underline{/36.87°}\text{ kVA}$$

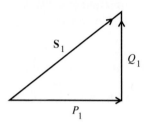

The power factor angle for the second load is (the angle is negative because the power factor is leading)

$$\theta_2 = \text{arc cos }(0.6) = -53.13°$$

The magnitude of the complex power is specified as 75 kVA, so

$$\mathbf{S}_2 = 75\underline{/-53.13°}\text{ kVA}$$

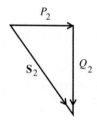

Adding the two complex powers gives the total complex power absorbed by the loads:

$$\mathbf{S} = \mathbf{S}_1 + \mathbf{S}_2 = 100\underline{/36.87°} + 75\underline{/-53.13°}$$
$$= 80 + j60 + 45 - j60$$
$$= 125 + j0 = 125\underline{/0°}\text{ kVA}$$

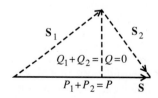

This gives a line current for the combined loads of

$$I_L = \frac{|\mathbf{S}|}{\sqrt{3}\,V_L} = \frac{125{,}000}{440\sqrt{3}} = 164 \text{ A} \quad \blacksquare$$

EXAMPLE 19.6

For the balanced 2400-V three-phase system shown in Fig. 19.12, let $\mathbf{V}_{ab}$ be the reference; that is, let $\mathbf{V}_{ab} = 2400\underline{/0^\circ}$ V. Assume a positive (abc) phase sequence.

(a) Calculate currents $\mathbf{I}_{AA_1}$, $\mathbf{I}_{AA_2}$, $\mathbf{I}_{A_2B_2}$, and $\mathbf{I}_{aA}$.

(b) Determine the complex power, real power, and reactive power absorbed by each load and the combined load. Also calculate the corresponding power factors.

FIGURE 19.12
A system with combined Δ and Y loads

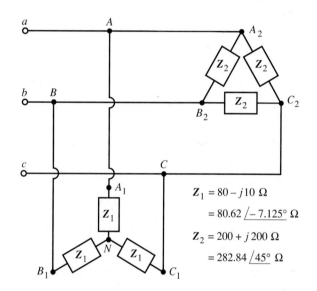

$$\mathbf{Z}_1 = 80 - j10 \ \Omega$$
$$= 80.62\ \underline{/-7.125^\circ}\ \Omega$$

$$\mathbf{Z}_2 = 200 + j\,200\ \Omega$$
$$= 282.84\ \underline{/45^\circ}\ \Omega$$

■ SOLUTION

(a) A straightforward way to calculate the requested line currents is

$$\mathbf{I}_{AA_1} = \frac{1}{\mathbf{Z}_1}\frac{\mathbf{V}_{AB}}{\sqrt{3}}(1\underline{/-30^\circ}) = \frac{1}{80 - j10}\frac{2400\underline{/0^\circ}}{\sqrt{3}}(1\underline{/-30^\circ})$$

$$= 17.19\underline{/-22.88^\circ} = 15.84 - j6.68 \text{ A}$$

$$\mathbf{I}_{AA_2} = \frac{1}{\mathbf{Z}_2}\sqrt{3}\,\mathbf{V}_{AB}(1\underline{/-30^\circ}) = \frac{1}{200 + j200}\sqrt{3}\,2400\underline{/0^\circ}(1\underline{/-30^\circ})$$

$$= 14.70\underline{/-75^\circ} = 3.80 - j14.20 \text{ A}$$

$$\mathbf{I}_{A_2B_2} = \frac{1}{\mathbf{Z}_2}\mathbf{V}_{AB} = \frac{1}{200 + j200}2400\underline{/0^\circ}$$

$$= 8.49\underline{/-45^\circ} = 6 - j6 \text{ A}$$

$$\mathbf{I}_{aA} = \mathbf{I}_{AA_1} + \mathbf{I}_{AA_2} = 15.84 - j6.68 + 3.80 - j14.20$$

$$= 19.64 - j20.88 = 28.66\underline{/-46.75^\circ} \text{ A}$$

(b) The individual complex powers are most easily found in the following manner:

$$S_1 = \frac{1}{Z_1^*} V_L^2 = \frac{1}{80 + j10} 2400^2 = 71{,}444 \underline{/-7.125^\circ}$$

$$= 70{,}892 - j8{,}862 \text{ VA}$$

$$S_2 = 3\frac{1}{Z_2^*} V_L^2 = 3\frac{1}{200 - j200} 2400^2 = 61{,}094 \underline{/45^\circ}$$

$$= 43{,}200 + j43{,}200 \text{ VA}$$

and the total complex power is

$$S = S_1 + S_2$$
$$= 114{,}092 + j34{,}338$$
$$= 119{,}148 \underline{/16.76^\circ} \text{ VA}$$

The real powers are

$$P_1 = \mathcal{R}e\,\{S_1\} = 70{,}892 \text{ W}$$
$$P_2 = \mathcal{R}e\,\{S_2\} = 43{,}200 \text{ W}$$
$$P = \mathcal{R}e\,\{S\} = 114{,}092 \text{ W}$$

The reactive powers are

$$Q_1 = \mathcal{I}m\,\{S_1\} = -8862 \text{ VAR}$$
$$Q_2 = \mathcal{I}m\,\{S_2\} = 43{,}200 \text{ VAR}$$
$$Q = \mathcal{I}m\,\{S\} = 34{,}338 \text{ VAR}$$

The power factors are

$$PF_1 = \cos\,(-7.125^\circ) = 0.99 \text{ leading}$$
$$PF_2 = \cos\,(45^\circ) = 0.707 \text{ lagging}$$
$$PF = \cos\,(16.76^\circ) = 0.96 \text{ lagging}$$

The complex powers could have been found first and the currents calculated from them. The other line currents and line-to-line currents can easily be found from the 120° symmetry property. ∎

The .AC analysis option of a circuit analysis program such as PSpice is very useful to analyze three-phase systems. The systems can be analyzed directly, or for a balanced system, all sources and loads can be converted to their equivalent Y. Then only one phase needs to be analyzed. The remaining voltages and currents can be obtained from the symmetry properties and by converting any equivalent Y connections back to the original Δ connection.

EXERCISES

13. For a balanced three-phase load connected to a positive-phase-sequence balanced three-phase source as shown below, if $I_{bB} = 5\underline{/20^\circ}$ A and $V_{BC} = 200\underline{/-150^\circ}$ V, find

(a) The complex power absorbed by the load
(b) The real power absorbed by the load

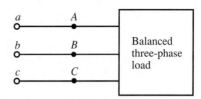

14. A balanced three-phase three-wire system contains a Δ-connected load with an impedance of $3 + j4\ \Omega$ per phase (line-to-line). A balanced Y-connected load with an impedance of $3 - j4\ \Omega$ per phase (line-to-neutral) is also connected in parallel with the Δ load. The line voltage is 440 V. Find:
    (a) The complex power absorbed by the Δ-connected load
    (b) The complex power absorbed by the Y-connected load
    (c) The complex power absorbed by the combined loads
    (d) The line current for the Δ load
    (e) The line current for the Y load
    (f) The line current for the combined loads

15. A three-phase 440-V (line-to-line) load has an input of 50 kVA with a power factor of 0.8 lagging. Find:
    (a) The complex power absorbed by the load
    (b) The line current
    (c) The kVAR rating of a lossless three-phase capacitor that must be connected in parallel with the motor to obtain a combined power factor of 0.9 leading

16. A 2400-V (line-to-line) three-phase motor is operated at rated voltage and has a mechanical output of 50 horsepower. The motor is operating with an efficiency of 90% and a power factor of 0.8 lagging. (1 hp = 0.7457 kW.) Find:
    (a) The complex power absorbed by the motor
    (b) The line current
    (c) The kVAR rating of a lossless three-phase capacitor that must be connected in parallel with the motor to obtain a combined power factor of 0.9 lagging
    (d) The line current required for the combined load

19.6 Power Measurements

We will now discuss how to measure average power in both single-phase and three-phase systems. Assume references for v and i as shown in Fig. 19.13. The instantaneous power absorbed by component X is

$$p = vi \tag{19.54}$$

Average power is calculated from

$$P = \lim_{T \to \infty} \frac{1}{T} \int_{-T/2}^{T/2} p\, dt \tag{19.55}$$

FIGURE 19.13
Device absorbing power

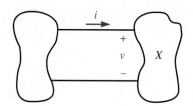

Average power is experimentally measured in the same manner, usually by an electromechanical *wattmeter* represented schematically in Fig. 19.14. A wattmeter has two coils: a current coil, represented by the horizontal coil, which produces a magnetic field proportional to the current, and a potential or voltage coil, represented by the vertical coil, which draws a small amount of current proportional to, and in phase with, the voltage. The interaction of the magnetic field established by the current coil with the magnetic field produced by the voltage coil causes a force, and thus a torque, proportional to the product vi. This torque tends to rotate the voltage coil, and thus the indicating needle, against the force of a spring. The mass of the moving coil prevents it from following the 120-Hz torque pulses produced by the 60-Hz voltage and current, with the result that the needle deflection is determined by the average of the product vi. Proper scale calibration gives a meter that reads average power. A more detailed description can be found in other sources.

The ($\pm$) mark on the current coil in the wattmeter symbol shown in Figure 19.14 indicates that the average power is calculated using the current into the ($\pm$) marked terminal. The ($\pm$) mark on the voltage coil indicates that the average power is calculated using the voltage with the ($+$) reference mark on this terminal and the ($-$) reference mark on the other voltage coil terminal. For an ideal wattmeter, the current coil is assumed to have zero impedance and the voltage coil is assumed to have infinite impedance. More accurate analysis may require a model with the appropriate impedance inserted in series with the current coil and in parallel with the voltage coil.

FIGURE 19.14
Measuring power absorbed
with a wattmeter

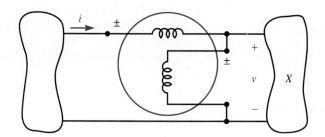

EXAMPLE 19.7

Assume an ideal wattmeter in Fig. 19.14. The voltage and current are

$$v = 120\sqrt{2} \cos (2\pi 60t + 15°) \text{ V}$$

$$i = 10\sqrt{2} \cos (2\pi 60t - 45°) \text{ A}$$

What is the wattmeter reading?

■ SOLUTION

$$\mathbf{V} = 120\underline{/15°} \text{ V}$$

$$\mathbf{I} = 10\underline{/-45°} \text{ A}$$

$$\mathbf{S} = \mathbf{VI}^* = 1200\underline{/60°} \text{ VA}$$
$$= 600 + j600\sqrt{3} \text{ VA}$$
$$= P + jQ$$

The wattmeter reading is

$$P = 600 \text{ W}$$

The same result could have been obtained from

$$P = \frac{1}{T} \int_0^T vi\, dt$$

$$= \frac{1}{1/60} \int_0^T 120\sqrt{2} \cos{(2\pi 60t + 15°)}10\sqrt{2} \cos{(2\pi 60t - 45°)}\, dt$$

$$= (60)1200 \int_0^{1/60} \left[\cos 60° + \cos{(2\pi 120t - 30°)}\right] dt$$

$$= 600 \text{ W} \quad ■$$

Residential customers are charged for the energy they consume. This total energy is recorded on an instrument called a watthour meter.

$$1 \text{ Wh} = (1 \text{ W})(1 \text{ h})$$

$$= \left(1\, \frac{\text{J}}{\text{s}}\right)(3600 \text{ s})$$

$$= 3600 \text{ J} \tag{19.56}$$

The principle of operation of a watthour meter can be found in many texts on ac circuits, power systems, and electrical machinery.

Average power measurements for three-phase three-wire circuits are only slightly more difficult than for single-phase circuits. Three wattmeters can be used, as shown in Fig. 19.15. The connection of point X will temporarily remain unspecified. The wattmeters will read a total absorbed power of

$$P = P_A + P_B + P_C$$
$$= \mathscr{R}e\,\{\mathbf{V}_{AX}\mathbf{I}_{aA}^* + \mathbf{V}_{BX}\mathbf{I}_{bB}^* + \mathbf{V}_{CX}\mathbf{I}_{cC}^*\}$$
$$= \mathscr{R}e\,\{(\mathbf{V}_{AO} + \mathbf{V}_{OX})\mathbf{I}_{aA}^* + (\mathbf{V}_{BO} + \mathbf{V}_{OX})\mathbf{I}_{bB}^* + (\mathbf{V}_{CO} + \mathbf{V}_{OX})\mathbf{I}_{cC}^*\}$$
$$= \mathscr{R}e\,\{\mathbf{V}_{AO}\mathbf{I}_{aA}^* + \mathbf{V}_{BO}\mathbf{I}_{bB}^* + \mathbf{V}_{CO}\mathbf{I}_{cC}^* + \mathbf{V}_{OX}(\mathbf{I}_{aA} + \mathbf{I}_{bB} + \mathbf{I}_{cC})^*\} \tag{19.57}$$

where point O is any reference point. Summation of the currents for a surface enclosing the load indicates

$$\mathbf{I}_{aA} + \mathbf{I}_{bB} + \mathbf{I}_{cC} = 0 \tag{19.58}$$

This permits us to write the total average power P absorbed by the load [Eq. (19.57)] as

$$P = \mathscr{R}e\,\{\mathbf{V}_{AO}\mathbf{I}_{aA}^* + \mathbf{V}_{BO}\mathbf{I}_{bB}^* + \mathbf{V}_{CO}\mathbf{I}_{cC}^*\} \tag{19.59}$$

FIGURE 19.15
Measuring three-phase
power with three wattmeters

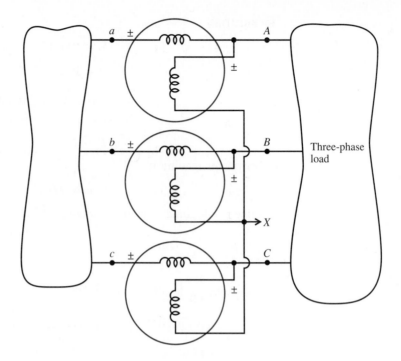

Reference points X and O were chosen arbitrarily. This implies that the sum of the powers indicated by the three meters is independent of the connection point X for a balanced or unbalanced system of either phase sequence.

Clearly, if the load is Y-connected, point X can be the load neutral N, and wattmeter A reads the power absorbed by the line-to-neutral impedance of phase A. An equivalent result holds for wattmeters B and C and line-to-neutral impedances in phases B and C. Thus the sum of the wattmeter readings P must be the power absorbed by the load. The value of P is independent of the location of X.

A useful result is obtained if point X is connected to point B. Wattmeter B must read zero because $\mathbf{V}_{BB} = 0$ and

$$P_B = \mathscr{R}e\,\{\mathbf{I}_{bB}\mathbf{V}_B\} = 0 \tag{19.60}$$

Thus *only two wattmeters* are required to read power for *any* three-phase three-wire circuit. A second useful result holds for *balanced positive (abc) phase sequence* three-phase circuits when power is measured by the two wattmeters A and C located as in Fig. 19.15 with X connected to B.

$$\begin{aligned}
P_A &= \mathscr{R}e\,\{\mathbf{V}_{AB}\mathbf{I}_{aA}^*\} \\
&= \mathscr{R}e\,\{\mathbf{V}_{AN}(\sqrt{3}\,\underline{/30^\circ})\mathbf{I}_{AN}^*\} \\
&= \mathscr{R}e\,\{\mathbf{V}_{AN}\mathbf{I}_{AN}^*(\sqrt{3}\,\underline{/30^\circ})\}
\end{aligned} \tag{19.61}$$

$$\begin{aligned}
P_C &= \mathscr{R}e\,\{\mathbf{V}_{CB}\mathbf{I}_{cC}^*\} \\
&= \mathscr{R}e\,\{-\mathbf{V}_{AN}(\sqrt{3}\,\underline{/30^\circ})(1\,\underline{/-120^\circ})[\mathbf{I}_{AN}(1\,\underline{/-240^\circ})]^*\} \\
&= \mathscr{R}e\,\{\mathbf{V}_{AN}\mathbf{I}_{AN}^*(\sqrt{3}\,\underline{/-30^\circ})\}
\end{aligned} \tag{19.62}$$

Thus
$$P_A + P_C = \mathscr{R}e\,\{\mathbf{V}_{AN}\mathbf{I}_{AN}^*(\sqrt{3}\underline{/30^\circ} + \sqrt{3}\underline{/-30^\circ})\}$$
$$= 3\mathscr{R}e\,\{\mathbf{V}_{AN}\mathbf{I}_{AN}^*\}$$
$$= P \qquad (19.63)$$

as was already known. We can also use the wattmeter readings to calculate the power factor angle:

$$\frac{P_A}{P_C} = \frac{\mathscr{R}e\,\{\mathbf{V}_{AN}\mathbf{I}_{AN}^*\sqrt{3}\underline{/30^\circ}\}}{\mathscr{R}e\,\{\mathbf{V}_{AN}\mathbf{I}_{AN}^*\sqrt{3}\underline{/-30^\circ}\}}$$

$$= \frac{V_{LN}I_L \cos{(\theta + 30^\circ)}}{V_{LN}I_L \cos{(\theta - 30^\circ)}}$$

$$= \frac{\cos{(\theta + 30^\circ)}}{\cos{(\theta - 30^\circ)}} = \frac{\cos 30^\circ \cos \theta - \sin 30^\circ \sin \theta}{\cos 30^\circ \cos \theta + \sin 30^\circ \sin \theta}$$

$$= \frac{\sqrt{3}\cos \theta - \sin \theta}{\sqrt{3}\cos \theta + \sin \theta} \qquad (19.64)$$

This can be solved for

$$\tan \theta = \frac{\sin \theta}{\cos \theta}$$

$$= \sqrt{3}\frac{P_C - P_A}{P_C + P_A} \qquad (19.65)$$

The tangent of θ is positive for a lagging power factor and negative for a leading power factor. This gives

$$\theta = \arctan\left(\sqrt{3}\frac{P_C - P_A}{P_C + P_A}\right) \qquad (19.66)$$

Thus, for a balanced positive sequence three-wire three-phase system, the two wattmeter readings provide not only the power absorbed by the load, but the power factor as well. A similar relation for θ can be found for a negative sequence system.

A low-power-factor load can result in a negative wattmeter reading. Most practical wattmeters do not indicate negative quantities. As a result, the current coil will have to be reversed for negative power readings. Low-power-factor loads will result in P being calculated as the difference of two numbers of nearly the same magnitude. This can lead to large percentage errors, so three wattmeters are often used to measure low-power-factor loads.

EXAMPLE 19.8

Two wattmeters are inserted in the lines of the network of Example 19.6 in the manner shown in Fig. 19.16. Find:
   (a) The reading of each wattmeter
   (b) The power absorbed by the load
   (c) The power factor of the load

FIGURE 19.16
Two wattmeters used to
measure three-phase power

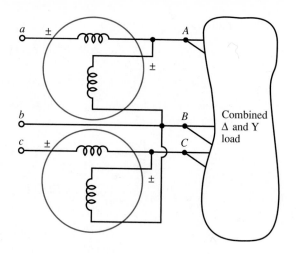

SOLUTION

(a) The currents and voltages required are available from Example 19.6. The wattmeter in line aA indicates

$$P_A = \mathscr{R}e\left\{\mathbf{V}_{AB}\mathbf{I}_{aA}^*\right\}$$
$$= \mathscr{R}e\left\{(2400\underline{/0°})(28.66\underline{/-46.75°})^*\right\}$$
$$= 47{,}130 \text{ W}$$

The wattmeter in line cC indicates

$$P_C = \mathscr{R}e\left\{\mathbf{V}_{CB}\mathbf{I}_{cC}^*\right\} = \mathscr{R}e\left\{(\mathbf{V}_{BC}\underline{/-180°})\mathbf{I}_{cC}^*\right\}$$
$$= \mathscr{R}e\left\{(2400\underline{/60°})(28.66\underline{/-46.75° - 240°})^*\right\}$$
$$= 66{,}960 \text{ W}$$

(b) The power absorbed by the combined load is

$$P = P_A + P_C = 47{,}130 + 66{,}960$$
$$= 114{,}090 \text{ W}$$

(c) From Eq. (19.65),

$$\theta = \arctan\left(\sqrt{3}\,\frac{(P_C - P_A)}{(P_A + P_C)}\right)$$
$$= 16.75°$$
$$\text{PF} = \cos\theta = 0.96 \text{ lag}$$

The combined load presents a lagging power factor because $\theta > 0$. ∎

REMEMBER

Two wattmeters can be used to measure the average power for a three-phase load. If the load is balanced, the wattmeter readings can also be used to calculate the power factor.

17. For the wattmeter connections shown in the following figure, the wattmeter readings are $P_1 = 3.2$ kW and $P_2 = 1.25$ kW.
    (a) Find the average power absorbed by the load.
    (b) If the power source is a positive phase sequence balanced three-phase source, and the three-phase load is balanced, find the load power factor.

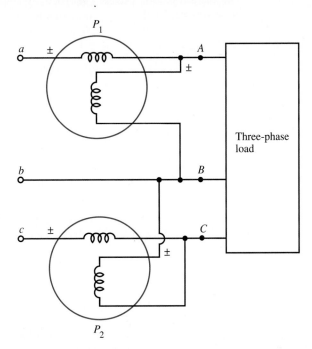

18. Two wattmeters are connected to a Y-connected load with phase impedance of $5\underline{/36.87°}$ Ω and a 440-V line voltage. The coils are connected as in Fig. 19.16. Calculate P_A, the power indicated by the wattmeter in line A, and P_C, the power indicated by the wattmeter in line C. Use P_A and P_C to calculate the total power and the power factor and indicate whether it is leading or lagging.

19.7 Instantaneous Power

The instantaneous power absorbed by a single linear load driven by a sinusoidal source is

$$p = vi = V_M \cos (\omega t + \phi) I_M \cos (\omega t + \phi - \theta)$$

$$= \frac{1}{2} V_M I_M \cos \theta + \frac{1}{2} V_M I_M \cos (2\omega t + 2\phi - \theta) \qquad (19.67)$$

where the passive sign convention was assumed. This instantaneous power consists of the sum of two terms, the first of which is the average power

$$P = \frac{1}{2} V_M I_M \cos \theta = V_{\text{rms}} I_{\text{rms}} \cos \theta \qquad (19.68)$$

The second term in Eq. (19.67) is a sinusoidally varying component at angular frequency 2ω. A single-phase three-wire system with balanced 120-V loads will have an instantaneous power twice this value and will thus still contain a sinusoidally varying term.

Any balanced polyphase system will have an instantaneous power that is constant. The following is a proof for a balanced positive-sequence three-phase load. The instantaneous power absorbed by such a load, written in terms of the rms line-to-neutral voltage V_{LN} and rms line current I_L, is

$$
\begin{aligned}
p &= v_{AN}i_{AN} + v_{BN}i_{BN} + v_{CN}i_{CN} \\
&= 2V_{LN}\cos(\omega t + \phi)I_L\cos(\omega t + \phi - \theta) \\
&\quad + 2V_{LN}\cos(\omega t + \phi - 120°)I_L\cos(\omega t + \phi - 120° - \theta) \\
&\quad + 2V_{LN}\cos(\omega t + \phi - 240°)I_L\cos(\omega t + \phi - 240° - \theta) \\
&= V_{LN}I_L[3\cos\theta + \cos(2\omega t + 2\phi - \theta) + \cos(2\omega t + 2\phi - \theta - 240°) \\
&\quad + \cos(2\omega t + 2\phi - \theta - 480°)] \\
&= V_{LN}I_L[3\cos\theta + \cos(2\omega t + 2\phi - \theta) + \cos(2\omega t + 2\phi - \theta - 240°) \\
&\quad + \cos(2\omega t + 2\phi - \theta - 120°)]
\end{aligned}
\tag{19.69}
$$

The three time-varying terms in Eq. (19.69) can be represented by three phasors that differ in phase by 120°. These terms thus sum to zero, leaving

$$
\begin{aligned}
p &= 3V_{LN}I_L\cos\theta \\
&= 3\frac{V_L}{\sqrt{3}}I_L\cos\theta \\
&= \sqrt{3}\,V_LI_L\cos\theta \\
&= P
\end{aligned}
\tag{19.70}
$$

The instantaneous power p is therefore equal to the average power P and is thus a constant. As the instantaneous power into a balanced three-phase load is constant, three-phase electric motors tend to produce less noise and vibration than single-phase motors of the same power.

REMEMBER

The instantaneous power absorbed by a balanced three-phase load is constant.

EXERCISE

19. A two-phase power system has three lines a, b, and n, with $v_{an} = V_m\cos\omega t$ and $v_{bn} = V_m\cos(\omega t - 90°)$. Calculate the instantaneous power supplied to a balanced two-phase load with an impedance of $\mathbf{Z}_{an} = \mathbf{Z}$ from line a to line n, and an impedance of $\mathbf{Z}_{bn} = \mathbf{Z}$ from line b to line n. Sketch the power absorbed by phase a, the power absorbed by phase b, and the total power as a function of t. For the sketch, assume $\mathbf{Z} = |\mathbf{Z}|\underline{/45°}$. Is the instantaneous power equal to the average power?

19.8　Unbalanced Three-Phase Systems

Only balanced three-phase circuits were considered in this chapter. We used symmetry to simplify power and current calculations. Unbalanced three-phase systems can be solved with the use of mesh-current or node-voltage analysis or with a computer program such as PSpice. Care must be exercised because models for balanced three-phase systems often rely on symmetry for simplification. For instance, mutual inductance between phases in a balanced system is usually modeled with the use of self-inductance only. The model will no longer be appropriate for unbalanced conditions. Unbalanced three-phase systems with significant impedances resulting from long transmission lines or electric motors and generators are most conveniently modeled using the method of *symmetrical components*. The method of symmetrical components is a specialized analysis tool beyond the scope of this text.

EXERCISE

20. A single wattmeter is used to measure the power absorbed by phase B of an *unbalanced* three-phase load, as shown in the following circuit. The balanced voltage source has a positive (abc) phase sequence, and $\mathbf{V}_{an} = 100\underline{/0°}$ V.
    (a) Find the average power read by the wattmeter.
    (b) Find the average power read by the wattmeter if the ($\pm$) terminal of the voltage coil is accidentally connected to line A rather than line B.
    (c) Repeat for a negative (acb) phase sequence.

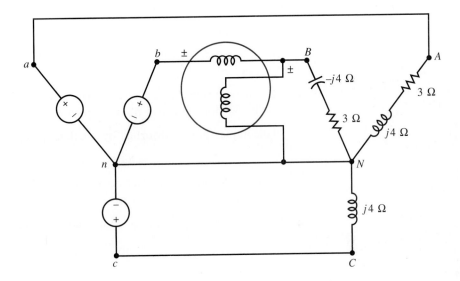

19.9　Summary

This chapter began with a discussion of the three-wire single-phase connection used to supply power to residences. We next described the generation of polyphase, and in particular three-phase, power. The Y and Δ connections were then discussed. We developed the relationships between phase voltages and currents and line voltages and currents in balanced three-phase systems. Symmetry properties were used to develop simple formulas for voltage, current, and average power calculations. We next showed how three-phase power could be measured using only two wattmeters, and followed this with a development that established that the instantaneous power delivered to a balanced three-phase load is a constant. A brief discussion of unbalanced systems concluded the chapter.

KEY FACTS

♦ Power is supplied to residences and many small businesses by 120/240-V three-wire single-phase service.

♦ Most power is generated and transmitted by balanced three-phase systems.

♦ Most industry is supplied by three-phase service.

♦ In power systems, voltage and current are specified in effective (rms) values unless otherwise stated.

♦ The line voltage V_L is used to specify the voltage of a power system.

♦ The line voltage is the phase voltage for a Δ-connected load.

♦ The magnitude of the line voltage is $\sqrt{3}$ times the magnitude of the line-to-neutral voltage.

♦ The line-to-neutral voltage is the phase voltage for a Y-connected load.

♦ For a positive phase sequence system, the line voltage leads the corresponding line-to-neutral voltage by 30°.

♦ The line current I_L is equal to the phase current for a Y-connected load.

♦ The magnitude of the line current is $\sqrt{3}$ times the phase current for a Δ-connected load.

♦ The phase current for a Δ-connected load leads the corresponding line current by 30° for a positive phase sequence system.

♦ The complex power absorbed by a balanced three-phase load is

$$\mathbf{S} = \sqrt{3}\,V_L I_L \underline{/\theta}$$

for both Δ- and Y-connected loads. The angle θ is the angle of the phase impedances.

♦ The complex power supplied to a Y-connected load is

$$\mathbf{S} = \frac{1}{\mathbf{Z}_{LN}^*}\,V_L^2$$

♦ The complex power supplied to a Δ-connected load is

$$\mathbf{S} = 3\,\frac{1}{\mathbf{Z}_{LL}^*}\,V_L^2$$

♦ The average power and power factor of a balanced three-phase system can be measured by two wattmeters.

♦ The instantaneous power p supplied to a balanced three-phase load by a balanced three-phase source is equal to the average power P and is thus constant.

Section 19.1

Refer to Fig. 19.17 for Problems 1 and 2. The line impedances are $\mathbf{Z}_s = j0.1 \, \Omega$, and the neutral impedance is $\mathbf{Z}_n = j0.2 \, \Omega$. The load represented by $\mathbf{Z}_1$ absorbs a complex power of $\mathbf{S}_1 = 240 + j0$ VA, and the load represented by $\mathbf{Z}_2$ absorbs a complex power of $\mathbf{S}_2 = 120 + j0$ VA when the load voltages are 120 V rms.

1. $\mathbf{Z}_3$ is an open circuit. What is the complex power absorbed by each load when connected as shown?
2. Calculate the complex power absorbed by each load if $\mathbf{Z}_3$ absorbs $\mathbf{S}_3 = 5760\underline{/36.87°}$ VA at 240 V rms. What percent change in the power absorbed by the loads occurs as a result of the presence of the line and neutral resistances?

For Problems 3 and 4, a single-phase inductive load of 8000 W has a power factor of 0.8. The line voltage is 240 V.

3. What is the line current?
4. Draw a diagram that indicates the complex power as a vector, and determine the reactive power rating of a lossless capacitor that is connected in parallel with the load to give an overall lagging power factor of 0.9. What is the line current for the combined load?

5. A single-phase inductive load of 100 kVA has a power factor of 0.8. The line voltage is 480 V. Draw a diagram that indicates the complex power as a vector, and determine the kVA rating of a load with a leading power factor of 0.2 that must be connected in parallel with the load to yield an overall lagging power factor of 0.9.
6. Your desk lamp is rated at 60 W and 120 V. Its power is supplied by the 60-Hz power system. Write the lamp voltage $v(t)$. Assume that the lamp is a linear resistance, and write the lamp current $i(t)$.

Section 19.2

7. Refer to the two-phase alternator of Fig. 19.2b. Points a' and b' are connected together. Write the voltage $v_{ab}(t)$ and the phasor voltage $\mathbf{V}_{ab}$.

Section 19.3

8. A balanced Y-connected load with a phase resistance of 12 Ω and an inductive reactance of 16 Ω is connected to a balanced three-phase source with a line voltage of 240 V. Find the line current and complex power absorbed by the load.
9. A balanced Y-connected three-phase resistance heater of total complex power $\mathbf{S} = P$ is sup-

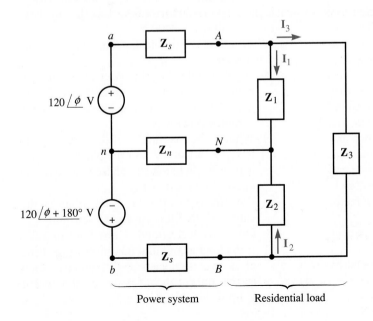

Power system   Residential load   **FIGURE 19.17**

plied power by a balanced three-wire three-phase system with an rms line voltage of V_L.
(a) Find the line current.
(b) If one wire is accidentally disconnected, is the load supplied single-phase, two-phase, or three-phase power?
(c) Find the line current and the heater power with the line disconnected (assume that the load is a linear resistance).

Section 19.4

10. A balanced Δ-connected load with a phase resistance of 12 Ω and an inductive reactance of 16 Ω is connected to a balanced three-phase source with a line voltage of 240 V. Find the line current and complex power absorbed by the load.

11. A balanced Δ-connected three-phase resistance heater of total complex power $\mathbf{S} = P$ is supplied power by a balanced three-wire three-phase system with an rms line voltage of V_L.
(a) Find the line current.
(b) If one wire is accidentally disconnected, is the load supplied single-phase, two-phase, or three-phase power?
(c) Find the line current and the heater power with the line disconnected (assume that the load is a linear resistance).
(d) Repeat (c) if a Y-connected heater is used.

Section 19.5

12. A balanced three-phase load is connected to a balanced three-phase power system. The line voltage is 480 V, and the line current is 10 A. The angle on the phase impedance of the load is 60°. Find the complex power $\mathbf{S}$, the real power P, and the reactive power Q absorbed by the load. Is the load inductive or capacitive?

13. A balanced three-phase load is connected to a transformer by three lines, each of which has a resistance of 1 Ω and an inductive reactance of 4 Ω. Measurements at the transformer end of the line indicate that the line voltage is 240 V, the line current is 10 A, and the power factor is 0.8 leading.
(a) Calculate the complex power delivered by the transformer.
(b) Calculate the complex power absorbed by the load.
(c) Calculate the line voltage at the load.

14. Two balanced three-phase loads are connected to a balanced three-phase line as shown in the following circuit. The loads are said to be connected in parallel. The line-to-line voltage at the load is known to be 240 V. Find:
(a) The complex power absorbed by the Y load
(b) The complex power absorbed by the Δ load
(c) The complex power absorbed by the combined loads
(d) The line current required for the combined load
(e) The voltage at the source end of the line

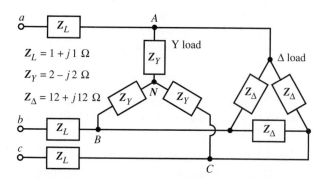

15. Repeat Problem 14 when the line-to-line voltage at the source end is known to be 240 V. [For (e), find the line-to-line voltage at the load.]

16. A balanced Δ-connected three-phase load absorbs a complex power of 100 kVA with a lagging power factor of 0.8 when the rms line-to-line voltage is 2400 V.
(a) What are the line-to-line impedances of the Δ load?
(b) What would be the equivalent line-to-neutral impedances required for a Y-connected load that would absorb the same complex power?
(c) What is the magnitude of the line current?
(d) If $\mathbf{V}_{ab} = 2400\underline{/0°}$ V and the phase sequence is positive, what is the phasor value of the line-to-line current $\mathbf{I}_{ab}$?
(e) For the equivalent Y load, what is the value of the line-to-neutral current $\mathbf{I}_{an}$? Does this correspond to any line current?
(f) If the load is an induction motor operating at an overall efficiency of 85 percent, what is the output horsepower of the motor (1 hp = 745.7 W)?

(g) What is the VAR of a three-phase capacitor bank connected in parallel with the load that will correct the power factor to 0.9 lagging? To unity power factor?

(h) What is the magnitude of the line current when the power factor is corrected to 0.9? What is the ratio of the line current after power factor correction to that before?

(i) Repeat (h) for the unity-power-factor correction. Can you say why industry might not correct their load to unity power factor? (Compare the decrease in line current per volt-ampere rating of the capacitance in each case.)

(j) Make a single phasor diagram of the line currents without power factor correction and with the power factor corrected to 0.9 and to 1.

Section 19.6

For Problems 17 and 18, a balanced three-phase load is connected to a balanced positive phase sequence source. Two wattmeters are connected as shown in Fig. 19.16.

17. The load absorbs a complex power of $\mathbf{S} = 5 + j20$ kVA. Find the two wattmeter readings.

18. The load absorbs a complex power of $\mathbf{S} = 5 - j20$ kVA. Find the two wattmeter readings.

19. Two wattmeters are connected to measure the power absorbed by the parallel loads of Problem 14. Assume a positive phase sequence. The meters are connected as in Fig. 19.16 and are at the load end of the line. Calculate P_A and P_C. Then find the power absorbed by the load and the power factor.

20. For Problem 19, if the phase sequence is actually negative, but the power factor calculations were made using P_A and P_C as if the phase sequence were positive, is $P_A + P_C$ still the correct value for the load power? Is the power factor calculated correctly? If not, what equation should be used?

21. Repeat Problem 19 if the wattmeters are connected at the source end of the line.

22. A wattmeter is connected so the current coil measures i_{aA} and the voltage coil measures v_{bc} for a balanced three-phase system. Relate the wattmeter reading P to the reactive power absorbed by the balanced load from a balanced positive phase sequence source.

23. Repeat Problem 22 for a negative phase sequence source.

Section 19.8

24. An unbalanced three-phase Δ load is connected to a balanced positive sequence three-phase voltage source as in Fig. 19.8 with a line-to-line voltage of 240 V. The line-to-line impedances $\mathbf{Z}_{AB} = 8 + j6\,\Omega$, $\mathbf{Z}_{BC} = 8 - j6\,\Omega$, and $\mathbf{Z}_{CA} = 10\,\Omega$.

(a) Find $\mathbf{I}_{aA}$, $\mathbf{I}_{bB}$, and $\mathbf{I}_{cC}$.

(b) If the impedance of each line connecting the 240-V source to the load is $j2\,\Omega$, recalculate the line currents. The suggested method is to convert the Δ of impedances to an equivalent Y. (The procedure is identical to that for the resistive Δ-Y conversion presented in Chapter 3. Then solve for $\mathbf{V}_{Nn}$ by the use of a single KCL equation. The currents can then be found by the use of KVL and Ohm's law.

25. Repeat Problem 24 for a negative phase sequence source.

Comprehensive Problems

26. A 2400-V three-phase motor operated at rated voltage has a rated output of 100 hp, a full load efficiency of 90 percent, and a lagging power factor of 0.8 at full load. When started, the motor will initially require four times the rated current, and the power factor can be approximated by zero. The motor is connected in parallel with a 100-kW balanced three-phase resistance heater. What is the line current for the combined load when the motor is started?

27. A three-phase motor absorbs 10 kVA at a lagging power factor of 0.6 and a line voltage of 208 V.

(a) What is the line current?

(b) What size capacitor bank, in kilovolt-amperes, must be connected in parallel with the load to reduce the line current for the combined load to 20.82 A?

28. A three-phase motor absorbs 100 kW of real power at a power factor of 0.707 lagging and a line voltage of 480 V.

(a) Find the complex power and line current.

(b) A three-phase capacitor bank, which can be assumed lossless, is connected in parallel with the motor and draws 25 kVA. For the combined load, find the complex power absorbed, the power factor, and the resulting line current.

29. Replace the line-to-line impedances of the Δ-connected load of Example 19.6 by equivalent line-to-neutral impedances in a Y connection. Then make a single-phase circuit for the line-a-to-neutral phase. Solve this for $I_{AA_1} I_{AA_2}$, and I_{aA}. Calculate the complex power absorbed by each single-phase load and the combined single-phase complex power. Next, find the three-phase complex powers by multiplying these quantities by 3 and compare your results with those obtained in Example 19.6.

30. Three voltmeters, each with arbitrary impedance, are connected in a Y to the lines of a balanced positive phase sequence three-wire system. The meter connected to line a reads value A V, the one connected to line b reads B V, and the one connected to line c reads C V. Show that the line-to-line voltage can be found from the expression

$$V_L = \left\{ \frac{1}{2} [A^2 + B^2 + C^2] + \frac{1}{2} [6(A^2B^2 \right.$$

$$\left. + B^2C^2 + C^2A^2) - 3(A^4 + B^4 + C^4)]^{1/2} \right\}^{1/2}$$

$$= \left\{ \frac{1}{2} [A^2 + B^2 + C^2] + \frac{1}{2} [3(A^2 + B^2 + C^2)^2 \right.$$

$$\left. - 6(A^4 + B^4 + C^4)]^{1/2} \right\}^{1/2}$$

This problem arises in efficiency tests of three-phase motors.

31. The following circuit can be used to check for the phase sequence of a balanced set of three-phase voltages. When the phase sequence is abc, as shown in Fig. 19.4, lamp A will be brighter than lamp B. When the phase sequence

is acb, as shown in Fig. 19.7, lamp B will be brighter than A. Calculate $\mathbf{V}_{AN}$ and $\mathbf{V}_{CN}$ for both phase sequences to verify that this is true. Assume that $\mathbf{V}_{ab} = 120\underline{/0°}$ V.

32. The following circuit can be used to check for the phase sequence of a balanced three-phase voltage. Assume that the voltmeter is ideal (draws no current) and indicates rms values and that the line-to-line voltage is 208 V. Find the voltmeter reading for an abc phase sequence and an acb phase sequence. In practice, the voltmeter is often replaced by a small neon lamp and series resistance. The brightness of the lamp is used to indicate the phase sequence.

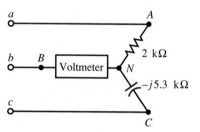

Power circuit problems are often worked on a normalized or per unit basis to simplify the arithmetic. Problems 33 through 39 use the per unit method. The per unit quantity is calculated from

$$\text{Per unit quantity} = \frac{\text{Actual quantity}}{\text{Base quantity}}$$

A similar procedure is often used in electronics, where $V_{\text{Base}} = 1$ V and $Z_{\text{Base}} = 1$ kΩ. In this case, the resulting per unit currents are numerically equal to their value in milliamperes. The quantities given in the following circuit for Problems 33 through 39 are in per unit values. The problem can be worked as though the units were volts, amperes, voltamperes, and ohms.

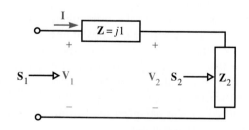

33. If $\mathbf{V}_1 = 1\underline{/0°}$ and $\mathbf{Z}_2 = 1.2 - j1.6$, find $\mathbf{I}$, $\mathbf{V}_2$, $\mathbf{S}_1$, and $\mathbf{S}_2$ in per unit values.

34. If $\mathbf{V}_2 = 1\underline{/0°}$ and $\mathbf{Z}_2 = 1.2 - j1.6$, find $\mathbf{I}$, $\mathbf{V}_1$, $\mathbf{S}_1$, and $\mathbf{S}_2$ in per unit values.

35. If $\mathbf{V}_1 = 1\underline{/0°}$ and $\mathbf{Z}_2 = 1.2 + j1.6$, find $\mathbf{I}$, $\mathbf{V}_2$, $\mathbf{S}_1$, and $\mathbf{S}_2$.

36. If $\mathbf{V}_1 = 1\underline{/0°}$ and $\mathbf{S}_1 = 0.3 - j0.4$, find $\mathbf{I}$, $\mathbf{V}_2$, $\mathbf{Z}_2$, and $\mathbf{S}_2$.

37. If $\mathbf{V}_1 = 1\underline{/0°}$ and $\mathbf{S}_2 = 0.3 - j0.4$, find $\mathbf{I}$, $\mathbf{V}_2$, $\mathbf{Z}_2$, and $\mathbf{S}_1$.

38. If $P_2 = 0.3$, Q_2 is unknown, $\mathbf{V}_1 = 1\underline{/0°}$ and $|\mathbf{V}_2| = 1$, find $\mathbf{V}_2$, $\mathbf{I}$, $\mathbf{Q}_2$, and $\mathbf{Z}_2$.

39. Two identical systems are connected together. The per unit equivalent circuit looking into system 1 from the connection is $\mathbf{V}_1 = 1\underline{/0°}$ and $\mathbf{Z}_1 = j1$, and that looking into system 2 is $\mathbf{V}_2 = 1\underline{/0°}$ and $\mathbf{Z}_2 = j1$. The voltage at the interconnection is $\mathbf{V}$. This is depicted in the model shown in the following circuit. Under these conditions, $\mathbf{V} = 1\underline{/0°}$ and no power is delivered to system 2. The operator of system 1 wants to deliver real power to system 2.

(a) The operator of system 1 raises voltage $\mathbf{V}_1$ to $1.2\underline{/0°}$. Find $\mathbf{V}$, $\mathbf{I}$, and the complex power $\mathbf{S} = \mathbf{VI}^* = P + jQ$ delivered to system 2 at the interconnection.

(b) The operator changes voltage $\mathbf{V}_1$ to $1\underline{/30°}$. Find $\mathbf{V}$, $\mathbf{I}$, and the complex power $\mathbf{S} = P + jQ$ delivered to system 2.

(c) Do the results surprise you? Why? This is one reason phase-shifting transformers, mentioned in Chapter 18, are used.

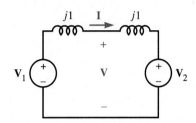

APPENDIXES

APPENDIX

A

Linear Equations and Determinants

If the number of simultaneous linear equations that we must solve is large, we usually use matrix notation as an aid in organizing and solving the equations.

A.1 Simultaneous Equations

We will now show how to write a set of simultaneous linear equations as a single matrix equation.

The *matrices*

$$\mathbf{v} = [v_j] = \begin{bmatrix} v_1 \\ v_2 \\ v_3 \\ v_4 \end{bmatrix} \tag{A.1}$$

and

$$\mathbf{s} = [s_i] = \begin{bmatrix} s_1 \\ s_2 \\ s_3 \\ s_4 \end{bmatrix} \tag{A.2}$$

are four rows by one column or 4×1 matrices (*column vectors*). Matrix

$$\mathbf{T} = [T_{ij}] = \begin{bmatrix} T_{11} & T_{12} & T_{13} & T_{14} \\ T_{21} & T_{22} & T_{23} & T_{24} \\ T_{31} & T_{32} & T_{33} & T_{34} \\ T_{41} & T_{42} & T_{43} & T_{44} \end{bmatrix} \begin{array}{l} \leftarrow \text{Row 1} \\ \\ \\ \leftarrow \text{Row 4} \end{array} \tag{A.3}$$

Column 1 ⎯⎯⎯   Column 4

is a four-row by four-column or a 4×4 matrix. Consider the set of equations

$$T_{11}v_1 + T_{12}v_2 + T_{13}v_3 + T_{14}v_4 = s_1 \tag{A.4}$$

$$T_{21}v_1 + T_{22}v_2 + T_{23}v_3 + T_{24}v_4 = s_2 \tag{A.5}$$

$$T_{31}v_1 + T_{32}v_2 + T_{33}v_3 + T_{34}v_4 = s_3 \tag{A.6}$$

$$T_{41}v_1 + T_{42}v_2 + T_{43}v_3 + T_{44}v_4 = s_4 \tag{A.7}$$

In matrix form this is written as

$$\mathbf{T}\mathbf{v} = \mathbf{s} \tag{A.8}$$

Equation (A.4) contributes the first row of matrix $\mathbf{T}$ and $\mathbf{s}$,

$$\begin{bmatrix} T_{11} & T_{12} & T_{13} & T_{14} \\ & & & \\ & & & \\ & & & \end{bmatrix} \cdot \begin{bmatrix} v_1 \\ v_2 \\ v_3 \\ v_4 \end{bmatrix} = \begin{bmatrix} s_1 \\ \\ \\ \end{bmatrix} \tag{A.9}$$

Equation (A.5) contributes the second row,

$$\begin{bmatrix} & & & \\ T_{21} & T_{22} & T_{23} & T_{24} \\ & & & \\ & & & \end{bmatrix} \cdot \begin{bmatrix} v_1 \\ v_2 \\ v_3 \\ v_4 \end{bmatrix} = \begin{bmatrix} \\ s_2 \\ \\ \end{bmatrix} \tag{A.10}$$

and so on. The pattern seems obvious. We can write equation n as

$$s_i = \sum_{j=1}^{4} T_{ij}v_j \qquad i = 1, 2, 3, 4 \tag{A.11}$$

The complete set of equations written as a single matrix equation is

$$\begin{bmatrix} T_{11} & T_{12} & T_{13} & T_{14} \\ T_{21} & T_{22} & T_{23} & T_{24} \\ T_{31} & T_{32} & T_{33} & T_{34} \\ T_{41} & T_{42} & T_{43} & T_{44} \end{bmatrix} \cdot \begin{bmatrix} v_1 \\ v_2 \\ v_3 \\ v_4 \end{bmatrix} = \begin{bmatrix} s_1 \\ s_2 \\ s_3 \\ s_4 \end{bmatrix} \tag{A.12}$$

EXAMPLE A.1

Write the following set of equations as a single matrix equation.

$$v_1 + 3v_2 + 2v_3 = 130$$

$$6v_1 + 5v_2 + 4v_3 = 280$$

$$9v_1 + 8v_2 + 7v_3 = 460$$

■ SOLUTION

$$\begin{bmatrix} 1 & 3 & 2 \\ 6 & 5 & 4 \\ 9 & 8 & 7 \end{bmatrix} \cdot \begin{bmatrix} v_1 \\ v_2 \\ v_3 \end{bmatrix} = \begin{bmatrix} 130 \\ 280 \\ 460 \end{bmatrix}$$ ■

EXAMPLE A.2

Write the following set of equations as a single matrix equation.

$$5i_1 + 6\frac{d}{dt}i_1 - 6\int i_3\,dt = -100\cos 6t$$

$$-24\frac{d}{dt}i_1 + 8\frac{d}{dt}i_2 + 4i_2 + 16\frac{d}{dt}i_3 = 0$$

$$24\frac{d}{dt}i_1 - 6\int i_1\,dt - 8\frac{d}{dt}i_2 - 16\frac{d}{dt}i_3 + 6\int i_3\,dt = 25$$

■ SOLUTION

$$\begin{bmatrix} 5 + 6\dfrac{d}{dt} & 0 & -6\int dt \\[2mm] -24\dfrac{d}{dt} & 8\dfrac{d}{dt} + 4 & 16\dfrac{d}{dt} \\[2mm] 24\dfrac{d}{dt} - 6\int dt & -8\dfrac{d}{dt} & -16\dfrac{d}{dt} + 6\int dt \end{bmatrix} \cdot \begin{bmatrix} i_1 \\ i_2 \\ i_3 \end{bmatrix} = \begin{bmatrix} -100\cos 6t \\ 0 \\ 25 \end{bmatrix}$$ ■

Solution of this matrix equation requires some knowledge of differential equations and will not be attempted in this appendix. When only resistive elements are present, the equations are algebraic. The solution of simultaneous linear algebraic equations is considered in the next section.

A.2 Determinants and Cramer's Rule

There are several ways in which systems of linear algebraic equations can be solved. These methods include simple substitution, Cramer's rule, and numerical algorithms, such as gaussian elimination, that are well suited for the digital computer. Although Cramer's rule is not efficient for the solution of a large number of simultaneous linear equations, the formulation is valuable for the derivation of certain network properties and we present it here.

The *determinant* of the $n \times n$ square matrix $\mathbf{a} = [a_{ij}]$, denoted by

$$\Delta = |\mathbf{a}|$$
$$= |a_{ij}| \tag{A.13}$$

is defined for any n in terms of sums of products of elements of $\mathbf{a}$.

The definition for arbitrary n, while compact, is not particularly convenient and will not be presented. For $n = 2$, the definition yields

$$\Delta = \begin{vmatrix} a_{11} & a_{12} \\ a_{21} & a_{22} \end{vmatrix} = a_{11}a_{22} - a_{12}a_{21} \tag{A.14}$$

and for $n = 3$ the definition yields

$$\Delta = \begin{vmatrix} a_{11} & a_{12} & a_{13} \\ a_{21} & a_{22} & a_{23} \\ a_{31} & a_{32} & a_{33} \end{vmatrix} = \begin{aligned} a_{11}a_{22}a_{33} + a_{12}a_{23}a_{31} + a_{13}a_{21}a_{32} \\ - a_{13}a_{22}a_{31} - a_{12}a_{21}a_{33} - a_{11}a_{23}a_{32} \end{aligned} \tag{A.15}$$

These relationships can be remembered if we recognize that the determinant for $n = 2$ is obtained from the product of the two terms on the diagonal line with negative slope minus the product of the two terms on the diagonal line with positive slope as shown below.

$$\begin{vmatrix} a_{11} & a_{12} \\ a_{21} & a_{22} \end{vmatrix}$$

$$\Delta = -a_{12}a_{21} \quad + a_{11}a_{22} \tag{A.16}$$

Similarly, for $n = 3$ the six terms are as shown:

$$\begin{aligned} & -a_{11}a_{23}a_{32} \\ \Delta = & -a_{12}a_{21}a_{33} \\ & -a_{13}a_{22}a_{31} \end{aligned} \quad \begin{vmatrix} a_{11} & a_{12} & a_{13} \\ a_{21} & a_{22} & a_{23} \\ a_{31} & a_{32} & a_{33} \end{vmatrix} \quad \begin{aligned} & +a_{13}a_{21}a_{32} \\ & +a_{12}a_{23}a_{31} \\ & +a_{11}a_{22}a_{33} \end{aligned} \tag{A.17}$$

We will now show an alternative method for evaluating the determinant of a matrix of order greater than two. If the matrix $\mathbf{a}_{ij}$ is the matrix formed by removing row i and column j from matrix $\mathbf{a}$, the scalar

$$\Delta_{ij} = (-1)^{i+j}|\mathbf{a}_{ij}| \tag{A.18}$$

is called the *cofactor* of element a_{ij}.

EXAMPLE A.3

Determine the cofactors Δ_{12}, Δ_{22}, and Δ_{32} of the matrix

$$\mathbf{a} = \begin{bmatrix} 1 & 3 & 2 \\ 6 & 5 & 4 \\ 9 & 8 & 7 \end{bmatrix}$$

■ SOLUTION

$$\Delta_{12} = (-1)^{1+2} \begin{vmatrix} 6 & 4 \\ 9 & 7 \end{vmatrix} = - \begin{vmatrix} 6 & 4 \\ 9 & 7 \end{vmatrix}$$

$$= -[(6)(7) - (4)(9)] = -6$$

$$\Delta_{22} = (-1)^{2+2} \begin{vmatrix} 1 & 2 \\ 9 & 7 \end{vmatrix} = \begin{vmatrix} 1 & 2 \\ 9 & 7 \end{vmatrix}$$

$$= [(1)(7) - (2)(9)] = -11$$

$$\Delta_{32} = (-1)^{3+2} \begin{vmatrix} 1 & 2 \\ 6 & 4 \end{vmatrix} = - \begin{vmatrix} 1 & 2 \\ 6 & 4 \end{vmatrix}$$

$$= -[(1)(4) - (2)(6)] = 8 \quad ■$$

The determinant Δ of the matrix $\mathbf{a}$ can be calculated from

---

Laplace's Expansion

$$\Delta = |\mathbf{a}| = \sum_{j=1}^{n} a_{ij}\Delta_{ij} \tag{A.19}$$

$$= \sum_{i=1}^{n} a_{ij}\Delta_{ij} \tag{A.20}$$

The sum in Eq. (A.19) is called the *Laplace expansion* of Δ about its ith row, and the sum in Eq. (A.20) is called the Laplace expansion of Δ about its jth column. We can expand the determinant about any row or column of matrix $\mathbf{a}$.

EXAMPLE A.4

Evaluate

$$\Delta = \begin{vmatrix} 1 & 3 & 2 \\ 6 & 5 & 4 \\ 9 & 8 & 7 \end{vmatrix}$$

■ SOLUTION

Expand the determinant about the first column, as in Eq. (A.20):

$$\Delta = a_{11}\Delta_{11} + a_{21}\Delta_{21} + a_{31}\Delta_{31}$$

$$\Delta_{11} = (-1)^{1+1}\begin{vmatrix} 5 & 4 \\ 8 & 7 \end{vmatrix} = (35 - 32) = 3$$

$$\Delta_{21} = (-1)^{2+1}\begin{vmatrix} 3 & 2 \\ 8 & 7 \end{vmatrix} = -(21 - 16) = -5$$

$$\Delta_{31} = (-1)^{3+1}\begin{vmatrix} 3 & 2 \\ 5 & 4 \end{vmatrix} = (12 - 10) = 2$$

Equation (A.20) gives the determinant:

$$\Delta = 1(3) + 6(-5) + 9(2)$$
$$= -9 \quad \blacksquare$$

We will now see how to use determinants to solve simultaneous linear equations. We can write the set of linear equations

$$\begin{aligned} a_{11}x_1 + a_{12}x_2 + \cdots + a_{1n}x_n &= b_1 \\ a_{21}x_1 + a_{22}x_2 + \cdots + a_{2n}x_n &= b_2 \\ &\vdots \\ a_{n1}x_1 + a_{n2}x_2 + \cdots + a_{nn}x_n &= b_n \end{aligned} \tag{A.21}$$

in matrix notation as

$$\begin{bmatrix} a_{11} & a_{12} & \cdots & a_{1n} \\ a_{21} & a_{22} & \cdots & a_{2n} \\ \vdots & & & \\ a_{n1} & a_{n2} & \cdots & a_{nn} \end{bmatrix} \cdot \begin{bmatrix} x_1 \\ x_2 \\ \vdots \\ x_n \end{bmatrix} = \begin{bmatrix} b_1 \\ b_2 \\ \vdots \\ b_n \end{bmatrix} \tag{A.22}$$

or more abstractly as

$$\mathbf{ax} = \mathbf{b} \tag{A.23}$$

If $\Delta^{(j)}$ is defined as the determinant of the matrix obtained by replacing the jth column of $\mathbf{a}$ by vector $\mathbf{b}$, the value of the variable x_j is given by

Cramer's Rule

$$x_j = \frac{1}{\Delta}\Delta^{(j)} \qquad j = 1, 2, \ldots, n \tag{A.24}$$

Alternatively, if $\Delta^{(j)}$ is expanded about column j, we can write Cramer's rule as

$$x_j = \frac{1}{\Delta}\sum_{i=1}^{n} b_i\Delta_{ij} \qquad j = 1, 2, \ldots, n \tag{A.25}$$

If $\Delta = 0$, the equations are linearly dependent, and a unique solution does not exist. The determinant $\Delta^{(j)}$ can, of course, be expanded about any convenient row or column, as can Δ. An $n \times n$ square matrix with n linearly independent rows (rank n), and thus with $\Delta \neq 0$, is said to be *nonsingular*.

EXAMPLE A.5

Solve the set of equations

$$x_1 + 3x_2 + 2x_3 = 13$$
$$6x_1 + 5x_2 + 4x_3 = 28$$
$$9x_1 + 8x_2 + 7x_3 = 46$$

by the use of Cramer's rule.

■ SOLUTION

First write the equations in matrix notation:

$$\begin{bmatrix} 1 & 3 & 2 \\ 6 & 5 & 4 \\ 9 & 8 & 7 \end{bmatrix} \cdot \begin{bmatrix} x_1 \\ x_2 \\ x_3 \end{bmatrix} = \begin{bmatrix} 13 \\ 28 \\ 46 \end{bmatrix}$$

From Example A.4,

$$\Delta = -9$$
$$\Delta_{11}^{(1)} = \Delta_{11} = 3$$
$$\Delta_{21}^{(1)} = \Delta_{21} = -5$$
$$\Delta_{31}^{(1)} = \Delta_{31} = 2$$

Then

$$x_1 = \frac{1}{\Delta} \Delta^{(1)} = \frac{1}{-9} \begin{vmatrix} 13 & 3 & 2 \\ 28 & 5 & 4 \\ 46 & 8 & 7 \end{vmatrix}$$

$$= \frac{1}{-9} \left[13\Delta_{11}^{(1)} + 28\Delta_{21}^{(1)} + 46\Delta_{31}^{(1)} \right]$$

$$= \frac{1}{-9} \left[13(3) + 28(-5) + 46(2) \right]$$

$$= \frac{1}{-9} (-9) = 1$$

$$x_2 = \frac{1}{\Delta} \Delta^{(2)} = \frac{1}{-9} \begin{vmatrix} 1 & 13 & 2 \\ 6 & 28 & 4 \\ 9 & 46 & 7 \end{vmatrix}$$

$$= \frac{1}{-9} \left[13\Delta_{12}^{(2)} + 28\Delta_{22}^{(2)} + 46\Delta_{32}^{(2)} \right]$$

$$\Delta_{12}^{(2)} = \Delta_{12} = (-1)^{1+2} \begin{vmatrix} 6 & 4 \\ 9 & 7 \end{vmatrix} = -(42 - 36) = -6$$

$$\Delta_{22}^{(2)} = \Delta_{22} = (-1)^{2+2} \begin{vmatrix} 1 & 2 \\ 9 & 7 \end{vmatrix} = (7 - 18) = -11$$

$$\Delta_{32}^{(2)} = \Delta_{32} = (-1)^{3+2} \begin{vmatrix} 1 & 2 \\ 6 & 4 \end{vmatrix} = -(4 - 12) = 8$$

$$x_2 = \frac{1}{-9} \left[13(-6) + 28(-11) + 46(8) \right]$$

$$= \frac{1}{-9} (-18) = 2$$

$$x_3 = \frac{1}{\Delta} \Delta^{(3)} = \frac{1}{-9} \begin{vmatrix} 1 & 3 & 13 \\ 6 & 5 & 28 \\ 9 & 8 & 46 \end{vmatrix}$$

$$= \frac{1}{-9} \left[13\Delta_{13}^{(3)} + 28\Delta_{23}^{(3)} + 46\Delta_{33}^{(3)} \right]$$

$$\Delta_{13}^{(3)} = \Delta_{13} = (-1)^{1+3} \begin{vmatrix} 6 & 5 \\ 9 & 8 \end{vmatrix} = (48 - 45) = 3$$

$$\Delta_{23}^{(3)} = \Delta_{23} = (-1)^{2+3} \begin{vmatrix} 1 & 3 \\ 9 & 8 \end{vmatrix} = -(8 - 27) = 19$$

$$\Delta_{33}^{(3)} = \Delta_{33} = (-1)^{3+3} \begin{vmatrix} 1 & 3 \\ 6 & 5 \end{vmatrix} = (5 - 18) = -13$$

$$x_3 = \frac{1}{-9} \left[13(3) + 28(19) + 46(-13) \right]$$

$$= \frac{1}{-9} (-27) = 3 \quad \blacksquare$$

Determinants of higher order than three can be evaluated by repeated expansion with cofactors, so no determinant of higher order than two need be evaluated by direct application of the definition.

Cramer's rule is seldom used to solve numerical problems involving more than three simultaneous equations. Computer programs based on gaussian elimination, or a related technique, are used.

■ PROBLEMS ■■■

Section A.1

1. Write the equations

$$3v_1 + v_2 + 3v_3 = 14$$
$$2v_1 - 2v_2 - v_3 = -5$$
$$v_1 + v_2 + 2v_3 = 9$$

as a matrix equation of the form

$$\mathbf{av} = \mathbf{i}$$

2. Write the equations

$$\frac{d^2}{dt^2} v_1 - 2v_2 = 5 \cos t$$

$$-3 \frac{d}{dt} v_1 + 4 \int v_2 \, dt = e^{-4t}$$

as a single matrix equation.

Section A.2

3. Evaluate the determinant of matrix **a** given

$$\mathbf{a} = \begin{bmatrix} 3 & 1 & 3 \\ 2 & -2 & -1 \\ 1 & 1 & 2 \end{bmatrix}$$

(a) By multiplying the elements along the diagonals as shown in Eq. (A.17)
(b) By Laplace's expansion about the first column
(c) By Laplace's expansion about the second column

(d) By Laplace's expansion about the third column
(e) By Laplace's expansion about the second row

4. Use Cramer's rule to solve for the voltages of Problem 1.

5. Use Cramer's rule to solve for the currents:

$$\begin{bmatrix} 6 & -2 & -3 \\ -2 & 7 & -4 \\ -3 & -4 & 9 \end{bmatrix} \cdot \begin{bmatrix} i_1 \\ i_2 \\ i_3 \end{bmatrix} = \begin{bmatrix} -7 \\ 0 \\ 16 \end{bmatrix}$$

Complex Arithmetic

There are two standard analytical forms for a complex number, the *rectangular* and the *exponential* forms. In addition to these analytical forms, there is a standard notation for complex numbers called the *polar* form.

B.1 Rectangular Form

The *rectangular form* of a complex number z is

$$z = x + jy \qquad \text{(B.1)}$$

in which x and y are real numbers, and the *imaginary unit j* is defined by $j^2 = -1$. In Eq. (B.1), the real number x is called the *real part* (or *real component*) of z, and can be identified by the *real part operator* $\mathcal{R}e\{\cdot\}$ as follows:

$$x = \mathcal{R}e\{z\} \qquad \text{(B.2)}$$

Similarly, the real number y is called the *imaginary part* (or *imaginary component*) of z and can be identified by the *imaginary part operator* $\mathcal{I}m\{\cdot\}$ as

$$y = \mathcal{I}m\{z\} \qquad \text{(B.3)}$$

If the real part, x, of z is nonzero and the imaginary part, y, is zero, then z is said to be *purely real*. If the imaginary part, y, of z is nonzero, and the real part, x, is zero, then z is said to be *purely imaginary*. For example, 3.1415 is purely real, and $-j7$ is purely imaginary. If both the real and the imaginary parts of z are zero, then z is said to be zero, and is denoted by $z = 0$.

Since it takes two real numbers to specify a complex number, it is convenient to display a complex number as a point in a plane, called the complex plane, as shown in Fig. B.1a. Each point in the complex plane corresponds to exactly one complex number. The abscissa in this representation is the coordinate of possible values of the real part, x. The ordinate* is the coordinate of j times possible values of the imaginary part, y. An alternative representation of the same complex number is shown in Fig. B.1b. In this figure, a directed line segment is drawn from the origin to the value of z, forming a vector in the complex plane. With some abuse of terminology[†], it will occasionally be convenient to refer to the complex number z as "the vector z." The projection of this vector onto the abscissa gives the value of the real part of z, whereas the projection onto the ordinate gives j times the value of the imaginary part of z.

FIGURE B.1
Graphical representation of a complex number z: (*a*) as a point; (*b*) as a vector. The length of the vector, $|z|$, equals the magnitude of z. The angle of the vector, $\theta = ang\{z\}$, equals the argument (angle) of z

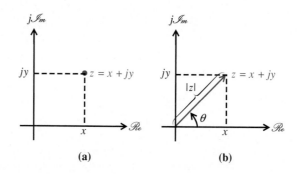

(a) (b)

---

* Mathematics texts correctly use y as the ordinate rather than jy. The convention to label the ordinate jy is, unfortunately, prevalent in electrical engineering. Despite the notational problem, we will use the prevalent convention.

† Complex numbers are actually not vectors in the mathematical sense. They can be multiplied and divided by other complex numbers. Vectors cannot be multiplied or divided by other vectors.

B.2   Exponential Form

The *exponential form* of a complex number z involves its *magnitude* (or *modulus*), denoted by $|z|$, and its *argument* (or *angle in radians*), denoted by $\theta = arg\{z\}$. These terms are defined in Fig. B.1b and can be seen to correspond to the length and the angle, respectively, of the vector z. It can readily be appreciated from Fig. B.1b that

$$x = |z| \cos \theta \tag{B.4}$$

and

$$y = |z| \sin \theta \tag{B.5}$$

so that

$$
\begin{aligned}
z &= x + jy \\
&= |z| \cos \theta + j|z| \sin \theta \\
&= |z|(\cos \theta + j \sin \theta) \\
&= |z|e^{j\theta}
\end{aligned}
\tag{B.6}
$$

Equation (B.6) follows from Euler's identity, which was proved in Section 7.3:

$$e^{j\theta} = \cos \theta + j \sin \theta \tag{B.7}$$

The series of equalities in Eqs. (B.4) through (B.6) shows the relationship between the rectangular form and the exponential form:

$$z = |z|e^{j\theta} \tag{B.8}$$

From Eqs. (B.4) through (B.6) (or directly from Fig. B.1b), we find further that

$$|z| = \sqrt{x^2 + y^2} \tag{B.9}$$

and

$$\theta = \tan^{-1} \frac{y}{x} \tag{B.10}$$

Equation (B.10) contains a hidden trap. Consider the complex numbers $z_1 = -1 + j1$ and $z_2 = 1 - j1$. By sketching z_1 and z_2 as vectors in the complex plane, we find that the angles are, respectively, $\theta_1 = arg\{-1 + j1\} = 135°$ and $\theta_2 = arg\{1 - j1\} = -45°$. On the other hand, Eq. (B.10) yields $\theta_1 = \tan^{-1} \frac{1}{-1}$ and $\theta_2 = \tan^{-1} \frac{-1}{1}$. The trap lies in the fact that, although *algebraically* $\frac{1}{-1}$ does equal $\frac{-1}{1}$, θ_1 certainly does not equal θ_2. Therefore, when evaluating Eq. (B.10), remember to correctly account for the signs of y and x.

The angle θ is undefined for $x = y = 0$. Length is never complex or negative. Therefore, regardless of the value of z, its magnitude is a nonnegative real number,

$$|z| \geq 0 \tag{B.11}$$

The factor $e^{j\theta}$ in Eq. (B.8) is a *unit* vector with the angle θ in the complex plane. This observation makes it obvious that, for example, $j = e^{j\pi/2}$, $e^{j\pi} = -1$, and $e^{j2\pi} = 1$, and so forth. [The student should confirm this using Eq. (B.7).]

The exponential representation contains an inherent ambiguity, since the addition of 2π radians to the angle of z does not change the value of z. (Rotation of the vector z of Fig. B.1b by 2π radians does not change the vector. Equivalently, $e^{j(\theta + 2\pi)} = e^{j\theta}e^{j2\pi} = e^{j\theta}$.) This ambiguity is sometimes removed by *defining* the angle of a complex number to lie in some 2π range, such as $0 \le \theta < 2\pi$ or $-\pi < \theta \le \pi$. If the argument of z, θ, is written as $\theta = \theta_0 + 2\pi k$, where $-\pi < \theta_0 \le +\pi$ and $k = 0, \pm 1, \ldots$, then θ_0 is called the *principal argument* of z and is denoted by $\mathscr{A}rg\,\{z\}$.

B.3 Polar Form

The *polar* form of a complex number z is simply the pair of its polar coordinates $|z|$ and θ of Fig. B.1b. There are two conventional notations used for the polar form,

$$z = |z|\underline{/\theta} \tag{B.12}$$

and

$$z = |z|\underline{/z} \tag{B.13}$$

Thus both $\underline{/\theta}$ and $\underline{/z}$ are sometimes used to denote the angle of z. Since (B.12) and (B.13) are definitions of notation rather than analytical formulas, the exponential or the rectangular forms are used in the analytical derivation of this book.

B.4 Arithmetic Operations

The rule of addition (or subtraction) of complex numbers is identical to the rule of addition (or subtraction) of vectors. Figure B.2 illustrates the addition of two complex numbers z_1 and z_2 as vector addition in the complex plane. The vector z_2 can be translated from the origin to the tip of vector z_1 when forming the sum. Translation of a vector in the complex plane does not change its value, since by definition a vector is a directed line segment. (The vector z_2 is completely specified by its length and angle or by the extent of its real and imaginary components. These quantities are invariant to a translation of the vector.)

It is more convenient to use the rectangular form than the polar form when we add complex numbers. The sum of two complex numbers $z_1 = x_1 + jy_1$ and $z_2 = x_2 + jy_2$ is then clearly seen to be a complex number $z_3 = x_3 + jy_3$ whose

FIGURE B.2
Addition of complex numbers

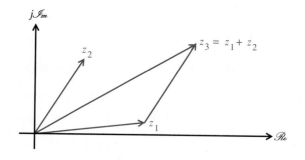

real component is the sum of the real components of z_1 and z_2 and whose imaginary component is the sum of the imaginary components of z_1 and z_2:

$$
\begin{aligned}
z_1 + z_2 &= x_1 + jy_1 + x_2 + jy_2 \\
&= x_1 + x_2 + j(y_1 + y_2) \\
&= x_3 + jy_3 \\
&= z_3
\end{aligned}
\tag{B.14}
$$

where $x_3 = x_1 + x_2$ and $y_3 = y_1 + y_2$.

The product of two complex numbers

$$
\begin{aligned}
z_1 z_2 &= |z_1| e^{j\theta_1} |z_2| e^{j\theta_2} \\
&= |z_1| |z_2| e^{j(\theta_1 + \theta_2)}
\end{aligned}
\tag{B.15}
$$

has a magnitude that is the product of the magnitudes of z_1 and z_2 and an angle that is the sum of the angles of z_1 and z_2. Thus, for any two complex numbers z_1 and z_2,

$$
|z_1 z_2| = |z_1| |z_2|
\tag{B.16}
$$

and
$$
\arg\{z_1 z_2\} = \arg\{z_1\} + \arg\{z_2\}
\tag{B.17}
$$

Alternatively, using the rectangular form,

$$
\begin{aligned}
z_1 z_2 &= (x_1 + jy_1)(x_2 + jy_2) \\
&= (x_1 x_2 - y_1 y_2) + j(x_1 y_2 + x_2 y_1)
\end{aligned}
\tag{B.18}
$$

Similarly, the quotient of two complex numbers is (for $z_2 \neq 0$)

$$
\begin{aligned}
\frac{z_1}{z_2} &= \frac{|z_1| e^{j\theta_1}}{|z_2| e^{j\theta_2}} \\
&= \frac{|z_1|}{|z_2|} e^{j(\theta_1 - \theta_2)}
\end{aligned}
\tag{B.19}
$$

Therefore

$$
\left| \frac{z_1}{z_2} \right| = \frac{|z_1|}{|z_2|}
\tag{B.20}
$$

and
$$
\arg\left\{ \frac{z_1}{z_2} \right\} = \arg\{z_1\} - \arg\{z_2\}
\tag{B.21}
$$

Alternatively, the division can be performed using the rectangular form as follows:

$$
\begin{aligned}
\frac{z_1}{z_2} &= \frac{x_1 + jy_1}{x_2 + jy_2} \\
&= \frac{x_1 + jy_1}{x_2 + jy_2} \cdot \frac{x_2 - jy_2}{x_2 - jy_2} \\
&= \frac{x_1 x_2 + y_1 y_2 + j(y_1 x_2 - y_2 x_1)}{x_2^2 + y_2^2} \\
&= \frac{x_1 x_2 + y_1 y_2}{x_2^2 + y_2^2} + j \frac{y_1 x_2 - y_2 x_1}{x_2^2 + y_2^2}
\end{aligned}
\tag{B.22}
$$

The procedure used to remove the imaginary factor in the denominator as shown is called *rationalizing the denominator*.

Observe that the exponential form yields more insight into the nature of complex multiplication and division than does the rectangular form. The polar form is also easier to apply in analytical expressions involving several factors of complex quantities. However, the rectangular form is sometimes more convenient to use when we evaluate complex expressions numerically.

B.5   The Complex Conjugate

The *complex conjugate* of a complex number z, denoted by z^*, is obtained by replacing the unit j by $-j$, which is the second root to $j^2 = -1$. For example, if $z = 3 + j2$, then $z^* = 3 - j2$. Replacement of the number j by $-j$ causes a reflection of z about the real axis, as shown in Fig. B.3. The rectangular form of the the complex conjugate of $z = x + jy$ is, in general,

$$z^* = x - jy \tag{B.23}$$

The polar form of the complex conjugate of $z = |z|e^{j\theta}$ is, in general,

$$z^* = |z|e^{-j\theta} \tag{B.24}$$

Therefore, complex conjugation reverses the sign of the imaginary part of z while leaving the real part of z unchanged. Equivalently, complex conjugation reverses the sign of the angle of z while leaving the magnitude of z unchanged.

FIGURE B.3
The number z and its complex conjugate z^*

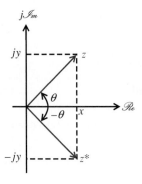

Another important observation is that for any complex number z,

$$z + z^* = 2\mathscr{R}e\left\{z\right\} \tag{B.25}$$

and

$$z - z^* = 2j\mathscr{I}m\left\{z\right\} \tag{B.26}$$

These relationships are illustrated in Fig. B.4, and are easily verified with the use of the rectangular form. It is also useful to recognize that

$$zz^* = |z|^2 \tag{B.27}$$

which is easily verified with the use of either the rectangular or the exponential form. An application of Eq. (B.27) arises in rationalizing a denominator, as illustrated in Eq. (B.22). The numerator and the denominator were both multiplied by z_2^*, rendering the denominator a positive real number.

FIGURE B.4
Illustration of Eqs. (B.25) and (B.26)

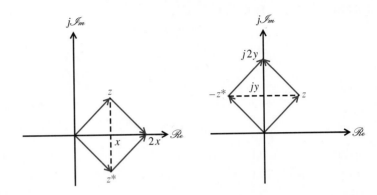

An important property of the operation of conjugation is that it commutes with respect to addition, multiplication, and division. The student should verify that

$$(z_1 + z_2)^* = z_1^* + z_2^* \tag{B.28}$$

$$(z_1 z_2)^* = z_1^* z_2^* \tag{B.29}$$

and

$$\left(\frac{z_1}{z_2}\right)^* = \frac{z_1^*}{z_2^*} \tag{B.30}$$

These relations make it easy to conjugate complex algebraic expressions. Simply replace j by $-j$ wherever it appears. For example,

$$\left[\frac{(3 + j)e^{j2}}{(4 - 5j)(7j)} + 2j\right]^* = \frac{(3 - j)e^{-j2}}{(4 + 5j)(-7j)} - 2j$$

B.6 Boldface Notation

Boldface notation is often used in electrical engineering to denote certain (but not all) complex quantities. The quantities that most often appear in boldface are current and voltage phasors $\mathbf{I}$ and $\mathbf{V}$, respectively; impedance and admittance $\mathbf{Z}$ and $\mathbf{Y}$, respectively; transfer function $\mathbf{H}$; and complex power $\mathbf{S}$. We use a subscript m to denote the magnitude of a phasor: $V_m = |\mathbf{V}|$ and $I_m = |\mathbf{I}_m|$. We often use boldface notation to denote complex voltage and current waveforms, $\mathbf{v}(t)$ and $\mathbf{i}(t)$ respectively.

B.7 Summary

The principal definitions and properties developed in this section are summarized in Table B.1.

TABLE B.1 Summary of complex arithmetic

NAME	FIGURE	DEFINITION	PROPERTIES
Rectangular form	B.1	$z = x + jy$	$x = |z|\cos\theta,\ y = |z|\sin\theta$
Exponential form	B.1	$z = |z|e^{j\theta}$	$|z| = \sqrt{x^2 + y^2},\ \theta = arg\,\{z\} = \tan^{-1}(y/x)$
Polar form		$z = |z|\underline{/\theta}$ or $z = |z|\underline{/z}$	
Addition	B.2	$z_1 + z_2 = (x_1 + x_2) + j(y_1 + y_2)$	—
Subtraction	—	$z_1 - z_2 = (x_1 - x_2) + j(y_1 - y_2)$	—
Multiplication	—	$z_1 z_2 = |z_1||z_2|e^{j(\theta_1 + \theta_2)}$	$|z_1 z_2| = |z_1||z_2|$ $arg\,\{z_1 z_2\} = arg\,\{z_1\} + arg\,\{z_2\}$
Division	—	$z_1/z_2 = [|z_1|/|z_2|]e^{j(\theta_1 - \theta_2)}$	$|z_1/z_2| = |z_1|/|z_2|$ $arg\,\{z_1/z_2\} = arg\,\{z_1\} - arg\,\{z_2\}$
Real part operator	B.1	$\mathcal{R}e\,\{z\} = x$	$\mathcal{R}e\,\{z_1 \pm z_2\} = \mathcal{R}e\,\{z_1\} \pm \mathcal{R}e\,\{z_2\}$
Imaginary part operator	B.1	$\mathcal{I}m\,\{z\} = y$	$\mathcal{I}m\,\{z_1 \pm z_2\} = \mathcal{I}m\,\{z_1\} \pm \mathcal{I}m\,\{z_2\}$
Conjugate operator	B.3	$z^* = x - jy = |z|e^{-j\theta}$	$(z_1 \pm z_2)^* = z_1^* \pm z_2^*$ $(z_1 z_2)^* = z_1^* z_2^*,\ (z_1/z_2)^* = z_1^*/z_2^*$
—	B.4	—	$z + z^* = 2\mathcal{R}e\,\{z\},\ z - z^* = 2j\mathcal{I}m\,\{z\}$

PROBLEMS

1. Plot the following eight numbers as vectors in the complex plane: $z_1 = 3$, $z_2 = 1\underline{/\pi}$, $z_3 = 1\underline{/\pi/2}$, $z_4 = -2j$, $z_5 = 3e^{j\pi/6}$, $z_6 = 1 - j$, $z_7 = 2e^{j3\pi/4}$, $z_8 = 3\underline{/-510°}$.

2. Express each of the following eight numbers in (a) exponential form and (b) polar form. (Write the argument of each number, θ, so that $-\pi < \theta \le \pi$.) $z_1 = 1$, $z_2 = j$, $z_3 = -1$, $z_4 = -j$, $z_5 = 1 + j$, $z_6 = 1 - j$, $z_7 = -1 - j$, $z_8 = -1 + j$.

3. Express each of the following seven numbers in rectangular form: $z_1 = 1\underline{/0°}$, $z_2 = 2e^{j\pi}$, $z_3 = 3e^{-j\pi/2}$, $z_4 = 10\underline{/-450°}$, $z_5 = -3e^{j\pi/4}$, $z_6 = 7\underline{/30°}$, $z_7 = -(2\underline{/60°})$.

4. (a) Illustrate $z_1 + z_2 + z_3$ graphically as the sum of vectors in the complex plane, where $z_1 = 1$, $z_2 = j$, and $z_3 = 2\underline{/135°}$.
   (b) Find $|z_1 + z_2 + z_3|$ and $arg\,\{z_1 + z_2 + z_3\}$.

5. Draw z_1, z_2, and $z_1 z_2$ as vectors in the complex plane, where $z_1 = 2\underline{/30°}$ and $z_2 = -4 + 4j$. Specify both the polar and the rectangular coordinates of $z_1 z_2$.

6. Draw z as an arbitrary vector in the complex plane, showing its length $|z|$ and angle θ and its rectangular coordinates x and y. Draw $1/z$ on the same plane and indicate its length and angle and its rectangular coordinates.

7. Evaluate each of the following expressions, putting your answers in (i) rectangular form, (ii) exponential form, and (iii) polar form.

   (a) $\dfrac{3\underline{/60°}}{1 + 3j}$ (b) $\dfrac{1 + j}{-1 - 3j}$

   (c) $\dfrac{e^{2j} + 1}{e^{j} - 1}$ (d) $\dfrac{e^{2j} + 1}{e^{j} + 1}$

(e) $\dfrac{(1-j)(4e^{j45°})}{1+3j+3\underline{/60°}}$

(f) $\dfrac{2\underline{/45°}+8\underline{/135°}}{(4\underline{/90°})(2\underline{/20°})}$

(g) $(5\underline{/22°})^{-1}+(6\underline{/185°})^{-1}$

(h) $\dfrac{4\underline{/65°}+(4+j)^{-1}}{17+e^{j60°}}$

8. Verify Eqs. (B.25) and (B.26).

9. Verify Eq. (B.27).

10. Verify Eqs. (B.28), (B.29), and (B.30).

11. Find the magnitudes and angles of the following expressions. (All quantities are real except j.)

(a) $\dfrac{1}{1+j\omega RC}$

(b) $\dfrac{V_m\underline{/\phi}}{j\omega RC+1}$

(c) $\dfrac{1+j\omega RC}{R-\omega^2 RLC+j\omega L}$

(d) $\dfrac{(a+bj)(c+dj)}{(e+fj)(h+ij)}$

12. Find the real and the imaginary parts of each of the following expressions. Put your answers in the simplest form possible. (All quantities are real except j.)

(a) $\dfrac{V_m\underline{/\phi}}{1+j\omega RC}e^{j\omega t}$

(b) $\dfrac{(1+j\omega RC)V_m e^{j\phi}}{R-\omega^2 RLC+j\omega L}e^{j\omega t}$

(c) $\dfrac{j\omega_1 L(A_1\underline{/\phi_1})}{2j\omega_1 L+R}e^{j\omega_1 t}+\dfrac{j\omega_2 L(A_2\underline{/\phi_2})}{2j\omega_2 L+R}e^{j\omega_2 t}$

13. Find the roots of the following equations and plot them as points in the complex plane.
(a) $s^2+4s+8=0$
(b) $s^2+6s+8=0$
(c) $s^8=1$
(d) $s^7=1$

14. Shade the regions in the complex (z) plane that satisfy
(a) $\mathscr{R}e\,\{z\}\geq 1$
(b) $\mathscr{I}m\,\{z\}\geq 1$
(c) $|z|\leq 1$
(d) $|z-1|=3$

15. Under what conditions does $|z_1+z_2|=|z_1|+|z_2|$?

16. Under what conditions does $|z_1+z_2|=|z_1|-|z_2|$?

17. (a) Take the real part of $e^{j(\theta+\phi)}$ to show that $\cos(\theta+\phi)=\cos\theta\cos\phi-\sin\theta\sin\phi$.
(b) Take the imaginary part of $e^{j(\theta+\phi)}$ to show that $\sin(\theta+\phi)=\cos\theta\sin\phi+\sin\theta\cos\phi$.

18. Show that $z_1 z_2=0$ if and only if at least one of the numbers z_1, z_2 is zero.

19. Show that

$$|z_1+z_2+\cdots+z_n|\leq|z_1|+|z_2|+\cdots+|z_n|$$

When does the equality hold?

20. Let n be any integer. Express the nth roots of unity in (a) exponential form and (b) rectangular form.

21. Evaluate each of the following expressions, putting your answers in (i) rectangular form, (ii) exponential form, and (iii) polar form.
(a) $j^{\sqrt{j}}$
(b) j^e
(c) $\cos(3+2j)$
(d) $\sin(3+2j)$
(e) $1+z+z^2+\cdots$, where $|z|<1$. (Hint: Let S equal the infinite series and subtract zS from S to get a closed form for S.)
(f) $1+z+z^2+\cdots+z^N$, where N is an integer.

22. Let $V(s)=1/(s-s_0)$ where s and s_0 are complex numbers. Show that, in general, $V^*(s)\neq V(s^*)\neq V^*(s^*)$.

Elementary Mathematical Formulas

The following standard formulas may be helpful in solving certain circuit problems.

C.1 Quadratic Formula

The roots of $ax^2 + bx + c = 0$ are given by

$$x = \frac{-b \pm \sqrt{b^2 - 4ac}}{2a} = -\frac{b}{2a} \pm \sqrt{\left(\frac{b}{2a}\right)^2 - \frac{c}{a}}$$

C.2 Trigonometric Formulas

$$\sin(-x) = -\sin x$$

$$\cos(-x) = \cos x$$

$$\sin x = \cos(x - 90°)$$

$$\cos x = \sin(x + 90°)$$

$$\cos^2 x + \sin^2 x = 1$$

$$\cos^2 x - \sin^2 x = \cos 2x$$

$$2 \sin x \cos x = \sin 2x$$

$$\sin(x + y) = \sin x \cos y + \cos x \sin y$$

$$\cos(x + y) = \cos x \cos y - \sin x \sin y$$

$$A \cos x + B \sin x = \sqrt{A^2 + B^2} \cos\left(x - \arctan\frac{B}{A}\right)$$

$$\sin x \sin y = \frac{\cos(x - y) - \cos(x + y)}{2}$$

$$\sin x \cos y = \frac{\sin(x + y) + \sin(x - y)}{2}$$

$$\cos x \cos y = \frac{\cos(x + y) + \cos(x - y)}{2}$$

C.3 Plane Triangle Formulas

Let A, B, C denote the angles of any plane triangle and a, b, c denote the corresponding opposite sides. The following apply

Law of sines

$$\frac{a}{\sin A} = \frac{b}{\sin B} = \frac{c}{\sin C}$$

Law of cosines

$$a^2 = b^2 + c^2 - 2bc \cos A$$

Law of tangents

$$\frac{b - c}{b + c} = \frac{\tan\frac{1}{2}(B - C)}{\tan\frac{1}{2}(B + C)}$$

C.4 Taylor Series

$$e^x = 1 + x + \frac{1}{2!}x^2 + \frac{1}{3!}x^3 + \cdots$$

$$\sin x = x - \frac{1}{3!}x^3 + \frac{1}{5!}x^5 - \cdots$$

$$\cos x = 1 - \frac{1}{2!}x^2 + \frac{1}{4!}x^4 - \cdots$$

C.5 Euler's Relations

$$e^{jx} = \cos x + j \sin x \qquad \text{where } j^2 = -1$$

$$\cos x = \frac{e^{jx} + e^{-jx}}{2}$$

$$\sin x = \frac{e^{jx} - e^{-jx}}{2j}$$

C.6 Miscellaneous Series and Integrals

$$\frac{1}{1-z} = 1 + z + z^2 + z^3 + \cdots \qquad \text{for } |z| < 1$$

$$\frac{1 - z^{n+1}}{1-z} = 1 + z + z^2 + \cdots + z^n \qquad \text{for any } z \neq 1, \text{ where } n \text{ is an integer}$$

$$\int_{-\infty}^{\infty} e^{-a^2x^2}\,dx = \frac{\sqrt{\pi}}{a} \qquad \text{for } a > 0$$

C.7 L'Hôpital's Rule

If $f(a) = g(a) = 0$, then

$$\lim_{x \to a} \frac{f(x)}{g(x)} = \lim_{x \to a} \frac{f'(x)}{g'(x)}$$

provided that $g'(a) \neq 0$. The primes denote derivatives with respect to x.

ANSWERS TO EXERCISES

CHAPTER 1

1. 2 mm **2.** 15 km **3.** 50 ms **4.** 400 ns **5.** 27 MW **6.** 7 ms
7. Consistent **8.** Inconsistent **9.** Consistent **10.** Inconsistent
11. Consistent, but lowercase preferable **12.** -6.25×10^3 electrons
13. $-20e^{-2t}$ A **14.** $-20 \sin 2t$ A **15.** $5(1 - e^{-2t})$ C **16.** $5 \sin 2t$ C
17. 12.5 V **18.** 4 C, 2 A **19.** Active **20.** Passive **21.** Active **22.** Passive
23. Passive **24.** $-\frac{1}{12}$ A **25.** $\frac{5}{12}$ A **26.** $\frac{36}{5}$ V **27.** $\frac{1}{3}$ A **28.** $-\frac{2}{3}$ A
29. 200 W, 200 W **30.** $100[\cos(\pi/3) + \cos(2\pi120t - \pi/3)]$ W, 50 W **31.** -96 W
32. -24 W **33.** 480 W **34.** -480 W **35.** 120 W **36.** Yes

CHAPTER 2

1. 5 **2.** 4 **3.** -9 A **4.** 14 A **5.** 8 **6.** 7 **7.** 6 A **8.** 5 **9.** 3
10. 3 **11.** 6 **12.** 9 V **13.** 14 V **14.** -12 V **15.** -12 V **16.** -16 V
17. -4 V, 2 V, -2 V, 9 V, -20 V, -7 V, 3 V, -9 V **18.** 20 V **19.** 0 V
20. 20 V **21.** -10 V **22.** 30 A **23.** 20 A **24.** -10 A **25.** 20 A
26. -30 A, 45 V, 900 W **27.** -8 A, -4 V, -40 W **28.** -20 A, 10 V, 200 W
29. -10 A, 20 V, -100 W **30.** KVL is violated **31.** KVL is violated
32. KCL is violated **33.** KCL is violated **34.** 6 V, 30 A **35.** -100 V, -10 A
36. 120 V, 0 A **37.** 1 V, -120 A **38.** 80 Ω **39.** 10 V, 2 Ω
40. 24 V, 12 A, 288 W **41.** 10 V, 5 A, 50 W **42.** -18 V, -54 A, 972 W
43. 42 V, 7 A, 294 W **44.** -63 V, -7 A, 441 W **45.** -75 V, -5 A, 375 W
46. -24 V, -2 A, 48 W **47.** 24 V, 3 A, 72 W **48.** 192 V, 24 A, 4608 W
49. -50 A **50.** -20 V **51.** 9 V if i_L is constant, 9.68 V if load is resistive
52. $-120\pi \sin 2\pi60t$ A **53.** $-18,850 \sin 2\pi120t$ W **54.** $25 + 25 \cos 2\pi120t$ J
55. $22e^{-3t}$ V, $t > 0$ **56.** $-132e^{-3t}$ A, $t > 0$ **57.** $24e^{-3t}$ A, $t > 0$
58. $-108e^{-3t}$ A, $t > 0$ **59.** Integrate Eq. (2.22) **60.** $-120\pi \sin 2\pi60t$ V
61. $-18,850 \sin 2\pi120t$ W **62.** $25 + 25 \cos 2\pi120t$ J **63.** $22e^{-3t}$ A, $t > 0$
64. $2e^{-3t}$ A, $t > 0$ **65.** $-132e^{-3t}$ V, $t > 0$ **66.** $24e^{-3t}$ V, $t > 0$
67. Integrate Eq. (2.35) **68.** $-1200 \sin 3t$ V, $-600 \sin 3t$ V
69. $-1125 \sin 3t$ V, 0 V, $B - \frac{25}{2} \cos 3t$ A **70.** $\frac{5}{6} \sin 3t$ A, $5 \cos 3t$ V
71. $\frac{8}{9} \sin 3t$ A, 0 V **72.** $(L_1 + M)di_y/dt - Mdi_x/dt$, $-(L_2 + M)di_y/dt + L_2di_x/dt$

73. $-\dfrac{L_2}{L_1L_2 - M^2} \displaystyle\int_{-\infty}^{t} v_y\, d\tau + \dfrac{L_2 + M}{L_1L_2 - M^2} \int_{-\infty}^{t} v_x\, d\tau;$

$\dfrac{L_1 + M}{L_1L_2 - M^2} \displaystyle\int_{-\infty}^{t} v_x\, d\tau - \dfrac{M}{L_1L_2 - M^2} \int_{-\infty}^{t} v_y\, d\tau$ **74.** 0.417   **75.** 6 H

76. $76 - \sqrt{10016}\, \cos(4t - 1.53)$ J   **77.** $76 - \sqrt{10016}\, \cos(4t + 1.53)$ J

CHAPTER 3

1. 12 A, 8 A, 24 A, -5 A, -39 A, 1440 W, 960 W, 2880 W, -600 W, -4680 W

2. -12 A, -8 A, -24 A, -5 A, 49 A

3. $2\cos 2t$ A, $-320\sin 2t$ A, $-5 + 320\sin 2t$ A, $-\cos 2t$ A, $5 - \cos 2t$ A, -5 A

4. $8\dfrac{d}{dt}v + \dfrac{1}{20}v + 10 + 32\displaystyle\int_0^t v\, d\tau = 2\cos 2t,\ t > 0$ **5.** 1 Ω, e^{2t} V

6. 15.75 Ω, $15.75e^{2t}$ V   **7.** 2 H, $4e^{2t}$ V   **8.** 79 F, $6.329e^{2t}$ mV

9. 8 Ω, 15 H, $57e^{2t}$ A   **10.** 2 H, 28 F, $20{,}250e^{2t}$ A   **11.** 30 A, -20 A   **12.** 3 A, -15 A

13. 12 V, 24 V, 8 V, -5 V, -39 V, 1440 W, 2880 W, 960 W, -600 W, -4680 W

14. -120 A, -12 V, -24 V, -8 V, -5 V, 49 V

15. $2\cos 2t$ V, $-320\sin 2t$ V, $-5 + 320\sin 2t$ V, $-\cos 2t$ V, $-\cos 2t + 5$ V, -5 V

16. $8\dfrac{d}{dt}i + \dfrac{1}{20}i + 10 + 32\displaystyle\int_0^t i\, d\tau = 2\cos 2t,\ t > 0$ **17.** 14 Ω, $\frac{74}{7}e^{3t}$ A

18. 20 Ω, $\frac{37}{5}e^{3t}$ A   **19.** 15 H, $\frac{148}{45}e^{3t}$ A   **20.** 15 H, $\frac{148}{45}e^{3t}$ A   **21.** 3 F, $1332e^{3t}$ A

22. 1.1461 F, $508.85e^{3t}$ A   **23.** 8 V   **24.** 12 V, -72 V   **25.** 3 V, -24 V

26. 2 V, 1 V   **27.** 24 V, 4 V   **28.** -56 V, $\frac{160}{3}$ V   **29.** 2 Ω, 10 Ω, 5 Ω

30. $R/3$ **31.** 230 Ω, 184 Ω, 115 Ω   **32.** $3R$ **33.** 9 A, 9 A   **34.** 16 Ω, 4 A

CHAPTER 4

1. -12 V, 36 V, -48 V, 8 A   **2.** 18 V, 36 V, -18 V, 3 A   **3.** 70 V, -105 V, 175 V

4. $\frac{1}{4}v_1 + \frac{1}{2}v_2 = -35$ **5.** 120 V, 70 V, 50 V

6. $\begin{bmatrix} \dfrac{1}{4} + \dfrac{1}{10}\displaystyle\int_{-\infty}^{t} d\tau & -\dfrac{1}{10}\displaystyle\int_{-\infty}^{t} d\tau \\[3mm] -\dfrac{1}{10}\displaystyle\int_{-\infty}^{t} d\tau & 2\dfrac{d}{dt} + \dfrac{1}{10}\displaystyle\int_{-\infty}^{t} d\tau \end{bmatrix} \cdot \begin{bmatrix} v_1 \\ v_2 \end{bmatrix} = \begin{bmatrix} 35 \\ -70 \end{bmatrix}$

7. $\begin{bmatrix} \dfrac{3}{5} & -\dfrac{1}{2} & \dfrac{11}{20} & -\dfrac{1}{2} \\[2mm] -\dfrac{1}{2} & -\dfrac{3}{2} & \dfrac{3}{2} & 1 \\[2mm] 1 & 0 & -1 & 0 \\[2mm] 0 & 1 & 0 & -1 \end{bmatrix} \cdot \begin{bmatrix} v_1 \\ v_2 \\ v_3 \\ v_4 \end{bmatrix} = \begin{bmatrix} 0 \\ 0 \\ 10 \\ 40 \end{bmatrix},$ $\begin{bmatrix} \dfrac{23}{20} & -1 \\[2mm] 1 & -\dfrac{1}{2} \end{bmatrix} \cdot \begin{bmatrix} v_1 \\ v_2 \end{bmatrix} = \begin{bmatrix} -\dfrac{29}{2} \\[2mm] 55 \end{bmatrix},$

146.47 V, 182.94 V, 136.47 V, 142.94 V   **8.** 14.65 A, 1.765 A, -16.41 A

9. -100 V, 10 V, 20 V, -220 W   **10.** 219 V, 20 V, 0 V, -796 W

11. 230 V, -1 V, 20 V, -1016.4 W   **12.** $\begin{bmatrix} \dfrac{3}{5} & -\dfrac{1}{2} & \dfrac{11}{20} & -\dfrac{1}{2} \\[2mm] 1 & 0 & -1 & 0 \\[2mm] 0 & 1 & 0 & 0 \\[2mm] 0 & 0 & 0 & 1 \end{bmatrix} \cdot \begin{bmatrix} v_1 \\ v_2 \\ v_3 \\ v_4 \end{bmatrix} = \begin{bmatrix} 0 \\ 10 \\ 140 \\ 100 \end{bmatrix}$

13. 109.13 V, 140 V, 99.13 V, 100 V   **14.** $\begin{bmatrix} C\dfrac{d}{dt} + \dfrac{1}{R} & 0 \\[2mm] -\alpha & 1 \end{bmatrix} \cdot \begin{bmatrix} v_1 \\ v_2 \end{bmatrix} = \begin{bmatrix} i_s \\ 0 \end{bmatrix}$

15.
$$\begin{bmatrix} \dfrac{13}{12} & -1 & 0 & 0 \\[2mm] -1 & \dfrac{4}{3} & \dfrac{5}{6} & -\dfrac{1}{2} \\[2mm] \dfrac{1}{4} & -1 & 1 & 0 \\[2mm] 0 & 0 & 0 & 1 \end{bmatrix} \cdot \begin{bmatrix} v_1 \\ v_2 \\ v_3 \\ v_4 \end{bmatrix} = \begin{bmatrix} 14 \\ 0 \\ 0 \\ -6 \end{bmatrix}$$

16.
$$\begin{bmatrix} \dfrac{13}{12} & -1 \\[2mm] -\dfrac{29}{24} & \dfrac{13}{6} \end{bmatrix} \cdot \begin{bmatrix} v_1 \\ v_2 \end{bmatrix} = \begin{bmatrix} 14 \\ -3 \end{bmatrix}$$

17. 24 V, 12 V, 6 V, −6 V, 48 W

18.
$$\begin{bmatrix} \dfrac{1}{3} + \dfrac{1}{6} & -\dfrac{1}{6} \\[2mm] -\dfrac{1}{6} & \dfrac{1}{6} + \dfrac{1}{10} + \dfrac{1}{15} \end{bmatrix} \cdot \begin{bmatrix} v_1 \\ v_2 \end{bmatrix} = \begin{bmatrix} 8 - 1 \\ 1 \end{bmatrix}, \quad$$ 18 V, 12 V

19.
$$\begin{bmatrix} \dfrac{1}{4} + \dfrac{1}{10} & -\dfrac{1}{10} \\[2mm] -\dfrac{1}{10} & \dfrac{1}{2} + \dfrac{1}{10} \end{bmatrix} \cdot \begin{bmatrix} v_1 \\ v_2 \end{bmatrix} = \begin{bmatrix} 35 \\ -70 \end{bmatrix}, \quad$$ 70 V, 105 V

20.
$$\begin{bmatrix} \dfrac{1}{R_3} + \dfrac{1}{R_4} + \dfrac{L_2}{L_1 L_2 - M^2} \int_{-\infty}^{t} d\tau & \dfrac{M - L_2}{L_1 L_2 - M^2} \int_{-\infty}^{t} d\tau & -\dfrac{1}{R_4} - \dfrac{M}{L_1 L_2 - M^2} \int_{-\infty}^{t} d\tau \\[3mm] \dfrac{M - L_2}{L_1 L_2 - M^2} \int_{-\infty}^{t} d\tau & \dfrac{1}{R_1} + \dfrac{L_1 + L_2 + 2M}{L_1 L_2 - M^2} \int_{-\infty}^{t} d\tau & \dfrac{M - L_1}{L_1 L_2 - M^2} \int_{-\infty}^{t} d\tau \\[3mm] -\dfrac{1}{R_4} - \dfrac{M}{L_1 L_2 - M^2} \int_{-\infty}^{t} d\tau & \dfrac{M - L_1}{L_1 L_2 - M^2} \int_{-\infty}^{t} d\tau & \dfrac{1}{R_3} + \dfrac{1}{R_4} + \dfrac{L_1}{L_1 L_2 - M^2} \int_{-\infty}^{t} d\tau \end{bmatrix} \cdot \begin{bmatrix} v_1 \\ v_2 \\ v_3 \end{bmatrix} = \begin{bmatrix} \dfrac{1}{R_3} v_s \\ 0 \\ 0 \end{bmatrix}$$

21. −12 A, 36 A, −48 A, 8 V **22.** 18 A, 36 A, −18 A, 3 V

23. 70 A, −105 A, 175 A **24.** $\frac{1}{4}i_1 + \frac{1}{2}i_2 = -35$ **25.** 120 A, 70 A, 50 A

26.
$$\begin{bmatrix} \dfrac{1}{4} + 10 \int_{-\infty}^{t} d\tau & -10 \int_{-\infty}^{t} d\tau \\[3mm] -10 \int_{-\infty}^{t} d\tau & \dfrac{1}{2}\dfrac{d}{dt} + 10 \int_{-\infty}^{t} d\tau \end{bmatrix} \cdot \begin{bmatrix} i_1 \\ i_2 \end{bmatrix} = \begin{bmatrix} 35 \\ -70 \end{bmatrix}$$

27.
$$\begin{bmatrix} \dfrac{3}{5} & -\dfrac{1}{2} & \dfrac{11}{20} & -\dfrac{1}{2} \\[2mm] -\dfrac{1}{2} & -\dfrac{3}{2} & \dfrac{3}{2} & 1 \\[2mm] 1 & 0 & -1 & 0 \\[2mm] 0 & 1 & 0 & -1 \end{bmatrix} \cdot \begin{bmatrix} i_1 \\ i_2 \\ i_3 \\ i_4 \end{bmatrix} = \begin{bmatrix} 0 \\ 0 \\ 10 \\ 40 \end{bmatrix}, \quad \begin{bmatrix} \dfrac{23}{20} & -1 \\[2mm] 1 & -\dfrac{1}{2} \end{bmatrix} \cdot \begin{bmatrix} i_1 \\ i_2 \end{bmatrix} = \begin{bmatrix} -\dfrac{29}{2} \\[2mm] 55 \end{bmatrix},$$

146.47 A, 182.94 A, 136.47 A, 142.94 A **28.** 14.647 V, 1.7647 V, −16.412 V

29. −100 A, 10 A, 20 A, −220 W **30.** 219.05 A, 20 A, −796.2 W

31. −100 A, 10 A, 20 A, −220 W **32.** 109.3 A, 140 A, 99.13 A, 100 A

33.
$$\begin{bmatrix} \dfrac{13}{12} & -1 & 0 & 0 \\[2mm] -1 & \dfrac{4}{3} & \dfrac{5}{6} & -\dfrac{1}{2} \\[2mm] \dfrac{1}{4} & -1 & 1 & 0 \\[2mm] 0 & 0 & 0 & 1 \end{bmatrix} \cdot \begin{bmatrix} i_1 \\ i_2 \\ i_3 \\ i_4 \end{bmatrix} = \begin{bmatrix} 14 \\ 0 \\ 0 \\ -6 \end{bmatrix}$$

34.
$$\begin{bmatrix} \dfrac{13}{12} & -1 \\[2mm] -\dfrac{29}{24} & \dfrac{13}{16} \end{bmatrix} \cdot \begin{bmatrix} i_1 \\ i_2 \end{bmatrix} = \begin{bmatrix} 14 \\ -3 \end{bmatrix}$$

35. 24 A, 12 A, 6 A, −6 A, 48 W   **36.**
$$\begin{bmatrix} \dfrac{1}{4}+\dfrac{1}{10} & -\dfrac{1}{10} \\[2ex] -\dfrac{1}{10} & \dfrac{1}{2}+\dfrac{1}{10} \end{bmatrix}\cdot\begin{bmatrix} i_1 \\ i_2 \end{bmatrix}=\begin{bmatrix} 35 \\ -70 \end{bmatrix}$$

37.
$$\begin{bmatrix} 4\dfrac{d}{dt}+3 & -6\dfrac{d}{dt} & -2 \\[2ex] -6\dfrac{d}{dt} & 12\dfrac{d}{dt}+5+\dfrac{1}{3}\displaystyle\int_{-\infty}^{t}d\tau & -8\dfrac{d}{dt}-\dfrac{1}{3}\displaystyle\int_{-\infty}^{t}d\tau \\[2ex] -2 & -8\dfrac{d}{dt}-\dfrac{1}{3}\displaystyle\int_{-\infty}^{t}d\tau & -8\dfrac{d}{dt}+9+\dfrac{1}{3}\displaystyle\int_{-\infty}^{t}d\tau \end{bmatrix}\cdot\begin{bmatrix} i_1 \\ i_2 \\ i_3 \end{bmatrix}=\begin{bmatrix} v_s \\ 0 \\ 0 \end{bmatrix}$$

38. 24 V, 12 V, 6 V, −6 V   **39.** 12 A   **40.** 10 V, 20 V, 30 V, 40 V
41. 0 V, 13.33 V, −6.667 V, 0 V, −2.222 A

CHAPTER 5

1. The control equation can be shown to be linear and time invariant. The proof follows that contained in Section 5.1. However, a dependent source is a four-terminal equation, and both input and output voltages and currents must be considered. Consider specifically the case for a voltage-controlled voltage source (refer to Fig. 2.11). The source is described by

$$\begin{bmatrix} v_{cd} \\ i_{ab} \end{bmatrix}=\begin{bmatrix} k & 0 \\ 0 & 0 \end{bmatrix}\cdot\begin{bmatrix} v_{ab} \\ i_{cd} \end{bmatrix}$$

This is of the form

$$\mathbf{Y}=\mathbf{T}\cdot\mathbf{X}$$

Matrix **T** is time invariant, and matrix multiplication is a linear transformation. Therefore, VCVS is a linear time-invariant component. Similar results hold for the other three dependent sources.   **2.** Not linear

3. $v_1(t)=\dfrac{1}{C}\displaystyle\int_{-\infty}^{t}i(\tau)\,d\tau$ and $v_2(t)=\dfrac{1}{C}\displaystyle\int_{-\infty}^{t}i(\tau-t_0)\,d\tau$ by a change in variables:

$v_2(t)=\dfrac{1}{C}\displaystyle\int_{-\infty}^{-t-t_0}(v)\,dv=v_1(t-t_0)\Rightarrow$ time invariant.   **4.** Use $v_{ab}=V_0$

5. Use $i_{ab}=I_0$ **6.** $-480+160-56=-376$ V   **7.** $6-\frac{24}{5}-2=-\frac{4}{5}$ V
8. 2 V   **9.** $\frac{25}{3}$ A, $\frac{35}{3}$ A
10.
$$\begin{bmatrix} C_1\dfrac{d}{dt}+\dfrac{1}{R_1}+\dfrac{1}{L}\displaystyle\int_{-\infty}^{t}d\tau & -\dfrac{1}{L}\displaystyle\int_{-\infty}^{t}d\tau \\[2ex] -\dfrac{1}{L}\displaystyle\int_{-\infty}^{t}d\tau & C_2\dfrac{d}{dt}+\dfrac{1}{R_2}+\dfrac{1}{L}\displaystyle\int_{-\infty}^{t}d\tau \end{bmatrix}\cdot\begin{bmatrix} v_1 \\ v_2 \end{bmatrix}=\begin{bmatrix} \dfrac{1}{R_1}v_a \\[2ex] \dfrac{1}{R_2}v_b \end{bmatrix}$$

11. A 40-V source in series with a 10-Ω resistance or a 4-A source in parallel with a 10-Ω resistance

12. A 200-V source in series with a 24-Ω resistance or a $\frac{25}{3}$-A source in parallel with a 24-Ω resistance   **13.** Same as Problem 11   **14.** Same as Problem 12
15. A 75-V source in series with a $\frac{75}{2}$-Ω resistance. A 2-A source in parallel with a $\frac{75}{2}$-Ω resistance.   **16.** 40 V, 10 Ω   **17.** 200 V, 24 Ω   **18.** 75 V, $\frac{75}{2}$ Ω
19. 0 Ω, 0 W, 1440 W   **20.** 10 Ω, 360 W, 360 W   **21.** 2 Ω, 18 W   **22.** $\frac{75}{2}$ Ω
23. −2 V   **24.** 36 V   **25.** The same as Fig. 5.31b   **26.** The same as Fig. 5.31a

27.

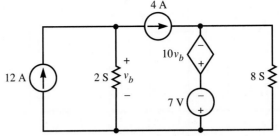

28. Reciprocity holds

29. Reciprocity holds   **30.** Reciprocity holds

CHAPTER 6

1. 8.203   **2.** 9.091   **3.** 10   **4.** 9.8 Ω **5.** 9.09 Ω **6.** 10 Ω **7.** 60650.17 Ω
8. 100 kΩ **9.** -9.01 **10.** $\frac{1}{11}$ (α is negative)   **11.** -9.01
12. $v_o = -R_3(v_1/R_1 + v_2/R_2)$, $v_o = -(v_1 + v_2)$
13. Source 1 sees a resistance of R_1. Source 2 sees a resistance of R_2. The output circuit is two VCVS in series. One has a value of $-(R_3/R_1)v_1$, and the other a value of $-(R_3/R_2)v_2$ with the $(+)$ references toward the output terminal and the $(-)$ references toward ground.
14. $v_o = \dfrac{R_2 + R_1}{R_1} \cdot \dfrac{R_b}{R_a + R_b} v_2 - \dfrac{R_2}{R_1} v_1$, $v_o = \dfrac{R_2}{R_1}(v_2 - v_1)$
15. Source v_1 sees a resistance of R_1 in series with a VCVS of value v_b. Source v_2 sees resistances R_a and R_b in series. The voltage across R_b is v_b. The output circuit consists of two VCVS in series one with value v_b and the other with value $-R_2/R_1\, v_1$, with $(+)$ reference marks toward the output terminal.
16. $v_0 = -\dfrac{1}{RC}\displaystyle\int_{-\infty}^{t} v_s\, d\tau$ **17.** $v_0 = -RC\dfrac{d}{dt}v_s$
18. $v_0 = \dfrac{(R_1 + R_2)(R_3 + R_4) + R_3R_4}{R_1R_4} v_{s2} - \dfrac{R_2(R_3 + R_4) + R_3R_4}{R_1R_4} v_{s1}$
19. Source s_1 sees R_1 in series with a VCVS of value v_{s2}. Source v_{s2} sees an open circuit. The output circuit consists of a VCVS of value A_1v_{s1} in series with a VCVS of value
A_2v_{s2} where $A_1 = -\dfrac{R_2(R_3 + R_4) + R_3R_4}{R_1R_4}$ and $A_2 = -\dfrac{(R_1 + R_2)(R_3 + R_4) + R_3R_4}{R_1R_4}$

CHAPTER 7

1. $2u(t - 2) - 2u(t - 5)$ **2.** $u(t) + u(t - 1) + 3u(t - 3)$
3. $10u(t - 1) - 5u(t - 2) - 5u(t - 3) = 10\Pi(t - \frac{3}{2}) + 5\Pi(t - \frac{5}{2}) = 5\Pi(t - \frac{3}{2}) + 5\Pi[(t - 2)/2]$
4. $3t[u(t) - u(t - 1)] = 3t\Pi(t - \frac{1}{2})$
5.

6.

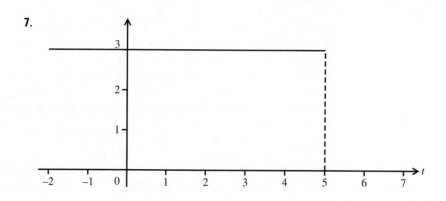

7.

8. $2t\,u\,(t-1)$

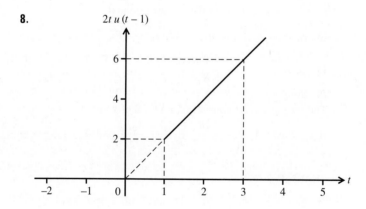

9. $2\,(t-1)\,u\,(t-1)$

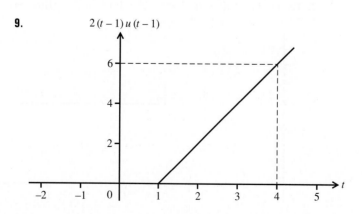

10.

11. $(t-2)u(t-2)$ **12.** $(t+\frac{1}{2})u(t+\frac{1}{2}) - (t-\frac{1}{2})u(t-\frac{1}{2})$ **13.** 1   **14.** 1

15. $\cos 2\pi t$ **16.** 1   **17.** 0.819   **18.** 0.995   **19.** 0.812   **20.** 0.912

21. $0.995 + j0.0998$ **22.** $0.812 + j0.584$ **23.** 1.1513   **24.** 0.001707   **25.** 0

26. b, c, and d   **27.** Periodic   **28.** Aperiodic   **29.** Periodic   **30.** 0

31. $\frac{1}{2}(V_{max} + V_{min})$ **32.** 5 A   **33.** 0   **34.** 4 A   **35.** 4 V   **36.** V_m **37.** $\frac{1}{2}I_m$

38. 4 A   **39.** 3.30 V   **40.** V_m **41.** $1/\sqrt{3}\,V_m$ **42.** 5 A   **43.** $\sqrt{17}$

44. Energy   **45.** Energy   **46.** Power   **47.** Show that $\lim\limits_{T\to\infty} \dfrac{1}{T}\displaystyle\int_{-T/2}^{T/2} v_1(t)v_2(t)\,dt = 0$

48. 4.95 V   **49.** 3.54 V   **50.** 4 A   **51.** 4.53 A   **52.** 2.61 A

CHAPTER 8

1. $(p + \frac{1}{4})i = 5e^{-2t}$ **2.** $(p^2 + 6p + 8)i = 0$ **3.** $(p^2 + 5p + 6)v = p(24e^{-3t})$

4. $(p^2 + 9p)i = 20e^{-t}$ **5.** $(p^2 + 6p + 9)v = 0$ **6.** $(p^2 + 9)i = 36$ **7.** $Ae^{-\frac{1}{4}t}$

8. $A_1e^{-2t} + A_2e^{-3t}$ **9.** $A_1e^{(-2+j1)t} + A_2e^{(-2-j1)t}$ **10.** $A_0 + A_1e^{-9t}$

11. $10e^{-20t}$ A   **12.** $(p+2)v = 0$ **13.** $(p^2 + 5p + 6)v = 0$ **14.** $(p^2 + 2p)v = 0$

15. $(A_0 + A_1t)e^{-3t}$ **16.** $A_0 + A_1t + A_2e^{-2t}$ **17.** $(A_0 + A_1t + A_2t^2)e^{-t}$

18. $(p^2 + 4p + 4)i = 0$ **19.** $(A_0 + A_1t)e^{-2t}$ **20.** $(12 - 42t)e^{-2t}$ A for $t > 0$

21. $(p^2 + 12p + 36)v = 0$ **22.** $(p^3 + 21p^2 + 144p + 320)v = 0$ **23.** $(p^3 + 2p^2)v = 0$

24. $(p^3 + 3p^2 + 3p + 1)v = 0$ **25.** $10e^t$ V for $t > 0$, unstable

26. $4 - 4e^{-4t}$ V for $t > 0$ **27.** $10e^{-2/3t}$ V for $t > 0$ **28.** $\frac{1}{2}e^{-(t-5)/5}$ A for $t > 5$

29. $10e^{-1/6t}$ A for $t > 0$ **30.** $-60e^{-72(t-5)}$ V for $t > 5$ **31.** $e^{-2t}(A\cos t + B\sin t)$

32. $A\cos 3t + B\sin 3t$ **33.** $(p^2 + 9)v = 0$ **34.** $(p^2 + 9)v = 0$

35. $(p^2 + 4p + 29)v = 0$ **36.** $(p^4 + 4p^3 + 78p^2 + 148p + 1369)v = 0$

37. $(p^3 + 5p^2 + 44p + 40)v = 0$ **38.** $(p^5 + 16p^3)v = 0$

39. $-11/6e^{-t} + 35/6e^{-25t}$ A for $t > 0$ **40.** $(4 - 60t)e^{-5t}$ A for $t > 0$

41. $e^{-3t}(4\cos 4t - 13\sin 4t)$ A for $t > 0$ **42.** $-544/3e^{-2t} + 736/3e^{-8t}$ V for $t > 0$

43. $(64 - 1216t)e^{-4t}$ V for $t > 0$ **44.** $e^{-3t}(64\cos\sqrt{7}t - 1152/\sqrt{7}\sin\sqrt{7}t)$ V for $t > 0$

45. CIRCUIT FILE FOR EXERCISE 8.45
```
R10    1    0    20
C10    1    0    0.025    IC=4
.TRAN    0.05    1.5    UIC
.PLOT    TRAN    V(1)
.END
```

46. CIRCUIT FILE FOR EXERCISE 8.46
```
L01    0    1    0.015625    IC=8
R12    1    2    3.125
C20    2    0    0.01    IC=16
.TRAN    0.002    0.048    UIC
.PLOT    TRAN    I(L01)
.END
```

47. CIRCUIT FILE FOR EXERCISE 8.47

```
LO1    0    1    0.015625    IC=8
R12    1    2    1
C20    2    0    0.01         IC=16
.TRAN    0.002    0.096    UIC
.PLOT    TRAN    I(LO1)
.END
```

CHAPTER 9

1. $-12e^{-6t}$ V   **2.** 6 V   **3.** $-\frac{3}{2}e^{-6t}$ V   **4.** $6te^{-2t}$ V   **5.** $-6e^{-3t}$ A

6. $4te^{-2t}$ V   **7.** $25t^2e^{-5t}$ V   **8.** $-333.33e^{-5t}$ A   **9.** $1000te^{-2t}$ A   **10.** 500 A

11. $1000/7e^{5t}$ A   **12.** $250e^{2t}$ A   **13.** $-6e^{-3t}$ V   **14.** $4te^{-2t}$ V   **15.** $25t^2e^{-5t}$

16. 4 A   **17.** $4 + 12e^{-t} + 36te^{-3t}$ **18.** $24 - 12e^{-2t}$ V   **19.** $2 + 5.238e^{2t}$ V

20. $20 \cos (3t + 8.13°)$ A   **21.** $4.851 \cos (2t - 75.96°)$ V   **22.** $16.971 \cos (5t - 135°)$ V

23. $-10 \cos 3t$ A $= 10 \cos (3t \pm 180°)$ A   **24.** $50t \cos 2t$ A

25. $A_1 + A_2e^{-t} + 5t - te^{-t}$ V   **26.** $A_1e^{-t} + A_2e^{-9t} + 0.485 \cos (2t - 75.96°)$ V

27. $e^{-t}(A \cos 6t + B \sin 6t) + 9.965 \cos (6t - 85.2°)$ V

28. $(A_0 + A_1t + A_2t^2)e^{-t} + 40$ V   **29.** $18, -13$ **30.** $7.136, -87.669$

31. $150 + e^{-t}(150 \cos 3t - 350 \sin 3t)$ V for $t > 0$ **32.** $24e^{-6t} - 12e^{-3t}$ A for $t > 0$

33. 2 A, 0 V, 2 A, 64 V, $6 - 4e^{-2t}$ A for $t > 0$ **34.** Same as Exercise 33

35. 48 V for $t < 0$, $88e^{-5t} - 40e^{-8t}$ V for $t > 0$ **36.** $40te^{-5t}$ A for $t > 0$

37. $-0.28e^{-2t} + 1.6 \cos (1.5t - 36.87°)$ A for $t > 0$

38. $(3.81 + 15.60t)e^{-5t}$ A for $t > 0$ **39.** $5(1 - e^{-24t})$ A for $t > 0$

40. $16 - 14e^{-t}$ A for $t > 0$ **41.** $16 - 5e^{-4t} + 5e^{-36t}$ A for $t > 0$, overdamped for both cases   **42.** $24(1 - e^{-1/3t})$ V for $t > 0$ **43.** Same as Exercise 42

44. $(12 - 22t)e^{-2t} - 9$ V for $t > 0$ **45.** $(10 + 22t)e^{-2t} - 9$ A for $t > 0$

46. $(-9 - 18t)e^{-2t} - 18$ V for $t > 0$, $(-9 - 18t)e^{-2t} - 18$ V for $t > 0$

47. INPUT FILE FOR EXERCISE 9.47

```
R30    3    0    0.33333
R20    2    0    1
C10    1    0    1    IC=3
L12    1    2    1    IC=1
V31    3    1    DC    12
.TRAN    0.1    3    UIC
.PLOT    TRAN    V(1), I(L12)
.END
```

48. INPUT FILE FOR EXERCISE 9.48

```
R30    3    0    0.33333
R20    2    0    1
C10    1    0    1    IC=3
L12    1    2    1    IC=1
V31    3    1    DC    12
.TRAN    0.1    3    UIC
.PROBE
.END
```

49. INPUT FILE FOR EXERCISE 9.49
```
R30     3    0    0.33333
R20     2    0    1
C10     1    0    1
L12     1    2    1
V34     3    4    DC     36
V41     4    1    PULSE(0,-12)
.TRAN   0.1  3
.PLOT   TRAN   V(1)   V(2)
.END
```

50. INPUT FILE FOR EXERCISE 9.50
```
R30     3    0    0.33333
R20     2    0    1
C10     1    0    1
L12     1    2    1
V34     3    4    DC     36
V41     4    1    PULSE(0,-12)
.TRAN   0.1  3
.PROBE
.END
```

51. CIRCUIT FILE FOR EXERCISE 9.51
```
R32     3    2    12
R21     2    1    8
L10     1    0    2
V30     3    0    DC     240
*SWITCH MODEL
V40     4    0    PULSE(2,-1)
S21     2    1    4    0    SMOD
.MODEL   SMOD   VSWITCH   (RON=1E-3   ROFF=1E6)
R40     4    0    1
*SWITCH MODEL COMPLETE
.TRAN   0.0001   0.4
.PROBE
.END
```

CHAPTER 10

4. (a) $\mathbf{I} = 10\underline{/20°}$ mA, $\mathbf{V} = 25\underline{/-65°}$ V   (b) $\mathbf{Z} = \mathbf{V}/\mathbf{I} = 2.5\underline{/-85°}$ kΩ

5. $v_p(t) = 15.8 \cos(2t + 63.4°)$ V

7. From $v = Ri$ we have $\mathscr{A}(p) = 1$ and $\mathscr{B}(p) = R$. Therefore, Eq. (10.17a) yields $\mathbf{Z} = R$.

8. From $v = L\dfrac{d}{dt}i$ we have $\mathscr{A}(p) = 1$ and $\mathscr{B}(p) = Lp$. Therefore, Eq. (10.17a) yields $\mathbf{Z} = j\omega L$.

9. By substituting $\mathbf{I}e^{j\omega t}$ and $\mathbf{V}e^{j\omega t}$ into $i = C\dfrac{d}{dt}v$ we obtain $\mathbf{I}e^{j\omega t} = j\omega C\mathbf{V}e^{j\omega t}$. Solve for $\mathbf{V}/\mathbf{I}$ to obtain $\mathbf{Z} = 1/j\omega C$.

10.

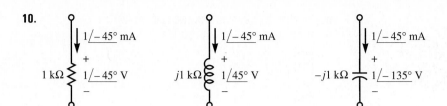

11. $v_R(t) = \cos(1000t - 45°)$ V, $v_L(t) = \cos(1000t + 45°)$ V, $v_C(t) = \cos(1000t - 135°)$ V

14. (a) 3 S   (b) 4 S   (c) 0.12 S   (d) -0.16 S   (e) capacitive

15. $\mathbf{Z} = R + j\left(\omega L - \dfrac{1}{\omega C}\right)$ (a) R (b) $\omega L - \dfrac{1}{\omega C}$ (c) $\dfrac{R}{R^2 + (\omega L - 1/\omega C)^2}$

(d) $-\dfrac{\omega L - 1/\omega C}{R^2 + (\omega L - 1/\omega C)^2}$ (e) $\dfrac{R}{R^2 + (\omega L - 1/\omega C)^2} - j\dfrac{\omega L - 1/\omega C}{R^2 + (\omega L - 1/\omega C)^2}$

16. Connect R_1 and R_2 in series. (a) The driving-point impedance is $\mathbf{Z} = R_1 + R_2$. The driving-point admittance is $\mathbf{Y} = 1/(R_1 + R_2)$. This illustrates that $\mathbf{Y} = \mathbf{Z}^{-1}$ for driving-point impedance and admittance.
(b) Measure transfer impedance and admittance as shown.

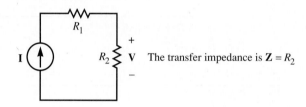

The transfer impedance is $\mathbf{Z} = R_2$

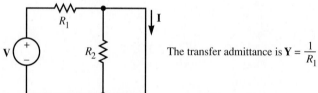

The transfer admittance is $\mathbf{Y} = \dfrac{1}{R_1}$

This illustrates that $\mathbf{Y} \neq \mathbf{Z}^{-1}$ for transfer impedance and admittance.
18. $j\omega L/(R + j\omega L)$ **19.** $2.07 \cos(1000t + 65.6°)$ mA

CHAPTER 11

1. $\mathbf{I} = 6\underline{/30°}$ **2.** $\mathbf{V} = 5\underline{/-150°}$ **3.** $\mathbf{I} = 4\underline{/-60°}$ **4.** $\mathbf{V} = 3\underline{/120°}$
6. $I_{m1} = \sqrt{I_{m2}^2 + I_{m3}^2 - 2I_{m2}I_{m3}\cos\phi}$, where $\phi = \underline{/\mathbf{I}_3} - \underline{/\mathbf{I}_2}$ **8.** $(p^2 + p + 1)v_0 = v_s$
9. Eq. (11.28) follows immediately from Eq. (11.27), Eq. (11.29) follows immediately from Eq. (11.26) with $N = 3$.
11. (a) $\mathbf{V}/\mathbf{I} = R + j\omega L \,\|\, R = R(R + j2\omega L)/(R + j\omega L)$
(b) $\mathbf{V}_0/\mathbf{V} = (R \,\|\, j\omega L)/(R + R \,\|\, j\omega L) = j\omega L/(R + j2\omega L)$
(c) $\mathbf{I}_1/\mathbf{V} = (\mathbf{I}_1/\mathbf{V}_0)(\mathbf{V}_0/\mathbf{V}) = 1/(R + j2\omega L)$

(d) $\mathbf{I_2}/\mathbf{V} = (\mathbf{I_2}/\mathbf{V_0})(\mathbf{V_0}/\mathbf{V}) = j\omega L/(R^2 + j2\omega LR)$ (e) $\mathbf{I_1}/\mathbf{I} = R/(R + j\omega L)$

(f) $\mathbf{I_2}/\mathbf{I} = j\omega L/(R + j\omega L)$

12. $\lim\limits_{\omega \to \infty} \dfrac{\mathbf{V}}{\mathbf{I}} = R_2$. This result makes sense because the impedances of the inductance and capacitance approach ∞ (open circuit) and 0 (short circuit), respectively, as $\omega \to \infty$.

13. $\begin{bmatrix} -j & -j & 0 \\ \hline -j & 4+j & -4+6j \\ \hline 0 & -4 & 4-j \end{bmatrix} \begin{bmatrix} \mathbf{V_1} \\ \mathbf{V_2} \\ \mathbf{V_3} \end{bmatrix} = \begin{bmatrix} j \\ 0 \\ 0 \end{bmatrix}$

14. $\begin{bmatrix} 1.5 & -1 \\ 2-j & j \end{bmatrix} \begin{bmatrix} \mathbf{V_1} \\ \mathbf{V_2} \end{bmatrix} = \begin{bmatrix} 1.5\underline{/30°} \\ 3\underline{/-60°} \end{bmatrix}, \begin{bmatrix} \mathbf{V_1} \\ \mathbf{V_2} \end{bmatrix} = \begin{bmatrix} 0.728\underline{/-74.05°} \\ 2.06\underline{/-119.1°} \end{bmatrix}$

15. $\begin{bmatrix} 1+4j & -1 & -4j \\ \hline 1 & 2-2j & 2j \\ \hline -2-4j & 2j & 6j \end{bmatrix} \begin{bmatrix} \mathbf{I_1} \\ \mathbf{I_2} \\ \mathbf{I_3} \end{bmatrix} = \begin{bmatrix} j \\ 0 \\ 0 \end{bmatrix}$

16. $\begin{bmatrix} \dfrac{1}{j\omega C} + j\omega L_1 & -j\omega(L_1 - M) & 0 \\ \hline -j\omega(L_1 - M) & j\omega(L_1 + L_2 - 2M) & R \\ \hline \dfrac{3}{j\omega C} & -1 & 1 \end{bmatrix} \begin{bmatrix} \mathbf{I_1} \\ \mathbf{I_2} \\ \mathbf{I_3} \end{bmatrix} = \begin{bmatrix} 0 \\ \mathbf{V_s} \\ 0 \end{bmatrix}$

17. $i(t) = A \cos(\omega_1 t + \phi_A) + B \sin(\omega_2 t + \phi_B)$, where $A = V_m/\sqrt{R^2 + \omega_1^2 L^2}$, $\phi_A = \phi_\mathbf{v} - \arctan(\omega_1 L/R)$, $B = \omega_2 L I_m/\sqrt{R^2 + \omega_2^2 L^2}$, $\phi_B = \phi_1 + 90° - \arctan(\omega_2 L/R)$

18. $v_0(t) = B_1 \cos(\omega_1 t + \phi_1) + B_2 \cos(\omega_2 t + \phi_2) + B_2 \cos(\omega_3 t + \phi_3)$, where

$$B_1 = A_1 \omega_1 L / \sqrt{R^2(1 - \omega_1^2 LC)^2 + \omega_1^2 L^2}, \quad \phi_1 = 90° - \arctan\left\{\frac{\omega_1 L}{R(1 - \omega_1^2 LC)}\right\},$$

$$B_2 = A_2 R \sqrt{R^2(1 - \omega_2^2 LC)^2 + \omega_2^2 L^2}, \quad \phi_2 = -\arctan\left\{\frac{\omega_2 L}{R(1 - \omega_2^2 LC)}\right\},$$

$$B_3 = A_3 R \omega_3^2 LC / \sqrt{R^2(1 - \omega_3^2 LC)^2 + \omega_3^2 L^2}, \quad \phi_3 = 180° - \arctan\left\{\frac{\omega_3 L}{R(1 - \omega_3^2 LC)}\right\}$$

19.

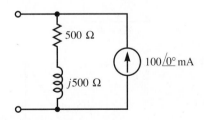

20. $\mathbf{V}_{oc} = \mathbf{I}_{sc} = 0$, $Z_{EQ} = \dfrac{j\omega RL}{R - r + j\omega L}$ **21.** $\mathbf{V}_{oc} = \mathbf{I}_{sc} = 0$, $Z_{EQ} = \dfrac{10\sqrt{2}}{99}\underline{/135°}\ \Omega$

22.

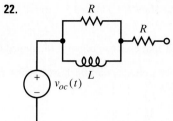

(a)

where $v_{oc}(t) = 10e^{-\frac{Rt}{L}}u(t)$ V

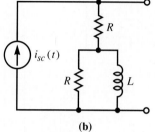

(b)

where $i_{sc}(t) = \dfrac{5}{R}e^{-\frac{Rt}{2L}}u(t)$ A

(b) $(L_p + R)v_{ab}(10) = 10L \int (t) - (2RL_p + R^2)i(t)$

23.

FREQ	VM(1)	VP(1)	VM(2)	VP(2)
6.366E-01	1.000E+01	9.268E-16	4.472E+00	6.344E+01

FREQ	VM(3)	VP(3)	IM(C12)	IP(C12)
6.366E-01	8.944E+00	-2.657E+01	2.236E+00	6.344E+01

24.

FREQ	VM(1)	VP(1)	VM(2)	VP(2)
1.592E+00	9.847E+01	7.856E+00	3.419E+01	-1.183E+02

FREQ	VM(3)	VP(3)	IM(R41)	IP(R41)
1.592E+00	2.417E+01	-1.633E+02	2.736E+00	-7.966E+01

CHAPTER 12

2. (a) 0.707 lagging   (b) 353.6 W   **5.** 500 VA   **9.** 353.6 VAR

12. (a) $500\underline{/45°}$ VA   (b)

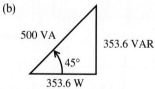

(c) 353.6 W, 353.6 VAR, 500 VA   **13.**

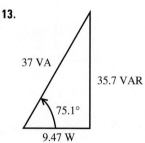

14. $18.5\underline{/-30.2°}$ VA **15.** 2.5 kW **16.** $0.5 - j0.5\,\Omega$ **17.** 10 kW
18. $-j0.5\,\Omega$ **19.** 3.53 W

CHAPTER 13

1. $\mathbf{H}(jR/L) = \dfrac{j}{1+j} = (1/\sqrt{2})\underline{/45°}$

2. $v_o(t) = 2.997 \times 10^{-2} \cos(3.14 \times 10^3 t + 99.4°) + 2.9997 \cos(3.14 \times 10^7 t + 27.57°)$

3. $\frac{1}{2}$ W

4. (a) $\mathbf{H}(j\omega) = \dfrac{\omega RC}{\sqrt{1 + (\omega RC)^2}}\,\left|\dfrac{\pi}{2} - \text{arc tan}\,(\omega RC)\right.$. The plots are identical to those in
Fig. 13.2 with R/L replaced by $1/(RC)$. (b) $1/(RC)$
(c) Stopband: $0 \le \omega \le 1/(RC)$, passband: $1/(RC) < \omega < \infty$

5. (a) $\mathbf{H}(j\omega) = \dfrac{1}{\sqrt{1 + (\omega LR)^2}}\,\underline{/-\text{arc tan}\,(\omega LR)}$. The plots are identical to those in
Fig. 13.5 with $1/(RC)$ replaced with R/L. (b) R/L
(c) Stopband: $R/L < \omega < \infty$, passband: $0 \le \omega \le R/L$

6. $v_o(t) = 10 + 6.965 \cos(0.1t - 5.71°) + (5/\sqrt{2})\cos(t - 45°) + 0.2985 \cos(10t - 84.3°)$ V

7. $\alpha = 100\ \text{s}^{-1}$, $\omega_d = 994.99$ rad/s

8. $v_o(t) = 10 + 7.02 \cos(50t - 0.574°) + 25 \cos(1000t - 90°) + 0.99 \cos(2000t - 172.4°)$

10. $\mathbf{H}(j\omega) = \dfrac{1}{j\omega - s_1}$, where $s_1 = -1/(RC) = -1$.

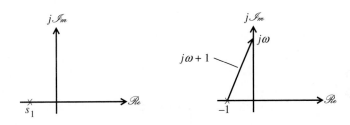

The amplitude and phase characteristics are given in Fig. 13.5 with $RC = 1$ s.

11. See Fig. 13.17. **12.** $\mathscr{A}(p)\mathbf{V}_o e^{st} = \mathscr{B}(p)\mathbf{V}_s e^{st}$. Perform the differentiations to obtain
$\mathscr{A}(s)\mathbf{V}_o e^{st} = \mathscr{B}(s)\mathbf{V}_s e^{st}$. Multiply both sides by e^{-st} to obtain the algebraic, frequency-domain equation $\mathscr{A}(s)\mathbf{V}_o = \mathscr{B}(s)\mathbf{V}_s$. Solve for $\mathbf{V}_o$ to obtain Eq. (13.31)

13. (a) $\mathscr{A}(p)v = \mathscr{B}(p)i$ where $\mathscr{A}(p) = \dfrac{1}{R}p + \dfrac{1}{L}$ and $\mathscr{B}(p) = p$,

$\mathbf{Z}(s) = \mathscr{B}(s)/\mathscr{A}(s) = sLR/(sL + R)$ (b) $v_p(t) = 12.7e^{-3000t} \cos(2000t + 31.3°)$ mV

14. $v_p(t) = 7.07 \cos(1000t + 65°)$ mV **15.** $v_p(t) = 15e^{-3000t}$ mV

17. (a)

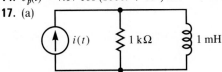

(b)

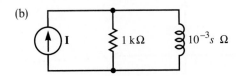

(c) $Z(s) = R \parallel sL = \dfrac{sL}{1 + sL/R} = \dfrac{10^{-3}s}{1 + 10^{-6}s}$ (d) $v_p(t) = 3.54 \cos{(10^6 t + 45°)} \, kV$

18. $\displaystyle\sum_{n=1}^{N} i_n(t) = \sum_{n=1}^{N} I_{mn}e^{\sigma t}\cos{(\omega t + \phi_n)} = \sum_{n=1}^{N} \mathcal{R}e\,\{I_n e^{st}\} = \mathcal{R}e\left\{\sum_{n=1}^{N} I_n e^{st}\right\} = \mathcal{R}e\left\{\left(\sum_{n=1}^{N} I_n\right)e^{st}\right\}$.

The equation $\displaystyle\sum_{n=1}^{N} I_n = 0$ is therefore a sufficient condition for $\displaystyle\sum_{n=1}^{N} i_n(t) = 0$

19. (a) $H(s) = 1/(1 + s)$ (b) (c) $s_1 = -1$

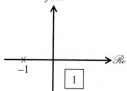

(d) $v_{oc}(t) = Ae^{-t}$ (e) $v_{op}(t) = 7.07 \cos{(t - 25°)} \, V$

20. (a) $H(s) = s/(1 + s)$ (b) (c) $s_1 = -1$

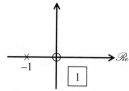

(d) $v_{oc}(t) = Ae^{-t}$ (e) $v_{op}(t) = 7.07 \cos{(t + 65°)} \, V$

21. (a), (b) By substituting Ie^{st} and Ve^{st} into the equation, differentiating as indicated and solving for V, we obtain $V = Z(s)I$ where $Z(s) = (s^2 + 7s)/(s^2 + 15s + 50)$

(c) $Z(s) = \dfrac{s(s - s_{z1})}{(s - s_{p1})(s - s_{p2})}$, where $s_{z1} = -7$, $s_{p1} = -5$, $s_{p2} = -10$

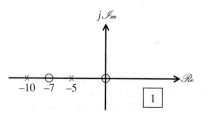

(d) $Z(j5) = 0.544\underline{/54.0°}$ (e) $V = 5.44\underline{/54.0°}$ (f) $v_p(t) = 5.44 \cos{(5t + 54.0°)}$
(g) $v_c(t) = A_1 e^{-5t} + A_2 e^{-10t}$

22. (a), (b) As in Exercise (21), we obtain $V = Z(s)I$ where
$Z(s) = (s^2 + 6s + 34)/(s^2 + 15s + 50)$

(c) $Z(s) = \dfrac{(s - s_{z1})(s - s_{z2})}{(s - s_{p1})(s - s_{p2})}$, where $s_{z1} = -3 + j5$. $s_{z2} = -3 - j5$, $s_{p1} = -5$, $s_{p2} = -10$

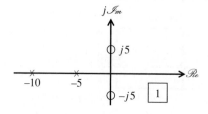

(d) $\mathbf{Z}(j5) = 0.396\underline{/1.73°}$ (e) $\mathbf{V} = 3.96\underline{/1.73°}$ (f) $v_p(t) = 3.96\cos(5t + 1.73°)$

(g) $v_c(t) = A_1 e^{-5t} + A_2 e^{-10t}$

23. For the rubber sheet analogy, we expect that, for $\omega > 0$, $|\mathbf{H}(j\omega)|$ will equal zero for $\alpha = 3$ and have peaks near $\omega = 2$ and $\omega = 4$.

24. (a)

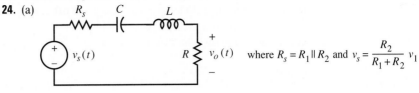

where $R_s = R_1 \parallel R_2$ and $v_s = \dfrac{R_2}{R_1 + R_2} v_1$

where $R_s = R_1 \parallel R_2$ and $v_s = \dfrac{R_2}{R_1 + R_2} v_1$

(b) $\dfrac{\mathbf{V}_o}{\mathbf{V}_1} = \dfrac{R_2}{R_1 + R_2} \cdot \dfrac{\mathbf{V}_o}{\mathbf{V}_s} = \dfrac{R_2}{R_1 + R_2} \cdot \dfrac{R}{R + R_s} \cdot \dfrac{s(R + R_s)/L}{(s - s_{p1})(s - s_{p2})}$, where s_{p1} and s_{p2} are given by Eqs. (13.50) through (13.52) where $\alpha = (R + R_s)/L$ and $\omega_0 = 1\sqrt{LC}$. By simplifying, we obtain $\mathbf{H}(j\omega) = \dfrac{R_2}{R_1 + R_2} \dfrac{sR/L}{(s - s_{p1})(s - s_{p2})}$ (c) $BW = 2\alpha = (R + R_s)/L$

25. (a) $\dfrac{sL}{s^2 RLC + sL + R}$

(b) Interchanging $R \leftrightarrow R^{-1}$, $L \leftrightarrow C$ and $\mathbf{I} \leftrightarrow \mathbf{V}$ we see that $\mathbf{H}_G(s)$ and $\mathbf{H}_R(s)$ of Eq. (13.49) are duals.

(c) $|\mathbf{H}_R(j\omega)|$ has a peak when $\omega = \omega_0 = 1\sqrt{LC}$. The expression $1\sqrt{LC}$ is unchanged when L and C are interchanged. By duality therefore, $|\mathbf{H}_R(j\omega)|$ has a peak when $\omega = \omega_0 = 1\sqrt{LC}$.

(d) The formula $BW = 1/RC$ is obtained by interchanging $R \leftrightarrow R^{-1}$ and $L \leftrightarrow C$ in Eq. (13.68).   **26.** $R = 1$ kΩ

27. $\mathbf{Y}(j\omega) = \dfrac{1}{R} + j\left(\omega C - \dfrac{1}{\omega L}\right)$. Solve $\mathcal{Im}\{\mathbf{Y}(j\omega)\} = 0$ to obtain $\omega_{upf} = \omega_0 = 1/\sqrt{LC}$. By interchanging $L \leftrightarrow C$ we obtain the dual result for the series RLC circuit, namely $\omega_{upf} = \omega_0 = 1/\sqrt{LC}$.

28. Since the susceptance $\left(\omega C - \dfrac{1}{\omega L}\right)$ equals zero for $\omega = \omega_{upf} = \omega_0$, no current can enter the enclosed part of the circuit when $\omega = \omega_0$. Therefore, $\mathbf{I}_R = \mathbf{I}_s$ and $\mathbf{V}_o = \mathbf{I}_s R$. This leads immediately to $\mathbf{I}_C = j\omega_0 CR\mathbf{I}_s$ and $\mathbf{I}_L = -\mathbf{I}_C = -j\omega_0 CR\mathbf{I}_s$

29. (a) 20π krad/s   (b) 8 kΩ, 0.127 H, 20 pF

30. (a) 20π krad/s   (b) 1 Ω, 15.9 mF, 1.59 μH   **31.** 7.43   **32.** 5.49

34. (a) For $\omega = 0$ the inductance acts like short-circuit and the capacitance acts like an open circuit.

(b) For $\omega = \infty$, the capacitance acts like a short-circuit and the inductance acts like an open circuit.

(c) For $\omega = 0$, the capacitance acts like an open circuit. For $\omega = \infty$, the inductance acts like an open circuit.

35. $\left(\dfrac{1}{2}\right)\dfrac{1}{5s + 1}$

36. Straight-line amplitude approximation: 0 dB low-frequency asymptote. 6 dB/octave downward corner at 0.05 rad/s. Upward corner to level off at 0.1 rad/s. Corrections for actual characteristic: ± 3 dB at corners.

37. Straight-line amplitude approximation: -3 dB low-frequency asymptote. 6 dB/octave upward corner at 0.1 rad/s. Downward corner to level off at 0.2 rad/s. Corrections for actual characteristic: ± 3 dB at corners.

38. Straight-line amplitude approximation: low-frequency asymptote has slope.
6 dB/octave and passes through -20 dB at $\omega = 0.01$. Downward corner to level off at
0.05 rad/s.
Corrections for actual characteristic: 3 dB at corner.
39. Straight-line amplitude approximation: -6 dB/octave straight line passing through
0 dB at $\omega = 10^3$ rad/s.
Corrections for actual characteristic: none.

CHAPTER 14

3. $1 + \displaystyle\sum_{n=1}^{5} \frac{2}{\sqrt{1 + (2\pi n f_1 RC)^2}} \cos\left[2\pi n f_1 t - \text{arc tan}\,(2\pi n f_1 RC)\right]$

5. Period $T = 1$ ms, $a_0 = 3$, $a_1 = 5$, $b_3 = 7$. All other coefficients equal zero.

6. (c) $a_0 = \dfrac{1}{T},\ a_n = \dfrac{2}{T},\ b_n = 0$ for $n = 1, 2, 3, \ldots$

(d) $v(t) = \dfrac{1}{T}\left[1 + 2\cos\,(2\pi f_1 t) + 2\cos\,(2\pi 2 f_1 t) + 2\cos\,(2\pi 3 f_1 t) + \cdots\right]$

7. The only difference between the Fourier coefficients of $v'(t)$ and $v(t)$ is the dc coefficient
$a_0 = A$ for $v'(t)$ and $a_0 = 0$ for $v(t)$.

8. (a) $v_1(t)$ and $v_2(t)$ (b) $v_3(t)$ and $v_4(t)$ (c) $v_2(t)$ and $v_4(t)$

9. $b_n = (-1)^{(n-1)/2}\dfrac{8}{(\pi n)^2}$ for $n = 1, 3, 5, \ldots$. All other coefficients equal zero.

10. $b_n = \dfrac{2A}{n\pi}(-1)^{n+1}$ for $n = 1, 2, 3, \ldots\ a_n = 0$ for $n = 0, 1, 2, \ldots$

11. (a) $A_n = \dfrac{8}{(\pi n)^2},\ \phi_n = (-1)^{(n+1)/2}\,90°$, for $n = 1, 3, 5, \ldots$ (b) $\tfrac{1}{3}$ W (c) 99.77%

12. $V_0 = \dfrac{A\tau}{T},\ V_n = \dfrac{A\tau}{T}\dfrac{\sin\,(n\pi f_1\tau)}{n\pi f_1\tau}$ for $n \neq 0$

13. (a) $V_n = \dfrac{1}{T}\displaystyle\int_T v(t)e^{-j2\pi n f_1 t}\,dt = \dfrac{1}{T}\int_T v(t)\cos\,(2\pi n f_1 t)\,dt - j\dfrac{1}{T}\int_T v(t)\sin\,(2\pi n f_1 t)\,dt$

If $v(t)$ is an even function of t, then $v(t)\sin\,(2\pi n f_1 t)$ is an odd function of t, and the
second integral on the right-hand side is zero. The first integral is an even function of n.
This integral is real when $v(t)$ is real.

14. $V_n = 1/T$ for $n = 0, \pm 1, \pm 2, \ldots$ **15.** $V_n = \begin{cases} j\dfrac{4}{\pi^2 n^2}(-1)^{(n+1)/2} & \text{for } n \text{ odd} \\[2mm] 0 & \text{for } n \text{ even} \end{cases}$

16. $V_n = \begin{cases} \dfrac{A}{4} - j\dfrac{2A}{\pi n}; & n = \pm 1 \\[2mm] -j\dfrac{2A}{\pi n}; & n = \pm 3, \pm 5, \ldots \\[2mm] \dfrac{A}{\pi(1 - n^2)}(-1)^{n/2}; & n \text{ even} \end{cases}$ **17.** (b) $G_n = \dfrac{A\tau}{T}\cdot\dfrac{1 - e^{-T/\tau}}{1 - (j2\pi n\tau/T)}$

18. (b) $(-1)^n V_n$ where V_n is the coefficient in FS8

19. When we differentiate the triangular wave of Fig. 14.1, we get an even symmetric
square wave having amplitude $A = 4/T$. The complex Fourier series coefficient is
$\dfrac{8}{\pi n T}(-1)^{(n-1)/2} = \dfrac{2A}{\pi n}(-1)^{(n-1)/2}$. This disagrees with FS1 because the square wave in FS1

has odd symmetry. **20.** $V_{on} = \dfrac{\sin\,(0.1 n\pi)}{0.1 n\pi(1 + j0.1 n\pi)}$ **21.** $V_{on} = \left(\dfrac{j0.1\pi n}{1 + j0.1\pi n}\right)\left(\dfrac{\sin\,(0.1 n\pi)}{0.1 n\pi}\right)$

22. $V_{on} = \dfrac{\sin(0.1n\pi)}{0.1n\pi\left[1 + j\left(n - \dfrac{9}{n}\right)\right]}$ **23.** None **24.** None **25.** None

26. Numbers 1 and 3 **27.** Number 2 **28.** Number 3 **29.** Number 3

CHAPTER 15

1. $V(f) = 1$ **2.** $V(f) = 1/(b - j2\pi f)$ **3.** $\dfrac{a + b}{ab + (2\pi f)^2 + j(b - a)2\pi f}$ **4.** 1

5. $A\cos 2\pi f_0 t$ **6.** $2W\,\text{sinc}\,(2Wt)$ **7.** (a), (b) follow directly from Eq. (15.1)

(c) The Fourier operator is linear. **8.** $B\sin 2\pi f_0 t$ **11.** $\dfrac{\alpha + j2\pi f}{\alpha^2 + (2\pi)^2(f^2 - f_c^2) + j4\pi f}$

12. (a) $V_x(f) = \dfrac{A}{2}\{\delta(f - f_c) + \delta(f + f_c)\} + \dfrac{Aa}{2}\{V(f - f_c) + V(f + f_c)\}$

(b) The density spectrum of the AM waveform contains impulses at $\pm f_c$.
13. $\delta(f) - e^{-j6\pi f}$
14. (a) Since $v_x(t) = v(t) + v(t)\cos 2\pi 2f_c t$ then $V_x(f) = V(f) + \frac{1}{2}V(f - 2f_c) + \frac{1}{2}V(f + 2f_c)$.
The low-pass filter will "pass" the $v(t)$ term and "reject" the $v(t)\cos 2\pi 2f_c t$ term.
(b) Perfect recovery of $v(t)$ occurs if $H(f)V_x(f) = V(f)$.

15. (a) Since $V_C(f) = \dfrac{1}{1 + j\omega RC}$ then $v_C(t) = \dfrac{1}{RC}e^{-t/RC}u(t)$.

(b) $v_R(t) = v_S(t) - v_C(t) = \delta(t) - \dfrac{1}{RC}e^{-t/RC}u(t)$

16.–18. May all be established directly from the definition $x(t) * y(t) = \int_{-\infty}^{+\infty} x(\lambda)y(t - \lambda)\,d\lambda$.
Alternatively, we can use Fourier transforms. For example, to establish the answer to
Exercise 16, we note that $x_1(t) * x_2(t) = F^{-1}\{X_1(f)X_2(f)\} = F^{-1}\{X_2(f)X_1(f)\} = x_2(t) * x_1(t)$.
19.–22. May all be established directly from the definition of convolution or by Fourier

transforms. **23.** $\dfrac{d}{dt}\delta(t), j2\pi f$ **24.** $u(t), \dfrac{1}{2}\delta(f) + \dfrac{1}{j2\pi f}$ **25.** $\delta(t - t_0), e^{-j2\pi ft_0}$
26. Violates Dirichlet condition 1 **27.** Satisfies all three Dirichlet conditions
28. Violates Dirichlet conditions 1 and 3 **29.** Violates Dirichlet condition 1
30. Violates Dirichlet condition 1 **31.** Violates Dirichlet condition 3
32. Satisfies all three Dirichlet conditions **33.** Violates Dirichlet condition 1
34. Violates Dirichlet condition 1 **35.** Violates Dirichlet conditions 1 and 2
36. The condition of Plancherel's theorem is satisfied by the signals of Exercises 27,
29, and 32.

CHAPTER 16

1. e^{-st_0}/s, $\mathcal{R}e\,\{s\} > 0$ **2.** $(1 - e^{-st_0})/s$, $\mathcal{R}e\{s\} > 0$ **3.** $\dfrac{1}{s + 3} + \dfrac{1}{s + 5}$, $\mathcal{R}e\,\{s\} > -3$

4. $\dfrac{1}{s + 3} + \dfrac{1}{s - 5}$, $\mathcal{R}e\,\{s\} > 5$ **5.** $1/s^2$, $\mathcal{R}e\,\{s\} > 0$ **6.** $\delta(t) + u(t) + e^{-2t}u(t)$

7. $1/s^2$ **8.** e^{-st_0}/s^2 **9.** $(1 + st_0)e^{-st_0}/s^2$ **10.** $\dfrac{5}{(s - 7)^2} + \dfrac{1\underline{/45^\circ}}{s - j\omega_0} + \dfrac{1\underline{/-45^\circ}}{s + j\omega_0}$

11. $(1 - e^{-s\tau})/s$ **12.** $u(t - 1)$ **13.** $e^{(-10 + j30)t}u(t)$ **14.** $e^t(\cos t - \sin t)u(t)$
15. $(2e^{-3t} - e^{-2t})u(t)$ **16.** Just set $\alpha = 0$ **17.** Just set $\alpha = 0$
18. Set $K_0 = 1$, $\sigma_0 = -\alpha$, and $\omega_0 = \omega_d$ in LT9.
19. Set $K_0 = j$, $\sigma_0 = -\alpha$, and $\omega_0 = \omega_d$ in LT9. **20.** Initial value: 0. Final value: 1
21. Initial value: 1, Final value: 0 **22.** Initial value: 0, Final value: 0

23. $\dfrac{0.5j}{s+j} - \dfrac{0.5j}{s-j}$ **24.** $\dfrac{1}{(s+1)^2} - \dfrac{1}{s+1}$ **25.** $1 + \dfrac{5}{s+3} - \dfrac{10}{s+4}$ **26.** $s^2 + s + \dfrac{1}{s+2}$

27. $s^2 - 4s + 20 - \dfrac{239}{s+2} + \dfrac{923}{s+3} - \dfrac{259}{s+4}$ **28.** $1 - \dfrac{1}{s+5} - \dfrac{8}{(s+1)^2 + 9}$

29. $\dfrac{a}{s-s_a} - \dfrac{a}{s-s_b}$ where $a = \dfrac{k_a k_b}{s_a - s_b}$ **30.** $\dfrac{k_a k_b}{(s-s_a)^2}$ **31.** $\dfrac{1/2}{s - j\omega_0} + \dfrac{1/2}{s + j\omega_0}$

32. $\dfrac{1/2j}{s - j\omega_0} - \dfrac{1/2j}{s + j\omega_0}$ **33.** $\dfrac{1/2}{s + \alpha - j\omega_0} + \dfrac{1/2}{s + \alpha + j\omega_0}$ **34.** $\dfrac{1/2j}{s + \alpha - j\omega_d} - \dfrac{1/2j}{s + \alpha + j\omega_d}$

35. $e^{-t}u(t)$ V **36.** $(10 - 16e^{-t})u(t)$ V **37.** $[1 - v_R(0^-)]e^{-tR/L}u(t)$

38. $v_L(t) = \dfrac{A\omega_0}{\sqrt{(R/L)^2 + \omega_0^2}} \cos\left[\omega_0 t + \phi + 90° - \tan^{-1}\left(\dfrac{\omega_0 L}{R}\right)\right]u(t) +$

$\left[\dfrac{(AR/L)}{\sqrt{(R/L)^2 + \omega_0^2}} \cos\left[\phi + 180° - \tan^{-1}\left(\dfrac{\omega_0 L}{R}\right)\right] - v_R(0^-)\right]e^{-(R/L)t}u(t)$

39. $v_0(t) = \dfrac{1}{C} e^{-\alpha t}u(t)\left[\cos(\omega_0 t) - \dfrac{\alpha}{\omega_0}\sin(\omega_0 t)\right]$ where $\alpha = \dfrac{1}{2RC}$ and $\omega_0^2 = \dfrac{1}{LC} - \left(\dfrac{1}{2RC}\right)^2$

40. $8e^{-t}u(t)$ A **41.** $4e^{-t}u(t)$ A **42.** $4e^{-t}u(t) + 4e^{-(t-1)}u(t-1)$ A

43. $[9.464e^{-t} + 1.464\cos(t) - 5.464\sin(t)]u(t)$

44. $v_1 = 10 + 2.5e^{-0.25t}$ V; $t \geq 0$, $v_2 = -2.5e^{-0.25t}$ V; $t \geq 0$

45. Forced response: $-10u(t) + 5tu(t) + 10e^{-0.5t}u(t)$ V

Stored-energy response: $-3e^{-0.5t}u(t)$ V

Particular response: $-10u(t) + 5tu(t)$ V

Complementary response: $7e^{-0.5t}u(t)$ V

46. Forced response:

$\dfrac{A}{\sqrt{1 + (\omega_0 RC)^2}}\left[\cos(\omega_0 t + \phi - \tan^{-1}\omega_0 RC) - \cos(\phi - \tan^{-1}\omega_0 RC)e^{-t/RC}\right]u(t)$

Stored-energy response: $v_0(0^-)e^{-t/RC}u(t)$

Particular response: $\dfrac{A}{\sqrt{1 + (\omega_0 RC)^2}}\cos(\omega_0 t + \phi - \tan^{-1}\omega_0 RC)u(t)$

Complementary response: $\left[v_0(0^-) - \dfrac{A}{\sqrt{1 + (\omega_0 RC)^2}}\cos(\phi - \tan^{-1}\omega_0 RC)\right]e^{-t/RC}u(t)$

47. $\left\{\dfrac{A}{R} + \left[\dfrac{A}{R_1 + R_2} - \dfrac{A}{R}\dfrac{R}{R_2} i_L(0^-)\right]e^{-tR/L}\right\}u(t)$, where $R = R_1 \parallel R_2$

Forced response: $\left\{\dfrac{A}{R_1} + \left[\dfrac{A}{R_1 + R_2} - \dfrac{A}{R_1}\right]e^{-tR/L}\right\}u(t)$

Stored energy response: $-\dfrac{R}{R_1} i_L(0^-)e^{-tR/L}u(t)$

Particular response: $\dfrac{A}{R_1}u(t)$

Complementary response: $\left[\dfrac{A}{R_1 + R_2} - \dfrac{A}{R_1} - \dfrac{R}{R_1} i_L(0^-)\right]e^{-tR/L}u(t)$

48. (a) no (b) yes (c) yes **49.** (a) no (b) yes (c) yes

50. (a) no (b) no (c) no **51.** (a) no (b) yes (c) yes

52. (a) no (b) no (c) no **53.** 2, 0, 0, 0, none

CHAPTER 17

1. and 2.
$$Z = \begin{bmatrix} \dfrac{33}{12} & 3 \\ \dfrac{29}{14} & 7 \end{bmatrix} \Omega, \ \mathbf{V}_{oc} = 0$$

3. and 4. $Z = \begin{bmatrix} 1 & 1 \\ 1 & 2+j \end{bmatrix} \Omega, \ \mathbf{V}_{oc} = \begin{bmatrix} 5\underline{/0°} \\ 10\underline{/0°} \end{bmatrix}$ V

5. and 6.
$$\mathbf{Y} = \begin{bmatrix} \dfrac{7}{12} & -\dfrac{1}{4} \\[2mm] -\dfrac{29}{60} & \dfrac{11}{20} \end{bmatrix} \text{S, } \mathbf{I}_{sc} = \mathbf{0}$$

7. and 8.
$$\mathbf{Y} = \begin{bmatrix} \dfrac{3}{2} - \dfrac{1}{2}j & -\dfrac{1}{2} + \dfrac{1}{2}j \\[2mm] -\dfrac{1}{2} + \dfrac{1}{2}j & \dfrac{1}{2} - \dfrac{1}{2}j \end{bmatrix} \text{S, } \mathbf{I}_{sc} = \begin{bmatrix} \dfrac{5\sqrt{2}}{2}\underline{/45°} \\[2mm] \dfrac{5\sqrt{2}}{2}\underline{/-45°} \end{bmatrix}$$
9. Yes

14. Set $\mathbf{h}_{11} = 0\ \Omega$, $\mathbf{h}_{21} = \mathbf{h}_{12} = 0$, $\mathbf{h}_{22} = 0$ S, $\mathbf{V}_{1ocsc} = 0$ V, $\mathbf{I}_{2ocsc} = 0$ A, and $\mathbf{h}_{21} = \beta$ in Fig. 17.23 to obtain Fig. 17.29.

15.
$$\mathbf{H} = \begin{bmatrix} \left(\dfrac{3}{5} + \dfrac{1}{5}j\right)\Omega & \dfrac{2}{5} - j\dfrac{1}{5} \\[2mm] -\dfrac{2}{5} + j\dfrac{1}{5} & \left(\dfrac{2}{5} - \dfrac{1}{5}j\right)\text{S} \end{bmatrix}, \begin{bmatrix} \mathbf{V}_{1ocsc} \\[2mm] -\mathbf{I}_{2ocsc} \end{bmatrix} = \begin{bmatrix} \sqrt{5}\underline{/63.4°}\text{ V} \\[2mm] 2\sqrt{5}\underline{/153.4°}\text{ A} \end{bmatrix}$$

16. Draw a single large enclosing networks A, I, and II. Draw another large circle enclosing networks IV and B. Apply KCL at the surface of the circles to show that network III is a two port.   **17.** -5

18. Use Fig. 17.36b where $\mathbf{Z} = \begin{bmatrix} j\omega L_1 & j\omega M \\ j\omega M & j\omega L_2 \end{bmatrix}$

Use Fig. 17.36d where $\mathbf{Y} = \begin{bmatrix} \dfrac{j\omega L_2}{|\mathbf{Z}|} & -\dfrac{j\omega M}{|\mathbf{Z}|} \\[2mm] -\dfrac{j\omega M}{|\mathbf{Z}|} & \dfrac{j\omega L_1}{|\mathbf{Z}|} \end{bmatrix}$, where $|\mathbf{Z}| = \omega^2(L_1 L_2 - M^2)$

Use Fig. 17.36f where $\mathbf{H} = \begin{bmatrix} \dfrac{|\mathbf{Z}|}{j\omega L_2} & \dfrac{M}{L_2} \\[2mm] -\dfrac{M}{L_2} & \dfrac{1}{j\omega L_2} \end{bmatrix}$

19. By inspection, obtain $\mathbf{Y} = \begin{bmatrix} 2 & -1 \\ -1 & 2 \end{bmatrix} \text{S, } \mathbf{I}_{sc} = \begin{bmatrix} -12 \\ +12 \end{bmatrix}$ A. (Use Tables 17.4 and 17.5 where $|\mathbf{Y}| = 3$ S^2 and Figs. 17.36a, c, and e.)

21. $\mathbf{H} = \begin{bmatrix} R_1 & 0 \\ -h & R_0^{-1} \end{bmatrix}$. Find the $\mathbf{z}$ and $\mathbf{y}$ parameters from Table 17.4 where $|\mathbf{H}| = R_1/R_0$, and use Figures 17.36b, d, and f as in Exercise 18.

CHAPTER 18

1. Replace voltages and currents with phasor values. Replace self- and mutual-inductances with $j\omega$ times the inductance.   **2.** $\begin{bmatrix} \mathbf{V}_1 \\ \mathbf{V}_2 \end{bmatrix} = \begin{bmatrix} j\omega L_1 & j\omega M \\ j\omega M & j\omega L_2 \end{bmatrix} \cdot \begin{bmatrix} \mathbf{I}_1 \\ \mathbf{I}_2 \end{bmatrix}$

3. $\begin{bmatrix} \mathbf{I}_1 \\ \mathbf{I}_2 \end{bmatrix} = \begin{bmatrix} \dfrac{1}{j\omega} \cdot \dfrac{L_2}{L_1 L_2 - M^2} & -\dfrac{1}{j\omega} \cdot \dfrac{M}{L_1 L_2 - M^2} \\[2mm] -\dfrac{1}{j\omega} \cdot \dfrac{M}{L_1 L_2 - M^2} & \dfrac{1}{j\omega} \cdot \dfrac{L_1}{L_1 L_2 - M^2} \end{bmatrix} \cdot \begin{bmatrix} \mathbf{V}_1 \\ \mathbf{V}_2 \end{bmatrix}$ **4.** $j\omega \dfrac{L_1 L_2 - M^2}{L_2}$

5. $\begin{bmatrix} \mathbf{Z}_s + R_1 + j\omega L_1 & j\omega M \\ j\omega M & \mathbf{Z}_L + R_2 + j\omega L_2 \end{bmatrix} \cdot \begin{bmatrix} \mathbf{I}_1 \\ \mathbf{I}_2 \end{bmatrix} = \begin{bmatrix} \mathbf{V}_s \\ 0 \end{bmatrix}$

6. $\begin{bmatrix} \mathbf{Z}_s + R_1 + j\omega L_1 & j\omega M \\ j\omega M & R_2 + j\omega L_2 \end{bmatrix} \cdot \begin{bmatrix} \mathbf{I}_1 \\ \mathbf{I}_2 \end{bmatrix} = \begin{bmatrix} \mathbf{V}_1 \\ \mathbf{V}_2 \end{bmatrix}$

7. $R_1 + j\omega L_1 + \dfrac{\omega^2 M^2}{R_2 + R_L + j\omega L_2}, \dfrac{j\omega M R_L}{(R_1 + j\omega L_1)(R_2 + R_L + j\omega L_2) + \omega^2 M^2}$

8. $-\dfrac{j\omega M}{R_2 + R_L + j\omega L_2}$ **9.** $\dfrac{(\omega M)^2 R_L}{[(R_1 + j\omega L_1)(R_2 + R_L + j\omega L_2) + \omega^2 M^2][R_2 + R_L + j\omega L_2]}$

10. 1 H, 39 H, 1 H **11.** 0 H (a short circuit), 38 H, 2 H **12.** -1 H, 37 H, 3 H

13. -4 H, 34 H, 6 H **14.** $\frac{79}{39}$ H, 79 H, 79 H **15.** 2 H, ∞ H (an open circuit), 38 H

16. $\frac{71}{37}$ H, -71 H, $\frac{71}{3}$ H **17.** $\frac{22}{17}$ H, -22 H, $\frac{22}{3}$ H **18.** 1 Ω, $j20\omega$ Ω, $j10\omega$, 2 Ω, $j40\omega$ Ω

19. $L_{\ell 1} = 12.93$ H, $L_{M1} = 7.07$ H, $L_{\ell 2} = 25.9$ H, $n_2/n_1 = \sqrt{2}$

20. $L_{\ell 1} = 12.93$ H, $L_{M2} = 14.14$ H, $L_{\ell 2} = 25.9$ H, $n_2/n_1 = \sqrt{2}$

21. Dot marks are at the top on both sides of the ideal transformer with turns ratio n_1/n_2.

22. Dot marks are at the top on both sides of the ideal transformer with turns ratio n_1/n_2. **23.** 12 A, 72 V, 4 A, 24 V, -12 A, 30 Ω

24. 12 A, 72 V, 4 A, -24 V, 12 A, 30 Ω **25.** 32.7 V, 0.723 Ω, -12 A **26.** $\frac{8}{3}$ A, $\frac{4}{3}$ A

27. $\frac{8}{3}$ A, $\frac{4}{3}$ A **28.** $\frac{8}{3}$ A, $\frac{4}{3}$ A **29.** $\frac{24}{17}$ A, $-\frac{12}{17}$ A **30.** 8 V, 4 V, 6 V

31. 8 V, 4 V, 6 V **32.** 8 V, 4 V, 6 V **33.** $-\frac{24}{11}$ V, $\frac{36}{11}$ V, 6 V **34.** 576 Ω, 48.2 Ω

35. 144 Ω, 12.04 Ω

CHAPTER 19

1. $10.9\underline{/-90°}$ A, $10.9\underline{/90°}$ A, $21.8\underline{/-90°}$ A, $109\underline{/0°}$ V, $109\underline{/180°}$ V, 0 V

2. $18.95\underline{/90°}$ A, $6.32\underline{/90°}$ A, 0 A, $189.5\underline{/0°}$ V, $63.2\underline{/180°}$ V, $50.53\underline{/180°}$ V

3. $18.5\underline{/78.1°}$ A, $6.18\underline{/78.1°}$ A, $24.7\underline{/-11.9°}$ A, $185\underline{/-11.9°}$ V, $61.8\underline{/168°}$ V, $49.5\underline{/168°}$ V **4.** $3394\underline{/45°}$ V, $3394\underline{/-75°}$ V, $3394\underline{/-195°}$ V or $3394\underline{/45°}$ V, $3394\underline{/165°}$ V, $3394\underline{/-75°}$ V **5.** $208\underline{/-120°}$ V, $208\underline{/-240°}$ V **6.** $120\underline{/-30°}$ V, $120\underline{/-150°}$ V, $120\underline{/-270°}$ V **7.** $6\underline{/-45°}$ A, $6\underline{/-165°}$ A, $6\underline{/-285°}$ A

8. $2163\underline{/15°}$ VA **9.** 208 V **10.** $12\underline{/0°}$ A, $12\underline{/-120°}$ A, $12\underline{/-240°}$ A

11. $20.78\underline{/-30°}$ A, $20.78\underline{/-150°}$ A, $20.78\underline{/-270°}$ A **12.** $2880\underline{/15°}$ VA

13. $5\underline{/20°}$ A, $1.732\underline{/160°}$ KVA, -1.628 kW

14. $116.2\underline{/53.13°}$ kVA, $38.72\underline{/-53.13°}$ kVA, $111.7\underline{/33.69°}$ kVA, 152.4 A, 50.8 A, 146.5 A

15. $50\underline{/36.87°}$ kVA, 65.61 A, -49.37 kVAR **16.** $51.8\underline{/36.87°}$ kVA, 12.46 A, -11.006 kVAR, 11.07 A **17.** 1.95 kW, 0.310 leading

18. 8.78 kW, 22.19 kW, 30.98 kW, 0.8 lagging

19. $[V_m^2/(2|\mathbf{Z}|)][\cos(\underline{/\mathbf{Z}}) + \cos(2\omega t - \underline{/\mathbf{Z}})]$ W, $[V_m^2/(2|\mathbf{Z}|)][\cos(\underline{/\mathbf{Z}}) - \cos(2\omega t - \underline{/\mathbf{Z}})]$ W, $V_m^2/|\mathbf{Z}| \cos(\underline{/\mathbf{Z}})$, $p = P_{av}$

20. 1200 W, 786 W, 1200 W, 1200 W, -1986 W

ANSWERS TO SELECTED PROBLEMS

CHAPTER 1

1. 2 A   **5.** arc tan $t + 1.107$ C, $-2 \le t \le 2$ **9.** 80.1 nA
11. (a) $-96,488$ C/gram atomic weight   (b) -894.4 C/gram   **13.** 12 V   **14.** 2 V
18. 0   **22.** $200 \cos 2\pi t$ W; 0   **26.** $100(\cos 2\pi t + \cos 6\pi t)$ W; 0
29. -144 W, 72 W, 48 W, -12 W, 36 W   **33.** 60 W   **37.** 10 C   **40.** 1000 C
42. 1,879,000 J   **45.** $5.04   **48.** 15 A

CHAPTER 2

1. 1 A   **5.** 7 A   **7.** -5 A   **11.** 2 A, 4 A, 3 A   **15.** 1 V   **19.** -5 V
23. -3 V   **28.** 4 A, 39 V, 108 A, -1620 W   **32.** 2 V, -4 V, 2 V, 3 V
34. -9 V, -24 V, -4 A, 2 A, 3.667 Ω   **36.** $-0.03 \sin 300t$ A
41. $-0.0048 \sin 200t$ A for $t > 0$; $203 - 200 \cos 200t$ V for $t \ge 0$
44. $3.25 - 1.25e^{-400t}$ A for $t \ge 0$ **48.** $-48e^{-2t}$ V for $t > 0$; $122 - 120e^{-2t}$ A for $t > 0$
49. $-40e^{-2t}$ V, $60e^{-2t}$ V, $28e^{-2t}$ V for $t > 0$
52. $100e^{-2t}$ V, $90e^{-2t}$ V, $15.5 - 10.5e^{-2t}$ A, $-4e^{-2t}$ A for $t > 0$ **56.** 0.408
60. Dot mark by terminal c **61.** 15 V, 120 V, -6 A, 720 W
64. $-22e^{-2t}$ A, $29e^{-2t}$ A, $-23e^{-2t}$ A, $-186e^{-2t}$ V, $142e^{-2t}$ V, $-266e^{-2t}$ V, $-442e^{-2t}$ V
for $t > 0$ **65.** 5 Ω in series with a 10-V source
68. The model consists of a voltage controlled voltage source with a 2-Ω resistance in
series with the dependent source with control equation $-10v_{ab}$. Terminals b and d are
connected.   **69.** 7.9 kW, 7.29 Ω, 32.9 A, $0.79

CHAPTER 3

1. $60 \cos 2t$ V, $10 \cos 2t$ A, $6 \cos 2t$ A, $4 \cos 2t$ A, $180(1 + \cos 4t)$ W, $120(1 + \cos 4t)$ W,
$-300(1 + \cos 4t)$ W   **5.** 12 V, -30 V, 42 V, 3 A, -2 A, -7 A   **7.** $4 \cos 25t$ A
12. 4 Ω   **13.** 6 H   **14.** 37 F
18. 10 A, 384 J, 48 J, 10 A, 400 J, No, 0 A, $\frac{4}{3}$A, $-\frac{4}{3}$A   **22.** 6 A, -18 A   **26.** 18 V
33. 24 V, 6 A, 108 W, 144 W, 180 W, -288 W, -144 W   **37.** -1.1 kV

40. $112 \cos 2t$ V   **42.** $18\,\Omega$ **43.** 37 H   **44.** 6 F   **49.** $L_1 + L_2 - 2M$
52. -72 A   **56.** $2.4\,\Omega$ **60.** $0\,\Omega$ **62.** $3.6\,\Omega$ **65.** $3\,\Omega$ **68.** $\frac{3}{4}R$
72. (a) -3000 W   (b) -600 W   **76.** 36 V, 2 A   **82.** 1.2 A
86. -10 V, 5 V, 5 V, $-\frac{1}{3}$ A, $\frac{1}{3}$ A, 0 A   **90.** $v/4$ **94.** $3\,\Omega$, 9 W, 3 W, 12 W, -24 W
98. 20 kΩ   **104.** $40\,\Omega$, unique   **106.** $2.24\,\Omega$ or $10.258\,\Omega$, not unique
110. 40-V source   **113.** Approximately 6 V and 1 A   **116.** 4 A

CHAPTER 4

2. 20 V, 70 V, 10 V, -1 A, 4 A, -2 A, -3 A, 5 A, -10 A, 1 A   **3.** 1 V
4. 50 A   **7.** 10.41 V, 20.41 V, 24.58 V   **8.** 2 V, 9 V, -2 A
9. 3 V, 7.5 V, -2.857 A   **11.** 67.13 V, -147.92 V, 66.34 V, -40 V
17. (a) 99.01   (b) -8.167, 0.0983 A, $10.17\,\Omega$ **22.** 7.692 A
27. $4e^{-t}$ V, -10 V, $-96e^{-t}$ V, -5 V, $15 + 92e^{-t}$ V, $-9e^{-t}$ V, $-5 - 87e^{-t}$ V for $t > 0$
28. -1 V, 4 V, -2 V, -3 V, 5 V, -10 V, 1 V   **29.** 1 A   **30.** 12 A
33. 10.41 A, 20.41 A, 24.58 A   **36.** 67.13 A, -147.92 A, 66.34 A, -40 A
43. 7.692 V   **48.** Node voltages   **52.** Mesh currents   **56.** Mesh currents
80. 9.818 V   **84.** 48.33 A   **85.** 179.17 V   **87.** 22.5 V
91. 11.43 V, 1.428 V, -3.57 V

CHAPTER 5

3. -8 V   **6.** -2 A, 1 A, 4 A   **11.** 24 V, $12\,\Omega$, 2 A   **14.** 28 V, $6\,\Omega$, $\frac{14}{3}$ A
15. 3 V, $-10\,\Omega$ **20.** 2 V, $\frac{1}{2}\,\Omega$ **22.** 120 V, $10\,\Omega$ **24.** $12\,\Omega$, 12 W
26. $266\,\Omega$, 157.99 W   **28.** $0\,\Omega$ **31.** 10 V
37. The network of Fig. 4.18 is the dual of the network of Fig. 4.7.
42. $50\,\Omega$, 19.5, -0.975, 19.0125   **44.** $\beta \geq 999$ **46.** 6 V
48. -240 V, $6\,\Omega$, -32.33 V, $0.8847\,\Omega$, $0.37321\,\Omega$ **49.** -13.3, 2 kΩ
50. 421.96 kΩ, -6.667

CHAPTER 6

1. $10\,\Omega$, $25\,\Omega$, 5000, no   **2.** -8.057, $1193.3\,\Omega$, $10.704\,\Omega$, yes

6. $\dfrac{d}{dt} v_o + \dfrac{1}{(A + 1)RC} v_o = -\dfrac{A}{(A + 1)RC} v_s,\ v_o = -\dfrac{1}{RC} \displaystyle\int_{-\infty}^{t} v_s\, d\tau$

7. $\dfrac{d}{dt} v_o + (A + 1)\dfrac{R_a}{L} v_o = -A\dfrac{d}{dt} v_s,\ v_o = -\dfrac{L}{R_a}\dfrac{d}{dt} v_s$

10. The input circuit is a resistance $R_{in} = [(1 + A)R_1 + R_2]/(1 + A)$. The output circuit
is a VCVS with gain $G = -A/[(1 + A)R_1 + R_2]$.   **14.** $C\dfrac{d}{dt} v_o + \dfrac{1}{R_b} v_o = -\dfrac{1}{R_a} v_s$

19. $v_o = -(v_{ref}R_x)/R_{ref}$ **20.** $i = v_s/R$, the output voltage is limited.   **22.** $2v_s^2$ mW

27. $v_o = \left(1 + \dfrac{R_b}{R_a} + \dfrac{2R_b}{R_c}\right)(v_{s1} - v_{s2})$ **36.** $\left(\dfrac{R_4}{R_2} + \dfrac{R_4}{R_f} - \dfrac{R_3}{R_1}\right)\dfrac{R_f}{R_3 + R_4} v_{ref}$

43. $v = Rg^{-1}(-Ri)$ where $v_L = g(i_L)$ and $i_L = g^{-1}(v_L)$; $v = Ri_L$

44. $v = -Ri$, a negative resistance   **45.** $v = -\dfrac{R^2}{L}\displaystyle\int_{-\infty}^{t} i\, d\tau$, a negative capacitance

46. $v = -R^2 C\dfrac{d}{dt} i$, a negative inductance

CHAPTER 7

1. $10e^{-(t-2)/4}[u(t-2) - u(t-6)]$ **4.** $\sum_{n=0}^{\infty} u(t-n)$ **7.** $5(1 - e^{-2t})u(t)$ A

11. $(1/n!)t^n u(t)$ **14.** $2\delta(t)$ **16.** $\sum_{n=-\infty}^{\infty} (-1)^n \delta(t-n)$ **19.** $18u(t-2)$ V   **21.** 3

24. $\mathscr{R}e\{100e^{j(50t+\pi/4)}\}$ **27.** $5e^{j2\pi6t} + 5e^{-j2\pi6t}$ **31.** $\frac{1}{2}e^{-3t}(e^{j2\pi(t-0.25)} + e^{-j2\pi(t-0.25)})$

34. 70.71   **38.** $10e^{-3t}(\cos 4t + j \sin 4t)$ **42.** 1.257 rad, 72°   **45.** 0.942 rad, 54°

48. 347.2 μs   **51.** 2.083 ms   **54.** 1.042 ms   **56.** Continuous

60. Discontinuous at $t = (n + \frac{1}{2})\pi$, n an integer   **64.** $\sqrt{5}$ **67.** Energy

70. Orthogonal   **73.** 7.071 A   **76.** 6.667 A   **78.** $P = 62.5$ W   **79.** $W = 0.25$ J

83. $W = 16$ J   **86.** (a) Periodic pulse train   (b) Period $= T$ (c) τ/T A

(d) τ/T A   (e) $\sqrt{\tau/T}$ A   **88.** (a) 1.1107   (b) 9.003 V

CHAPTER 8

1. -7 **4.** $-3, -5, -5$ **7.** Ae^{3t} **10.** $A_1 + A_2e^{-t}$ **13.** $-320e^{-20t}$ V

16. $A_1e^{-t} + A_2e^{-2t}$ **19.** 14 V   **22.** Unstable   **25.** Marginally stable

29. $A > -1$ **32.** $10e^{-20t}$ V   **35.** $10e^{-1/2t}$ A   **38.** $80e^{-20t}$ A

39. $e^{-t}(A \cos 3t + B \sin 3t)$, stable   **41.** $A \cos 2t + B \sin 2t$, marginally stable

45. $20e^{-4t} \cos 2t$ **49.** $R = \sqrt{2}\,\Omega$ **52.** $R = 4\sqrt{5}\,\Omega$ **54.** 157.114

57. CIRCUIT FILE FOR PROBLEM 8.57

```
R10     1    0    10
L12     1    2    5      IC=80
E02     0    2    1    0    9
R20     2    0    13
.TRAN    0.005    0.15    UIC
.PLOT    TRAN    I(L12)
.END
```

61. CIRCUIT FILE FOR PROBLEM 8.61

```
R20     2    0    12
R21     2    1    6
L21     2    1    1      IC=0
C10     1    0    0.05555556    IC=36
.TRAN    0.04    1    UIC
.PLOT    TRAN    V(1), I(L21)
.END
```

63. $40 \cos 1/4t - 20 \sin 1/4t$ V, $10 \cos 1/4t + 20 \sin 1/4t$ A, 2000 J   **64.** $20e^{-100t}$ V

66. $10e^{-10t}$ V   **69.** $9/5e^{-2t} + 21/5e^{-12t}$ A

70. $e^{-t}(8 \cos t - 4 \sin t)$ A, $e^{-t}(4 \cos t - 12 \sin t)$ A   **72.** $10e^{-2t}$ A

73. $5 + 5e^{-2t}$ V, $15 - 15e^{-2t}$ V

CHAPTER 9

1. $-6e^{-2t}$ **4.** $16/3e^{-3t}$ **7.** $25t^2e^{-5t}$ **10.** 36   **13.** $(t - \frac{1}{12})u(t)$ A

16. 24 for $t > 0$ **19.** $12t^2e^{-3t}$ for $t > 0$ **22.** $10te^{-10t}u(t)$ V

25. $450t^2e^{-5t}u(t)$ A   **28.** $3 - 2e^{-2t} + 45/17e^{2t}$

30. $0.5604 \cos (14t + 71.69°)$, steady state

35. $9.6e^{-t} \cos (4t - 53.13°)$, steady state   **39.** $25 - 5e^{-4t}$ V

42. $-3.2625e^{-t} + 21.3625e^{-9t} - 0.1 \cos 3t$ V

44. $112.64e^{-6t} + 8e^{-3t} \cos (4t - 23.13°)$ V

47. $71/3e^{-2t} + \sqrt{2}/3 \cos (2t - 45°)$ V **50.** $90 - 30e^{-4t}$ V

52. $10 + 5/2(e^{-10t} - e^{-50t})u(t)$ A **55.** $15 + 12{,}000e^{-100t} - 12{,}000e^{-150t}$ V

58. $102 - 2e^{-2t}$ V **60.** $100 + e^{-3t}(100 \cos 4t + 75 \sin 4t)$ V

62. $\frac{3}{2} - 1/2e^{-3t}$ V **64.** $9 \cos (4t - 53.13°)$ V for $t < 0$, $27/5e^{-3t}$ V for $t > 0$

69. (a) 12 V, 0 A (b) 12 V, $-12/5$ A (c) Ae^{-12t} (d) 0 (e) $12e^{-12t}$ V

74. (a) 6 A, 0 V (b) 6 A, -72 V (c) Ae^{-4t} (d) 0 (e) $6e^{-4t}$ A

83. (a) 0 V, 0 A, 0 V, 6 A (b) 0 V, -6 A, 0 V, 6 A (c) 0 V, 0 A, 0 V, 0 A

(d) $e^{-3t}(A \cos 4t + B \sin 4t)$ (e) $-450e^{-3t} \sin 4t$ V

103. INPUT FILE FOR PROBLEM 9.103 **113.** $100(1 - e^{-t})$ V, 0.1 F

```
I1      0   1    DC    20
R10     1   0    0.25
R12     1   2    0.16667
C12     1   2    1    IC=1
C20     2   0    1    IC=5
.TRAN   0.02     3    UIC
.PROBE
.END
```

114. $100(1 - e^{-10t})$ V, 0.01 F **122.** $C = 0.0002725$ F, $4.489 \cos (10t - 20.6°)$ A

132. (a) $(R^3C^3p^3 + 6R^2C^2p^2 + 5RCp + 1)v_o = (6R^2C^2p^2 + 5RCp + 1)v_s$

(b) $V_s(1 - 0.0597e^{s_1t} + 0.280/e^{s_2t} - 1.2204e^{s_3t})v(t)$ V, where $s_1 = -0.30798 \ s^{-1}$,

$s_2 = -0.64310 \ s^{-1}$, $s_3 = -5.04892 \ s^{-1}$

CHAPTER 10

4. $v_c(t) = Ae^{s_1t}$ and $s_1 = -\dfrac{1}{(1 - \alpha)RC}$. For $\alpha < 1$, $v_c(t) \to 0$ as $t \to \infty$. For $\alpha = 1$, $v_c(t) = 0$

for all t. For $\alpha > 1$, $v_c(t) \to \infty$ as $t \to \infty$

5. $A(p)e^{j\omega t}$ denotes that $A(p)$ *operates* on $e^{j\omega t}$. $A(j\omega)e^{j\omega t}$ is the *result* of the operation.
Consider for example, $A(p) = p$. Then $pe^{j\omega t} = j\omega e^{j\omega t}$ but $p \neq j\omega$.

7. (a) $Z = \dfrac{\omega^2 RL^2}{R^2 + (\omega L)^2} + j\dfrac{\omega R^2 L}{R^2 + (\omega L)^2}$ (b) 1 Mrad/s **8.** $Cpv = (RCp + 1)i$

9. $j\omega CV = (1 + j\omega RC)\mathbf{I}$ **10.** $Z = (1 + j\omega RC)/j\omega C$ **11.** $\mathbf{v}(t) = \mathbf{V}e^{j\omega t}$ where $\mathbf{V} = \mathbf{ZI}$

13. $v_p = 100\sqrt{2} \cos (1000t - 25°)$ V

17. (a) $\underbrace{\left(3\dfrac{d}{dt} + 2\right)(v_R(t) + jv_I(t))}_{\mathscr{A}(p)\mathbf{v}(t)} = \underbrace{3\dfrac{d}{dt}v_R(t) + 2v_R(t)}_{\mathscr{A}(p)v_R(t)} + j\underbrace{\left[3\dfrac{d}{dt}v_I(t) + 2v_I(t)\right]}_{j\mathscr{A}(p)v_I(t)}$

The real part of the left-hand side is $\mathscr{R}e\{\mathscr{A}(p)\mathbf{v}(t)\}$, and the real part of the right-hand
side is $\mathscr{A}(p)v_R(t)$. Therefore, $\mathscr{R}e\{\mathscr{A}(p)\mathbf{v}(t)\} = \mathscr{A}(p)\,\mathscr{R}e\{\mathbf{v}(t)\}$ for $\mathscr{A}(p) = 3p + 2$

(b) $\underbrace{\left(\sum_{k=0}^{n} a_k p^k\right)(v_R(t) + jv_I(t))}_{\mathscr{A}(p)\mathbf{v}(t)} = \underbrace{\sum_{k=0}^{n} a_k p^k v_R(t)}_{\mathscr{A}(p)v_R(t)} + j\underbrace{\sum_{k=0}^{n} a_k p^k v_I(t)}_{j\mathscr{A}(p)v_I(t)}$,

so $\mathscr{R}e\{\mathscr{A}(p)\mathbf{v}(t)\} = \mathscr{A}(p)\,\mathscr{R}e\{\mathbf{v}(t)\}$ for every real $\mathscr{A}(p)$. (c) Same derivation as in (b).

(d) We can take the real part of the equation $\mathscr{A}(p)\mathbf{v}(t) = \mathscr{B}(p)\mathbf{i}(t)$ to obtain
$\mathscr{R}e\{\mathscr{A}(p)\mathbf{v}(t)\} = \mathscr{R}e\{\mathscr{B}(p)\mathbf{i}(t)\}$. By (b) and (c) the above becomes
$\mathscr{A}(p)\,\mathscr{R}e\{\mathbf{v}(t)\} = \mathscr{B}(p)\,\mathscr{R}e\{\mathbf{i}(t)\}$. To obtain Eq. (10.21) let $\mathbf{i}(t) = \mathbf{I}e^{j\omega t}$ and $\mathbf{v}(t) = \mathbf{V}e^{j\omega t}$

24. From Table 10.1, $R(\omega)G(\omega) = R^2(\omega)/[R^2(\omega) + X^2(\omega)]$. This equals 1 if and
only if $X(\omega) = 0$.

28. $v_o(t) = \dfrac{\omega L}{\sqrt{R^2 + \left(\omega L - \dfrac{1}{\omega C}\right)^2}}\, V_m \cos\left[\omega t + \phi + 90° - \arctan\left(\dfrac{\omega L}{R} - \dfrac{1}{\omega R C}\right)\right]$

33. $v_o(t) = \mathbf{H}(j\omega)\mathbf{V}_s e^{j\omega t}$ where $\mathbf{H}(j\omega)$ is determined in Problem 32.

38. (a) $(1 + h)Lp^2 i_o + Rpi_o + (1 + h)i/C = pv_s$

(b) $\mathbf{H}(j\omega) = j\omega C/[(1 + h) - (1 + h)\omega^2 LC + j\omega RC]$

(c) The steady-state response to $v_s(t) = A \cos(\omega t + \phi)$ is $i_o(t) = |\mathbf{H}(j\omega)|A \cos(\omega t + \phi + \underline{/\mathbf{H}(j\omega)})$ where $|\mathbf{H}(j\omega)| = \omega C/\sqrt{[(1 + h) - (1 + h)\omega^2 LC]^2 + (\omega RC)^2}$ and

$\underline{/\mathbf{H}(j\omega)} = \dfrac{\pi}{2} - \arctan\{\omega RC/[(1 + h) - (1 + h)\omega^2 LC]\}$ **44.** $\mathbf{H}(j10^6) = -10{,}000$

47. $t_o = (3 + 4k)\pi/4$ ms where k is an integer.   **62.** $v(t) = 6 \cos(t + 53°) + 2 \cos 2t$ V

CHAPTER 11

2. $i(t) = \sqrt{2} \cos(\omega t + 45°)$ **8.** $2.77\underline{/4.98°}$ A   **10.** $(4p + 2)i = pv$

14. $\mathbf{Z} = \dfrac{R + r - \omega^2 rLC + j\omega(rRC + L)}{1 + j\omega rC}$ **16.** $(rCp + 1)v = (rLCp^2 + rRCp + Lp + R + r)i$

21. $\mathbf{Z}_{in} = \dfrac{\sqrt{5} - 1}{2}\,\Omega$ **25.** $\left(\dfrac{L}{R}p + 1\right)v_3 = v_2$

30. $\begin{bmatrix} j\omega C + \dfrac{1}{j\omega L} + \dfrac{1}{R} & -\dfrac{1}{R} \\[2mm] -\dfrac{1}{R} & j\omega C + \dfrac{1}{j\omega L} + \dfrac{1}{R} \end{bmatrix} = \begin{bmatrix} \dfrac{\mathbf{V}_s}{j\omega L} \\[2mm] j\omega C\mathbf{V}_s \end{bmatrix}$

32. $\begin{bmatrix} -j & 3 \\ 2j & 1 \end{bmatrix}\begin{bmatrix} \mathbf{V}_1 \\ \mathbf{V}_2 \end{bmatrix} = \begin{bmatrix} 18 - j5 \\ 5 + j10 \end{bmatrix}$, $\begin{bmatrix} \mathbf{V}_1 \\ \mathbf{V}_2 \end{bmatrix} = \begin{bmatrix} 5.02\underline{/4.9°} \\ 5.86\underline{/0°} \end{bmatrix}$

35. (a) $\begin{bmatrix} \mathbf{Z}_a^{-1} + \mathbf{Z}_b^{-1} & 0 & 0 \\ 0 & \mathbf{Z}_1^{-1} + \mathbf{Z}_2^{-1} & -\mathbf{Z}_2^{-1} \\ -A & A & 1 \end{bmatrix}\begin{bmatrix} \mathbf{V}_{(+)} \\ \mathbf{V}_{(-)} \\ \mathbf{V}_0 \end{bmatrix} = \begin{bmatrix} \mathbf{Z}_a^{-1}\mathbf{V}_{s2} \\ \mathbf{Z}_1^{-1}\mathbf{V}_{s1} \\ 0 \end{bmatrix}$,

(b) $\mathbf{V}_0 = \dfrac{A}{1 + A\beta}\left[-\dfrac{\mathbf{Z}_2}{\mathbf{Z}_1 + \mathbf{Z}_2}\mathbf{V}_{s1} + \dfrac{\mathbf{Z}_b}{\mathbf{Z}_a + \mathbf{Z}_b}\mathbf{V}_{s2}\right]$, where $\beta = \mathbf{Z}_1/(\mathbf{Z}_1 + \mathbf{Z}_2)$

(c) $\mathbf{V}_0 = -\dfrac{\mathbf{Z}_2}{\mathbf{Z}_1}\mathbf{V}_{s1} + \dfrac{\mathbf{Z}_b(\mathbf{Z}_1 + \mathbf{Z}_2)}{\mathbf{Z}_1(\mathbf{Z}_a + \mathbf{Z}_b)}\mathbf{V}_{s2}$ **42.** $\begin{bmatrix} 1 + j10 & -j90 \\ -j90 & 400 + j1960 \end{bmatrix}\begin{bmatrix} \mathbf{I}_1 \\ \mathbf{I}_2 \end{bmatrix} = \begin{bmatrix} 10\underline{/0°} \\ 0 \end{bmatrix}$

46. $\begin{bmatrix} j\omega L_1 & j\omega M \\ j\omega M & j\omega L_2 \end{bmatrix}\begin{bmatrix} \mathbf{I}_1 \\ \mathbf{I}_2 \end{bmatrix} = \begin{bmatrix} \mathbf{V}_1 \\ \mathbf{V}_2 \end{bmatrix}$ **47.** $\begin{bmatrix} j\omega(L_a + L_b) & j\omega L_c \\ j\omega L_c & j\omega(L_b + L_c) \end{bmatrix}\begin{bmatrix} \mathbf{I}_1 \\ \mathbf{I}_2 \end{bmatrix} = \begin{bmatrix} \mathbf{V}_1 \\ \mathbf{V}_2 \end{bmatrix}$

48. $L_a = L_1 - M$, $L_b = L_2 - M$, $L_c = M$ **50.** $i(t) = 4.17 \sin(2t + 33.7°) - 2.5$ A

54. $v_o(t) = 5 \cos(t + 10°) + 2.5 \sin(2t + 33.7°) + 10$ V

57.

$1\,\Omega$

63. If $\mathbf{V}_s \neq 0$, we obtain an impossible theoretical circuit when we short the input terminals. Thus $\mathbf{I}_{sc}$ is not defined for this circuit.   **65.** $\sqrt{20} \cos(1000t - 71.6°)$ A

71. Thévenin equivalent

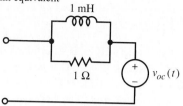

where $v_{oc}(t) = 5\sqrt{2} \cos(1000t + 45°) + 0.4\sqrt{5} \sin(2000t - 26.6°)$ V

Norton equivalent

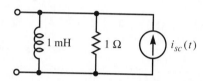

where $i_{sc}(t) = -10 \cos 1000t - 10 \sin 2000t$ A

72.

FREQ	VM(1)	VP(1)	VM(2)	VP(2)
1.000E+03	2.000E-03	0.000E+00	1.000E-20	0.000E+00
FREQ	VM(3)	VP(3)	VM(4)	VP(4)
1.000E+03	3.183E-09	-9.000E+01	1.000E-20	0.000E+00
FREQ	VM(5)	VP(5)		
1.000E+03	3.183E-04	9.000E+01		

74.

FREQ	VM(1)	VP(1)	VM(2)	VP(2)
1.000E+03	1.000E+00	0.000E+00	1.000E+00	0.000E+00
FREQ	VM(3)	VP(3)	VM(4)	VP(4)
1.000E+03	9.970E-01	0.000E+00	1.000E+00	0.000E+00
FREQ	VM(5)	VP(5)	IM(V12)	IP(V12)
1.000E+03	9.817E+00	1.800E+02	1.030E-04	0.000E+00

Z(INPUT)=V(1)/I(V12)=9709 OHMS AT 0 DEGREES

75.

FREQ	VM(1)	VP(1)	VM(2)	VP(2)
1.000E+03	1.000E+00	0.000E+00	9.950E-01	5.713E+00
FREQ	VM(3)	VP(3)	VM(4)	VP(4)
1.000E+03	9.920E-01	5.713E+00	9.950E-01	5.713E+00
FREQ	VM(5)	VP(5)	IM(C12)	IP(C12)
1.000E+03	9.769E+00	-1.743E+02	1.025E-04	5.713E+00

C12=1/(970.9*2*PI*1000)=0.1693E-6
Z(INPUT)=V(1)/I(C12)=9756 OHMS AT -5.713 DEGREES

79. $Z = \sqrt{L/C}$ **82.** 20 V   **86.** 46.9 A

CHAPTER 12

2. 819 W   **8.** $\omega = 1/\sqrt{LC}$ **10.** 3.2 Ω, 4.97 mF

13. (a) 7.07 kW   (b) -7.07 kVAR   (c) 118 A   (d) $V_{\text{eff}}^2 \left[\dfrac{1}{R} + j\omega C \right]$

(e) 2.04 Ω, 1.3 mF   **19.** (a) $\frac{1}{3}\underline{/45°}$ Ω   (b) $150\underline{/45°}$ VA   (c) 150 VA

23. (a) 283 mA   (b) 62.5 V rms, 77.7 V rms   (c) PF $\simeq 1$ (d) $24.01\underline{/1.72°}$ VA

28. (a) 2 H   (b) 2 W   **32.** $1 - j\,\Omega$ **36.** (a) $1 - j\,\Omega$ (b) 12.5 W

40. $28.6 + j0.18$ kVA   **41.** 28.6 W   **42.** 119 A   **43.** PF $\simeq 1$

CHAPTER 13

3. $\mathbf{H}(j\omega) = \left(\dfrac{R}{R + R_s} \right) \left(\dfrac{\omega L}{\sqrt{R_p^2 + \omega^2 L^2}} \right) \Big/\dfrac{\pi}{2} - \text{arc tan } (\omega L/R_p)$, where $R_p = R \,\|\, R_s$

6. (a) 0   (b) $\mathbf{V}_o = (K\underline{/-\omega_1\tau})\mathbf{V}_1$ for $\omega_1 < \omega_0$. Therefore
$v_o(t) = KV_{1m} \cos \left[\omega_1(t - \tau) + \phi_{V1} \right]$
(c) $v_o(t) = 6 + 2 \cos (300t + 2.81°) + 4 \sin (800t - 45.8°)$ **7.** 1.9 kΩ, 10 μF

13. (a) 1 kΩ
(b) The admittance of the parallel LC part of the circuit, $j(\omega C - 1/\omega L)$, equals zero for $\omega = 1/\sqrt{LC}$. Therefore, when $\omega = 1/\sqrt{LC}$, $\mathbf{I}_R = \mathbf{I}_s = \mathbf{V}_s/(r + R) = 5\underline{/0°}$ mA and
$$\mathbf{I}_L = -\mathbf{I}_C = -j\omega C \left(\dfrac{R}{r + R} \right) \mathbf{V}_s = 5\underline{/-90°}\ \text{A}$$

14. For $\omega = 0$, the inductance and capacitance behave like short and open circuits, respectively. Therefore, $\mathbf{I}_L = \mathbf{I}_s = 10\underline{/0°}$ mA and $\mathbf{I}_C = \mathbf{I}_R = 0$ mA.

15. For $\omega = \infty$, the inductance and capacitance behave like open and short circuits, respectively. Therefore, $\mathbf{I}_C = \mathbf{I}_s = 10\underline{/0°}$ mA and $\mathbf{I}_L = \mathbf{I}_R = 0$ mA.

17. (a) Let p denote differentiation with respect to t. Take the real parts of both sides of the equation $p^2\mathbf{v} + 15p\mathbf{v} + 50\mathbf{v} = p^2\mathbf{i} + 7\mathbf{i}$ to obtain
$\mathscr{R}e\,\{p^2\mathbf{v} + 15p\mathbf{v} + 50\mathbf{v}\} = \mathscr{R}e\,\{p^2\mathbf{i} + 7\mathbf{i}\}$. Since $\mathscr{R}e\,\{\cdot\}$ is a linear operator, this becomes
$p^2\,\mathscr{R}e\,\{\mathbf{v}\} + 15p\,\mathscr{R}e\,\{\mathbf{v}\} + 50\mathscr{R}e\,\{\mathbf{v}\} = p^2\,\mathscr{R}e\,\{\mathbf{i}\} + 7\mathscr{R}e\,\{\mathbf{i}\}$, which is the equality sought.
(b) Same derivation as (a), except that $\mathscr{R}e\,\{\ \}$ is replaced with $\mathscr{I}m\,\{\ \}$.

22. (a) $(2p + 1)v = (4p + 1)i$ (b) $v(t) = -0.13e^{-0.5t}\,u(t) + 4.17e^{-4t} \cos (3t + 16.9°)u(t)$

28. $\dfrac{1 + sR_2C}{1 + sC(R_1 + R_2)}$ **35.** $\pm j/\sqrt{LC}$ **36.** (a) $-\dfrac{1}{RC}$ (b) 0

39. (a) $-\alpha \pm j\omega_d$ where $\alpha = \dfrac{R}{2L}$ and $\omega_d = \sqrt{\dfrac{1}{LC} - \left(\dfrac{R}{2L} \right)^2}$ (b) 0 and ∞

44. (a) $\dfrac{s + 2}{s + 1}$ (b) -2 Np/s   (c) -1 Np/s   (d) Ae^{-t} (e) Ae^{-2t}

45. (a) $\dfrac{LCs^2 + RCs + 1}{RLCs^2 + sL}$

(b)

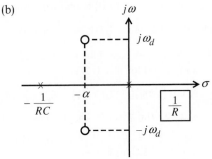

(c) $e^{-\alpha t}(A\cos\omega_d t + B\sin\omega_d t)$

50. (a) (i) 0   (ii) $KV_{1m}\cos\left[\omega_1(t-\tau)+\phi_{v1}\right]$

(b) $v_o(t) = 4\sin(800t - 45.8°) + 17\cos(1100t - 18°)$

54. (a) $\sqrt{\dfrac{1}{LC} - \left(\dfrac{R}{L}\right)^2}$ (b) $\dfrac{L}{RC}$

(c) From Eq. (12.24), $W_{C,ave} = \frac{1}{4}C|V_C|^2$ where V_C is the phasor voltage across C. From Eq. (12.20) $W_{L,ave} = \frac{1}{4}L|I_L|^2$ where I_L is the phasor current through L. Substituting $V_C = (R + j\omega_{upf}L)I_L$ where ω_{upf} is given in (a) we find that $W_{C,ave} = W_{L,ave}$.

63. (a) Second order pole at -0.7 Np/s (assumed). First order pole at -100 Np/s. Second order pole at -500 Np/s (assumed)

(b) Second order zero at 0 Np/s. First order zero at -10 Nps

(c) $H(s) = \dfrac{s^2\left(\dfrac{s}{10}+1\right)}{\sqrt{0.7}\left(\dfrac{s}{0.7}+1\right)^2\left(\dfrac{s}{100}+1\right)\left(\dfrac{s}{500}+1\right)^2}$ (assumed)

(d) The answers labeled "assumed" are not unique because the corner frequencies 0.7 rad/s and 500 rad/s could have been caused by first-order complex poles. In addition, we have assumed that the circuit is minimum phase.

78. (a) $H(s) = 1/(\tau s + 1)$ where $\tau = 1$ ms

(b) $H_2(s) = (\tau s + 1)/(\tau^2 s^2 + 3\tau s + 1)$, $H_3(s) = 1/(\tau^2 s^2 + 3\tau s + 1)$

81. (a) $H(s) = (-\tau_1\tau_2 s^2 + 1)/[\tau_1\tau_2 s^2 + (\tau_1 + \tau_2)s + 1]$, where $\tau_1 = L/R$ and $\tau_2 = RC$

(b) $(\tau_1\tau_2\omega^2 + 1)/\sqrt{(1 - \tau_1\tau_2\omega^2)^2 + (\tau_1 + \tau_2)^2\omega^2}$ (c) arc tan $\{(\tau_1 + \tau_2)\omega/(1 - \tau_1\tau_2\omega^2)\}$

82. $RC = L/R$ **83.** 4 mH, 250 μF

84. $H(s) = \dfrac{sC_1R_2(sC_2R_3 + 1)}{s^2C_1C_2[R_1R_2 + R_1R_3 + R_2R_3] + s[C_1(R_1 + R_2) + C_2(R_2 + R_3)] + 1}$

85. (a) 0   (b) $\frac{2}{7}$

92. (a) $H(s) = \left(\dfrac{R_2}{R_1 + R_2}\right)\left(\dfrac{R_1C_1s + 1}{R_pC_ps + 1}\right)$, where $R_p = R_1 \parallel R_2$ and $C_p = C_1 + C_2$

(b) $1/(1 + \alpha)$ (Typically 0.1)   (c) R_2C_2/R_1 or C_2/α (d) $(R_1 + R_2)\left\|\dfrac{1}{sC_p}\right.$

CHAPTER 14

2. $T = 1$ ms, $a_0 = 10$, $a_3 = 7$, $a_4 = \dfrac{5}{\sqrt{2}}$, $b_4 = -\dfrac{5}{\sqrt{2}}$. The remaining coefficients equal zero.

5. The mean square value of $v(t)$ is equal to the squared length of the vector $\mathbf{v}$.

9. (b) No

(c) The only difference between $v(t)$ and $v'(t)$ is that $v'(t)$ contains a nonzero coefficient a_0.

10. (b), (c) See Problem 9   **12.** $A_0 = \dfrac{A}{T}$, $A_n = \dfrac{2A}{T}$, and $\phi_n = 0$ for $n = 1, 2, \ldots$

14. (τ/T) 100%   **16.** $a_0 = A$, $a_n = b_n = 0$ for $n = 1, 2, \ldots$

17. $a_0 = 1/T$, $a_n = 2/T$ and $b_n = 0$ for $n = 1, 2, \ldots$

19. $v(t - t_d) = A_o + \sum\limits_{n=1}^{\infty} A_n\cos\left[2\pi nf_1(t - t_d) + \phi_n\right] = A_o + \sum\limits_{n=1}^{\infty} A_n\cos(2\pi nf_1 t + \phi_n - 2\pi nf_1 t_d)$

20. $V_n = 1$ for $|n| \le 5$ and zero otherwise   **26.** (b) Yes   **36.** 13.7 μF

37. 0.255H, 39.7 pF   **40.** $\dfrac{1}{\pi}$ V   **41.** $\dfrac{1}{\pi} + \dfrac{1}{2}\cos(2000\pi t)$ V

CHAPTER 15

1. $2W \operatorname{sinc}[2W(t - t_d)]$ **2.** (a) $\dfrac{\tau}{1 + j2\pi f \tau}(1 - e^{-j2\pi f T})$

(b) $|V(f)| = \dfrac{\tau \sqrt{[1 - e^{-T/\tau} \cos 2\pi f T]^2 + [e^{-T/\tau} \sin 2\pi f T]^2}}{\sqrt{1 + (2\pi f \tau)^2}}$,

$\underline{/V(f)} = \arctan\left\{ \dfrac{e^{-T/\tau} \sin 2\pi f T}{1 - e^{-T/\tau} \cos 2\pi f T} \right\} - \arctan(2\pi f \tau)$

5.–7. $V(f) = \tau \operatorname{sinc}(f\tau)$, $v_T(t) = \displaystyle\sum_{n=-\infty}^{+\infty} V_n e^{j2\pi n f_1 t}$ where $f_1 = 1/T$ and $V_n = \dfrac{\tau}{T} \operatorname{sinc}\left(\dfrac{n\tau}{T}\right)$.

Let $T = 5\tau$, 2τ, and τ as requested.   **10.** 0   **12.** (a) Use FT15 with $V_n = 1/T$.

(b) Each function has a Fourier transform having the same functional form as the original function.

(c) There are an infinite number. Start with any even function $g(t)$ and let $G(f) = \mathscr{F}\{g(t)\}$. Form the new function $v(t) = g(t) + G(t)$. Then $V(f) = g(f) + G(f) = v(f)$ has the same functional form as $v(t)$.

13. The function $g(t) = u(t)$. Therefore $G(f) = \dfrac{1}{2}\delta(f) + \dfrac{1}{j2\pi f}$ and

$$\mathscr{F}\left\{ \int_{-\infty}^{t} v(\lambda)\, d\lambda \right\} = G(f)V(f) = \dfrac{1}{2}V(f)\delta(f) + V(f)\dfrac{1}{j2\pi f} = \dfrac{1}{2}V(0)\delta(f) + V(f)\dfrac{1}{j2\pi f}$$

14. 1 J

18. $\displaystyle\int_{-\infty}^{+\infty} x(t)y^*(t)\, dt = \int_{-\infty}^{+\infty} \left\{ \int_{-\infty}^{+\infty} X(f)e^{j2\pi f t}\, df \right\} \left\{ \int_{-\infty}^{+\infty} Y(\xi)e^{j2\pi \xi t}\, d\xi \right\}^* dt$

$\displaystyle = \int_{-\infty}^{+\infty} \left\{ \int_{-\infty}^{+\infty} X(f)e^{j2\pi f t}\, df \right\} \left\{ \int_{-\infty}^{+\infty} Y^*(\xi)e^{-j2\pi \xi t}\, d\xi \right\} dt$

$\displaystyle = \int_{-\infty}^{+\infty} \int_{-\infty}^{+\infty} X(f)Y^*(\xi) \left\{ \int_{-\infty}^{+\infty} e^{j2\pi(f-\xi)t}\, dt \right\} df\, d\xi$

$\displaystyle = \int_{-\infty}^{+\infty} \int_{-\infty}^{+\infty} X(f)Y^*(\xi)\delta(f - \xi)\, df\, d\xi$

$\displaystyle = \int_{-\infty}^{+\infty} X(f)Y^*(f)\, df$

19. $\dfrac{E_0(w\tau)}{E_0(\infty)} = \dfrac{2}{\pi} \tan^{-1}(2\pi W\tau)$

28. (a) $H(f)$ (b) $A \cos 2\pi f_s t$ (c) 0   (d) $2W \operatorname{sinc}(2Wt)$

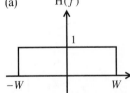

(e) $\operatorname{sinc}(2f_a t)$ (f) $(f_b/f_a)\operatorname{sinc}(2f_a t)$ (g) $v_s(t)$

CHAPTER 16

1. $\dfrac{1}{s - a} e^{-s t_0}$, pole at $s = a$, abscissa of convergence $= a$

2. $\dfrac{V}{s - s_0}$, pole at $s = \sigma_0 + \omega_0$, abscissa of convergence $= \sigma_0$

3. $\dfrac{(\frac{1}{2})V}{s - s_0} + \dfrac{(\frac{1}{2})V^*}{s - s_0^*}$ where $s_0 = \sigma_0 + j\omega_0$ and $\mathbf{V} = V_m\underline{/\phi_0}$, abscissa of convergence $= \sigma_0$,

poles at $s = \sigma_0 \pm j\omega_0$

7. Since $v(t)$ is causal, $V(f) = \mathscr{F}\{v(t)\} = \int_{0^-}^{\infty} v(t)e^{-j2\pi ft}\,dt$. But

$V(s) = \mathscr{L}\{v(t)\} = \int_{0^-}^{\infty} v(t)e^{-st}\,dt$. By substituting $j2\pi f$ for s, we see that $V(j2\pi f) = V(f)$

9. $\dfrac{1}{s}[e^{st/2} - e^{-st/2}]e^{-sto}$

18. $\dfrac{e^{sT}}{e^{sT} - 1}\,V_0(s)$, the abscissa of convergence equals the largest element of $\{\sigma_0, 0\}$

21. $\lim\limits_{s \to 0} \dfrac{sA}{s^2 + \omega_0^2} = 0$. This is incorrect because $A\cos(\omega_0 t)u(t)$ has no final value.

25. Since $D(s) = (s - s_1)Q(s)$ then $D'(s) = \dfrac{d}{ds}[(s - s_1)Q(s)] = Q(s) + (s - s_1)\dfrac{d}{ds}Q(s)$

Therefore $D'(s_1) = Q(s_1)$ and $k_{11} = (s - s_1)\dfrac{N(s)}{D(s)}\Big|_{s=s_1} = \dfrac{N(s_1)}{Q(s_1)} = \dfrac{N(s_1)}{D'(s_1)}$

28. $[2e^{-3(t-4)} - e^{-2(t-4)}]u(t)$ **30.** $[-\tfrac{25}{4} + 25t + \tfrac{105}{4}e^{-4t}]u(t)$ **43.** $10e^{-100t}u(t)$ V

44. $10u(t) - e^{-50t}u(t)\left[10\cos(\sqrt{997500}\,t) + \dfrac{50}{\sqrt{997500}}\sin(\sqrt{997500}\,t)\right]$

47. $3.161u(t - 0.001) - 3.161e^{-2000(t-0.001)}u(t - 0.001)$ V **51.** 10 V

54. (b) $H(s) = \dfrac{s}{s + 1/\tau}$ where $\tau = \dfrac{L}{(\beta + 1)R}$ (d) $i = 3e^{-t/\tau}$ mA for $t \geq 0$. When we

substitute $\beta = 0$, -1, and -2, this becomes $3e^{-tR/L}$ mA, 3 mA, and $3e^{tR/L}$ mA, respectively.

55. (b) $Y(s) = \dfrac{sC}{\tau} \cdot \dfrac{1}{s + 1/\tau}$ where $\tau = (\beta + 1)RC$

(d) $i = \dfrac{2C}{\tau}e^{-t/\tau}$ A for $t \geq 0$. When we substitute $\beta = 0$ and $\beta = -2$, this becomes

$\dfrac{2}{R}e^{-t/RC}$ A and $-\dfrac{2}{R}e^{+t/RC}$ A, respectively. Take limit $\beta \to -1$ to get $2C\delta(t)$ A as the

current for $\beta = -1$. **62.** $[5 + 6e^{-4000t}]u(t)$ V

68. (a) $0.234888\cos(2t + 3.366°)u(t) + 0.299e^{-2t}\cos(3t - 141.6°)$

(b) $(p^2 + 4p + 13)y = (p + 2)x$ where y and x denote output and input, respectively.

71. (a) $i = 0.8$ mA for $t < 0$. Since $i > 0$, switch S2 is closed.

(b) $(12 - 24e^{-t/\tau})u(t)$ V where $\tau = 15$ ms (c) 10.4 ms

CHAPTER 17

3. $\mathbf{Z} = \begin{bmatrix} \mathbf{Z}_A \parallel (\mathbf{Z}_A + \mathbf{Z}_B) + \mathbf{Z}_C & \dfrac{\mathbf{Z}_A^2}{2\mathbf{Z}_A + \mathbf{Z}_B} + \mathbf{Z}_C \\[2ex] \dfrac{\mathbf{Z}_A^2}{2\mathbf{Z}_A + \mathbf{Z}_B} + \mathbf{Z}_C & \mathbf{Z}_A \parallel (\mathbf{Z}_A + \mathbf{Z}_B) + \mathbf{Z}_C \end{bmatrix}$, $\mathbf{V}_{oc} = 0$

4. $\mathbf{V}_{oc} = \mathbf{V}_{2oc} + \mathbf{z}_{21}\dfrac{\mathbf{V}_s - \mathbf{V}_{1oc}}{\mathbf{Z}_s + \mathbf{z}_{11}}$, $\mathbf{Z}_{Th} = \mathbf{z}_{22} - \dfrac{\mathbf{z}_{12}\mathbf{z}_{21}}{\mathbf{Z}_s + \mathbf{z}_{11}}$ **7.** No

10. $\mathbf{y}_{11} = \mathbf{y}_{22} = \dfrac{1}{\mathbf{Z}_1 + \mathbf{Z}_1 \parallel \mathbf{Z}_2} + \dfrac{1}{\mathbf{Z}_A + \mathbf{Z}_A \parallel \mathbf{Z}_B}$,

$\mathbf{y}_{12} = \mathbf{y}_{21} = -\left(\dfrac{\mathbf{Z}_2}{\mathbf{Z}_1 + \mathbf{Z}_1}\right)\dfrac{1}{\mathbf{Z}_1 + \mathbf{Z}_1 \parallel \mathbf{Z}_2} - \left(\dfrac{\mathbf{Z}_B}{\mathbf{Z}_A + \mathbf{Z}_B}\right)\dfrac{1}{\mathbf{Z}_A + \mathbf{Z}_A \parallel \mathbf{Z}_B}$

13. $\mathbf{I}_{sc} = \begin{bmatrix} 7.26\underline{/65.6°} \\ 4.28\underline{/24.5°} \end{bmatrix}$ A **19.** $i_2 = -\beta_2\beta_1 i_1$

29. Replace each of networks A and B of the series connection illustrated in Problem 28 with their z-parameter equivalents of Fig. 17.36a. The terminal equations of the resulting connection are given by KVL as: $V_1 = (z_{11A} + z_{11B})I_1 + (z_{12A} + z_{12B})I_2 + V_{1ocA} + V_{1ocB}$, $V_2 = (z_{21A} + z_{21B})I_1 + (z_{22A} + z_{22B})I_2 + V_{2ocA} + V_{2ocB}$ and may be written as

$$\begin{bmatrix} V_1 \\ V_2 \end{bmatrix} = \begin{bmatrix} z_{11A} + z_{11B} & z_{12A} + z_{12B} \\ z_{21A} + z_{21B} & z_{22A} + z_{22B} \end{bmatrix} \begin{bmatrix} V_1 \\ V_2 \end{bmatrix} + \begin{bmatrix} V_{1ocA} + V_{1ocB} \\ V_{2ocA} + V_{2ocB} \end{bmatrix}$$

Let us define sums of matrices as follows

$$\begin{bmatrix} z_{11A} & z_{12A} \\ z_{21A} & z_{22B} \end{bmatrix} + \begin{bmatrix} z_{11B} & z_{12B} \\ z_{21B} & z_{22B} \end{bmatrix} = \begin{bmatrix} z_{11A} + z_{11B} & z_{12A} + z_{12B} \\ z_{21A} + z_{21B} & z_{22A} + z_{22B} \end{bmatrix}$$

and

$$\begin{bmatrix} V_{1ocA} \\ V_{2ocA} \end{bmatrix} + \begin{bmatrix} V_{1ocB} \\ V_{2ocB} \end{bmatrix} = \begin{bmatrix} V_{1ocA} + V_{1ocB} \\ V_{2ocA} + V_{2ocB} \end{bmatrix}$$

Then the z-parameter matrix of the series network equals the sum of the z-parameter matrices of networks A and B. Notice also that a similar result applies for the open-circuit voltage matrices.

31. The solution to this problem is the dual of that of Problem 29.

32. $\dfrac{V_o}{V_s} = \dfrac{z_{21}Z_L}{(z_{11} + Z_s)(z_{22} + Z_L) - z_{12}z_{21}}$ **33.** $\dfrac{V_2}{V_s} = \dfrac{V_o}{V_s} = \dfrac{-y_{21}Y_s}{(y_{11} + Y_s)(y_{22} + Y_L) - y_{12}y_{21}}$

34. $\dfrac{V_2}{V_s} = \dfrac{V_o}{V_s} = \dfrac{-h_{21}Z_L}{Z_L(h_{11} + Z_s)(h_{22} + Y_L) + h_{21}h_{12}}$

CHAPTER 18

1. 6.25 Ω, 0.1115 H, 0.0531 H   **2.** 3.75 Ω, 0.0453 H, 0.0531 H   **3.** 0.747

4. $30\underline{/53.13°}$ Ω $= 18 + j24$ Ω   **11.** 8 H, 38 H, 2 H

12. 0 H (a short circuit), 6 H, 2 H   **15.** $\frac{198}{19}$ H, $\frac{198}{4}$ H, 198 H

16. 2 H, ∞H (an open circuit), 6 H   **19.** 15 mH, 10 μF, 3 mH

24. $M \leq L_1$ and $M \leq L_2$ **27.** $k \leq \sqrt{L_1/L_2}$ and $k \leq \sqrt{L_2/L_1}$

28. $R_1 = 10$ Ω, $X_{\ell 1} = 21.72$ Ω, $X_{m1} = 28.28$ Ω, $R_2 = 4$ Ω, $X_{\ell 2} = 10.86$ Ω, $n_2/n_1 = 1/\sqrt{2}$

32. $R_1 = 1$ Ω, $X_{\ell 1} = 9$ Ω, $X_{m1} = 1$ Ω, $R_2 = 2$ Ω, $X_{\ell 2} = 36$ Ω, $n_2/n_1 = 2$

35. $1.644\underline{/-9.46°}$ A, $32.88\underline{/9.46°}$ VA, $14.704\underline{/107.1°}$ V, $7.352\underline{/-72.9°}$ V

38. $2\underline{/-23.13°}$ A, $4\underline{/-23.13°}$ A, $-\frac{8}{5}\underline{/-23.13°}$ A, $-\frac{16}{3}\underline{/-23.13°}$ A

39. $n_2/(n_1 + n_2)$, $-(n_1 + n_2)/n_2$, $[(n_1 + n_2)/n_2]^2 Z_L$ **42.** $1\underline{/0°}$ A, $2\underline{/0°}$ A

44. $2\underline{/0°}$ V, $-2\underline{/0°}$ V   **50.** 11.52 kΩ, 720 Ω

51. 1.267 μF, 2.044 mH, 5.05 μF, $3.095\underline{/0.49°}$ Ω, 0.408   **54.** $\sqrt{6}$, 2.566

58. $(j\omega L_2 n_2^2 Z_s)/[j\omega L_2(n_1 + n_2)^2 + n_2^2 Z_s]$, $Z_s + [(n_1 + n_2)/n_2]^2[j\omega L_2 Z_L/(j\omega L_2 + Z_L)]$

61. $(n_2/n_1)^2 R_s/2$. $2R_s + (2n_1/n_2)^2 R_L$

66. $L_{\ell 1} = 50.25$ mH, $L_{m1} = 4.950$ H, $L_{\ell 2} = 0.1256$ mH   **71.** $30\underline{/53.13°}$ Ω

76. Connect the 109-V supply to the 120-V coil. Connect the dot-marked end of the 120-V coil to the unmarked end of the 12-V coil. Rated coil voltages and currents are not exceeded. Insulation of the 12-V coil to the core might be a problem.

77. $21.3\underline{/-20.51°}$ V, $0.5196\underline{/-20.91°}$ A   **78.** $21.71\underline{/-19.01°}$ V, $0.5367\underline{/-17.95°}$ A

CHAPTER 19

1. 239.98 W, 119.99 W   **3.** 41.7 A   **4.** $10\underline{/36.87°}$ KVA, -2.125 KVAR, 37.04 A

6. $120\sqrt{2} \cos(2\pi 60t + \phi_v)$ V, $0.5\sqrt{2}\cos(2\pi 60t + \phi_v)$ A   **10.** 20.8 A, $2.88\underline{/53.13°}$ kVA

12. $8.31\underline{/60°}$ kVA, 4.16 kW, 7.2 kVAR

14. $20.4\underline{/-45°}$ kVA, $10.18\underline{/45°}$ kVA, $22.8\underline{/-18.43°}$ kVA, 54.8 A, 323 V

17. -3.27 kW, 8.27 kW   **22.** $Q = \sqrt{3}P_M$ **26.** 102.5 A   **27.** 27.8 A, -3.5 kVAR

31. $103.7\underline{/-48.4°}$ V, $27.8\underline{/71.6°}$ V, $27.8\underline{/-11.6°}$ V, $103.7\underline{/-108.4°}$ V

33. $0.745\underline{/26.6°}$ pu, $1.49\underline{/-26.6°}$ pu, $0.745\underline{/-26.6°}$ pu, $1.111\underline{/-53.13°}$ pu

36. $0.5\underline{/53.13°}$ pu, $1.082\underline{/-16.09°}$ pu, $2.165\underline{/-69.22°}$ pu, $0.54\underline{/-69.22°}$ pu

38. $1\underline{/-17.46°}$ pu, $0.304\underline{/-8.73°}$ pu, -0.0461 pu, $3.295\underline{/-8.73°}$ pu

39. (a) 1.1 pu, $-j0.1$ pu, $j0.11$ pu;   (b) $0.966\underline{/15°}$ pu, $0.259\underline{/15°}$ pu, $0.250 + j0.067$ pu;
(c) Shifting the phase allows system 1 to deliver real power to system 2.

APPENDIX A

1. $\begin{bmatrix} 3 & 1 & 3 \\ 2 & -2 & -1 \\ 1 & 1 & 2 \end{bmatrix} \cdot \begin{bmatrix} v_1 \\ v_2 \\ v_3 \end{bmatrix} = \begin{bmatrix} 14 \\ -5 \\ 9 \end{bmatrix}$ **2.** $\begin{bmatrix} \dfrac{d^2}{dt^2} & -2 \\ -3\dfrac{d}{dt} & 4\int dt \end{bmatrix} \cdot \begin{bmatrix} v_1 \\ v_2 \end{bmatrix} = \begin{bmatrix} 5\cos t \\ e^{-4t} \end{bmatrix}$

3. -2 for all parts   **4.** 1 V, 2 V, 3 V   **5.** 1 A, 2 A, 3 A

APPENDIX B

7. (a) (i) $0.929 - j0.190$ (ii) $0.948e^{-j11.6°}$ (iii) $0.948\underline{/-11.6°}$
(c) (i) $0.539 - j0.989$ (ii) $1.127e^{-j61.4°}$ (iii) $1.127\underline{/-61.4°}$
(e) (i) $0.376 - j0.842$ (ii) $0.922e^{-j65.9°}$ (iii) $0.922\underline{/-65.9°}$
(g) (i) $0.0193 - j0.0604$ (ii) $0.0634e^{-j72.2°}$ (iii) $0.0634\underline{/-72.2°}$

11. (a) $\dfrac{1}{\sqrt{1 + (\omega RC)^2}}$ and $-\arctan(\omega RC)$ (b) $\dfrac{V_m}{\sqrt{1 + (\omega RC)^2}}$ and $\phi - \arctan(\omega RC)$

(c) $\sqrt{\dfrac{1 + (\omega RC)^2}{(R - \omega^2 RLC)^2 + (\omega L)^2}}$ and $\arctan(\omega RC) - \arctan\left[\dfrac{\omega L}{R - \omega^2 RLC}\right]$

(d) $\sqrt{\dfrac{(a^2 + b^2)(c^2 + d^2)}{(e^2 + f^2)(h^2 + i^2)}}$ and $\arctan(b/a) + \arctan(d/c) - \arctan(f/e) - \arctan(i/h)$

13. (a) $-2 + j2$ and $-2 - j2$ (b) -2 and -4 (c) $e^{j2\pi n/8}$ for $n = 0, 1, 2, \ldots, 7$
(d) $e^{j2\pi n/7}$ for $n = 0, 1, 2, \ldots, 6$ **15.** $\underline{/z_2} = \underline{/z_1}$

19. We can demonstrate the inequality by sketching z_1, $i = 1, 2, \ldots, n$ as vectors
and examining the vector sum $z_1 + z_2 + \cdots + z_n$. Inspection shows that the length
$|z_1 + z_2 + \cdots + z_n|$ is less than or equal to the sum of the lengths $|z_1| + |z_2| + \cdots + |z_n|$,
with equality if and only if $\underline{/z_1} = \underline{/z_2} = \cdots = \underline{/z_n}$.

21. (a) There are two roots. Let $\beta = \dfrac{\pi}{2\sqrt{2}}$. Then the two roots are given by

(i) $e^{-\beta}\cos\beta + je^{-\beta}\sin\beta$ and $e^{\beta}\cos\beta - je^{\beta}\sin\beta$ (ii) $e^{-\beta}e^{j\beta}$ and $e^{\beta}e^{-j\beta}$
(iii) $e^{-\beta}\underline{/\beta}$ and $e^{\beta}\underline{/-\beta}$ (b) (i) $\cos(e\pi/2) + j\sin(e\pi/2)$ (ii) $e^{je\pi/2}$ (iii) $1\underline{/e\pi/2}$
(c) (i) $-3.7245 - j0.5119$ (ii) $3.759e^{-j172.7°}$ (iii) $3.759\underline{/-172.2°}$

INDEX

The Standard for Circuit Simulation
FREE SOFTWARE

PSpice is a circuit simulator package used to calculate the behavior of electrical circuits. Class instructors can receive complimentary evaluation versions for *both* the IBM-PC and Macintosh by submitting a request on company or educational letterhead to:

> Product Marketing Dept.
> MicroSim Corporation
> 20 Fairbanks
> Irvine, CA 92718

Duplication of the diskettes for your students is encouraged.

PSpice is a registered trademark of MicroSim Corporation.

THE GREEK ALPHABET

α	A	Alpha
β	B	Beta
γ	Γ	Gamma
δ	Δ	Delta
ε	E	Epsilon
ζ	Z	Zeta
η	H	Eta
θ	Θ	Theta
ι	I	Iota
κ	K	Kappa
λ	Λ	Lambda
μ	M	Mu
ν	N	Nu
ξ	Ξ	Xi
o	O	Omicron
π	Π	Pi
ρ	P	Rho
σ	Σ	Sigma
τ	T	Tau
υ	Υ	Upsilon
ϕ	Φ	Phi
χ	X	Chi
ψ	Ψ	Psi
ω	Ω	Omega